Energy Design for Tomorrow

Energy Design für morgen

For Christa Daniels

Klaus Daniels
Ralph E. Hammann

Energy Design for Tomorrow

Energy Design für morgen

Edition Axel Menges

Content

Part 3

Inhalt

Teil 3

Preface

Vorwort

I'd put my money on solar energy...
I hope we don't have to wait till oil and coal run out before we tackle that.

Thomas Edison, in conversation with Henry Ford and Harvey Firestone

March 1931

I'd put my money on solar energy...
I hope we don't have to wait till oil and coal run out before we tackle that.

Thomas Edison, im Gespräch mit Henry Ford und Harvey Firestone

März 1931

The new millennium is now almost a decade old. That which was begun with great festive optimism was soon followed by the events of September 11, 2001, which led to deep political and security distortions on a global scale that are now being more and more dominated by a linked set of existential questions with far-reaching consequences for the future existence of mankind. As a result of this recent discussion, the issue of global climate change has risen to prominence worldwide.

Among the many problems humanity will have to address in the 21 st century, three that ought to be accorded utmost priority, because of an increasing world population that should total 10 billion people by the end of the first century of the new millennium, are the following[1]:

- providing access to clean drinking water,
- securing healthy and sufficient nutrition,
- assuring adequate health care.

It is self-evident that questions concerning the securing of a stable and sustainable energy supply for this rise in population are of great importance. Equally, the consequences of changing attitudes with regard to energy-related topics, which provide the backdrop of this book, need to be studied intensively.

The future increase in global primary energy demand,[2] as predicted by the International Energy Agency (IEA), and the so-called "Reference Scenario 2006" published by the same agency, should be the basis for a realistic discussion.[2]

Currently, the majority of energy generation is derived from the transformation of non-renewable, fossil resources with a finite availability, such as crude oil, natural gas, coal, and uranium. Undoubtedly, there are clear indicators that allow for a relatively concrete estimation of availability of such natural resources, and it becomes in some cases astonishingly clear – on a historical scale – how short such time spans will be.

Das neue Millennium ist fast eine Dekade alt. Was in vielen Metropolen der Welt mit großem festlichem Optimismus begonnen wurde und kurze Zeit später durch die Ereignisse des 11. September 2001 zu einschneidenden, global wirkenden und politischen Verwerfungen geführt hat, wird aktuell mehr und mehr dominiert durch eine verknüpfte Kette von existenziellen Fragestellungen mit erheblichen und weltweiten Auswirkungen. Dabei avancierte „Klimawandel" zu einem der weltweit meist verbreiteten Begriffe.

Probleme, die die Menschheit des 21. Jahrhunderts als Ganzes mit größter Priorität zu lösen hat, sind:

- sauberes Trinkwasser,
- gesunde und ausreichende Ernaehrung,
- der Zugang zu adäquater medizinischer Versorgung

für eine prognostizierte Weltbevölkerung von 10 Milliarden Menschen zum Ausgang des ersten Jahrhunderts des neuen Millenniums[1].

Direkten Bezug zum genannten Zuwachs an globaler Population haben selbstverständlich die Aspekte einer globalen Sicherung der Energieversorgung, und die Konsequenzen einer veränderten Grundanschauung zu allen energiewirtschaftlichen Fragen, wie sie Thema dieses Buches sind.

Dem Anstieg des primären Energieverbrauchs,[2] dargestellt durch Prognosen der Internationalen Energiebehörde (IEA) und dem von ihr herausgegebenen globalen Energieverbrauch-Referenz-Szenarios 2006 ist Rechnung zu tragen.

Der maßgebliche Anteil bei der Energieerzeugung wird derzeit noch aus nicht erneuerbaren Ressourcen, wie Erdöl, Erdgas, Kohle und Uran gedeckt. Diese Rohstoffe sind endlich, und neuere Indikatoren erlauben eine relative klare Abschätzung, wie historisch kurz die Zeiträume sind, für die sie noch verfügbar bleiben.

In this book, the authors criticize the questionable prevailing thinking of depleting such valuable resources simply for energy generation. In addition, the related consequences of increases in carbon dioxide output into the atmosphere as a result of fossil fuel consumption are discussed. Scientists united in the Intergovernmental Panel on Climate Change (IPCC) strongly suggest that this output of the greenhouse gas of CO_2 on a global scale ought to be limited to no more than 500 ppm if the observed increase in global temperature is to be limited to 2 °C.[3]

However, the purpose of this book is far from painting an inevitable doomsday scenario. On the contrary, it is conceived as the most actual contribution to a vision of transformation of current "business-as-usual" practices in energy supply in general and their use in buildings in particular towards a truly sustainable future, employing renewable forms of energy.

The authors describe here, with great enthusiasm and in detail, already-successful new forms of sustainable, non-fossil energy-generating technologies that may serve as alternatives to current practice, including thermal solar power, solar electricity, wind and hydroelectric power, the generation of electricity with the help of the ocean's waves and tides, or forms that are still in a state of experimentation. The book presents sustainable energy systems that can be used either locally or in connection with long-distance or large-area regional utility grids, as well as technologies that are integrated into the building itself, such as building-integrated photovoltaics (BIPV), building-integrated thermal solar (BITS), and building-integrated wind power (BIWP). Furthermore, this book makes it possible for the first time to make clear comparisons between a wide range of otherwise sometimes confusing national energy data. Here, such data were converted and calculated into comparable units of kW, MW, kWh, and MWh, respectively.

Es wird in dieser Publikation nicht nur darauf verwiesen, dass es extrem fragwürdig ist, wertvolle fossile Rohstoffe zum überwiegenden Anteil zur Energieerzeugung zu benutzen, sondern es werden die Konsequenzen, die mit dieser Form der Energieerzeugung zusammenhängen, angesprochen – die Erhöhung des CO_2-Ausstoßes. Eine Limitierung dieses CO_2-Outputs auf 500 ppm wird von Wissenschaftlern, verbunden im Intergovernmental Panel on Climate Change (IPCC), gefordert, wenn die globale Temperaturerhöhung auf 2 °C beschränkt werden soll.[3]

Der Tenor der vorliegenden Publikation ist jedoch weit entfernt von der Verbreitung einer Endzeitstimmung. Vielmehr ist das Buch als aktueller Beitrag und Vision einer Transformationsstrategie hin zu einer nachhaltigen Energieversorgung mit erneuerbaren Ressourcen zu verstehen.

Die Autoren stellen hier mit großem Enthusiasmus die erfolgreichen oder in Zukunft erfolgversprechenden, nicht-fossilen Alternativen zur derzeitigen Praxis der Energiebereitstellung, wie solare Wärmekraft, solare Elektrizitätserzeugung, Technologien zur Energieerzeugung aus der Kraft des Meeres und Windkraft, im Detail vor. Es werden dabei solche Technologien besprochen, die in der Lage sind, entweder lokal oder überregional vernetzt erneuerbare Energie zu erzeugen, oder die individuell in Gebäuden zum Einsatz kommen können, wie zum Beispiel building-integrated photovoltaics (BIPV), building-integrated thermal solar (BITS) oder building-integrated wind power (BIWP). Ferner erlaubt dieses Buch zum ersten Mal, Vergleiche zu Energieverbrauch und Erzeugung für ein breites Spektrum von Ländern zu treffen. Um dies zu ermöglichen, wurden umfangreiche, und in vielen Fällen verwirrende, länderbezogene Einzeldaten in einheitliche und vergleichbare Werte nach Kilowatt, Megawatt bzw. Kilowattstunden, Megawattstunden pro Jahr errechnet und diesem Buch zu einem besseren Verständnis beigegeben.

1 World Population to 2300, UN Report 2004 (Mid-report)
2 2006 World Energy Outlook (Source: IEA, ISBN 92-64-10989-7)
3 Intergovernmental Panel on Climate Change, Summary for Policymakers, Emission scenarios, 2000, ISBN: 92-9169-113-5

1 World Population to 2300, UN Report 2004 (Mid-report)
2 2006 World Energy Outlook (Source: IEA, ISBN 92-64-10989-7)
3 Intergovernmental Panel on Climate Change, Summary for Policymakers, Emission scenarios, 2000, ISBN: 92-9169-113-5

How is such a focus on energy and its generation and consumption related to the built environment and architecture?

On a global average, building-related activities such as construction, operation, and maintenance consume more than 40 % of a country's energy. Without any doubt, this is a significant percentage, and in some cases it is more than the percentage needed for a national economy's transportation or industrial sectors. Furthermore, buildings are characterized by a much longer lifespan than that of cars, household appliances, or other consumer products (which, nevertheless, need to achieve definite efficiency increases). Errors made today in the conception of the built environment, as a result of either a lack of expertise or misunderstood design philosophies, will place long-term environmental burdens not just on the owners and users of buildings but also on society in general.

Hence, the book introduces a palette of intelligent and appealing sustainable design solutions, including renewable energy systems that strive not only to reduce energy consumption within the structures but also to provide highly comfortable environments. With great interest, the authors continue to observe the potential of such advanced-concept integrations utilizing renewable energy and to investigate how renewable energy may serve as a catalyst in the initial design phase of any architectural project. Based on the essential and careful observation of individual locations and climates, such strategies provide designers with a "natural way" to generate an architectural parti and the subsequent building design.

Was hat ein solcher Fokus auf Energie, ihre Erzeugung und ihren Verbrauch mit der gebauten Umwelt und mit Architektur zu tun?

In einem globalen Mittel verbrauchen gebäudebezogene Aktivitäten wie Konstruktion, Bewirtschaftung und Unterhalt von Gebäuden mehr als 40 % der Energie eines Landes. Dies ist ein erheblicher Prozentsatz und in vielen Ländern größer als der Anteil für Transport und für die Erzeugung industrieller Produkte. Gebäude sind zudem charakterisiert durch eine erheblich längere Lebensdauer als Automobile, Geräte des täglichen Lebens und anderer Bereiche einer nationalen Ökonomie, die auch in Zukunft weiterhin deutliche Effizienzsteigerungen erzielen müssen. Fehler, die heute entweder durch Unkenntnis oder falsch verstandenen Design-Individualismus bei der Konzipierung und der Realisierung unserer gebauten Umwelt gemacht werden, führen auf lange Sicht zu erheblichen Belastungen nicht nur der entsprechenden Bauherren und Nutzer, sondern der Gesellschaft als Ganzes.

Aus diesem Grund stellt das Buch eine Reihe hoch interessanter Beispiele von unterschiedlichen baulichen Strukturen und Nutzungen vor, bei denen versucht wird, nicht nur wertvolle Energie einzusparen, sondern auf intelligente Weise hochkomfortable Räume bei gleichzeitiger Integration von Technologien zur Nutzung erneuerbarer Energien zu schaffen. Mit großem Interesse verfolgen die Autoren weiterhin die gestalterischen, architektonischen Möglichkeiten, die sich aus einer sorgfältigen Analyse des jeweiligen Gebäudestandorts und der Integration von Technologien zur Erzeugung von Energie sozusagen "auf natürliche Weise" für den planenden Architekten ergeben.

Such a complex undertaking as presented in this publication, with more than 500 images, diagrams, and tables, most of which were generated specifically for this book, cannot be achieved without qualified and financial support from colleagues, partners and colleges. The authors want to thank especially:

Klaus Betz of Imtech Engineering, Hamburg, Germany;
Carlo Baumschlager and Dietmar Eberle of Baumschlager-Eberle Architects, Lochau, Austria;
Dirk U. Hindrichs and Winfried Heusler of Schueco International KG, Bielefeld, Germany;
Michael Küpper, Klaus G. Peter, Jacob Platzer and Klaus Daniels – HL-Technik Engineering Partner GmbH, Munich, Germany;
Thomas Wetter of HL Technik AG, Zurich, Switzerland;
Siegfried Timmler of TTC Technology GmbH, Recklinghausen, Germany;
College of Fine and Applied Arts, University of Illinois, Urbana-Champaign, Illinois, USA.

The graphical design of the book was provided by Riemer Design, Munich, Germany.

Klaus Daniels
Ralph Hammann

Munich, Champaign
December 2008

Eine solch umfassende Publikation mit über 500 Bildern und Graphiken, die zum Teil für dieses Buch eigens erstellt wurden, ist nicht ohne inhaltliche und finanzielle Unterstützung von Kollegen, Partnern und Hochschulen möglich. Die Autoren danken aus diesem Grund insbesondere:

Klaus Betz – Firma Imtech, Hamburg, Deutschland;
Carlo Eberle und Dietmar Baumschlager – Büro Baumschlager-Eberle, Lochau, Österreich;
Dirk. U. Hindrichs und Winfried Heusler – Firma Schüco, Bielefeld, Deutschland;
Michael Küpper, Klaus G. Peter, Jacob Platzer und Klaus Daniels – Firma HL-Technik Engineering Partner GmbH, München, Deutschland;
Thomas Wetter – Firma HL-Technik AG, Zürich, Schweiz;
Siegfried Timmler – Firma TTC, Recklinghausen, Deutschland;
College of Fine and Applied Arts, University of Illinois, Urbana-Champaign, Illinois, USA.

Die graphische Bearbeitung des vorliegenden Buches wurde durch Riemer Design in München erstellt.

Klaus Daniels
Ralph Hammann

München, Champaign
Dezember 2008

Introduction

Einleitung

The challenges for humanity that we face in the course of the 21 st century are staggering: dramatic increases in population for at least some areas of the globe, increasingly asymmetrical wealth distribution, scarcity of some important natural resources – and a brutal struggle for their exploitation and distribution, the destruction of the natural support basis of our existence that has become more and more prevalent and clear, and climate changes, to which we were already alerted during the decades of the 70 s and 80 s of the past century.

Out of this overwhelming kaleidoscope of problems facing the human race, only a minute spectrum will be addressed, especially those that are related to sustainability.

In 1987, the United Nations Brundtland Commission (formally, the World Commission on Environment and Development or WCED) defined a sustainable development as follows: "Sustainable development is development that meets the needs of the present without compromising the ability of future generations to meet their own needs."

Sustainable development rests on three major principles – ecology, economy, and social justice – of which ecology seems currently to take priority in the public discourse, possibly as a result of concerns about climate change and depleted natural resources.

Nature is organized according to the principles of nutrients and metabolisms, clearly avoiding all notions of waste. A fruit tree in nature produces flowers and fruit, pollinated by insects, and its abundant flowers are not worthless – after falling to the ground, they enhance the quality of the soil with the help of numerous organisms and microbes. Nothing is wasted.

Wherever we are on the planet, animals and humans exhale carbon dioxide, which in turn is absorbed by plants and used for their growth. Nitrogen is present in all living organisms in proteins, nucleic acids, and other molecules. It is a component of animal waste and is essential nutrient for all plants and organisms that convert it, together with the nutrients of the earth, such as carbon, hydrogen, and oxygen, into oxygen in a perpetual life cycle. Waste, if it occurs, becomes nutrients in return.

This biological life cycle has been sustained for millions of years on a planet with an abundance of magnificent species. Humans are the only species in this system that extracts large quantities of nutrients but only rarely returns them into the cycle in a usable form.

Die Herausforderungen des einundzwanzigsten Jahrhunderts durch Bevölkerungsexplosion in verschiedensten Regionen der Welt, wachsende Instabilität infolge asymmetrisch verteilten Wohlstands, Rohstoffverknappung und Verteilungskämpfe sowie die Zerstörung unserer natürlichen Lebensgrundlagen dämmern uns mehr und mehr, angetrieben durch das zunehmende Erkennen der klimatischen Konsequenzen, vor denen bereits in den 70er und 80er Jahren des letzten Jahrhunderts gewarnt wurde.

Nur einen kleinen Teil der erkennbaren Probleme werden die nachfolgenden Ausführungen behandeln – insbesondere die, die sich mit dem Thema Nachhaltigkeit oder auch Sustainability auseinandersetzen.

1987 definierte die Brundtland-Kommission eine Entwicklung als nachhaltig, wenn sie den Bedürfnissen der heutigen Generationen entspricht, ohne die Möglichkeit künftiger Generationen zu gefährden, ihre eigenen Bedürfnisse zu befriedigen.

Nachhaltige Entwicklung steht auf drei Säulen: Ökologie, Ökonomie und Soziales, von denen derzeit die ökologische Dimension vor dem sich abzeichnenden Klimawandel und den knapper werdenden Ressourcen den breitesten Raum in der öffentlichen Diskussion einnimmt.

Die Natur funktioniert nach einem System von Nährstoffen und Metabolismen, in dem kein Abfall vorkommt. Ein Obstbaum produziert viele Blüten und Früchte, damit ein neuer Baum keimen und wachsen kann. Der Überfluss an Blüten ist nicht etwa wertlos, sondern dient nach dem Herabfallen auf den Boden dazu, zahlreiche Organismen und Mikroorganismen zu ernähren, wobei gleichzeitig die Bodenbeschaffenheit verbessert wird.

Überall auf der Welt atmen Tiere und Menschen Kohlendioxid aus, das die Pflanzen aufnehmen und für ihr Wachstum nutzen. Stickstoff aus den Abfällen wird von Mikroorganismen, Tieren und Pflanzen in Protein umgewandelt, und die wichtigsten Nährstoffe der Erde wie Kohlenstoff, Wasserstoff, Sauerstoff und Stickstoff durchlaufen einen Kreislauf und werden immer wieder neu verwendet. Abfall ist somit wieder Nahrung.

Dieses zyklische biologische System lässt seit Jahrmillionen einen Planeten mit einer prächtigen Vielfalt gedeihen. Lediglich der Mensch ist die einzige Spezies, die dem Boden große Mengen an Nährstoffen entzieht, die zum Teil für biologische Prozesse gebraucht werden, die jedoch nur selten in brauchbarer Form zurückgeführt werden.

Following the early advances of industry, the natural balance of materials has begun to shift toward imbalance. Man extracts materials in large quantities from the surface or the outer crust of the planet, then modifies, synthesizes, and treats them in a fashion that, in most cases, renders their safe return to their origin prohibitive. In addition to the biological metabolism of the biosphere, a second metabolism of the "techno-sphere" needs to attract our attention in order to bring peace to the process of industry. Here, we especially need to focus on the need to intelligently "deconstruct" an overwhelming number of industrial products for the sake of sustainability.

It is a question of design as to how we can achieve the benign return of industrially manufactured products and materials to the metabolism of the techno-sphere. They need to become "nutrients" for the new. In this regard, the problem of widespread contamination has to be primarily alleviated under all circumstances.

A "technical nutrient" is a material or product that is conceived and constructed in such a way that it allows its return into a technical life cycle – i.e., the technical metabolism from which it originates. A new demand for harmless "deconstruction" is raised, in addition to the concept of recycling. The design of a product needs to be conceived in such a way that the product can be easily disassembled and deconstructed, with its separate components returned to a positive techno-sphere cycle. Manufacturers of materials and products ought to be responsible for the return of their products, which they need to prepare for future reuse. This will eventually also result in designs that rely upon manufacturing without the use of hazardous components and substances.

Architects, civil engineers, and engineers of building systems need to redefine their roles as "nutrient managers". It has to be understood that the materials challenge in constructing buildings far exceeds the challenges of energy consumption and supply. It is therefore also important to remember the term "ecologically correct building construction", the science that deals with the interaction between living organisms, the material world, the metabolism and energy balance of the biosphere, and the products we utilize for buildings.

Currently in the U.S., certification of a wide range of building materials and products based on the Greenguard Environmental Institute's product certification program for low-emitting interior building materials, furnishings, and finish systems is gaining momentum.

Mit dem Aufkommen der Industrie hat sich das natürliche Gleichgewicht der Materialien auf der Erde verschoben. Der Mensch nimmt sich Substanzen von der Erdoberfläche oder aus der Erdkruste und bereitet sie auf, ändert und synthetisiert sie zu riesigen Mengen von Material, das dem Boden nicht wieder gefahrlos zugeführt werden kann. Neben dem biologischen Metabolismus oder der Biosphäre (Kreisläufe der Natur) muss uns insbesondere der zweite – technische – Metabolismus (Technosphäre) interessieren, um die Kreisläufe der Industrie, zu denen auch der Abbau mancher technischer Materialien zählt, wieder in Ordnung zu bringen.

Mit dem richtigen Design sollten weitestgehend alle von der Industrie hergestellten Produkte und Materialien den Metabolismen zugeführt werden können, um „Nahrung" für etwas Neues zu liefern. Dies setzt voraus, dass Kontaminationen unbedingt vermieden werden.

Ein „technischer Nährstoff" ist somit ein Material oder Produkt, dass so konstruiert ist, dass es in den technischen Kreislauf zurückkehren kann – in den industriellen Metabolismus -, dem es entstammt. Hieraus entsteht die Forderung nach Abcyceln anstatt Recyceln. Ein Design von Erzeugnissen als Produkt beinhaltet, das Produkt so zu konzipieren, dass es zerlegt werden kann. Hersteller von Produkten sollten die Verpflichtung haben, die ursprünglich erworbenen Materialien zurückzunehmen und wiederzuverwenden. Hieraus resultiert, dass ein Design zu entwickeln ist, das gänzlich ohne gefährliche Stoffe auskommt.

Architekten, Bauingenieure und Ingenieure für technische Anlagen müssten sich als kreative und intelligente „Nährstoffmanager" verstehen. Die materielle Problematik beim Bau von Häusern ist weitaus größer als das Energieproblem. Insofern sollte man sich des Begriffs des ökologisch richtigen Bauens wieder erinnern – der Wissenschaft, die sich mit den Wechselbeziehungen zwischen den Organismen und der unbelebten und der belebten Umwelt befasst sowie mit dem Stoff- und Energiehaushalt der Biosphäre und ihren Untereinheiten.

In den USA nimmt zurzeit eine eigene Greenguard-Zertifizierung für das Ausgasungsverhalten von Produkten einen breiten Raum ein. Wichtiger wäre jedoch, die Gesamtbilanz von Energie und Rohstoffen so zu bewerten, dass hieraus resultierend Materialien zu gestalten sind wie ein Nährstoff, da falsch verstandenes Recycling zusätzliche Umweltbelastungen mit sich bringt. Ein zukünftiges, gutes Produktdesign bzw. Baudesign sollte die nächste Nutzung des Objekts bereits berücksichtigen – ein so genanntes „design for reincarnation". Solche Gebäude zum Beispiel beinhalten eine gezielte Auswahl von Materialien, die sich vor allem gut trennen lassen.

More important would be the development of a total budget of primary materials and embedded energies of materials, with the goal of developing materials as "nutrients". Recycling – if understood in the wrong way – will only add to our environmental problems. "Design for reincarnation" could be a future guiding principle by which all designs for materials used in buildings try to anticipate a new utilization of the built environment. Such buildings would then be composed of material assemblies with the goal of their easy and safe deconstruction and reverse engineering.

To exemplify the task, we can use the basic, or primitive, recycling technologies of the steel sector may be used, in which valuable non-ferrous metals are often lost in an unsophisticated recycling process. Concrete and cement are "enhanced" with harmful additives that contaminate the atmosphere and prevent a return to the "nutrient cycle".

"Eco-effectiveness" could be a new philosophy. By this we mean not minimizing the ecological footprint but increasing it, deepening it, and creating a wet area that allows for other organisms to thrive.

In Germany, building waste currently constitutes approximately 70 % of all waste. In particular, the so-called "multi-component" building composites are almost impossible to deconstruct into their initial materials. Insulation materials and other interior components represent the bulk of all building construction-related waste. Typically, the lesser challenges are steel, aluminum, glass, and wood, if they are available in single varietals.

German automobile manufacturers may serve as an advanced example of the future requirements of the building sector. Under pressure from the German legislature, they will be obliged to take back any of their products at the end of their useful lifespan. This has stimulated not only an extensive recycling program but also an intense design effort – a pre-engineering for future de-construction, so to speak.

According to the German engineer Werner Sobek, future buildings should be characterized as follows:

- zero energy: In total, they will not require energy for their annual operation.
- zero emissions: They will not emit any harmful substances.
- zero waste: All materials will be completely recyclable.

Beispielsweise lässt sich feststellen, dass im Baustahl wertvolle Buntmetalle durch Primitivrecycling verloren gehen und in Zement und Beton schädliche Additive und problematische Zusatzstoffe eingesetzt werden, die eine Kontamination der Biosphäre verursachen und somit einen echten „Nährstoffkreislauf" verhindern. Öko-Effektivität bedeutet letztendlich, nicht den ökologischen Fußabdruck zu minimieren, sondern einen großen Abdruck zu machen, der gleichzeitig ein Feuchtgebiet ist, das Lebensraum für andere Lebewesen bietet.

Festzustellen ist in Deutschland zurzeit, dass der Anteil der Baumassenabfälle am gesamten Abfallaufkommen ca. 70 Prozent ausmacht, wobei vom einstmals Gebauten gerade die Mehrkomponentenbauteile in ihre Einzelwerkstoffe kaum zu zerlegen sind. Vor allem Dämm- und Ausbaustoffe bereiten mengenmäßig die größten Probleme, unkritisch sind dabei sortenrein vorliegende Bauteile aus Stahl, Aluminium, Holz und Glas.

Mit der Ankündigung einer Rücknahmeverpflichtung im Automobilbau ist eine Forschung und Entwicklung eingetreten hin zum recyclinggerechten Konstruieren, d.h. zum methodisch vorstrukturierten Zusammen- und Auseinanderbauen von Komponenten.

Gemäß Prof. Werner Sobek sollten sich Gebäude dadurch auszeichnen, dass sie

- für ihren Betrieb in der Jahressumme keine Energie benötigen (zero energy),
- keine schädlichen Emissionen abgeben (zero emission),
- vollkommen reziklierbar sind (zero waste).

Das Triple-zero-Konzept erfordert in Bezug auf die Durchsetzung des Nachhaltigkeitsaspekts in der gebauten Umwelt einen weltweiten Umdenkungsprozess. Politik, Wissenschaft und Industrie bereiten die Einführung des Nachhaltigkeitsaspekts in der gebauten Umwelt vor, seine Umsetzung wird im Wesentlichen in den Händen von Architekten und Ingenieuren liegen, somit in Händen von Menschen, die bis heute in der Regel noch über keine durchgreifenden Konzepte für Konzeption, Konstruktion und Gestaltung dieser „nachhaltigen Architektur" verfügen. Sicher nicht mehr gefragt sein wird in Zukunft die Antwort des weltweiten architektonischen Schaffens auf den Bericht des Club of Rome – der Postmodernismus – Dekonstruktivismus – Superdutch – Blob und andere Stilrichtungen, die in der Regel alle ökologischen Probleme schlichtweg ignorieren.

Vor dem Hintergrund des Klimawandels müssen Investoren wie Nutzer gleichermaßen den grundsätzlichen Bedarf an überbauter Fläche hinterfragen und den Umgang mit natürlichen Ressourcen effizienter

This so-called "Triple-Zero Concept" requires a sustainable rethinking process on a global basis. Governments, the sciences, and industry are currently in the process to preparing for such a definition of sustainability. When it comes to the built environment, the responsibility is mostly in the hands of architects and engineers – in other words, specialists who up to now have been ill-prepared for such a radical readjustment of their practices. In the past, the global response of the design profession to concerns addressed by the Club of Rome were stylistics exercises such as Deconstructivismn, Super-Dutch, "Blobs", and other styles. The idea was to simply ignore issues of ecology and sustainability.

Faced with the backdrop of climate change and resource depletion, both investors and users need to contemplate their requirements for built surface area. They also need to devise a way to work more efficiently with the available resources, which means, in addition to the design of more energy-efficient, long-lasting, and sustainable buildings, the need for adaptive re-use of real estate. For investors, all of this will be of interest when savings in variable costs are capable of covering the added initial investment costs. Market economics for energy cost will certainly play a decisive role as well. An increase in image value for sustainable buildings can be observed already, but it remains to be seen whether such ecological image gains can be turned into economic gains. Without doubt, future clients in the real estate market will consider, in addition to the economic aspects of a project, the ecological framework as well. Long-term investors will have to solve the equation of the competing goals of economy, ecology, and technology to develop a long-term strategy to remain successful in a challenging market.

gestalten. Das bedeutet in erster Linie, vermehrt energieeffiziente und umweltfreundliche Gebäude zu planen und zu bauen und Bestandsimmobilien zu modernisieren. Von besonderem Interesse aus Investorensicht wird, wenn das Einsparpotenzial bei den variablen Kosten die zusätzlichen Investitionskosten deckt. Beides hängt in hohem Maß vom Marktpreis für Energie ab und ob sich der Reputationsgewinn durch ein Bekenntnis zu nachhaltigem Wirtschaften für Mieter wie Vermieter gleichermaßen in Marktanteile ummünzen lässt. Es ist davon auszugehen, dass zukünftig die Mieter großer Immobilienflächen zusätzlich zu ökonomischen Gesichtspunkten verstärkt auch ökologische Faktoren in ihre Entscheidungen einbeziehen. Ein langfristig agierender Investor tut gut daran, die bisher divergierenden Zielkonflikte zwischen Ökonomie, Technik und Ökologie zugunsten einer Langfriststrategie aufzulösen.

Part 1

The greenhouse effect

Fossil energy sources, renewable energy

Energy cost

Energy consumption of selected world regions (IEA 2004)

Detailed view: Germany – case study

Detailed view: Switzerland

Detailed view: France

Global energy scenarios

Teil 1

Der Greenhouse-Effekt

Fossile Energieträger/erneuerbare Energien

Energiekosten

Energieverbräuche ausgewählter Regionen (IEA – 2004)

Detailbetrachtung Bundesrepublik Deutschland – ein Beispiel

Detailbetrachtung Schweiz

Detailbetrachtung Frankreich

Energieszenarien, weltweit

1 The greenhouse effect

1 Der Greenhouse-Effekt

The problem of the future availability of fossil fuels is currently being overshadowed by the discussion of how their use may contribute to global climate change. In this discussion, two opposing camps of scientists and climate researchers come to very different conclusions. One group, led by the Intergovernmental Panel on Climate Change (IPCC), is convinced that climate change is a result of a rise in carbon dioxide levels in the atmosphere, while another group of scientits disputes this view and suggests that climate change is a natural-occurring process as a result of increased solar activity.

The authors of this book are reluctant to position themselves on the side of either of those camps because they are not climate scientists and thus are unable to fully fathom the complex causalities of the issue.

As a result, the authors will present both positions equally weighted.

The fact that the concentration of carbon dioxide gas in the atmosphere, which was at around 280 ppm in the year 1800, has increased to levels of around 350 ppm today cannot be disputed. The increase of the surface temperature of the Earth is currently approximately 0.8 K (Kelvin).

Historically, times of colder temperatures on earth have always given way to periods of warmer temperatures without a clear association to CO_2 levels. Calculations made by the Niels Bohr Institute of Copenhagen allow the conclusion that sudden temperature increases of 7 – 10 K every 1,500 years are natural chaotic fluctuations of the climate system. What also can be derived from these studies is that so far no correlation exists between rising CO_2 levels and the global temperature increase.

The analysis of ice cores drilled deep into glaciers has revealed a cycle of glacials and interglacials that have occurred throughout Earth's history. Currently, we are in a warm period of an ice age, as seen at glaciers at least on one of the Earth's poles. Warm periods are typically brief periods in a glacial period.

Between 8000 and 6000 B.C., temperatures in the northern hemisphere were significantly warmer than they are today, but at the same time the CO_2 concentration in the atmosphere passed through a minimum of around 260 ppm. When this level started to rise in the centuries that followed – without human intervention – the temperature, interestingly, started to decrease.

Die Problematik der zeitlichen Verfügbarkeit fossiler Brennstoffe wird zurzeit total überlagert von der Diskussion des Klimawandels. Bei dieser Diskussion stehen sich zwei Lager von Wissenschaftlern und Klimaforschern gegenüber, wobei eine Gruppe – angeführt von IPCC – die Meinung vertritt, dass der Klimawandel durch die CO_2-Erhöhung ausgelöst wurde und die zweite Gruppe von Wissenschaftlern dieses bestreitet und die Meinung vertritt, dass die Klimaveränderungen auf erhöhte Sonnenaktivitäten zurückzuführen ist.

Die Autoren können sich in diese Diskussion nicht einbringen, da sie keine Klimaforscher sind und die detaillierten Zusammenhänge nur zum Teil nachvollziehen können.

Insofern werden beide Ansichten berücksichtigt und kurz beschrieben.

Dass die weltweite CO_2-Konzentration in der Atmosphäre, die vor ca. 1800 bei ca. 280 ppm lag, nunmehr auf ca. 350 ppm gestiegen ist, ist unbestreitbar, der Anstieg der Erdoberflächentemperatur beträgt zurzeit ca. 0,8 K (Kelvin).

Kalt- und Warmzeiten haben sich im Laufe der Erdgeschichte unablässig abgelöst. Dabei gab es keinen Zusammenhang zwischen Temperatur und CO_2 in der Atmosphäre. Berechnungen des Niels-Bohr-Instituts, Kopenhagen, lassen vermuten, dass abrupte Wärmeeinbrüche von 7 – 10 K ca. alle 1.500 Jahre zufällige Erscheinungen sind, chaotische Fluktuationen des Klimasystems selbst. Feststellbar ist auch, dass zumindest bisher der CO_2-Gehalt in der Atmosphäre keinen Einfluss auf Temperaturveränderungen hatte.

Wie man aus der Auswertung von Bohrkernen weiß, lösten Eis- und Warmzeiten einander ab. Zurzeit befinden wir uns in der Warmzeit eines Eiszeitalterns, das durch Gletscher an mindestens einem der Pole gekennzeichnet ist. Warmzeiten innerhalb eines Eiszeitalters sind im Regelfall Kurzepochen.

Zwischen 8000 und 6000 v. Chr. war es auf der Nordhalbkugel deutlich wärmer als heute, während der CO_2-Gehalt der Atmosphäre ein Minimum von etwa 260 ppm durchlief und danach ohne jeden menschlichen Einfluss anstieg, wobei gleichzeitig interessanterweise die Temperatur sank.

Zur Zeit Christi Geburt herrschte wiederum eine kleine Warmzeit – die Züge der Römer berichten davon. Etwa 400 Jahre später setzte eine Völkerwanderung vom Norden in den Süden ein – eine Reaktion auf eine beginnende Kaltzeit, die etwa 500 Jahre dauerte und etwa 800 n. Chr. endete. Ca. 200 Jahre später besiedelten Norweger zum Teil Grönland (Grünland) und Nordamerika. Ab 1300 begann eine neue Kaltzeit, die

Around the time of the birth of Jesus of Nazareth, a brief warm period was observed, as documented in the records of Roman military campaigns. Around 400 years later, mass migration from the North to the South of Europe began in response to the beginning cold period that lasted around 500 years, until about 800 A.D. Approximately 200 years later, Norwegians settled and cultivated parts of Greenland and North America. Then, in 1300, another cold period began that resulted, between approximately 1500 and 1700, in extremely low temperatures in Central Europe, with summer highs not greater than 15 °C. Although the temperature in these periods changed dramatically from year to year, CO_2 levels remained nearly constant. Any increase in global temperatures at the time was prevented by the eruption of the Laki volcano in Iceland in 1783, which ejected several hundred million tons of dust and gas into the atmosphere. As a result, the Sun's rays were reflected back into space, and the Earth experienced subsequent drops in temperature.

In contrast to the advocates of the theory that links rising CO_2 levels to a rise in global atmospheric temperature, scientists of international repute point out a correlation between the temperature increase and solar activity and its influence on the formation of clouds. Never in the past 1,000 years was solar activity as strong as it was during the 20th century.

On the other side of the argument, the U.N. Intergovernmental Panel on Climate Change (IPCC), which published its findings in a report in April 2007, concluded that the increase in global temperatures can – with a 90 % probability – be linked to human activity. It was said that CO_2, a color- and odorless gas that dissolves in water, plays the decisive role. At a temperature of 20 °C, one cubic meter of water accepts 0.5 g of CO_2, yet at a temperature of 0 °C the amount it accepts is 1 g. Therefore, if water is warmed up it will release CO_2 back into the atmosphere, while if it cools it will absorb carbon dioxide again. Because about two-thirds of the surface of the Earth is covered with water, only a slight increase in water temperature will release large amounts of the gas into the atmosphere. A reversal of this process is impossible due to saturation limits.

To a large degree, plant materials take the most advantage of an increase in CO_2 levels in the atmosphere because they act as the largest sink for this gas. If global temperature rises and carbon dioxide levels increase, plants increase their rate of growth, an effect that surpasses the influence of the bodies of the Earth's water in absorbing CO_2. It should also be noted that CO_2 concentration in the atmosphere amounts to only a few percent, whereas water vapor exceeds the climate gas by a multitude. Clouds, formed out of the water vapor contained in the Earth's atmosphere, influence the temperature more than

mit tiefsten Temperaturen in Mitteleuropa um 1500 – 1700 (höchste Sommertemperatur ca. 15 °C) festzustellen war. Obwohl sich in dieser Zeit die Außentemperaturen von Jahr zu Jahr drastisch veränderten, blieb der CO_2-Gehalt nahezu konstant. Ein Anstieg der weltweiten Temperaturen ab ca. 1800 wurde durch die Explosion des isländischen Vulkans Laki (1783) verhindert. Infolge des Ausstoßes von mehreren hundert Millionen Tonnen Staub und Gas in die Atmosphäre wurden Sonnenstrahlen ins Weltall rückreflektiert, und die Erdoberfläche kühlte sich ab.

Im Gegensatz zu den Verfechtern der CO_2-Temperaturerhöhungs-Theorie führen anerkannte Wissenschaftler die Temperaturerhöhungen auf der Erde, wie sie zurzeit festgestellt werden, auf Änderungen der Aktivität der Sonne und ihren Einfluss auf die Wolkenbildung zurück. Die Sonnenaktivität des zwanzigsten Jahrhunderts war in den vergangenen tausend Jahren nicht so stark wie heute.

Das IPCC (Intergovernmental Penal on Climate Change) verkündete im April 2007, dass ein Temperaturanstieg auf der Erde mit 90prozentiger Sicherheit durch Menschen gemacht ist. Hierbei soll das CO_2, ein farb- und geruchsloses Gas, das sich in Wasser gut löst, die wesentliche Rolle spielen. Bei 20 °C nimmt ein Kubikmeter Wasser 0,5 g CO_2 auf, bei 0 °C jedoch 1 g. Erwärmt sich also das Wasser, so gibt es CO_2 an die Erdatmosphäre ab. Kühlt es sich ab, nimmt es wiederum CO_2 auf. Da die Erde zu ca. zwei Dritteln von Wasser bedeckt ist, werden bereits bei einer geringen Erwärmung des Wassers große Mengen an CO_2 freigesetzt. Der umgekehrte Weg ist wegen der Sättigungsgrenzen nicht möglich.

Der Anstieg des CO_2 in der Atmosphäre kommt im Wesentlichen Pflanzen zugute, da diese der größte CO_2-Senker sind. Wenn es auf der Erde wärmer wird und in der Atmosphäre ein höherer CO_2-Pegel besteht, beschleunigt sich das Wachstum, und ihr Einfluss übersteigt den des Wassers. Festzustellen ist, dass CO_2 nur einen Anteil von wenigen Prozent haben soll, während Wasserdampf das Klimagas um ein Vielfaches übertrifft. Wolken, gebildet durch Wasserdampf, beeinflussen die Temperaturen stärker als üblicherweise bekannt, wobei das IPCC selbst feststellt, dass eine Verdoppelung des CO_2-Gehalts einen Temperaturanstieg von ca. 0,7 °C zur Folge hat.

Zurzeit leben auf der Erde ca. 6.7 Milliarden Menschen. Sie erzeugen eine CO_2-Emission per anno von ca. 2,45 Milliarden Tonnen. Alle Autos dieser Welt emittieren im gleichen Zeitraum ca. 2,1 Milliarden Tonnen CO_2. Insgesamt fügen alle durch menschliches Handeln erzeugten CO_2-Emissionen nur 1 – 4 Prozent zu den natürlichen CO_2-Emissionen hinzu, was mit hoher Wahrscheinlichkeit keine Klimakatastrophe auslöst.

typically known today. Even the IPCC recognizes the fact that a doubling of CO_2 concentration only raises the temperature around 0.7 °C.

The Earth's population is currently about 6.7 billion, and we produce roughly 2.45 billion tons of CO_2 emissions annually. Automobiles alone currently emit 2.1 billion tons of CO_2 emissions per year. However, all human activity combined now contributes only 1 – 4 % to the natural carbon emission – an increase that in all probability will not cause a climatic collapse.

1.1 The greenhouse effect: in detail

The Sun and the Earth's atmosphere are the two main factors related to the development of the greenhouse effect, as seen in **Figure 1**. Around 70 % of the rays of the Sun penetrate the Earth's atmosphere, with the remainder being reflected back into space. The Sun's energy is absorbed by the oceans and land masses and is later returned as long-wave infrared radiation back to the atmosphere. The captured heat energy in the atmosphere is absorbed by gases and water vapor and radiated back as well. This process is typically described as the greenhouse effect, without which the Earth's temperature would be around 33 K lower than it is today. An increase of greenhouse gas concentration and water vapor causes an increase in global temperature.

The greenhouse effect observed on Earth exists to some degree on all planets in our solar system. On Venus, high concentrations of carbon dioxide let surface temperatures rise to over 400 °C. On Earth, on the other hand, greenhouse gas concentrations are in a constant flux that is influenced by not only plants but also humans, animals, and the oceans. Thus, our temperatures are balanced to comfortable levels.

1.1 Der Treibhauseffekt

Die Sonne und die Erdatmosphäre sind die Hauptfaktoren für die Entstehung des Treibhauseffekts, **Bild 1**. Ca. 70 Prozent der einfallenden Sonnenstrahlung dringen in die Erdatmosphäre ein, die verbleibenden 30 Prozent werden ins Weltall reflektiert. Die Sonnenenergie, welche die Atmosphäre durchdringt, wird von den Ozeanen und Landmassen aufgenommen und schließlich in Form von Wärme (langwellige Infrarotstrahlung) erneut in die Atmosphäre abgegeben. Die in der Erdatmosphäre eingeschlossene Wärme wird von Gasen und Wasserdampf aufgenommen und wiederum abgestrahlt. Dieser Prozess wird als der so genannte Treibhauseffekt bezeichnet. Ohne diesen Effekt wäre die Temperatur an der Erdoberfläche heute um ca. 33 K niedriger. Mit zunehmender Konzentration der Treibhausgase sowie der Wasserdampfmengen nimmt die Erwärmung auf der Erde zu.

Der auf der Erde bestehende Treibhauseffekt besteht auf annähernd allen Planeten unseres Sonnensystems. Auf dem Planeten Venus lassen hohe Kohlendioxidkonzentrationen die Oberflächentemperaturen auf über 400 °C ansteigen. Da sich auf der Erde die Treibhausgase in einem ständigen Kreislauf befinden, an dem Pflanzen, Menschen und Tiere wie auch Meere beteiligt sind, hat sich ein angenehmes Klima eingestellt.

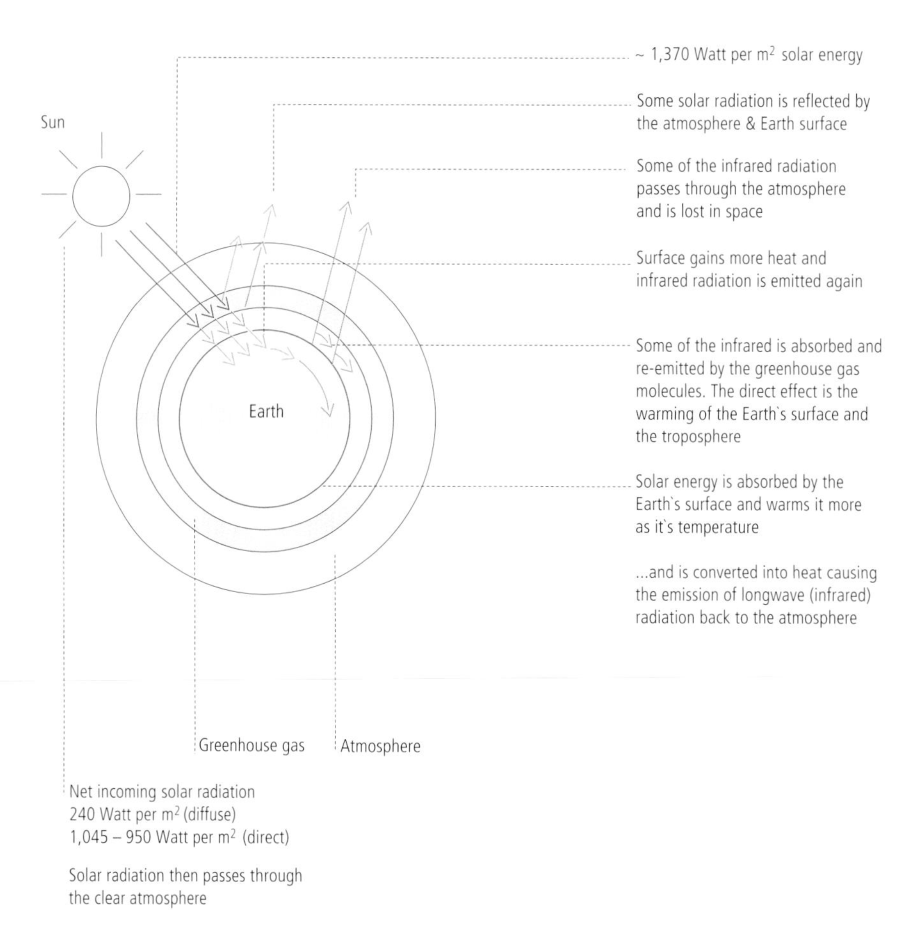

According to the reports of the IPCC, the temperature increase on Earth is mainly a result of increases in greenhouse gas levels. The current rise of greenhouse gas concentrations in the atmosphere, mainly carbon dioxide and methane, are the result of human activities such as deforestation, the burning of fossil fuels, and agriculture. The greenhouse gases emitted by chimneys, car exhausts, dump pits, and agro-industrial complexes play a role in the processes determining the Earth's temperature. Without a doubt, human activity has caused an imbalance in such processes. There is also no dispute about the fact that there is a change of the atmospheric composition since the industrial revolution. Human activity, in general, is seen as being responsible for the change in surface temperature.

Greater surface temperatures cause the melting of glaciers, the retreat of snow and ice caps, and an increase in sea water levels. At an average surface temperature change of 0.6 K (with a margin of error of ±0.2 K) in the 20th century the levels of the oceans increase by 1 – 2 mm per year.

Gemäß den Darstellungen des IPCC wird der Temperaturanstieg auf der Erde primär festgemacht an einem deutlichen Anstieg der Treibhausgase. Die zurzeit steigende Konzentration der Treibhausgase in der Atmosphäre (insbesondere Kohlendioxid und Methan) ist im Wesentlichen eine Folge menschlicher Aktivitäten, bei der vorwiegend die Abholzung von Wäldern, Verbrennung fossiler Brennstoffe und die Landwirtschaft eine große Rolle spielen. Die Treibhausgase (aus Schornsteinen, Auspuffanlagen, landwirtschaftlichen Betrieben, Mülldeponien usw.) spielen demnach eine Rolle in den natürlichen Abläufen, welche die Erdtemperatur bestimmen. Unzweifelhaft haben menschliche Eingriffe das bis dato bestehende Gleichgewicht gestört. Unbestreitbar ist auch die Veränderung des Erdklimas seit der industriellen Revolution. Die von Menschen verursachten Emissionen werden als die Ursache erachtet, die für höhere Oberflächentemperaturen sorgen.

Höhere Oberflächentemperaturen führen infolge schmelzender Gletscher und des Rückgangs von Eis- und Schneeflächen zu einem steigenden Meeresspiegel. Bei einer durchschnittlichen Temperaturerhöhung an der Erdoberfläche im 20. Jahrhundert um ca. 0,6 K (Fehlertoleranz ± 0,2 K) haben sich die Meeresspiegel um jährlich ca. 1 – 2 mm erhöht.

Figure 1
The greenhouse effect

Source: Greenpeace

Bild 1
Der Greenhouse-Effekt

Quelle: Greenpeace

"Factor Four" is a concept that was introduced in 1995 in a book of the same name written by Ernst Ulrich von Weizsäcker. In the publication, von Weizäcker simulates the effect of CO_2 concentrations for an energy consumption of 300 x 10^9 MWh/a. **Figure 2** shows the changes in CO_2 emissions since 1900 as well as CO_2 concentrations in the atmosphere. Parallel to these changes, the Earth's surface temperature is depicted. According to von Weizäcker – and the IPCC – a warm period can only be prevented by a lowering of the carbon dioxide concentration.

When it comes to lowering CO_2 concentration, the emissions of different nations on Earth play different roles, of course. **Figure 3**, which shows CO_2 emissions in tons per capita and per year, suggests that some individual countries need to play a more decisive role in the reduction of this gas than others.

Zum Zeitraum 1955 wurde von Ernst Ulrich von Weizsäcker im Buch „Faktor Vier" eine Simulation des CO_2-Effekts bei einem Energieverbrauch von 300 x 109 MWh/a dargestellt. **Bild 2** zeigt die Veränderung der CO_2-Emissionen seit 1900 sowie die CO_2-Konzentration in der Atmosphäre. Parallel hierzu dargestellt ist der Temperaturwechsel, d.h. der Anstieg der Temperatur an der Erdoberfläche. Gemäß E.U. von Weizsäcker und den letzten Aussagen des IPCC kann die Warmzeit nur dadurch aufgehalten werden, dass der CO_2-Pegel gesenkt wird.

Bei der Senkung des CO_2-Pegels spielen selbstverständlich die CO_2-Emissionen in einzelnen Ländern der Welt eine unterschiedliche Rolle. **Bild 3** zeigt die CO_2-Emissionen in Tonnen pro Kopf und Jahr und weit aus, dass einzelne Länder eine hervorragende Rolle bei der Senkung spielen müssen.

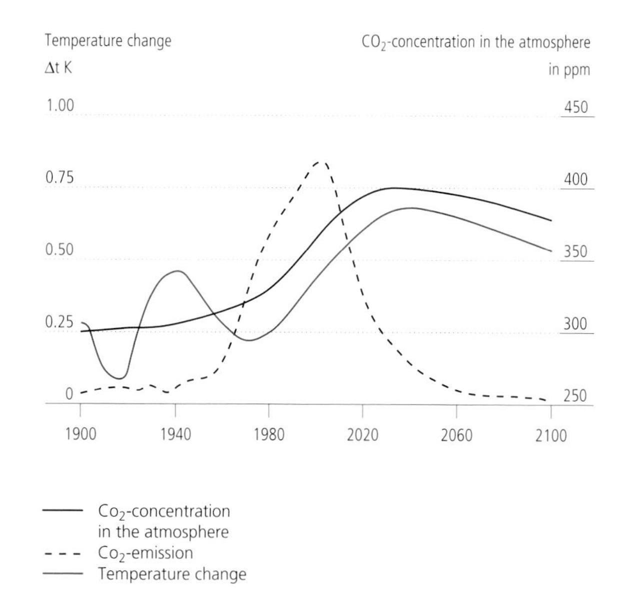

Figure 2
Simulation of the CO_2 effect for an energy consumption of 300 x10^9 MWh/a
Solar and nuclear power use, as of 1995

Source: E.U. v. Weizsäcker, Faktor Four

Bild 2
Simulationen des CO_2-Effektes bei einem Energieverbrauch von 300 x10^9 MWh/a
Solar- und Nuklearanwendung, Stand 1995

Quelle: E.U. v. Weizsäcker, Faktor Vier

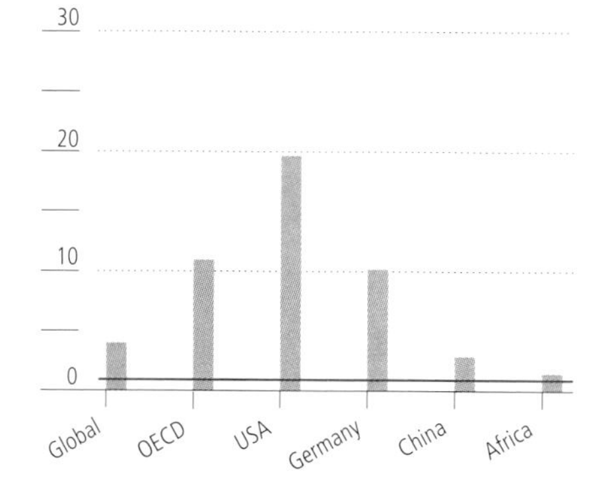

Figure 3
CO_2 emissions in tons per capita and year

Source: Energy Policy, 35/2007

Bild 3
CO_2 Emissionen in t pro Kopf und Jahr

Quelle: Energy Policy, 35/2007

1.2 CO_2 reduction, 2 K scenario

As a result of the findings of the IPCC, several European countries have agreed that a temperature increase of more than 2 K should not be tolerated, and a resulting CO_2 concentration should be limited to not more than 450 ppm. This, in turn, means that CO_2 emissions, according to the newest model simulations, should be decreased in 2050 to 10 Gt/a (Gigatons, 1 billion tons), which constitutes a reduction of today's projected emission levels by 60 %, or, as shown in **Figure 3**, to a level of one ton per capita per year. The development of global primary energy supply by means of primary energy, and under the assumption of a 2 K scenario until 2050, is shown in **Figure 4.1**. In order to achieve such a reduction of primary energy use, great advances in the efficiency of appliances and machines will need to be achieved, and a much greater use of renewable energy sources will have to be pursued. Then, as the diagram shows, fossil fuels such as gas, oil, and coal, as well as nuclear power, will play an increasingly minor role, with nuclear energy from 2030 onwards ceasing to supply any power at all. This hypothesis seems to be premature because it can be assumed that the current uranium reserves in several countries will be utilized before this energy generation method will be abandoned.

1.2 CO_2-Reduzierung, 2 K-Szenario

Basierend auf den Resultaten des IPCC hat eine Reihe europäischer Länder einen Konsens dahingehend gefunden, dass eine mittlere Temperaturerhöhung von 2 K nicht überschritten werden soll und hieraus resultierend eine CO_2-Konzentration unter 450 ppm liegen muss. Dies setzt wiederum voraus, dass der CO_2-Ausstoß nach den bestehenden Modellberechnungen auf 10 Gt/a (Gigatonnen) im Jahr 2050 reduziert werden muss oder sollte, was einer Reduktion vom derzeitigen, hochgerechneten Emissionslevel um ca. 60 Prozent entspricht. Dieser Anspruch liegt der Darstellung in Bild 3 zugrunde, die Begrenzung der CO_2-Emissionen pro Kopf und Jahr auf ca. 1 Tonne. Eine Entwicklung der globalen Primärenergieversorgung unter dem 2 K-Szenario ist in **Bild 4.1** bis zum Jahr 2050 dargestellt. Diese Darstellung weist aus, dass zur Erreichung der Reduzierung der Primärenergieverbräuche einmal eine deutliche Effizienzverbesserung Energie verbrauchender Geräte und Maschinen erfolgt. Weiterhin zeigt das Bild, dass in hohem Maß erneuerbare Energien zur Energiebereitstellung eingesetzt werden. Fossile Brennstoffe wie Erdgas, Erdöl, Kohle und Nuklearenergie spielen eine zunehmend geringere Rolle, wobei bei diesem Szenario die Nuklearenergie ab 2030 nicht mehr in Erscheinung tritt. Diese Annahme dürfte verfehlt sein, da zumindest damit zu rechnen ist, dass die noch derzeit bestehenden Uranvorräte in einzelnen Ländern aufgebraucht werden.

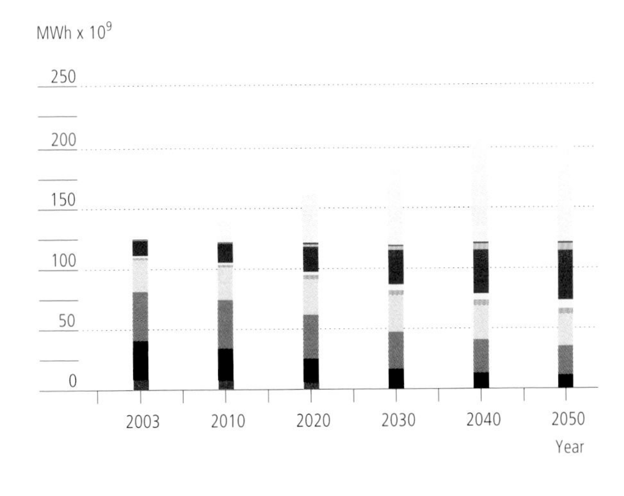

Figure 4.1
Development of the global primary energy supply according to the 2 K scenario.

Source: Elsevier, Energy Policy, International Journal 35/2007

Bild 4.1
Entwicklung der globalen Primärenergieversorgung im 2 K Szenario.

Quelle: Elsevier, Energy Policy, International Journal 35/2007

Figure 4.2 shows a reduction scenario for CO_2 emissions under the assumption introduced in **Figure 4.1**. As can be seen, the largest contributing factor in the reduction of carbon emissions will be the increased efficiency of systems. According to **Table 1**, the use of various energy fuels on the one hand, and the implementation of new technologies on the other (as shown in **Figure 5**), will both play important roles.

Bild 4.2 zeigt die Reduzierung der CO_2-Emissionen unter dem Szenario gemäß **Bild 4.1**. Wie das Bild ausweist, wird der größte Effekt durch die Verbesserung der Effizienz erzielt. Dabei eine Rolle spielen wird einerseits der Einsatz der verschiedensten Brennstoffe gemäß **Tabelle 1** und zum anderen der Einsatz verschiedenster Technologien, wie in **Bild 5** ausgewiesen.

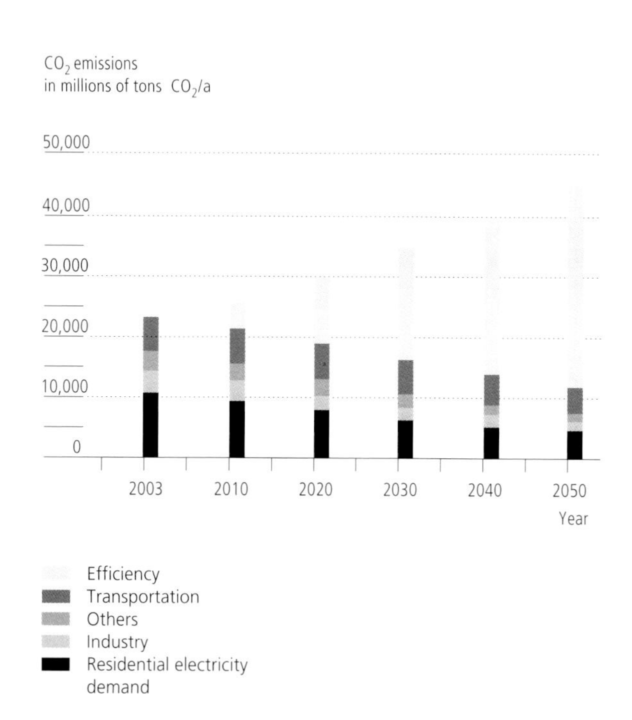

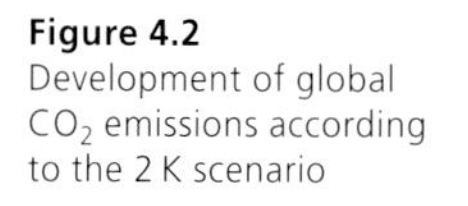

Figure 4.2
Development of global CO_2 emissions according to the 2 K scenario

Source: Elsevier, Energy Policy, International Journal 35/2007

Bild 4.2
Entwicklung der globalen CO_2-Emissionen unter dem 2 K Szenario

Quelle: Elsevier, Energy Policy, International Journal 35/2007

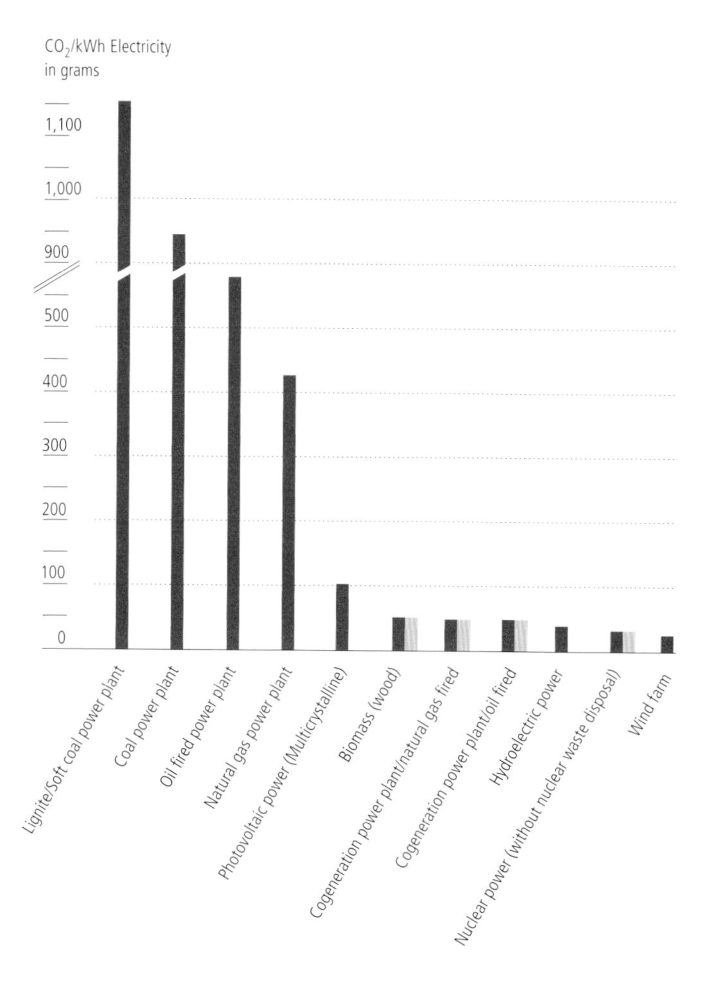

Figure 5
CO_2 emissions in grams per kilowatt hour produced electricity for various power plant technologies

Source: Der Spiegel, 12/2007

Bild 5
CO_2-Emissionen pro Kilowattstunde erzeugtem Strom, in Gramm bei verschiedenen Kraftwerkstechnologien

Quelle: Der Spiegel, 12/2007

		g CO_2/kWh$_{End}$
Fuel	Heating oil	303
	Natural gas	249
	Liquefied gas	263
	Coal	439
	Lignite (Brown coal)	452
	Wood scrap	35
	Fire wood	6
	Wood pellets	42
Electricity	Electricity mix	647
	Heat-electricity mix	932
	Heat-electricity coal	1036
	Photovoltaic, localized	234
	Wind power (Coastal wind farms and transmission)	19
Local and remote district heating	Remote district heating 70% combined-heat-power plant (CHP)	217
	Remote district heating 35% combined-heat-power plant (CHP)	312
	Remote district heating 0% combined-heat-power plant (CHP)	408
	Local district heating, 70% gas-fired combined-heat-power plant (CHP)	- 68
	Local district heating, 35% gas-fired combined-heat-power plant (CHP)	129
	Local district heating, 0% gas-fired combined-heat-power plant (CHP)	325

Distribution chain from generation of final energy to delivery in building including material supply for thermal energy utilities, without support energy in buildings
Local supply units with gas-fired CHP plus natural gas peak demand boiler
Remote district heating by coal-fired condensation heat and power plants plus oil-fired peak demand boiler

Table 1
CO_2 emissions for various energy sources and technologies

Source: Institut Wohnen und Umwelt GmbH (IWU), Darmstadt, Germany 09.01.2006/GEMIS 4.3.

		g CO_2/kWh$_{End}$
Brennstoffe	Heizöl EL	303
	Erdgas H	249
	Flüssiggas	263
	Steinkohle	439
	Braunkohle	452
	Holzhackschnitzel	35
	Brennholz	6
	Holz-Pellets	42
Strom	Strom-mix	647
	Heizstrom-mix	932
	Heizstrom-Steinkohle	1036
	PV-Strom (erzeugernah)	234
	Wind (Park Küste und Verteilung)	19
Fern- /Nahwärme	Fernwärme 70% KWK	217
	Fernwärme 35% KWK	312
	Fernwärme 0% KWK	408
	Nahwärme/Gas BHKW 70% KWK	- 68
	Nahwärme/Gas BHKW 35% KWK	129
	Nahwärme/Gas BHKW 0% KWK	325

Vorgelagerte Kette für die Endenergie bis Übergabe im Gebäude inkl. Materialaufwand für Wärmeerzeuger, ohne Hilfsenergie im Haus
Nahwärmeversorgung durch Erdgas-BHKW (= Anteil KWK) + Erdgas-Spitzenkessel
Fernwärmeversorgung durch Steinkohle-Kondensationskraftwerk (= Anteil KWK) + Heizöl-Spitzenkessel

Tabelle 1
CO_2-Emissionen verschiedener Energieträger und Technologien

Quelle: Institut Wohnen und Umwelt GmbH (IWU), Darmstadt, Deutschland 09.01.2006/GEMIS 4.3.

For Germany, **Figure 6.1** shows the aim of reduction of carbon emissions from 1990 onward, according to the so-called governmental "Leitszenario 2006". **Figure 6.2** shows the amounts of avoided emissions as a result of renewable energy use. The avoidance of CO_2 emissions is distributed across all fuel types, including oil (transportation), thermal, and electrical energy, and requires, before 2050, the accelerated use of renewable energy sources.

Whether reduced CO_2 emissions or the 2 K scenario will play the decisive role is of lesser importance for the authors of this book, who are not experts in the field of climate research. We are in no position to make a final judgment on whether, and to what extent, changes in climate are a result of changes in greenhouse gas emissions. What seems, on the other hand, to be of great significance is the fact that the global discussion of the 2 K scenario has led to a significant shift in current thinking in the direction of increased efficiencies, decreases in energy losses, and an accelerated use of renewable technologies. This will result in the very important circumstance of saving remaining fossil fuel reserves for future use. Again, we should remember the definition of sustainability put forth by the Brundtland commission: "Sustainable development is development that meets the needs of the present without compromising the ability of future generations to meet their own needs".

Beispielhaft für die Bundesrepublik Deutschland ist in **Bild 6.1** der angestrebte Rückgang der CO_2-Emissionen ab 1990 unter dem bekannten Leitszenario 2006 dargestellt. **Bild 6.2** weist vermiedene CO_2-Emissionen durch den Einsatz erneuerbarer Energien aus. Die Vermeidung von CO_2-Emissionen durch erneuerbare Energien verteilt sich auf die verschiedensten Ansprüche wie Kraftstoffe (Transport), Wärme und elektrische Energie und setzt voraus, dass tatsächlich bis zum Jahr 2050 in erheblichem Umfang erneuerbare Energien zum Einsatz kommen.

Ob nun der verringerte CO_2-Ausstoß oder das 2 K-Szenario wirklich eine Rolle spielt, ist für die Autoren dieses Buches nicht wesentlich, da sie keine Klimaforscher sind, die in der Lage wären zu beurteilen, ob tatsächlich die Treibhausgase für die Klimaveränderungen verantwortlich sind. Erheblich ist jedoch, dass durch die weltweite Diskussion um das 2 K-Szenario ein breites Umdenken eingesetzt hat, einerseits Wirkungsgrade deutlich zu verbessern, Verluste deutlich zu reduzieren und auf erneuerbare Energien zu setzen, um damit einen erheblichen Effekt dahingehend zu bewirken, fossile Brennstoffe auch noch in Zukunft zur Verfügung zu haben. Erinnern wir uns nochmals an die Nachhaltigkeit gemäß der Definition der Brundtland-Kommission: „... was den Bedürfnissen der heutigen Generation entspricht, ohne die Möglichkeit künftiger Generationen zu gefährden, ihre eigenen Bedürfnisse zu befriedigen."

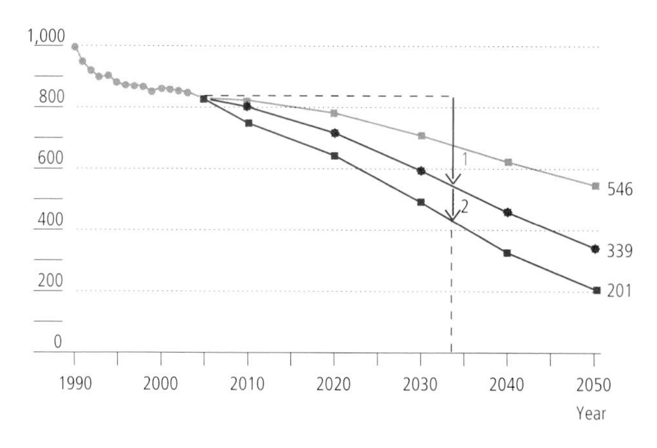

Figure 6.1
CO_2 emissions beginning since 1990 (temperature adjusted) Leitszenario 2006
Total contribution of renewable energies since 2005 and contribution in CO_2 reductions by improved efficiencies since 2005

Source: Bundesministerium für Umwelt, Naturschutz und Reaktorsicherheit, Germany Leitstudie 2007 "Ausbaustrategie Erneuerbare Energien"

Bild 6.1
CO_2- Emissionen ab 1990 (temperaturbereinigt), Leitszenario 2006 (Gesamtbeitrag der erneuerbaren Energien ab 2005 und Beitrag zusätzlicher Effizienz ab 2005 zur CO_2- Minderung)

Quelle: Bundesministerium für Umwelt, Naturschutz und Reaktorsicherheit, Leitstudie 2007 "Ausbaustrategie Erneuerbare Energien"

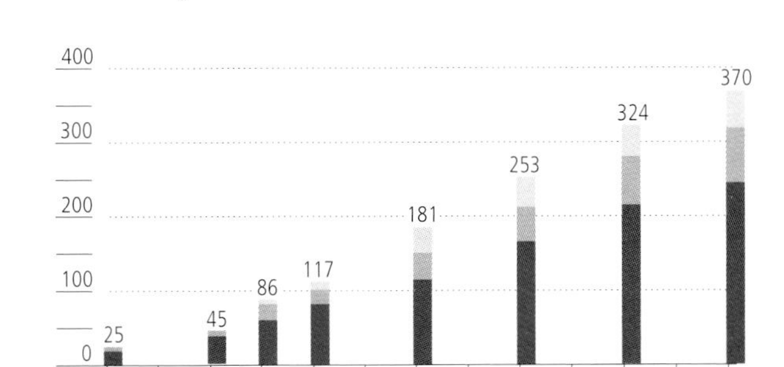

Figure 6.2
Avoided CO_2 emissions by renewable energy use

Source: Bundesministerium für Umwelt, Naturschutz und Reaktorsicherheit, Germany, Leitstudie 2007 "Ausbaustrategie Erneuerbare Energien"

Bild 6.2
Durch erneuerbare Energien vermiedene CO_2-Emissionen

Quelle: Bundesministerium für Umwelt, Naturschutz und Reaktorsicherheit, Leitstudie 2007 "Ausbaustrategie Erneuerbare Energien"

2 Fossil energy sources, renewable energy

2 Fossile Energieträger/erneuerbare Energien

The projection of global energy consumption up to the year 2100 is shown in **Figure 7**. The increase in consumption, on the one hand, is a result of the increase in world population from around 6.5 billion today to 9.5 – 10 billion; on the other hand, it is a result of the fact that those countries with the most rapid increase in population, such as China and India, also have the greatest incentive to elevate their living standards to those previously reserved for the so-called developed nations of Europe and North America.

Der global geschätzte Primärenergieverbrauch bis zum Jahr 2100 ist in **Bild 7** dargestellt. Der Anstieg des Primärenergieverbrauchs ergibt sich einmal durch die Bevölkerungszunahme von zurzeit ca. 6,5 Milliarden auf ca. 9,5 – 10 Milliarden Menschen auf der Erde und weiterhin dadurch, dass gerade die Nationen mit großer Bevölkerung (Indien, China usw.) ihre Länder auf einen Standard entwickeln, der vergleichbar dem der europäischen Länder und den Ländern Nordamerikas ist.

2.1 Prognosis of energy consumption

Figures 8.1 and 8.2 show in detail the course of energy consumption since 1959 and the projections for 2100.

From the data in **Figure 8.1**, we can see that in 2008, on a global basis, heavy reliance on the known fossil fuel resources is common. Coal, oil, gas, and nuclear energy play the paramount roles in supplying us with energy.

Photovoltaic technology, geothermal, solar thermal, the use of biomass, and the power generated by wind and hydroelectric power are just beginning to serve the global energy market.

Over the course of the decades to come, the percentage of renewable energy use will have to be increased significantly, and these energy sources will become the most prevalent for the target year of 2100. The primary sources will probably be energy generated from solar radiation, from wind, and as a result of biomass processing. In comparison with those renewable sources, the power generated from hydroelectric plants, together with the power from the waves of the ocean, or tidal power plants, will play only a minor role.

2.1 Prognostizierte Energieverbräuche

Bild 8.1 und 8.2 zeigen im Detail die Entwicklung der Energieverbräuche seit 1959 bis hochgerechnet 2100.

Aus **Bild 8.1** lässt sich für den Zeitraum 2008 gut ablesen, dass wir weltweit noch im Wesentlichen auf bekannte fossile Brennstoffe setzen. Kohle, Erdöl, Erdgas und Kernkraft spielen noch die wesentliche Rolle in der Energieerzeugung.

Photovoltaik, Geothermie, Solarthermie, Einsatz von Biomasse, Wind- und Wasserkraft beginnen gerade mit noch sehr kleinen Anteilen den Energiemarkt zu bedienen.

Der Anteil der erneuerbaren Energien nimmt in den nächsten Jahren zwangsläufig deutlich zu und muss im Jahr 2100 die wesentliche Rolle spielen. Dabei im Vordergrund steht die Erzeugung von Energie durch Sonne, Wind und Biomasse. Die Erzeugung von Energie durch Wasserkraft sowie Wellen- und Gezeitenkraftwerke spielt eher eine untergeordnete Rolle.

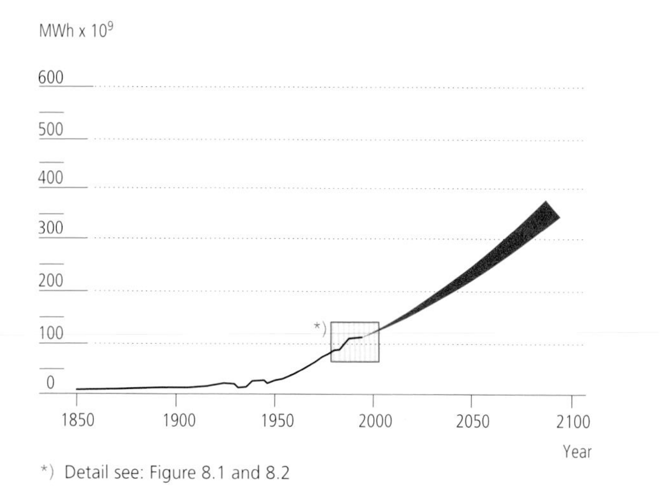

Figure 7
Estimated global primary energy consumption up to 2100

Bild 7
Globaler geschätzter Primärenergieverbrauch bis 2100

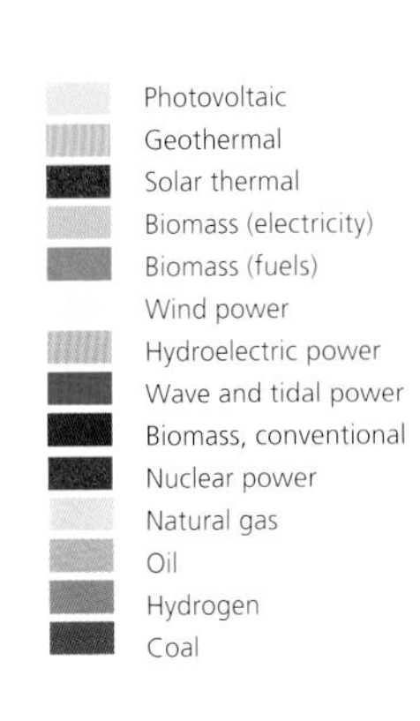

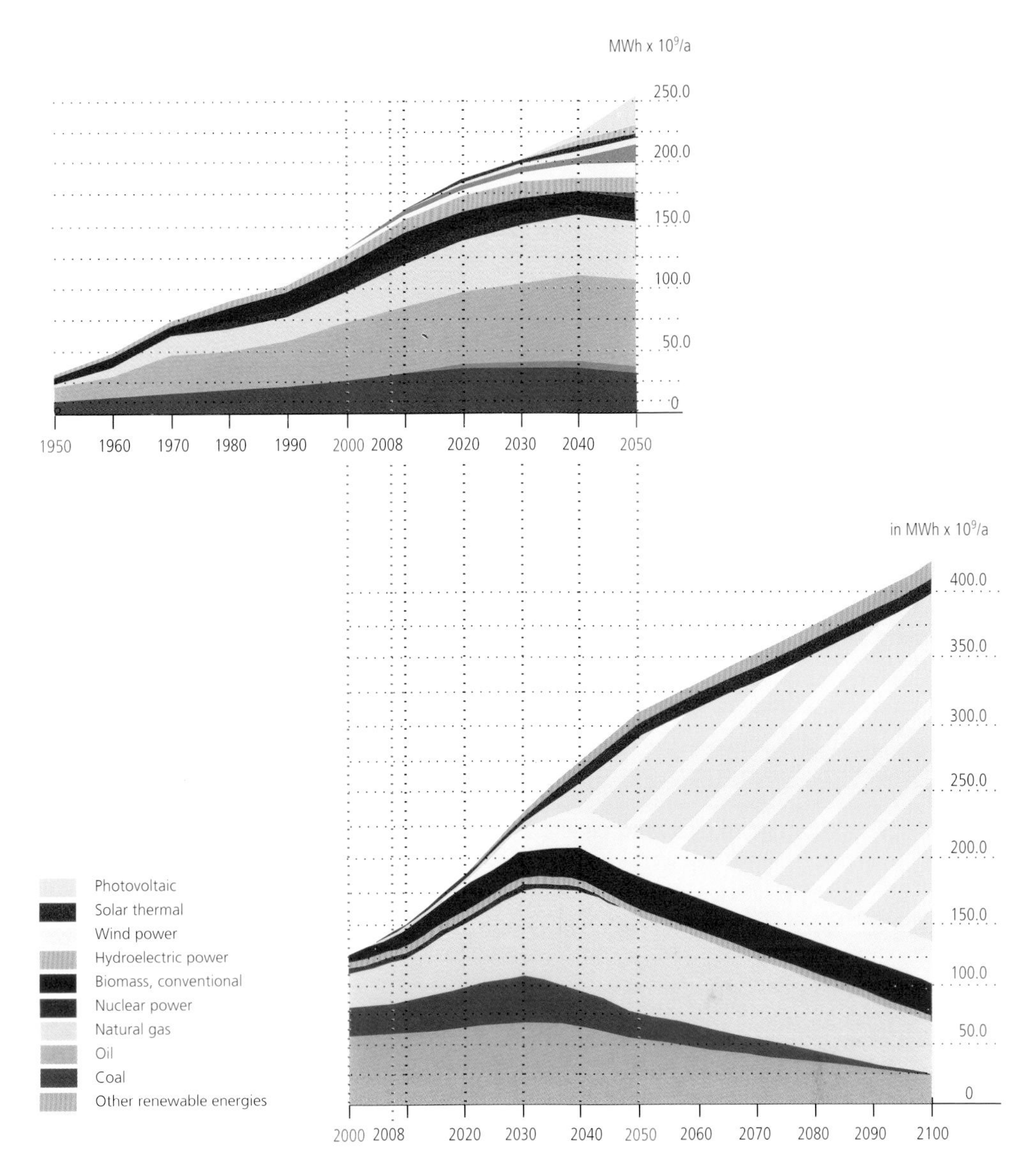

Figure 8.1
Estimated primary energy consumption and supply by various energy sources

Source: Spiegel-Spezial, 5/2006

Bild 8.1
Geschätzter Primärenergiebedarf und -deckung durch verschiedene Energieträger

Quelle: Spiegel-Spezial, 5/2006

Figure 8.2
Estimated primary energy consumption and supply by various energy sources

Source:
www.solarwirtschaft.de, 2007

Bild 8.2
Geschätzter Primärenergiebedarf und -deckung durch verschiedene Energieträger

Quelle:
www.solarwirtschaft.de, 2007

2.2 Primary energy offered by nature

Nature presents us with an abundant amount of energy (see **Figures 9.1, 9.2.**) The theoretically available energy of 1,524,240 x 10^9 MWh/a as a result of solar radiation is approximately 3,630 times larger than the prognosis for a future global primary energy demand of 420 x 10^9 MWh/a.

Even if, conservatively, we assume an average coefficient of performance of solar technology of only 15 %, the available solar energy exceeds consumption by a factor of 5.44.

In contrast, hydroelectric power – that is, energy generated with the help of water reservoirs – is only capable of meeting the energy demand to a degree of 10 %.

Much higher is still is the potential to generate energy out of biomass. If pursued resolutely, the biomass globally available at this moment would be sufficient to satisfy energy requirements, if high coefficients of performance in their conversion into usable energy are implemented.

Tidal and wave energy, if implemented rigorously, could also theoretically satisfy global energy requirements, and the same can be said for wind energy conversion, if we were determined to implement wind farms in all global regions with strong wind conditions.

Figure 9.2 shows a possible scenario for all countries. The assumption here is that, due to the limited availability of surface area, solar radiation is only being used on 10 % of land in countries with strong solar gains.

The utilization of hydropower will reach a possible maximum implementation stage in the next 50 to 100 years.

Since biomass power generation obviously has to be seen in conjunction with the nutritional requirements of an expanding world population, its role cannot be as prominent as previously conceived.

Tidal and wave power generation are in the infancy of their development, and so a reliable prognosis is difficult to make at this point. It is assumed that the degree to which energy can be generated out of the forces of ocean waves and tides will be far less than the diagram in **Figure 9.2** suggests.

2.2 Primärenergieangebot der Natur

Die Natur hält ein unendlich großes Angebot an Energien bereit, **Bilder 9.1/9.2**. Das theoretische Angebot der Solarenergie mit 1.524.240 x 10^9 MWh/a ist ca. 3.630 mal größer als der prognostizierte Primärenergiebedarf der Erde mit ca. 420 x 10^9 MWh/a.

Selbst unter der Berücksichtigung eines lediglich mittleren Wirkungsgrades von ca. 15 Prozent überschreitet das Angebot der Sonne den Verbrauch um das 5,44-fache.

Wasserkraft, d.h. Kraftwerke in Verbindung mit Stauseen, können maximal nur ca. 10 Prozent des Energiebedarfs decken.

Deutlich größer ist der Anteil der Energieerzeugung auf Basis Biomasse. Bei konsequenter Nutzung der weltweit verfügbaren Biomasse könnte bereits der Energiebedarf gedeckt werden, setzt man hohe Wirkungsgrade bei der Umwandlung voraus.

Wellen- und Meeresenergie könnte allein wiederum bei konsequenter Anwendung den Weltenergiebedarf decken. Gleiches gilt auch in Bezug auf die Windkraft, hier vorausgesetzt, dass in allen windstarken Regionen der Erde konsequent Windfarmen installiert werden.

In **Bild 9.2** ist ein denkbares Szenario für alle Länder der Welt dargestellt, wobei nunmehr davon ausgegangen wurde, dass lediglich 10 Prozent der Solarstrahlung aufgrund der Flächenverfügbarkeit in sonnenreichen Gegenden genutzt wird.

Die Nutzung von Wasserkraft (Hydropower) wird voraussichtlich in den nächsten 50 – 100 Jahren auf einen möglichen Endzustand ausgebaut.

Der Einsatz von Biomasse ist in Verbindung zu sehen mit der Beschaffung von Nahrungsmitteln einer wachsenden Weltbevölkerung und wird daher nicht die hervorragende Rolle spielen können wie denkbar.

Der Einsatz von Wellen- und Meeresenergie steht noch am Anfang, so dass eine Prognose hier sehr schwierig ist. Allgemein davon ausgegangen wird, dass der Umfang des Energiebeitrags durch Wellen- und Meeresenergie deutlich kleiner ist als in **Bild 9.2** dargestellt.

If we assume that approximately 2 % of all theoretically available wind energy can technically be converted into electric energy with the help of large wind power plants, the energy gained from wind would be the second largest renewable source, after solar energy.

As suggested at the beginning of this chapter, **Figure 9.2** illustrates just one possible scenario for any given country or region. Depending upon the specifics of locally available renewable resources, the individual scenarios per country or region may differ significantly.

Geht man davon aus, dass ca. 2 Prozent der verfügbaren (technisch nutzbaren) Windenergie tatsächlich zum Einsatz kommen, d.h. entsprechend viele, große Windenergieanlagen stationiert werden, könnte die Windkraft neben der Solarenergie die wesentliche Rolle spielen.

Wie bereits eingangs gesagt, ist das in **Bild 9.2** gezeigte Szenario nur ein denkbares Szenario, das für ein Land oder eine Region zutreffen könnte. Je nach Land und seinem spezifischen, natürlichen Angebot werden die Lösungsansätze bei der Nutzung erneuerbarer Energien ganz unterschiedlich ausfallen.

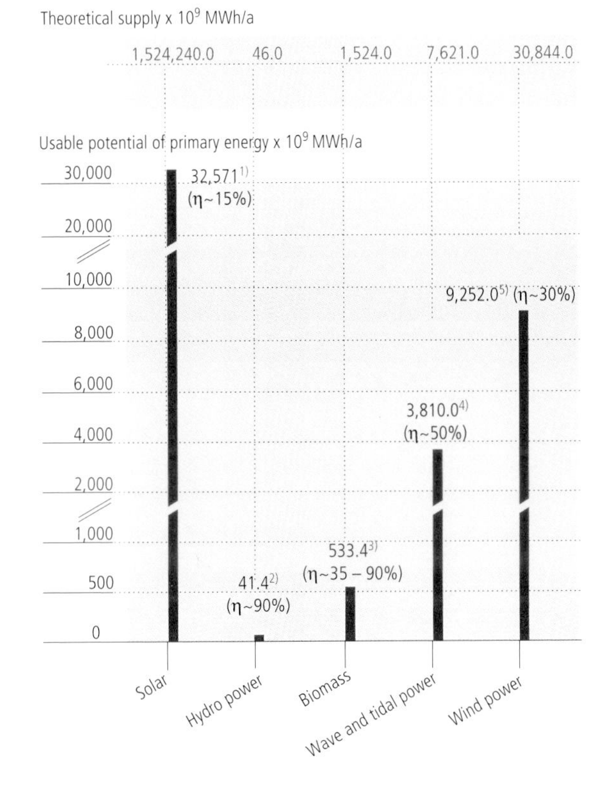

Bild 9.1
Theoretische Nutzung und Primärangebot erneuerbarer Energien

Quelle: Bundesverband Erneuerbarer Energien (BEE), Deutschland, Eurec. Agency/Eurosolar, 2007

Figure 9.1
Theoretical utilization and primary energy supply by renewable energies

Source: Bundesverband Erneuerbarer Energien (BEE), Deutschland, The European Renewable Energy Research Centres Agency (Eurec)/Eurosolar, 2007

% = maximum performance of technical plants
1) usable potential if land surface area is completely utilized, approximately 145 x 106 km2
2) maximum hydroelectric power utilization
3) complete surface area utilization for biomass generation
4) estimated value, plants currently under development
5) maximum development of appropriate wind farm locations h in %, plant coefficient of performance
η in % per plant and related coefficient of performance

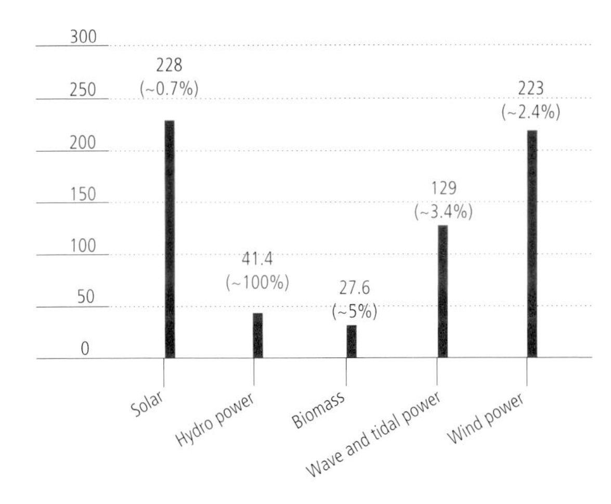

Figure 9.2
Potential global renewable energy supply scenario (all countries)

Depending on availability of natural resources at specific locations significant deviations may exist.

% = percentage of technically usable potential or area

Source: Bundesverband Erneuerbarer Energien (BEE), Deutschland, The European Renewable Energy Research Centres Agency Eurec/Eurosolar, 2007

Bild 9.2
Denkbares Scenario für alle Länder der Welt

(je nach Angebot der natürlichen Ressourcen je Standort ergeben sich massive Abweichungen)

% Angaben = prozentuale Anteile des techn. nutzbaren Potenzials oder Flächen

Quelle: Bundesverband Erneuerbarer Energien (BEE), Deutschland, Eurec. Agency/Eurosolar, 2007

2.3 Availability of fossil energy sources (oil, gas, coal, uranium)

In light of the assumption that the global energy economy will have to undergo a significant reconfiguration process, it remains interesting to see how much time remains for such changes. A look at currently available fossil fuel reserves is necessary.

Figure 10.1 shows both the currently known and assumed reserves of oil in the various regions of the globe.

The Middle East specifically, with countries such as Saudi-Arabia, Iran, and Iraq, harbors by far the greatest reserves, although it seems that they are in decline and are approaching the end of their life span.

Figure 10.2 shows the yearly resource exploration between 1930 and 2010. Juxtaposed is the oil production for the same period, which is equal to the approximate consumption, and the reserves that have been set up by individual economies. As seen in this diagram, the annual production of oil for the past 30 years has continuously declined, which contrasts with a significant increase in consumption over the same period. Currently, the demand is covered only by reserves that were identified prior to 1980 and that are currently being exploited. In order to extend known reserves further into the future, the replacement of fossil fuel energy by renewable energy is mandatory.

2.3 Verfügbarkeit fossiler Energieträger (Öl/Gas/Kohle/Uran)

Bezüglich des sich abzeichnenden Umbaus der weltweiten Energiewirtschaft ist von großem Interesse festzustellen, wie viel Zeit noch bleibt, um die entsprechenden Umstellungen durchzuführen. Hierzu ist es notwendig, sich ein Bild davon zu machen, welche fossilen Brennstoffe noch vorhanden sind.

Bild 10.1 zeigt die zurzeit bekannten und vermuteten Reserven von Öl in den verschiedensten Regionen der Welt.

Der Mittlere Osten – hier insbesondere die Länder Saudi Arabien, Iran und Irak – besitzen mit hohem Abstand die größten Ölreserven, die jedoch offensichtlich ihrem Ende entgegengehen.

In Bild 10.2 sind dargestellt die jährlichen Förderungen sowie Explorationen der Jahre 1930 – 2010. Dem gegenübergestellt ist die Ölproduktion, gleich zu setzen mit dem Circa-Verbrauch und den in einzelnen Ländern gebildeten Reserven. Wie die Darstellung zeigt, geht die jährliche Förderung von Öl seit ca. 30 Jahren kontinuierlich zurück, während der Verbrauch steigt. Insofern wird zurzeit der Verbrauch durch die Reserven gedeckt, die vor ca. 1980 bereits festgestellt wurden und zurzeit ausgebeutet werden. Um die Reserven weiter strecken zu können, ist es unerlässlich, einen zunehmenden Anteil an Primärenergie durch erneuerbare Energien zu ersetzen.

Figure 10.2
Global fossil fuel reserves

Source: SZ Wissen, 1/2005

Bild 10.2
Weltweite fossile Brennstoffreserven

Quelle: SZ Wissen, 1/2005

Yearly production
Explorations (partly based on assumptions)
Oil production, equal to approximate consumption

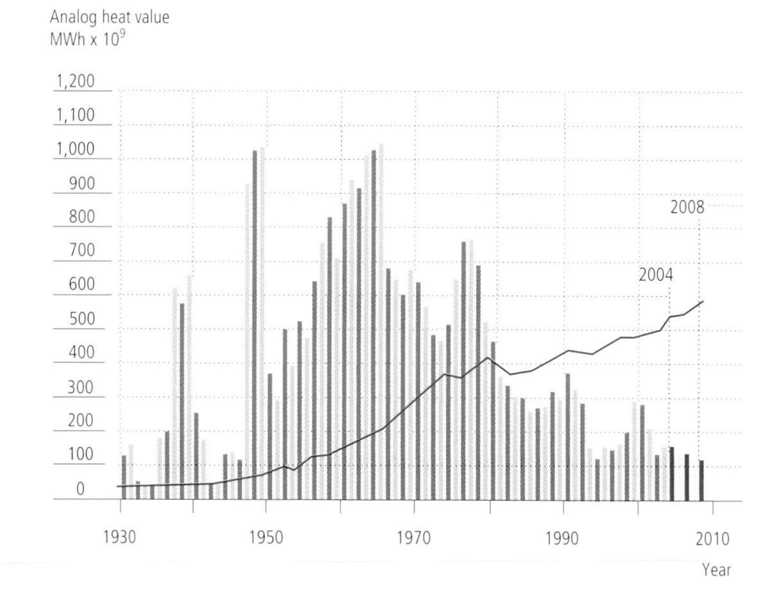

MWh x 10^9
1,200
1,100
1,000
900
800
700
600
500
400
300
200
100
0
62.1
89.7
178.25
170.2
226.55
1,163.8
Asia-Pacific
North America
Africa
South and Central America
Europe and Eurasia
Middle East

Figure 10.1
Currently known and assumed energy amount of oil reserves

Source: Oil & Gas Journal BP, USGS, IAEA

Bild 10.1
Zurzeit bekannte und vermutete Reserven von Öl (Energiemenge)

Quelle: Oil & Gas Journal BP, USGS, IAEA

Figure 11 shows the currently known and predicted global reserves of gas. Again, the regions of Europe, Eurasia, and especially Russia and the Middle East are of primary importance.

The currently known and predicted reserves of coal are distributed mainly across regions of Asia (China), Africa, Europe, and Eurasia, as well as the Middle East. Coal reserves are several times as great as those of oil and gas, which of course results in a longer period of availability and utilization (**Figure 12**.)

In **Bild 11** dargestellt sind die zurzeit bekannten und vermuteten Reserven von Gas, wobei hier die Regionen Europa und Eurasia (insbesondere Russland) und wiederum der Mittlere Osten im Vordergrund stehen.

Die zurzeit bekannten und vermuteten Reserven von Kohle verteilen sich im Wesentlichen auf die Region Asien (China), Afrika, Europa und Eurasia sowie wiederum auf den Mittleren Osten. Sie betragen ein Mehrfaches dessen an Öl und Gas, so dass eine größere Verfügbarkeit (Zeitraum) besteht, **Bild 12**.

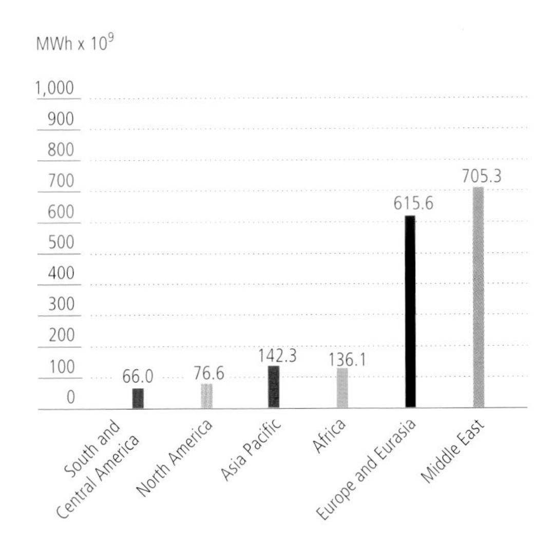

Figure 11
Currently known and assumed energy amount of natural gas reserves

Source: Oil & Gas Journal
BP, USGS, IAEA

Bild 11
Zurzeit bekannte und vermutete Reserven von Gas (Energiemenge)

Quelle: Oil & Gas Journal
BP, USGS, IAEA

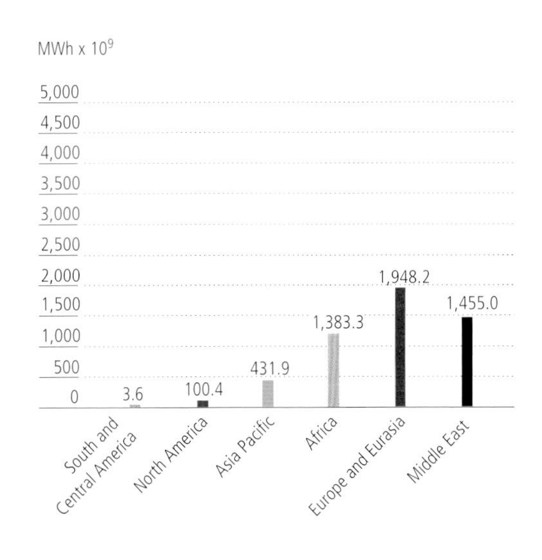

Figure 12
Currently known and assumed energy amount of coal reserves

Source: Oil & Gas Journal
BP, USGS, IAEA

Bild 12
Zurzeit bekannte und noch vermutete Reserven von Kohle (Energiemenge)

Quelle: Oil & Gas Journal
BP, USGS, IAEA

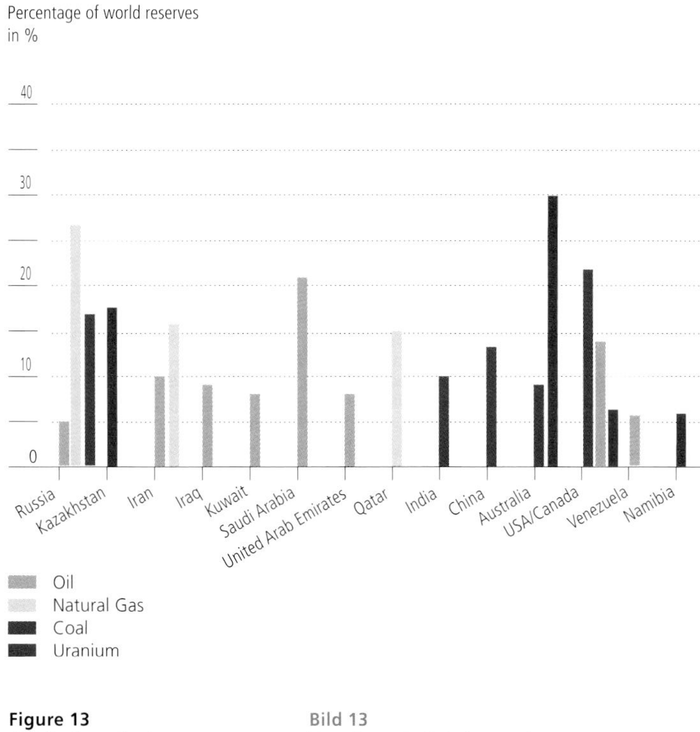

Figure 13
Distribution of primary energy resources

(Shown are countries with more than 5 % of the global resource. Smaller percentages of additional energy resources are omitted.)

Source: Oil & Gas Journal, BP, USGS, IAEA

Bild 13
Verteilung der Primärenergieträger

(Länder mit mehr als 5 % Anteil an den weltweiten Reserven des jeweiligen Rohstoffes. Geringe Anteile weiterer Energieträger nicht berücksichtigt)

Quellen: Oil & Gas Journal, BP, USGS, IAEA

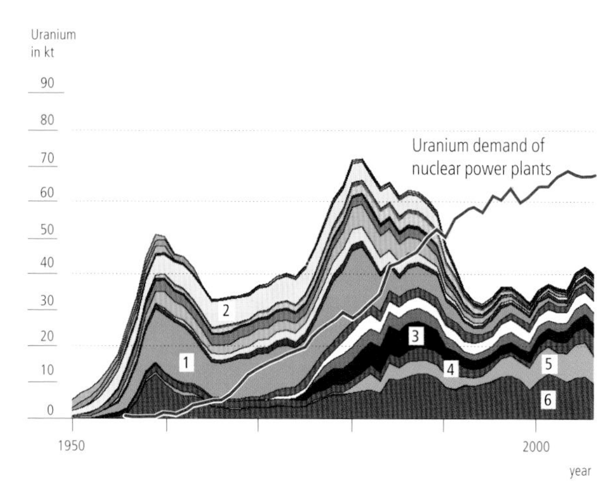

Figure 14
Uranium production for various nations in the years 1950-2006 and approximate uranium demand of nuclear power plants.
(There will be a supply gap of raw uranium which needs to be compensated by currently known reserves. The supply gap remains in effect and will cause uranium to be in short supply already starting in the year 2013. If by then no new additional mine explorations yield uranium the first reactors may be forced to terminate energy production.)

Source: Sonnenenergie, Heft 9/10, 2007
Author APSO Deutschland e.V.

Bild 14
Uranförderung der einzelnen Länder von 1950 – 2006 sowie Uranbedarf der Kernreaktoren.

(Die Versorgungslücke,die durch Lagerbestände ausgeglichen werden muss, wird keineswegs kleiner. Im Jahr 2013 wird ein großer Teil der Vorräte nicht mehr verfügbar sein. Wenn bis dahin neue Minen nicht zusätzliches Uran auf den Markt bringen können, werden vermutlich ab 2013 die ersten Reaktoren aus Brennstoffmangel abgeschaltet werden müssen.)

Quelle: Sonnenenergie, Heft 9/10, 2007
Autor APSO Deutschland e.V.

Figure 13 presents the distribution of primary energy sources among countries that have a greater than 5 % share of globally available raw energy. In principle, these are the 14 countries of the world that possess the lion's share of global resources. Just identifying which countries are on this roster makes clear what consequences future distribution wars might entail.

Up to a few years ago, it was assumed that nuclear energy generated from uranium would play a major role in power generation. However, as already noted for oil, the reserves of uranium are nearing their end, and many of today's nuclear power plants are being operated by reserves that were discovered and exploited before 1990 (**Figure 14**).

The amount of theoretically available fossil resources is a result of consumption predictions and the known and assumed reserves.

With the help of **Table 2**, we can identify the projected consumption until 2100 and the assumed reserves; it is intentional that two sources are used here.

In **Bild 13** dargestellt ist die Verteilung der Primärenergieträger auf die Länder, die einen höheren Anteil an den weltweiten Reserven des jeweiligen Rohstoffs besitzen (>5 %). Im Wesentlichen 14 Länder der Erde besitzen den Löwenanteil der Primärenergieträger, und aus dieser Darstellung lässt sich bereits entnehmen, um welche Länder es sich handelt, wenn es um den Verteilungskampf fossiler Brennstoffe geht.

Bis vor weinigen Jahren ist man noch davon ausgegangen, dass Uran (Kernenergie) für Atomkraftwerke eine besondere Rolle spielen wird. Wie beim Erdöl lässt sich gemäß **Bild 14** wiederum gut ablesen, dass die Uranreserven ihrem Ende entgegengehen und zum heutigen Zeitpunkt viele Atomkraftwerke aus den Reserven bedient werden, die vor 1990 entdeckt und erschlossen wurden.

Die theoretische Verfügbarkeit der fossilen Brennstoffe ergibt sich einerseits aus den geschätzten Verbräuchen und andererseits aus geschätzten und bekannten Reserven.

According to data by the German solar power industry (which, quite naturally, has its own business interest in such predictions), the availability of gas stretches into the future for another 29 – 53 years, oil for another 32 – 37 years, and coal for a period of about 330 – 410 years.

In contrast, according to BP's Statistical Review of World Energy (2007), gas will still be available for another 66 years, oil for approximately 45 years, and coal for 180 years. The percentage of uranium, as shown in the case of the other energy sources in 10^9 MWh, is marginal, but the reprocessing of nuclear elements will play a role in extending the availability of nuclear power further, and a period of 30 – 40 years of uranium availability can be assumed.

In **Tabelle 2** sind die prognostizierten Verbräuche bis 2100 und die geschätzten Reserven gegenübergestellt, wobei hier bewusst zwei Quellen zur Betrachtung herangezogen wurden.

Gemäß den Darstellungen der deutschen Solarwirtschaft mit ihren eigenen Interessen ergibt sich eine Verfügbarkeit für Gas von 29 – 53 Jahren, Öl für ca. 37 – 32 Jahre und Kohle für ca. 330 – 410 Jahre.

Ermittelt man die Verfügbarkeit in Jahren gemäß den Statistiken der BP, „Statistical Review of World Energy" (Zeitraum 2007), so ergeben sich Verfügbarkeiten in Gas von ca. 66 Jahren, Öl von ca. 45 Jahren und Kohle von ca. 180 Jahren. Der Anteil an Uran, an-gegeben wie bei den anderen Primärenergieträgern in 10^9 MWh, nimmt dabei eine sehr kleine Rolle ein, wobei hier jedoch die Wiederaufbereitung von Kernbrennstoffen insofern eine Rolle spielt, als dadurch der verfügbare Zeitraum länger wird und ebenfalls ca. 30 – 40 Jahre betragen wird.

Consumption estimate ($x10^9$ MWh/10a)	**Natural gas**	**Coal**	**Oil**	**Uranium**
2000 – 2010	350	250	600	
– 2020	470	300	620	
– 2030	650	350	680	
– 2040	780	340	700	
– 2050	850	250	610	
– 2060	800	190	550	
– 2070	720	150	450	
– 2080	650	120	400	
– 2090	580	50	350	
– 2100	500	10	270	
Total consumption Σ_{100a} ($x10^9$ MWh/100a)	**6,350**	**2,010**	**5,230**	
Average Σ_{1a} ($x10^9$ MWh/a)				
(www.Solarwirtschaft.de)	**63.50**	**20.10**	**52.30**	
(BP Statistical Review)	**26.2***	**29.4***	**41.8***	
Estimated reserves ($x10^9$ MWh)				
(www.Solarwirtschaft.de)	**ca. 1,839**	**ca. 8,279**	**ca. 1,925**	
(BP Statistical Review)	**ca. 1,741.9***	**ca. 5,322.4***	**ca. 1,890.6***	**ca. 68.0***
Availability in years under average consumption, 2000 – 2100				
(www.Solarwirtschaft.de)	**ca. 29 a**	**ca. 412 a**	**ca. 36,8 a**	
(BP Statistical Review)	**ca. 66.4 a***	**ca. 181 a***	**ca. 45.2 a***	

Geschätzte Verbräuche ($x10^9$ MWh/10a)	**Gas**	**Kohle**	**Öl**	**Uran**
2000 – 2010	350	250	600	
– 2020	470	300	620	
– 2030	650	350	680	
– 2040	780	340	700	
– 2050	850	250	610	
– 2060	800	190	550	
– 2070	720	150	450	
– 2080	650	120	400	
– 2090	580	50	350	
– 2100	500	10	270	
Verbrauchssumme, Σ_{100a} ($x10^9$ MWh/100a)	**6.350**	**2.010**	**5.230**	
Durchschnittl., Σ_{1a} ($x10^9$ MWh/a)				
(www.Solarwirtschaft.de)	**63,50**	**20,10**	**52,30**	
(BP Statistical Review)	**26,2***	**29,4***	**41,8***	
Geschätzte Reserven ($x10^9$ MWh)				
(www.Solarwirtschaft.de)	**ca. 1.839**	**ca. 8.279**	**ca. 1.925**	
(BP Statistical Review)	**ca. 1.741,9***	**ca. 5.322,4***	**ca. 1.890,6***	**ca. 68,0***
Verfügbarkeit in Jahren bei ø Verbrauch, 2000 – 2100				
(www.Solarwirtschaft.de)	**ca. 29 a**	**ca. 412 a**	**ca. 36,8 a**	
(BP Statistical Review)	**ca. 66,4 a***	**ca. 181 a***	**ca. 45,2 a***	

Table 2
Reserves and consumption of fossil fuels

Source:
www.solarwirtschaft.de
*) Source: BP Statistical Review of World Energy, June 2007

Tabelle 2
Reserven und Verbräuche von fossilen Rohstoffen

Quelle:
www.solarwirtschaft.de
*) Quelle: BP Statistical Review of World Energy, June 2007

3 Energy cost

3 Energiekosten

3.1 Changes in cost of fossil fuels

Research by the International Monetary Fund (IMF) has documented a direct link between fluctuations in economic performance and access to energy resources by the world's economies. According to these studies, energy consumption increases with a change from an agrarian economy to a more industrialized society and then to a consumer culture. Or, to put it differently, an increase in gross domestic product is achieved by a constant rate of energy consumption.

Production of energy to satisfy the current and future demands of our societies will have to change because it is indisputable that our natural resources are approaching the end of their life span.

In his remarkable book "Twilight in the Desert: The Coming Saudi Oil Shock and the World Economy," the author Matthew R. Simmons presents not only the notion that world oil reserves are already on a steady decline but also states that, as a result of the depletion, we will, in the near future, have to live with a price per barrel of oil of around U.S.$200 – 250. If true, this will most certainly lead to significant warping of the entire world economy. In a striking interview, the author details clearly that worldwide the easily accessible, exploitable oil has been exhausted nearly completely. Governmental leaders around the globe, however, have yet to respond resolutely to this condition. According to internal studies of Aramco, Saudi Arabia as early as the 1980s had to lower its exploration quota in order to protect the oil fields and prevent them from being destroyed prematurely. The expected massive cost increases for oil (**Figure 15**) are a result of the fact that it will become increasingly more difficult to exploit oil depots such as oil sands or high-sulfur oil. Sophisticated and time-consuming methods of extraction will need to be employed, which, consequently, will increase production cost and the price of the commodity. Simmons also shows that oil production in Russia is currently declining and that Mexico's and China's reserves are close to collapse. The production of offshore oil from fields in the North Sea has also sunk by 25 % since 1999.

"The next massive oil shock can hit us any day."

Since a large percentage of oil is used worldwide for transportation alone, corrections of the current way of consumption must first start there. For the U.S., especially, this would necessitate a complete reconfiguration of the work environment and the habits of consumption. Today's technology renders it unnecessary that hundreds of millions of people drive their individual automobiles to work on a daily basis, and

3.1 Kostenveränderungen fossiler Energieträger

Forschungen des Internationalen Währungsfonds (IWF) haben gezeigt, dass es eine direkte Verbindung zwischen wirtschaftlichen Schwankungen und dem Zugang zu Energien gibt. Dabei hat sich herausgestellt, dass durch die Veränderung eines Landes oder einer Region, von einer Agrarwirtschaft über die Industrialisierung zur Konsumgesellschaft, der Energieverbrauch gleichermaßen steigt, d.h. das Bruttoinlandsprodukt steigt bei gleichem Energieverbrauch.

Die Produktion von Energie für die heutige Gesellschaft muss und wird sich ändern, da unbestreitbar ist, dass die fossilen Ressourcen ihrem Ende entgegengehen.

Matthew R. Simmons hat in seinem bemerkenswerten Buch „Wenn der Wüste das Öl ausgeht" über den Rückgang der Ölreserven in Saudi-Arabien berichtet und zudem in einem Interview „Es kann uns täglich treffen" sehr markant beschrieben, dass nicht nur die Ölreserven dem Ende entgegengehen, sondern wir in den kommenden Jahren mit einem Ölpreis von ca. 200 – 250 USD pro Barrel rechnen müssen, was zu massiven Verwerfungen der Wirtschaft führen kann. In diesem Interview beschreibt er, dass das beste, leicht abbaubare Öl nahezu komplett verbraucht ist – jedoch offensichtlich die Regierungen der westlichen Industrieländer darauf bis heute nur bedingt reagieren. Nach Studien interner Aramco-Unterlagen konnte er unschwer feststellen, dass Saudi-Arabien bereits in den frühen 80er Jahren seine Förderquoten senken musste, um seine Ölfelder nicht vorzeitig zu zerstören. Die zu erwartenden massiven Preisschübe im Ölbereich, vergleiche **Bild 15**, ergeben sich daraus, dass in Zukunft vermehrt schwefelhaltiges Öl, Ölsande und andere, schwer abbaubare Produkte genutzt werden müssen und sich hierdurch die Produktionskosten zwangsläufig massiv erhöhen. Simmons stellt zudem fest, dass selbst in Russland die Ölproduktion sinkt, Mexiko und China annähernd am Rande eines Förderkollapses stehen und weiterhin die Produktion von Öl aus der Nordsee seit 1999 um ein Viertel gesunken ist.

„Der nächste, massive Ölschock kann uns jeden Tag treffen."

Da ein hoher Anteil des Öls weltweit im Straßenverkehr verbraucht wird, muss ein massives Gegensteuern gegen den Verbrauch im Wesentlichen an dieser Stelle beginnen. Gerade für die USA bedeutet dies eine notwendige, komplette Neugestaltung der Arbeitswelt und der Konsumgewohnheiten. Heutige Technologien machen es überflüssig, dass täglich hunderte Millionen

it is unnecessary that our food is transported thousands of miles only to be deep frozen, later thawed out, and then "cooked" in a microwave oven – only to be without taste at the end.

Where the transportation sector is concerned, oil consumption will be directly affected by drastically increased fuel costs at the pump, and highly efficient cars with engines such as the so-called "Three-liter car"*) will be in high demand indeed.

Figure 16 shows the projections for the cost of oil, gas, and coal for the coming decades. If we wish to maintain our current high standards of living in the industrialized countries at reasonable cost, countermeasures will have to be developed with the goal of lowering consumption. Since buildings in Europe and elsewhere typically use around 40 % of a country's entire energy consumption, important work lies ahead for all of us in the building design and construction industry.

Menschen mit dem Auto zur Arbeit fahren, und ebenso unnötig ist es, Lebensmittel tausende von Meilen zu transportieren, tiefzukühlen und nach dem Auftauen mit der Mikrowelle festzustellen: Es schmeckt nach nichts.

In Bezug auf den Ölverbrauch durch den Straßenverkehr wird sich sehr schnell eine Regelung dann ergeben, wenn die Benzinpreise auf ein Mehrfaches dessen steigen, was man zurzeit bezahlt. Das „Drei-Liter-Auto"*) wird im Falle entsprechender Rohölkostensteigerungen sofort nachgefragt.

Bild 16 zeigt die vermutlichen Preisentwicklungen bei Öl, Erdgas und Steinkohle in den nächsten Jahrzehnten. Will man den gehobenen Standard in Zukunft noch bezahlbar machen, müssen Gegenreaktionen schnellstens eingeleitet werden, die darin bestehen, die Verbräuche zu reduzieren. Da Gebäudestrukturen in Mitteleuropa mehr als 40 Prozent der Energien verbrauchen, tut sich hier ein weites Feld auf, an dem es zu arbeiten gilt.

Figure 15
Price of crude oil
(West Texas Intermediate WTI)

Source: Federal Reserve,
Spiegel Special, Nr. 5/2006

Bild 15
Rohölpreis
(West Texas Intermediate)

Quelle: Federal Reserve,
Spiegel Special, Nr. 5/2006

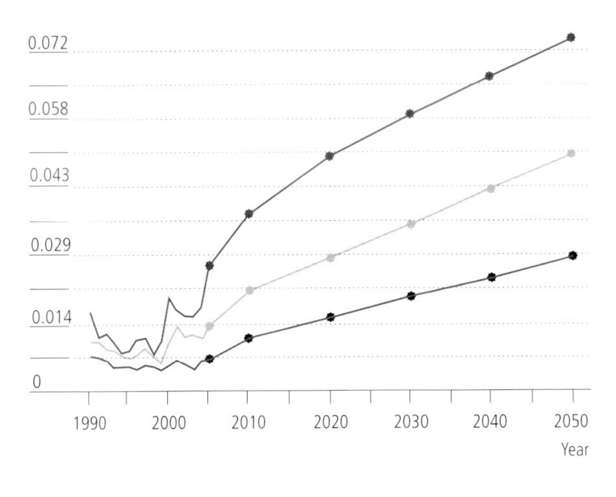

Figure 16
Estimate of future price of oil, natural gas and coal including added CO_2 tax.

Source: Bundesministerium für Umwelt, Naturschutz und Reaktorsicherheit, Germany. Leitstudie 2007 "Ausbaustrategie Erneuerbare Energien"

Bild 16
Schätzungen der zukünftigen Preise für Öl, Erdgas und Steinkohle (mit CO_2 Aufschlag)

Quelle: Bundesministerium für Umwelt, Naturschutz und Reaktorsicherheit, Leitstudie 2007 "Ausbaustrategie Erneuerbare Energien"

*) The term "Three-Liter-Car" was coined in the late 1990 s of the last century and describes a car emitting less than 90 g CO_2/km according to the EU regulation RL93/116/EU. This amount of emission is equal to a fuel consumption of approximately 70 miles per gallon of diesel fuel (US) or 61 miles per gallon (US) of unleaded.

*) Der Begriff „Drei-Liter-Auto" wird nach europäischem Steuerrecht (RL93/116/EG) seit 1996 mit einer Kohlendioxidemission von 90 g CO_2/km verbunden. Das entspricht einem Streckenverbrauch von etwa 3,4 l/100 km Diesel oder 3,8 l/100 km Benzin.

3.2 Cost tendencies for renewable energy sources

Without doubt, in the near future fossil fuels will need to be replaced by renewable energy sources. This consensus has led to an intensified search, particularly during the past decade, for ways to become independent of fossil fuel. Of interest is a view of how future cost predictions for electrical energy will develop. **Figure 17.1** shows the projected cost development for renewable, electricity-generating energy sources up to the year 2050.

In addition to the traditional generating costs for electricity, the generating cost of thermal energy will be of great interest because it represents a large percentage of total consumption. **Figure 17.2** shows the estimated cost for thermal solar collectors, the use of geothermal energy (shallow and deep geothermal), heat generation by biomass in small power plants, and, finally, large biomass district-heat power plants.

A comparison of the costs of fossil fuel sources with renewable energy sources is shown in **Figures 16, 17.1** and **17.2**. It can be seen that, at least for now, the first are still significantly cheaper than the latter.

Whether the prognosis above will have value in the foreseeable future depends, among other factors, upon trends in development and changes in the real consumption of primary energy.

According to a study by the international management consulting firm McKinsey & Company published in Febr. 2008, the annual global investment necessary to improve energy efficiency significantly needs to be about $170 billion (U.S. Dollars) or £ 87 billion British Pounds. A profit of $29 billion per year was estimated as a result of this investment, which is equal to a 17 % return. In the study, the necessary spending was distributed across the following areas:

- Residential approx. 40 billion $ p.a.
- Commercial approx. 22 billion $ p.a.
- Industrial approx. 83 billion $ p.a.
- Transportation approx. 25 billion $ p.a.

The necessary investment distribution across geographical regions would be as follows:

- USA approx. 38 billion $ p.a.
- China approx. 28 billion $ p.a.
- Remaining World approx. 104 billion $ p.a.

3.2. Kostenentwicklungen erneuerbarer Energien

Die fossilen Energieträger müssen in der Zukunft durch erneuerbare Energien ersetzt werden. Diese Erkenntnis hat dazu geführt, dass im letzten Jahrzehnt eine teilweise intensive Entwicklung eingesetzt hat, mehr natürliche Ressourcen zu nutzen und sich von den fossilen Energieträgern unabhängiger zu machen. Insofern ist von Interesse festzustellen, wie zukünftige Kostenentwicklungen zum Beispiel bei den Stromgestehungskosten aussehen werden. **Bild 17.1** zeigt die voraussichtliche Kostenentwicklung stromerzeugender, erneuerbarer Energietechnologien bis zum Jahr 2050.

Neben den Stromgestehungskosten sind die Wärmegestehungskosten ebenfalls von großem Interesse, da sie einen hohen Anteil am Gesamtenergieverbrauch ausmachen. **Bild 17.2** zeigt die voraussichtlichen Kostenentwicklungen von Kollektoren (solarthermische Wärmeerzeugung), Nutzung der Erdwärme (untiefe und tiefe Geothermie), Heizung durch Biomasse in Kleinanlage sowie Biomasse-Heizwerke zur Versorgung größerer städtischer Einheiten (Fernwärme).

Vergleicht man die zu erwartenden Übergangspreise zur Erzeugung von Energien in € /kWh gemäß **Bild 16** und **Bild 17.1/17.2**, so lässt sich feststellen, dass zumindest nach den Prognosen die Erzeugung der Energien durch fossile Brennstoffe noch deutlich billiger ist.

Ob sich die Prognosen, wie vor dargestellt, bewahrheiten werden, dürfte wohl davon abhängen, wie sich der Primärenergieverbrauch in Zukunft tatsächlich verändert.

Nach einer im Februar 2008 vorgestellten Studie der Fa. McKinsey müssen Investkosten zur deutlichen Steigerung der Energieeffizienz mit ca. 170 Mrd. (Milliarden) US-Dollar pro Jahr veranschlagt werden (ca. £ 87 Mrd. Pfund), wobei ein Gewinn von ca. 29 Mrd. $ pro Jahr entstehen würde (17 Prozent Verzinsung). Die Investitionen müssten eingesetzt werden für:

- Wohnen ca. 40 Mrd. $/a
- Handel, Gewerbe ca. 22 Mrd. $/a
- Industrie ca. 83 Mrd. $/a
- Transport ca. 25 Mrd. $/a

Die Investitionen würden sich in etwa verteilen auf Regionen:

- USA ca. 38 Mrd. $/a
- China ca. 28 Mrd. $/a
- Restliche Welt ca. 104 Mrd. $/a

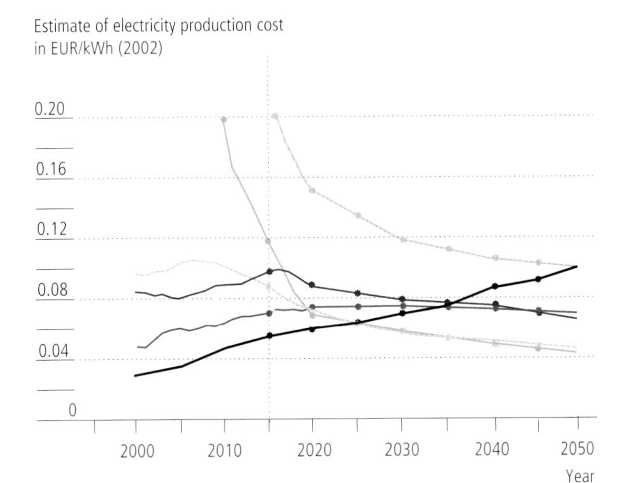

Figure 17.1
Future electricity production cost under the assumption of most advanced fossil fuel power plant technology and renewable energy generation until 2050.
(Averages of the total renewable energy mix according to the Leitszenario 2006)

(Value of currency as of 2002; interest rate 6 %/a; shown are average values of several individual technologies)

Source: Bundesministerium für Umwelt, Naturschutz und Reaktorsicherheit, Germany. Leitstudie 2007 "Ausbaustrategie Erneuerbare Energien"

Bild 17.1
Zukünftige Stromgestehungskosten bei Einsatz modernster fossiler Kraftwerke und der stromerzeugenden EE-Technologien bis 2050
(Mittelwerte des gesamten EE-Mixes im Leitszenario 2006)

(Geldwert 2002; realer Zinssatz 6 %/a; jeweils Mittelwerte mehrerer Einzeltechnologien)
EE – erneuerbare Energien

Quelle: Bundesministerium für Umwelt, Naturschutz und Reaktorsicherheit, Leitstudie 2007 "Ausbaustrategie Erneuerbare Energien"

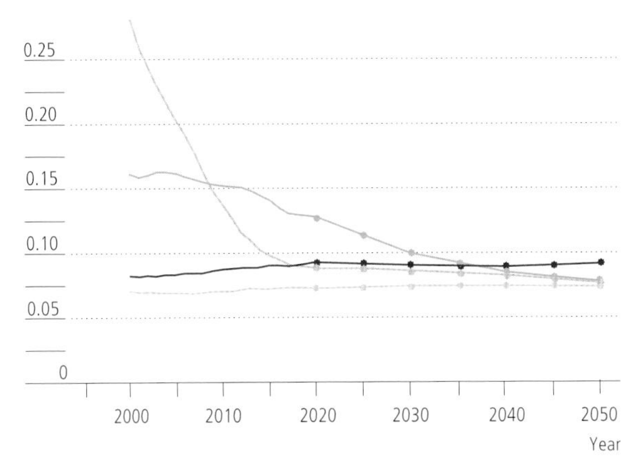

Figure 17.2
Estimate of thermal energy production cost for solar collectors, geothermal energy, individual biomass heating units and biomass district heating plants
(Average values of several individual technologies)

Source: Bundesministerium für Umwelt, Naturschutz und Reaktorsicherheit, Germany. Leitstudie 2007 "Ausbaustrategie Erneuerbare Energien"

Bild 17.2
Kostenentwicklung der Wämegestehungskosten durch Kollektoren, Erdwärme, Biomasse-Einzelheizungen und Biomasse-Heizwerken
(jeweils Mittelwerte mehrerer Einzeltechnologien)

Quelle: Bundesministerium für Umwelt, Naturschutz und Reaktorsicherheit, Leitstudie 2007 "Ausbaustrategie Erneuerbare Energien"

With such annual investments, the CO_2 level could be stabilized at around 550 ppm, resulting in a reduction of oil consumption of around 64 million barrels per day. Only less than 1 % of the countries GDP would have to be reserved for this goal.

The study by McKinsey researchers resulted in the interesting finding that the most energy-consuming facets of world energy consumption can be seen in the heavy industry of China on the one hand and the residential sector in the U.S. on the other. These sectors deserve high priority due to their very significant energy demands.

The study suggests that energy efficiency (lower consumption with increased coefficients of performance) is the best, cheapest, and quickest way to show positive energy-saving results. Since the study expects a return of 17 % annually, it seems that we should encourage large investors to take on the task without delay. The question therefore remains why we wait with the implementation. Postponing such intervention will only result in vastly greater expenses in the future.

Durch diese jährlichen Investitionen könnte nach der McKinsey-Studie der CO_2-Pegel bei 550 ppm gestoppt werden. Weniger als 1 Prozent der Bruttosozialprodukte der Länder (ca. 0,4 Prozent) würden zurzeit reichen, um Einsparungen von ca. 64 Millionen Barrel Öl pro Tag zu erzielen.

Als besonders energieträchtig bezüglich des Verbrauchs hat die Studie festgestellt, dass die Schwerindustrie Chinas und die Wohnhäuser Nordamerikas von besonderer Bedeutung sind. Beide Bereiche verbrauchen besonders viel Energie und sind somit die, in denen massive Verbesserungen schnellstens angegangen werden sollten.

Die Steigerung der Effizienz (geringerer Verbrauch bei deutlich höheren Wirkungsgraden) wäre gemäß Studie der schnellste, billigste und sinnvollste Weg, um zu einem Erfolg zu kommen. Bei einem Profit von 17 Prozent pro Jahr müssten die großen Investoren sofort starten – worauf warten wir, denn warten heißt, in Zukunft deutlich mehr zu bezahlen.

4 Energy consumption of selected world regions

(Excerpt from International Energy Agency Statistics 2004)

4 Energieverbräuche ausgewählter Regionen

(Auszug aus Statistik der International Energy Agency, 2004)

4.1 Region: Europe

On the basis of data provided by the International Energy Agency (IEA) in their 2004 report, **Figures 18 – 31** show the distribution of primary energy use, consumption of renewable energy, and the percentages of energy consumption. We should begin this review by mentioning that the data provided by this international agency sometimes differs significantly from the detailed data provided by the individual member countries (see Chapter 5, France, Germany, and Switzerland). In addition, IEA data in some instances do not cover the same time period as the data presented in **Chapter 5**.

Nevertheless, the graphs shown here allow for an overall trend analysis and prognosis, although their depicted absolute values may need to be adjusted further. Although the IEA claims to base their statistical data on information directly gathered from the member countrys' sources, it seems recommendable to collect data from those countries directly, if more detailed research is desired.

What is most unexpected is that the agency's data show the percentage of losses from the conversion of primary energy to final energy – the energy offered at the individual building – as only 5 %, although such losses are typically in the range of 35 %.

Germany is a country that uses a variety of fossil fuels, such as coal, oil, gas, and nuclear energy, to satisfy its primary energy needs. In addition to the traditional energy sources, 5.3 % of the total energy is recruited from renewable sources. The largest consumers of energy can are the industrial sector, dwellings, and the commercial sector (**Figures 18.1 – 18.3**).

Switzerland shows that two major energy sources play a deciding role in the country's energy mix, which are hydroelectric power and nuclear energy conversion. The structure of energy consumption is similar to the one in Germany (**Figures 19.1 – 19.3**).

4.1 Region Europa

Für eine Reihe europäischer Länder sind auf der Basis der Statistiken der IEA (International Energy Agency, 2004) die Primärenergieverbräuche, Verbräuche aus erneuerbaren Energien und Verteilung des Energieverbrauchs jeweils dargestellt, **Bilder 18 bis 31**. Dabei vorauszuschicken ist, dass die von der IEA erstellten Statistiken nicht mit den detailliert erstellten Statistiken einzelner Länder gemäß Kapitel 5 (Bundesrepublik Deutschland, Schweiz, Frankreich) übereinstimmen, sondern zum Teil erhebliche Abweichungen aufweisen. Hinzu kommt, dass die für das Jahr 2004 erhobenen Daten der IEA nicht den gleichen zeitlichen Rahmen treffen wie die Betrachtungen gemäß **Kapitel 5**.

Insofern zeigen die Bilddarstellungen und Auswertungen für die verschiedensten Regionen einerseits zwar eine Tendenz, jedoch andererseits unter Umständen zu korrigierende absolute Werte. Daher ist es sicher für jedes Land und für die Einzelbetrachtung verschiedener Länder vonnöten, die jeweils im Land erarbeiteten Statistiken neuesten Datums auszuwerten.

Auffällig ist insbesondere in den Datenerhebungen der IEA, dass die Energieverluste nur in etwa um 5 Prozent liegen, während die Differenzen zwischen Primärenergie und Endenergie (Energieangebot am energieverbrauchenden Gebäude) im Bereich um 35 Prozent anzusiedeln sind.

Die Bundesrepublik Deutschland ist ein Land, das seinen Primärenergieverbrauch aus den verschiedensten Quellen des fossilen Angebots deckt (Kohle, Erdöl, Erdgas, Kernenergie). Neben den traditionellen Energiequellen wird ein Anteil von ca. 5,3 Prozent durch erneuerbare Energien bereitgestellt. Die größten Verbraucher der benötigten Energien liegen im Bereich Industrie, Wohnen und Gewerbe, **Bilder 18.1 – 18.3**.

Die Schweiz zeigt zwei Bereiche der Energieversorgung, die besonders hervorzuheben sind. Es handelt sich hierbei einmal um den Einsatz von Kernenergie und zum zweiten um den Einsatz von Wasserkraft. Die Energieverbrauchsstruktur ist vergleichbar der der Bundesrepublik Deutschland, Bilder **19.1 – 19.3**.

Source: OECD/International Energy Agency (IEA), 2007

Figures
18.1 – 31.3
37.1 – 56.3
61.1 – 68.3

Quelle: OECD/International Energy Agency (IEA), 2007

Bilder
18.1 – 31.3
37.1 – 56.3
61.1 – 68.3

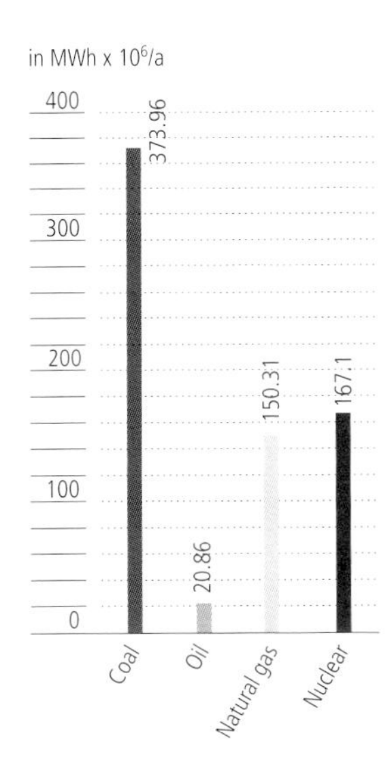

Figure 18.1
Primary energy consumption, Germany

Bild 18.1
Primärenergieverbrauch, Deutschland

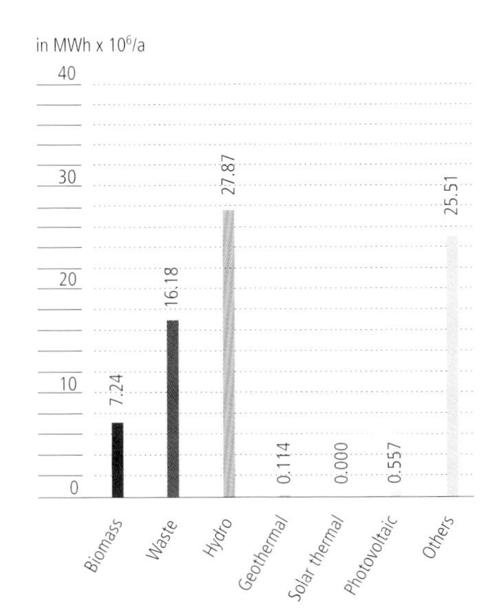

Figure 18.2
Renewable energy consumption, Germany

Bild 18.2
Verbrauch aus erneuerbaren Energien, Deutschland

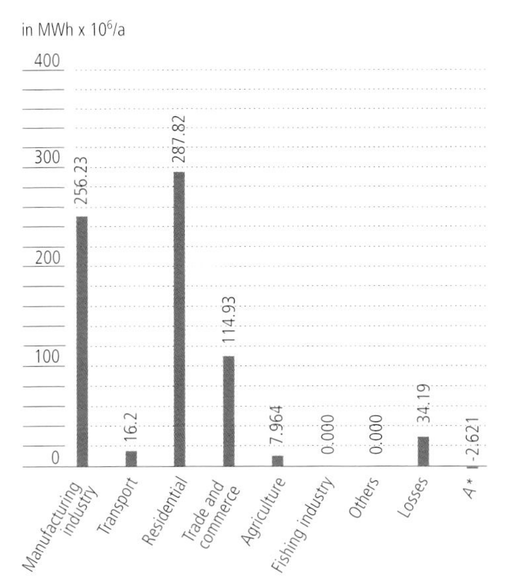

A* Electric energy import/export balance

Figure 18.3
Distribution of energy consumption, Germany

Bild 18.3
Verteilung des Energieverbrauchs, Deutschland

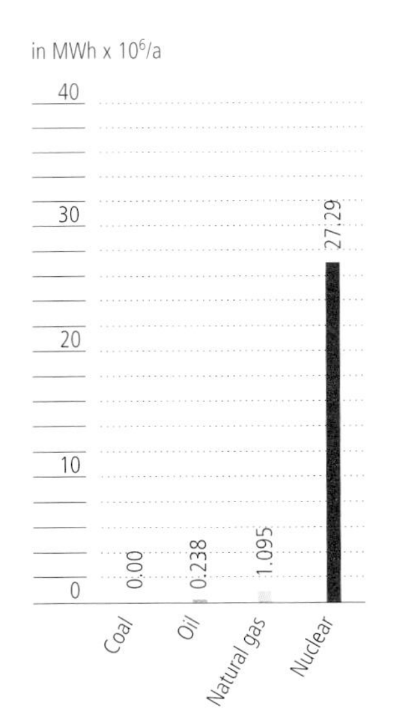

Figure 19.1
Primary energy consumption, Switzerland

Bild 19.1
Primärenergieverbrauch, Schweiz

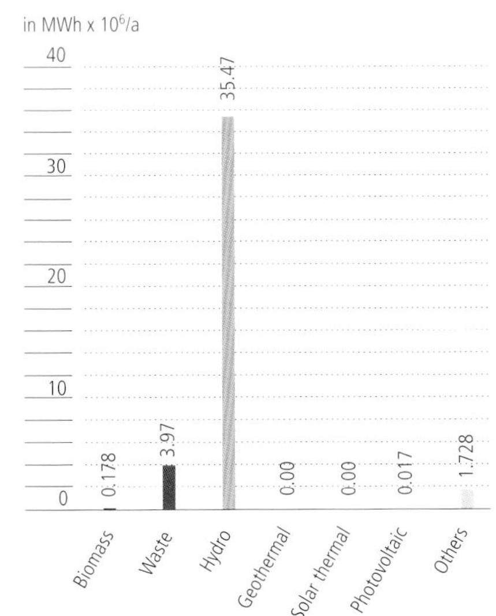
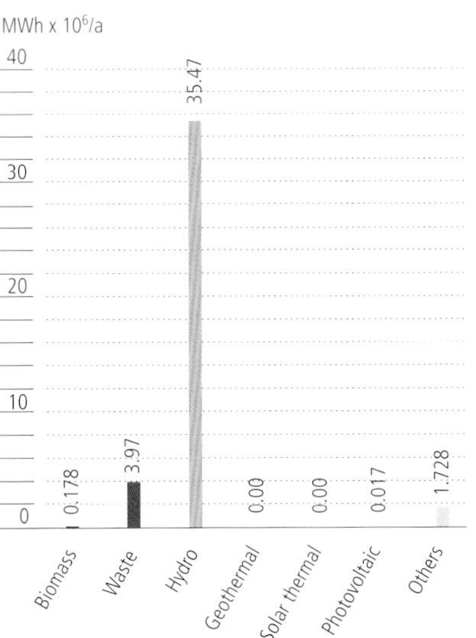

Figure 19.2
Renewable energy consumption, Switzerland

Bild 19.2
Verbrauch aus erneuerbaren Energien, Schweiz

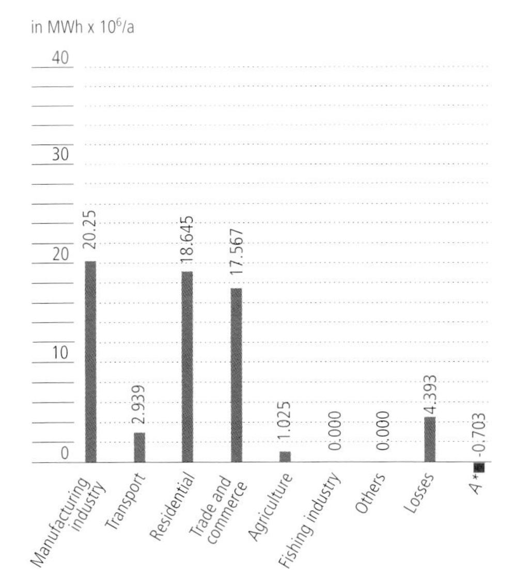

A* Electric energy import/export balance

Figure 19.3
Distribution of energy consumption, Switzerland

Bild 19.3
Verteilung des Energieverbrauchs, Schweiz

Austria's energy structure is comparable to Germany's except for the absence of nuclear power. Hydroelectric power and energy generated from biomass are the two major sources of Austria's renewable energy contribution. The distribution of energy consumption is similar to the countries presented above (**Figures 20.1 – 20.3**).

Österreich zeigt in Bezug auf die Darstellung der Primärenergie eine ähnliche Struktur wie Deutschland, jedoch ohne Kernenergie. Der Einsatz erneuerbarer Energien ergibt sich primär aus der Nutzung von Wasserkraft (Wasserkraftwerke) und dem Einsatz von Biomasse. Die Verteilung des Endenergieverbrauchs entspricht wiederum derjenigen in den bisher besprochenen Ländern, Bilder **20.1 – 20.3**.

Primary energy demand in the UK is supplied by the four usual fossil fuel sources. According to IEA data, the country's renewable energy remains very small. Energy consumption distribution, on the other hand, is similar to other highly developed European countries (**Figures 21.1 – 21.3**).

Der Primärenergiebedarf Englands, **Bilder 21.1 – 21.3** wird durch die vier bekannten Energieträger fossiler Energien gedeckt, wobei der Einsatz erneuerbarer Energien zumindest nach der Statistik der IEA einen sehr geringen Umfang ausmacht. Die Struktur der Energieverteilung entspricht wiederum einem typischen industrialisierten Land, ähnlich Deutschland, Schweiz und Österreich.

Norway is the European country that is in a position to satisfy its energy demand largely from renewable sources, mainly hydroelectric power. Due to this favorable condition, energy can be offered at fairly inexpensive rates, which leads to a high per-capita energy-consumption factor. The distribution of consumer sectors is typical for a post-industrial country (**Figure 22.1 – 22.3**).

Norwegen stellt sich als Land dar (**Bilder 22.1 – 22.3**), das in hohem Maß in der Lage ist, seinen Energiebedarf aus erneuerbaren Energien – hier insbesondere Wasserkraft – zu decken. Der Anteil der erneuerbaren Energien am gesamten Energieverbrauch ist außerordentlich hoch (Wasserkraft) und hat letztlich dazu geführt, dass die elektrische Energie einerseits sehr kostengünstig angeboten wird und daher andererseits traditionell umfänglich eingesetzt wird (hoher elektrischer Energiebedarf pro Kopf). Die Verteilung des Energieverbrauchs entspricht wiederum dem eines typischen nachindustriellen Landes.

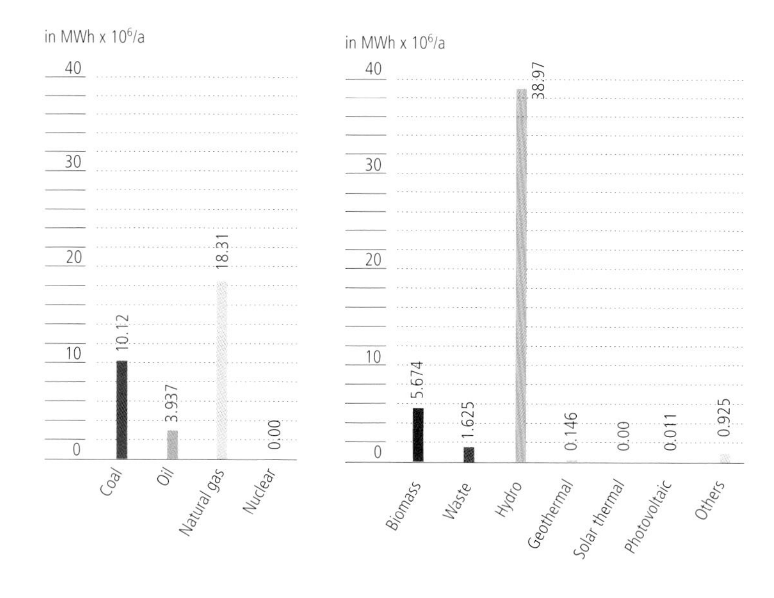

Figure 20.1
Primary energy consumption, Austria

Bild 20.1
Primärenergieverbrauch, Österreich

Figure 20.2
Renewable energy consumption, Austria

Bild 20.2
Verbrauch aus erneuerbaren Energien, Österreich

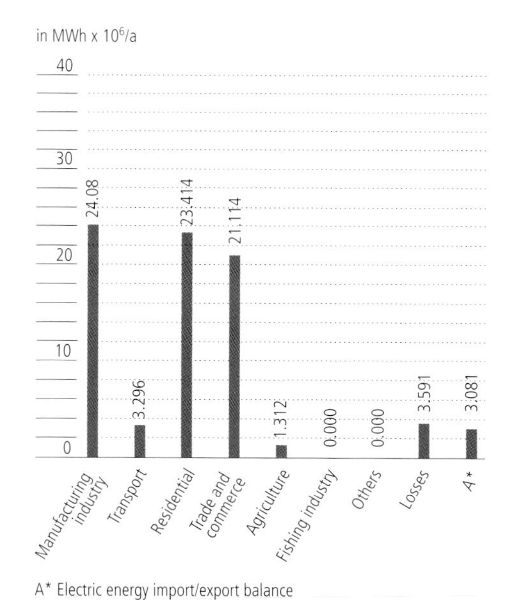

A* Electric energy import/export balance

Figure 20.3
Distribution of energy consumption, Austria

Bild 20.3
Verteilung des Energieverbrauchs, Österreich

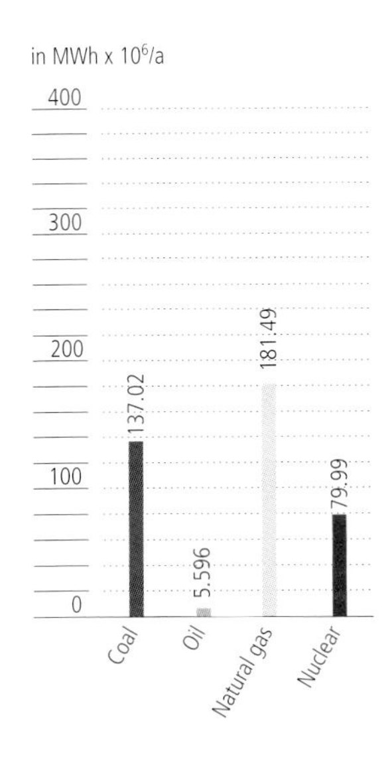

Figure 21.1
Primary energy consumption, United Kingdom (UK)

Bild 21.1
Primärenergieverbrauch, England

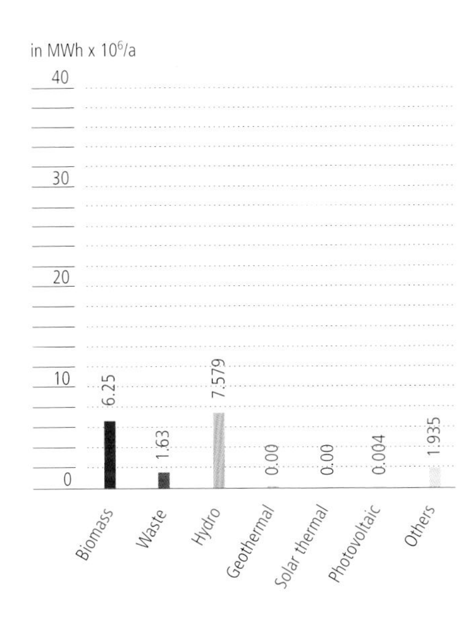

Figure 21.2
Renewable energy consumption, United Kingdom (UK)

Bild 21.2
Verbrauch aus erneuerbaren Energien, England

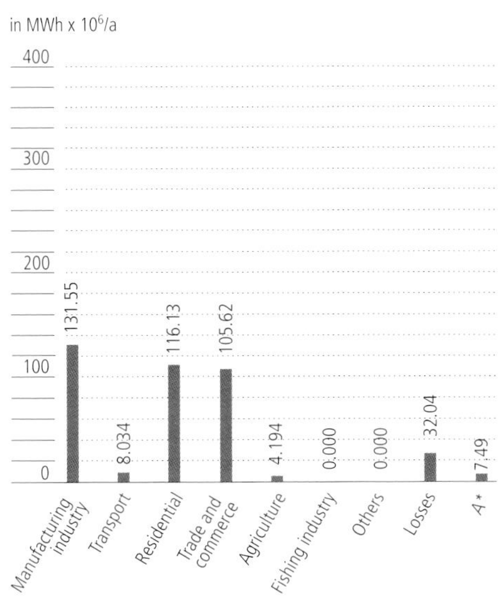

A* Electric energy import/export balance

Figure 21.3
Distribution of energy consumption, United Kingdom (UK)

Bild 21.3
Verteilung des Energieverbrauchs, England

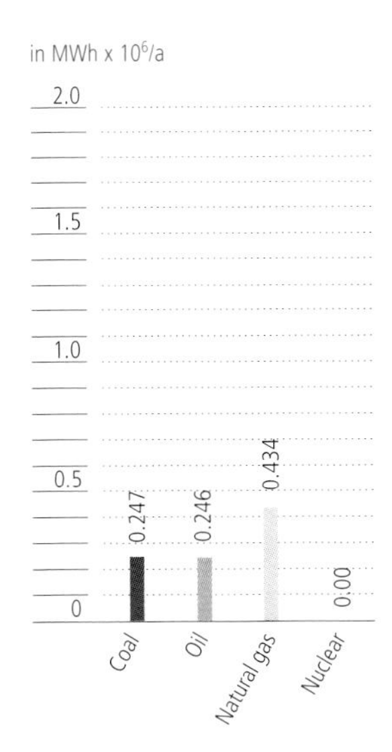

Figure 22.1
Primary energy consumption, Norway

Bild 22.1
Primärenergieverbrauch, Norwegen

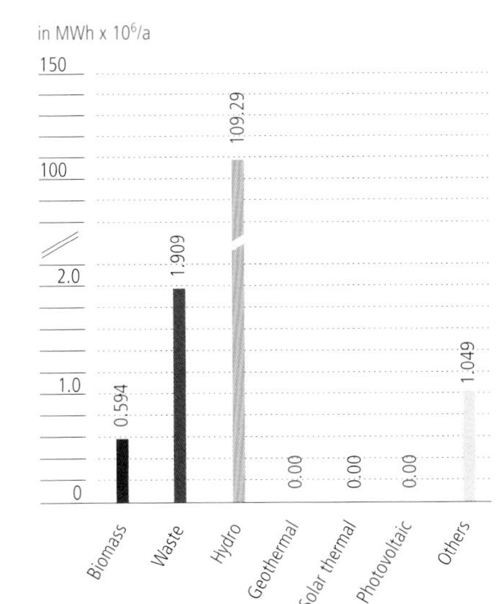

Figure 22.2
Renewable energy consumption, Norway

Bild 22.2
Verbrauch aus erneuerbaren Energien, Norwegen

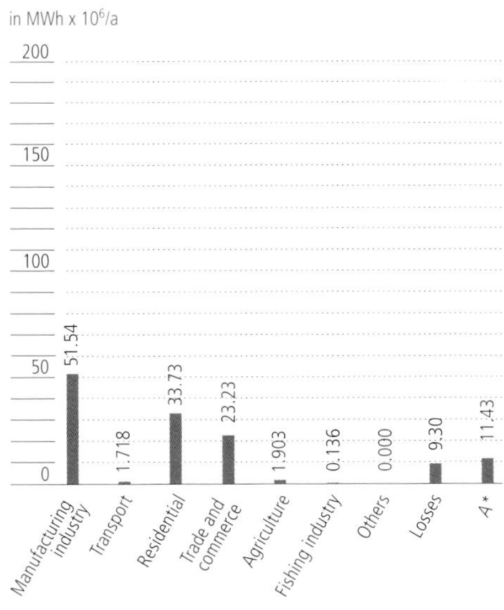

A* Electric energy import/export balance

Figure 22.3
Distribution of energy consumption, Norway

Bild 22.3
Verteilung des Energieverbrauchs, Norwegen

Sweden, in stark contrast to Norway, uses nuclear power to a large extent to provide energy. Yet, energy generated from biomass, waste, and hydroelectric power are also well represented in the energy palette of this country, and they could in fact substitute for more traditional sources easily if their implementation were to double in size. Again, distribution across consumer sectors is typical for a post-industrial economy (**Figures 23.1 – 23.3**).

Schweden deckt im Gegensatz zu Norwegen einen hohen Anteil seines Primärenergiebedarfs durch Kernenergie. Der Anteil erneuerbarer Energien durch Biomasse, Abfallstoffe und Wasserkraft ist nach dieser Statistik ein sehr hoher und könnte bei Verdoppelung der erneuerbaren Energien die traditionellen Energieträger weitestgehend ablösen. Die Verteilung des Energieverbrauchs entspricht wiederum der der nachindustriellen Länder, **Bilder 23.1 – 23.3**.

Finland, illustrated in **Figures 24.1 – 24.3**, gains its energy, according to the IEA data, to a large degree from coal, gas, and nuclear energy. In contrast to Norway and Sweden, the use of renewable energy is significantly lower. It is also worth mentioning that industrial consumers are around 50 % more significantly represented in the distribution than in other comparable countries.

Finnland, **Bilder 24.1 – 24.3**, setzt zur Deckung des Primärenergieverbrauchs gemäß Statistik der IEA in großem Umfang Kohle, Erdgas und Kernenergie ein. Der Einsatz erneuerbarer Energien ist im Gegensatz zu Norwegen und Schweden deutlich geringer. Auffällig in Bezug auf die Verteilung des Energieverbrauchs ist, dass offensichtlich die Industrie einen wesentlichen Anteil (ca. 50 Prozent) des gesamten Energiebedarfs ausmacht.

Iceland can be described as a country with almost perfect conditions when it comes to energy supply. More than 99 % of all energy consumed on the island is derived from renewable sources, with geothermal energy and hydroelectric power being the dominant sources. The distribution of energy consumer areas is very similar to Norway (**Figures 25.1 – 25.3**).

Island kann in Bezug auf den Einsatz fossiler Brennstoffe als geradezu paradiesisches Land bezeichnet werden. Gemäß der IEA-Statistik werden im Wesentlichen mehr als 99 Prozent aller Energieverbräuche durch Wasserkraft und Geothermie gedeckt (erneuerbare Energien). Die Verteilung des Energieverbrauchs zeigt wiederum eine ähnliche Struktur wie zum Beispiel Norwegen, **Bilder 25.1 – 25.3**.

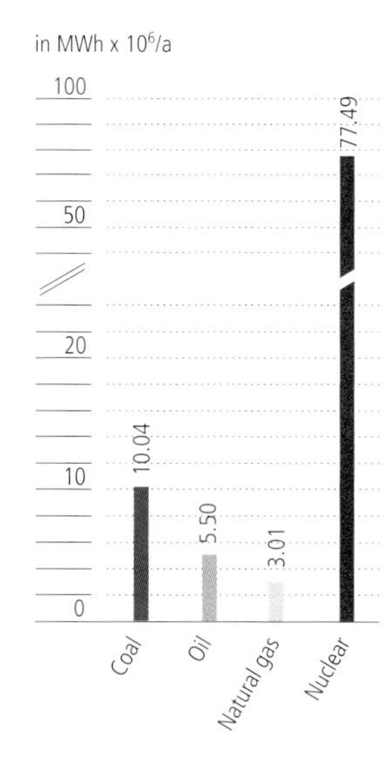

Figure 23.1
Primary energy consumption, Sweden

Bild 23.1
Primärenergieverbrauch, Schweden

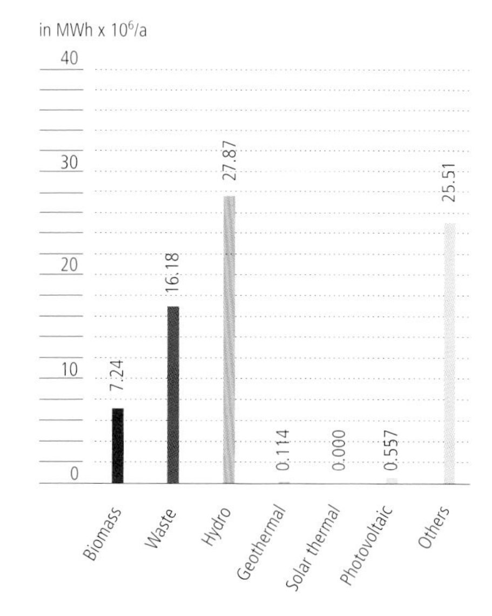

Figure 23.2
Renewable energy consumption, Sweden

Bild 23.2
Verbrauch aus erneuerbaren Energien, Schweden

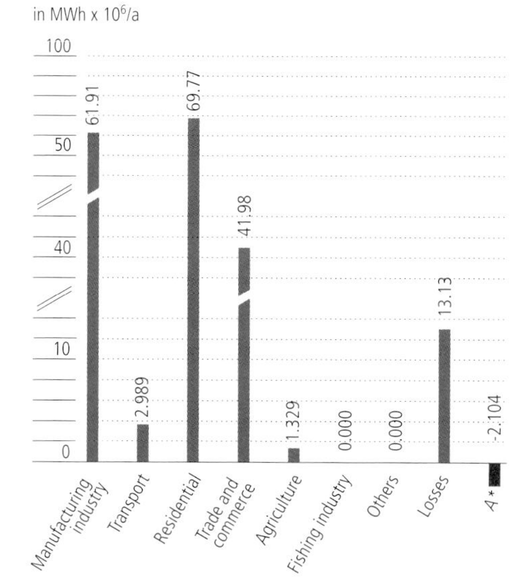

Figure 23.3
Distribution of energy consumption, Sweden

Bild 23.3
Verteilung des Energieverbrauchs, Schweden

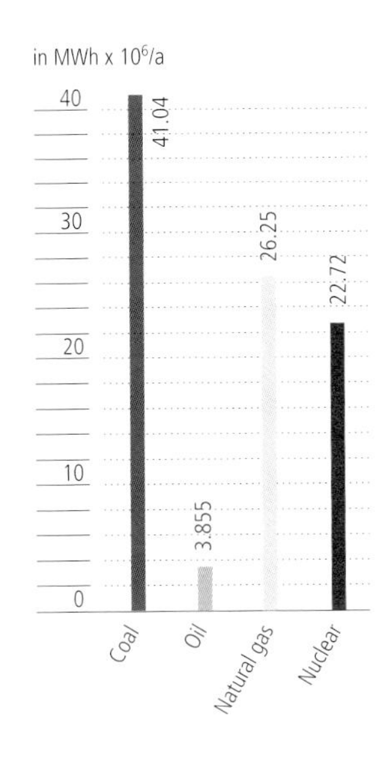

Figure 24.1
Primary energy consumption, Finland

Bild 24.1
Primärenergieverbrauch, Finnland

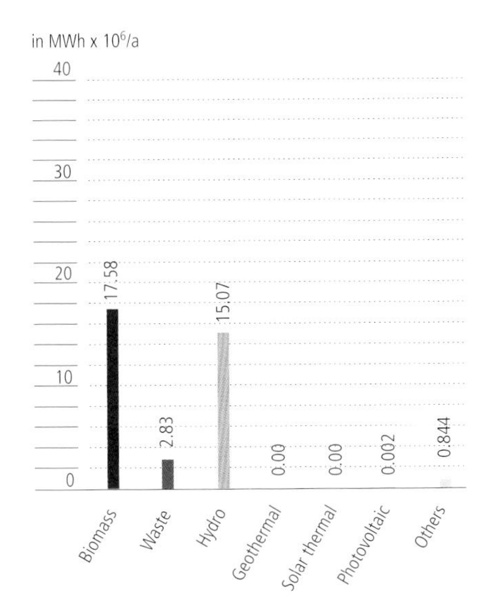

Figure 24.2
Renewable energy consumption, Finland

Bild 24.2
Verbrauch aus erneuerbaren Energien, Finnland

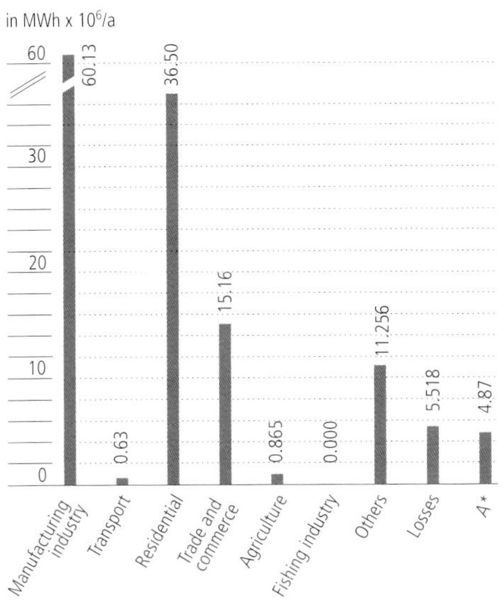

Figure 24.3
Distribution of energy consumption, Finland

Bild 24.3
Verteilung des Energieverbrauchs, Finnland

Figure 25.1
Primary energy consumption, Iceland

Bild 25.1
Primärenergieverbrauch, Island

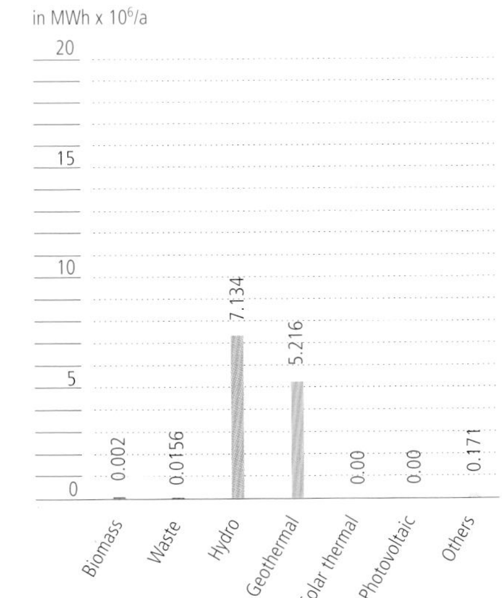

Figure 25.2
Renewable energy consumption, Iceland

Bild 25.2
Verbrauch aus erneuerbaren Energien, Island

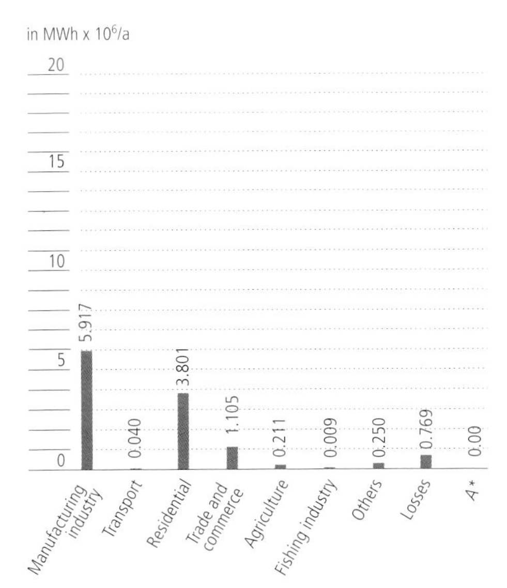

Figure 25.3
Distribution of energy consumption, Iceland

Bild 25.3
Verteilung des Energieverbrauchs, Island

Russia is one of the world's largest energy exporters. It uses all primary energy resources available in the country. The share of renewable energy is just 10 %, as a result, mainly, of the abundance of hydroelectric power sources. It is interesting that the country's greatest energy consumers are – according to the comprehensive data provided by the IEA – residential dwellings. The reason for this may be in the climatic conditions but also, it can be assumed, in their inefficient, energy-waste-promoting operation (**Figures 26.1 – 26.3**).

Russland als eines der großen Energie exportierenden Länder nutzt alle Ressourcen zur Deckung des Primärenergieverbrauchs, die im eigenen Land vorkommen. Der Anteil erneuerbarer Energien liegt im Bereich um ca. 10 Prozent und ergibt sich im Wesentlichen aus der Nutzung von Wasserkraft. Die Verteilung des Energieverbrauchs mit dem Schwerpunkt im Bereich Wohnen dürfte sich aufgrund der klimatischen Bedingungen wie gezeigt darstellen, unter Umständen auch dadurch, dass zum heutigen Zeitpunkt Wohnobjekte einen deutlich zu hohen Energieverbrauch aufweisen, **Bilder 26.1 – 26.3**.

Figures 27.1 – 27.3 show the current energy scenario for Spain. Around 20 % of the country's energy comes from renewable resources, with the remainder being distributed across all available fossil fuels. Consumer distribution is coherent with that of a post-industrial society.

Spanien (**Bilder 27.1 – 27.3**) setzt auf alle konventionellen Energieträger, wobei der Anteil aus erneuerbaren Energien (Biomasse, Abfallstoffe und Wasserkraft) in etwa einen Anteil um 20 Prozent ausmacht. Die Verteilung des Energieverbrauchs entspricht in ihrer Struktur wiederum der einer typischen nachindustriellen Gesellschaft.

Italy obtains its primary energy generation from coal, oil, and gas, with nuclear energy playing only a very minor role. Except for Iceland, Italy seems to be the country with the highest geothermal energy contribution of all of Europe. Waste products and hydroelectric power also complement the energy mix. Energy consumption is comparable to Spain's, a typical case of a post-industrial economy (**Figures 28.1 – 28.3**).

In Italien wird der Primärenergiebedarf durch Kohle, Erdöl und Erdgas gedeckt, Kernenergie spielt keine wesentliche Rolle. Neben Island scheint in Italien ein nennenswerter Anteil der erneuerbaren Energien aus geothermischen Nutzungen zu stammen. Einen deutlich höheren Beitrag liefern im Bereich der erneuerbaren Energien Wasserkraftwerke und die Nutzung von Abfallstoffen. Die Verteilung des Energieverbrauchs entspricht in etwa der Spaniens und ist wiederum typisch für ein nachindustrielles Land, **Bilder 28.1 – 28.3**.

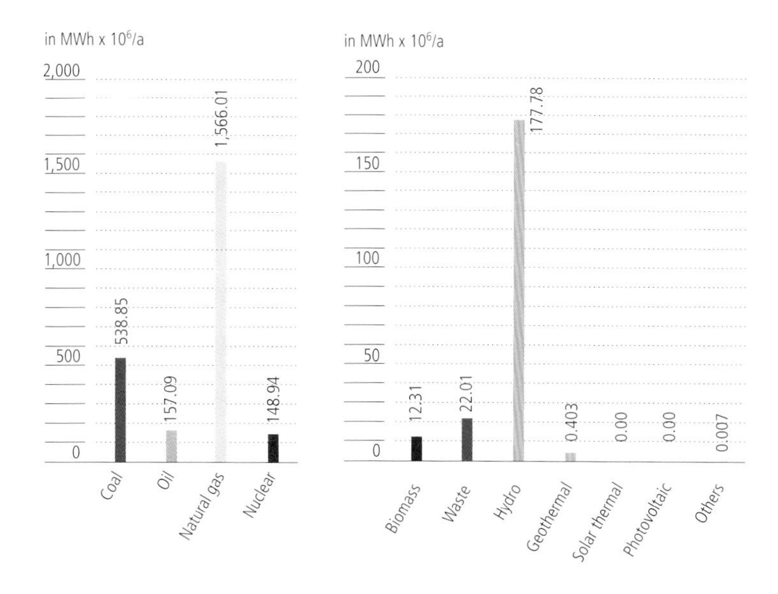

Figure 26.1
Primary energy consumption, Russia

Bild 26.1
Primärenergieverbrauch, Russland

Figure 26.2
Renewable energy consumption, Russia

Bild 26.2
Verbrauch aus erneuerbaren Energien, Russland

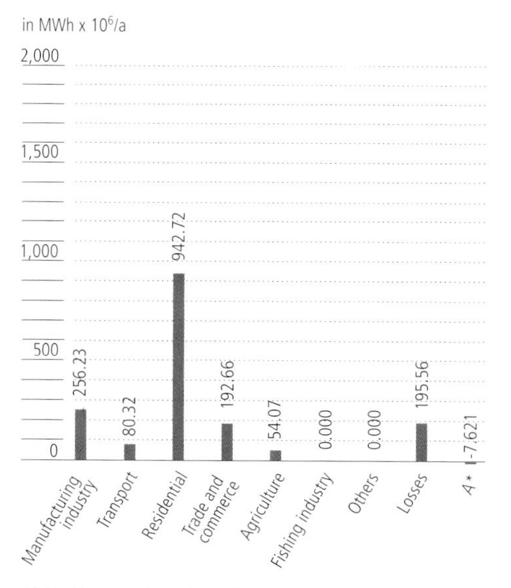

A* Electric energy import/export balance

Figure 26.3
Distribution of energy consumption, Russia

Bild 26.3
Verteilung des Energieverbrauchs, Russland

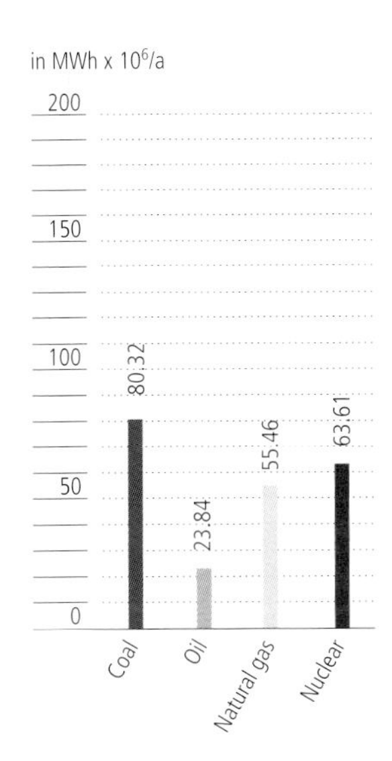

Figure 27.1
Primary energy consumption, Spain

Bild 27.1
Primärenergieverbrauch, Spanien

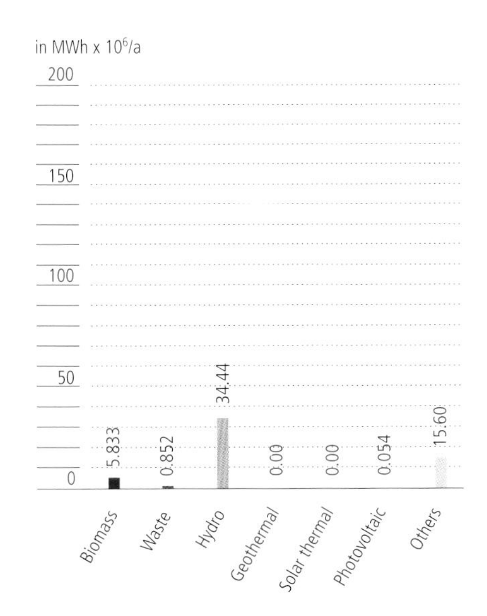

Figure 27.2
Renewable energy consumption, Spain

Bild 27.2
Verbrauch aus erneuerbaren Energien, Spanien

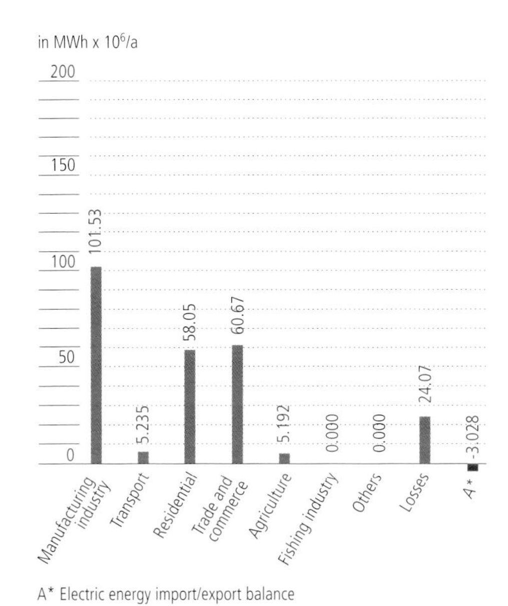

A* Electric energy import/export balance

Figure 27.3
Distribution of energy consumption, Spain

Bild 27.3
Verteilung des Energieverbrauchs, Spanien

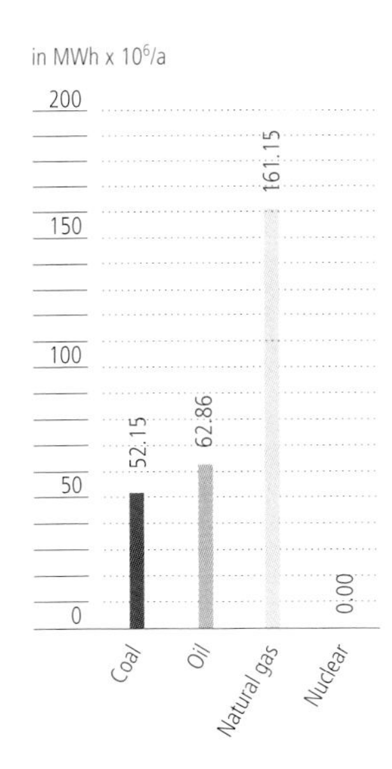

Figure 28.1
Primary energy consumption, Italy

Bild 28.1
Primärenergieverbrauch, Italien

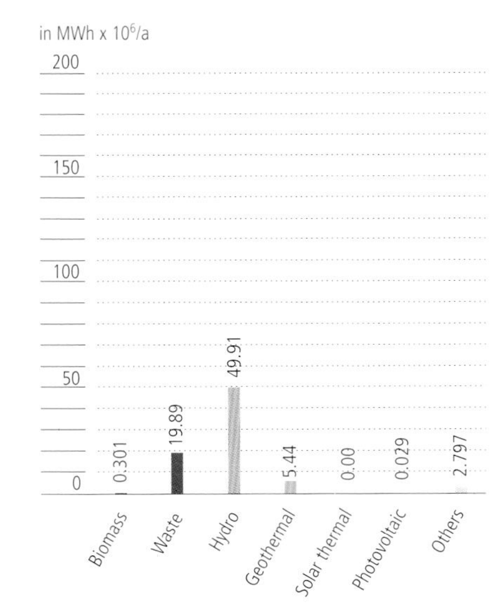

Figure 28.2
Renewable energy consumption, Italy

Bild 28.2
Verbrauch aus erneuerbaren Energien, Italien

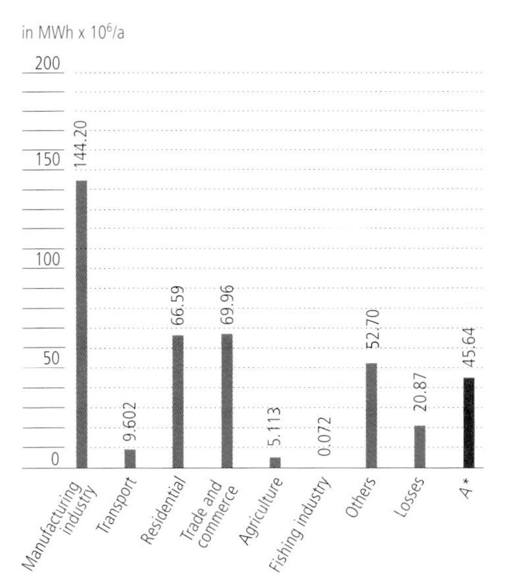

A* Electric energy import/export balance

Figure 28.3
Distribution of energy consumption, Italy

Bild 28.3
Verteilung des Energieverbrauchs, Italien

Greece uses all available classic fossil fuels and almost no nuclear energy, and renewable energy is derived from hydroelectric power and other non-fossil fuel sources. Consumer distribution is similar to that for other European countries, with two exceptions: the percentage of consumption for transport is comparatively small, but for agriculture, on the other hand, it is much greater than in other countries in Europe (**Figures 29.1 – 29.3**).

France, shown in **Figures 30.1 – 30.3**, supplies energy primarily with nuclear power, which represents approximately 85 % of the country's energy budget. Because of this somewhat unique condition, the country will have to confront a substantial reconfiguration of its entire energy sector in only a few decades. However, renewable energy does contribute to the mix with almost 9 %, the majority being provided by hydroelectric power. Energy consumption distribution is similar to that in comparable European countries.

A situation in contrast to France is found in Turkey, a country that uses, in essence, no nuclear energy to supply its demand for primary energy demand (**Figures 31.1 – 31.3**). Gas is the major primary energy source, followed by coal and oil. A very minor role is played by renewable energy sources, with the exception of hydroelectric power.
In this country, great potentials for the exploitation of solar and wind energy exist and need to be pursued more decisively.

As seen in **Figure 31.1**, a large percentage of energy is used for the industrial sector in Turkey, followed by residential uses, commerce, and the trades. Included in the last categories are the heightened consumption percentages that are contributed by the tourism sector. **Figure 31.3**, which allows a comparison with the image in **29.3** (Greece), displays a similar structure, although the industrial sector still makes a large contribution to consumption.

Griechenland nutzt die klassischen Energieträger Kohle, Erdöl und Erdgas und offensichtlich praktisch keine Kernenergie. Der Einsatz erneuerbarer Energien entstammt im Wesentlichen ausschließlich der Wasserkraft sowie anderen erneuerbaren Energiequellen. Die Verteilung des Energieverbrauchs zeigt wiederum die typisch europäische Tendenz, jedoch mit einem bemerkenswert kleinen Anteil für Transport und einem deutlich höheren Anteil für die Landwirtschaft, **Bilder 29.1 – 29.3**.

Frankreich, **Bilder 30.1 – 30.3**, deckt einen wesentlichen Teil seiner Primärenergie durch Kernenergie (ca. 85 Prozent der elektrischen Energie aus Kernkraftwerken) und steht somit in einigen Jahrzehnten vor einem erheblichen Umbau seiner Elektroenergieerzeugung. Der Einsatz erneuerbarer Energien macht in Frankreich ca. annähernd 9 Prozent aus und ergibt sich im Wesentlichen aus der Nutzung von Wasserkraft. Die Energieverteilung, d.h. Verteilung auf verschiedenste Verbrauchstrukturen, entspricht wiederum der eines nachindustriellen Landes und ist vergleichbar mit vielen anderen europäischen Ländern.

In der Türkei, **Bilder 31.1 – 31.3**, wird gemäß IEA im Gegensatz zu Frankreich praktisch keine Kernenergie eingesetzt, um den Primärenergieverbrauch zu decken. Der Einsatz von Erdgas stellt den größten Primärenergieträger dar, gefolgt von Kohle und Erdöl. Im Bereich der erneuerbaren Energien ist noch ein wesentlicher Anteil an Wasserkraft festzustellen, wobei weitere erneuerbare Energieträger eine sehr untergeordnete Rolle spielen und sich gerade auf dem Sektor der Windenergie und vor allem Solarenergie vieles entwickeln sollte.

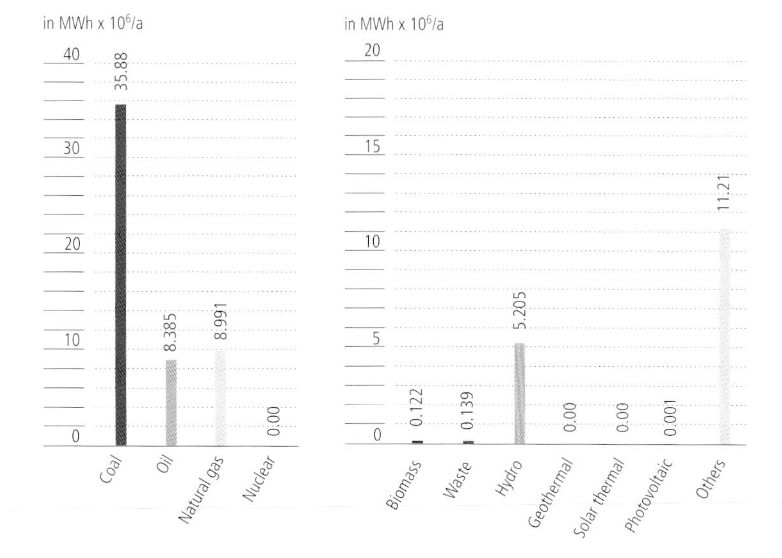

Figure 29.1
Primary energy consumption, Greece

Bild 29.1
Primärenergieverbrauch, Griechenland

Figure 29.2
Renewable energy consumption, Greece

Bild 29.2
Verbrauch aus erneuerbaren Energien, Griechenland

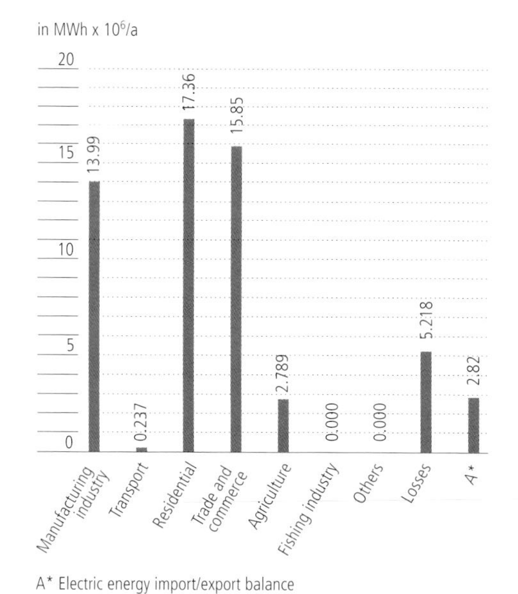

Figure 29.3
Distribution of energy consumption, Greece

Bild 29.3
Verteilung des Energieverbrauchs, Griechenland

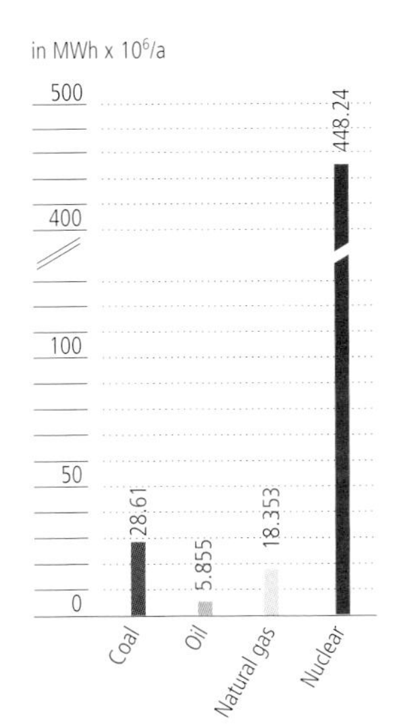

Figure 30.1
Primary energy consumption,
France

Bild 30.1
Primärenergieverbrauch,
Frankreich

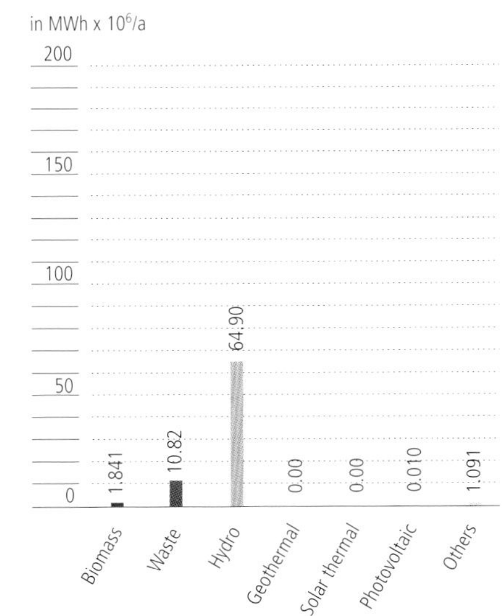

Figure 30.2
Renewable energy consumption,
France

Bild 30.2
Verbrauch aus erneuerbaren Energien,
Frankreich

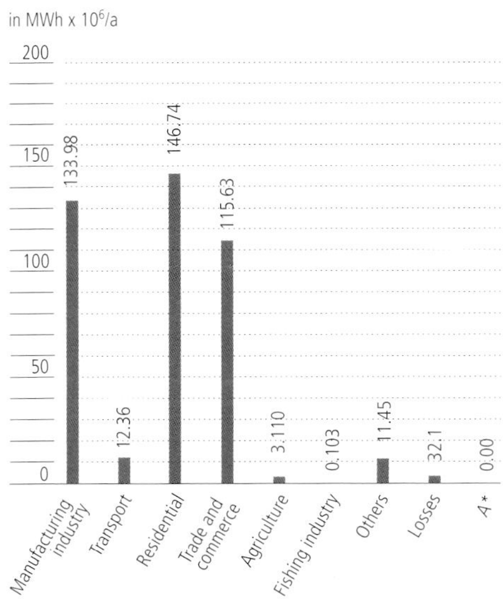

A* Electric energy import/export balance

Figure 30.3
Distribution of energy consumption,
France

Bild 30.3
Verteilung des Energieverbrauchs,
Frankreich

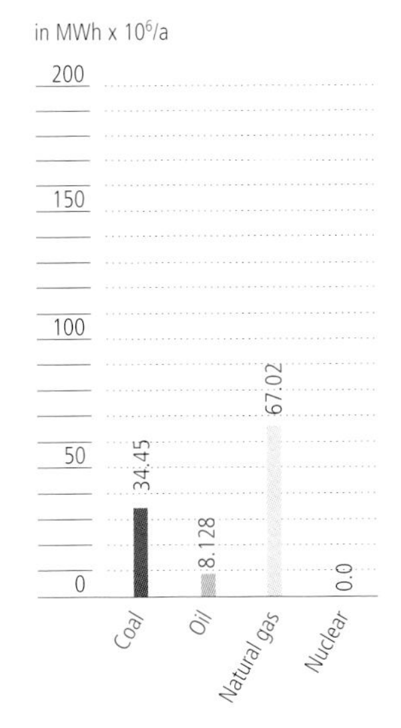

Figure 31.1
Primary energy consumption,
Turkey

Bild 31.1
Primärenergieverbrauch,
Türkei

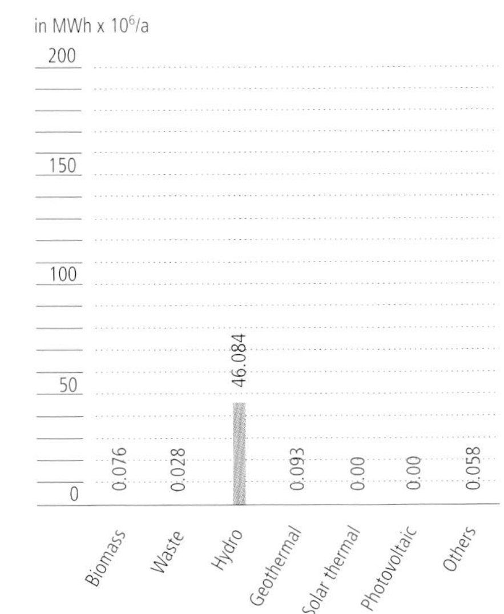

Figure 31.2
Renewable energy consumption,
Turkey

Bild 31.2
Verbrauch aus erneuerbaren Energien,
Türkei

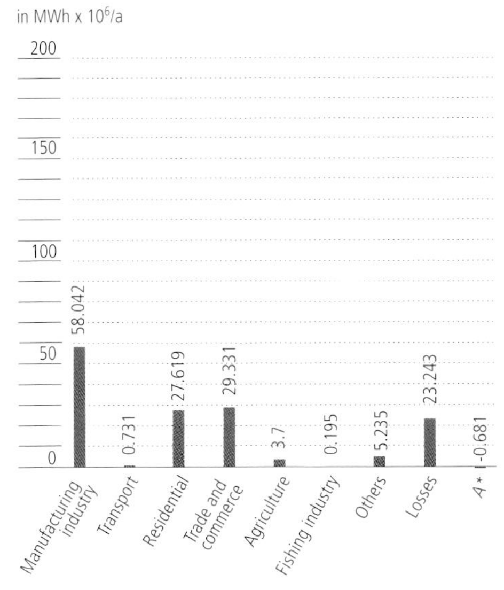

A* Electric energy import/export balance

Figure 31.3
Distribution of energy consumption,
Turkey

Bild 31.3
Verteilung des Energieverbrauchs,
Türkei

4.1.1 Use of renewable energy sources in Europe

In the aftermath of the first two energy crises in 1973 and 1978, Europe intensified the effort to become eventually and gradually independent of fossil fuels. Individual countries in the union approached the goal with different means, which have now shown different results.

Figure 32 presents the average time of solar incidence, a measure of the degree to which solar technology may be used in different regions.

Figure 33 gives the measure of how much photovoltaic power generation was installed in the year 2007 by the various countries. Germany has the greatest percentage of installed solar power devices in all of Europe – by a significant margin – as a result of regulatory intervention and tax incentives. The same is the case when it comes to wind power usage in Germany (**Figure 34**), although its role of leader in wind power may soon be lost to other European countries because of their much longer, wind-intensive coastlines. These could be Denmark, Spain, Portugal, the Netherlands, Greece, Sweden, Italy, the U.K., France, Estonia, Belgium, Finland, Lithuania, and Latvia.

4.1.1. Einsatz erneuerbarer Energien in Europa

In den europäischen Ländern ist im Wesentlichen nach den ersten Energiekrisen (1973/1978) das Bemühen entstanden, sich zumindest zum Teil von den fossilen Brennstoffen abzukoppeln und somit einen Teil der konventionellen Energieträger durch erneuerbare Energien zu ersetzen. Die einzelnen Ansätze wurden in den europäischen Ländern politisch unterschiedlich unterstützt und haben somit auch zu unterschiedlichen Ergebnissen geführt.

Bild 32 zeigt beispielhaft für Europa die durchschnittliche Sonnenscheindauer – ein Maßstab für den Einsatz von solartechnischen Anlagen zur Erzeugung von Energien.

Bild 33 weist die im Jahr 2007 installierte Photovoltaikleistung in einzelnen europäischen Ländern aus, wobei politisch gewollte Unterstützungsmaßnahmen dazu geführt haben, dass in Deutschland mit deutlichem Abstand die meisten photovoltaischen Anlagen installiert wurden. Gleiches gilt im Prinzip auch für die installierten Windleistungen durch Windkraftanlagen, **Bild 34**. Die Spitzenstellung, die Deutschland trotz seiner verhältnismäßig kurzen Küstenlinien einnimmt, dürfte sich in den nächsten Jahren zugunsten der Länder verändern, die über windreiche Küstenregionen verfügen. Hierzu gehören insbesondere Länder wie Dänemark, Spanien, Irland, Portugal, Niederlande, Griechenland, Schweden, Italien, Großbritannien, Frankreich, Estland, Belgien, Finnland, Litauen und Lettland.

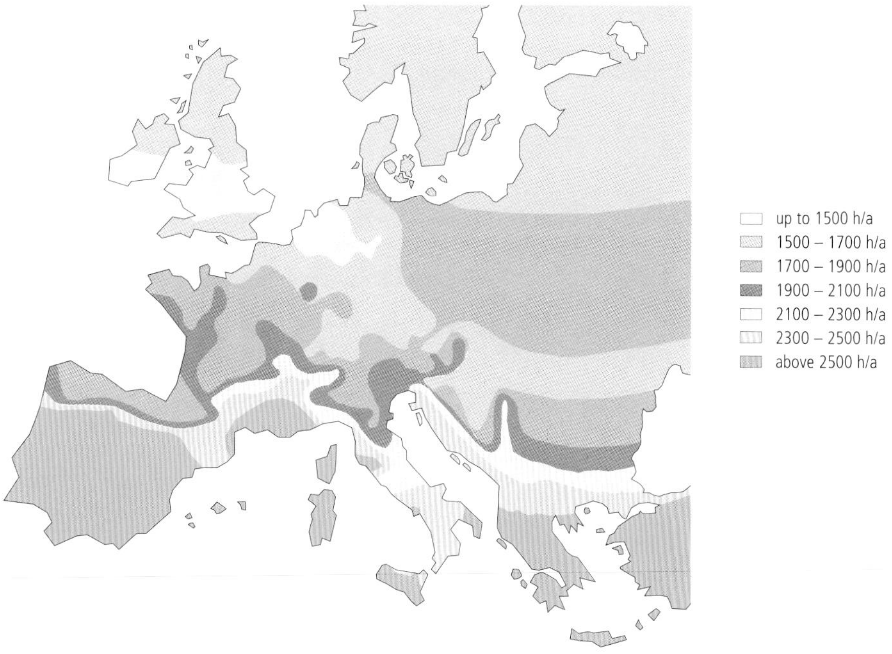

up to 1500 h/a
1500 – 1700 h/a
1700 – 1900 h/a
1900 – 2100 h/a
2100 – 2300 h/a
2300 – 2500 h/a
above 2500 h/a

Figure 32
Average annual sunshine hours in Central and Northern Europe

Source:
Low Tech-Light Tech-High Tech, Klaus Daniels
Birkhäuser Verlag, 1998

Bild 32
Durchschnittliche Sonnenscheindauer in Zentral- und Nordeuropa

Quelle:
Low Tech-Light Tech-High Tech, Klaus Daniels
Birkhäuser Verlag, 1998

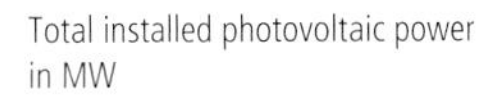

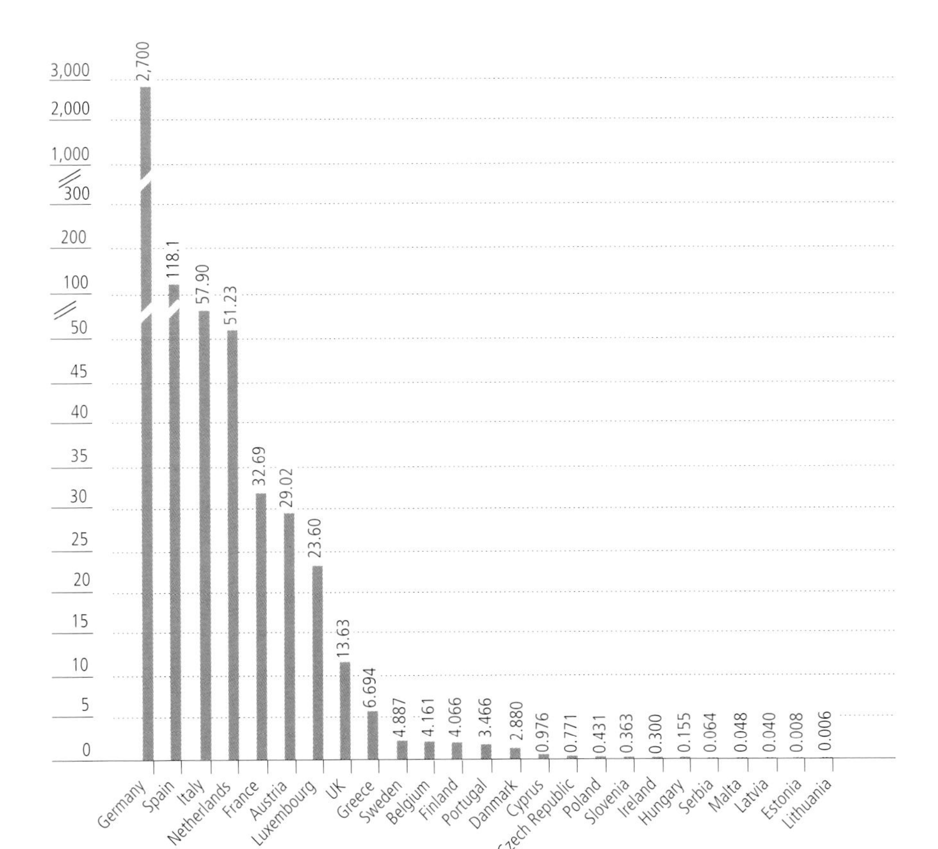

Figure 33
Total installed photovoltaic power in the European Union

Source: Eurobserv'ER 2007

Bild 33
Installierte Photovoltaik-Leistung in der Europäischen Union

Quelle: Eurobserv'ER 2007

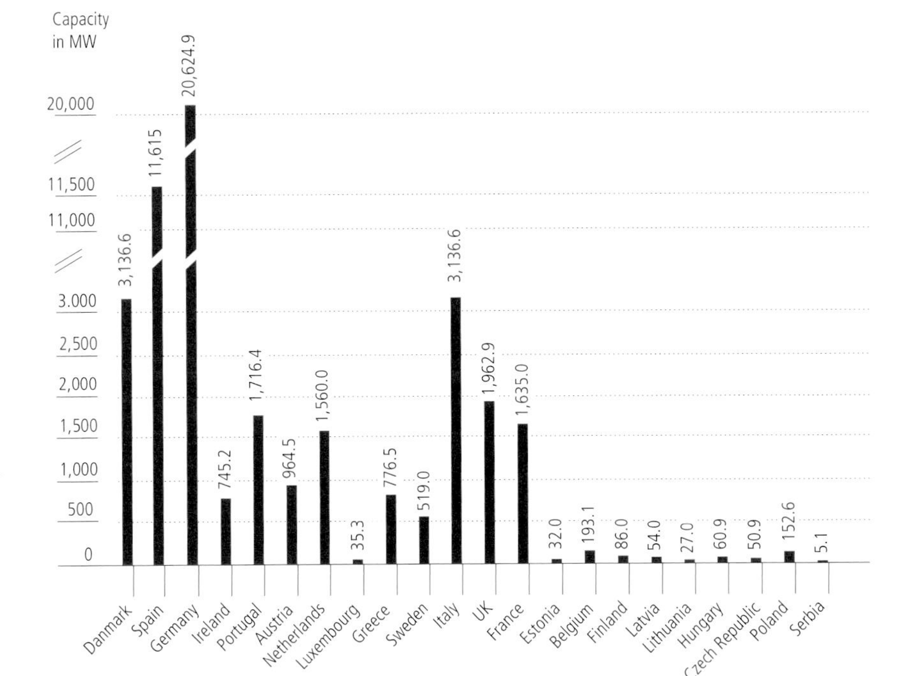

Figure 34
Installed wind power in MW, Europe 2006

Source: Sonnenenergie 05/2007

Bild 34
Installierte Windleistung in MW, Europa 2006

Quelle: Sonnenenergie 05/2007

The European leaders in biogas generation are Germany and the U.K., with Italy, Spain, and France far behind but with substantial potential (**Figure 35**).

Germany is also the leader in another sector of renewable energy generation. As shown in **Figure 36**, the country's per capita consumption of bioethanol, biodiesel, and other biogenic transportation fuels is greater than anywhere in the union, followed by similar consumption patterns in Austria, and Sweden.

Figures 33–36 allow an approximate estimate that the renewable energy generation from various sources in Europe is around 10 % of total energy consumption.

Bei der Produktion von Biogas in Europa liegen Deutschland und Großbritannien gemäß den neusten Statistiken deutlich im Vordergrund, während eine Vielzahl anderer europäischer Länder wie zum Teil Italien, Spanien, Frankreich noch ein erhebliches Potenzial heben können, **Bild 35**.

Gemäß **Bild 36** nimmt in Bezug auf den Einsatz von Biotreibstoffen die Bundesrepublik Deutschland wiederum eine Vorreiterrolle ein. Bezogen auf die Pro-Kopf-Verbräuche von Bioethanol, Biodiesel und anderen Biotreibstoffen liegen Österreich und Schweden in einer ähnlichen Größenordnung.

Gemäß den Darstellungen **Bild 33 bis 36** kann zurzeit sehr grob gerechnet ein Anteil erneuerbarer Energien von ca. 10 Prozent, bezogen auf den gesamten Primärenergiebedarf Europas, festgestellt werden.

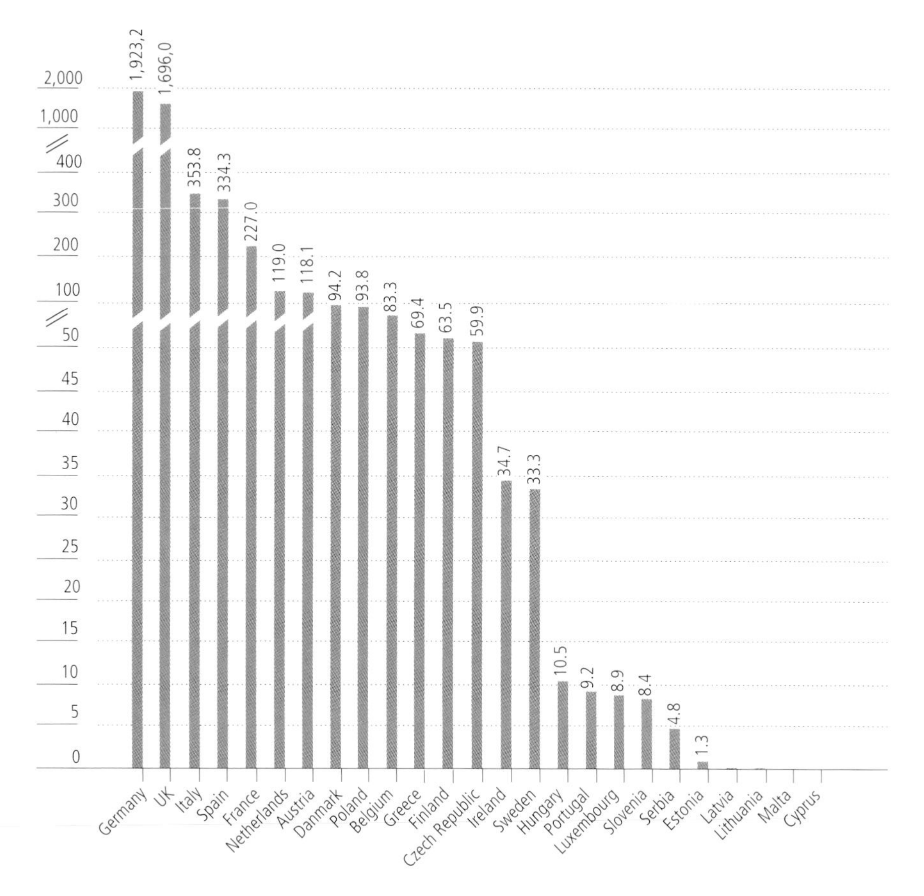

European Union total 5,346.7 kilo tons (including landfill gas, sewage treatment gas and others)

Figure 35
Primary energy production of biogas in the European Union 2006

Source: EurObserv'ER 2007

Bild 35
Primärproduktion von Biogas in der EU 2006

Quelle: EurObserv'ER 2007

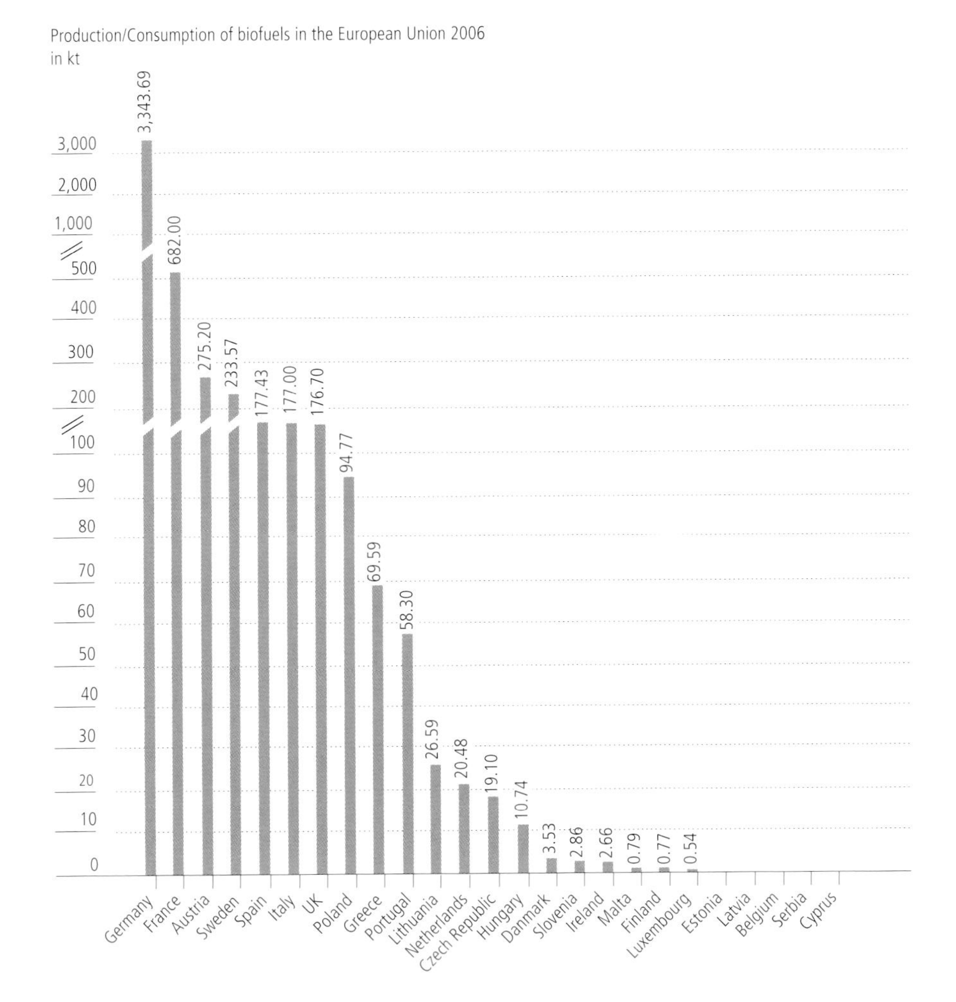

Figure 36
Consumption of biofuels in the European Union 2006

Source: EurObserv'ER 2007

Bild 36
Verbrauch von Biotreibstoffen in der EU 2006

Quelle: EurObserv'ER 2007

4.2 North and South American region

We selected some key countries as representatives of this continent, which are discussed in the statistics in **Figures 37–43**.

Canada uses, besides coal, oil, gas, and nuclear primary energy, a very significant amount of renewables to satisfy its energy demand. The contribution of hydroelectric power is particularly remarkable. However, except for biomass, other renewable technologies do not play a large role. In analyzing the consumer distribution, it is noteworthy that for a post-industrial country the percentage of energy used in the industrial sector is very high, while the percentage of energy used for transportation on the other hand is low (**Figures 37.1–37.3**).

The United States of America is the country with the world's highest energy consumption per capita—similar to, but greater by a large margin than such economies as Russia and China. The U.S. relies on all available raw fossil fuel energy potentials, with the nuclear sector being a major contributor (**Figures 38.1–38.3**).

4.2 Region Nord- und Südamerika

Aus der Region Nord- und Südamerika wurden nur einige wesentliche Schlüsselländer in die Betrachtung einbezogen, die in den Statistiken, **Bilder 37** bis **43**, dargestellt sind.

Kanada nutzt neben Kohle, Erdöl und Erdgas sowie Kernenergie einen außerordentlich hohen Anteil an erneuerbaren Energien, um seinen Energiebedarf zu decken. Dabei fällt der hohe Anteil an Wasserkraft auf. Neben der Biomasse als erneuerbarer Energieträger spielen zurzeit weitere Techniken der erneuerbaren Energien keine wesentliche Rolle. Bei der Verteilung des Energieverbrauchs ist auffällig, dass der Anteil für die Industrie in Bezug auf ein nachindustrielles Land sehr stark zu Buche schlägt, dafür der Anteil für Transporte sehr gering scheint, **Bilder 37.1 – 37.3**.

Die Vereinigten Staaten als das Land mit dem höchsten Energiebedarf pro Kopf und einem ähnlich großen Energiebedarf wie China und bedingt Russland setzt auf alle vier Bereiche der fossilen Brennstoffe, wobei wiederum Kernenergie einen sehr hohen Anteil einnimmt, **Bilder 38.1 – 38.3**.

The U.S. on the other hand, uses an amount of renewable energy that should definitely not be underestimated, largely because of the availability of hydroelectric power, which contributes around 15 – 18 % of total energy, with the state of California being the major producer. The distribution of energy consumption is similar to any post-industrial nation, although the relatively low percentage shown in the graph for energy used in the transportation sector may actually be somewhat higher.

Mexico, documented in **Figures 39.1 – 39.3**, is highly dependent upon energy from oil and gas and relies only marginally on coal. Nuclear power amounts to about 5 % of the total energy mix. With a 20 % contribution, renewable energy from biomass, hydroelectric power, and geothermal applications completes the palette of primary energy sources. The distribution across user sectors represents that of a developed industrial nation.

Der Bereich der erneuerbaren Energien spielt in den USA zurzeit infolge der Nutzung von Wasserkraft eine nicht zu unterschätzende Rolle. Der Gesamtumfang des Einsatzes erneuerbarer Energien liegt bei grob geschätzten ca. 15 – 18 Prozent, wobei ein wesentlicher Teil in Kalifornien dargestellt wird. Die Verteilung des Energieverbrauchs entspricht wiederum der Struktur eines nachindustriellen Landes mit jedoch einem erstaunlich geringen Anteil für Transporte, der vermutlich deutlich höher liegt.

Mexiko, **Bilder 39.1 – 39.3**, ist offensichtlich in hohem Umfang von Erdöl und Erdgas abhängig, bedingt von Kohle. Die Kernenergie macht in Bezug auf die Darstellung des Primärenergieverbrauchs in etwa einen Anteil von 5 Prozent aus. Der Anteil der erneuerbaren Energien aus Biomasse, Wasserkraft und Geothermie ergänzt mit ca. 20 Prozent die Deckung des gesamten Primärenergiebedarfs. Die Verteilung der Energieverbräuche entspricht in etwa der Kanadas und somit der eines entwickelten Industrielandes.

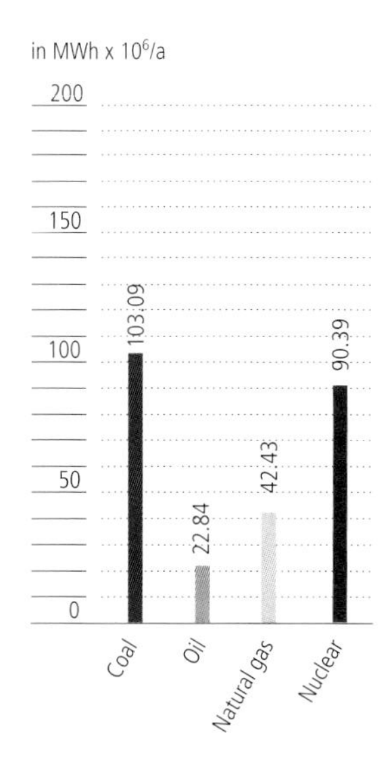

Figure 37.1
Primary energy consumption, Canada

Bild 37.1
Primärenergieverbrauch, Kanada

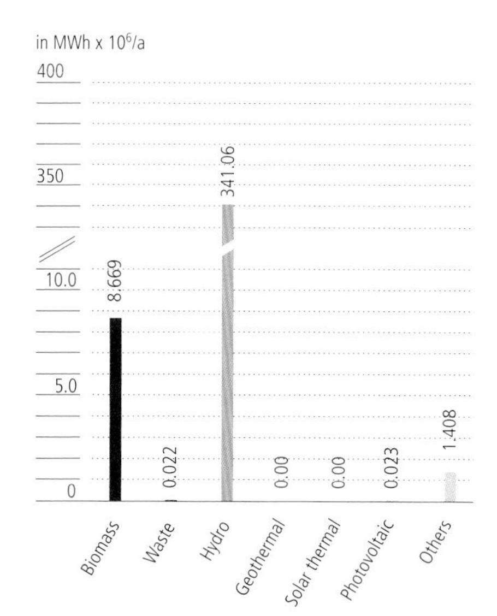

Figure 37.2
Renewable energy consumption, Canada

Bild 37.2
Verbrauch aus erneuerbaren Energien, Kanada

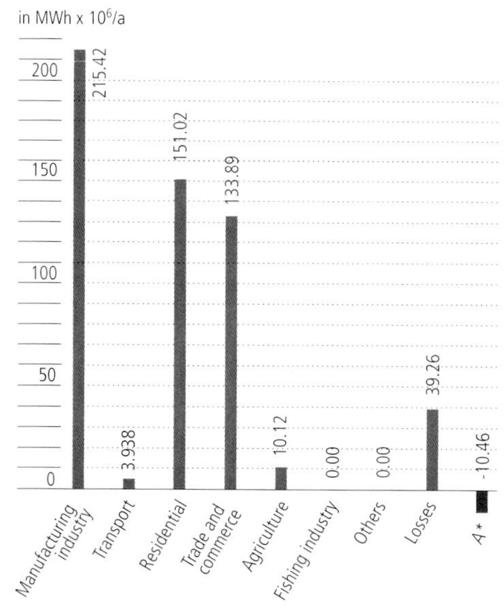

Figure 37.3
Distribution of energy consumption, Canada

Bild 37.3
Verteilung des Energieverbrauchs, Kanada

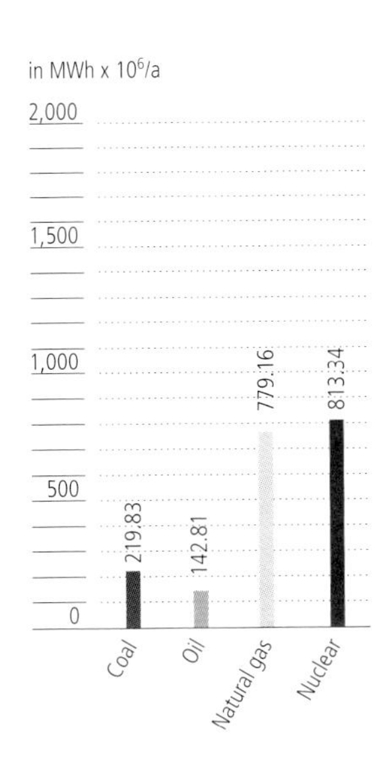

Figure 38.1
Primary energy consumption,
USA

Bild 38.1
Primärenergieverbrauch,
USA

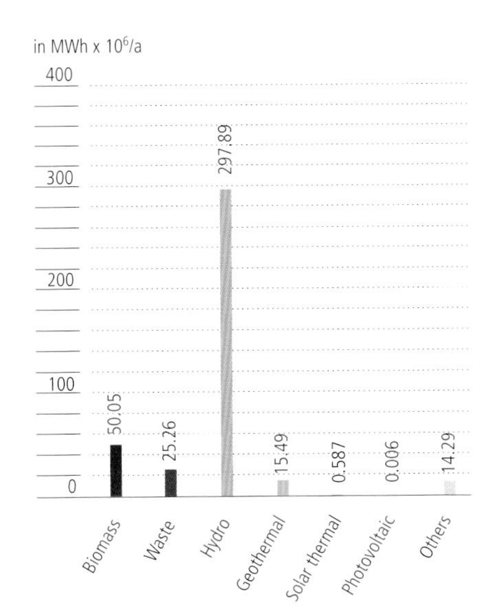

Figure 38.2
Renewable energy consumption,
USA

Bild 38.2
Verbrauch aus erneuerbaren Energien,
USA

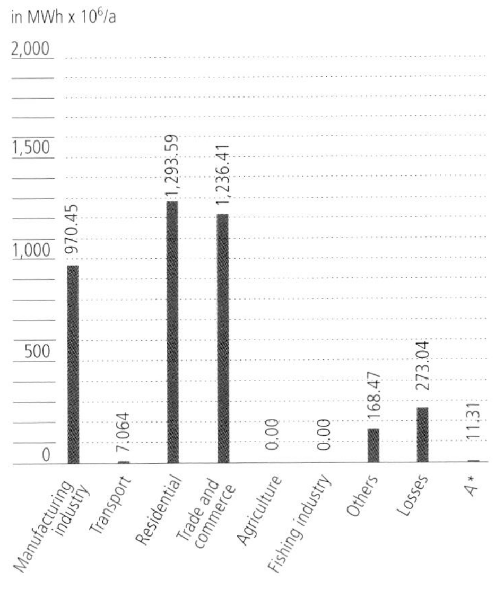

A* Electric energy import/export balance

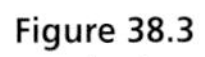

Figure 38.3
Distribution of energy consumption,
USA

Bild 38.3
Verteilung des Energieverbrauchs,
USA

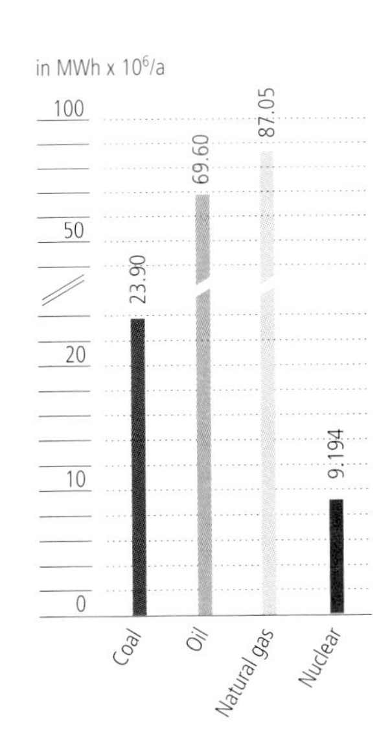

Figure 39.1
Primary energy consumption,
Mexico

Bild 39.1
Primärenergieverbrauch,
Mexiko

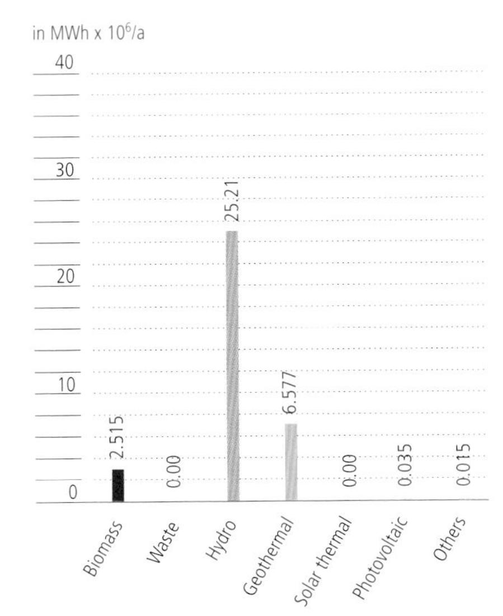

Figure 39.2
Renewable energy consumption,
Mexico

Bild 39.2
Verbrauch aus erneuerbaren Energien,
Mexiko

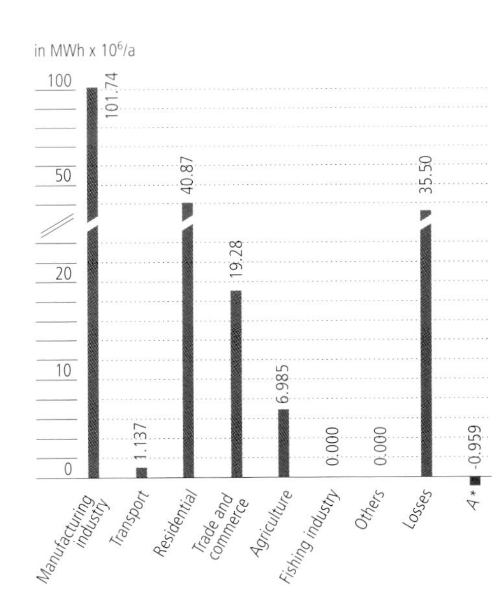

A* Electric energy import/export balance

Figure 39.3
Distribution of energy consumption,
Mexico

Bild 39.3
Verteilung des Energieverbrauchs,
Mexiko

Venezuela is one of the major oil exporting nations and supplies the majority of its energy from this source and from oil production-associated gas. However, renewable energy is used to some degree, with hydroelectric power being the major source. The consumer distribution is similar to that of Mexico, and for both countries it is notable that transmission and other losses on the path from primary to final energy are very high (**Figures 40.1 – 40.3**).

Brazil has almost a balanced energy mix, with the percentages for the use of coal, oil, gas, and nuclear energy being very similar. Yet, compared with the country's renewable energy, mainly hydroelectric power, these sources seem to pale in comparison. Another substantial contribution to renewable energy is the use of biomass. With regard to electrical energy supply, Brazil seems to be entirely independent of fossil fuel power generation. The distribution of energy across consumer types is similar to that of other developed industrial nations, but with a higher percentage of use by agro-industrial sector (**Figures 41.1 – 41.3**).

In contrast to Brazil, Argentina seems to be betting on the use of gas as its main energy source. Gas plays a decidedly important role in the country's primary energy supply. Another important contributor is hydroelectric power, and a small percentage is provided by biomass energy. Energy consumption is distributed across consumers much as in the cases of Brazil and Venezuela, with the main consumer being, again, the industrial sector (**Figures 42.1 – 42.3**).

Venezuela als Erdöl exportierendes Land deckt seinen Primärenergiebedarf primär aus den im Land vorkommenden Erdöl- und Erdgas-Lagerstätten. Ergänzt wird die Energiebereitstellung durch einen Anteil aus erneuerbarer Energie infolge Wasserkraft. Die Verbrauchsstruktur entspricht in etwa der Mexikos, wobei in beiden Ländern auffällig ist, dass im Verhältnis zu den Energieverbräuchen die Verluste im Zuge der Energiebereitstellung sehr hoch sind, **Bilder 40.1 – 40.3**.

Brasilien besitzt eine annähernd ausgewogene Energiebereitstellung durch Kohle, Erdöl, Erdgas und Kernenergie. Diese Primärenergieträger machen jedoch nur einen geringen Anteil dessen aus, was durch Wasserkraft (erneuerbare Energie) dargestellt wird. Zudem erfolgt ein wesentlicher Beitrag durch Biomasse. Bezogen auf den elektrischen Energiebedarf dürfte Brasilien völlig unabhängig von den fossilen Brennstoffen sein. Die Energieverteilung entspricht der eines entwickelten Industrielandes mit einem höheren Landwirtschaftsanteil, **Bilder 41.1 – 41.3**.

Argentinien setzt im Gegensatz zu Brasilien offensichtlich massiv auf den Einsatz von Erdgas, das eine dominierende Rolle bei der Primärenergiebereitstellung spielt. Eine weitere dominierende Rolle spielt der Einsatz von Wasserkraft bei einem geringen Anteil von Biomasse zur Erzeugung von Energien aus erneuerbaren Energieträgern. Die Verteilung der Energieverbräuche verhält sich ähnlich wie die Brasiliens und Venezuelas, dominierender Verbraucher ist hierbei wieder die Industrie, **Bilder 42.1 – 42.3**.

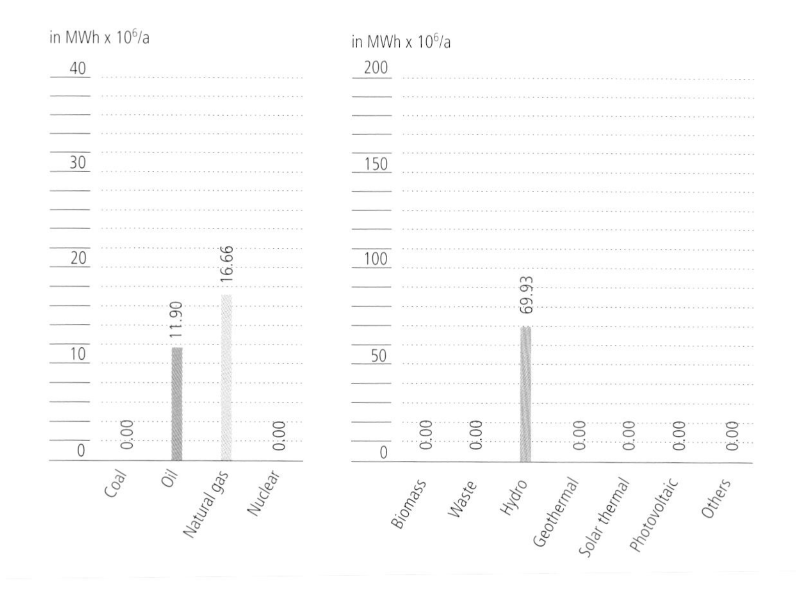

Figure 40.1
Primary energy consumption, Venezuela

Bild 40.1
Primärenergieverbrauch, Venezuela

Figure 40.2
Renewable energy consumption, Venezuela

Bild 40.2
Verbrauch aus erneuerbaren Energien, Venezuela

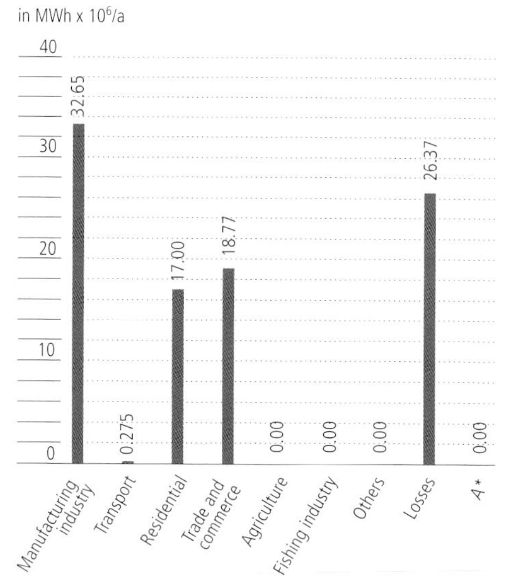

A* Electric energy import/export balance

Figure 40.3
Distribution of energy consumption, Venezuela

Bild 40.3
Verteilung des Energieverbrauchs, Venezuela

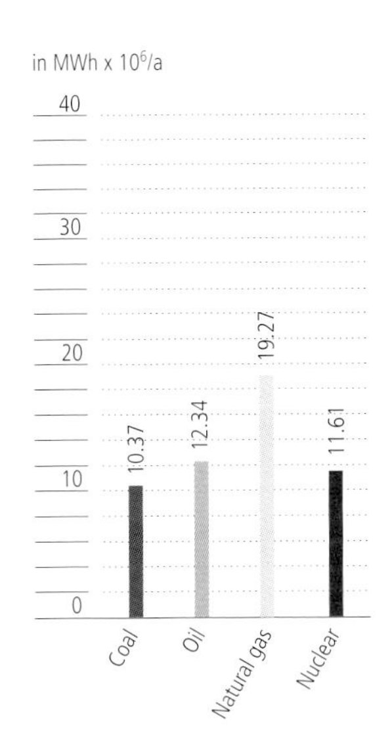

Figure 41.1
Primary energy consumption,
Brazil

Bild 41.1
Primärenergieverbrauch,
Brasilien

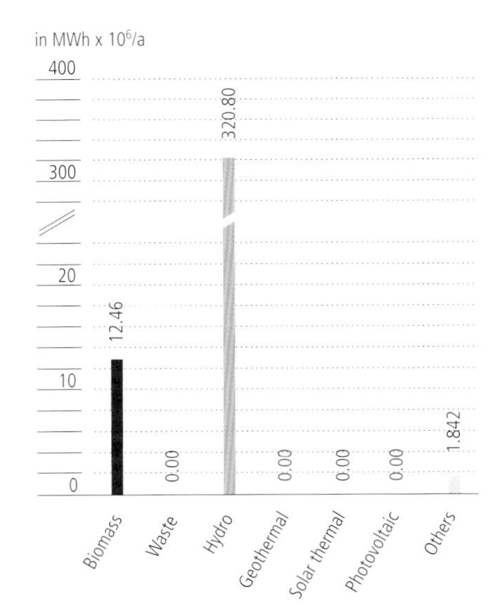

Figure 41.2
Renewable energy consumption,
Brazil

Bild 41.2
Verbrauch aus erneuerbaren Energien,
Brasilien

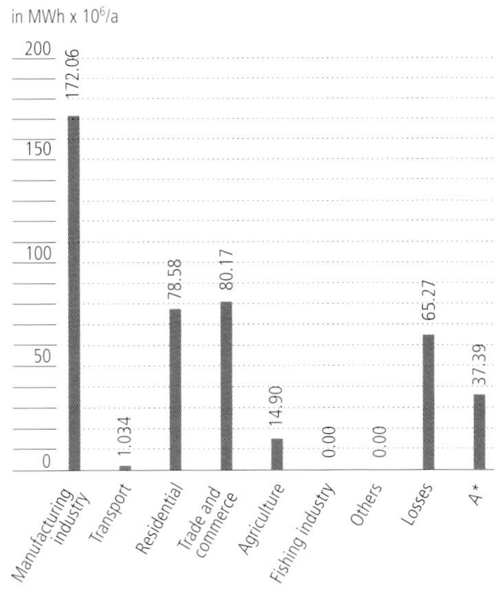

A* Electric energy import/export balance

Figure 41.3
Distribution of energy consumption,
Brazil

Bild 41.3
Verteilung des Energieverbrauchs,
Brasilien

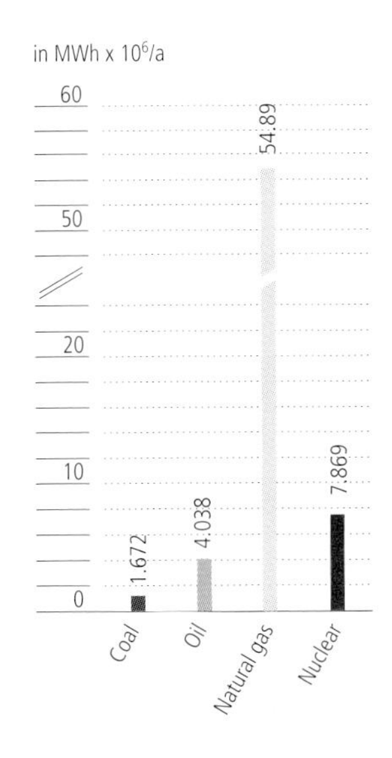

Figure 42.1
Primary energy consumption,
Argentina

Bild 42.1
Primärenergieverbrauch,
Argentinien

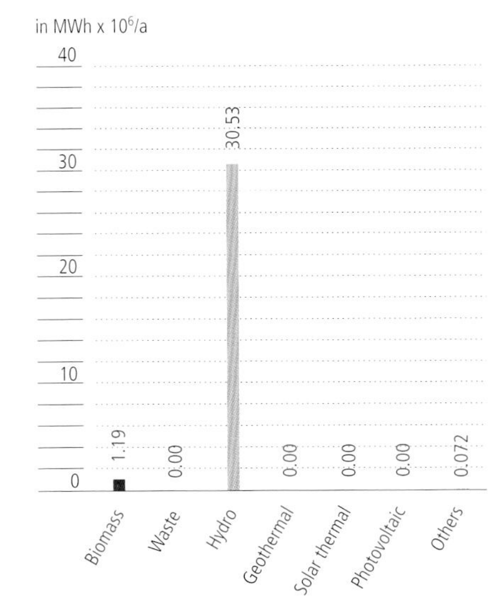

Figure 42.2
Renewable energy consumption,
Argentina

Bild 42.2
Verbrauch aus erneuerbaren Energien,
Argentinien

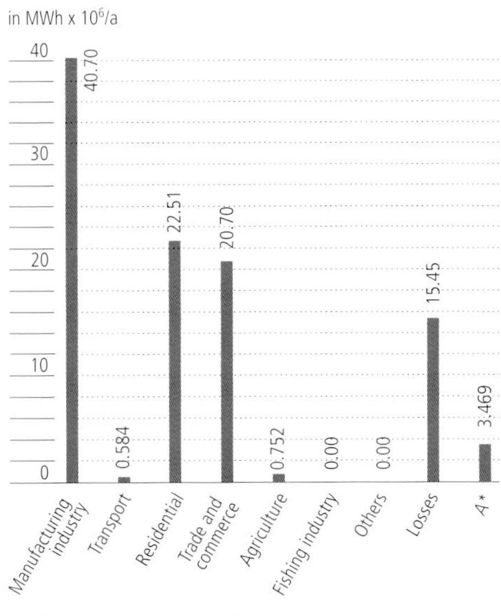

A* Electric energy import/export balance

Figure 42.3
Distribution of energy consumption,
Argentina

Bild 42.3
Verteilung des Energieverbrauchs,
Argentinien

In the images **43.1 – 43.3**, the primary energy consumption and the related consumer sectors are shown for Chile. Similar to Argentina, Chile uses gas to a large degree, followed in magnitude by coal. As for the other two fossil resources, oil is mainly consumed in the transportation sector, and nuclear energy is not represented in the country's energy supply structure at all.

Hydroelectric power is Chile's major renewable energy source, with biomass being a distant second. All other renewable sources are too small to have any statistical impact. For Chile, with its very extended coastline, wind power could be a significant factor for tomorrow's energy supply. In addition, energy gained from the sea, such as tidal or wave energy could be successfully introduced. With the availability of such renewable sources, future energy shortages for the country are unlikely. When analyzing Chile's energy consumer distribution, the large percentage of use by industry is notable, especially when compared with Argentina. High consumption rates can be noted as well for the country's residential sector, commerce, and trade. Energy transmission losses are remarkably small, which seems to indicate that energy generation plants such as district heating plants and other power plants and consumers are located within close proximity, especially when compared with Argentina.

In den **Bildern 43.1 – 43.3** sind die Primärenergieverbräuche aus fossilen Brennstoffen sowie erneuerbarer Energien und die Verteilung derselben in Chile ausgewiesen. Wie in Argentinien setzt Chile in hohem Umfang Erdgas ein, gefolgt von Kohle. Erdöl kommt im Wesentlichen dem Verkehr zugute, während Kernenergie im Gegensatz zu Argentinien keine Rolle spielt.

Die Wasserkraft ist das wesentliche Element des Energiebeitrags aus erneuerbaren Energien, gefolgt in weitem Abstand von Biomasse. Alle weiteren Energiebereitstellungen aus erneuerbaren Energien sind offensichtlich so gering, dass sie keinerlei Rolle spielen. Gerade Chile mit seiner unendlich langen Küste kann in Zukunft in hohem Maß seine Energie durch Windenergie und Gezeiten- und Wellenenergie bereitstellen, so dass sich langfristig kein Energieengpass ergeben dürfte. Bei der Verteilung des Energieverbrauchs sticht der Energiebedarf der Industrie besonders hervor und ist prozentual deutlich größer als der Argentiniens. Ein hoher Energiebedarf besteht auch im Bereich Wohnen, Handel und Gewerbe. Die Energieverluste durch Übertragungen über lange Strecken sind in Chile erstaunlich gering, vergleicht man die Übertragungsverluste mit denen Argentiniens. Insofern ist davon auszugehen, dass die notwendigen Energieerzeugungseinrichtungen (Kraftwerke, Heizwerke usw.) jeweils dicht an den Verbrauchszentren liegen.

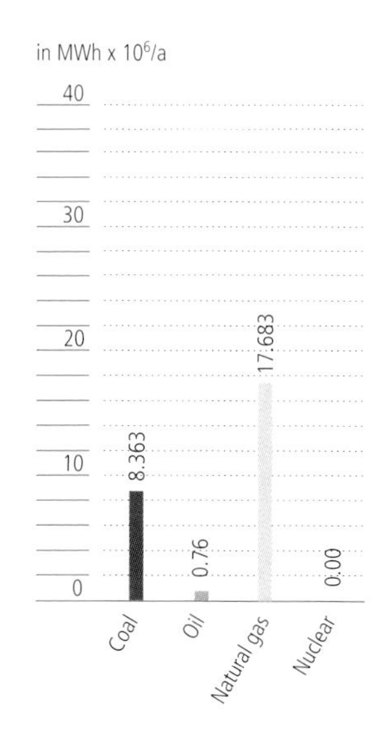

Figure 43.1
Primary energy consumption, Chile

Bild 43.1
Primärenergieverbrauch, Chile

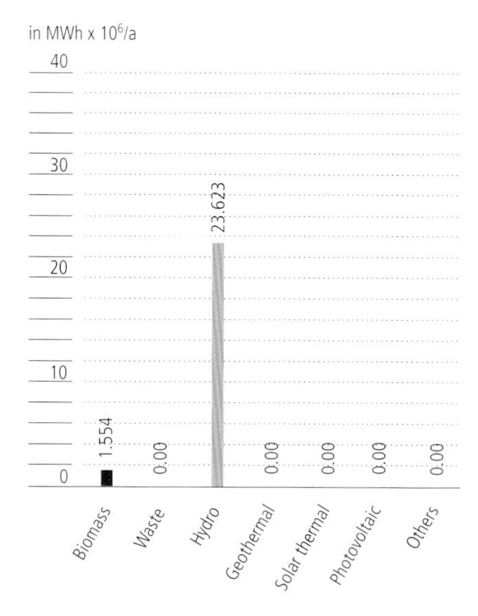

Figure 43.2
Renewable energy consumption, Chile

Bild 43.2
Verbrauch aus erneuerbaren Energien, Chile

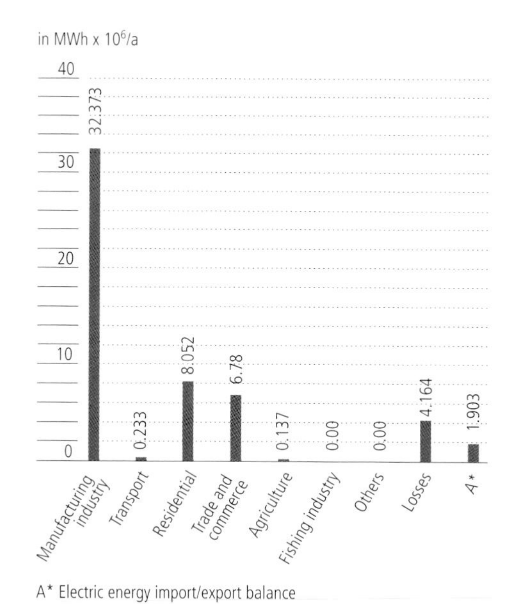

Figure 43.3
Distribution of energy consumption, Chile

Bild 43.3
Verteilung des Energieverbrauchs, Chile

4.3 Africa and Middle East region

The countries of this region, as shown in **Figures 44 – 56**, could greatly benefit in their energy supplies from the abundance of solar energy available in the region. If we consider the consequences of a looming energy crisis, the percentage of solar-based energy generation in these countries would certainly be much greater than documented in IEA 2004 statistics.

Morocco provides final energy by using coal and oil, and to a small degree hydroelectric power. The country's major consumers are the industrial sector and residential dwellings (**Figures 44.1 – 44.3**).

4.3 Region Afrika und Vorderer Orient

Die Länder Nordafrikas und insbesondere des Vorderen Orients, dargestellt in den Diagrammen der **Bilder 44** bis **56**, zeichnen sich durch ein hohes Angebot an Solarenergie aus. Insofern wäre zu erwarten, dass bei entsprechender zeitgemäßer Entwicklung und Erkennung der auf die Welt zukommenden Energiekrise der Einsatz von Solarenergie eine bereits deutlich größere Rolle spielen würde, als sich gemäß den Statistiken der IEA 2004 zeigt.

Marokko deckt seine Primärenergie im Wesentlichen durch Kohle und Erdöl, in geringem Umfang durch Wasserkraft. Die Energieverbrauchsstruktur zeigt den höchsten Bedarf im Bereich der Industrie und im Wohnen, **Bilder 44.1 – 44.3**.

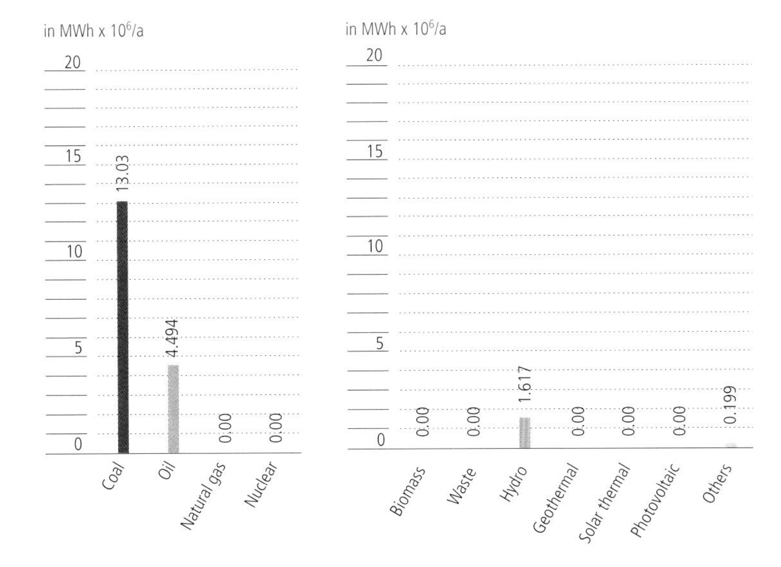

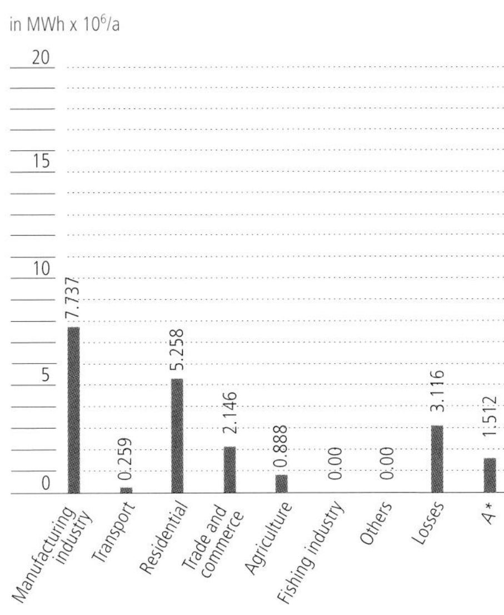

A* Electric energy import/export balance

Figure 44.1
Primary energy consumption, Morocco

Bild 44.1
Primärenergieverbrauch, Marokko

Figure 44.2
Renewable energy consumption, Morocco

Bild 44.2
Verbrauch aus erneuerbaren Energien, Marokko

Figure 44.3
Distribution of energy consumption, Morocco

Bild 44.3
Verteilung des Energieverbrauchs, Marokko

In contrast to Morocco, Algeria mainly uses gas as its energy source and oil to only a small degree; the contributions of hydroelectric power and renewable energy sources are insignificant. The greatest share in consumption is the residential sector (for air-conditioning and potentially water treatment and desalination of sea water), which dominates other shown consumer sectors (**Figures 45.1 – 45.3**).

Tunisia uses mainly oil and gas for energy generation, and only insignificant amounts of hydroelectric power. The main consumer sector is industry, with a balanced mix shown for the sectors of residential, commerce, and trade (**Figures 46.1 – 46.3**).

Libya is an oil-exporting nation and relies on its own locally produced fossil fuel resources such as oil and gas. Renewable energy seems to be still absent from the statistics. Residential consumers are dominant in the distribution, followed by the percentages for transmission losses and the use of energy by industry (**Figures 47.1 – 47.3**).

Im Gegensatz zu Marokko setzt offensichtlich Algerien im Wesentlichen auf den Energieträger Gas mit kleinen Anteilen an Erdöl und annähernd vernachlässigbarem Kleinanteil im Bereich Wasserkraft. Der größte Energieverbrauch entsteht im Wohnbereich (Klimatisierung, unter Umständen Wasseraufbereitung) durch Meerwasser-Entsalzungsanlagen und dominiert alle weiteren Verbräuche, **Bilder 45.1 – 45.3**.

Tunesien setzt bezüglich der Darstellung der Primärenergieverbräuche auf Erdöl und Erdgas und in geringem Umfang auf Wasserkraft. Die Verteilung des Energieverbrauchs zeigt eine Dominanz im Bereich Industrie und einen ausgewogenen Bedarf bei Wohnen, Handel und Gewerbe, **Bilder 46.1 – 46.3**.

Libyen als Erdöl exportierendes Land setzt bezüglich der Deckung des Primärenergiebedarfs auf die heimischen Energieträger Erdöl und Erdgas, wobei erneuerbare Energien zurzeit offensichtlich noch keine Rolle spielen. Die Verteilung des Energieverbrauchs zeigt eine deutliche Dominanz im Wohnbereich, gefolgt von Energieverlusten und Energieverbräuchen im Bereich Industrie, **Bilder 47.1 – 47.3**.

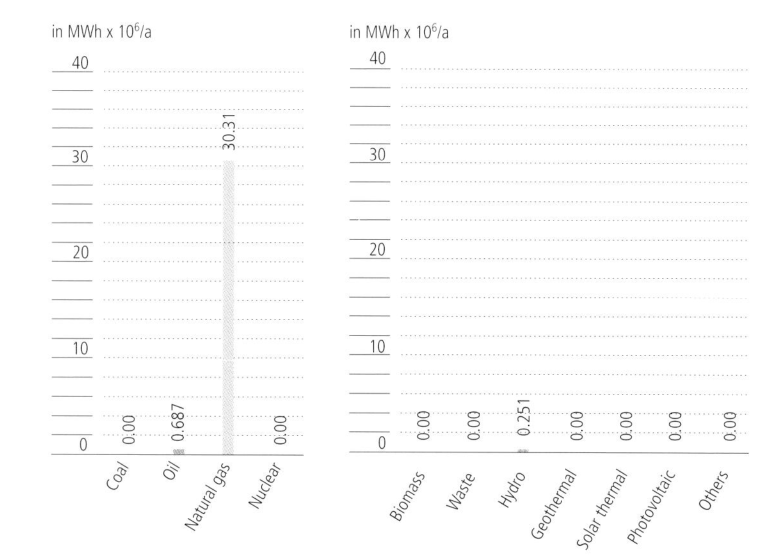

Figure 45.1
Primary energy consumption, Algeria

Bild 45.1
Primärenergieverbrauch, Algerien

Figure 45.2
Renewable energy consumption, Algeria

Bild 45.2
Verbrauch aus erneuerbaren Energien, Algerien

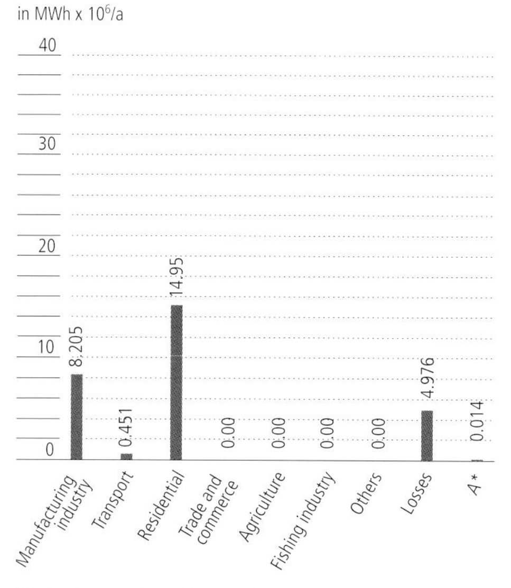

A* Electric energy import/export balance

Figure 45.3
Distribution of energy consumption, Algeria

Bild 45.3
Verteilung des Energieverbrauchs, Algerien

in MWh x 10⁶/a
20
15
10
5
0
0.00
1.08
11.79
0.00
Coal
Oil
Natural gas
Nuclear

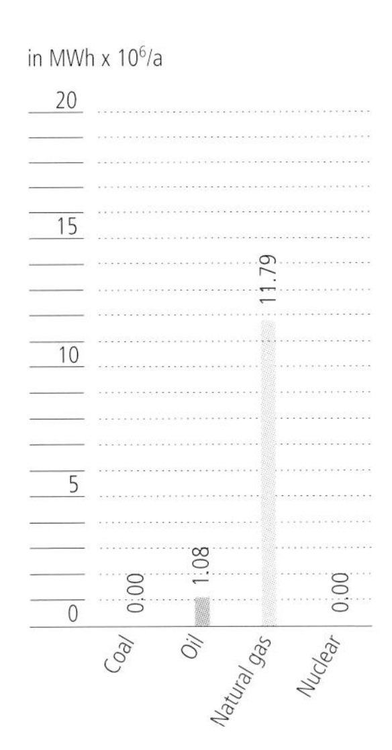

Figure 46.1
Primary energy consumption, Tunisia

Bild 46.1
Primärenergieverbrauch, Tunesien

in MWh x 10⁶/a
0.20
0.15
0.10
0.05
0
0.00
0.00
0.154
0.00
0.00
0.00
0.044
Biomass
Waste
Hydro
Geothermal
Solar thermal
Photovoltaic
Others

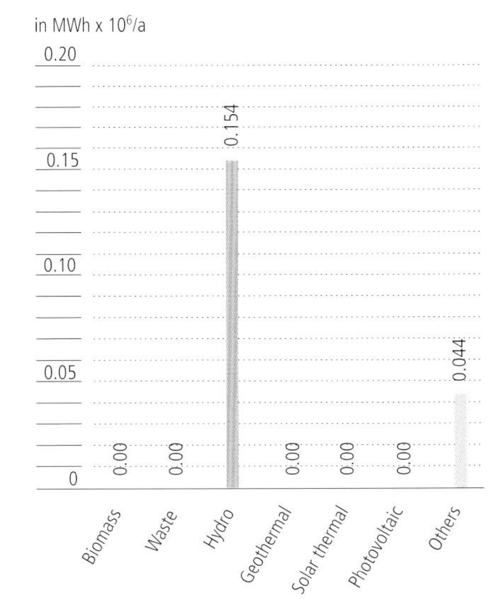

Figure 46.2
Renewable energy consumption, Tunisia

Bild 46.2
Verbrauch aus erneuerbaren Energien, Tunesien

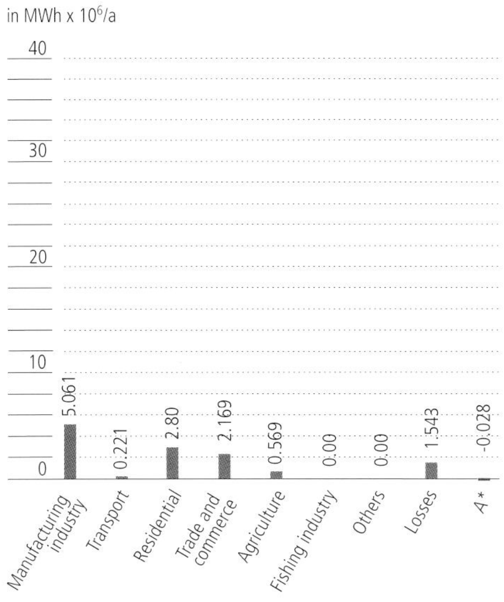

in MWh x 10⁶/a
40
30
20
10
0
5.061
0.221
2.80
2.169
0.569
0.00
0.00
1.543
-0.028
Manufacturing industry
Transport
Residential
Trade and commerce
Agriculture
Fishing industry
Others
Losses
A*

A* Electric energy import/export balance

Figure 46.3
Distribution of energy consumption, Tunisia

Bild 46.3
Verteilung des Energieverbrauchs, Tunesien

in MWh x 10⁶/a
40
30
20
10
0
0.00
16.30
3.907
0.00
Coal
Oil
Natural gas
Nuclear

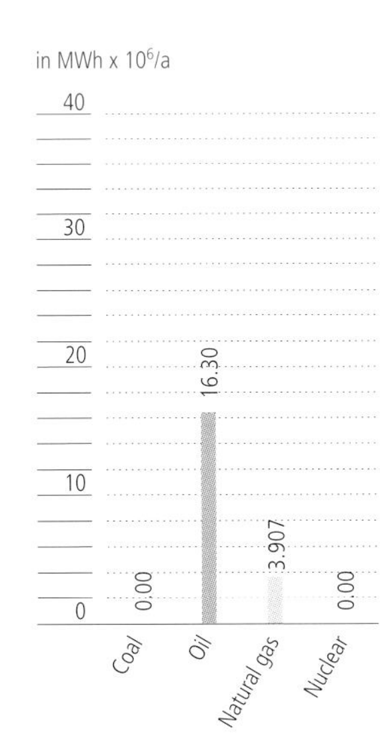

Figure 47.1
Primary energy consumption, Libya

Bild 47.1
Primärenergieverbrauch, Libyen

in MWh x 10⁶/a
40
30
20
10
0
0.00
0.00
0.00
0.00
0.00
0.00
0.00
Biomass
Waste
Hydro
Geothermal
Solar thermal
Photovoltaic
Others

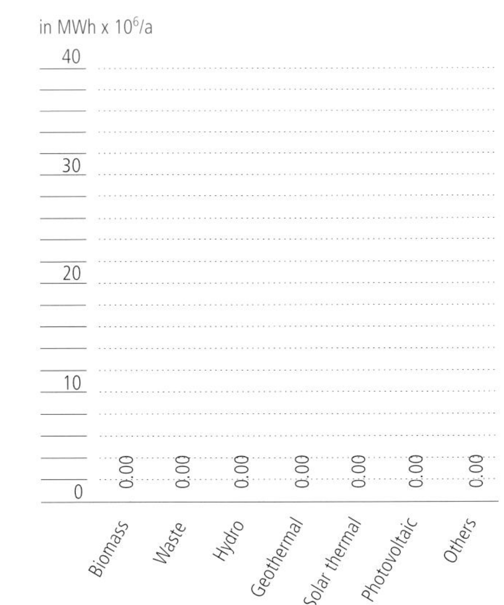

Figure 47.2
Renewable energy consumption, Libya

Bild 47.2
Verbrauch aus erneuerbaren Energien, Libyen

in MWh x 10⁶/a
40
30
20
10
0
2.907
0.00
7.742
0.00
1.410
0.00
0.00
5.74
0.00
Manufacturing industry
Transport
Residential
Trade and commerce
Agriculture
Fishing industry
Others
Losses
A*

A* Electric energy import/export balance

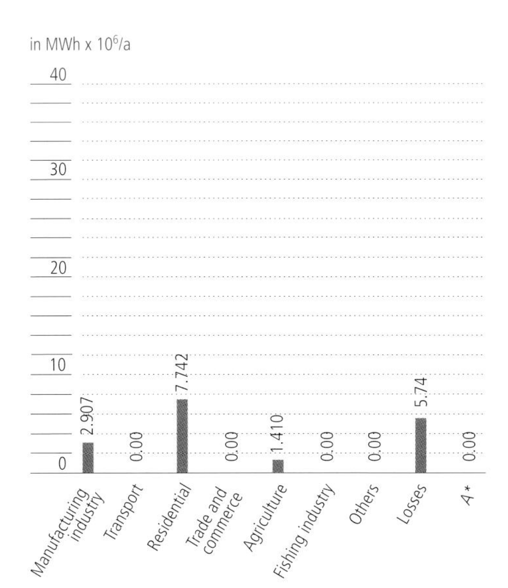

Figure 47.3
Distribution of energy consumption, Libya

Bild 47.3
Verteilung des Energieverbrauchs, Libyen

Egypt gains its primary energy from gas and oil (50 %), followed by renewable resources, mainly the hydroelectric power generated by the Aswan reservoir, which is considered the largest hydroelectric power compound in Africa. Distribution of consumption shows two dominate sectors, industry and residential, with commerce and trade also being important areas. As shown in **Figure 48.3**, Egypt is able to export a small amount of its energy to neighboring countries.

Nigeria belongs to the group of the world's largest oil producers. To meet its own power demand, the country uses mainly gas and hydroelectric power, with oil making only a minor contribution. Energy is finally consumed by the residential sector and the service industry, followed by the percentages for transmission losses (**Figures 49.1 – 49.3**).

South Africa provides energy first with coal and second with nuclear power. About 2 % of the country's energy comes from renewable sources. However, it is notable that, besides the use of biomass, solar thermal energy is well represented. Energy consumption in the country is largely dominated by the industrial sector, which is twice as large as the consumption of any of the other sectors. A small amount of energy is exported to South Africa's neighbors (**Figures 50.1 – 50.3**).

Ägypten erzeugt seine Primärenergie im Wesentlichen durch Erdgas, gefolgt von Erdöl (ca. 50 Prozent) bei einem Anteil von ca. 10 Prozent an erneuerbaren Energien durch Wasserkraft (insbesondere Assuan-Staudamm). Die Verteilung des Energieverbrauchs zeigt zwei dominierende Bereiche, Industrie und Wohnen und drei ähnlich große Energieabnehmer (Handel und Gewerbe, Verluste und sonstige Energieverbraucher), **Bilder 48.1 – 48.2**. Wie **Bild 48.3** ausweist, exportiert Ägypten in geringem Umfang elektrische Energie in benachbarte Länder.

Nigeria als ein großes Erdöl exportierendes Land nutzt in geringem Umfang Erdöl und in hohem Umfang Erdgas sowie Wasserkraft zur Erzeugung seiner Energien. Der größte Energieverbraucher dabei ist der Wohn- und Dienstleistungsbereich, gefolgt von Verlusten zur Übertragung von Energien, **Bilder 49.1 – 49.3**.

Südafrika stellt seine Energieverbräuche primär durch Kohle und sekundär durch Kernenergie im Bereich der fossilen Brennstoffe dar. Ca. 2 Prozent der notwendigen Energien wird durch erneuerbare Energien beigesteuert, wobei bemerkenswert ist, dass neben Biomasse vor allem auch Solarthermie eine Rolle spielt. Der Energieverbrauch des Landes wird deutlich dominiert durch den Verbrauch der Industrie. Dieser ist etwa doppelt so groß wie alle sonstigen Energieverbraucher. In geringem Umfang wird elektrische Energie in benachbarte Regionen exportiert, **Bilder 50.1 – 50.3**.

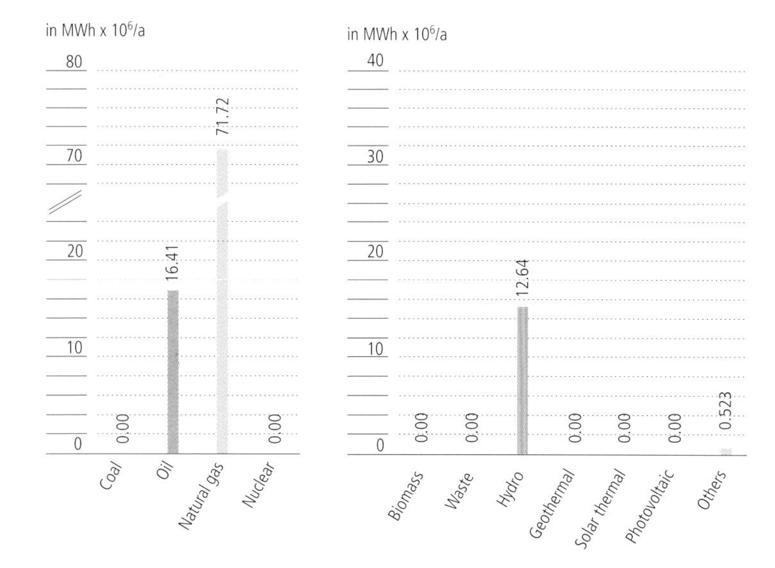

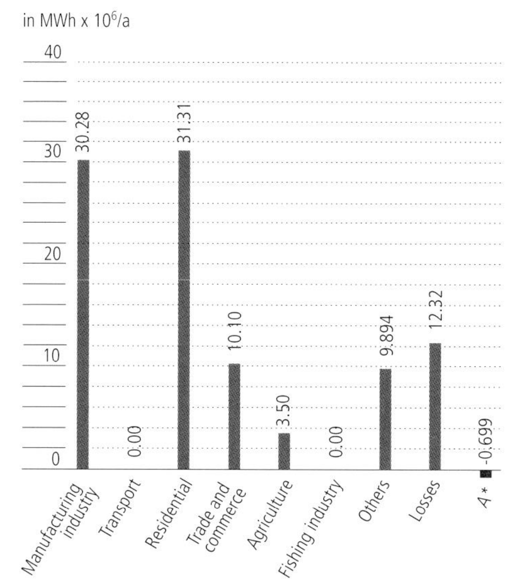

Figure 48.1
Primary energy consumption, Egypt

Bild 48.1
Primärenergieverbrauch, Ägypten

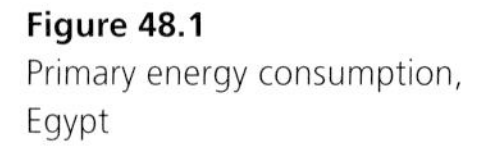

Figure 48.2
Renewable energy consumption, Egypt

Bild 48.2
Verbrauch aus erneuerbaren Energien, Ägypten

Figure 48.3
Distribution of energy consumption, Egypt

Bild 48.3
Verteilung des Energieverbrauchs, Ägypten

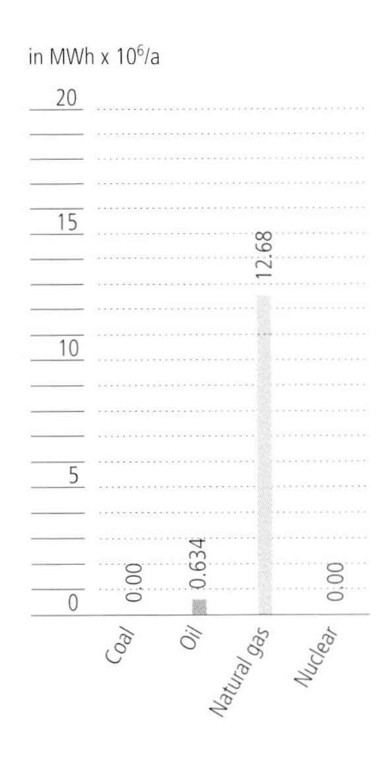

Figure 49.1
Primary energy consumption,
Nigeria

Bild 49.1
Primärenergieverbrauch,
Nigeria

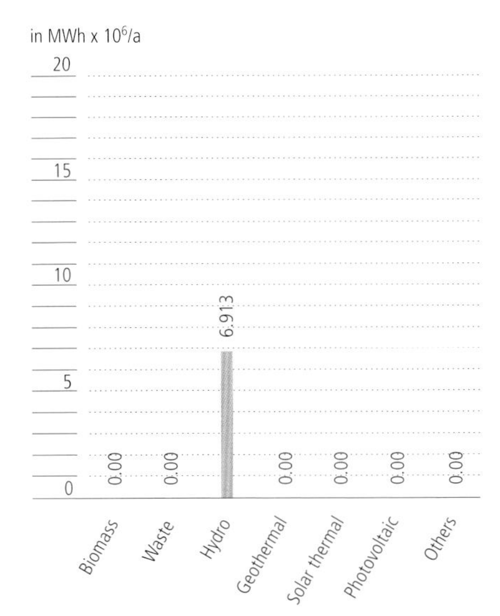

Figure 49.2
Renewable energy consumption,
Nigeria

Bild 49.2
Verbrauch aus erneuerbaren Energien,
Nigeria

A* Electric energy import/export balance

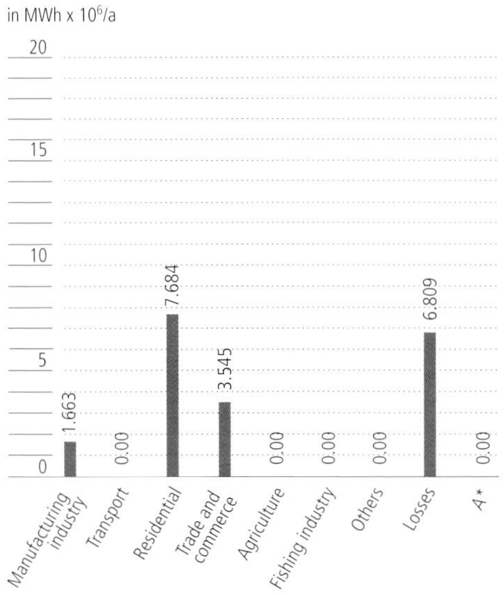

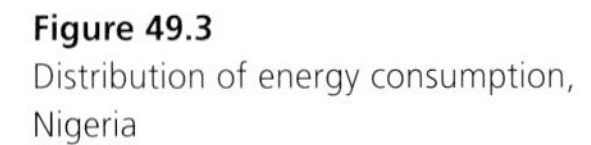

Figure 49.3
Distribution of energy consumption,
Nigeria

Bild 49.3
Verteilung des Energieverbrauchs,
Nigeria

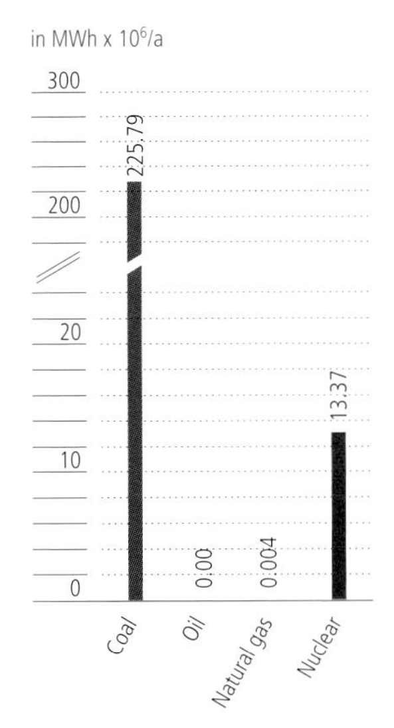

Figure 50.1
Primary energy consumption,
South Africa

Bild 50.1
Primärenergieverbrauch,
Südafrika

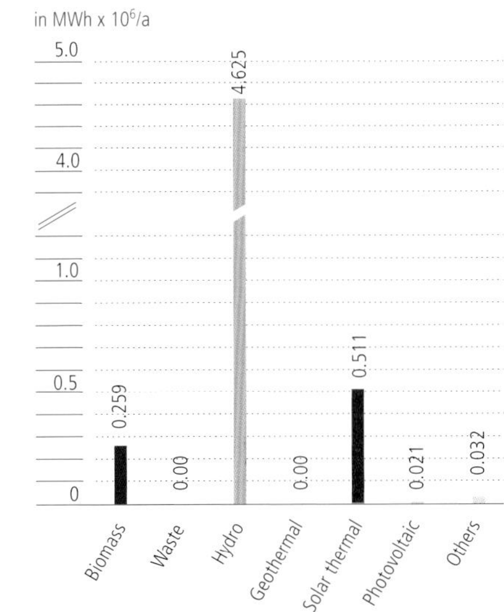

Figure 50.2
Renewable energy consumption,
South Africa

Bild 50.2
Verbrauch aus erneuerbaren Energien,
Südafrika

A* Electric energy import/export balance

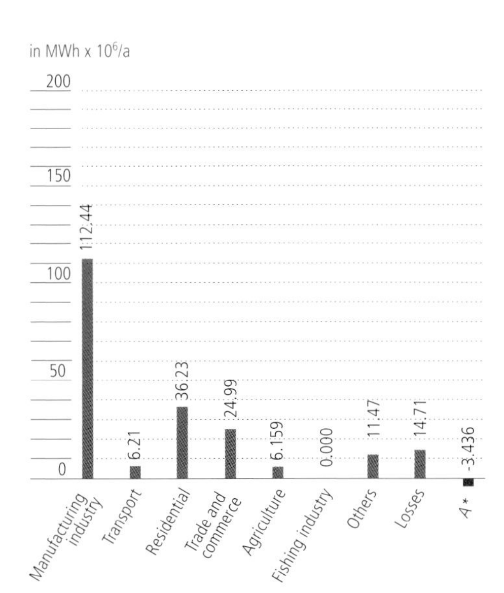

Figure 50.3
Distribution of energy consumption,
South Africa

Bild 50.3
Verteilung des Energieverbrauchs,
Südafrika

For all countries in the Middle East, it can be said that generally gas and oil are the major sources of energy generation.

According to IEA 2004 data, the energy requirements of Iran are satisfied by oil to a smaller degree, but mainly by gas (**Figures 51.1 – 51.3**). A less significant role is played by hydroelectric power. As in other comparable countries, the distribution of energy consumers shows the main sector to be industry, followed closely by residential, trade, and commerce. Transmission losses play an important role, as well the agricultural sector.

Iraq is an oil-exporting nation and supplies its energy mainly from that resource, followed insignificantly by other sources such as hydroelectric power. Due to a current lack of reliable data, no conclusions with regard to consumption can be made (**Figures 52.1 – 52.3**).

The United Arab Emirates – and among this group especially the member country Abu Dhabi – export oil and gas as well. The UAE supply the local energy demand with their own resources. Renewable energy is too small to be recorded by the IEA report and will only be developed into a significant sector in the future. Distributions of consumer sectors show, in principle, the areas of residential dwelling, commerce, and the trade sector; transmission losses are small, as are the contributions of agriculture and industry (**Figures 53.1 – 53.3**).

Im Bereich des Vorderen Orients wird augenfällig, dass in den entsprechenden Ländern Erdgas und Erdöl die dominierende Rolle spielen.

Der Iran deckt gemäß den Statistiken der IEA 2004, bilder **51.1 – 51.3**, seinen Energiebedarf im Wesentlichen durch Erdöl und vor allem Erdgas, in geringem Umfang durch Wasserkraft. Bei der Verteilung der Energieverbräuche zeigt sich wie üblich als größter Einzelverbraucher die Industrie, dicht gefolgt vom Bereich Wohnen. Handel und Gewerbe sowie Verluste machen jeweils einen nennenswerten Anteil an den Energieverbräuchen aus, wobei zudem ein nennenswerter Energiebedarf im Bereich der Landwirtschaft festgestellt werden kann.

Das ebenfalls Erdöl exportierende Land Irak deckt fast ausschließlich seinen Energiebedarf durch Erdöl und in sehr geringen Umfang durch Wasserkraft. Infolge fehlender Daten kann die Verteilung des Energieverbrauchs nicht genau dargestellt werden, **Bilder 52.1 – 52.3**.

Die Vereinigten Arabischen Emirate, hier insbesondere Abu Dhabi, exportieren ebenfalls Erdöl und Erdgas und decken ihren Bedarf selbst in ihrer Region mit den heimischen Primärenergieträgern. Ein Verbrauch an erneuerbaren Energien dürfte sich im nennenswerten Umfang erst in der Zukunft einstellen und ist zurzeit noch so gering, dass er durch die Statistik der IEA nicht erfasst werden konnte.

Die Verteilung des Energieverbrauchs liegt in den Vereinigten Arabischen Emiraten im Wesentlichen im Bereich Wohnen, Handel und Gewerbe. Erst mit deutlichem Abstand folgen die Bereiche Industrie und Landwirtschaft sowie Verluste, **Bilder 53.1 – 53.3**.

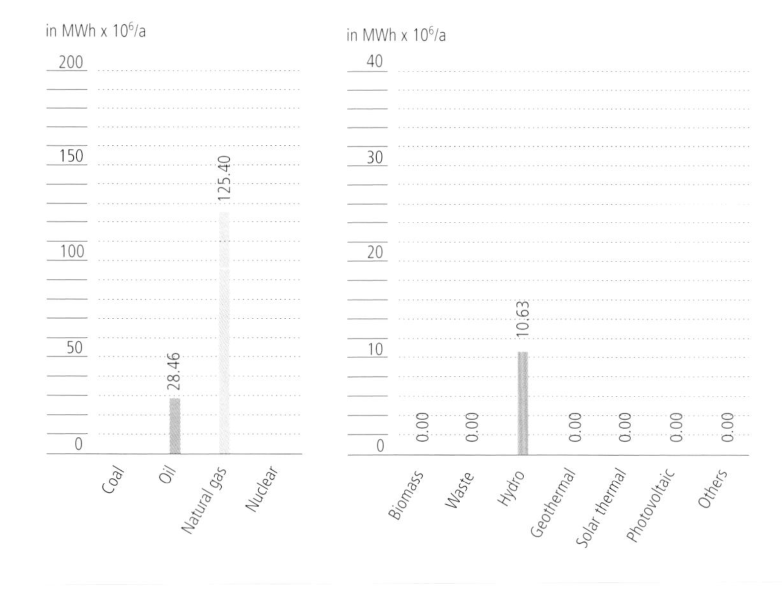

Figure 51.1
Primary energy consumption,
Iran

Figure 51.2
Renewable energy consumption,
Iran

Bild 51.1
Primärenergieverbrauch,
Iran

Bild 51.2
Verbrauch aus erneuerbaren Energien,
Iran

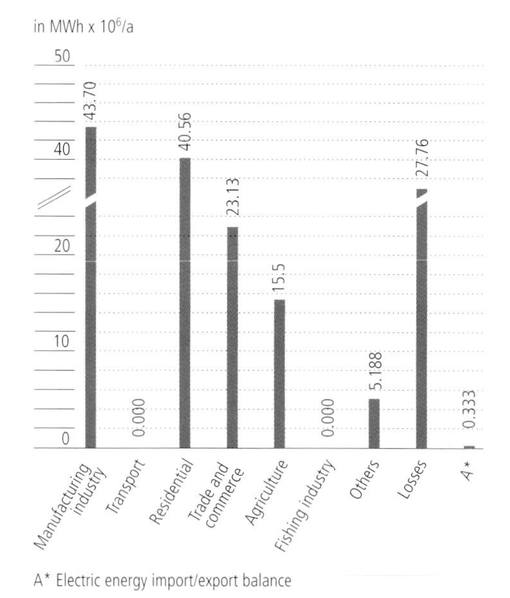

Figure 51.3
Distribution of energy consumption,
Iran

Bild 51.3
Verteilung des Energieverbrauchs,
Iran

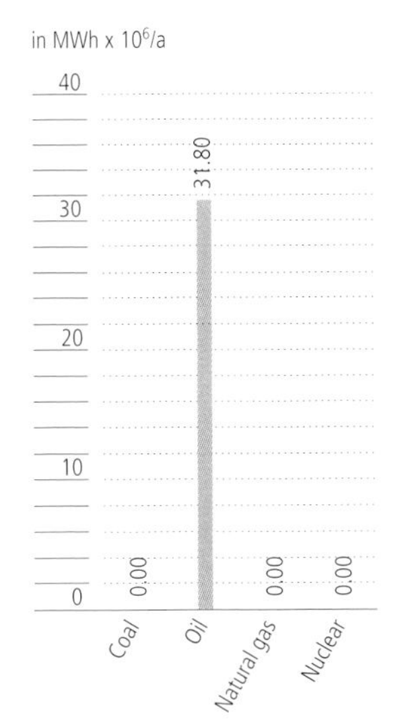

Figure 52.1
Primary energy consumption,
Iraq

Bild 52.1
Primärenergieverbrauch,
Irak

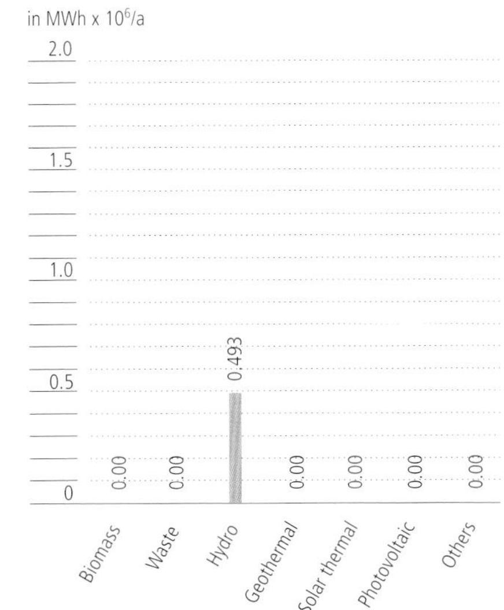

Figure 52.2
Renewable energy consumption,
Iraq

Bild 52.2
Verbrauch aus erneuerbaren Energien,
Irak

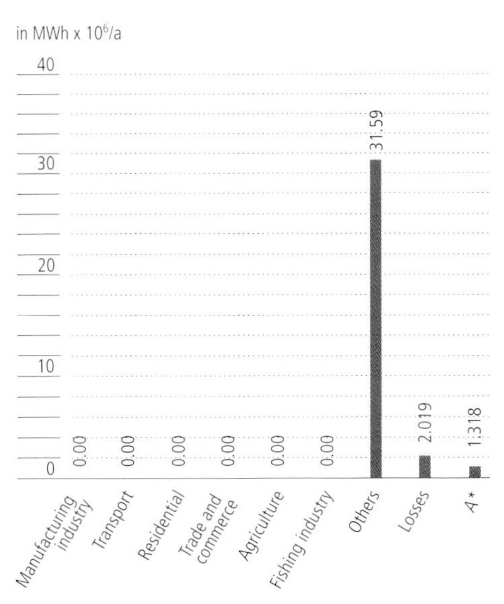

A* Electric energy import/export balance

Figure 52.3
Distribution of energy consumption,
Iraq

Bild 52.3
Verteilung des Energieverbrauchs,
Irak

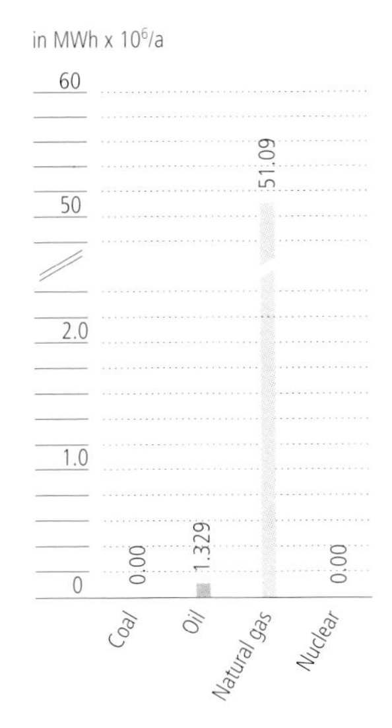

Figure 53.1
Primary energy consumption,
UAE

Bild 53.1
Primärenergieverbrauch,
UAE

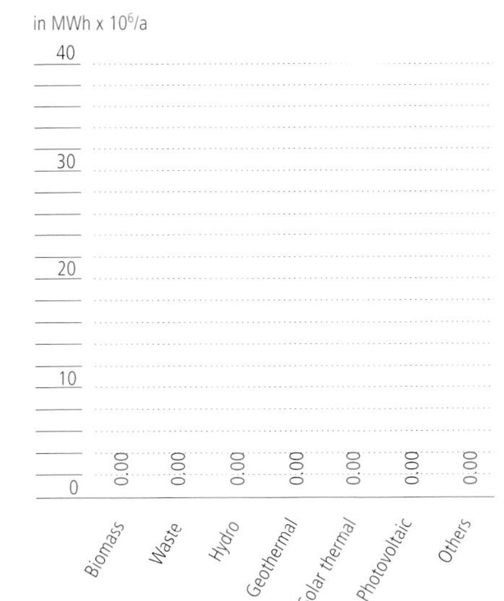

Figure 53.2
Renewable energy consumption,
UAE

Bild 53.2
Verbrauch aus erneuerbaren Energien,
UAE

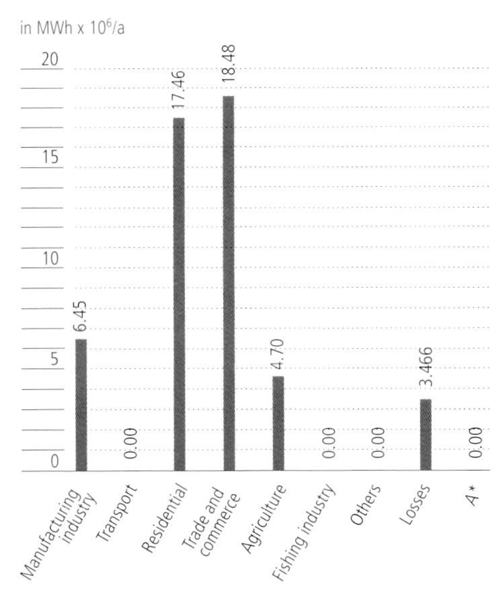

A* Electric energy import/export balance

Figure 53.3
Distribution of energy consumption,
UAE

Bild 53.3
Verteilung des Energieverbrauchs,
UAE

Saudi Arabia is the world's foremost oil- and gas-exporting country, and its primary energy demand is supplied by these resources. Renewable energy sources are obviously too small to be recorded by IEA data. Energy is mainly used for electric power generation for air-conditioning of residential and commercial buildings. Energy consumption in the area of industry is astonishingly small (**Figures 54.1 – 54.3**).

Oman, shown in **Figures 55.1 – 55.3**, is an oil-exporting economy as well, and it supplies its own demand from the same sources – gas and oil. As seen previously in the countries of this region, the percentage of renewable sources is too small to be documented by the International Energy Agency. The consumption structure is similar to that of Saudi Arabia, with the majority of energy being used in the residential, trade, and commerce sectors.

Yemen satisfies its energy demand with oil, and other, renewable, sources are not documented. The residential sector and agriculture are the two main consumers, and transmission losses are significant (**Figures 56.1 – 56.3**).

Saudi-Arabien als das wesentliche Öl und Gas exportierende Land der Welt deckt seinen Primärenergieverbrauch aus seinen heimischen Quellen. Eine Bedarfsdeckung durch erneuerbare Energien ist offensichtlich so gering, dass hierzu entsprechende Angaben seitens des IEA fehlen. Ein wesentlicher Anteil des Energieverbrauchs wird durch die elektrische Energieversorgung und insbesondere Kühlung von Wohnobjekten sowie Handels- und Gewerbeimmobilien genutzt. Der Anteil des Energieverbrauchs für Industrie ist für dieses Land erstaunlich gering, **Bilder 54.1 – 54.3**.

Der Oman, **Bilder 55.1 – 55.3**, ein ebenfalls Erdöl exportierendes Land, deckt seinen Energiebedarf wiederum aus heimischen Quellen, d.h. Erdöl und Erdgas. Wiederum feststellbar ist, dass der Beitrag zum Energieverbrauch durch erneuerbare Energien offensichtlich so gering ist, dass er von der IEA nicht erfasst wurde. Die Verbrauchsstruktur des Oman gleicht annähernd der Saudi-Arabiens, d.h. der wesentliche Teil der Energieverbräuche wird genutzt zum Betreiben von Wohnbereichen, Handel und Gewerbe.

Der Jemen deckt seinen Energiebedarf im Wesentlichen nur aus Erdöl und ergänzt dieses Angebot durch keine nennenswerten Mengen an erneuerbaren Energien. Infolgedessen fehlen hier wiederum entsprechende statistische Angaben. Wohnen und Landwirtschaft sind die dominierenden Größen bei der Verteilung des Energieverbrauchs, gefolgt von Übertragungsverlusten, **Bilder 56.1 – 56.3**.

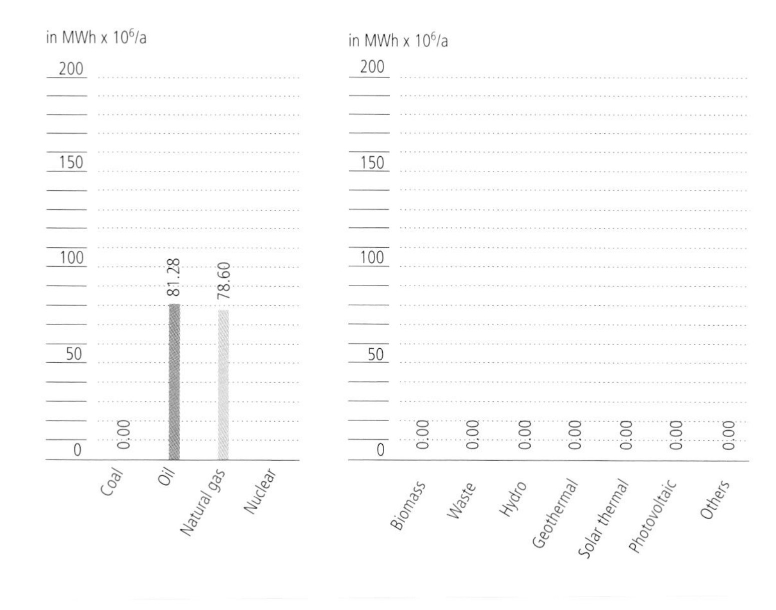

Figure 54.1
Primary energy consumption, Saudi Arabia

Bild 54.1
Primärenergieverbrauch, Saudi-Arabien

Figure 54.2
Renewable energy consumption, Saudi Arabia

Bild 54.2
Verbrauch aus erneuerbaren Energien, Saudi-Arabien

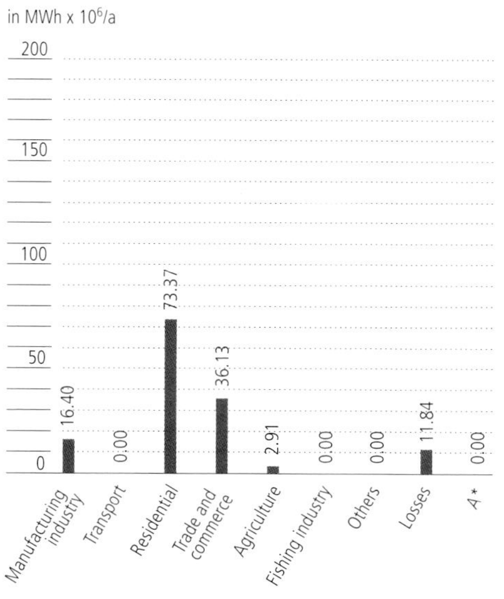

A* Electric energy import/export balance

Figure 54.3
Distribution of energy consumption, Saudi Arabia

Bild 54.3
Verteilung des Energieverbrauchs, Saudi-Arabien

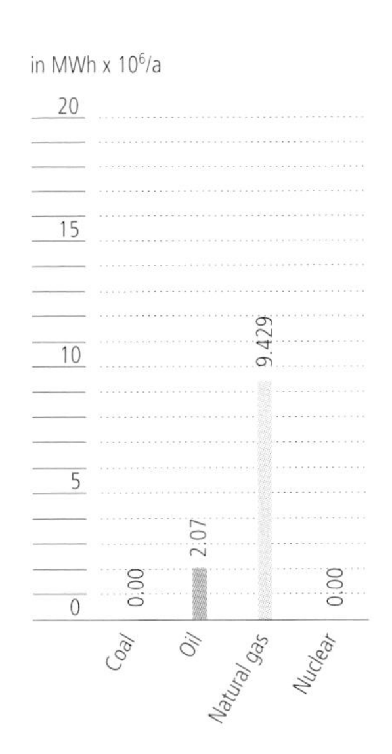

Figure 55.1
Primary energy consumption, Oman

Bild 55.1
Primärenergieverbrauch, Oman

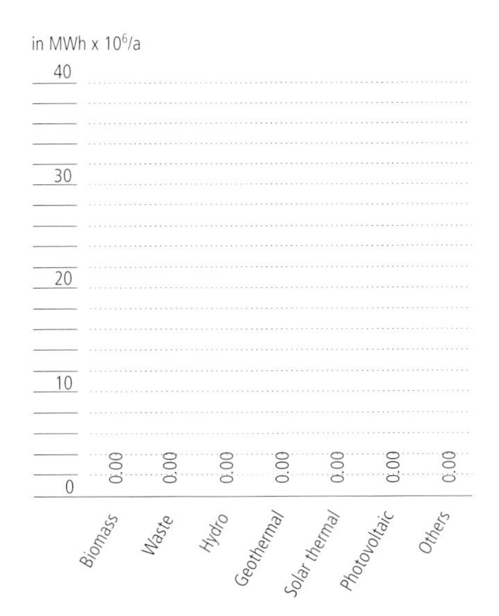

Figure 55.2
Renewable energy consumption, Oman

Bild 55.2
Verbrauch aus erneuerbaren Energien, Oman

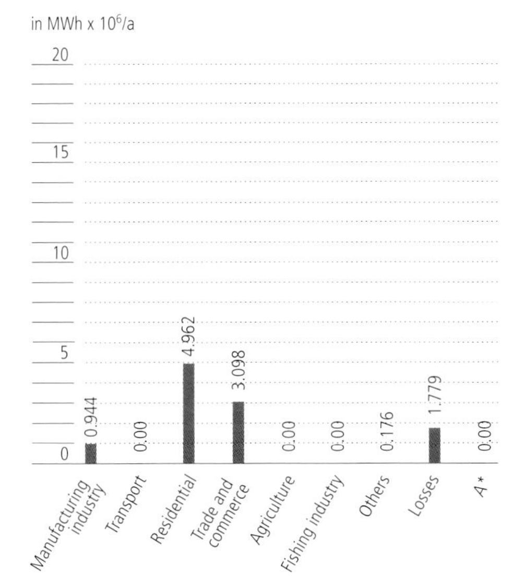

Figure 55.3
Distribution of energy consumption, Oman

Bild 55.3
Verteilung des Energieverbrauchs, Oman

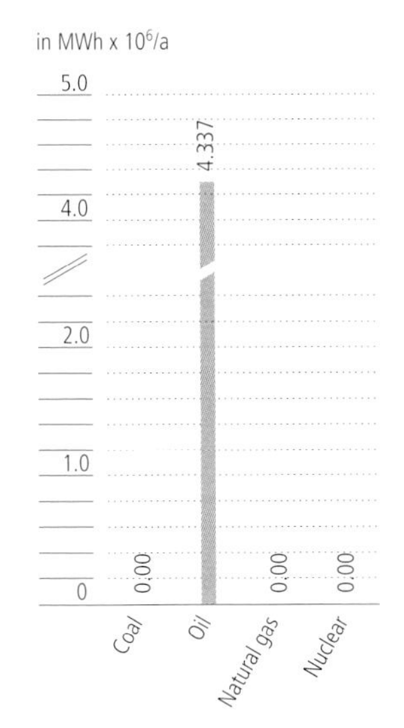

Figure 56.1
Primary energy consumption, Yemen

Bild 56.1
Primärenergieverbrauch, Jemen

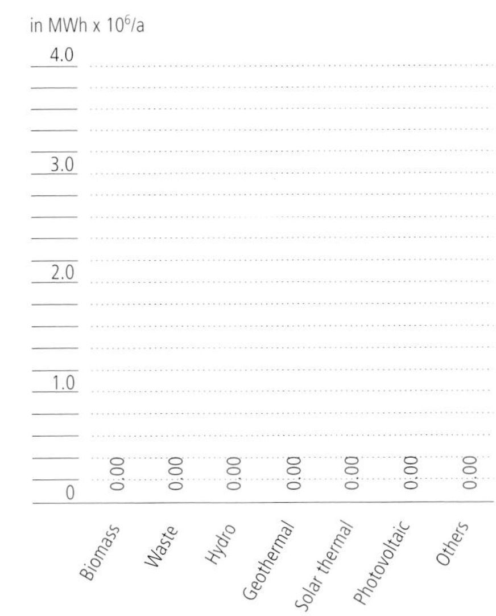

Figure 56.2
Renewable energy consumption, Yemen

Bild 56.2
Verbrauch aus erneuerbaren Energien, Jemen

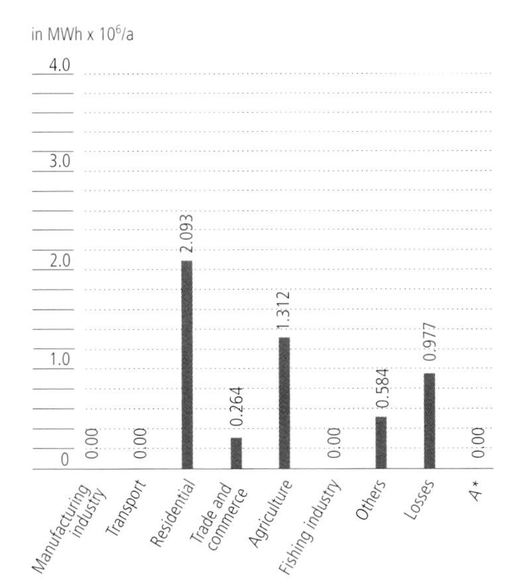

Figure 56.3
Distribution of energy consumption, Yemen

Bild 56.3
Verteilung des Energieverbrauchs, Jemen

In April 2005, the German Aerospace Center, under contract by the German Federal Ministry for the Environment, Nature Conservation, and Nuclear Safety, conducted an in-depth study of the Mediterranean region of Europe that was aimed at the potential replacement of fossil fuel energy supplies with renewable sources.

The countries studied were Portugal, Spain, Italy, Greece, Morocco, Algeria, Tunisia, Libya, Egypt, Israel, Jordan, Lebanon, Syria, Turkey, Iraq, Iran, Saudi Arabia, Kuwait, Bahrain, Qatar, United Arab Emirates, Oman, and Yemen. **Figure 57** shows the grid of electric high-voltage power supply in the Mediterranean region as well as a desired replacement by renewable sources per country. **Figure 58** presents the electric power generation in these countries, and **Figure 59** shows the renewable power that is recommended for implementation by 2050. The resulting land area necessary, as well as the coefficients of performance and the available energies, are presented in **Table 3**.

Figure 60 shows the CO_2 emissions in millions of tons per year that result from the conventional energy production out of fossil fuels for all of the above-listed countries and the savings of emissions if a switch to renewable sources could be implemented.

Im April 2005 wurde eine Studie für die mediterrane Region und den Vorderen Orient durch das Deutsche Aerospace Center (DLR) im Auftrag des Bundesministeriums für Umwelt-, Naturschutz und Reaktorsicherheit vorgestellt, die den langfristigen Umbau der Energieversorgung hin zu erneuerbaren Energien aufzeigte.

Die in die Projektion einbezogenen Länder waren Portugal, Spanien, Italien, Griechenland, Marokko, Algerien, Tunesien, Libyen, Ägypten, Israel, Jordanien, Libanon, Syrien, Türkei, Irak, Iran, Saudi-Arabien, Kuwait, Bahrain, Katar, Vereinigte Arabische Emirate, Oman und Jemen. **Bild 57** zeigt eine transmediterrane Hochspannungsversorgung und gleichzeitig eine erste Angabe, in welchen Ländern und Regionen erneuerbare Energien zum Einsatz kommen sollen, um die Abhängigkeit von fossilen Brennstoffen aufzulösen. **Bild 58** stellt die elektrische Energieerzeugung in den aufgeführten Ländern, **Bild 59** die zu installierende Kapazität bis jeweils 2050 dar. Die hieraus resultierenden Landverbräuche, Wirkungsgrade und zur Verfügung stehenden Energien sind in **Tabelle 3** ausgewiesen.

Bild 60 weist die CO_2-Emissionen zur Erzeugung der elektrischen Energien in Millionen Tonnen pro Jahr für alle in Betracht kommenden Länder (wie zuvor dargestellt) aus und gleichzeitig die Einsparung von CO_2-Emissionen gegenüber einem Betrieb auf Basis fossiler Brennstoffe.

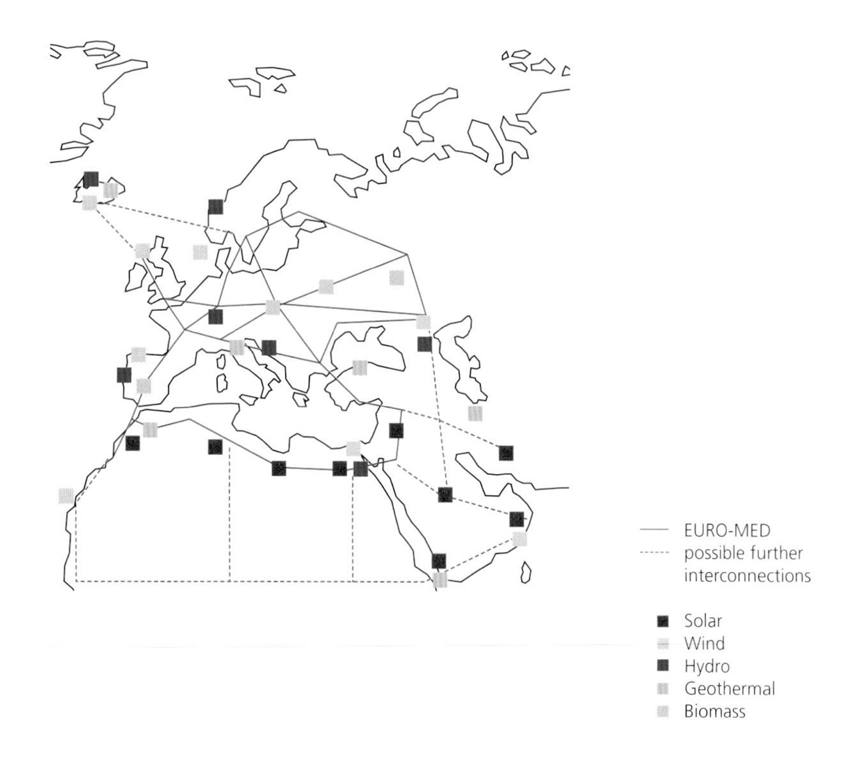

Figure 57
Potential of a trans mediterranean power grid connecting consumers with the most appropriate sites of renewable energy generation according to the European Union (EU) and the Middle East and North Africa (MENA) association agreement. (MENA)

Source, Report by: Concentrating Solar Power for the Mediterranean Region (MED-CSP)

Bild 57
Potenzielles trans-mediterranes Verteilungsnetz zur Nutzung erneuerbarer Energien und Vernetzung von Verbraucherzentren und best-geeigneten Standorten gemäß Europäischer Union und Naher Osten Vereinigung (MENA)

Quelle:
Concentrating Solar Power for the Mediterranean Region MED-CSP

Figure 58
Annual electricity generation of analyzed countries according to CG/HE scenario*

*CG/HE Closing 25 % of the per Capita GDP Gap with the U.S. by 2050/High Efficiency of the Power Sector

Source: Concentrating Solar Power for the Mediterranean Region (MED-CSP)

Bild 58
Jährliche Elektrizitätserzeugung der analysierten Länder gemäß CG/HE Szenarium.

Quelle: Concentrating Solar Power for the Mediterranean Region (MED-CSP)

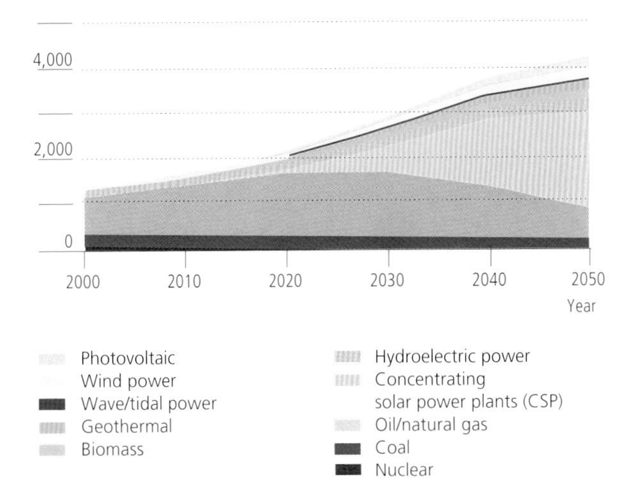

Figure 59
Installed power and peak load of analyzed countries of CG/HE scenario*

*CG/HE: Closing 25 % of the per Capita GDP Gap with the U.S. by 2050/High Efficiency of the Power Sector

Source: Concentrating Solar Power for the Mediterranean Region (MED-CSP)

Bild 59
Installierte Leistung und maximaler Verbrauch der analysierten Länder gemäß CG/HE Szenarium

Quelle: Concentrating Solar Power for the Mediterranean Region (MED-CSP)

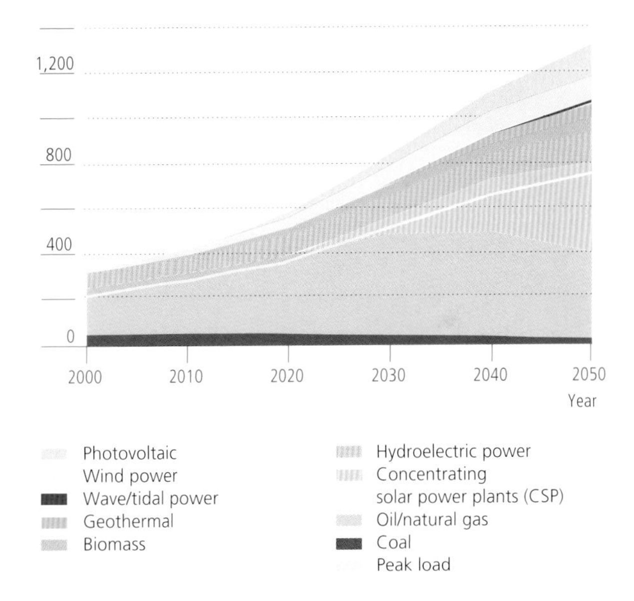

Figure 60
CO_2-Emissionen of electricity generation in million tons per year for all countries for the scenario CG/HE* and emissions that would occur in a Busines-as-usual Case (BAU)

*CG/HE: Closing 25 % of the per Capita GDP Gap with the U.S. by 2050/High Efficiency of the Power Sector

Source: Concentrating Solar Power for the Mediterranean Region (MED-CSP)

Bild 60
CO_2 Emissionen der Elektrizitätserzeugung in Millionen Tonnen aller Erzeugerländer und Emissionen die entstehen bei Beibehaltung eines Status quo.

Quelle:
Concentrating Solar Power for the Mediterranean Region MED-CSP

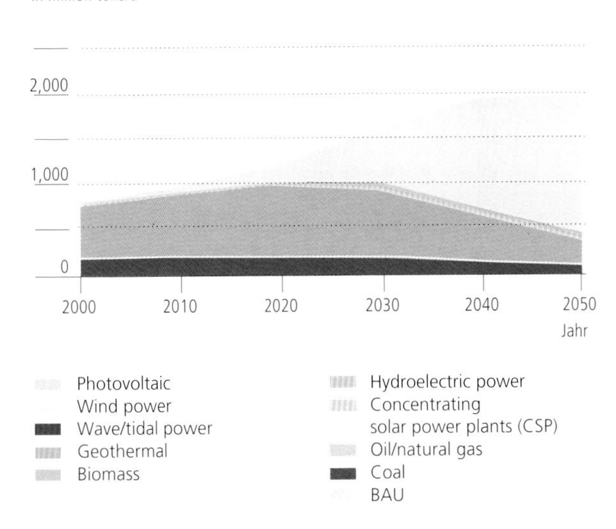

	Unit capacity	Capacity credit Land use	Capacity factor	Resource	Applications	Comment
Wind power	1 kW – 5 MW	46 km²/ 10^6 MWh/a	15 – 50 %	Kinetic energy of the wind	Electricity	Fluctuating, supply defined by resource
Photovoltaic	1 W – 5 MW	Only on roofs	15 – 25 %	Direct and diffuse irradiance on a fixed surface tilted with latitud angle	Electricity	Fluctuating, supply defined by resource
Biomass Agricultural resources Wood	1 kW – 25 MW	Fields 2 km²/ 10^6 MWh/a	40 – 60 %	Biogas from the decom-position of organic residues, solid residues and wood	Electricity and heat	Seasonal fluctuations by good storability, power on demand
Geothermal (Hot dry rock)	25 kW – 50 MW	2 – 8 km²/ 10^6 MWh/a	40 – 90 %	Heat of hot dry rocks in several 1000 meters depth	Electricity and heat	No fluctuations, power on demand
Hydropower	1 kW – 1000 MW	10 km²/ 10^6 MWh/a	10 – 90 %	Kinetic energy and pressure of water streams	Electricity	Seasonal fluctuation, good storability in dams, used also as pump storage for other sources
Solar chimney	100 kW – 200 MW		20 to 70 %	Direct and diffuse irradiance on a horizontal plane	Electricity	Seasonal fluctuations, good storability, base load power
Concentrating solar thermal power (CSP)	10 kW – 200 MW	6 – 10 km²/ 10^6 MWh/a	20 to 90 %	Direct and irradiance on a surface tracking the sun	Electricity and heat	Fluctuations are compensated by thermal storage and fuel, power on demand
Gas turbine	0.5 kW – 100 MW	Power Plant	10 – 90 %	Natural gas, fuel oil	Electricity and heat	Power on demand
Steam cycle	5 kW – 500 MW	Power Plant	40 – 90 %	Coal, lignite, fuel oil, natural gas	Electricity and heat	Power on demand
Nuclear	1000 MW	Power Plant	90 %	Uranium	Electricity and heat	Base load power
Desalination by CSP	10^6 – 10^9 m³ a	2 – 100 km²		Direct irradiance on a surface tracking the sun	Electricity and heat	Seasonal fluctuations, good storability, base load power

Table 3
Characteristics of various contemporary power technologies

Source: Concentrating Solar Power for the Mediterranean Region (MED-CSP), 2005

	Energieertrag/ Einheit	Energie-erzeugende Systeme	Kapazitäts-grad	Ressource	Energie-bereitstellung in Form von:	Bemerkung
Windkraft	1 kW – 5 MW	46 km²/ 10⁶ MWh/a	15 – 50 %	kinetische Energie durch Wind	elektrische Energie	Betreiben durch Windströmung
Photovoltaik	1 W – 5 MW	nur auf Dächern	15 – 25 %	direkte und diffuse Strahlungsenergie	elektrische Energie	Betreiben durch Solarstrahlung
Biomasse aus Landwirtschaft Wäldern	1 kW – 25 MW	Felder/Wälder 2 km²/ 10⁶ MWh/a	40 – 60 %	Biogas aus Pflanzenrückständen sowie Holz	elektrische Energie und Wärmeenergie	Dauerbetrieb durch Lagermöglichkeit
Tiefe Geothermie (Hot dry rock)	25 kW – 50 MW	2 – 8 km²/ 10⁶ MWh/a	40 – 90 %	Wärmeenergie aus heißem Gestein verschiedener Tiefe	elektrische Energie und Wärmeenergie	Dauerbetrieb durch Erdwärme
Wasserkraft	1 kW – 1000 MW	10 km²/ 10⁶ MWh/a	10 – 90 %	kinetische Energie und Druck von Wasserströmen	elektrische Energie	Dauerbetrieb infolge hoher Speicherkapazität, u.U. Pumpspeicherkraftwerk
Solarkraftwerk	100 kW – 200 MW		20 to 70 %	direkte und diffuse Strahlung auf ein Glaszelt	elektrische Energie	saisonaler Betrieb an klaren Tagen, Energiespeicherung in Netzen
Parabolspiegelsysteme, konzentrierend (CSP)	10 kW – 200 MW	6 – 10 km²/ 10⁶ MWh/a	20 to 90 %	direkte und diffuse Strahlungsenergie (nachgeführte Parabolspiegelsysteme)	elektrische Energie und Wärmeenergie	Betrieb bei Solarstrahllung, Wärmespeicherung im System, u.U.zusätzlicher Einsatz von Gas, Kohle od. Öl
Gasturbine	0.5 kW – 100 MW	Kraftwerk	10 – 90 %	Erdgas	elektrische Energie und Wärmeenergie	Energieerzeugung nach Notwendigkeit
Dampfkreislauf	5 kW – 500 MW	Kraftwerk	40 – 90 %	Kohle, Braunkohle, Erdöl, Erdgas	elektrische Energie und Wärmeenergie	Energieerzeugung nach Notwendigkeit
Nuklearenergie	1000 MW	Kraftwerk	90 %	Uran	elektrische Energie und Wärmeenergie	Grundversorgung für elektrischen Energieverbrauch
Entsalzung durch CSP	$10^6 – 10^9$ m³ a	2 - 100 km²		direkte und diffuse Strahlungsenergie (nachgeführte Parabolspiegelsysteme)	elektrische Energie und Wärmeenergie	Betrieb bei Solarstrahlung, Wärmespeicherung im System, u.U.zusätzlicher Einsatz von Gas, Kohle od. Öl

Tabelle 3
Vergleichbare Betrachtung verschiedener Systeme erneuerbarer Energien mit speziellen Daten

Quelle: Concentrating Solar Power for the Mediterranean Region (MED-CSP), 2005

4.4. Asia and Oceania region

Figures 61 – 68 present the current energy situation for selected important countries of the Asian and Oceanic region.

According to the IEA 2004 study, China covers its main energy demand by use of coal, and to a minor degree by hydroelectric power, oil, gas, and nuclear power. Biomass is an essential element in the power mix, but other renewables remain in the realm of an insignificant percentage. The distribution of China's energy consumption shows that the majority of the energy is consumed by the industrial sector. Because a large percentage of the required energy is supplied by coal, a long-term supply guarantee can be assumed. Nevertheless, there is no getting away from the fact that China needs to substantially lower its energy consumption, mainly in the sectors of industry and residential dwellings (**Figures 61.1 – 61.3**).

Hong Kong has historically taken a different path of development than the Chinese mainland, so a comparative analysis is of interest. Coal, and then gas, are the main sources of primary energy for Hong Kong, and renewable energy sources do not show up in the statistics. However, the distribution of energy consumption differs significantly from that of the Chinese mainland because consumption of energy by industrial end users plays almost no role, while values for trade and commerce are disproportionately high (**Figures 62.1 – 62.3**).

4.4 Region Asien und Ozeanien

In den **Bildern 61 – 68** sind einige wesentliche Länder des asiatischen und ozeanischen Raumes dargestellt.

Gemäß den Studien der IEA, 2004 deckt China seinen wesentlichen Primärenergiebedarf aus Kohle und geringeren Anteilen aus Wasserkraft, Erdöl, Erdgas und Kernenergie. Ein noch nennenswerter Anteil an der Energieerzeugung ist der Einsatz von Biomasse, während alle weiteren erneuerbaren Energien zurzeit der Erhebung so gering waren, dass sie nicht ausgewiesen wurden. Die Verteilung des Energieverbrauchs zeigt, dass der mit Abstand größte Anteil in den Industriebereich fließt. Nachdem sich feststellen lässt, dass für Kohle die größte Verfügbarkeit besteht, wäre zumindest langfristig die Energieversorgung auf dieser Basis gesichert. Gleichwohl wird China nicht darum herumkommen, seinen Energiebedarf im Bereich Industrie und Wohnen deutlich zu senken, **Bilder 61.1 – 61.3**.

Nachdem Hongkong als Teil Chinas historisch eine völlig andere Entwicklung genommen hat, ist ein Vergleich zu China von Interesse. Kohle, gefolgt von Erdgas sind die wesentlichen Energieträger zur Erzeugung von Primärenergien, während entweder die erneuerbaren Energien keinen nennenswerten Anteil bilden oder nicht erfasst werden konnten. Die Verteilung des Energieverbrauchs unterscheidet sich zu Restchina ganz erheblich, der Verbrauch von Energie für die Industrie spielt eher eine untergeordnete Rolle, während der für Handel und Gewerbe überproportional groß ist, **Bilder 62.1 – 62.3**.

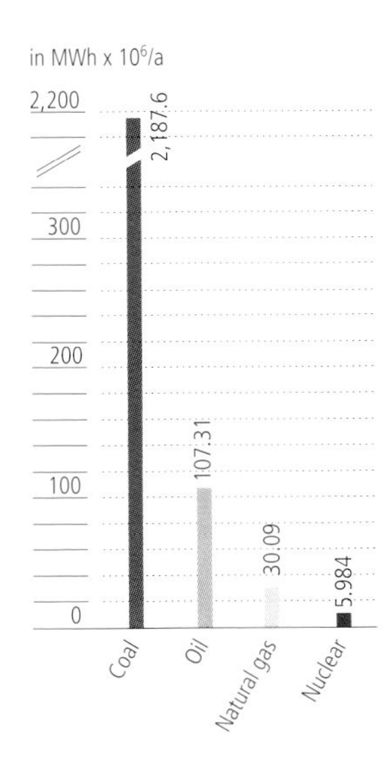

Figure 61.1
Primary energy consumption, China

Bild 61.1
Primärenergieverbrauch, China

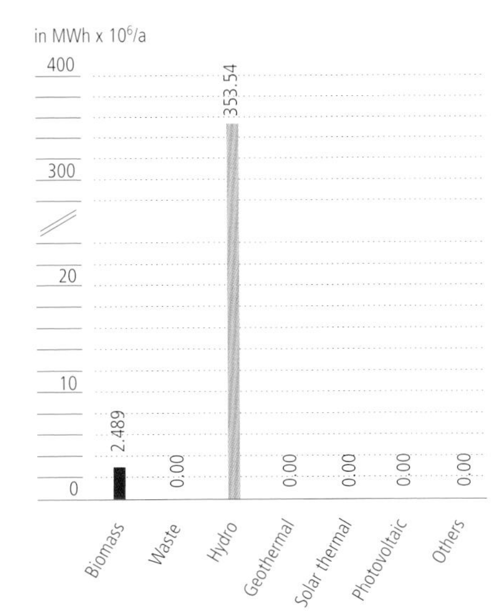

Figure 61.2
Renewable energy consumption, China

Bild 61.2
Verbrauch aus erneuerbaren Energien, China

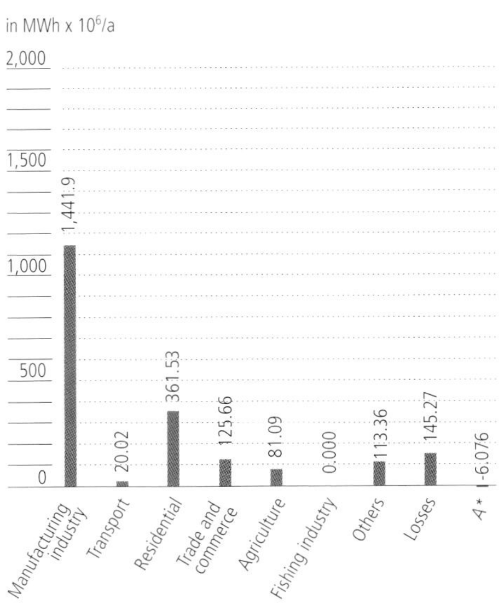

Figure 61.3
Distribution of energy consumption, China

Bild 61.3
Verteilung des Energieverbrauchs, China

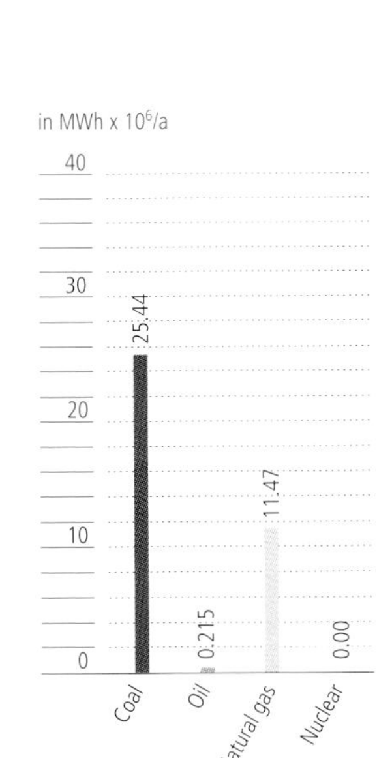

Figure 62.1
Primary energy consumption, Hong Kong

Bild 62.1
Primärenergieverbrauch, Hongkong

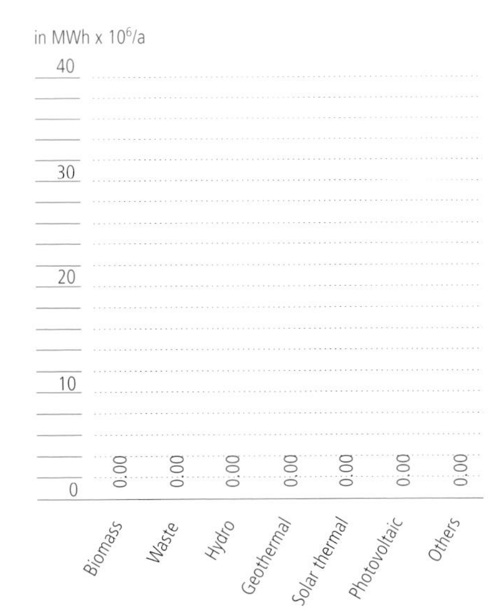

Figure 62.2
Renewable energy consumption, Hong Kong

Bild 62.2
Verbrauch aus erneuerbaren Energien, Hongkong

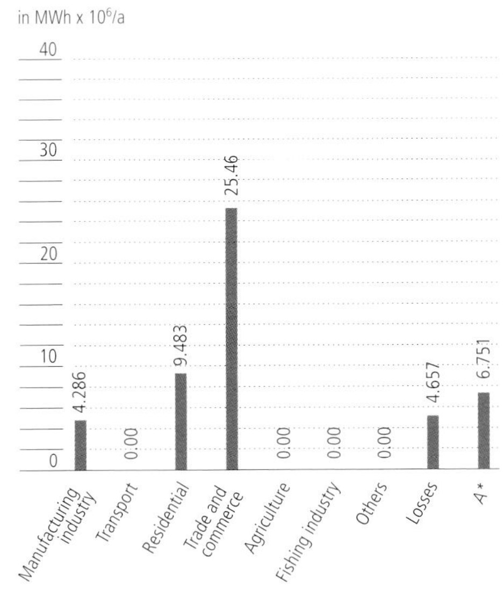

Figure 62.3
Distribution of energy consumption, Hong Kong

Bild 62.3
Verteilung des Energieverbrauchs, Hongkong

India's energy scenario is very similar to that of China, although the percentage of renewable energy used is significantly lower, as shown in **Figures 63.1 – 63.3**. The largest final energy consumer is the manufacturing sector, followed by very significant losses in the energy's transformation and transmission. Residential and agricultural users are as prominent proportionally as in China, but in absolute values they are much smaller.

Japan is a highly industrialized economy, of course, and it consumes any available energy source currently available, both fossil and renewable. The latter's contribution is commendably high, amounting to 11 %. Consumption is distributed across industry, residential, trade, and commerce, and other energy consumers are proportionally small (**Figures 64.1 – 64.3**).

South Korea is also a highly developed industrial country, and its energy data are similar to Japan's, yet with a much smaller percentage of renewable energy contribution to the overall demand mix. Consumption is distributed across the known categories and is very comparable to the Japanese case (**Figures 65.1 – 65.3**).

Indien zeigt eine ähnliche Primärenergieverbrauchsstruktur wie China, wobei der Anteil der erneuerbaren Energien deutlich geringer ist, **Bilder 63.1 – 63.3**. Wiederum wie in China ist der größte Energieverbraucher der Bereich der Industrie, gefolgt von Verlusten in der Energieerzeugung und Verteilung, die überproportional groß sind. Wohnen und Landwirtschaft sind zwei ähnlich große Verbraucher wie in China, liegen jedoch deutlich unter den Werten der Volksrepublik.

Japan als hochindustrialisiertes Land setzt gleichermaßen auf sämtliche Primärenergieträger, weist jedoch dabei einen hohen Anteil an erneuerbaren Energien aus – dieser beträgt ca. 11 Prozent. Die Verteilung des Energieverbrauchs betrifft im Wesentlichen die Bereiche Industrie, Wohnen, Handel und Gewerbe, während alle anderen Energieverbraucher deutlich kleiner sind, **Bilder 64.1 – 64.3**.

Südkorea als ebenfalls hochindustrialisiertes Land zeigt eine ähnliche Verbrauchsstruktur wie Japan, jedoch mit einem deutlich kleineren Anteil an erneuerbaren Energien. Die Verteilung des Energieverbrauchs für Industrie, Transport, Wohnen, Handel und Gewerbe usw. zeigt wiederum eine ähnliche Struktur wie Japan und ist demgemäß in etwa vergleichbar, **Bilder 65.1 – 65.3**.

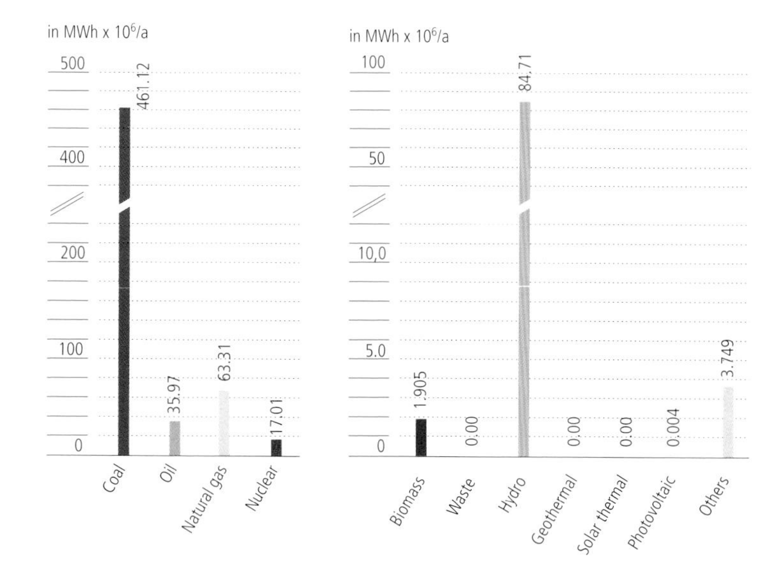

Figure 63.1
Primary energy consumption, India

Bild 63.1
Primärenergieverbrauch, Indien

Figure 63.2
Renewable energy consumption, India

Bild 63.2
Verbrauch aus erneuerbaren Energien, Indien

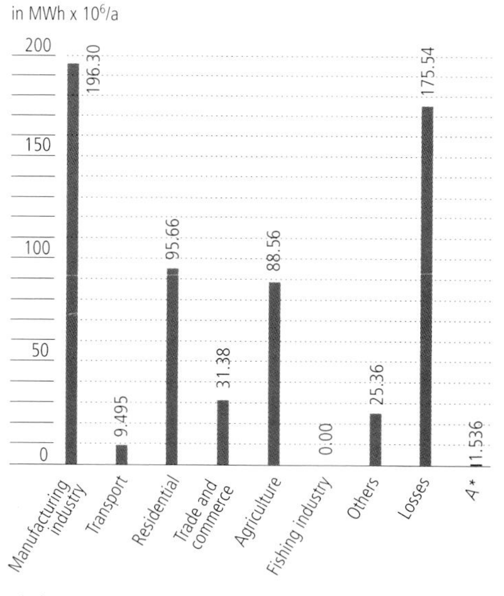

A* Electric energy import/export balance

Figure 63.3
Distribution of energy consumption, India

Bild 63.3
Verteilung des Energieverbrauchs, Indien

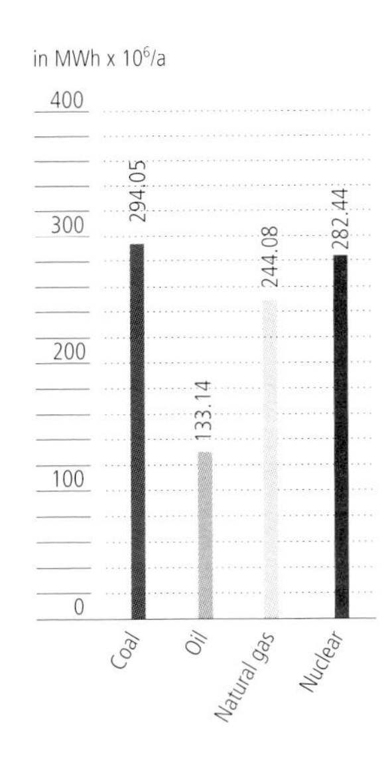

Figure 64.1
Primary energy consumption,
Japan

Bild 64.1
Primärenergieverbrauch,
Japan

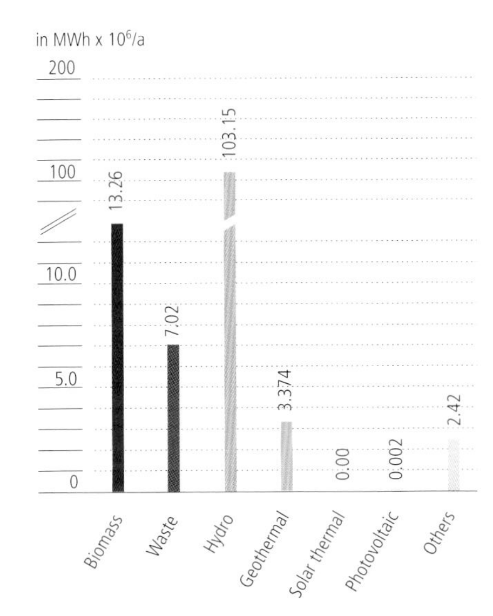

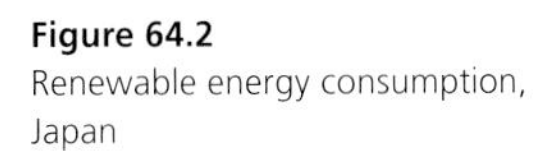

Figure 64.2
Renewable energy consumption,
Japan

Bild 64.2
Verbrauch aus erneuerbaren Energien,
Japan

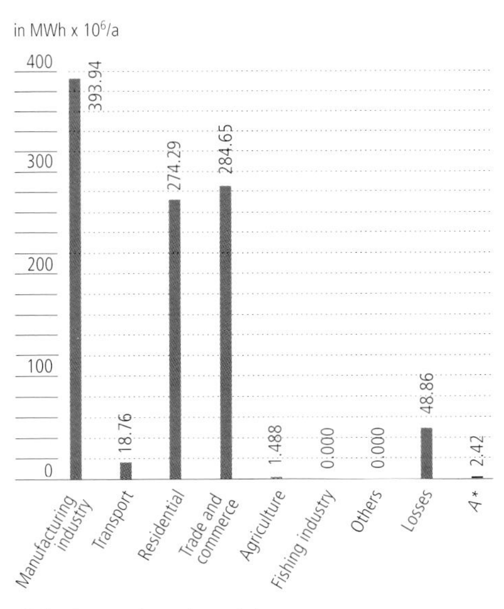

A* Electric energy import/export balance

Figure 64.3
Distribution of energy consumption,
Japan

Bild 64.3
Verteilung des Energieverbrauchs,
Japan

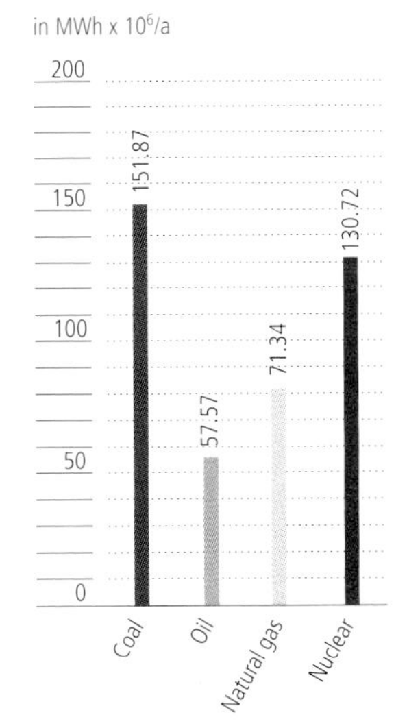

Figure 65.1
Primary energy consumption,
South Korea

Bild 65.1
Primärenergieverbrauch,
Südkorea

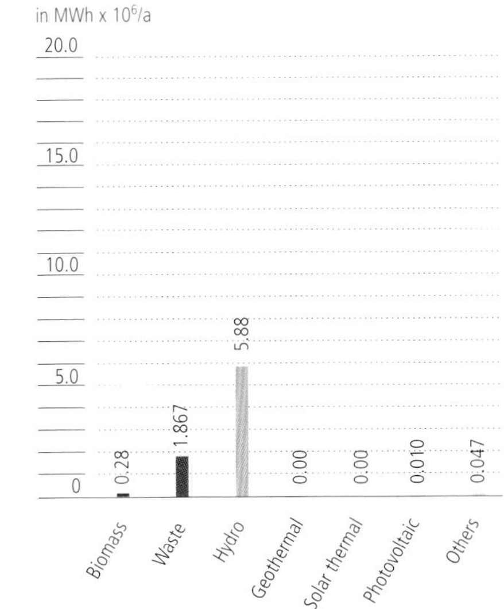

Figure 65.2
Renewable energy consumption,
South Korea

Bild 65.2
Verbrauch aus erneuerbaren Energien,
Südkorea

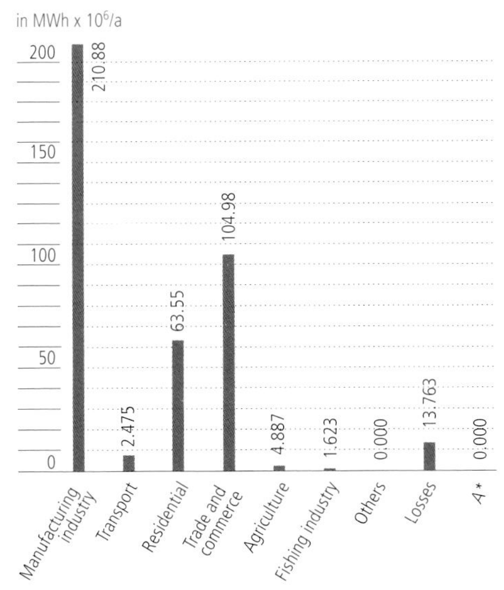

A* Electric energy import/export balance

Figure 65.3
Distribution of energy consumption,
South Korea

Bild 65.3
Verteilung des Energieverbrauchs,
Südkorea

Thailand, as a strongly aspiring economy, still counts mainly on the use of fossil resources to satisfying its growing demand for energy. Its use of renewables is still very small and limited to biomass and hydroelectric power. The consumer distribution is comparable to that in South Korea and Japan, with residential uses, industry, and commerce being the main consumers, **Figures 66.1 – 66.3**.

Australia, as seen in **Figures 67.1 – 67.3**, currently relies to a large degree on fossil fuels, yet the percentage of renewable energy sources is rising significantly. In order to replace its reliance on fossil fuels with renewables in the future, Australia has the potential to utilize its vast, unpopulated, yet solar-gain-intensive regions in the interior, north, and west of the country for solar thermal and photovoltaic applications. Energy consumption distribution is very similar to other countries in the Asian hemisphere, with industry, residential, and commerce being the main consumers.

New Zealand can be characterized to a large degree as an energy-autonomous country (**Figures 68.1 – 68.3**). The country employs only a small amount of coal or gas in the supply of its energy, the majority coming from hydroelectric power, biomass, and geothermal energy plants. Consumption patterns are similar to those in Australia, yet there is a larger percentage of losses than in that country.

Thailand als aufstrebende Industrienation setzt primär auf die fossilen Brennstoffe ohne Einsatz von Kernenergie bei einem gleichzeitig geringen Anteil an erneuerbaren Energien aus Wasserkraft und Biomasse. Die Verteilung der Energieverbräuche zeigt im Wesentlichen die gleiche Tendenz wie Südkorea oder Japan, Industrie, Wohnen, Handel und Gewerbe stehen im Vordergrund der Energieverbräuche, **Bilder 66.1 – 66.3**.

Australien setzt zurzeit wohl noch im Wesentlichen auf die fossilen Brennstoffe, **Bilder 67.1 – 67.3**, jedoch mit einem nennenswerten Anteil beim Einsatz erneuerbarer Energien. Aufgrund seiner menschenleeren und sonnenreichen Regionen im Norden wird Australien sicher in Zukunft massiv auf Solarthermie und Photovoltaik setzen, um sich von den fossilen Brennstoffen zu lösen. Die Verteilung der Energieverbräuche zeigt wiederum das charakteristische Bild vieler asiatischer Staaten – Industrie, Wohnen, Handel und Gewerbe nehmen den größten Anteil für sich in Anspruch.

Neuseeland scheint ein im Wesentlichen energieautarkes Land zu sein, **Bilder 68.1 – 68.3**, da es nur einen kleineren Anteil an Kohle und Erdgas bei der Energieerzeugung einsetzt, einen wesentlich größeren Anteil aus erneuerbaren Energien durch Wasserkraft, Geothermie und Biomasse. Die Verteilung des Energieverbrauchs zeigt wiederum eine vergleichbare Struktur wie die Australiens, jedoch mit einem etwas höheren Anteil an Verlusten.

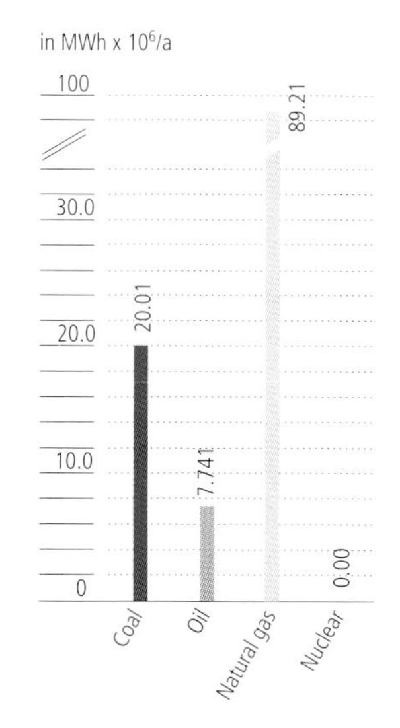

Figure 66.1
Primary energy consumption, Thailand

Bild 66.1
Primärenergieverbrauch, Thailand

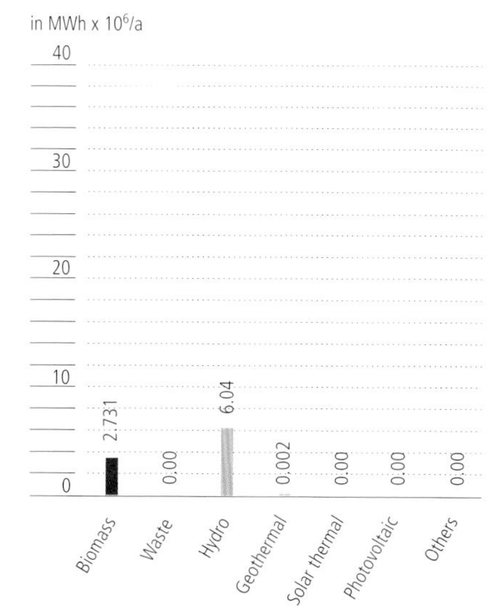

Figure 66.2
Renewable energy consumption, Thailand

Bild 66.2
Verbrauch aus erneuerbaren Energien, Thailand

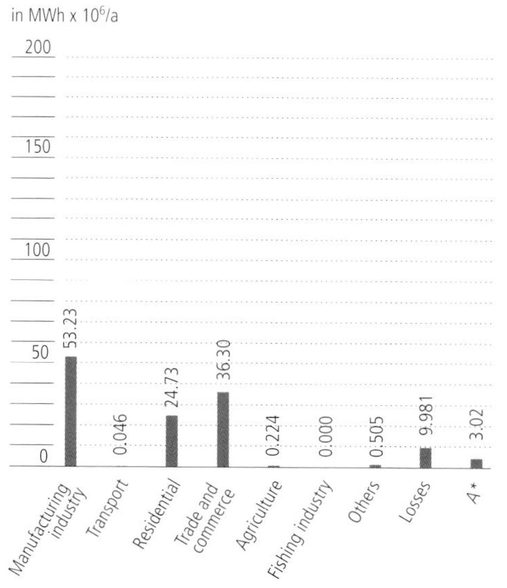

A* Electric energy import/export balance

Figure 66.3
Distribution of energy consumption, Thailand

Bild 66.3
Verteilung des Energieverbrauchs, Thailand

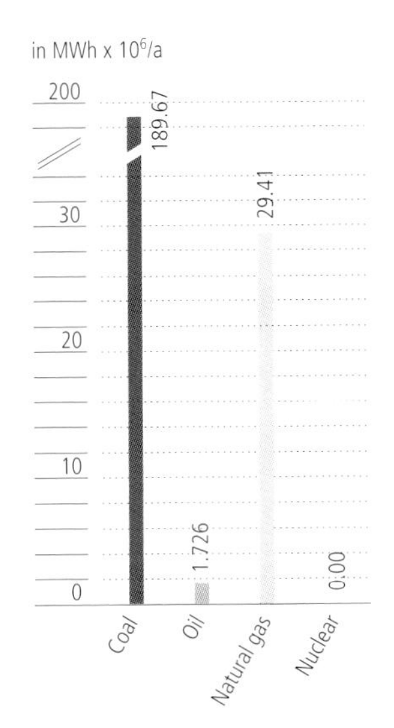

Figure 67.1
Primary energy consumption, Australia

Bild 67.1
Primärenergieverbrauch, Australien

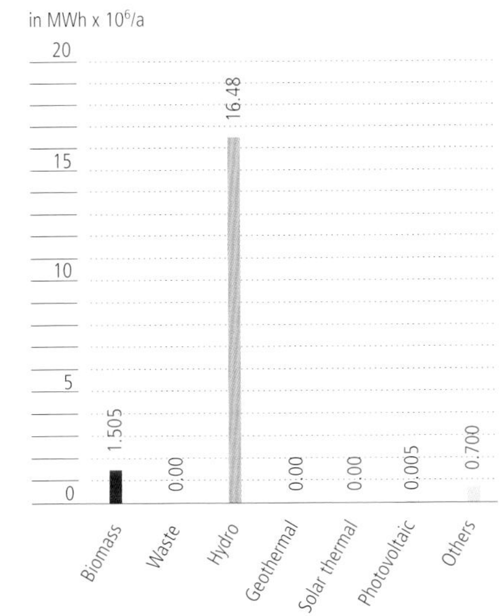

Figure 67.2
Renewable energy consumption, Australia

Bild 67.2
Verbrauch aus erneuerbaren Energien, Australien

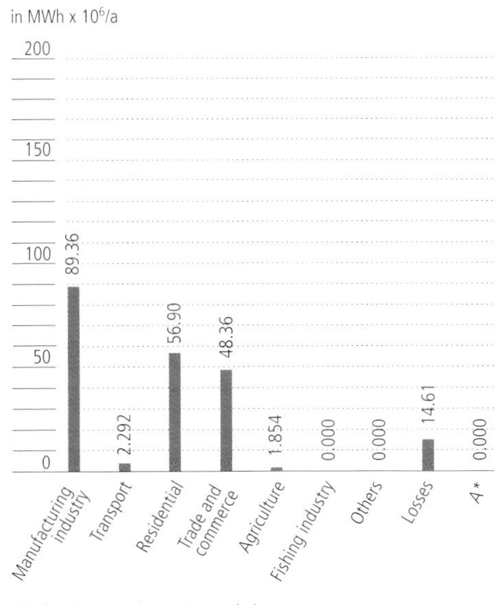

A* Electric energy import/export balance

Figure 67.3
Distribution of energy consumption, Australia

Bild 67.3
Verteilung des Energieverbrauchs, Australien

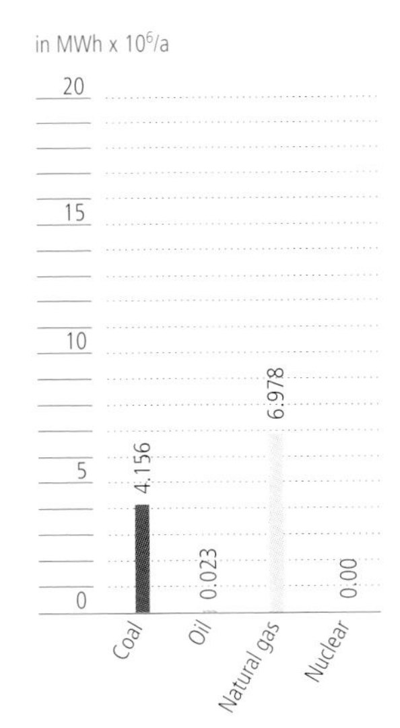

Figure 68.1
Primary energy consumption, New Zealand

Bild 68.1
Primärenergieverbrauch, Neuseeland

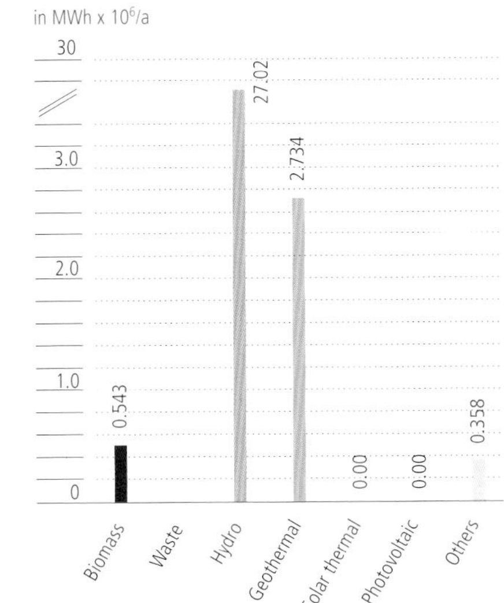

Figure 68.2
Renewable energy consumption, New Zealand

Bild 68.2
Verbrauch aus erneuerbaren Energien, Neuseeland

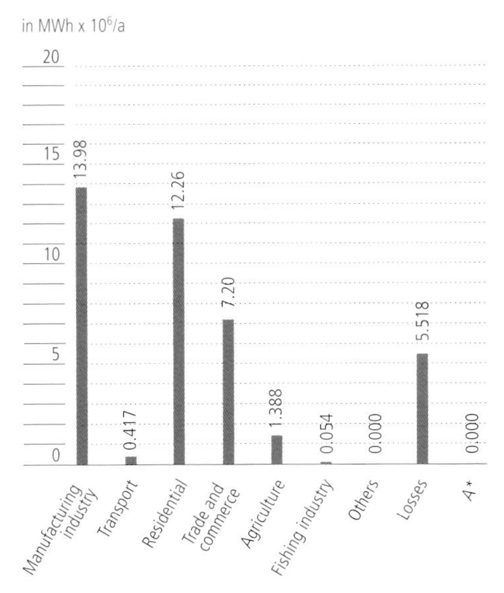

A* Electric energy import/export balance

Figure 68.3
Distribution of energy consumption, New Zealand

Bild 68.3
Verteilung des Energieverbrauchs, Neuseeland

Conclusion

In this first general analysis of the various regions of the globe, one result becomes apparent: renewable resource potentials are still almost completely undeveloped – unfortunately. This is particularly the case for solar thermal energy. **Figure 69** illustrates the regions in which such technologies could be successfully employed, on the basis of the great solar gains and the high temperatures available in these areas. Both heat and cooling energy, in addition to electrical energy, could be generated in these countries very successfully with solar thermal technology.

To satisfy the energy demand of all of the countries in North Africa, the Mediterranean region, and additionally some countries of the Middle Eastern region discussed earlier in this chapter, a comparatively small land area of only 0.15 % of the Saharan desert would need to be covered with solar thermal systems. Learning from that example, it would be a strategically wise and profitable move of solar-rich countries such as Australia and New Zealand to completely disengage from a looming distribution war for energy and consider a massive implementation of solar energy generating plants in the areas of their countries with an abundance of solar energy exposure.

Fazit

Bei einer ersten Grobanalyse der verschiedensten Regionen der Welt zeigt sich, dass der Einsatz erneuerbarer Energien, insbesondere der Solarthermie, noch völlig unterentwickelt ist. **Bild 69** deutet die Regionen an, in denen in hohem Maß insbesondere durch konzentrierende Solarsysteme thermische Energie auf hohem Temperaturniveau gewonnen werden kann, um hieraus sowohl elektrische Energie als auch Wärmeenergie und Kälteenergie zu erzeugen.

Um zum Beispiel die Region um das Mittelmeer mit den zuvor aufgeführten Ländern Nordafrikas, Südeuropas und zum Teil des Vorderen Orients mit elektrischer Energie auf Solarthermiebasis zu versorgen, würde es ausreichen, ca. 0,15 Prozent der Sahara mit entsprechenden solarthermischen Systemen zu überziehen. Gerade Australien und Neuseeland könnten sich bei entsprechender Politik sehr schnell vom Energieverteilungskampf losmachen, da sie in ihren eigenen Ländern über ausreichende natürliche Ressourcen verfügen.

Global radiation per annum

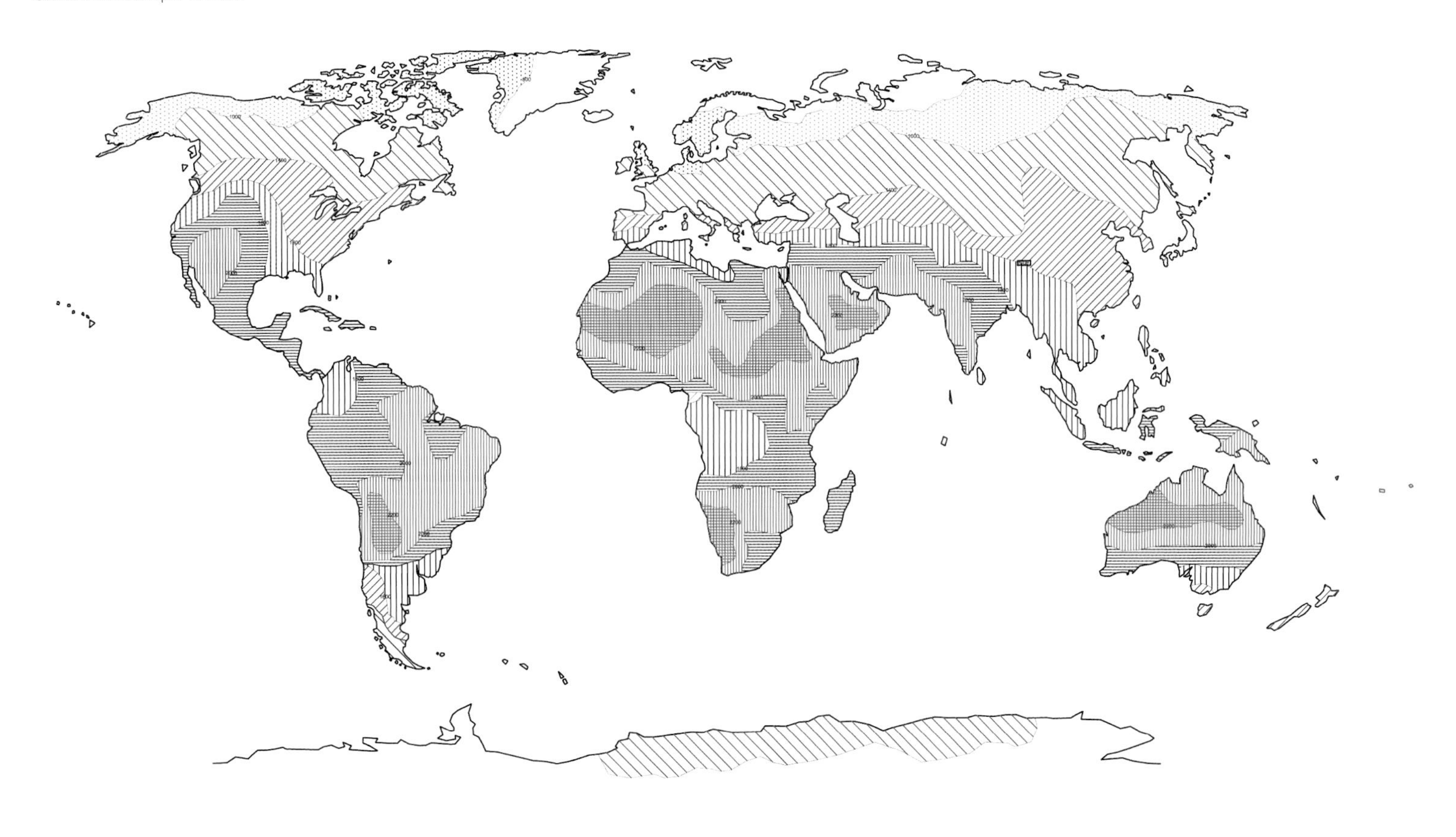

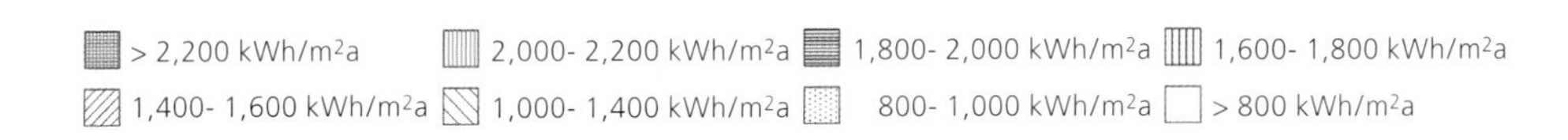

Figure 69
Distribution of global solar radiation

Regions with more than 1950 kWh/m²a are suited for solar production of electricity

Source: Hindrichs, Daniels, plusminus 20°/40° latitude Edition Axel Menges, 2007

Bild 69
Verteilung der Intensität der globalen Sonneneinstrahlung in kWh/m²a

Regionen mit mehr als 1950 kWh/m²a sind geeignet für Elektrizitätserzeugung mittels Photovoltaik.

Quelle: Hindrichs, Daniels, plusminus 20°/40° latitude Edition Axel Menges, 2007

5 Detailed view: Germany – case study

5 Detailbetrachtung Bundesrepublik Deutschland – ein Beispiel

Because the authors have access to detailed data concerning Germany's energy supply and demand, we use this country as a case study of the current state of the industry, the options available to us for development, and what a future energy sector might look like. It should be noted that the example of Germany may be useable by other countries only with some reservations due to other given parameters, but the overall conclusions concerning applicable concepts are still valid. The goal is to use the German case as a lesson to avoid future energy crunches and, with its help, to develop a whole new design of a future "energy landscape."

For European countries, one source of renewable energy that already contributes significantly to the energy mix is available in abundance: water. This observation cannot be generalized on a global basis, of course because in many countries the water supply for drinking and cooking purposes is less than reliable and often far from being secure. Yet, even for Europe we need to see how rising temperatures (a warm period) could lead to a situation in which hydroelectric power generation cannot be taken for granted and determine what alternative future potential exists for supplying power.

Figure 70 shows the water consumption (benchmark data) for various building types per liter/per capita and day. We can see that the use of the "food stuff" water amounts to approximately 100 liters/capita and day, although water consumption in hot and dry regions of the Earth can be twice that amount, or greater. **Figure 71** shows the water consumption for Germany under the assumption that besides high-grade potable water grey water or processed rainwater could be substituted. At least some appropriate applications are well served with grey water instead of the highest-grade water, and so the substantial effort required to process and provide potable water, in this case, could be eased.

Da den Autoren dieses Buches für die Bundesrepublik Deutschland genauste Zahlen und Analysen vorliegen, soll dieses Land beispielhaft einen Einblick darüber geben, wo wir zurzeit stehen, welche Potenziale zu heben sind und wie eine neue Energiewirtschaft aussehen könnte. Das Beispiel Deutschland kann selbstverständlich für andere Länder nur bedingt gelten und für manche nur lediglich eine Anleitung sein, wie man entsprechend vorgehen könnte und sollte, um die zukünftig zu erwartenden Energieengpässe auszugleichen bzw. eine völlig neue „Energielandschaft" zu entwickeln.

Ein für die europäischen Länder nicht entscheidendes Thema ist die Wasserversorgung, die jedoch in vielen Regionen der Welt sehr wohl entscheidend ist. Auch für Europa kann im Zuge der Temperaturerhöhungen (Warmzeit) die Wasserthematik, d.h. der Wasserverbrauch, ein deutlich schärferes Profil einnehmen als zurzeit, und insofern wird ein erster Hinweis darauf gegeben, welches Potenzial hier zu heben ist.

Bild 70 zeigt die Wasserverbräuche (Benchmarkzahlen) verschiedener Gebäudestrukturen in Deutschland in Liter pro Person und Tag. Grob zusammengefasst, lässt sich feststellen, dass der Wasserverbrauch in etwa bei 100 Liter pro Person und Tag liegt (im Wesentlichen das Lebensmittel Frischwasser), während gerade in heißen und trockenen Regionen der Welt der Wasserverbrauch zum Teil doppelt so hoch und höher ist. **Bild 71** zeigt die Struktur des Wasserverbrauchs in Deutschland mit der Möglichkeit, anstelle aufwendig aufbereiteten Trinkwassers sinnvollerweise auch aufbereitetes Regenwasser oder Grauwasser zum Einsatz zu bringen. Hierdurch würde sich der Aufwand zur Aufbereitung von Frischwasser massiv senken lassen, weitere Vorteile werden durch den Einsatz aufbereiteten Regenwassers eingekauft.

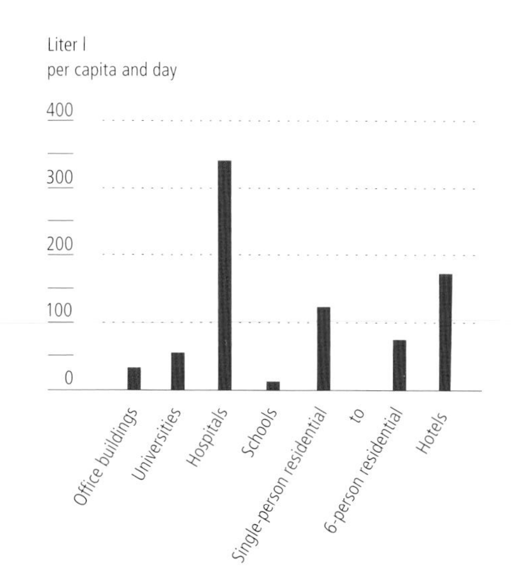

Figure 70
Water consumption for various building types, averages in liter/capita day

Source: Daniels/author

Bild 70
Wasserverbräuche in verschiedenen Gebäuden, Durchschnittswerte in Liter/Person+Tag

Source: Daniels/Autor

Europe also shows that, in addition to the valuable resource of water, other savings opportunities, such as in the electricity sector, exist as well. **Figure 72** depicts the annual electricity demand per household for the European member countries.

In those countries that keep the specific cost (€/kWh) for electricity extremely low, energy consumption is naturally high. It is prudent to discuss whether an electric consumption rate with a factor of 1:10 within the EU is still justifiable – how can it be that one country in the same climate region and therefore having similar weather conditions uses only one tenth as much electrical energy per capita and day as another country? For example, in the case of Norway, consumption of electricity happens on a grand scale because its energy is mostly derived from renewable sources. It would seem to make more sense for that country to have its surplus energy exported to other countries in order to reduce dependency on fossil fuel consumption.

Dass in Europa weitere erhebliche Potenziale zur Einsparung vorhanden sind, zeigt unter anderem auch das **Bild 72**, den jährlichen Strombedarf je Haushalt in den verschiedensten europäischen Ländern.

In den Ländern, in denen der spezifische Preis für elektrische Energie (z.B.€/kWh) außerordentlich niedrig ist, wird naturgemäß auch entsprechend viel verbraucht. Es ist sicher an der Zeit, darüber nachzudenken, ob ein Stromverbrauch im Verhältnis 1:10 noch vertretbar ist oder nicht, d.h. warum unter Umständen ein Land nur ein Zehntel dessen verbraucht wie ein anderes Land in derselben Region und unter ähnlichen Witterungsbedingungen. Selbstverständlich kann es sich ein Land Norwegen sehr viel leichter leisten, elektrische Energie in großem Umfang zu konsumieren, wenn diese Energie durch erneuerbare Energien dargestellt werden kann. Sinnvoller wäre es sicher jedoch, diese Energie zu exportieren und damit einen Beitrag zu leisten, den Einsatz von fossilen Brennstoffen zu reduzieren.

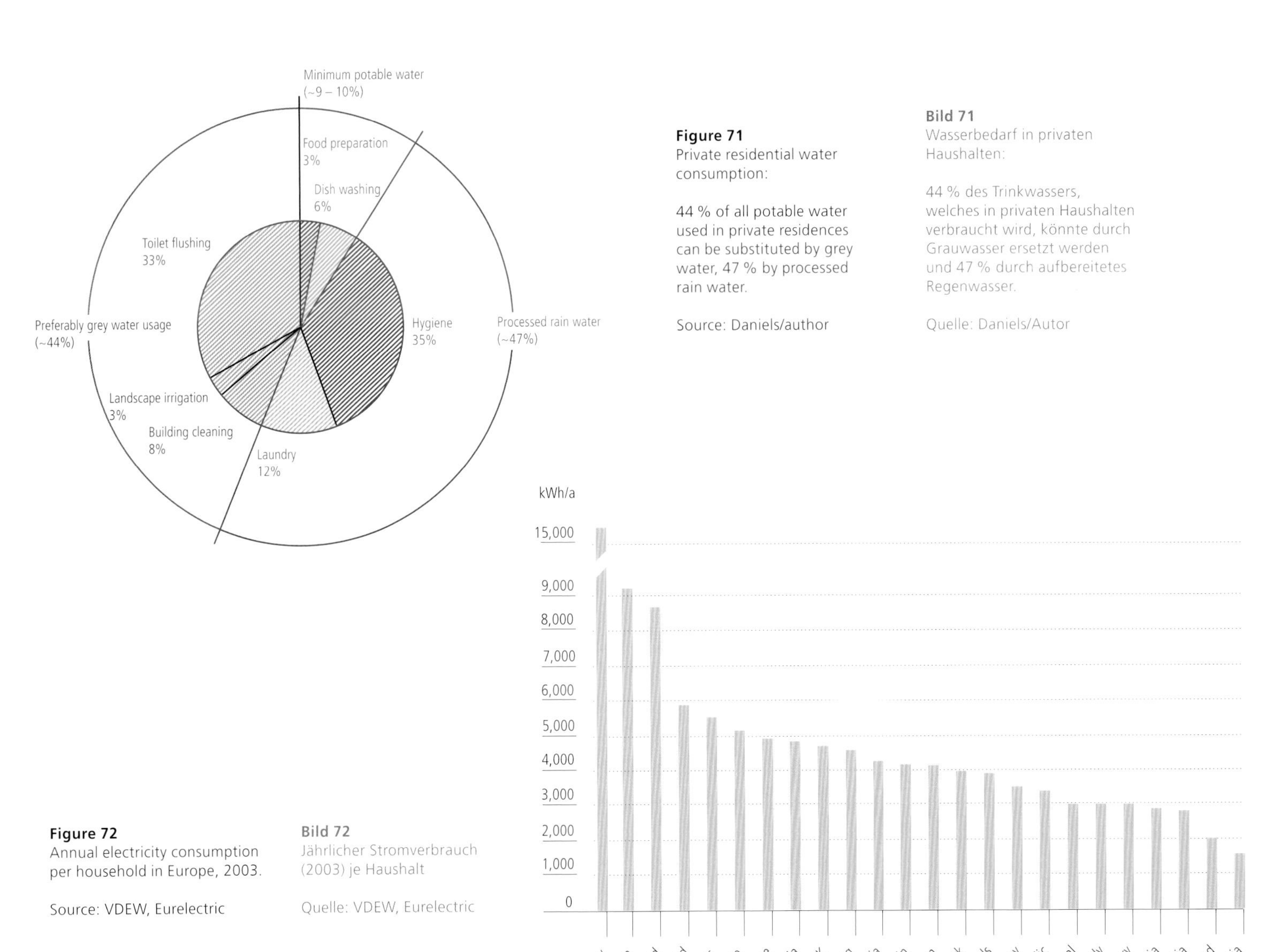

Figure 71
Private residential water consumption:

44 % of all potable water used in private residences can be substituted by grey water, 47 % by processed rain water.

Source: Daniels/author

Bild 71
Wasserbedarf in privaten Haushalten:

44 % des Trinkwassers, welches in privaten Haushalten verbraucht wird, könnte durch Grauwasser ersetzt werden und 47 % durch aufbereitetes Regenwasser.

Quelle: Daniels/Autor

Figure 72
Annual electricity consumption per household in Europe, 2003.

Source: VDEW, Eurelectric

Bild 72
Jährlicher Stromverbrauch (2003) je Haushalt

Quelle: VDEW, Eurelectric

5.1 Electricity and thermal consumption, 2006

Figure 73 shows that the average total energy consumption per household in Germany is around 37,000 kWh. Thermal energy use contributes approximately 52 % to the total amount, individual transportation another 35 %, and electrical energy consumption approximately 13 %. The energy consumption of the average German household is therefore twice as great as the target of the so-called "2,000-Watt-Society" that we introduce in Chapter 9.2. In other words, Germany would need to lower its energy consumption in a relatively brief period by half, meaning a reduction in thermal energy consumption of 50 %, as well as significant reductions in the use of gasoline and warm water and in the heating of buildings.

The graphs in Figures 74 and 75 were developed on the basis of Endenergie VDI 3807, the so-called benchmark goals put forward by the Society of German Engineers (VDI), a financially independent and politically unaffiliated, non-profit organization of 132,000 engineers and natural scientists. Figure 74 shows the consumption of electrical energy in kWh/m²a for various building types as minimum, maximum, and weighed-average values. Amplitudes between depicted minimal and maximum values are in some cases significant, which allows the conclusion that some dwellings are obviously being operated in a very inefficient manner. Similarly, Figure 75 presents thermal consumption data that shows great differences between minimal and maximum consumption as well. Buildings with such high consumption rates need to either terminate operation or, at the least, be completely refurbished to allow for adaptation to future energy-consumption standards.

The projected future increase in energy costs will most certainly add incentives for accelerated renovation or termination of some of the existing non-performing building stock of a country.

5.1 Strom- und Wärmeverbräuche, 2006

Bild 73 zeigt einen durchschnittlichen Endenergieverbrauch pro Haushalt in Deutschland mit insgesamt ca. 37.000 kWh. Dabei macht die Wärmeenergie einen Anteil von ca. 52 Prozent aus, der individuelle Transport ca. 35 Prozent und der elektrische Energiebedarf ca. 13 Prozent. Damit liegt der Gesamtenergieverbrauch eines durchschnittlichen deutschen Haushalts bei ca. dem Doppelten dessen, was in Kapitel 9.2 die 2000-Watt-Gesellschaft der Zukunft beanspruchen soll, oder umgekehrt, die Bundesrepublik Deutschland müsste in relativ kurzer Zeit ihren Gesamtenergieverbrauch auf die Hälfte senken, was heißen könnte, Reduzierung des Wärmeenergiebedarfs auf die Hälfte, Reduzierung des Benzinverbrauchs auf die Hälfte, Reduzierung des Warmwasserbedarfs oder der Heizleistung zur Erzeugung desselben ebenfalls auf 50 Prozent.

Auf der Basis von Benchmarkzahlen (Endenergie gemäß VDI 3807) wurde die Struktur der Bilder 74 und 75 erarbeitet. Das Bild 74 zeigt den Stromverbrauch in kWh/m²a für verschiedenste Gebäudestrukturen mit Minimal- und Maximalwerten sowie gewichteten Mittelwerten. Die Ausschläge zwischen Maximalverbräuchen und Minimalverbräuchen sind zum Teil extrem groß, woran zumindest erkennbar ist, dass eine Reihe von Häusern extrem unwirtschaftlich betrieben wird. Gleiches gilt auch für die in Bild 75 gezeigten Heizenergie-Verbrauchskennwerte, die wiederum extrem große Ausschläge ausweisen. Die jeweils angegebenen Minimalwerte zeigen an, was zu erreichen ist, Gebäude mit Maximalwerten müssten entweder total rückgebaut werden oder zumindest schleunigst erheblich saniert werden, um den Ansprüchen der Zukunft zu genügen.

Mit den zu erwartenden massiven Preissteigerungen in der Zukunft wird sicher die Thematik der Sanierung oder gegebenenfalls des Rückbaus beschleunigt.

Figure 73
Final energy consumption per household in Germany (100 % = 37,000 kWh)

Source: Nuclear power plant Gösgen

Bild 73
End-Energieverbrauch pro Haushalt, (100 % = 37.000 kWh)

Quelle: Kernkraftwerk Gösgen

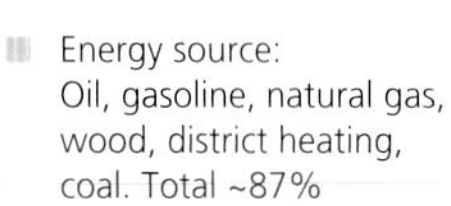

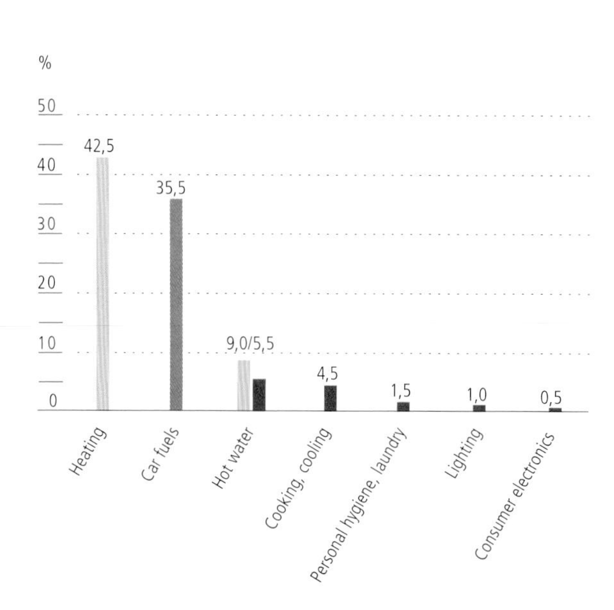

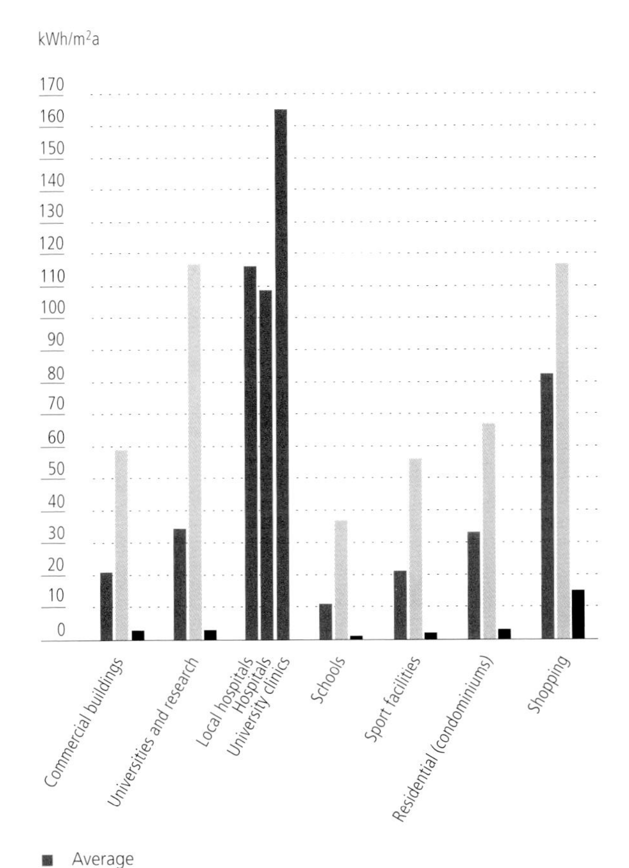

Figure 74
Electricity Consumption Value in kWh/(m²a).
Final Energy according: Verein Deutsche Ingenieure VDI 3807 (Society of German Engineers)

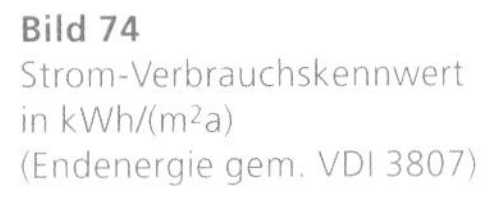

Bild 74
Strom-Verbrauchskennwert in kWh/(m²a)
(Endenergie gem. VDI 3807)

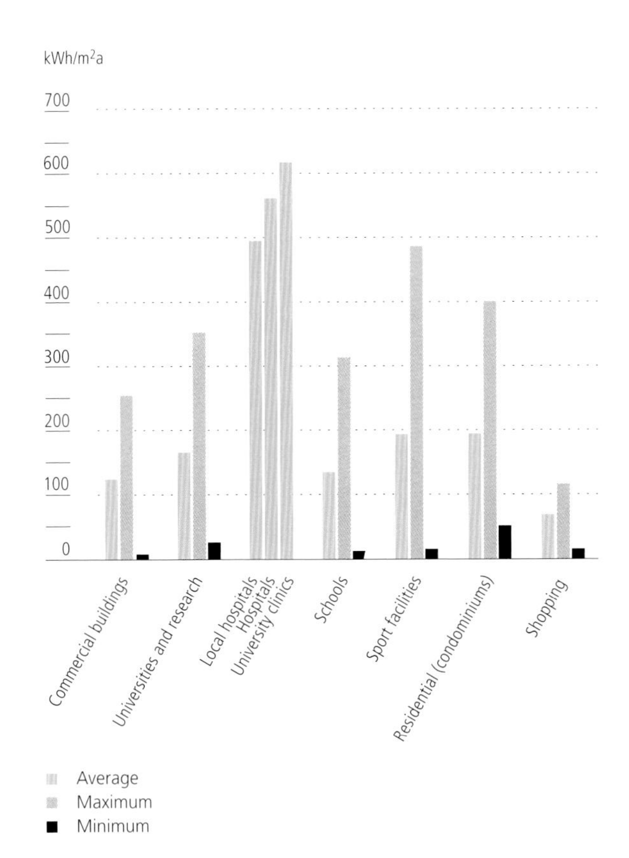

Figure 75
Heating Consumption Value in kWh/(m²a). Final energy according: Verein Deutsche Ingenieure VDI 3807 (Society of German Engineers)

Bild 75
Heizenergie-Verbrauchskennwert in kWh/m²a
(Endenergie gem. VDI 3807)

5.2 Primary energy consumption, Leitszenario 2006

Figure 76.1 depicts the typical composition of the consumption of primary energy and the percentage of renewable energy in Germany today. Renewable energy contributes up to 5.3 % and is far from the more lean consumption rate it could be. In 2006, the German government, represented by the Ministry of Energy, published the so-called "Leitszenario", a benchmark for the structure of future energy consumption up to the year 2050, as shown in Figure 76.2. The benchmark goals are, of course, to lower fossil fuel use and simoultaneously to increase renewable energy consumption, but also to reduce overall energy consumption from today's 4,000 MWh x 10^6 to 2,200 MWh x 10^6 in the year 2050.

5.2 Primärenergieverbräuche, Leitszenario 2006

Bild 76.1 zeigt die zurzeit typische Struktur des Primärenergieverbrauchs in Deutschland und dabei den Anteil erneuerbarer Energien. Der Anteil erneuerbarer Energien beträgt ca. 5,3 Prozent und entspricht selbstverständlich nicht dem, was wir haben sollten. Durch die Regierung der Bundesrepublik Deutschland wurde, vertreten durch das Bundesministerium für Umwelt, Naturschutz und Reaktorsicherheit, 2006 ein Leitszenario erstellt, das die Struktur des zukünftigen Energieverbrauchs bis 2050 beschreibt, Bild 76.2. Das Szenario weist aus, in welcher Form die fossilen Brennstoffe verringert werden sollen bei gleichzeitigem Anstieg des Einsatzes erneuerbarer Energien. Insgesamt jedoch sollen die Primärenergieverbräuche von zurzeit ca. 4.000 MWh x 10^6 gesenkt werden bis 2050 auf ca. 2.200 MWh x 10^6.

Figure 76.3, which allows us to compare the primary energy consumption per user,clearly shows that the highest contribution to energy consumption is made by losses in the processes of energy conversion and transmission. There is therefore an urgent need to increase efficiencies and performance and to reduce transmission losses for all systems involved in the production and supply chain.

Bild 76.3 zeigt ergänzend die derzeitige Primärenergieverteilung nach Verbrauchern, wobei mit Abstand der größte Verbrauch durch Verluste und Aufbereitung entsteht. Hieraus resultiert der Anspruch, sowohl die Aufbereitungs- wie auch die Transportverluste einerseits massiv zu senken und zum anderen die Wirkungsgrade energienutzender Anlagen deutlich zu erhöhen.

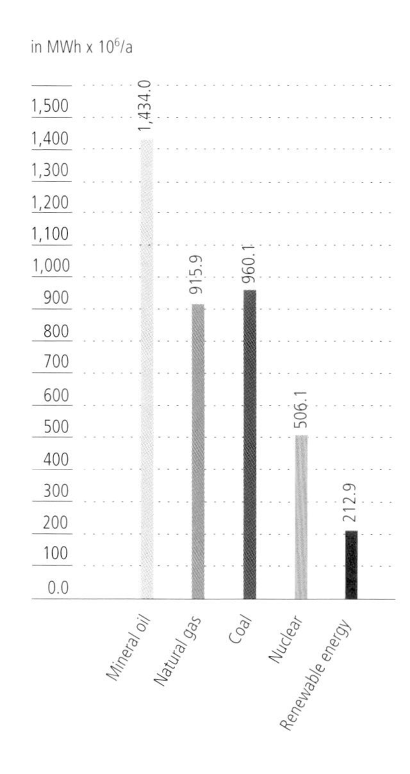

Figure 76.1
Pattern of primary energy consumption in Germany and utilization of renewable energy (2006)

Source: Arbeitsgemeinschaft Energiebilanz, Berlin

Bild 76.1
Struktur des Primärenergieverbrauchs in Deutschland und Einsatz der erneuerbaren Energien (2006)

Quelle: Arbeitsgemeinschaft Energiebilanz, Berlin

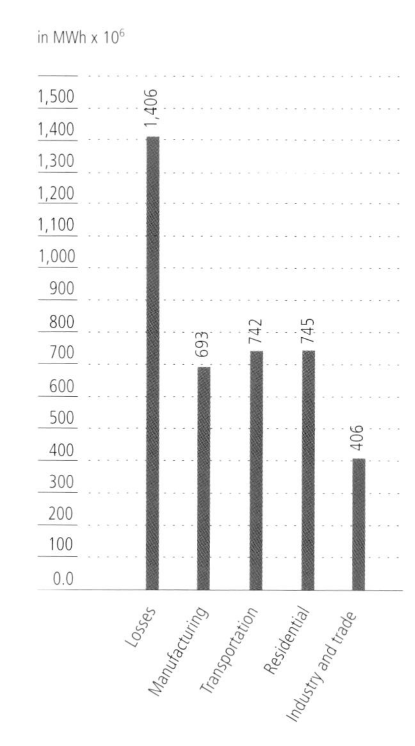

Figure 76.3
Distribution of primary energy consumption by consumer category in Germany

Source: Arbeitsgemeinschaft Energiebilanz, Berlin

Bild 76.3
Primär-Energieverteilung nach Verbrauchern

Quelle: Arbeitsgemeinschaft Energiebilanz, Berlin

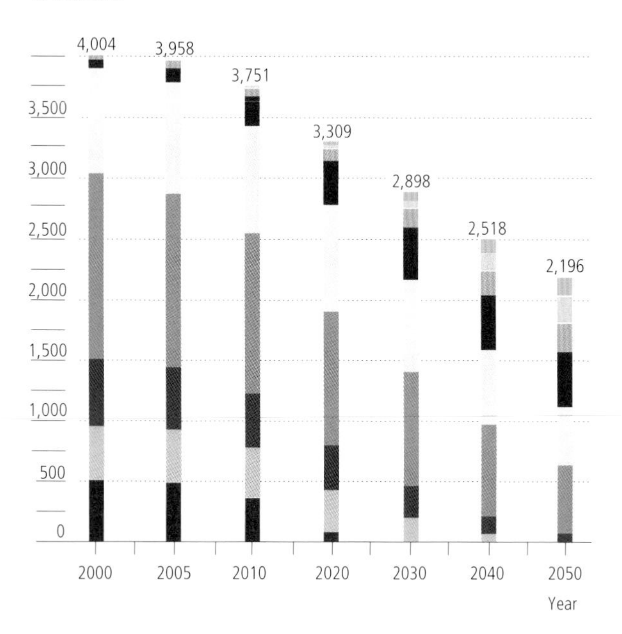

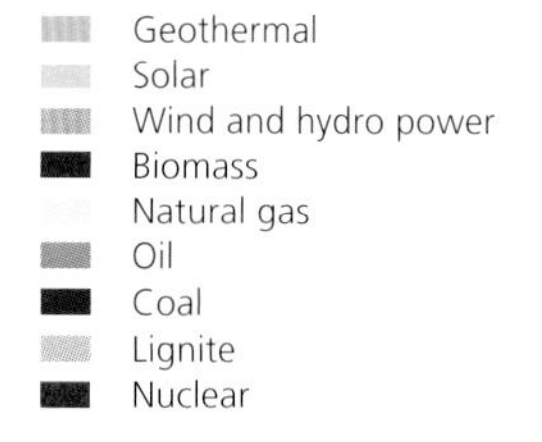

Figure 76.2
Pattern of primary energy consumption for various energy resources, according to Leitszenario 2006 (Coefficient of Performance Method)

Source: Bundesministerium für Umwelt, Naturschutz und Reaktorsicherheit, Germany. Leitstudie 2007 "Ausbaustrategie Erneuerbare Energien"

Bild 76.2
Struktur des Primärenergieverbrauchs im Leitszenario 2006 nach Energieträgern (Wirkungsgradmethode)

Quelle: Bundesministerium für Umwelt, Naturschutz und Reaktorsicherheit, Leitstudie 2007 "Ausbaustrategie Erneuerbare Energien"

5.3 Final energy consumption, Leitszenario 2006

If we subtract the percentage of energy that is typically lost during the conversion of primary energy – e.g., oil at oil fields, coal in coal mines, gas at its sources – we arrive at a unit called final energy, which is the amount of total energy that affects us in the operation of our buildings. Yet, even this final stage of the energy supply chain will undergo efficiency losses in performance: an incandescent light bulb, for example, receives 100 units of final energy but converts just 15 of those to light, with the rest – 85 %! – being lost as uncaptured thermal energy. **Figure 76.4** shows the degree to which we need to reduce final energy by the year 2050 by means of increases in efficiency, reduction in transformation losses, and the avoidance of non-energy consumption.

In order to reach the goal of fossil energy reduction, the need for increased implementation of renewable energy strategies is unavoidable. **Figure 76.5** shows how the percentage of renewable sources as part of the generation of final energy may be potentially increased.

5.3 Endenergieverbräuche, Leitszenario 2006

Zieht man vom Primärenergiebedarf (z.B. Primärenergie an der Quelle: Ölquelle, Bergwerk, Gasquelle usw.) die Energiemengen ab, die durch vorgelagerte Prozessketten außerhalb der Systemgrenze Gebäude bei der Gewinnung, Umwandlung und Verteilung der jeweils eingesetzten Brennstoffe entstehen, so erhält man beim Eintritt in das Gebäude die Endenergiemenge, d.h. die, auf die der Nutzer eines Gebäudes einen Einfluss hat. Die Endenergie wiederum wird nur zum Teil durch Wirkungsgradverluste tatsächlich genutzt (bestes Beispiel: Glühbirne Endenergie 100, Lichtenergie ca. 15 Prozent, Wärmeenergie ca. 85 Prozent). **Bild 76.4** zeigt, wie groß der Anteil der Endenergie in den Jahren bis 2050 sein soll und welche Reduzierungen erreicht werden sollen durch Effizienzsteigerungen, Vermeidung von Umwandlungsverlusten, Vermeidung von Nicht-Energieverbräuchen.

Zur Erreichung der Ziele bezüglich des Verbrauchs an fossilen Brennstoffen ist der Einsatz erneuerbarer Energien vermehrt angesagt. **Bild 76.5** zeigt, wie in den nächsten Jahren und Jahrzehnten die Endenergiemengen durch verschiedene erneuerbare Energiequellen gesteigert werden können.

Figure 76.4
Pattern of final energy consumption, non-energy consumption and transformation losses according to Leitszenario 2006.

Source:Bundesministerium für Umwelt, Naturschutz und Reaktorsicherheit, Germany. Leitstudie 2007 "Ausbaustrategie Erneuerbare Energien"

Bild 76.4
Struktur des Endenergieverbrauchs, nichtenergetischer Verbrauch undUmwandlungsverluste im Leitszenario 2006;

Quelle: Bundesministerium für Umwelt, Naturschutz und Reaktorsicherheit, Leitstudie 2007 "Ausbaustrategie Erneuerbare Energien"

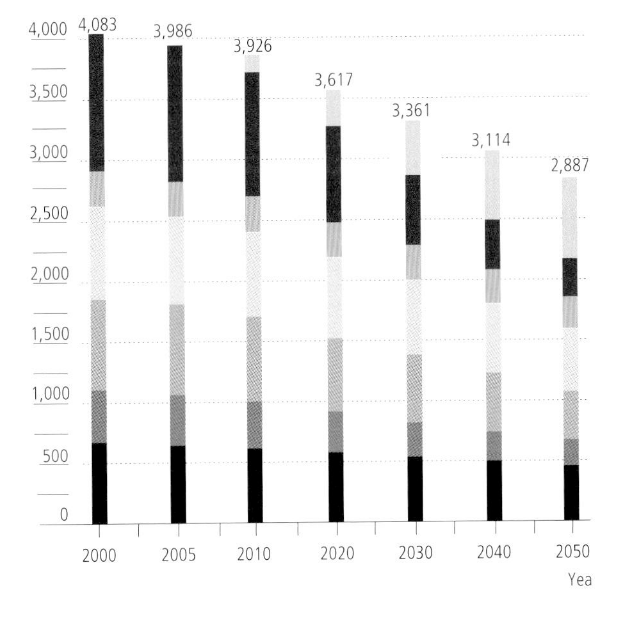

Figure 76.4

Figure 76.5
Final energy contribution by renewable energy resources until 2050, according to Leitszenario 2006.

Source: Bundesministerium für Umwelt, Naturschutz und Reaktorsicherheit, Germany, Leitstudie 2007 "Ausbaustrategie Erneuerbare Energien"

Bild 76.5
Endenergiebeitrag erneuerbarer Energien im Leitszenario 2006 nach Energiequellen bis zum Jahr 2050

Quelle: Bundesministerium für Umwelt, Naturschutz und Reaktorsicherheit, Leitstudie 2007 "Ausbaustrategie Erneuerbare Energien"

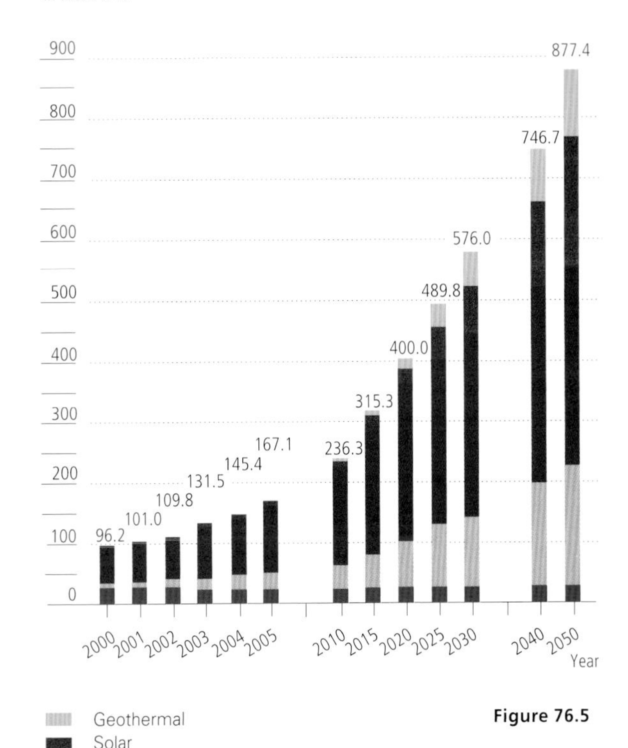

Figure 76.5

As a supplement, and in order to allow for a better further analysis, **Figures 77 – 78.2** are presented, which depict final energy consumption per industrial sector, for commerce, and for the service industry.

Zur Ergänzung der Informationen und zur besseren Analyse dienen die **Bilder 77 – 78.2**, die die Endenergieverbräuche für verschiedene Anwendungszwecke bzw. für verschiedene Verbrauchsbereiche wie Industrie, Gewerbe, Handel und Dienstleistungen zeigen.

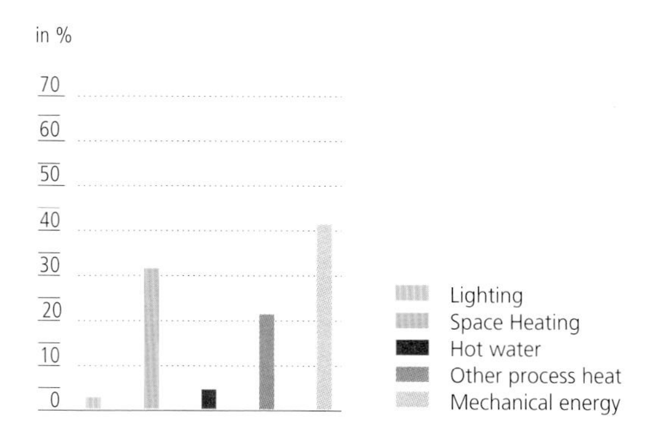

Figure 77
Final energy consumption by usage in Germany, year 2000.

Source: Arbeitsgemeinschaft Energiebilanz, Berlin

Bild 77
Endenergieverbrauch insgesamt nach Anwendungszwecken in Deutschland im Jahr 2000

Quelle: Arbeitsgemeinschaft Energiebilanz, Berlin

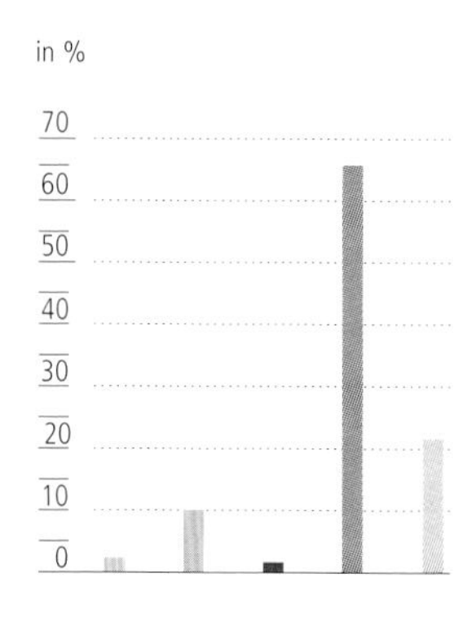

Figure 78.1
Final energy consumption in the industrial sector, Germany, 2000.

Source: Arbeitsgemeinschaft Energiebilanz, Berlin

Bild 78.1
Endenergieverbrauch im Sektor **Industrie** nach Anwendungszwecken im Jahre 2000

Quelle: Arbeitsgemeinschaft Energiebilanz, Berlin

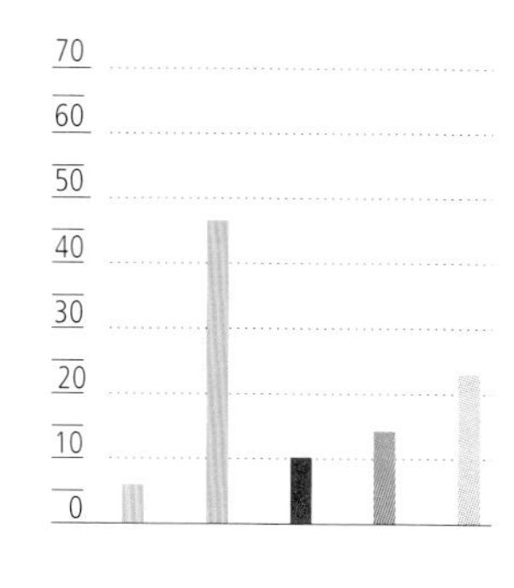

Figure 78.2
Final energy consumption per usage type for the manufacturing, trade and service industries in Germany, 2000

Source: Arbeitsgemeinschaft Energiebilanz, Berlin

Bild 78.2
Endenergieverbrauch im Sektor **Gewerbe, Handel, Dienstleistungen**nach Anwendungszwecken im Jahre 2000

Quelle: Arbeitsgemeinschaft Energiebilanz, Berlin

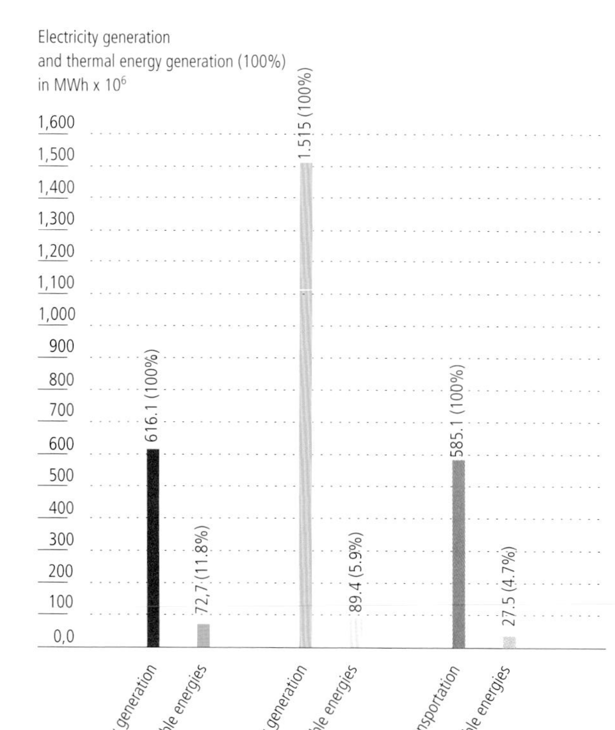

Figure 79
Percentage of renewable energies of total primary energy consumption in Germany 2006

Source: Arbeitsgemeinschaft Energiebilanz, Berlin

Bild 79
Anteile der erneuerbaren Energien am Gesamt-Primärenergieverbrauch (2006)

Quelle: Arbeitsgemeinschaft Energiebilanz, Berlin

Figure 80
Distribution of renewable energies in Germany
= 11,8 % (72,7 MWh x 10^6)
as part of total primary electricity generation, 2006

Source: BMU nach Arbeitsgruppe Erneuerbare Energien – Statistik (AGEE-Stat)

Bild 80
Verteilung der erneuerbaren Energien
= 11,8 % (72,7 MWh x 10^6)
an der Gesamtstromerzeugung (Primärenergie), Stand 2006

Quelle: BMU nach Arbeitsgruppe Erneuerbare Energien – Statistik (AGEE-Stat)

5.4 Renewable energy, Leitszenario 2006 Electricity generation

Figure 81
Development of electricity generation by renewable resources until 2020 according to the German Renewable Energies Act (EEG)

Source: Bundesministerium für Umwelt, Naturschutz und Reaktorsicherheit, Germany. Leitstudie 2007 "Ausbaustrategie Erneuerbare Energien".

Bild 81
Entwicklung der Stromerzeugung aus EE (Primärenergie) bis 2020 im Leitszenario unter EEG-Bedingungen

Quelle: Bundesministerium für Umwelt, Naturschutz und Reaktorsicherheit, Leitstudie 2007 "Ausbaustrategie Erneuerbare Energien"

If we break down the contribution of renewable energy as part of the total primary energy consumption up to the year 2006, we arrive at the graphical representation shown in **Figure 79**. Renewable energy contributes 11.8 % to the total electric energy generation, 5.9 % to thermal energy provision, and 4.7 % to the transportation sector (traffic). Obviously, all such contributions to the general power mix are far from being exhausted.

Figure 80 shows the distribution of renewable energy across total primary electric energy generation sources. Wind and hydroelectric power are the dominant contributing factors today, followed by biogas and biogenic solid fuels. Currently, only a small role is played by deep geothermal applications, and the area of photovoltaic power generation needs to be significantly improved. On the governmental, regulatory level, this can be achieved partly by reconsidering tax incentives for their implementation and operation.

Figure 81 depicts the development of electrical energy generation from renewable sources until the year 2020, according to the Leitszenario. It becomes evident that a further expansion of existing hydroelectric power supply is unlikely and that the major share of supply by renewable energy has to continue to come from wind power and biomass transformation.

5.4 Erneuerbare Energien, Leitszenario 2006 Stromerzeugung

Schlüsselt man die Anteile der erneuerbaren Energien am Gesamtprimärenergieverbrauch bis einschließlich 2006 auf, so stellt sich das in **Bild 79** ausgewiesene Szenario dar.

An der Gesamtstromversorgung beträgt der Anteil erneuerbarer Energien 11,8 Prozent, an der Gesamtwärmebereitstellung der Anteil erneuerbarer Energie 5,9 Prozent und am Straßenverkehr (Transport) 4,7 Prozent. Diese Anteile der erneuerbaren Energien an der Gesamtprimärenergieversorgung bzw. am Gesamtprimärenergieverbrauch sind weit weg von dem, was noch erreicht werden muss. **Bild 80** zeigt die Verteilung erneuerbarer Energien an der Stromerzeugung (Primärenergie).

Windenergie und Wasserkraft sind die dominierenden Größen, gefolgt von Biogas und biogenen Festbrennstoffen. Zurzeit noch keine Rolle spielt die tiefe Geothermie, und auch der Bereich der Photovoltaik muss sich deutlich erhöhen. Hierzu wäre es unter Umständen vonnöten, über einen erneuten steuerlichen Anreiz den Ausbau zu beschleunigen.

Bild 81 zeigt die Entwicklung der Stromerzeugung aus erneuerbaren Energien bis 2020 nach dem Leitszenario 2006. Hier wird deutlich, dass in Deutschland die Wasserkraft nicht mehr ausbaufähig ist und sowohl Windenergie wie auch Biomasse die wesentlichen Anteile in Zukunft übernehmen sollen.

Figure 80

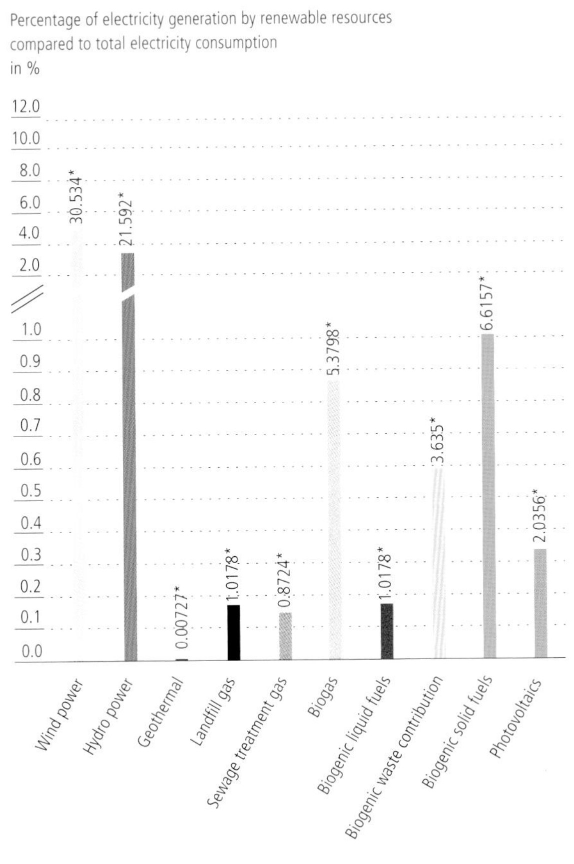

Figure 81

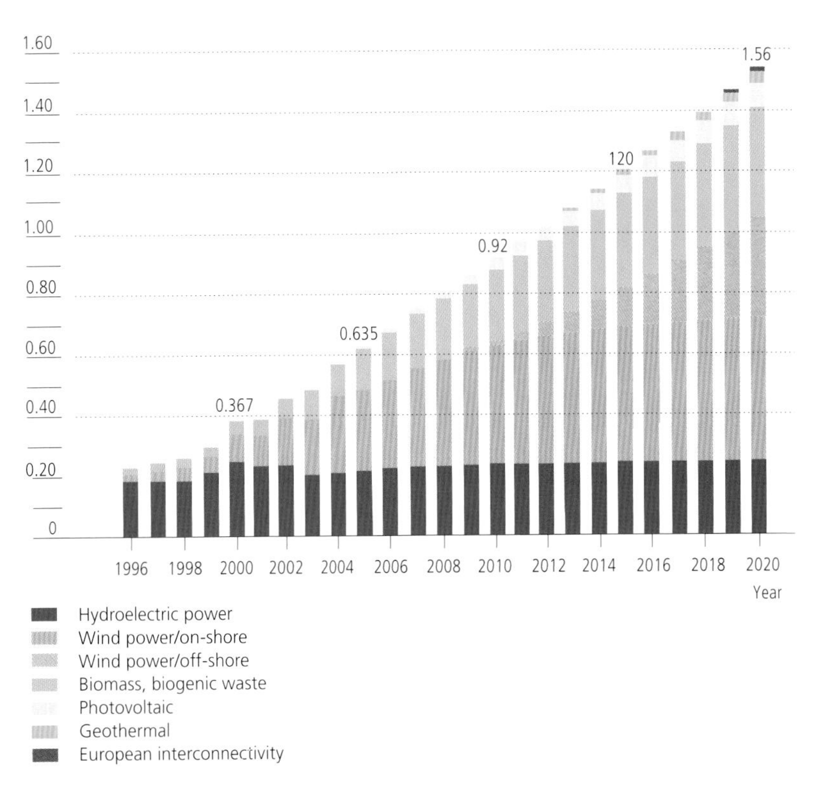

5.5 Renewable energy, Leitszenario 2006 Thermal energy generation

As documented in **Figure 79**, renewable energy in 2006 represents only 5.9 %, or 89.4 MWh x 10^6 – just a small part of thermal energy generation. It is derived mainly from biogenic solid fuel waste from households and industry, as well as liquid biogenic fuels or those in gaseous form from agricultural processes. Deep geothermal energy as an almost ideal energy source plays an entirely insignificant role thus far and needs further strengthening in the near future. Geothermal applications close to the surface of the earth, such as slinky collectors and energy piles, in connection with water-source heat pumps, are also not represented sufficiently yet, and this can be said for solar technologies as well. In all such cases of renewable energy applications, special incentives for their further distribution and use should be considered (**Figure 82**)

According to the Leitszenario 2006, the percentage of use of fossil heating oil, coal, and gas in the generation of thermal energy needs to be reduced significantly, **Figure 83**.

5.5 Erneuerbare Energien, Leitszenario 2006 Wärmebereitstellung

Wie aus **Bild 79** zu entnehmen, beträgt der Anteil der erneuerbaren Energie bei der Wärmebereitstellung ca. 5,9 Prozent = 89,4 MWh x 10^6 für den Zeitraum 2006. Wesentliche Elemente dabei sind biogene Festbrennstoffe aus Haushalten und Industrie sowie biogene flüssige und gasförmige Brennstoffe aus dem Bereich der Landwirtschaft. Die tiefe Geothermie als eine ideale Quelle der Wärmebereitstellung spielt eine völlig untergeordnete Rolle und ist in den nächsten Jahren deutlich zu entwickeln. Die oberflächennahe Geothermie, d.h. Flach- und Tiefensonden in Verbindung mit Wärmepumpen, sind ebenfalls noch deutlich zu gering vertreten, so dass hier wie auch bei der Solarthermie unter Umständen mit einem besonderen Anreiz die weitere Nutzung beschleunigt werden sollte, **Bild 82**.

Gemäß Leitszenario 2006 soll der Anteil an Heizöl massiv gesenkt werden, gleichermaßen der Anteil an Kohle und an Erdgas, **Bild 83**.

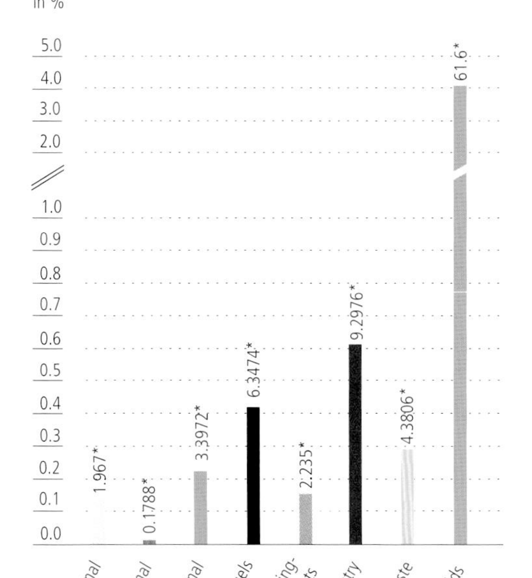

Figure 82
Distribution of thermal energy production (primary energy) from renewable resources = 5.9 % of total thermal energy demand (89.4 MWh x 10^6)

Source: Arbeitsgemeinschaft Energiebilanz, Berlin

Bild 82
Verteilung der Wärmebereitstellung (Primärenergie) aus erneuerbaren Energien, = 5.9 % (89,4 MWh x 10^6) des Gesamt-Wärmeenergiebedarfs

Quelle: Arbeitsgemeinschaft Energiebilanz, Berlin

The percentage of thermal energy produced by fossil fuels in long-distance and near-distance district heating plants remains comparable to the fossil-fuel-operated combined-heat-power plants (CHP). The Leitszenario goal is to increase the use of biomass, photovoltaic, and deep geothermal generation for thermal applications.

A further analysis of renewable energy, especially of geothermal applications, solar thermal, and biomass, is shown in **Figure 84**. According to this graph, the percentage of geothermally operated near-distance district heating will increase rapidly until the year 2050, but a much more accelerated increase in the development of such technologies would be even more desirable.

Der Wärmeenergieanteil aus Fern- und Nahwärme infolge fossiler Brennstoffe nimmt durchgängig einen Anteil ein, ähnlich dem der industriellen Kraftwärmekopplung auf Basis fossiler Energieträger. Der Einsatz von Biomasse wie auch die Anzahl von Kollektoren sollen einen breiteren Raum einnehmen, ähnlich wie der Ausbau der Erdwärme (tiefe Geothermie).

Eine weitere Aufschlüsselung des Einsatzes erneuerbarer Energien – hier insbesondere Geothermie, Solarthermie (Kollektoren) und Biomasse – zeigt **Bild 84**. Danach steigt der Anteil geothermischer Anlagen für den Nahwärmebereich deutlich bis 2050 an. Ein deutlich früherer Einsatz der Geothermie in noch größerem Umfang wäre aus Sicht des Autors notwendig und erstrebenswert.

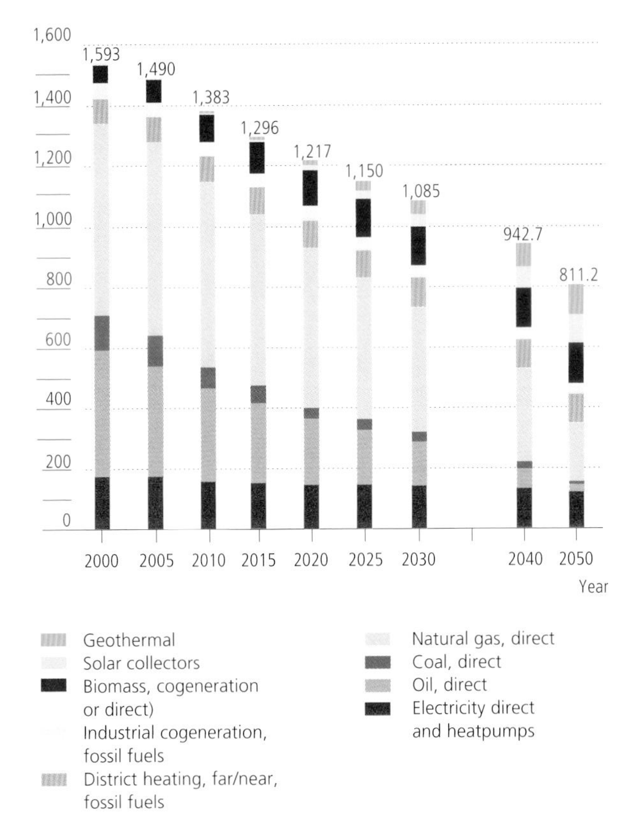

Figure 83
Various energy sources used for thermal production according to Leitszenario 2006

Source: Bundesministerium für Umwelt, Naturschutz und Reaktorsicherheit, Germany Leitstudie 2007 "Ausbaustrategie Erneuerbare Energien"

Bild 83
Energieeinsatz zur Wärmebereitstellung im Leitszenario 2006 nach Energieträgern

Quelle: Bundesministerium für Umwelt, Naturschutz und Reaktorsicherheit, Leitstudie 2007 "Ausbaustrategie Erneuerbare Energien"

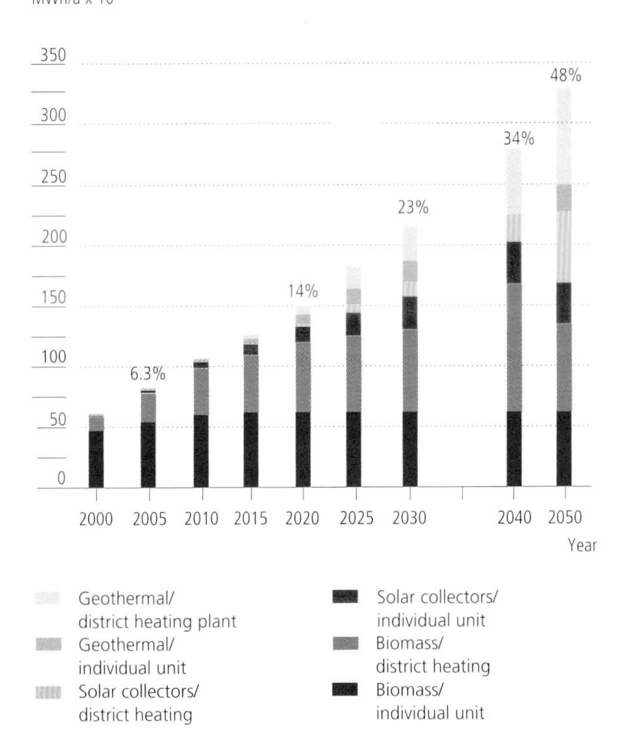

Figure 84
Contribution of renewable energy of total thermal energy demand according to Leitszenario 2006

Source: Bundesministerium für Umwelt, Naturschutz und Reaktorsicherheit, Germany, Leitstudie 2007 "Ausbaustrategie Erneuerbare Energien"

Bild 84
Beitrag erneuerbarer Energien zur Deckung des Wärmebedarfs im Leitszenario 2006

Quelle: Bundesministerium für Umwelt, Naturschutz und Reaktorsicherheit, Leitstudie 2007 "Ausbaustrategie Erneuerbare Energien"

With regard to the future use of biomass in Germany, Gerhard Hausladen and his team of scientists at the Technical University of Munich's Department of Building Climatics and Systems developed the following data and interdependencies:

Part 1

- Total surface area Germany: 35,709,200 ha
- Area covered by buildings: 4,652,100 ha
- Area covered by water: 827,900 ha

- Wooded areas used for construction and preservation: 7,500,000 ha
- Wooded areas used for energy today: 750,000 ha
- Wooded areas, future expansion potential: 1,750,000 ha

- Agricultural areas used for food supply: 17,000,000 ha
- Agricultural areas used for energy today: 1,980,000 ha

(Source: Federal Statistical Office, Germany; Reference year: 2006)

Part 2

	Area (ha)	Energy generation (50MWh/ha); (MWh)	Percentage of primary energy (%)
Wooded areas, total	10,000,000	(500,000,000)	12.7
Used for energy generation	2,500,000	125,000,000	3.2
Agricultural areas, total	19,000,000	(950.000.000)	24.1
Used for energy generation	2,000,000	100,000,000	2.5
Approximate total energy generation		**225,000,000**	

The agricultural surface area in Germany that is available for energy generation would be enough to provide 6 % of the country's primary energy requirements. Also, 2 % can be added to this amount if straw, industrial waste wood, animal waste and methane gas from sewage treatment plants and landfills were used.

Hinsichtlich des zukünftigen Einsatzes von Biomasse in Deutschland hat sich Gerhard Hausladen (Technische Universität München, Lehrstuhl für Bauklimatik und Haustechnik) mit seinem Team mit der Thematik auseinandergesetzt und folgende Relationen und Daten erarbeitet:

Teil 1

- Gesamtfläche Deutschland: 35.709.200 ha
- Bebaute Fläche: 4.652.100 ha
- Wasserfläche: 827.900 ha

- Waldfläche für Bauholz, Naturschutz: 7.500.000 ha
- Heutige Nutzung Wald für Energie: 750.000 ha
- Weiteres Potenzial Wald für Energie: 1.750.000 ha

- Ackerland für Ernährung: 17.000.000 ha
- Heutige Nutzung Ackerland für Energie: 1.980.000 ha

(Datenquelle: Statistisches Bundesamt, Bezugsjahr: 2006)

Teil 2

	Fläche (ha)	Energieertrag (50MWh/ha); (MWh)	Anteil an Primärenergie (%)
Waldfläche gesamt	10.000.000	(500.000.000)	12,7
davon für Energie	2.500.000	125.000.000	3,2
Landwirtschaftliche Fläche gesamt	19.000.000	(950.000.000)	24,1
davon für Energie	2.000.000	100.000.000	2,5
Ca.-Energieertrag total		**225.000.000**	

Durch die für Energie zur Verfügung stehende Fläche an Landwirtschaft und Wald können 6 Prozent des Primärenergieanteile in Deutschland gedeckt werden. Hierzu kommen noch gut 2 Prozent durch die Bereiche Stroh, Industrierestholz, tierische Exkremente sowie Klär- und Deponiegas.

The German Advisory Council on Global Change (WBGU), which consults for the Federal Government on energy policies, reports in a study the following potential of biomass source-related energy supply for Germany:

Demgegenüber zeigt eine andere Studie des wissenschaftlichen Beirats der Bundesregierung:

Biomass potential, Germany

	Energy produced (MWh)	Primary energy percentage of total (%)
- Wooded areas	171,000,000	4.3
- Agricultural areas	83,000,000	2.1
- Straw, animal waste Industrial waste wood Landfill/waste treatment gas	90,000,000	2.3
Total	344,000,000	

Biomassepotential in Deutschland

	Energieertrag (MWh)	Anteil an Primärenergie (%)
– Waldfläche	171.000.000	4,3
– Landwirtschaftliche Fläche	83.000.000	2,1
– Stroh, tierische Exkremente Industrieholz Klär- und Deponiegas	90.000.000	2,3
Gesamt	344.000.000	

Both studies show a certain range of fluctuation, a condition that seems to be common in such early prognostic data. However, the data can be used for guidance when further estimates are being delivered.

Beide Erhebungen zeigen die üblichen Bandbreiten von Prognosen, die wohl zurzeit noch nicht exakter erstellt werden können. Gleichwohl sind sie insofern interessant, als sie Anhaltswerte auch für weitere Abschätzungen bieten.

5.6 Renewable energy, Leitszenario 2006 Energy for the transportation sector

In **Figure 79**, the percentage of renewable energy in land transportation is shown as 4.7 %. This amount of renewable energy fuels, or 27.5 x 10^6 MWh/a, is being provided by biodiesel, vegetable oil, and bioethanol (see **Figure 85**.)

Figure 86 shows the aim of the Leitszenario 2006, and especially the degree to which fossil gasoline use will need to be reduced. The figure also shows the percentages of diesel fuel and kerosene. Thus, the goals of the Leitszenario remain basically unchanged because apparently no real alternatives to these fuels exist at this moment. The use of hydrogen gas for fuel cells is supposedly increasing slightly, according to the prognosis, while that of biofuels has increased significantly. Biofuels, however, show clearly varying efficiencies in comparison with diesel and fossil gasoline, and the uncertain yearly yield of agricultural fields results in greatly varied annual energy supplies from that renewable resource (**Figure 87**).

5.6. Erneuerbare Energien, Leitszenario 2006 Energieeinsatz im Verkehr

Gemäß **Bild 79** beträgt der Anteil erneuerbarer Energien im Straßenverkehr ca. 4,7 Prozent. Diese 4,7 Prozent = 27,5 x 10^6 MWh/a werden durch Biodiesel, Pflanzenöl und Bioethanol, siehe **Bild 85**, dargestellt.

Bild 86 zeigt das angestrebte Leitszenario 2006, wobei hier gut erkennbar ist, in welcher Form das Benzin deutlich reduziert werden soll, gleichermaßen Dieselkraftstoff. Kerosin bleibt in ähnlicher Größenordnung bestehen wie heute, da es hier offensichtlich zurzeit keine vergleichbaren Ersatzstoffe gibt. Der Einsatz von Erdgas (Brennstoffzellentechnik) soll leicht erhöht werden, der Einsatz von Biokraftstoffen deutlich. Die Biokraftstoffe zeigen je nach Art eine unterschiedliche Effizienz gegenüber Diesel bzw. Benzin, wobei der Jahresertrag eines Hektars Anbaufläche wiederum zu sehr unterschiedlichen Energiemengen führt, **Bild 87**.

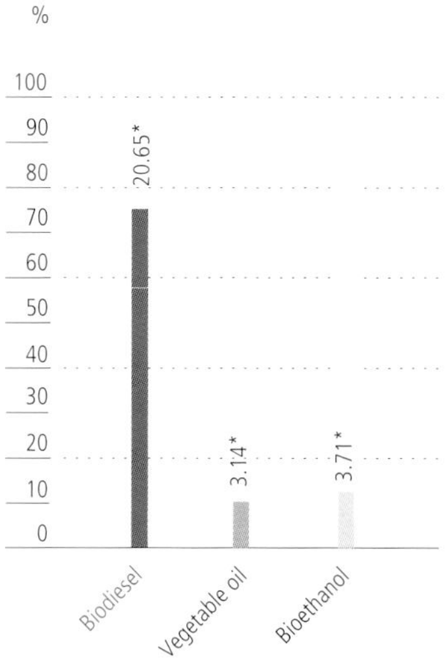

*) MWh x 10^6

Figure 85
Percentages and distribution of biogenic fuels in Germany (2006)

Source: BMU nach Arbeitsgruppe Erneuerbare Energien – Statistik (AGEE-Stat)

Bild 85
Struktur der biogenen Kraftstoffe in Deutschland (2006)

Quelle: BMU nach Arbeitsgruppe Erneuerbare Energien – Statistik (AGEE-Stat)

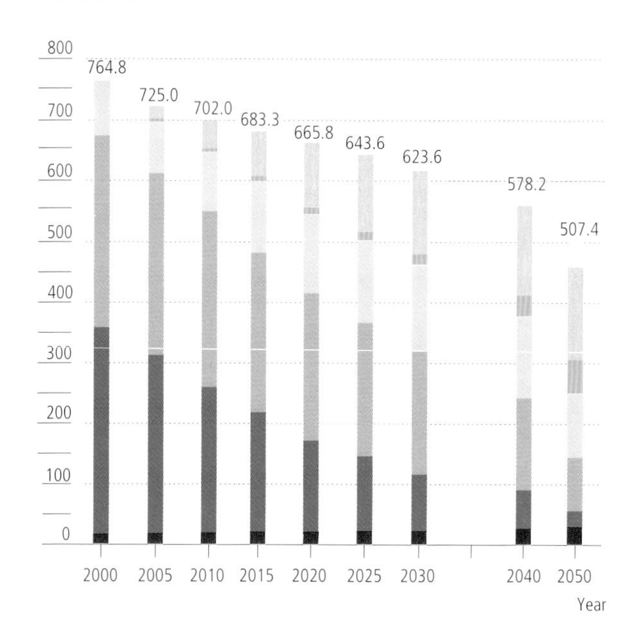

Figure 86
Energy distribution across fuel types used for transportation according to "Effizienzvariante" UBA 2006

Source: Bundesministerium für Umwelt, Naturschutz und Reaktorsicherheit, Germany, Leitstudie 2007 "Ausbaustrategie Erneuerbare Energien"

Bild 86
Energieeinsatz im Verkehr im Leitszenario 2006 nach Kraftstoffarten (in Anlehnung an die "Effizienzvariante" in UBA 2006)

Quelle: Bundesministerium für Umwelt, Naturschutz und Reaktorsicherheit, Leitstudie 2007 "Ausbaustrategie Erneuerbare Energien"

The use of renewable energy allows for a reduction in CO_2 emissions, as shown in **Figure 88**. It is apparent from this graph that electrical energy generation is more advanced in its development than the sectors of thermal energy supply and supply by renewable-based gasoline substitutes. In other words, thermal energy supply by renewable energy sources is in the early stage of its potential and needs to be promoted in order for it to provide a much higher percentage of the total energy supply.

All of the new technologies related to the generation of energy out of renewable sources are already an important economic factor that will have to gain further weight. First, analysis shows that Germany's local sales of renewables reached about 21.6 billion Euro, of which sales of biomass for energy generation amounted to the greatest contributing portion, with 38 %. Returns of solar plants resulted in a share of approximately 28 %, and wind power-based systems contributed approximately 26 %. Investments in renewable energy both for inland and export trade have created some 214,000 new jobs, including those indirectly related jobs in expert and other levels of the value chain.

The significant increase in job numbers when compared with previous years can be attributed largely to the demand for systems for thermal energy generation. In this sector, the large cost increases for oil and gas have certainly stimulated growth of the renewable sector significantly. Expected future turns of the energy price spiral will without doubt cause massive increases in the use of alternative conversion systems for renewable power generation.

Durch die Nutzung der erneuerbaren Energien können CO_2-Emissionen vermieden werden, wie sie in **Bild 88** dargestellt sind. Hier zeigt sich, dass die Stromerzeugung zurzeit noch einen deutlichen Vorsprung gegenüber der Wärmebereitstellung und der Bereitstellung von Kraftstoffen hat. Im Umkehrschluss kann somit festgestellt werden, dass die Wärmebereitstellung einen noch zu geringen Anteil an erneuerbarer Energie besitzt und massiv ausgebaut werden muss.

Die Entwicklung der Technologien zur Nutzung erneuerbarer Energien ist schon zurzeit ein interessanter Wirtschaftsfaktor, der zunehmend ausgebaut werden muss. Erste Analysen zeigen, dass der Inlandsumsatz im Jahr 2006 rund 21,6 Milliarden Euro betrug, wobei der umsatzstärkste Bereich mit 38 Prozent die energetische Nutzung von Biomasse war. Der Umsatz im Bereich Solarenergie betrug ca. 28 Prozent, der der Windenergie ca. 26 Prozent. Die Investitionen im Bereich erneuerbarer Energien (Inlandsanlagen, Export) führten dazu, dass zurzeit ca. 214.000 Arbeitsstellen bestehen – Beschäftigte in Deutschland unter Einbeziehung des Außenhandels und vorgelagerter Wertschöpfungsstufen.

Die deutliche Steigerung von Beschäftigten gegenüber den Vorjahren ist vor allem auf die große Nachfrage nach Anlagen zur Wärmeerzeugung zurückzuführen, ausgelöst durch die sich immer höher entwickelnden Öl- und Gaspreise. Insofern kann davon ausgegangen werden, dass die zu erwartenden Preisschübe im Bereich der fossilen Brennstoffe weitere massive Steigerungen bei der Entwicklung von Anlagen zum Einsatz erneuerbarer Energien auslösen werden.

Figure 88
Reduction of CO_2 emissions by renewable energy use in Germany 2006

Source: BMU Arbeitsgruppe Erneuerbare Energien – Statistik (AGEE-Stat); Zentrum für Sonnenenergie- und Wasserstoffforschung Baden-Württemberg (ZSW); Gutachten zur CO_2-Minderung im Stromsektor durch den Einsatz erneuerbarerEnergien, 2005, Fraunhofer Institut für System- und Innovationsforschung Karlsruhe; Öko-Institut Darmstadt

Bild 88
Vermiedene CO_2-Emissionen durch die Nutzung erneuerbarer Energien in Deutschland (2006)

Quelle: BMU nach Arbeitsgruppe Erneuerbare Energien – Statistik (AGEE-Stat); Zentrum für Sonnenenergie- und Wasserstoffforschung Baden-Württemberg (ZSW); Studie "Gutachten zur CO_2-Minderung im Stromsektor durch den Einsatz erneuerbarer Energien", 2005, Fraunhofer Institut für System- und Innovationsforschung Karlsruhe; Öko-Institut Darmstadt

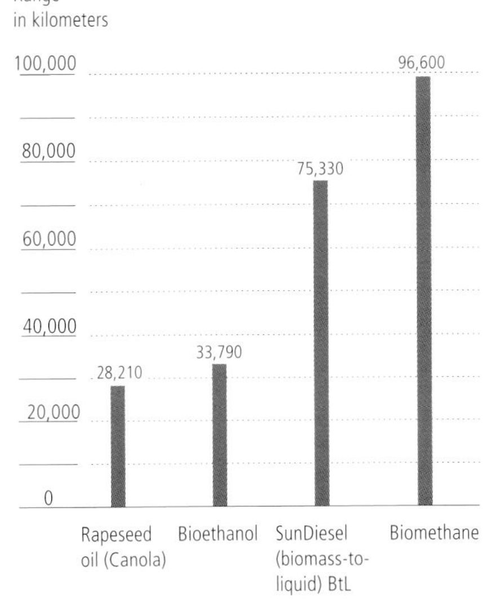

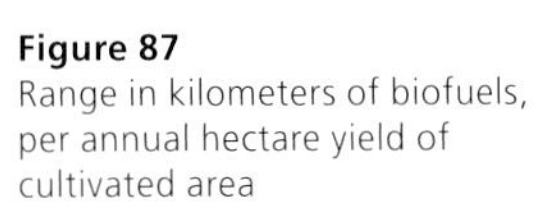
Figure 87
Range in kilometers of biofuels, per annual hectare yield of cultivated area

Source: Spiegel Spezial, 5/2006

Bild 87
Reichweiten mit Bio-Kraftstoffen aus dem Jahresertrag eines Hektars Anbaufläche

Quelle: Spiegel Spezial, 5/2006

	Yield per hectare	Efficiency compared to diesel	Efficiency compared to gasoline
Rapeseed oil	1.550 l	91%	–
Bioethanol	2.560 l	–	66%
SunDiesel	4.050 l	93%	–
Biomethane	3.560 l	–	140%

Million tons
70
60
50
40
30
20
10
0
67.1
20.7
8.7
Electricity
Thermal
Fuels
Water
Wind
Biomass
Photovoltaic/ Solar thermal
Geothermal
Biodiesel
Vegetable oil
Bioethanol

6 Detailed view Switzerland

6 Detail-betrachtung Schweiz

Energy supply by renewable sources: the Swiss approach

Switzerland, represented by the Swiss Federal Office of Energy (Bundesamt für Energie, BFE) publishes a yearly overview of national energy consump-tion data and how the energy is being provided. The total final energy consumption in Switzerland in 2005 was approximately 247.199 x 10^6 MWh, and the total primary energy demand was 314.44 x 10^6 MWh.

The amount of renewable energy in 2005 amounted to 38.5 x 10^6 MWh, or around 12.25 % of primary energy supply. **Figure 89** gives an overview of the sectors of Swiss primary generation consumption and the use of renewable energy for 2005, and **Figure 90** shows the distribution of energy consumption broken down by end-consumer categories (Source: Bundesamt für Energie, Bern).

The percentages of final energy consumption that consist of renewable energy sources **Figure 91**show clearly that in Switzerland hydroelectric power, as might be expected, plays a major role. If we compare the amount of total electricity generation with thermal energy and the nations's transportation-related energy requirements, we can see that the latter two are small – in fact, very small.

However, the percentages of total final energy consumption that consist of renewable energy are incomplete because shallow geothermal energy generation is not yet included in the data. Geothermal units, especially in their shallow form, which are very prevalent in Switzerland, are operated in conjunction with water-source heat pumps and are able to lower the cost of thermal energy generation for any given building significantly.

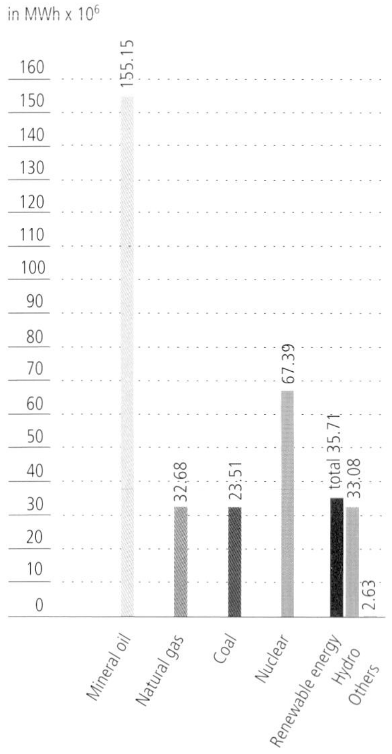

Energiebereitstellung – erneuerbare Energien –, der schweizerische Weg

Die Schweizerische Eidgenossenschaft, vertreten durch das Bundesamt für Energie (BFE), gibt jährlich einen Gesamtüberblick über die Gesamtenergieverbräuche und die Ressourcenbereitstellung. Der Endenergieverbrauch der Schweiz lag im Jahr 2005 bei insgesamt ca. 247,199 x 10^6 MWh, der Gesamtprimärenergiebedarf bei ca. 314,44 x 10^6 MWh.

Der Anteil der erneuerbaren Energien betrug in der Eidgenossenschaft ca. 38,5 x 10^6MWh (2005) und trug somit mit 12,25 Prozent zur Primärenergiebereitstellung bei. **Bild 89** gibt einen Überblick über die Struktur des Primärenergieverbrauchs der Schweiz und den Einsatz erneuerbarer Energien für den Zeitraum 2005. **Bild 90** zeigt die Energieverteilung nach Verbrauchern (Quelle: Bundesamt für Energie, Bern).

Der Anteil der erneuerbaren Energien am Endenergieverbrauch ist in **Bild 91** dargestellt. Hier wird sofort augenfällig, dass Wasserkraftwerke eine wesentliche Rolle im Bereich der erneuerbaren Energien spielen. Vergleicht man die Gesamtstromerzeugung zur Wärmebereitstellung und Energien für den Verkehr, so zeigt sich, dass die Anteile in den beiden letztgenannten Feldern eher gering bis sogar sehr gering sind.

Die Darstellung der Anteile der erneuerbaren Energien am Endenergieverbrauch sind jedoch insofern nicht vollständig, als insbesondere der Bereich der untiefen Geothermie in den einschlägigen Statistiken nicht ausgewiesen wurde. Die geothermischen Anlagen, insbesondere im Bereich der untiefen Geothermie, die in großer Anzahl in der Schweiz installiert sind, werden im Wesentlichen mit elektrisch betriebenen Wärmepumpen genutzt und senken die Wärmebereitstellung der entsprechenden Objekte.

in MWh x 10^6

Mineral oil 155.15
Natural gas 32.68
Coal 23.51
Nuclear 67.39
Renewable energy total 35.71
Hydro 33.08
Others 2.63

Figure 89
Primary energy consumption and renewable energy use (primary energy) in Switzerland (2005)

Source: Bundesamt für Energiewirtschaft, Bern, Switzerland

Bild 89
Struktur des Primärenergieverbrauchs und Einsatz der erneuerbaren Energien (Primärenergie) in der Schweiz (2005)

Quelle: Bundesamt für Energiewirtschaft, Bern, Schweiz

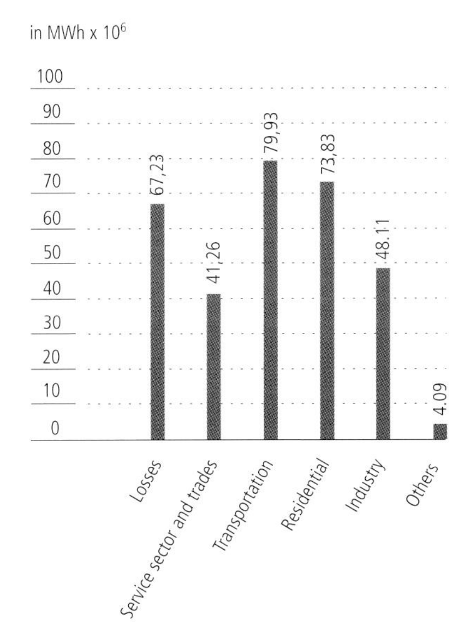

Figure 90
Energy distribution according to consumer type

Source: Bundesamt für Energiewirtschaft, Bern, Switzerland, 2006

Bild 90
Energieverteilung nach Verbrauchern

Quelle: Bundesamt für Energiewirtschaft, Bern, Schweiz, 2006

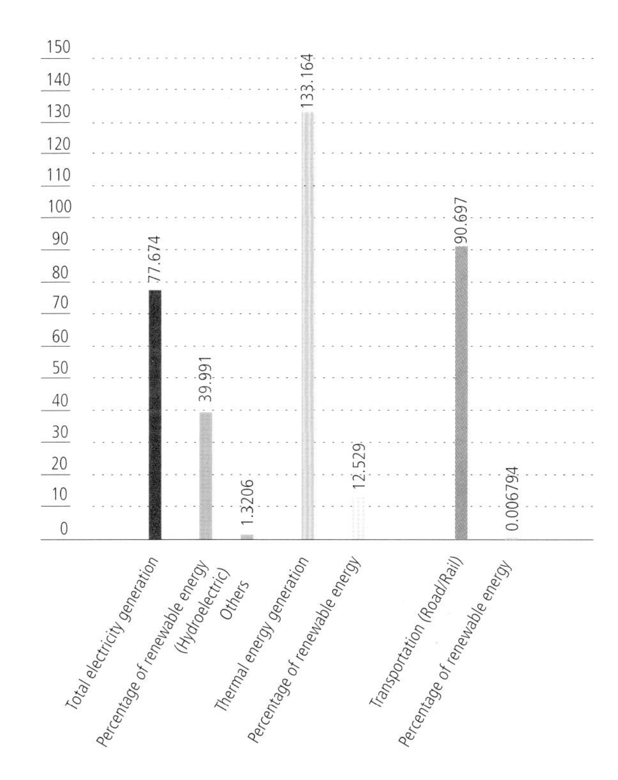

Figure 91
Percentage of renewable energy of total final energy consumption in Switzerland

Source: Bundesamt für Energiewirtschaft, Bern, Switzerland, 2006

Bild 91
Anteile der erneuerbaren Energien am Gesamt-Endenergieverbrauch in der Schweiz

Quelle: Bundesamt für Energiewirtschaft, Bern, Schweiz, 2006

Generation of electricity

Wind power

The low percentage of wind-generated power (generally, about 0.008 x 10^6 MWh) in Switzerland is due to insufficient prevailing wind speeds. Wind power clearly has no significant impact on overall energy generation.

Hydroelectric power

The majority of power generated from renewable sources in Switzerland is produced by hydroelectric power. As in Germany, the share of total power generation has remained nearly constant during the past few years.

Biomass

With a total of 0.078 x 10^6 MWh, biomass does not play a significant role in Switzerland's overall power supply. Yet, the demand for biomass is increasing, not only on the global market but also in Switzerland. Compact biomass, including biogenic waste, has the largest share (5.411 x 10^6 MWh). Liquid biomass, as well as biogenic gaseous fuels, is again of insignificant importance.

Stromerzeugung

Windenergie

Der Anteil der Windenergie ist in der Schweiz mangels ausreichender Windgeschwindigkeiten mit ca. 0,008 x 10^6 MWh eher sehr bescheiden und spielt somit keine wesentliche Rolle.

Wasserkraft

Die Stromerzeugung aus Wasserkraft ist wie in der Bundesrepublik Deutschland im Wesentlichen seit Jahren konstant und macht den überwiegenden Anteil der erneuerbaren Energien bei der Stromerzeugung aus.

Biomasse

Die Biomasse mit insgesamt ca. 0,078 x 10^6 MWh spielt bei der Stromerzeugung keine wesentliche Rolle. Die Nachfrage nach Biomasse ist, wie am Weltmarkt, auch in der Schweiz deutlich gestiegen.Feste Biomasse einschließlich biogener Abfall macht dabei den wesentlichen Anteil aus (5,411 x 10^6 MWh). Flüssige Biomasse sowie biogene gasförmige Brennstoffe spielen wiederum eine untergeordnete Rolle.

Photovoltaic

Photovoltaic power generation contributes 0.019 x 10^6 MWh to the total power mix. In other words, again as in Germany, it still plays a minor role, although a future increase can be expected.

Deep Geothermal power generation

The use of deep geothermal energy generation is also underdeveloped in Switzerland, as in Germany. Because statistical data by the BFE are not available, it can be assumed that this sector, especially when related to deep bore holes, plays an insignificant role.

Geothermal power generation

At the same time, despite their absence from the statistical data of the BFE, geothermal applications have a long tradition in Switzerland, and many buildings are equipped with such systems.

Thermal energy

As previously documented, thermal energy generation results in 5.522 x 10^6 MWh, and if we include solar thermal energy and shallow geothermal applications it supplies around 9.848 x 10^6 MWh of power and contributes around 4.3 % to the overall budget.

Solar thermal energy

Data in reference to solar thermal energy systems in Switzerland are, unfortunately, absent from the statistics of the BFE, although a great number of such systems exist throughout the country.

Bio fuels

Bio fuels, such as bio-diesel, bio-ethanol, and vegetable oils, are not documented in the government data currently available. It can be assumed that they play a role similar to that in the neighboring country of Germany.

Table 4 introduces the distribution of renewable energy sources that contribute to the overall power generation in this country.

Photovoltaik

Die Photovoltaik mit einem Anteil von 0,019 x 10^6 MWh/a spielt wie in Deutschland zurzeit noch eine untergeordnete Rolle und lässt sich in Zukunft deutlich anheben.

Tiefe Geothermie

Die Nutzung der tiefen Geothermie ist ähnlich unterentwickelt wie die in Deutschland. Da den einschlägigen Statistiken des Bundesamts für Energiewirtschaft keine Daten zu entnehmen sind, kann davon ausgegangen werden, dass dieser Bereich – insbesondere mit Tiefensonden – zurzeit noch keine Rolle spielt.

Geothermie

Der Einsatz oberflächennaher geothermischer Anlagen (Erdsonden/Grundwasser usw.) hat in der Schweiz eine lange Tradition und führt daher dazu, dass eine Vielzahl in Gebäuden installierter Anlagen zum Einsatz kommt, um im Wesentlichen auf Basis elektrisch betriebener Wärmepumpen Wärmeenergie bereitzustellen.

Wärmeenergie

Wie bereits festgestellt, beträgt der Anteil der Wärmeenergiebereitstellung durch erneuerbare Energien in der Schweiz lediglich 5,522 x 10^6 MWh bzw. unter Berücksichtigung der Solarthermie und oberflächennahen Geothermie ca. 9,848 x 10^6 MWh. Somit liegt der Anteil der erneuerbaren Energieträger bei ca. 4,3 Prozent.

Solarthermie

Leider sind genaue Angaben zu solarthermischen Anlagen den Statistiken des Bundesamtes für Energiewirtschaft nicht zu entnehmen, obwohl auch in der Schweiz solarthermische Anlagen in großem Umfang in Betrieb sind.

Biokraftstoffe

Zu Biokraftstoffen, d.h. dem Einsatz von Biodiesel, Bioethanol und Pflanzenöl, ist den einschlägigen Statistiken nichts zu entnehmen. Sie sind sicher in ähnlichem Umfang verbreitet wie in Deutschland.

Tabelle 4 zeigt die Aufschlüsselung der erneuerbaren Energien, die an der Strom- und Wärmeerzeugung beteiligt sind.

Electricity	(x10[6] MWh)
Wind power	0.008
Hydroelectric power	32.730
Total biomass	0.078
- Solid biomass, including biogenic waste	(0.055)
- Biogas	(0.023)
- Liquid biomass	(—)
Landfill and sewage treatment gas	0.153
Photovoltaic	0.019
Geothermal	(—)
∑ Electricity	**32.988**

Thermal energy	(x10[6] MWh)
Total biomass	5.522
- solid biomass, including biogenic waste	(5.411)
- liquid biomass	(0.104)
- biogenic gaseous fuels	(0.007)
Solar thermal energy	4.326
Deep geothermal energy	(—)
Near-surface geothermal energy	no ind.
∑ Thermal energy	**9.848**

∑ Final energy of renewable sources	**42.836**

Strom	(x10[6] MWh)
Windenergie	0,008
Wasserkraft	32,730
Biomasse, insg.	0,078
davon feste Biomasse, einschl. biogener Abfall	(0,055)
davon Biogas	(0,023)
davon flüssige Biomasse	(—)
Deponie- und Klärgas	0,153
Photovoltaik	0,019
Geothermie	(—)
∑ Strom	**32,988**

Wärme	(x10[6] MWh)
Biomasse, insg.	5,522
davon feste Biomasse, einschl. biogener Abfall	(5,411)
davon flüssige Biomasse	(0,104)
davon biogene gasförmige Brennstoffe	(0,007)
Solarthermie	4,326
tiefe Geothermie	(—)
oberflächennahe Geothermie	keine Ang.
∑ Wärme	**9,848**

∑ Endenergie aus erneuerbaren Energien	**42,836**

	MWh x 10[6]	**Percentage in %**
Residential	17.7	27.3
Agriculture	1.1	1.6
Industry/manufacturing	19.0	29.3
Service industry	15.3	23.5
Transportation	4.8	7.4
Transmission losses	4.3	6.7
Pumping storage	2.7	4.2
Total consumption	64.8	100.0

Percentage of renewable energy contribution according to consumer categories

	MWh x 10[6]	**Anteil in %**
Haushalt	17,7	27,3
Landwirtschaft	1,1	1,6
Industrie	19,0	29,3
Dienstleistung	15,3	23,5
Verkehr	4,8	7,4
Transportverlust	4,3	6,7
Pumpspeicherung	2,7	4,2
Totalverbrauch	64,8	100,0

Anteile erneuerbarer Energien nach Verbrauchern

Table 4
Contribution of renewable energy sources to energy supply in Switzerland, 2006

Source: Bundesamt für Energiewirtschaft, Bern, Switzerland

Tabelle 4
Beitrag der erneuerbaren Energien zur Energiebereitstellung in der Schweiz, 2006

Quelle: Bundesamt für Energiewirtschaft, Bern, Schweiz

7 Detailed view: France

7 Detailbetrachtung Frankreich

Energy supply: renewable energy, the French way

Each year the two French government departments of ecology and sustainable development (Ministère de l'Écologie, du Développement et de l'Aménagement durables) and the department of economic development (Ministère de l'Économie, des Finances et de l'Emploi) publish a comprehensive overview of energy sources and energy supply and demand for France. According to the 2007 report, the country's primary energy consumption was 2,412 x 10^6 MWh, of which 213 x 10^6 MWh were contributed by renewable sources. This amounts to 8.33 % of the total primary energy demand (**Figure 92**).

It has to be noted here that government records do not explicitly list France's thermal energy demand, so these figures were drawn as an estimated conclusion from the provided data. We therefore need to point out that the values for thermal energy display a certain imprecision compared with the other data.

Figure 92 lists an overview of France's primary energy supply structure, including the percentage of renewable energy supply. As seen in the diagram, and as commonly understood, France has for years relied heavily on nuclear energy, but it will have to leave this path entirely over the next 50 years. Consequently, the country will be faced with a rather substantial overhaul of large sectors of its energy supply.

Figure 93 lists the distribution of various energy sources broken down by end-consumer types. Similarly to the data from Germany – and in contrast to Switzerland – transmission losses amount to around 34 % of the total energy generated. Not yet included is the true cost of the nuclear energy generation sector, which will have a substantial impact in the future.

The percentage of renewable energy is shown in **Figure 94**. As a result of France's great hydroelectric power potential, the percentage of renewable energy in the overall power supply mix is significant (**Figure 95**.)

Energiebereitstellung – erneuerbare Energien –, der französische Weg

Das Ministerium für Ökologie und nachhaltige Entwicklung (Ministère de l'Écologie, du Développement et de l'Aménagement Durables) und das Ministerium für Ökonomie und Finanzen sowie Beschäftigung (Ministère de l'Économie, des Finances et de l'Emploi) geben jährlich einen Gesamtüberblick über die Gesamtenergieverbräuche und die Ressourcenbereitstellung aus. Gemäß der Ausgabe 2007 lag der Primärenergieverbrauch Frankreichs bei ca. 2.412 x 10^6 MWh bei einem Anteil erneuerbarer Energien von ca. 213 x 10^6 MWh. Dies entspricht einem Anteil von 8,33 Prozent des Primärenergiebedarfs, **Bild 92**.

Bezüglich der Erarbeitung vergleichbarer Diagramme und Zahlen muss darauf hingewiesen werden, dass der Wärmeenergiebedarf Frankreichs nicht ausgewiesen wurde, sondern durch Rückrechnung nur annähernd ermittelt werden konnte. Insofern haben die Zahlenangaben eine relative Unschärfe, was den Wärmesektor betrifft.

Bild 92 gibt einen Überblick über die Struktur des Primärenergieverbrauchs Frankreichs bei Angabe des Prozentsatzes erneuerbarer Energien. Wie die Diagrammstruktur ausweist, hat Frankreich bekanntermaßen seit Jahren auf den Einsatz der Nuklearenergie gesetzt und wird sich wohl in den nächsten 50 Jahren relativ total hiervon lösen müssen. Insofern steht Frankreich ein erheblicher Umbau der Energiewirtschaft bevor.

Bild 93 weist in etwa die Verteilung der Energien nach verschiedenen Verbraucherstrukturen aus. Wie bei der Bundesrepublik Deutschland sind im Gegensatz zur Schweiz auch in Frankreich die Energieverluste eine nennenswerte Größe und machen ca. 34 Prozent der primär eingesetzten Energie aus. Hierbei in keiner Form berücksichtigt sind die Folgen der Kernenergiewirtschaft, die u.U. noch erheblich zu Buche schlagen werden.

Der Anteil der erneuerbaren Energien am Endenergieverbrauch ist in **Bild 94** ausgewiesen. Der relativ hohe Anteil erneuerbarer Energien in Bezug auf die Gesamtstromerzeugung ergibt sich für Frankreich aus einem hohen Potenzial an Wasserkraft, **Bild 95**.

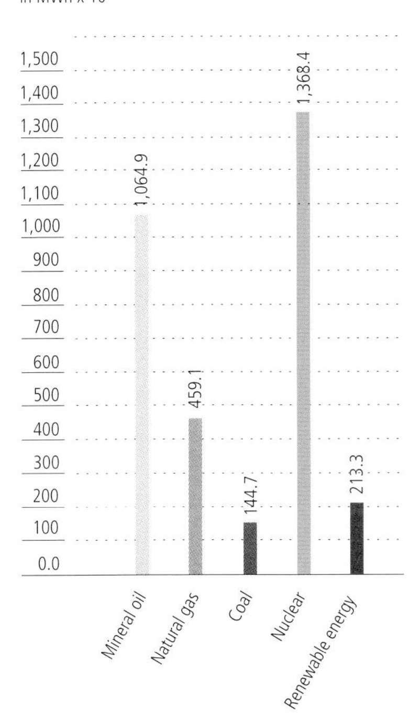

Figure 92
Primary energy consumption and renewable energy use in France

Source: Ministère de l'Ecologie, Développement et de l'Aménagement durables, 2007

Bild 92
Struktur des Primärenergieverbrauchs und Einsatz der erneuerbaren Energien in Frankreich

Quelle: Ministère de l'Ecologie, Développement et de l'Aménagement durables, 2007

Figure 93
Energy distribution according to consumer type

Source: Ministère de l'Ecologie, de l'Energie, du Développement durable et de l'Aménagement du territoire, 2007

Bild 93
Energieverteilung nach Verbrauchern

Quelle: Ministère de l'Ecologie, de l'Energie, du Développement durable et de l'Aménagement du territoire, 2007

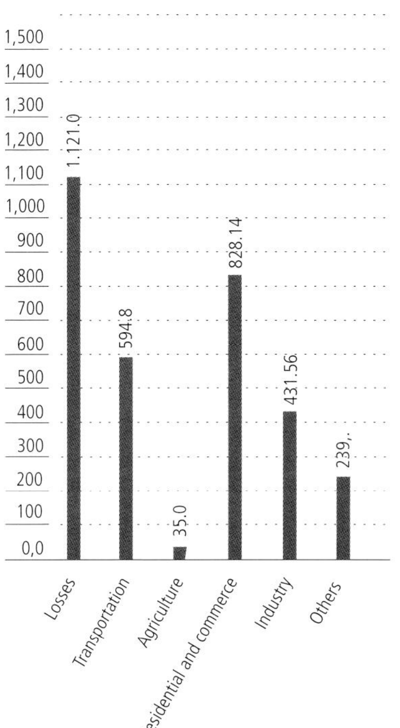

Figure 94
Percentage of renewable energy of total final energy consumption in France

Source: Ministère de l'Ecologie, Développement et de l'Aménagement durables, 2007

Bild 94
Anteile der erneuerbaren Energien am Gesamt-Endenergieverbrauch in Frankreich

Quelle: Ministère de l'Ecologie, Développement et de l'Aménagement durables, 2007

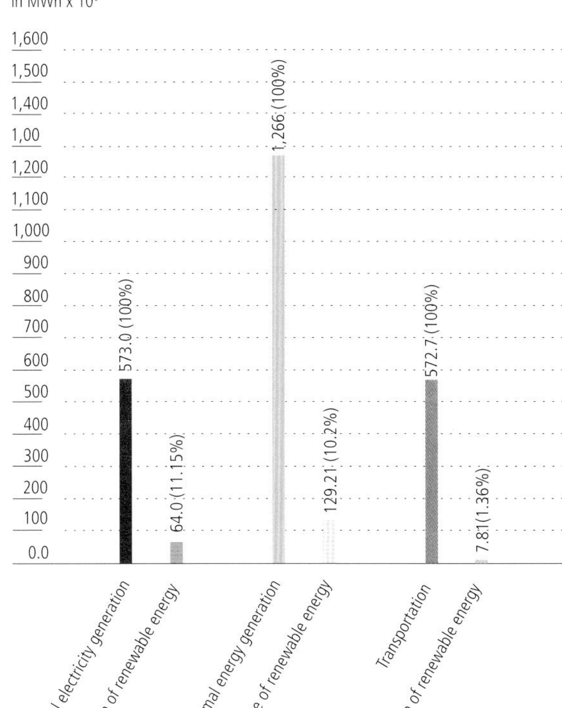

When it comes to thermal energy supply, **Figure 96**, we are able to identify a percentage of renewable energy of about 10 %, which is significantly greater than that of either Germany or Switzerland. This same condition applies equally to France's transportation sector.

Generation of Electric Energy

Wind power

Electric power generated by the use of wind turbines represents only approximately 3.26 % of the total energy generated and is therefore much less than that of Germany. Because France's coastline (with strong coastal winds) is much longer than that of its neighbor to the east, France clearly can take advantage of a still-untapped opportunity for future renewable energy generation from wind.

Hydroelectric power

By a wide margin, hydroelectric power is the largest contributing factor in the renewable power mix of France. However, because the likelihood of a significant increase in this resource sector is minimal, other renewable energy sources need to be investigated.

Biomass

Data for the role of biomass in France's energy sector were not obtainable, and thus assumptions needed to be made.

When the numbers were calculated, we established the hypothesis that mainly methane gas from landfill sites was used for electric energy generation and that all other biogenic sources were used to generate thermal energy.

Photovoltaic

With 0.36 x 10^6 MWh/a, power generation with the help of photovoltaic systems currently fails contribute a large percentage of France's renewable energy supply. This condition clearly is ready for a significant improvement, especially in light of the necessary future substitution of alternative fuels for nuclear-based electric energy generation in France's power supply.

Geothermal energy

Although it contributes a larger share than in the countries already reviewed (see **Figure 96**), power generation with geothermal systems is still in the early stages of development in France, as it is in other European countries. It amounts to only 1.51 x 10^6 MWh/a of power as part of the overall supply mix.

Im Bereich der Wärmebereitstellung, **Bild 96**, konnte annähernd eine anteilige erneuerbare Energieversorgung von ca. 10 Prozent ermittelt werden, die deutlich über dem Anteil in Deutschland und in der Schweiz liegt.Der Anteil erneuerbarer Energien im Bereich Verkehr liegt in Frankreich ebenfalls deutlich über dem Deutschlands und der Schweiz.

Stromerzeugung

Windenergie

Die Windenergie macht einen Anteil an der Stromerzeugung von ca. 3,26 Prozent aus und liegt damit deutlich unter dem Anteil der Bundesrepublik Deutschland. Da Frankreich eine deutlich längere Küstenlinie, bezogen auf die Größe des Landes, besitzt als Deutschland, besteht hier ein erhebliches Potenzial, um zukünftig einen höheren Stromanteil durch Windenergie darzustellen.

Wasserkraft

Der Anteil der Wasserkraft ist mit weitem Abstand der größte bei der Bereitstellung von elektrischer Energie durch erneuerbare Energien. Dieser Anteil dürfte sich jedoch kaum steigern lassen, so dass alle anderen Ressourcen vermehrt und verstärkt zum Einsatz kommen müssen.

Biomasse

Der Anteil der Biomasse an der Stromversorgung konnte nicht eindeutig festgestellt werden und musste demnach überschlägig ermittelt werden.

Bei der Erarbeitung der Zahlen wurde davon ausgegangen, dass im Wesentlichen Deponiegas zur Erzeugung elektrischer Energie zum Einsatz kommt, während alle anderen biogenen Anteile im Wesentlichen der Wärmeversorgung dienen.

Photovoltaik

Der Anteil der Photovoltaik mit ca. 0,36 x 10^6 MWh/a spielt offensichtlich zurzeit in Frankreich noch keine wesentliche Rolle und muss sicher massiv verstärkt werden, um einen hohen Anteil der nuklear bereitgestellten Elektroenergie zu übernehmen.

Geothermie

Die Geothermie mit ca. 1,51 x 10^6 MWh/a spielt zwar eine größere Rolle als in den zuvor dargestellten Ländern, ist jedoch vergleichsweise ähnlich unterentwickelt wie in den meisten anderen Ländern Europas, vergleiche **Bild 96**.

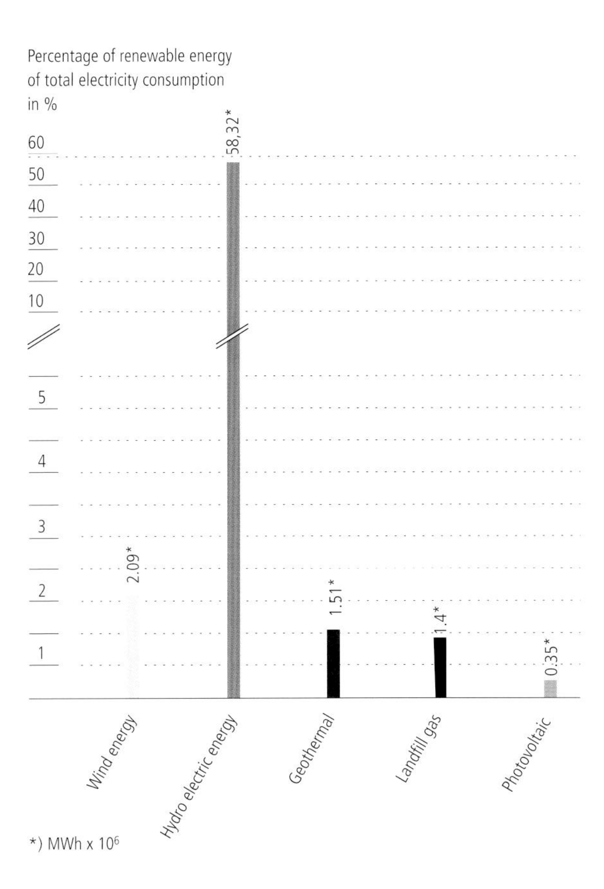

Figure 95
Distribution of electric energy production from renewable resources = 11.15 % of total electric energy demand (64.0 MWh x 10^6)

Source: Ministère de l'Ecologie, Développement et de l'Aménagement durables, 2007

Bild 95
Gesamtsumme Stromerzeugung aus erneuerbaren Energien, vom Gesamt-Bruttostromverbrauch = 11,15 % (64,0 MWh x 10^6)

Quelle: Ministère de l'Ecologie, Développement et de l'Aménagement durables, 2007

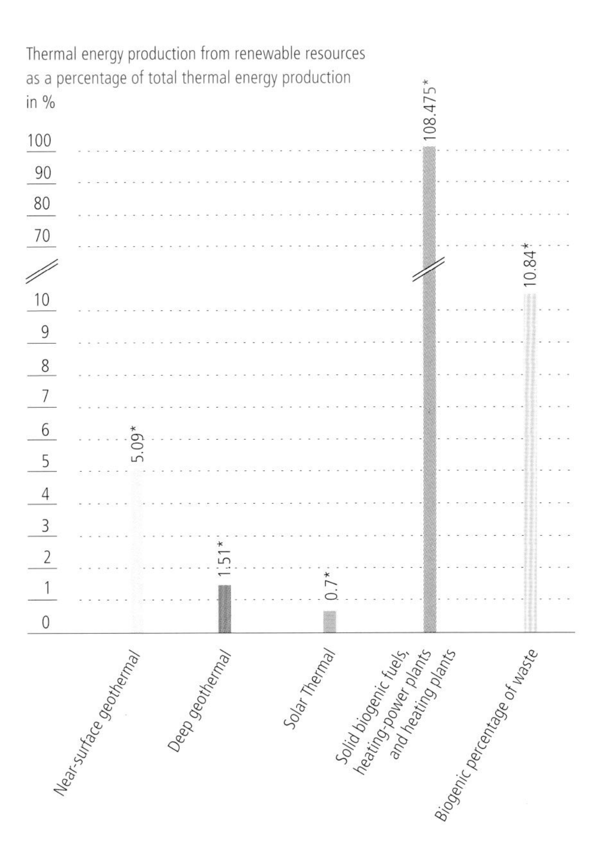

Figure 96
Distribution of thermal energy production from renewable resources = 10.21 %
of total thermal energy demand (89,4 MWh x 10^6)

Source: Ministère de l'Ecologie, Développement et de l'Aménagement durables, 2007

Bild 96
Gesamtsumme Wärmebereitstellung aus erneuerbaren Energien, von der Gesamt-Wärmebereitstellung = 10,21 %
(129,21 MWh x 10^6)

Quelle: Ministère de l'Ecologie, Développement et de l'Aménagement durables, 2007

Thermal energy

The supply of thermal energy by means of renewable energy sources comes to around 129 x 10^6 MWh/a, or 10.2 %. As already noted, government data in reference to thermal energy supply could not be obtained, so the following estimates are therefore approximations (**Figure 96**).

Biomass

We assumed that the percentage of solid biogenic fuels for use in thermal power plants and district heating units amounts to around 66.2 % of all renewable energy, or approximately 85,4 x 10^6 MWh. Because France has the largest forested regions in Europe, it can be concluded with certainty that the percentage of solid biomass can be increased if a political will for more renewable energy use takes hold.

The use of biogenic waste products amounts to 6.6 % of all renewables used for thermal energy supply, or 8.53 x 10^6 MWh. In this sector, as well, significant potential for increase exists.

Geothermal energy

Shallow geothermal applications – that is, the water-source heat pump operated using earth tubes and groundwater – contribute about 5.1 x 10^6 MWh energy, which is 3.94 % of all renewable energy supply and twice the amount of that type of surface-related energy than is produced in Germany.

The use of deep geothermal systems, on the other hand, represents only around 1.18 % of renewable energy, or 1.522 x 10^6 MWh, which is similarly insignificant as in Germany and Switzerland.

To compensate for the necessary shift from fossil fuels to renewable energy in the future, we must assume that the percentage of deep geothermal power plants will increase very substantially.

Solar thermal power

Solar thermal power systems currently contribute at a rate of 0.7 x 10^6 MWh to the total thermal power supply, or only 0.55 %. Since solar gains in the southern regions of France are much greater than in northern Europe, great potential for an increase in use of this renewable source does indeed exist.

Wärmeenergie

Die Bereitstellung von Wärmeenergie durch erneuerbare Energien nimmt in Frankreich eine Größenordnung von ca. 129 x 10^6 MWh/a ein. Somit beträgt der Anteil der Wärmeenergiebereitstellung durch erneuerbare Energien ca. 10,2 Prozent. Wie bereits festgestellt, konnten genaue Zahlen des Gesamtwärmeenergiebedarfs Frankreichs aus den einschlägigen staatlichen Darstellungen nicht entnommen werden. Insofern sind alle nachfolgenden Zahlen nur angenäherte Zahlen, vergleiche **Bild 96**.

Biomasse

Der Anteil biogener Festbrennstoffe für Heizkraft- und Heizwerke dürfte bei ca. 66,2 Prozent der erneuerbaren Energien für Wärmeenergie liegen (ca. 85,4 x 10^6 MWh). Da Frankreich das waldreichste Land Europas ist, kann davon ausgegangen werden, dass sich dieser Anteil deutlich erhöhen lässt, wenn konsequent der Einsatz von Biomasse betrieben wird.

Der Einsatz biogener Abfälle mit ca. 8,53 x 10^6 MWh nimmt einen Anteil von ca. 6,6 Prozent der erneuer baren Energien bei der Wärmebereitstellung ein. Auch hier können mit hoher Wahrscheinlichkeit die entsprechenden Anteile deutlich vergrößert werden.

Geothermie

Die oberflächennahe Geothermie (Wärmepumpenbetriebe mittels Sonden und Grundwasser) trägt einen Anteil von 5,1 x 10^6 MWh zur Wärmebereitstellung bei. Dies entspricht einem Anteil von ca. 3,94 Prozent der erneuerbaren Energien und liegt somit bei mehr als dem Doppelten der oberflächennahen Geothermieanlagen Deutschlands.

Der Einsatz tiefer Geothermie mit ca. 1,18 Prozent der erneuerbaren Energien (ca. 1,522 x 10^6 MWh) spielt eine ähnlich untergeordnete Rolle wie in Deutschland und der Schweiz.

Es ist sicher davon auszugehen, dass der Anteil der tiefen Geothermie in den nächsten Jahren massiv verstärkt wird, um den Einsatz fossiler Brennstoffe auszugleichen.

Solarthermie

Solarthermische Anlagen tragen zurzeit ca. 0,7 x 10^6 MWh zur Gesamtwärmebereitstellung bei. Dies entspricht einem Anteil von etwa 0,55 Prozent. Vergegenwärtigt man sich insbesondere, dass Südfrankreich einen deutlich höheren Solareintrag als Nordeuropa bietet, sind hier wiederum erhebliche Entwicklungsmöglichkeiten angezeigt.

Bio fuels

Liquid biogenic fuels were consumed in the year 2006 to generate thermal energy at a rate of 7.81 x 10^6 MWh, or 1.36 % of all fuels used in the transportation sector.

If we compare the use of biofuels in France with that of Germany (27 x 10^6 MWh versus 7.81 x 10^6 MWh), it becomes apparent that significant increases can and should be pursued.

For a better understanding, **Table 5** lists the contributions of renewable energy to France's entire energy supply for the year 2006, with the above-mentioned uncertainty about some data, because no comparative values for other countries were available at the time.

Electric energy	(x10^6 MWh)
Wind power	2.1
Hydroelectric power	57.0
Total biomass	4.08
-Solid biomass including biogenic waste	(1.4)
-Biogas	(2.68)
Photovoltaic power	0.35
Σ Electric energy	**63.53**
Thermal energy	(x10^6 MWh)
Total biomass	121.99
- Solid biomass	(108.47)
including biogenic waste	(10.84)
Solar thermal	0.7
Deep geothermal	1.51
Near-surface thermal	5.01
Σ Thermal energy	**129.21**
	(x10^6 MWh)
Σ Biofuels/biodiesel	**7.81**
Σ Final energy from renewable energy resources	**200.55**

Table 5
Contribution of renewable energy sources to energy supply in France, 2006

Source: Ministère de l'Écologie, du Dévelopement et de l'Aménagement durables, Ministère de l'Économie, des Finances et de l'Emploi.

Biokraftstoffe

Biogene, flüssige Kraftstoffe wurden im Jahr 2006 mit ca. 7,81 x 10^6 MWh eingesetzt. Dies entspricht in etwa einem Anteil von 1,36 Prozent der Kraftstoffe, die für den Straßenverkehr benötigt werden.

Vergleicht man den Einsatz der biogenen Kraftstoffe in Frankreich zu Deutschland (ca. 27 x 10^6 MWh zu 7,81 x 10^6 MWh), so wird deutlich, dass in Frankreich auf diesem Feld erheblich nachgerüstet werden kann und sollte.

Tabelle 5 listet zum besseren Nachvollziehen die Anteile der Beiträge der erneuerbaren Energien zur Energiebereitstellung in Frankreich im Jahr 2006 auf, wobei nochmals darauf hingewiesen werden muss, dass alle Zahlen eine gewisse Unschärfe besitzen, da kein absolut vergleichbares Zahlenmaterial zu den anderen Ländern vorlag.

Strom	(x10^6 MWh)
Windenergie	2,1
Wasserkraft	57,0
Biomasse, insg.	4,08
davon feste Biomasse, einschl. biogener Abfall	(1,4)
davon Biogas	(2,68)
Photovoltaik	0,35
Σ Strom	**63,53**
Wärme	(x10^6 MWh)
Biomasse, insg.	121,99
davon feste Biomasse,	(108,47)
einschl. biogener Abfall	(10,84)
Solarthermie	0,7
tiefe Geothermie	1,51
oberflächennahe Geothermie	5,01
Σ Wärme	**129,21**
	(x10^6 MWh)
Σ biologische Kraftstoffe	**7,81**
Σ Endenergie aus erneuerbaren Energien	**200,55**

Tabelle 5
Beitrag der erneuerbaren Energien zur Energiebereitstellung in Frankreich, 2006

Quelle: Ministère de l'Écologie, du Dévelopement et de l'Aménagement durables, Ministère de l'Économie, des Finances et de l'Emploi.

8 Global energy scenarios

8 Energieszenarien, weltweit

The course of the development of global energy supply and demand is characterized by three key factors:

- Population increase – that is, the number of people who will use energy in the future.
- Economic development as measured by gross domestic product (GDP), which typically directly influences energy consumption or the mentality of energy entitlement.
- Energy intensity, which is the measure of the energy efficiency of a nation's economy. It is calculated as units of energy per unit of GDP.

The following outlook is based on the assumption that world population will increase up to the year 2050 from today's 6.3 billion to 8.9 billion. **Figure 97** shows the increase of population for specific regions on Earth. According to the predictions of the International Energy Agency (IEA), which uses United Nations data, the rates of increase vary significantly across the various listed regions of the world. Great increases in population are being predicted for Africa, Latin America, South Asia, East Asia, and the Middle East. Countries in the geographical regions of Europe, North America, and the Pacific region will not see any significant changes, which will somewhat ease the overall conditions for future energy demand.

Die Entwicklung des globalen Energieszenarios ist geprägt durch drei wesentliche Schlüsselfaktoren

- Entwicklung der Bevölkerung, d.h. Anzahl Personen, die Energien verbrauchen
- Ökonomische Entwicklung, ausgedrückt durch das Bruttosozialprodukt (gross domestic product). Das Bruttosozialprodukt beeinflusst im Regelfall auch den Energieverbrauch bzw. den Anspruch an Energienutzung
- Energieintensität, d.h. wie viel Energie ist notwendig, um eine Einheit des Bruttosozialprodukts zu erzeugen.

Die nachfolgenden Betrachtungen zum weltweiten Energieszenario bauen darauf auf, dass sich die Weltbevölkerung bis 2050 von heute ca. 6,3 Milliarden auf 8,9 Milliarden vergrößert. **Bild 97** zeigt den Zuwachs von Menschen in einzelnen Regionen der Welt, wobei nach dem IEA-Referenzszenario, welches auf der Projektion der Weltbevölkerung gemäß United Nations basiert, die Zuwachsraten in den einzelnen Regionen der Welt sehr unterschiedlich sind. Besondere Zuwächse werden prognostiziert für Afrika, Südamerika, Südasien, Ostasien und den Mittleren Osten. In den Ländern Europas, Amerikas und im pazifischen Raum werden praktisch keine großen Zuwächse erwartet, was wiederum die Situationen in den entsprechenden Regionen hinsichtlich des zukünftigen Energiebedarfs entschärft.

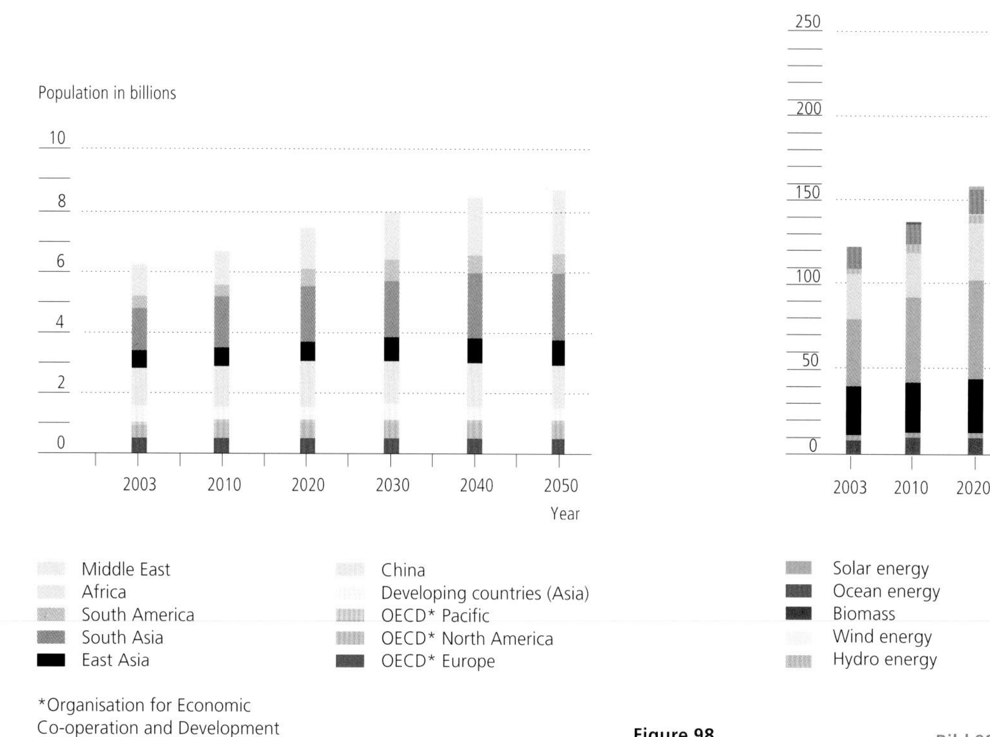

Figure 97
Global population increase estimate

Source: report global energy scenario, Greenpeace, 2007

Bild 97
Geschätzte Zunahme der Weltbevölkerung

Quelle: report global energy scenario, Greenpeace, 2007

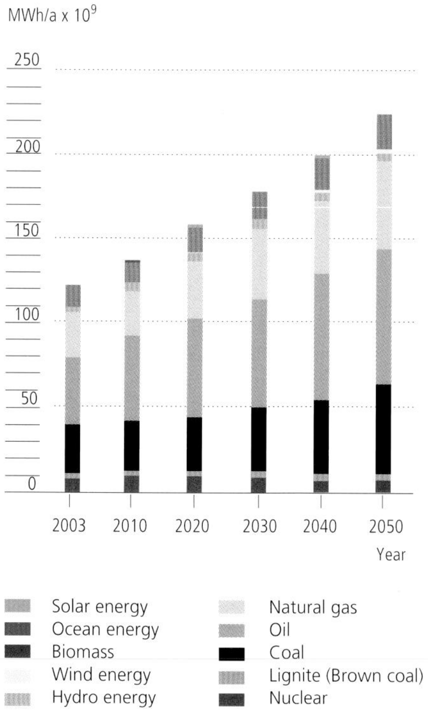

Figure 98
Development of primary energy consumption according to the reference scenario

Source: report global energy scenario, Greenpeace, 2007

Bild 98
Entwicklung des Primärenergieverbrauchs unter dem Referenz-Szenario

Quelle: report global energy scenario, Greenpeace, 2007

8.1 Primary energy and CO_2 emissions

If, on the one hand, we consider a scenario of generally increasing world population, and on the other hand a shift from currently developing to industrial nations, the graph in **Figure 99** is helpful. It shows the progression of primary energy consumption as a reference case. The increasing energy demand between 2003 and 2050, as shown in **Figure 98**, could then be satisfied only if renewable energy sources were used in a very significant manner. However, this would mean that drastic investments in such new technologies would have to be made immediately, a condition that does not exist due to the limited financial resources of a great number of countries. Greenpeace introduced the term of an Energy-(R)evolution, as we have already noted in previous chapters. Energy efficiencies will, according to all predictions, continue to increase, and according to **Figure 99** primary energy consumption should become stabilized by those improvements, although the dual use of both fossil fuels and renewable energy will still be prevalent.

The increased use of renewable sources of energy is supposed to decrease carbon emissions significantly, as shown in **Figure 100**, and it also presents the hope that this may have a positive effect on climate change.

The so-called reference case as well as the Energy-(R)-Evolution allows for a more detailed analysis.

8.1 Primärenergie und CO_2-Emissionen

Unter Berücksichtigung einer steigenden Weltbevölkerung sowie weiterhin unter Berücksichtigung der Entwicklung von Schwellenländern zu Industrienationen wurde die Entwicklung des Primärenergieverbrauchs als Referenzszenario, wie in **Bild 99** dargestellt, ermittelt. Der steigende Energiebedarf gemäß **Bild 98** von 2003 bis 2050 könnte nur dann gedeckt werden, wenn in erheblichem Umfang gegenüber dem derzeitigen Zustand erneuerbare Energien zum Einsatz kommen. Dies bringt allerdings mit sich, dass erhebliche Investitionsmittel eingesetzt werden können, die in vielen Ländern nicht vorhanden sind. Insofern wurde durch Greenpeace ein Energie-(R)Evolution-Szenario in ähnlicher Form geschrieben, wie wir es bereits in den vergangenen Kapiteln kennen gelernt haben. Durch insbesondere die Steigerung der Effizienz (Wirkungsgrad) beim Einsatz der verschiedensten Primärenergieträger soll sich gemäß **Bild 99** der Primärenergieverbrauch annähernd stabilisieren. Dabei werden sowohl fossile Energieträger wie auch erneuerbare Energien zum Einsatz kommen.

Durch den vermehrten Einsatz erneuerbarer Energien soll gleichzeitig der CO_2-Ausstoß gemäß **Bild 100** deutlich reduziert werden – in der Hoffnung, hierauf einen Einfluss auf den Klimawandel nehmen zu können.

Eine detailliertere Betrachtung zu den einzelnen Energieformen zeigt sowohl das Referenzszenario als auch das Energie-(R)Evolution-Szenario.

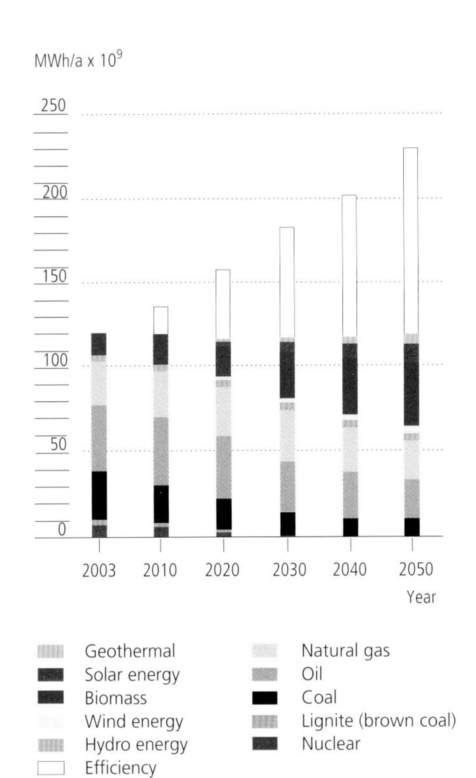

Figure 99
Development of primary energy consumption according to the Energy-(R)evolution scenario. (Efficiency: Reduction compared with the reference scenario)

Source: report global energy scenario, Greenpeace, 2007

Bild 99
Entwicklung des Primärenergieverbrauchs unter dem Energie-[R]Evolution-Szenario (Effizienz: Reduzierung verglichen mit demReferenzszenario)

Quelle: report global energy scenario, Greenpeace, 2007

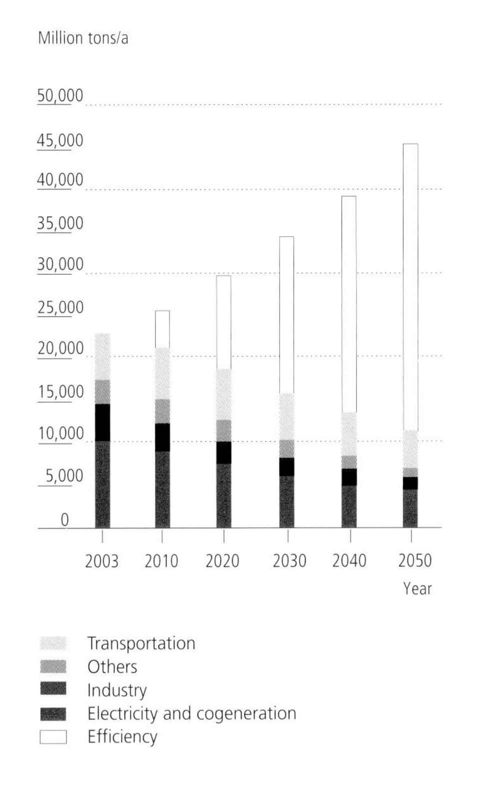

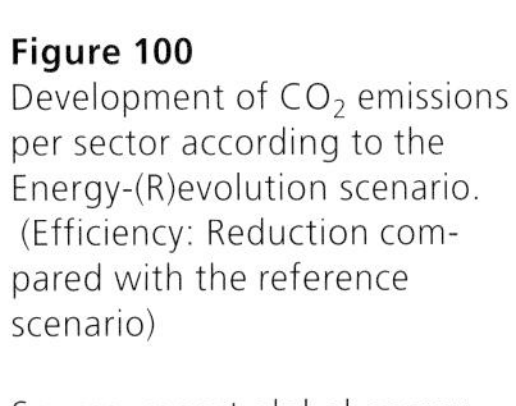

Figure 100
Development of CO_2 emissions per sector according to the Energy-(R)evolution scenario. (Efficiency: Reduction compared with the reference scenario)

Source: report global energy scenario, Greenpeace, 2007

Bild 100
Entwicklung der CO_2-Emissionen nach Sektoren unter dem Energie-[R]Evolution-Szenario (Effizienz: Reduzierung verglichen mit dem Referenzszenario)

Quelle: report global energy scenario, Greenpeace, 2007

8.2 Generation of electricity

Figure 101 shows the development of global electricity generation as documented in the reference case. A mix of fossil energy sources and renewables could, theoretically, satisfy the world's hunger for energy. Of great interest in this respect is the predicted ever-increasing use of coal, which is inextricably tied to the related issues of increasing global air pollution. Because these causal conditions should be reason to consider alternatives, **Figure 102** shows how intense efforts to increase efficiencies and reduce losses through transmission processes can be used as a counter-measure.

8.2 Stromerzeugung

Im **Bild 101** ist die Entwicklung der globalen Stromversorgung unter dem Referenzszenario dargestellt. Eine Mischung sowohl fossiler Energieträger wie auch erneuerbarer Energien könnte theoretisch den weltweiten Energiehunger stillen. Hierbei besonders auffällig ist der immer massiver werdende Einsatz von Kohle mit der daraus resultierenden zunehmenden Luftverschmutzung weltweit. Aus diesem Grund soll ein Szenario zur Stromversorgung gemäß **Bild 102** greifen, d.h. massive Bemühungen zur Steigerung der Wirkungsgrade und Verringerung der Transportverluste sowie ein verstärkter Einsatz von erneuerbaren Energien.

Figure 101
Development of global electricity supply according to the Reference Scenario

Source: report global energy scenario, Greenpeace 2007

Bild 101
Entwicklung der globalen Stromversorgung unter dem Referenz-Szenario

Quelle: report global energy scenario, Greenpeace, 2007

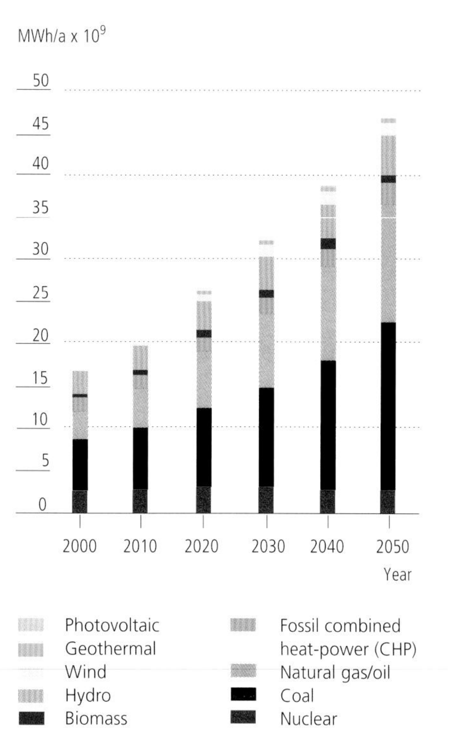

Figure 102
Development of global electricity supply according to the Energy-(R)evolution Scenario (Efficiency: Reduction compared to Reference Scenario)

Source: report global energy scenario, Greenpeace 2007

Bild 102
Entwicklung der Stromversorgung unter dem Energie-(R)Evolution-Szenario (Effizienz: Reduzierung verglichen mit dem Referenzszenario)

Quelle: report global energy scenario, Greenpeace, 2007

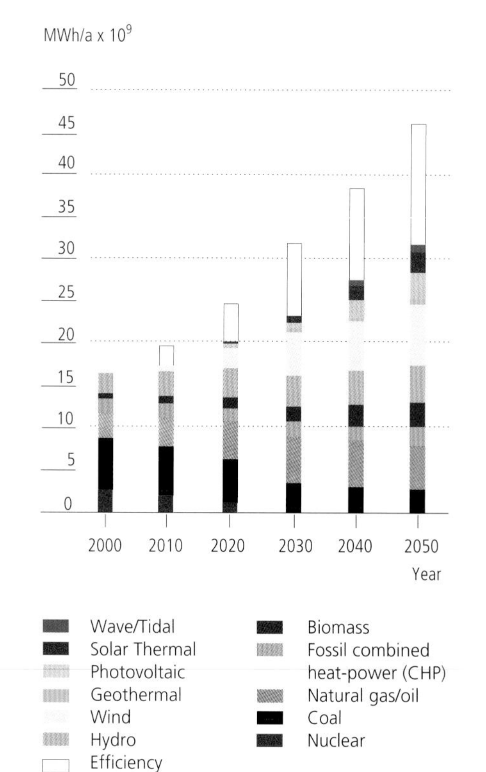

8.3 Generation of thermal energy

The development of future thermal energy supply is shown under the reference case scenario in **Figure 103**. Here, we can see the rather improbable case that a rapid acceleration of fossil fuel use will not be likely because resources will diminish. Only because of such an assumption are we more inclined to use the Energy-(R)evolution scenario, which, as shown, predicts a future duality of fossil and renewable energy sources with a gradual decrease of the first (**Figure 104**). As in the case of primary electrical energy supply, the efficiency of thermal energy supply needs to be pursued more aggressively, and this includes reduction of transmission losses and increased performance of all machinery and equipment involved.

8.3 Wärmeenergie-Erzeugung

Die Entwicklung der Wärmeversorgung unter dem Referenzszenario ist in **Bild 103** dargestellt. Diese Darstellung zeigt eine eher unwahrscheinliche Entwicklung, da eine rasante Zunahme des Einsatzes fossiler Rohstoffe eher nicht eintreten kann infolge der zunehmenden Verknappung der entsprechenden Energieträger. Schon aus diesem Grund muss das in **Bild 104** gezeigte Energie-(R)Evolution-Szenario greifen, d.h. die Verringerung fossiler Brennstoffe bei gleichzeitiger, deutlicher Erhöhung des Einsatzes erneuerbarer Energien. Wie bei der Primärenergiebereitstellung für elektrische Energie muss auch bei der Wärmeversorgung mit einer erheblichen Verbesserung der Effizienz gerechnet werden, d.h. massive Reduzierung der Verluste und Verbesserung von Wirkungsgraden maschineller Einrichtungen.

Figure 103
Development of heat supply according to the Reference Scenario

Source: report global energy scenario, Greenpeace 2007

Bild 103
Entwicklung der Wärmeversorgung unter dem Referenz-Szenario

Quelle: report global energy scenario, Greenpeace, 2007

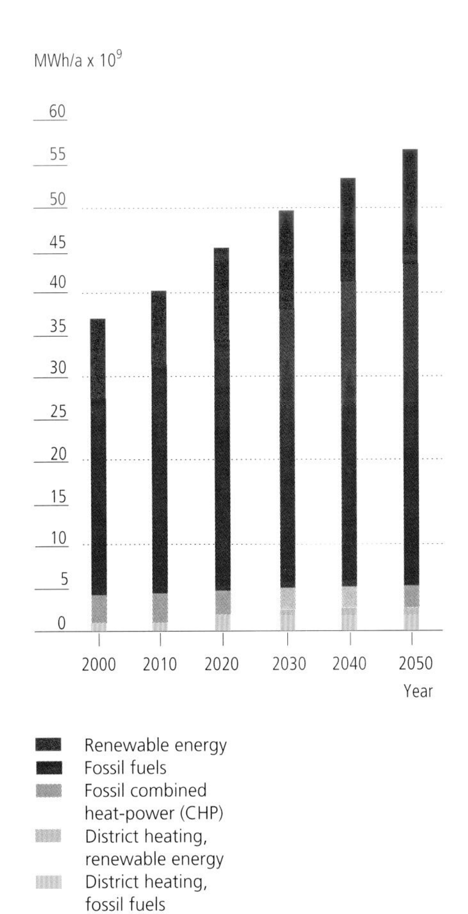

Figure 104
Development of heat supply according to the Energy-(R)evolution Scenario
(Efficiency: Reduction compared to Reference Scenario)

Source: report global energy scenario, Greenpeace 2007

Bild 104
Entwicklung der Wärmeversorgung unter dem Energie-(R)Evolution-Szenario. Einsparung durch Effizienzsteigerung.
(Effizienz: Reduzierung verglichen mit dem Referenzszenario)

Quelle: report global energy scenario, Greenpeace, 2007

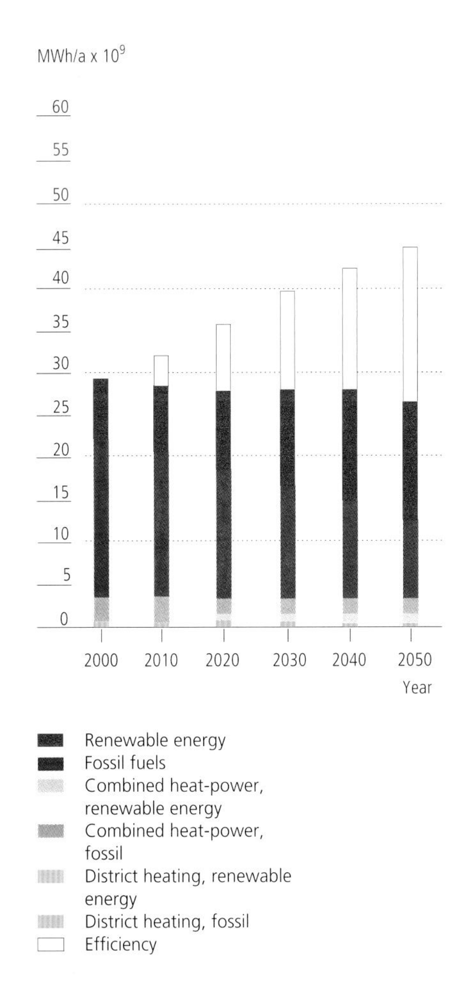

8.4 Energy-(R)evolution

Figures 105 and **106** describe the scenario recommended by Greenpeace, which is commonly known as the Energy-(R)evolution. In it, one can see the degree to which nuclear energy and energy from coal, oil, and brown coal will decline as percentages of the energy mix, and how gas usage, on the other hand, especially with the employment of combined heat and power plant generation (CHP), will increase until 2050.

In a parallel development, shown in **Figure 106**, hydroelectric power, wind power, photovoltaic applications on buildings, bio-energies, and geothermal energy are increasing. The use of geothermal power plants and energy from the oceans as tidal and wave power are completing the power mix. The largest role in the process of a stabilization of the global energy market will have to be played by the above-mentioned decreases in losses (primary and final energy use) and increases in performance.

It is self-evident that the prognostic scenarios shown in **Figures 105** and **106** need to be modified on a constant basis. Especially when it comes to geothermal energy and solar power plants, the authors expect more significant increases in contribution to the energy supply mix than shown in **Figure 106**.

8.4 Energie-(R)Evolution

Bilder 105 und **106** beschreiben das von Greenpeace empfohlene Szenario, beschrieben auch als Energie-(R)Evolution. Hier wird deutlich sichtbar, in welchem Umfang Nuklearenergie, Braunkohle, Steinkohle und Öl zurückgefahren werden und Gas als fossiler Energieträger insbesondere unter Einsatz von Kraft-Wärmekopplungen bis zum Jahr 2050 zunimmt.

Parallel soll gemäß **Bild 106** die Wasserkraft, Windenergie, Photovoltaik (an Gebäuden), Bioenergie und Geothermie zunehmen. Der Einsatz geothermischer Kraftwerke sowie die Nutzung von Meeresenergie (Gezeiten- und Wellenenergie) ergänzen den Energiemix. Den größten Anteil an der Stabilisierung der Energieverbräuche bei gleichzeitig steigender Bevölkerung und gleichzeitig steigendem Bedarf sollen die Verminderungen von Energieverlusten (Primärenergie-Endenergie) und die Steigerung von Wirkungsgraden herbeiführen.

Dass die in den **Bildern 105 und 106** prognostizierten Veränderungen selbstverständlich laufend angepasst und modifiziert werden müssen, bedarf keiner weiteren Erläuterung. Gerade im Bereich der Geothermie und der solarbetriebenen Kraftwerke sehen die Autoren deutliche Steigerungen gegenüber den in **Bild 106** gezeigten Werten.

Figure 105
Energy-(R)evolution
A world energy outlook with fuel energy, reduction through sustainable energy
(see figure 106)

Source: report global energy scenario, Greenpeace 2007

- Nuclear
- Lignite
- Coal
- Oil/Diesel
- Natural gas
- Combined Heat and Power

Bild 105
Die Energie-(Revolution)
Ausblick der Nutzung fossiler Brennstoffe, Reduktion durch den Einsatz erneuerbarer Energien (s. Bild 106)

Quelle: report global energy scenario, Greenpeace 2007

Very clearly, **Figures 107 – 109** seem to illustrate the implications as a result of use of renewable energy, such as solar, wind, and geothermal sources.

For individual countries and regions, the changes in absolute per-capita consumption are shown, and the land area requirements necessary to implement such renewable technologies are depicted as well.

Since deep geothermal applications only require insignificantly small land area for their installation, **Figure 109** does not include them in the graphic representation.

Sehr anschaulich dürften die **Bilder 107 – 109** sein, die den Einsatz erneuerbarer Energien (Solar/Wind/Geothermie) darstellen.

Für einzelne Länder und Regionen sind jeweils die Veränderungen der Energiemengen zwischen 2003 und 2050 (absolute Werte/Pro-Kopf-Verbräuche) ausgewiesen. Ebenfalls ausgewiesen sind die gesamten Landflächen, die notwendig werden, die Technologien der erneuerbaren Energien zu installieren.

Da die tiefe Geothermie keine entsprechenden oder nur unwesentlich kleine Landflächen benötigt, fehlen in **Bild 109** entsprechende Angaben.

Figure 106
Energy-(R)evolution
A sustainable world energy outlook

Source: report global energy scenario, Greenpeace 2007

- Hydropower
- Wind
- Photovoltaics
- Bioenergy
- Geothermal
- Solarthermal/Power Plants
- Ocean Energy
- Efficiency

Bild 106
Die Energie-(Revolution)
Zukünftige Energiebereitstellung durch erneuerbare Energien

Quelle: report global energy scenario, Greenpeace 2007

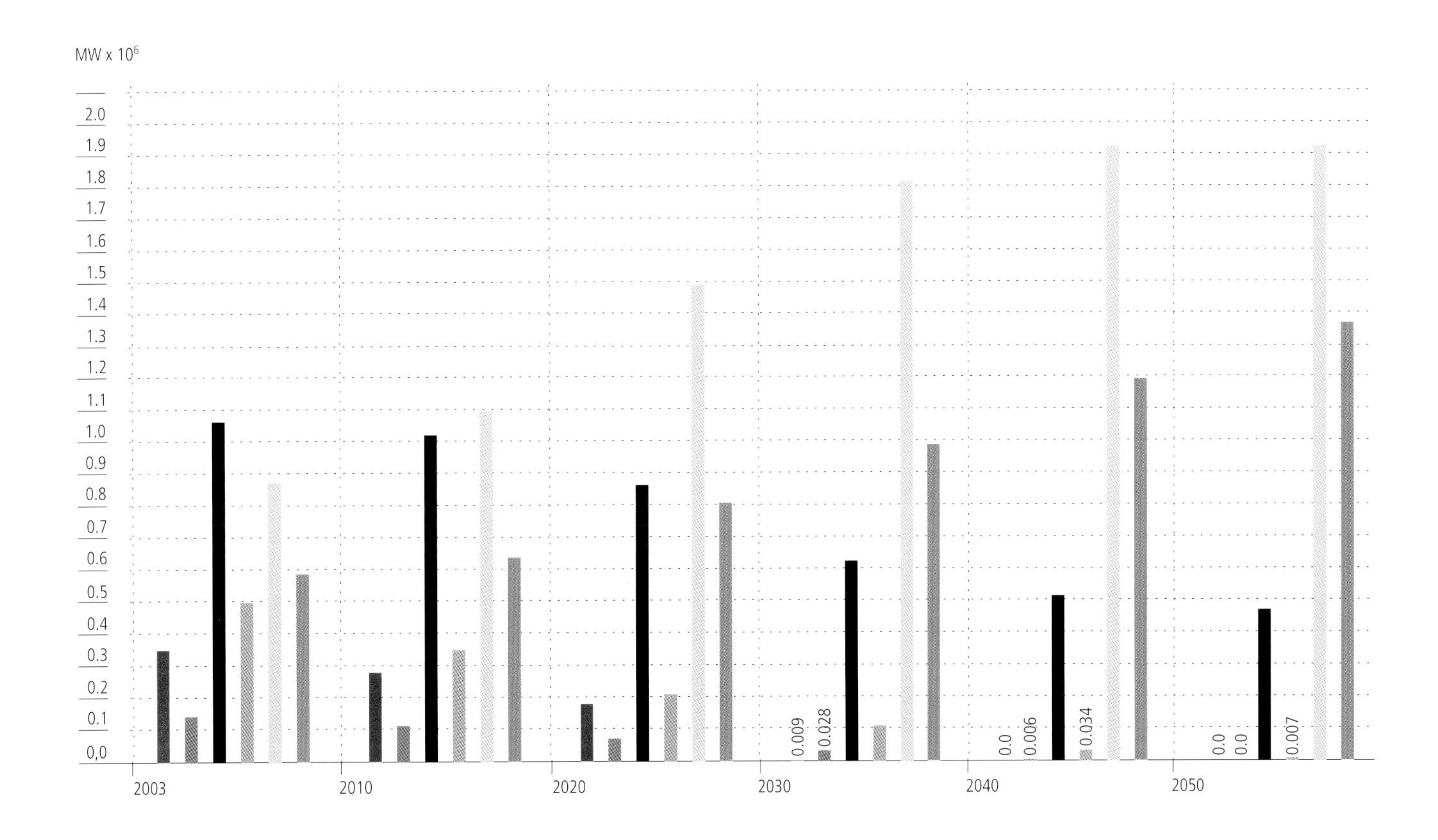
MW x 10^6
2.0
1.9
1.8
1.7
1.6
1.5
1.4
1.3
1.2
1.1
1.0
0.9
0.8
0.7
0.6
0.5
0.4
0.3
0.2
0.1
0,0
0.009
0.028
0.0
0.006
0.034
0.0
0.0
0.007
2003
2010
2020
2030
2040
2050

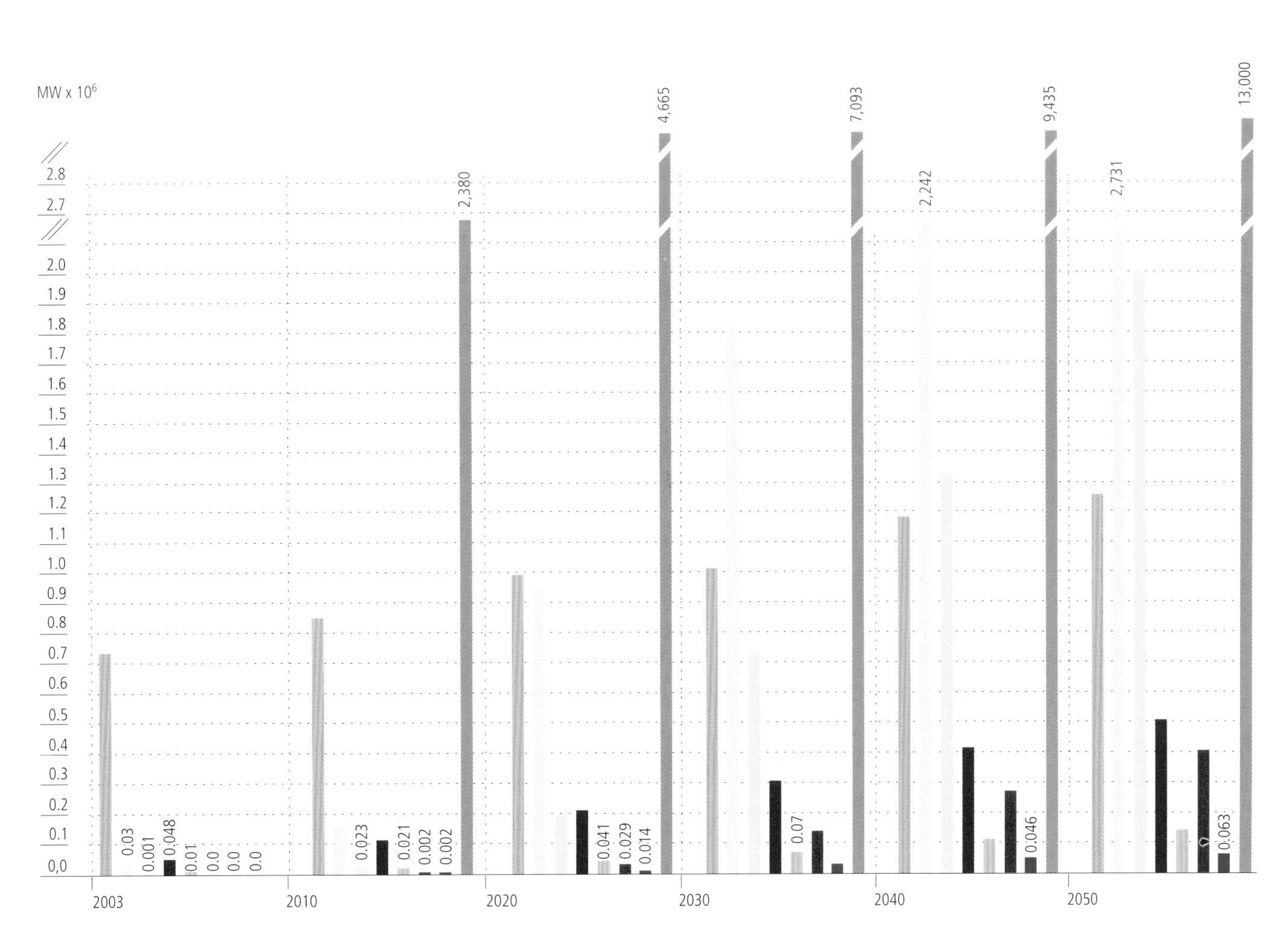
MW x 10^6
2.8
2.7
2.0
1.9
1.8
1.7
1.6
1.5
1.4
1.3
1.2
1.1
1.0
0.9
0.8
0.7
0.6
0.5
0.4
0.3
0.2
0.1
0,0
0.03
0.001
0.048
0.01
0.0
0.0
0.0
0.023
0.021
0.002
0.002
2,380
4,665
0.041
0.029
0.014
7,093
0.07
2,242
9,435
0.046
2,731
13,000
0.063
2003
2010
2020
2030
2040
2050

Figure 107
Renewable resource – solar
Solar reference scenario and the Energy-(R)evolution scenario (alternative scenario)

Source: Greenpeace

1

A	Reference scenario %	Reference scenario MWhx10⁶	Alternative scenario %	Alternative scenario MWhx10⁶
2003	0.05 M	15.85 H	–	–
2050	0.26 M	117.56	6.7	1,300

B	Reference scenario MWh	Alternative scenario MWh
2003	0.037	–
2050	0.201	2.22 M

Solar energy
OECD North America

2

A	Reference scenario %	Reference scenario MWhx10⁶	Alternative scenario %	Alternative scenario MWhx10⁶
2003	0.01	0.556	–	–
2050	0.03	4.45 L	7.4	619.4

B	Reference scenario MWh	Alternative scenario MWh
2003	0.001	–
2050	0.007	0.982

Solar energy
South America

A Production per Region (% of global share / MWhx10⁶)
B Production per Person (MWh)

H Highest
M Middle
L Lowest

3

A	Reference scenario %	Reference scenario MWhx10⁶	Alternative scenario %	Alternative scenario MWhx10⁶
2003	0.05 M	10.84	–	–
2050	0.25	64.77 M	6	851.2

B	Reference scenario MWh	Alternative scenario MWh
2003	0.02 M	–
2050	0.127 M	1.62

Solar energy
OECD Europe

4

A	Reference scenario %	Reference scenario MWhx10⁶	Alternative scenario %	Alternative scenario MWhx10⁶
2003	0.00 L	0.00 L	–	–
2050	0.17	35.31	14.9	1,823

B	Reference scenario MWh	Alternative scenario MWh
2003	0.00 L	–
2050	0.019	0.992

Solar energy
Africa

1
2

Bild 107
Erneuerbare Energie – Solar
Solar Referenz-Szenarium und Energie-(R)Evolution Szenarium (Alternatives Szenarium)

Source: Greenpeace

5

	Reference scenario		Alternative scenario	
A	%	MWhx10⁶	%	MWhx10⁶
2003	0.18 H	8.9	–	–
2050	0.32	34.75	38 H	2,124 H

B	MWh	MWh
2003	0.049 H	–
2050	0.098	6.0 H

Solar energy
Middle East

6

	Reference scenario		Alternative scenario	
A	%	MWhx10⁶	%	MWhx10⁶
2003	0.00 L	0.00 L	–	–
2050	0.46	162.4 H	8.1	1,716

B	MWh	MWh
2003	0.00 L	–
2050	0.115	1.22

Solar energy
China

7

	Reference scenario		Alternative scenario	
A	%	MWhx10⁶	%	MWhx10⁶
2003	0.00	0.278	–	–
2050	0.00	0.834	7.8	808.4

B	MWh	MWh
2003	0.001	–
2050	0.003 L	2.84

Solar energy
Transition economies

8

	Reference scenario		Alternative scenario	
A	%	MWhx10⁶	%	MWhx10⁶
2003	0.00 L	0.00 L	–	–
2050	0.17	33.64	12.2 M	1,266 M

B	MWh	MWh
2003	0.00 L	–
2050	0.015	0.572

Solar energy
South Asia

9

	Reference scenario		Alternative scenario	
A	%	MWhx10⁶	%	MWhx10⁶
2003	0.00 L	0.00 L	–	–
2050	0.39	65.33	5.5 L	491.2 L

B	MWh	MWh
2003	0.00 L	–
2050	0.073	0.552 L

Solar energy
East Asia

10

	Reference scenario		Alternative scenario	
A	%	MWhx10⁶	%	MWhx10⁶
2003	0.09	8.62 M	–	–
2050	0.85 H	110.37	11.5	755.9

B	MWh	MWh
2003	0.044	–
2050	0.604 H	4.14

Solar energy
OECD Pacific

Figure 108
Renewable resource – wind
Solar reference scenario
and the Energy-(R)evolution
scenario
(alternative scenario)

Source: Greenpeace

1

2

1

	Reference scenario		Alternative scenario	
A	%	MWhx10^6	%	MWhx10^6
2003	0.04 M	12.23	–	–
2050	0.73	330.26	7.7	1,501 H

B	MWh	MWh
2003	0.028	–
2050	0.563	2.56

Solar energy
OECD North America

2

	Reference scenario		Alternative scenario	
A	%	MWhx10^6	%	MWhx10^6
2003	0.01	0.278	–	–
2050	0.32	55.04	9.3 H	780.62 M

B	MWh	MWh
2003	0.001	–
2050	0.087	1.24 M

Solar energy
Latin America

A Production per Region (% of global share / MWhx10^6)
B Production per Person (MWh)

H Highest
M Middle
L Lowest

3

	Reference scenario		Alternative scenario	
A	%	MWhx10^6	%	MWhx10^6
2003	0.21 H	44.48 H	–	–
2050	2.1 H	545.44 H	7.2	1,016

B	MWh	MWh
2003	0.084 H	–
2050	1.07 H	1.99

Solar energy
OECD Europe

4

	Reference scenario		Alternative scenario	
A	%	MWhx10^6	%	MWhx10^6
2003	0.01	0.556	–	–
2050	0.14 L	28.91	1.1 L	133.16 L

B	MWh	MWh
2003	0.001	–
2050	0.016 L	0.072 L

Solar energy
Africa

Bild 108
Erneuerbare Energie – wind
Solar Referenz-Szenarium und
Energie-(R)Evolution Szenarium
(Alternatives Szenarium)

Source: Greenpeace

A	%	MWhx10^6	%	MWhx10^6
2003	0.00 L	0.00 L	–	–
2050	0.18	20.02 L	3.5	195.7

B	MWh	MWh
2003	0.00 L	–
2050	0.057	0.55

Solar energy
Middle East

A	%	MWhx10^6	%	MWhx10^6
2003	0.01	1.11	–	–
2050	0.53 M	189.32	5.7	1,201

B	MWh	MWh
2003	0.001	–
2050	0.135	0.853

Solar energy
China

A	%	MWhx10^6	%	MWhx10^6
2003	0.00 L	0.00 L	–	–
2050	0.21	38.64	7.7	800.64

B	MWh	MWh
2003	0.00	–
2050	0.136 M	2.82 H

Solar energy
Transition economies

8

	Reference scenario		Alternative scenario	
A	%	MWhx10^6	%	MWhx10^6
2003	0.05 M	3.61 M	–	–
2050	0.26 M	38.09	4.6	475.38

B	MWh	MWh
2003	0.003	–
2050	0.017	0.215

Solar energy
South Asia

9

	Reference scenario		Alternative scenario	
A	%	MWhx10^6	%	MWhx10^6
2003	0.00 L	0.00 L	–	–
2050	0.21	35.03	6.1 M	550.44

B	MWh	MWh
2003	0.00 L	–
2050	0.039	0.619

Solar energy
East Asia

10

	Reference scenario		Alternative scenario	
A	%	MWhx10^6	%	MWhx10^6
2003	0.02	1.67	–	–
2050	0.69	90.07 M	7.6	500.4

B	MWh	MWh
2003	0.009 M	–
2050	0.493	2.74

Solar energy
OECD Pacific

Figure 109
Renewable resource – geothermal
Solar reference scenario and the Energy-(R)evolution scenario (alternative scenario)

Source: Greenpeace

1

2

1

	Reference scenario		Alternative scenario	
A	%	MWhx10[6]	%	MWhx10[6]
2003	0.54 H	172.64 H	–	–
2050	0.78 H	353.06 H	5.5 M	1,059

B	MWh	MWh
2003	0.405 H	–
2050	0.602 H	1.81

Solar energy
OECD North America

2

	Reference scenario		Alternative scenario	
A	%	MWhx10[6]	%	MWhx10[6]
2003	0.31 M	16.96	–	–
2050	0.54	93.96 M	3.6	301.07

B	MWh	MWh
2003	0.038	–
2050	0.149	0.478

Solar energy
Latin America

A Production per Region (% of global share / MWhx10[6])
B Production per Person (MWh)

H Highest
M Middle
L Lowest

3

	Reference scenario		Alternative scenario	
A	%	MWhx10[6]	%	MWhx10[6]
2003	0.20	41.70 M	–	–
2050	0.61	157.63	8.6	1,221 H

B	MWh	MWh
2003	0.079 M	–
2050	0.310	2.40

Solar energy
OECD Europe

4

	Reference scenario		Alternative scenario	
A	%	MWhx10[6]	%	MWhx10[6]
2003	0.10	6.12	–	–
2050	0.05	9.45	2.26	272.72

B	MWh	MWh
2003	0.007	–
2050	0.005	0.148

Solar energy
Africa

Bild 109
Erneuerbare Energie – geothermal
Solar Referenz-Szenarium und Energie-(R)Evolution Szenarium (Alternatives Szenarium)

Source: Greenpeace

7

3

5

6

9

4

8

10

5

A	Reference scenario %	Reference scenario MWhx10^6	Alternative scenario %	Alternative scenario MWhx10^6
2003	0.00 L	0.00 L	–	–
2050	0.00 L	0.278 L	6.89	384.75
B		MWh		MWh
2003		0.00 L		–
2050		0.001 L		1.09 M

Solar energy
Middle East

6

A	Reference scenario %	Reference scenario MWhx10^6	Alternative scenario %	Alternative scenario MWhx10^6
2003	0.00	0.00	–	–
2050	0.06	21.13	0.12 L	25.85 L
B		MWh		MWh
2003		0.00		–
2050		0.015		0.018 L

Solar energy
China

7

A	Reference scenario %	Reference scenario MWhx10^6	Alternative scenario %	Alternative scenario MWhx10^6
2003	0.00	0.556	–	–
2050	0.30	55.88	7.8	814.54
B		MWh		MWh
2003		0.002		–
2050		0.196 M		2.87 H

Solar energy
Transition economies

8

A	Reference scenario %	Reference scenario MWhx10^6	Alternative scenario %	Alternative scenario MWhx10^6
2003	0.00	0.00	–	–
2050	0.17	33.92	4.02	411.44 M
B		MWh		MWh
2003		0.00		–
2050		0.015		0.186

Solar energy
South Asia

9

A	Reference scenario %	Reference scenario MWhx10^6	Alternative scenario %	Alternative scenario MWhx10^6
2003	2.00	124.0	–	–
2050	1.33	221.84	9.2 H	827.9
B		MWh		MWh
2003		0.199		–
2050		0.249		0.931

Solar energy
East Asia

10

A	Reference scenario %	Reference scenario MWhx10^6	Alternative scenario %	Alternative scenario MWhx10^6
2003	0.10	9.45	–	–
2050	0.38 M	48.93	2.81	181.81
B		MWh		MWh
2003		0.048		–
2050		0.268		0.995

Solar energy
OECD Pacific

Summary

Whether on a global or regional basis, one argument can be safely made, as follows:

- Regional activity needs to become global.
- The aim of achieving efficiency improvements, limiting transmission losses, and the improving performance need to be further pursued aggressively.
- Fossil energy sources need to become significantly more expensive to strengthen the incentive for change.
- Energy savings measures in the area of electric and thermal power generation and in all aspects related to fuels need to be fully exhausted.
- Developed nations with their greatly inflated energy requirements need to be part of an avant-garde supporting the development of alternative technologies of threshold nations.
- The waste of natural sources has to be stopped as fast as possible by the introduction and development of truly recyclable materials with an appropriate design.

According to research by the European Union, approximately 40 % of all energy used is consumed by the building sector for the construction, operation, and maintenance of buildings, including power for lighting, heating, cooling and electrical plug loads. The situation for the U.S. is comparable yet leads obviously to much greater absolute numbers. Aspiring countries in the rapidly developing Middle East and in parts of Asia already show a much greater energy demand per capita than even the U.S., mainly due to higher average temperatures and a philosophy of "everything-goes".

Since the authors of this publication are only capable of helping to improve conditions in the building sector when it comes to the built environment, the following chapters will show examples of possible implementation of new technology in buildings and discuss which reasonable classifications for energy-efficient structures can be made.

Zusammenfassung

Ob global oder regional – eins lässt sich sehr gut aus allem vorher Gesagten ableiten:

- Regionales Handeln muss zum gewollten globalen Szenario führen.
- Die angestrebten Energieeffizienzverbesserungen durch die Vermeidung von hohen Transportverlusten und unter Steigerung der Wirkungsgrade eingesetzte Energien müssen konsequent weiterverfolgt werden.
- Die fossilen Primärenergieträger müssen sich deutlich verteuern, um die Menschheit so unter Druck zu setzen, dass sie zum Handeln gezwungen wird.
- Energieeinsparungsmaßnahmen sowohl im Bereich des elektrischen Energiebedarfs wie auch des Wärmeenergiebedarfs und der Kraftstoffe müssen voll ausgeschöpft werden.
- Die in hohem Maß energieverzehrenden entwickelten Nationen müssen einerseits eine Vorreiterrolle übernehmen und andererseits den Schwellenländern bei der Entwicklung eigener Strategien unter die Arme greifen.
- Die Ressourcenverschwendung von Rohstoffen muss schnellstens durch tatsächlich recycelfähige Materialien und ein entsprechendes Design gestoppt werden.

Nach Recherchen der Europäischen Kommission entfallen in der EU auf den Bau und die Instandhaltung von Gebäuden (einschließlich Heizung, Kühlung, Beleuchtung und elektronische Ausstattung) ca. 40 Prozent des Verbrauchs aller Rohstoffe. In den USA ist die Situation bedingt vergleichbar – der Energiebedarf ist allerdings noch deutlich größer. Ähnlich verhält es sich auch zum Teil in den aufstrebenden Ländern des Nahen Ostens und Teilen Asiens, wo aufgrund der hohen Temperaturen und zum Teil der Einstellung „alles ist möglich" der Energiebedarf pro Kopf wiederum deutlich höher ist als selbst in den USA.

Da die Verfasser dieser Schrift lediglich mitwirken können, den Energieverbrauch von Gebäuden (weltweit) zu beeinflussen, soll nachfolgend dargestellt werden, welche Strategien bei Hochbauten zum Einsatz kommen können und an welchen Klassifizierungen energiesparender Häuser man sich sinnvollerweise orientieren sollte.

Part 2

Buildings for the post-fossil fuel era

Real estate of the future – post-fossil-fuel buildings

Small is beautiful! Exemplary small-scale projects

Zero-energy super-tall buildings: vision or illusion?

Teil 2

Postfossiles Bauen

Die Immobilie der Zukunft – postfossile Gebäude

Klein, aber fein, beispielhafte Bauten

Zero-energy super-tall buildings – vision oder Illusion

9 Buildings for the post-fossil fuel era

9 Postfossiles Bauen

Strategies for energy savings and protection of resources etc. in building design

The majority of energy-saving measures in buildings require greater initial investments. The return on investment (ROI) due to reduced operating cost is in the mid- to long-term range, and to a degree this explains why mostly large commercial enterprises take steps to minimize the energy consumption of their real estate. They are typically capable of affording higher initial investment cost in order to be able to take advantage of lower operating cost in the long run.

Strategien der Energie-Einsparung, Ressourcenschonung etc. bei Hochbauten

Ein Großteil der Energie-Sparoptionen bei Gebäuden erfordert im Regelfall höhere Investitionen, die sich erst mittel- und zum Teil langfristig durch Einsparungen bei den Energiekosten wieder bezahlt machen. Dies erklärt, warum eher gewerbliche Unternehmen Schritte einleiten, um ihren Energieverbrauch zu minimieren. Sie sind gewöhnlich in der Lage, höhere Anfangsinvestitionen zu tragen, um dann ihre Betriebskosten zu senken.

Topic	HQE®	BREEAM	LEED®	CASBEE-NC
Topic A Management	SME – Goal 3 – Construction site Goal 7 – Maintenance	M – Management	Organized according 4 topics	Q 2 – Service quality
Topic B Location	Goal 1 – Immediate environment	Use of soil and ecology T – Transport	SS – Sustainable sites	Q 3 – Ecology of site
Topic C Indoor environment	Goals: Comfort (8, 9, 10, 11) Health: (12, 13)	HW – Heath and wellbeing	IEQ – Indoor environmental quality	Q 1 – environmental quality
Topic D Consumption of resources	Goal 4 – Energy Goal 5, 14– Water Goal 2 – Materials Goal 6 – Waste	E – Energy W – Water MW – Materials, Waste	EA – Energy & atmosphere WE – Water efficiency MR – Materials & resources	LR1 – Energy LR2 – Resources and materials including water
Topic E Environmental impact	see: Goal 4 – energy	Pollution Use of soil, ecology	——	LR3 – Environment beyond site
Topic F Social-economical dimension	——	——	——	Q3 – Psychological comfort/ social dimension
Topic G Creativity/ Opening of system	——	——	Innovation & design	——

Figure 110
Comparison between various global sustainability standards and their environmental impact
Greenpeace, 2007

Bild 110
Vergleichende Analyse der Struktur verschiedener Methoden der Bewertung der Umweltverträglichkeit von Gebäuden

At the level of the individual residential homeowner, the higher initial cost can be prohibitive. This group seeks to find a compromise between efficiency improvements and initial investment cost. Nevertheless, it is wise to pursue available technology and energy efficiency improvements. Necessary additional financing is usually paid for by greatly reduced energy bills over the lifespan of the structure.

The current or future classification of building stock based on national rating standards, such as LEED® in the U.S., Energiepass in Germany, and others (**Figure 110**), in many cases results in an increased resale value of the real estate object. For example, it is apparent

Privatpersonen („Häuslebauer") meinen im Regelfall, sich die höheren Anfangsinvestitionskosten nicht leisten zu können, und versuchen daher, eher einen Kompromiss zwischen Effizienzverbesserung und Investitionsaufwand zu finden. Gleichwohl tun sie gut daran, im Notfall durch Kredite und gegebenenfalls staatliche Unterstützungen die erhöhten Investitionen vorzufinanzieren, um im laufenden Betrieb durch Energieeinsparungen die Kredite wieder zurückzuführen.

Durch die zukünftigen Klassifizierungen von Hochbauten (LEED, USA, Energiepass, Deutschland usw.) – **Bild 110** – kann im Regelfall ein erhöhter Werterhalt erreicht werden. Energiesparende Häuser dürften vor

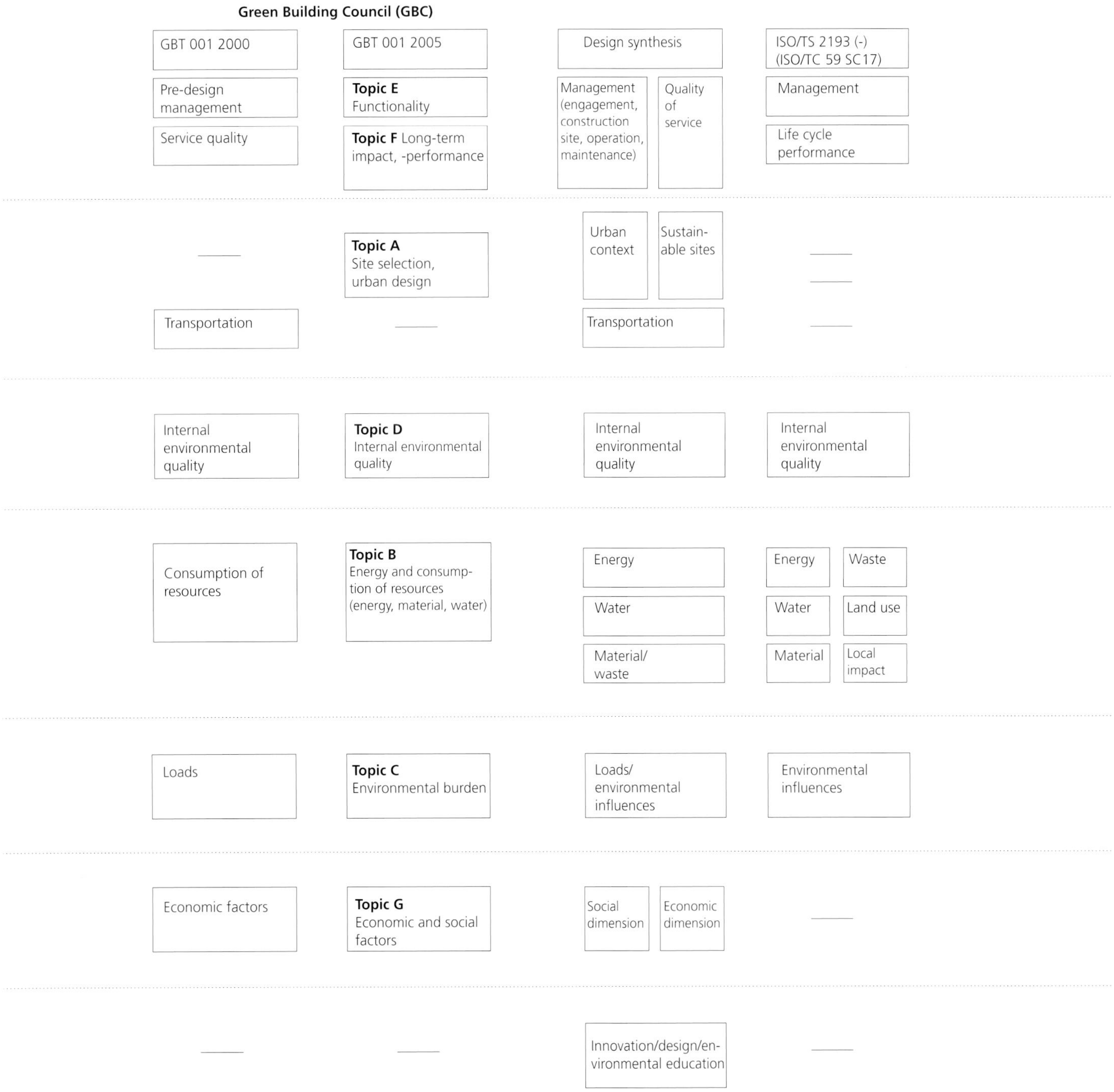

Green Building Council (GBC)			
GBT 001 2000	GBT 001 2005	Design synthesis	ISO/TS 2193 (-) (ISO/TC 59 SC17)
Pre-design management Service quality	**Topic E** Functionality **Topic F** Long-term impact, -performance	Management (engagement, construction site, operation, maintenance) Quality of service	Management Life cycle performance
—— Transportation	**Topic A** Site selection, urban design ——	Urban context Sustainable sites Transportation	—— —— ——
Internal environmental quality	**Topic D** Internal environmental quality	Internal environmental quality	Internal environmental quality
Consumption of resources	**Topic B** Energy and consumption of resources (energy, material, water)	Energy Water Material/waste	Energy Waste Water Land use Material Local impact
Loads	**Topic C** Environmental burden	Loads/environmental influences	Environmental influences
Economic factors	**Topic G** Economic and social factors	Social dimension Economic dimension	——
——	——	Innovation/design/environmental education	——

that energy-efficient structures with greatly reduced operating cost will gain market value in an environment of rising primary energy cost, a condition which will certainly be of great interest to institutional investors in the future.

The greater presence of real estate investment trusts (REITs) in today's financial markets provides many investors with a more accessible way to invest in commercial real estate. REITs function similarly to exchange-traded real estate funds, but technical and tax-related differences also exist. The REIT fund is a tax designation for a corporation investing in real estate that reduces or eliminates corporate income taxes. In return, REITs are required to distribute 90 % of their income, which may be taxable in the hands of the investors. The REIT structure was designed to provide a similar structure for investment in real estate as mutual funds provide for investment in stocks. REITs profit primarily from a reduction of operating cost, which improves the value of the portfolio and increases the net profit, especially since returns have to be paid out to stock holders.

When a person or an organization seeks investment opportunities in real estate, one of the first actions should be to identify any new regulations that have been introduced by the individual country to foster energy savings. In 2005, the U.S. passed the Energy Policy Act, which provides tax incentives for photovoltaic applications for residences and energy refurbishment of the existing building stock. In the European Union, including countries such as Norway and Switzerland, energy standards are planned to be revised every five years, and the currently most applicable energy standard has to be complied with in large-scale renovations and all new construction.

allem in Zukunft auch und insbesondere für institutionelle Anleger von besonderem Interesse sein, da bei steigenden Energiekosten insbesondere die Objekte nachgefragt werden, die mit geringeren Energieverbräuchen und Energiekosten auskommen.

Mit der Verbreitung von Real Estate Investment Trusts (REITs) hat sich der Zugang zu Anlageprodukten in Immobilien stark erweitert. REITs funktionieren ähnlich wie börsennotierte Immobilienfonds, wobei jedoch technische und steuerliche Unterschiede bestehen. Die REIT-Struktur soll ähnliche Investitionsstrukturen für Immobilien schaffen, wie sie Anlagefonds für Investitionen in Aktien bieten. Daher profitieren REITs primär über eine Senkung der Betriebskosten durch Verbesserungen der Energieeffizienz innerhalb ihrer Portfolios. Diese Verbesserungen haben positive Auswirkungen auf das Nettoergebnis, das sich bei REITs potenziell positiv auf die Bewertung auswirken kann, vor allem, weil REITs den Hauptanteil des Betriebsertrags an die Aktionäre ausschütten müssen.

Im Auge haben sollte man bei der Entwicklung von Immobilien zusätzlich, dass in einzelnen Ländern immer wieder neue Standards eingeführt werden, um Anreize zur Energiereduzierung zu schaffen. 2005 wurde in den USA der Energy Policy Act verabschiedet, der Anreize wie Steuervorteile beim Einsatz von Photovoltaikanlagen an Wohnhäusern oder für Sanierungen von Häusern schafft. Im Jahr 2006 wurde in den EU-Ländern einschließlich Norwegen und Schweiz beschlossen, mindestens alle fünf Jahre eine Aktualisierung der Energienormen vorzunehmen, wobei bei größeren Sanierungen und Neubauten die neuen Energieeffizienzstandards einzuhalten sind (Gesetz).

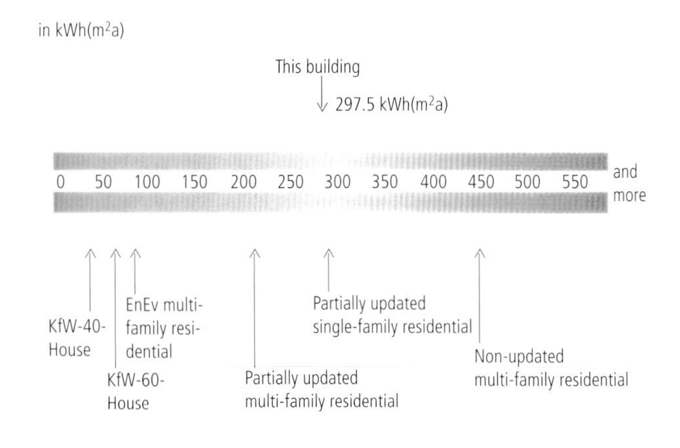

KfW KfW Bank Group, Germany
EnEV EnEV German Energy Saving Law

to Figure 110
The German Energy Pass for buildings

Source: dena, Deutsche Energie Agentur

zu Bild 110
Der deutsche „Energiepass"

Quelle: dena, Deutsche Energie Agentur

Organizational structure of the Green Building Councils

The World Green Building Council, founded in 1998 in Nagoya, Japan, is the foremost global, not-for-profit organization working to move the property industry toward sustainability. The World GBC supports world-wide definition of standards technologies, products, and projects to help reduce energy consumption in the building sector.

The organization's vision statement is:
The World GBC not-for-profit organization is working to transform the property industry toward sustainability through its members, the national Green Building Councils. It provides a federation union of national Green Building Councils whose common goal is the sustainable transformation of the global property industry.

It wants to be the primary global voice for green building issues, and it represents no less than 60 % of the global property industry through the national Green Building Councils.

The U.S. Green Building Council (USGBC), for instance, is a non-profit organization founded in 1994 and committed to expanding sustainable building practices. USGBC is composed of organizations from across the building industry that are working to advance structures that are environmentally responsible, profitable, and healthy places to live and work. Members include building owners and end-users, real estate developers, facility managers, architects, designers, engineers, general contractors, subcontractors, product and building system manufacturers, government agencies, and nonprofits. National Green Building Councils are financed by structured membership dues and are widely supported by the building industry.

The World Green Building Council consists today of the nine member countries that have their own national chapters (Australia, India, Japan, Canada, Mexico, New Zealand, Taiwan, The United Arab Emirates, and the United States), as well as other countries that are currently in the process of establishing their own national chapters.
(further information: www.worldgbc.org)

Various national rating systems and their assessment

Under the umbrella of the World Green Building Council, several national building rating and certification systems have been developed, including the following:

Die Organisationsform der Green Building Councils

Die Gründung des World Green Building Council (World-GBC) wurde erstmalig 1998 in Nagoya, Japan, angekündigt, um die Arbeit der nationalen GBCs zu koordinieren und voranzutreiben. Der World-GBC unterstützt weltweit die Entwicklung der Standards, Technologien, Produkte und Projekte. Er gilt ebenfalls als unpolitisches globales Forum für die Nachhaltigkeitsdiskussionen im Baubereich.

Die Vision des World-GBC lautet:
„The peak global not-for-profit organization is working to transform the property industry towards sustainability through its members national GBCs"

Als gemeinnützige Organisation verfolgt der World-GBC das Ziel, als weltweit führende Einrichtung über die nationalen Mitglieder die Immobilienbranche in Richtung Nachhaltigkeit voranzubringen.

Der World-GBC stellt als direkte Ziele seiner Gründung dar, technisches Wissen und Fortschritt der landesspezifischen Informationen über nachhaltiges Bauen zu verbreitern. Eines der Hauptziele des World-GBC ist es zudem, Mitglieder auf der ganzen Welt, eingeschlossen der Entwicklungsländer, zu werben.

Der World Green Building Council (World-GBC) stellt außer Untersstützung und Förderung auch Richtlinien für die Gründung weiterer Green Building Councils zur Verfügung. In den Richtlinien wird angeregt, dass die Mitgliedschaft die nationale geographische Umgebung und die Vielzahl der Akteure reflektieren sollte. Der US-GBC beispielsweise hat momentan mehr als 7200 Mitglieder, mit steigender Tendenz. Die bestehenden nationalen GBCs finanzieren sich durch individuell gestaffelte Mitgliedsbeiträge und werden von der Bauindustrie gut angenommen und unterstützt.

Der World-GBC besteht momentan aus neun Mitgliedsländern mit eigenen Green Building Councils (Australien, Indien, Japan, Kanada, Mexiko, Neuseeland, Taiwan, USA, Vereinigte Arabische Emirate) und weiteren Ländern, die aktiv an der Gründung ihres eigenen GBC arbeiten.
(mehr Informationen: www.worldgbc.org)

Bewertungssysteme verschiedener Länder und deren Erfolgsfaktoren

In den letzten Jahren wurden weltweit verschiedene erfolgreiche Bewertungssysteme für Gebäude entwickelt, und unter dem Dachverband des World-GBC verbreitet. Zu diesen gehören:

- BREEAM, the Building Research Establishment Assessment Method (U.K.)
- CASBEE, the Comprehensive Assessment System for Building Environmental Efficiency (Japan)
- HQE, The Haute Qualité Environnementale or HQE® (High-Quality Environmental Standard) (France)
- LEED® Leadership in Energy & Environmental Design (U.S.)
- LEED® in Canada
- Green Star (Australia)
- Green Star NZ (New Zealand)
- TGBRS TERI'S (The Energy and Resources Institute) Green Building Rating System (India)

Many of these rating systems build upon earlier established standards such as Green Star, LEED, and BREEAM, which are adapted to national specifics and have been further developed in the process. Discussions about the relative practical value of individual rating systems exist, but a consensus about the "best" rating system cannot be found because comparability is difficult to establish due to national differences. While architects, planners, and engineers are concerned with these issues, various approaches exist to make buildings sustainable, so the goal is to evaluate the results and provide transparency for investors, owners, and users.

The following chapter illustrates the four important rating systems for buildings. A Table in the Appendix provides an overview.

BREEAM (United Kingdom)

The Building Research Establishment Ltd. Environmental Assessment Method, or BREEAM, controls, distributes, and develops the rating system. Rating categories are available for the building typologies of office buildings, industrial structures, schools, prisons, residential structures, hospitals, single-family residential buildings (so-called Ecohomes and EcohomesXB), and existing residential buildings and retail establishments.

The system presents an ecological award after a process of building performance evaluations based on a set of criteria that investigate a building's compatibility with its environment on local, global, and internal levels. Points are given for the several criteria.

Architects and designers are thus encouraged to consider valuation criteria at an early stage of the design process in order to score BREEAM value points, which will, in the end, determine the classification of the finished building.

- BREEAM Building Research Establishment Assessment Method (England)
- CASBEE Comprehensive Assessment System for Building Environmental Efficiency (Japan)
- HQE Haute Qualité Environnementale (Frankreich)
- LEED Leadership in Energy & Environmental Design (USA)
- LEED Canada (Kanada)
- Green Star (Australien)
- Green Star NZ (Neuseeland)
- TGBRS TERI'S (The Energy and Resources Institute) Green Building Rating System (Indien)

Viele Bewertungssysteme bauen auf anderen, vorausgegangenen Systemen auf (z.B. Green Star und LEED auf BREEAM), führen diese fort oder wurden länderspezifisch angepasst. Über den relativen Nutzen, den die verschiedenen Bewertungssysteme erzielen, wird in Anwendung und Wissenschaft viel diskutiert. Ein Konsens über das „beste System" existiert verständlicherweise nicht, da die Vergleichbarkeit oft nicht gegeben und auch nicht Ziel der Anwendung der Systeme ist. Aus Sicht der Planer und Architekten gibt es unzählige Wege, die gewünschte Wirkung zu erzielen, und viele Arten, die erzeugte Wirkung zu messen und zu bewerten, um sie den Investoren, Eigentümern und Gebäudenutzern darstellen zu können.

In den folgenden Absätzen werden vier Bewertungssysteme für Gebäude kurz beschrieben. Eine Übersichtstabelle im Anhang stellt die Systeme im Überblick dar.

BREEAM (England)

BREEAM wurde von BRE (Building Research Establishment Ltd.) entworfen, kontrolliert, verbreitet und weiterentwickelt. Das BREEAM-Bewertungsschema ist verfügbar für Büros, Industrie, Schulen, Gerichte, Gefängnisse, Mehrfamilienhäuser, Krankenhäuser, Häuser (Ökohäuser), bestehende Siedlungen und Wohnhäuser.

BREEAM vergibt ein ökologisches Gütesiegel nach der Prüfung der Gebäudeperformance hinsichtlich einer Reihe von ökologischen Kategorien. Diese bewerten die Auswirkungen des Gebäudes auf seine Umwelt auf globaler, regionaler, lokaler und innenräumlicher Ebene. Für bestimmte Leistungskriterien werden vordefinierte Punktzahlen vergeben.

Die Architekten und Planer werden somit ermutigt, so früh wie möglich die Kategorien und Leistungskriterien zu beachten, um ihre Chance auf eine gute BREEAM-Bewertung zu erhöhen.

Points are obtainable in each category, and combinations of points are possible. A weighting of various ecological criteria allows the addition of several points to a total score. After assessing buildings using its criteria, BREEAM provides an overall score that will fall within a band providing either a "Pass", "Good", "Very Good", or "Excellent" rating. A certificate is also awarded that can be used by the owner as a marketing tool. Around 65,000 buildings have been awarded certificates so far, and another 270,000 are currently registered.

CASBEE (Japan)

CASBEE was developed according to a policy that the system should be structured to award high assessments to superior buildings, thereby enhancing incentives for designers and others. It was felt that the system should be as simple as possible and be applicable to buildings in a wide range of applications,and that it should take into consideration issues and conditions that are specific to Asia in general and to Japan in particular.

CASBEE is composed of four assessment tools corresponding to a building's life cycle. The so-called "CASBEE Family" is the collective name for these four tools and for a set of expanded tools for specific purposes. The CASBEE assessment tools are CASBEE for Pre-design, CASBEE for New Construction, CASBEE for Existing Buildings, and CASBEE for Renovation, which are intended to serve as guidance at each stage of the design process. Each tool is intended for a separate purpose and target user and is designed to accommodate a wide range of uses (offices, schools, apartments, etc.) in the evaluated buildings.

Haute Qualité Environnementale or HQE® (France)

The Haute Qualité Environnementale or HQE® (High-Quality Environmental Standard) is a standard for green building in France that is based on the principles of sustainable development first set out at the 1992 Earth Summit. The standard is controlled by the Paris-based Association pour la Haute Qualité Environnementale (ASSOHQE). Demand for certification of buildings according to HQE seems to be increasing. The rating procedure has the goals of introducing clients and investors to sustainable concepts and of offering suggestions for the optimization of buildings. It is an entirely voluntary process geared toward introducing ecology into the logic of builders.

The HQE method was first applied for the evaluation of buildings, mainly residential projects, in 1994, and since 1997 a complete set of criteria have been developed and structured by the HQE Association.

Punkte werden grundsätzlich in jeder Kategorie vergeben, dabei kommt es aber auch auf die Kombination an. Unterschiedliche ökologische Gewichtungen ermöglichen das Zusammenfügen der Punkte zu einer Gesamtbewertungspunktzahl. Die erreichte Punktzahl wird in Form einer allgemeinen Wertung ausgedrückt und in Klassen von „Ausgezeichnet" über „Sehr gut" und „Gut" bis „Durchschnittlich" eingeteilt. Das Gebäude kann auf dieser Skala eingeordnet werden, und ein Zertifikat kann vom Eigentümer zu Werbungszwecken genutzt werden. Bisher wurden ca. 65.000 Gebäude nach BREEAM zertifiziert und 270.000 registriert.

CASBEE (Japan)

Das JSBC (Japan Sustainable Building Consortium) hat CASBEE (Comprehensive Assessment System for Building Environmental Efficiency) entwickelt, welches anwendungsbezogen die ökologische Performance von Gebäuden klassifiziert:

CASBEE wurde nach folgendem Leitfaden entworfen:

1. Das System soll so strukturiert werden, dass hohe Bewertungen von übergeordneten Gebäuden erzielt werden, um somit den Anreiz der Architekten und anderer zu steigern.
2. Das Bewertungssystem soll so einfach wie möglich sein.
3. Das System soll für eine Vielfalt von Gebäuden anwendbar sein.
4. Das System soll Aufgaben und Probleme speziell für Japan und Asien mit einbeziehen.

CASBEE ist aus vier Bewertungstools, die dem Lebenszyklus eines Gebäudes entsprechen, zusammengesetzt: CASBEE für Entwurf, Neubau, existierende Gebäude und Erneuerungen. Jedes Tool ist für eine separate Anwendung mit eigenem Nutzungsziel vorgesehen und wurde für verschiedene Anwendungsfälle (Bürogebäude, Schulen, Wohnungen, usw.) entworfen. CASBEE wird in naher Zukunft in jedem Lebensabschnitt eines Gebäudes landesweit genutzt, um die Nachhaltigkeit im Bausektor zu stärken.

HQE (Frankreich)

Die wachsenden ökologischen Anforderungen in Angeboten und Baugesuchen in Frankreich zeigen, dass die Nachfrage nach der Anwendung der „HQE-Praktiken" bei der Gebäudeentwicklung steigt. Das französische HQE (Haute Qualité Environnementale) entspricht den gebräuchlichen Regeln, die die Erfahrungen mit HQE reflektieren.

With regard to the ecological management of building projects, the sustainable design of buildings is reviewed in 14 steps, ranging from site, construction type, and operative management all the way to comfort and health of the occupants after completion of the project.

The impact of building materials on the overall sustainability of the project is assessed based on the Environmental Product Declaration (EPD), which honors material selection according to ecological criteria.
An EPD is defined as "quantified environmental data for a product with pre-set categories of parameters based on the ISO 14040 series of standards, but not excluding additional environmental information".
HQE works in three phases: project contract, design, and execution on the construction site. Aspects of indoor air quality (IAQ) as well as emissions are part of the evaluation. Currently, 25 completed buildings have been certified by HQE, including 8 public buildings and 17 in the private sector.

LEED® (United States)

The Leadership in Energy and Environmental Design (LEED®) Green Building Rating System, developed by the U.S. Green Building Council, provides a suite of standards for environmentally sustainable construction. It is a voluntary standard with the goal of creating „healthier" buildings that use resources more effectively in their construction and operation.

Members of the U.S. Green Building Council represent all sectors of the building industry, and the LEED® rating system is continuously being developed with the goals of promoting integrated design practice, providing ecological leadership in the building industry, and fostering competition among sustainable ideas. There are 13,500 member organizations, including corporations, governmental agencies, nonprofits, and others from throughout the building industry.

LEED® and the USGBC strive for education of the client base and the architects, designers, and engineers involved, and they seek a complete market transformation. Three different ways to include LEED® principles into the building process can characterized as follows:

- LEED® may be used as a guiding principle for the design team during schematic design to implement sustainable design principles.
- LEED® ratings are a way to provide information to clients and other involved parties to prove that sustainable principles, materials, and methods were incorporated into the project.
- Both the U.S. and the Canadian Green Building Council are entitled to certify buildings.

Die Vorgehensweise des HQE hat das Ziel, die Auftraggeber (Bauherren, Bauträger) zu einem ökologischen Umdenken und Optimieren zu führen. Es ist eine vollkommen freiwillige Vorgehensweise, um die Ökologie in die Logik der Beteiligten einzubeziehen.

Die HQE-Methode wird seit 1994 bei Bauprojekten (Wohnbau und sonstigem Gebäudebau) getestet und wurde 1997 endgültig von der HQE-Association strukturiert. Das Schema hat 2 Dimensionen:

1. Das ökologische Management von Bauprojekten
2. Den nachhaltigen Gebäudeentwurf, der über 14 betroffene Aspekte bewertet wird von „Grundstück und Konstruktion", „Management" (operative Phase) bis hin zu „Komfort" und „Gesundheit" der Nutzer.

Kenntnisse über die ökologischen Auswirkungen der Bauprodukte basieren auf einer EPD (Environmental Product Declaration) und die Auswahl wird mit in die ökologischen Kriterien aufgenommen.Das Gleiche wird bei gesundheitlichen Aspekten (z.B. Emissionen und Innenluft) erforderlich. HQE deckt drei Phasen ab: Auftrag, Entwurf und Ausführung. Die Audits werden am Ende der drei Phasen durchgeführt. Im Moment sind 25 Projekte zertifiziert, davon 17 private und 8 öffentliche Gebäude.

LEED (USA)

LEED ist das Bewertungsverfahren, das vom US-GBC entwickelt wurde, um die Nachhaltigkeit beim Gebäudeentwurf abzuschätzen und die Ziele der Nachhaltigkeit einzubeziehen. Es ist ein freiwilliger nationaler Standard, basierend auf einer allgemeinen Abstimmung. LEED zielt darauf ab, Gebäude „gesünder" und ressourcenwirksamer zu entwickeln und zu betreiben.

Mitglieder des US-GBC, die alle Bereiche der US-amerikanischen Bauindustrie repräsentieren, entwickelten LEED auf Basis von BREEAM und fahren kontinuierlich mit der Weiterentwicklung fort. Die Ziele von LEED sind, mit einem herkömmlichen standardisierten Bewertungssystem die Nachhaltigkeit zu definieren, für integrierte, ganzheitliche Entwurfspraktiken zu werben, die ökologische Führungsrolle der Bauindustrie widerzuspiegeln, Wettbewerb im nachhaltigen Bauen anzuregen, das Bewusstsein der Konsumenten in Bezug auf den Nutzen nachhaltiger Gebäude zu erhöhen und um den Markt Richtung Nachhaltigkeit zu verändern. Das LEED-System kann auf drei verschiedene Arten genutzt werden, um die Nachhaltigkeit eines Gebäudeentwurfs zu verbessern:

Chapter programs

USGBC has over 72 local organizing groups and chapters throughout the U.S. that provide local green building resources, education, and leadership opportunities. Local chapter members can connect with green building experts in their area, develop local green building strategies, and tour green building projects.

The rating system is structured according to building typology:

- The LEED for New Construction rating system was first released in 2000.
- LEED for Commercial Interiors and Existing Buildings became available in 2004 to address the tenant market and the operations and maintenance of existing buildings.
- LEED for Core & Shell became available in July 2006 for spec developments.
- LEED for Homes was launched in December 2007.

And, as the last addition to the family of rating and certification programs:

- LEED for Neighborhood Development, Retail, and Healthcare is currently in a pilot test.

LEED® principles and the rating systems can be implemented in all 50 states of the U.S. , but the majority of certified projects are in California. As of February 2008, a total of 9,867 projects were registered, including 6,223 new structures, 926 existing buildings, and 1,272 core and shell buildings; 1,283 buildings were certified, 952 of those being new construction.

Summary of international rating systems

With regard to the existing rating systems, especially the U.S. LEED systems, the following can be said.

- Concentration: With the LEED rating system, the U.S. Green Building Council was successful in channeling the efforts of the American green building movement and bundling it into one significant sustainable-building initiative.
- Simplicity: Internationally successful rating systems ought to be simple in their use, and the results of a building evaluation need to be communicated in a clear way.
- Business models: The activities of the not-for-profit U.S. Green Building Council has provided a basis for a great variety of related business models for players in the sustainable building industry.
- Demand: Exemplary, high-profile projects that have received certification have created a strong and growing market demand for certification of all projects.

- LEED kann als Entwurfsleitfaden für das Planungsteam gelten, um ökologische Kriterien in den Gebäudeentwurf einzubeziehen.
- LEED-Bewertungsberichte sind Mittel, den Kunden und anderen Interessierten zu zeigen, dass in den Entwurf ökologische Kriterien einbezogen wurden.
- Ein Gebäudeentwurf kann vom amerikanischen oder kanadischen Green Building Council zertifiziert werden.

Die LEED-Richtlinien wurden in allen 50 Staaten der USA übernommen (am häufigsten in Kalifornien angewandt). Seit dem Beginn im Jahr 2000 wurden mehr als 1400 Projekte mit verschiedenen Beurteilungsstufen registriert. Bis jetzt wird der Markt von Verwaltungs- und Regierungsbauten und dem gemeinnützigen Sektor dominiert (ca. 75 Prozent aller LEED registrierten Projekte).

Am häufigsten werden Bürogebäude, Schulen und Universitätsgebäude mit LEED bewertet. Das Bewertungsschema erlaubt die Einteilung in „zertifiziert", „Silber", „Gold" oder „Platin-Auszeichnung".

Erfolgsfaktoren der Bewertungssysteme

Zusammengefasst lassen sich aus den existierenden Bewertungssystemen, allen voran LEED in den USA, folgende Erfolgsfaktoren ableiten:

- Kanalisieren: Der US-GBC hat die Green-Building-Bewegung in den USA mit Hilfe ihres im Markt etablierten Bewertungsinstruments LEED zusammengeführt und die Aktivitäten kanalisiert.
- Einfachheit: Die erfolgreichen Bewertungssysteme sind einfach anzuwenden und die Ergebnisse der Bewertung sind einfach zu kommunizieren.
- Geschäftsmodelle: Um die Aktivitäten der non-profit Organisation US-GBC ist ein Markt entstanden, der verschiedenste Geschäftsmodelle für Akteure im Baubereich eröffnet hat.
- Nachfrage: Leuchtturmprojekte haben die Nachfrage nach einer Zertifizierung mittels der Bewertungsinstrumente stark stimuliert.

Die Richtlinien umfassen unter anderem auch Energiezertifizierungsprogramme für alle Gebäude, die neu gebaut, vermietet oder verkauft werden oder öffentlich zugänglich sind. Hierdurch erhalten potenzielle Käufer Informationen, um den baulichen und energetischen Zustand eines Gebäudes zu bewerten, einschließlich zu erwartender Kosten und eventuell nötiger Renovierungen und Modernisierungen.

Tabelle 6 gibt abschließend eine Übersicht typischer Energieeinsparstandards Mitteleuropas auszugsweise wieder.

There are also energy certifications available for new construction and for buildings that are on the market for sale or lease. Potential buyers or users now have access to information about a building with respect to its energy consumption or the condition of its construction. The expected cost for energy consumption of real estate, and a building's need for renovation and modernization can also be evaluated prior to signing a contract.

Table 6 shows an excerpt of energy saving standards currently available in Middle Europe.

Table 6
Various energy saving standards in European countries

Source: Energieatlas, Nachhaltige Architektur, Hegger, Fuchs, Stark, Zeumer, Birkhäuser Verlag, 2007

Standard		Reference	Limits	Notes
EnEV Residential	D	Primary energy demand for heating, ventilation, potable water	depending on compactness	Minimum required by German law
EnEV Non-Residential	D	Primary energy demand for heating, ventilation, potable wate lighting	In reference to base case buildings	Minimum required by German law
KfW 60 Building	D	Primary energy value according EnEV	max. 60 kWh/m²a	Required to secure funding
KfW 40 Building	D	Primary energy value according EnEV	max. 40 kWh/m²a	Required to secure funding
Minergie Building	CH	Weighted energy index (final energy): Heating, ventilation, potable wate air-conditioning	max. 42 kWh/m²a	Additional requirements, i.e. skin of the building, mechanical ventilation, investment cost
Minergie Plus Building	CH	Weighted energy index (final energy): Heating, ventilation, potable wate air-conditioning	25–30 kWh/m²a	Additional requirements, i.e. air-tightness, installed heat power, heating demand, electricity demand
Klimahaus	I	Heat energy demand	max. 50 kWh/m²a	Klimahaus A: max. 30 kWh/m²a Klimahaus Gold: max. 10 kWh/m²a
HQE	F	Weighted energy index (primary energy), heat energy	max. 50 kWh/m²a	Various additional requirements, case-by-case evaluation
Passivhaus	D	heat energy demand	max. 15 kWh/m²a	Secondary requirement: Primary energy for heating, potable water and electricity max. 120 kWh/m²a

Tabelle 6
Übersicht verschiedener Energiespar-Strandards in Mitteleuropa

Quelle: Energieatlas, Nachhaltige Architektur Hegger, Fuchs, Stark, Zeumer Birkhäuser Verlag, 2007

Standard		Bilanzierungsebene	Grenzwerte	Bemerkungen
EnEV Wohngebäude	D	Primärenergiebedarf für Heizung, Lüftung und Trinkwarmwasser	in Abhängigkeit von der Kompaktheit	gesetzliche Mindestanforderung in Deutschland
EnEV Nichtwohngebäude	D	Primärenergiebedarf für Heizung, Lüftung, Trinkwasser, Kälte, Beleuchtung	in Abhängigkeit von Referenzgebäuden	gesetzliche Mindestanforderung in Deutschland
KfW 60 Haus	D	Primärenergiekennwert nach EnEV	max. 60 kWh/m²a	Nachweis für Fördermittel
KfW 40 Haus	D	Primärenergiekennwert nach EnEV	max. 40 kWh/m²a	Nachweis für Fördermittel
Minergiehaus	CH	gewichtete Energiekennzahl (Endenergie): Heizung, Lüftung, Trinkwarmwasser, Klimatisierung	max. 42 kWh/m²a	weitere Anforderungen: z. B. Gebäudehülle, mechanische Lüftung, Kosten
Minergie Plus Haus	CH	gewichtete Energiekennzahl (Endenergie): Heizung, Lüftung, Trinkwarmwasser, Klimatisierung	25–30 kWh m²a	weitere Anforderungen: Luftdichte, installierte Wärmeleistung, Heizwärmebedarf, Strombedarf
Klimahaus	I	Heizenergiebedarf	max. 50 kWh/m²a	Klimahaus A: max. 30 kWh/m²a Klimahaus Gold: max. 10 kWh/m²a
HQE	F	gewichtete Energiekennzahl (Primärenergie) Heizenergiebedarf	max. 50 kWh/m²a	diverse Nebenforderungen gemäß Einzeldarstellung
Passivhaus	D	Heizwärmebedarf	max. 15 kWh/m²a	Nebenanforderung: Primärenergiebedarf für Heizung, Trinkwarmwasser und Haushaltsstrom max. 120 kWh/m²a

9.1. Classification of building construction according to energy savings (Energy Savings Regulation for Buildings, or Energieeinsparverordnung für Gebäude (EnEV), Germany)

The Energy Savings Regulation (EnEV) is the major regulatory instrument of the government to lower energy consumption in the building sector in Germany. With its help, sustainable living conditions and a reduction of carbon dioxide emissions are being strengthened. The EnEV, which is based on the Energy Savings Law of 1976, ties the diverse German regulations concerning insulation, heating systems, and energy consumption in general together in one instrument.

The new holistic method of combining different measures is supposed to support integrative design, in which building systems are truly conceived early and together with the architectural design. By fostering high standards for detailing and detail execution on the construction site, it promotes high quality in the built environment. In addition, the law is aimed at providing high efficiencies of the integrated building systems.

EnEV compliance is mandatory for all new construction, with a distinction being made between buildings with normal or low interior temperatures. **Figure 111** shows, in the example of a low-energy building, exemplary deviations based on user habits and interior temperatures.

9.1 Klassifizierung energiesparender Häuser (EnEV, Deutschland)

Die Energieeinsparverordnung (EnEV) in Deutschland stellt die wesentliche ordnungspolitische Komponente zur Minderung des Energieverbrauchs im Gebäudebereich dar. Hiermit soll ein wichtiger Beitrag zur Reduzierung der Schadstoffemissionen und zur Daseinsvorsorge geleistet werden. Auf der Grundlage der Energieeinspargesetze aus dem Jahr 1976 werden die bisher diesen Bereich behandelnden Wärmeschutz- und Heizungsanlagenverordnungen in Deutschland zu einer Verordnung, der EnEV, zusammengefasst.

Die ganzheitliche Betrachtung und Einbeziehung der Anlagen- und Bautechnik soll die erforderliche, integrative energetische Planung fördern. Durch eine gute Detailplanung und gute Detailausführung soll eine verbesserte Qualität der Baukonstruktion erreicht werden. Darüber hinaus wird in dem Nachweisverfahren der Verordnung die Effizienz einer guten Gebäudeanlagentechnik deutlich herausgestellt.

Der Gültigkeitsbereich der EnEV beinhaltet neu zu errichtende Gebäude, wobei zwischen Gebäuden mit normalen und niedrigen Innentemperaturen unterschieden wird. **Bild 111** zeigt beispielhaft Abweichungen bei einem Niedrigenergiegebäude bei verschiedenen Nutzerverhalten und Innentemperaturen.

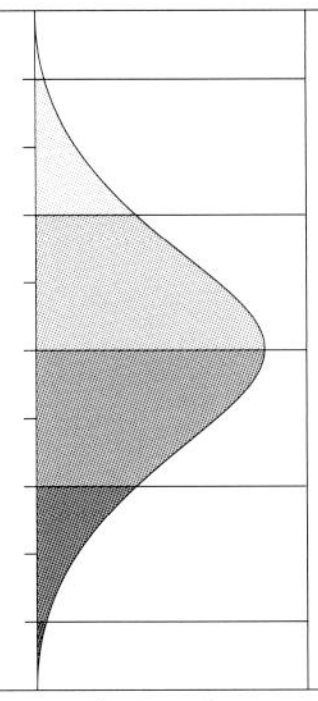

User behavior	Interior temperature	Energy demand for hot water	Air changes	Deviation from EnEV[1]	Electricity demand incl. lighting	Heat energy from appliances
Extreme Savers 10 % Quantil	17.0 °C	5.0 kW/m²a	0.4 /h	14.5 %	10 kWh/m²a	2.3 kWh/m²a
Savers 30 % Quantil	18.5 °C	10.1 kW/m²a	0.6 /h	2.5 %	15 kWh/m²a	3.4 kWh/m²a
Average 50 % Quantil	19.5 °C	15.1 kW/m²a	1.1 /h	14.2 %	20 kWh/m²a	4.5 kWh/m²a
Squanderer 70 % Quantil	21.0 °C	25.2 kW/m²a	1.5 /h	34.7 %	30 kWh/m²a	6.8 kWh/m²a
Extreme Squanderer 90 % Quantil	23. °C	35.3 kW/m²a	2.0 /h	59.4 %	40 kWh/m²a	9.0 kWh/m²a

[1] Deviations of a low energy house (A_n = 363.52 m²) with an annual primary energy demand of 63.1 kWh / m²a

Figure 111
Affect of user habits on energy consumption in residential buildings

Source: Energieatlas, Nachhaltige Architektur, Hegger, Fuchs, Stark, Zeumer, Birkhäuser Verlag, 2007

Bild 111
Einfluss des Nutzerverhaltens auf den Energiebedarf von Wohngebäuden

Quelle: Energieatlas, Nachhaltige Architektur Hegger, Fuchs, Stark, Zeumer Birkhäuser Verlag, 2007

According to the EnEV, the primary energy demand of non-residential and residential buildings is defined as a function of their surface-to-volume ratio (S/V ratio), and compliance with this value is required. **Figures 112** and **113** show the limited primary energy demand in kWh/m²a for nonresidential and residential construction. As a special requirement for residential buildings, the EnEV defines the limits for heat energy used for warm water generation, again as a function of the S/V ratio.

On the basis of the maximum allowable annual primary energy consumption for a specific building project, the regulation now allows the definition of the maximum allowable heat energy, and as a result the definition of U-values, or the measure of a material's or material assembly's heat-conducting properties, for all enclosing building elements such as exterior walls, windows, roofs, slabs-on-grade, etc.

Gemäß EnEV wird der Primärenergiebedarf nach Nicht-Wohngebäuden und Wohngebäuden in Abhängigkeit des Oberflächen-/Volumenverhältnisses (A/V-Verhältnis) beschrieben, der jeweils entsprechend einzuhalten ist. Die **Bilder 112** und **113** zeigen den limitierten Primärenergiebedarf in kWh/m²a (Nicht-Wohngebäude) bzw. kWh/m²a (Wohngebäude). Bei den Wohngebäuden zusätzlich definiert ist die Begrenzung der Heizleistung für die Aufbereitung von Warmwasser wiederum in Abhängigkeit des A/V-Verhältnisses.

Auf der Basis des maximal erlaubten Jahres-Wärmeenergiebedarfs (Primärenergiebedarfs) ermitteln sich unter Berücksichtigung der Primärenergiefaktoren die maximal erlaubten Heizleistungen und wiederum hieraus resultierend die notwendigen Wärmedurchgangskoeffizienten für Außenwände, Fenster, Dächer, bodenberührende Flächen usw..

Primary energy demand
(Thermal energy)kWh/m²a

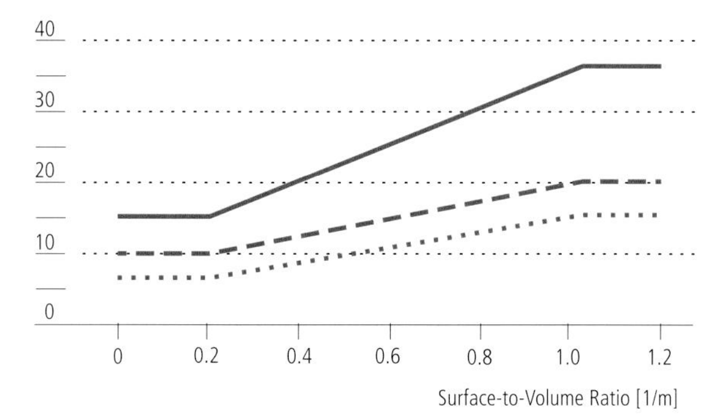

Primary energy demand Q_p
Plant Installation Factor e_p
Annual heat demand Q_H
Annual hot-water demand Q_W

$Q_p = e_p{}^* (Q_H + Q_W)$

*) dependent on primary energy source, area of building in m² and plant design (thermal energy generation)

— EnEV 2007
-- EnEV 2009
··· Electric energy

Primary energy demand (Thermal, distribution and transportation energy)in kWh/m²a

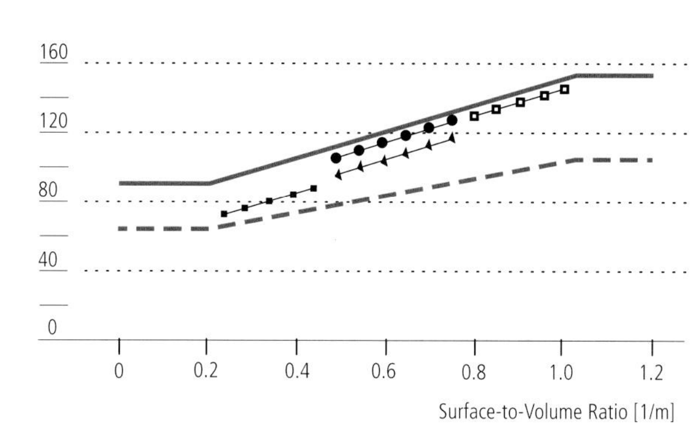

Primary energy demand hot-water
- ▪ Large-scale multi-unit residential
- ◂ Multi-unit residential and large inner and end townhouse units
- ● Small inner and end townhouse units
- ▫ Freestanding single-family residential

— EnEV 2007
-- EnEV 2009

Figure 112
Primary thermal energy demand for **non-residential buildings** as a function of Surface-to-Volume Ratio according DIN 4701-10

Source: EnEV (2007/2009)

Bild 112
Anforderungsgröße Primärenergiebedarf" für **Nichtwohngebäude**, in Abhängigkeit vom A/V_e-Verhältnis (nach DIN 4701-10)

Quelle: EnEV (2007/2009)

Figure 113
Primary thermal energy demand for hot-water generation with various technologies for **residential buildings** as a function of Surface-to-Volume Ratio according DIN 4701-10

Source: EnEV (2007/2009)

Bild 113
Anforderungsgröße "Primärenergiebedarf" für **Wohngebäude** mit unterschiedlicher Warmwasserbereitung in Abhängigkeit vom A/V_e-Verhältnis (nach DIN 4701-10)

Quelle: EnEV (2007/2009)

The calculation of the annual primary energy use according to the German norm DIN V 4701-10) takes into consideration (1) the losses of the installed building systems, and (2) the energy gains from the surrounding environment. As a result, the norm describes the efficiency of the selected building technology with so-called "Aufwandzahlen" (effort factors) – i.e., a cost-benefit ratio. With the help of the primary energy factors, as shown in **Table 9**, and with respect to the individual technical system and the energy source used, an effort factor is calculated. If this factor is multiplied by the sum of heat energy used for heating and warm water generation, the result is the amount of yearly primary energy consumption. **Figure 114** explains these conditions graphically.

Die Berechnung des Jahres-Primärenergiebedarfs (gemäß DIN V 4701-10) berücksichtigt die Verluste der Anlagentechnik und Wärmegewinne aus der Umwelt. Hieraus erfolgt eine Beschreibung der energetischen Effizienz des Gesamtanlagensystems über Aufwandzahlen, das Verhältnis von Aufwand zu Nutzen. Unter Berücksichtigung von Primärenergiefaktoren, wie z.B. in **Tabelle 9** dargestellt, wird je nach Anlagentechnik und eingesetztem Energieträger eine Anlagenaufwandszahl gebildet. Die Multiplikation der Anlagenaufwandszahl mit der Summe aus Heizwärme- und Warmwasserwärmebedarf ergibt den Jahres-Primärenergiebedarf. **Bild 114** beschreibt nochmals die Zusammenhänge.

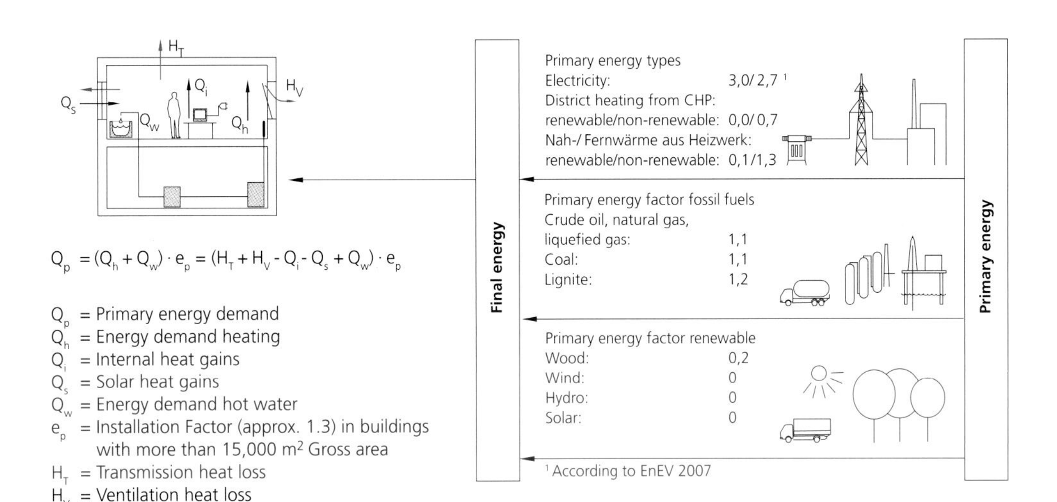

Figure 114
Assessment of primary and final energy and depiction of an Energy Pass with target values

Source: Energieatlas, Nachhaltige Architektur, Hegger, Fuchs, Stark, Zeumer, Birkhäuser Verlag, 2007

Bild 114
Bewertung der Primär- und Endenergie sowie Darstellung einer Auswertung (Energiepass) mit anzustrebenden Werten

Quelle: Energieatlas, Nachhaltige Architektur, Hegger, Fuchs, Stark, Zeumer Birkhäuser Verlag, 2007

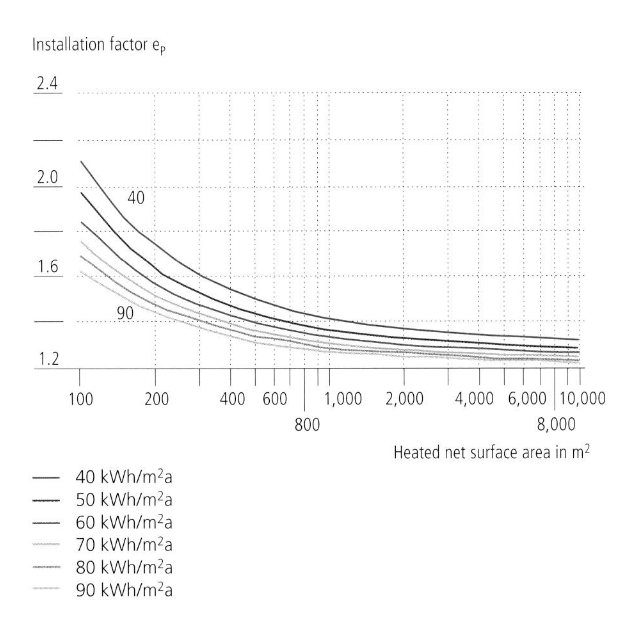

to Figure 114
Examples of various Installation Factors e_P as a function of heated net surface area and annual heat demand q_h.

Source: Taschenbuch für Heizungs- und Klimatechnik, Recknagel, Sprenger, Schramek Oldenbourg, 2007

zu Bild 114
Beispielhafte Auftragung der Anlagen-Aufwandszahl e_P in Abhängigkeit der beheizten Gebäudenutzfläche und dem Jahres-Heizwärmebedarf q_h.

Quelle: Taschenbuch für Heizungs- und Klimatechnik, Recknagel, Sprenger, Schramek Oldenbourg, 2007

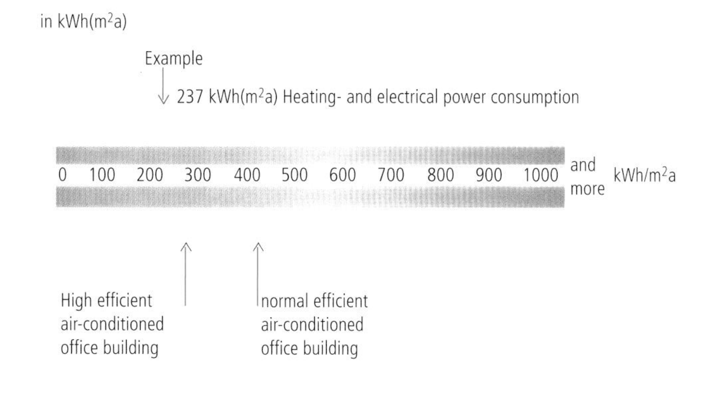

to Figure 114
Example: the German energy passport – for an office building

Source: dena,
Deutsche Energie Agentur

zu Bild 114
Beispiel: der deutsche „Energiepass" – für ein Verwaltungsgebäude

Quelle: dena,
Deutsche Energie Agentur

to Figure 114
Detailed analysis of energy consumption for heating, cooling, ventilation, lighting and hot water generation in buildings.

Source: energy passport for an office building, example. (Year of construction: 2006)

zu Bild 114
Detailanalyse über den unterschiedlichen Energiebedarf für Heizung, Kühlung, Lüftung, Licht und Warmwasser

Quelle: Energiepass eines Verwaltungsgebäudes, beispielhaft (erbaut 2006)

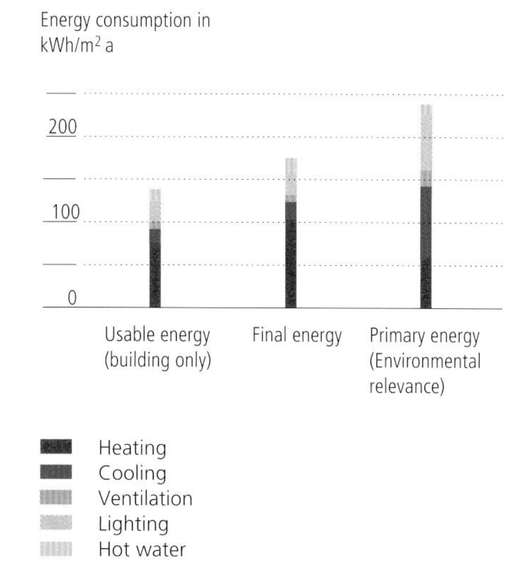

Kreditanstalt für Wiederaufbau (KfW), Germany KfW Development Bank

The KfW Banking Consortium is a financial institution of the German Federal Government that provides loans for new construction and energy-saving measures for existing construction; long-term, low-interest financing for the construction of new KfW Energy-Saving Houses 40 (buildings with a minimum of 40 % energy savings) and 60 (with a minimum of 60 % energy savings) and so-called passive houses; and financing for the installation of heating systems based on renewable energy sources or heating technologies that lead to considerable savings in new buildings.

Minergie® and Minergie-P® Buildings

Minergie® of Switzerland is a voluntary standard for the building industry that is designed to promote reduced energy consumption, to reduce emissions, to increase the use of renewable energy sources, and to improve the overall quality of life. The Minergie® standard accomplishes these goals by defining threshold values for energy consumption. Various methods to achieve the standard are feasible, but the building has to be seen as an integrated system – an approach that is very similar to the German EnEV with its requirements for building enclosure and building systems. Minergie® requires that the heating, ventilation, and warm water supply systems incorporate intelligent configurations that allow for the greatest reduction in energy consumption.

KfW-60-/40-Haus, Deutschland

Die Kreditanstalt für Wiederaufbau ist ein bundeseigenes Bankeninstitut, das Fördermittel bei Neubauten und Sanierungen vergibt. Die Ausgabe von Fördermitteln ist jedoch daran gebunden, dass Wärmeenergieverbräuche von entweder maximal 60 kWh/m²a bzw. 40 kWh/m²a nicht überschritten werden. Hieraus resultiert die zuvor angegebene Kurzbezeichnung.

Minergie- und Minergie-P-Haus

Der Minergiestandard der Schweiz ist ein freiwilliger Baustandard, der den rationellen Energieeinsatz und die breite Nutzung erneuerbarer Energien bei gleichzeitiger Verbesserung der Lebensqualität und Senkung der Umweltbelastung ermöglicht. Bei Minergiegebäuden wird das Ziel, Grenzwerte im Energieverbrauch zu erreichen, definiert. Dabei ist es möglich, verschiedene Wege zu gehen, wobei besonders wichtig ist, dass das ganze Gebäude als integrales System betrachtet wird – somit ähnlich der EnEV in Bezug auf die Gebäudehülle und die Gebäudetechnik. Bei der Heizung, Lüftung und Warmwasseraufbereitung sind sinnvolle Kombinationen gefragt, um mit minimalstem Heizenergieverbrauch auszukommen.

The Minergie-P® standard describes and qualifies those buildings that use even less energy than Minergie® requires. Quality requirements with regard to comfort and economy are defined as well. Thus, a building that strives for classification according to the Minergie-P® standard has to be designed from the very first stages and in all of its building parts for optimization of all operating processes.

Additional insulation layers will not suffice to reach the stringent standards – what is required is an overall concept for system optimization and ease of use of all elements involved. **Figure 115** shows the differences between Minergie® and Minergie-P®.

Der Standard Minergie-P® bezeichnet und qualifiziert Objekte, die einen noch tieferen Energieverbrauch als Minergiegebäude anstreben. Analog zum Minergie stellt auch Minergie-P® hohe Anforderungen an das Komfortangebot und die Wirtschaftlichkeit. Ein Haus, das den sehr strengen Anforderungen von Minergie-P® genügen soll, ist als Gesamtsystem und in all seinen Teilen konsequent auf dieses Ziel hin zu planen und zu bauen und zudem im Betrieb zu optimieren.

Dabei genügen nicht allein zusätzliche Wärmedämmschichten, sondern eine einfache Bedienbarkeit des Gebäudes und der technischen Einrichtungen ist zwingend erforderlich. **Bild 115** stellt den Unterschied zwischen einem Minergie® und einem Minergie-P®-Gebäude dar.

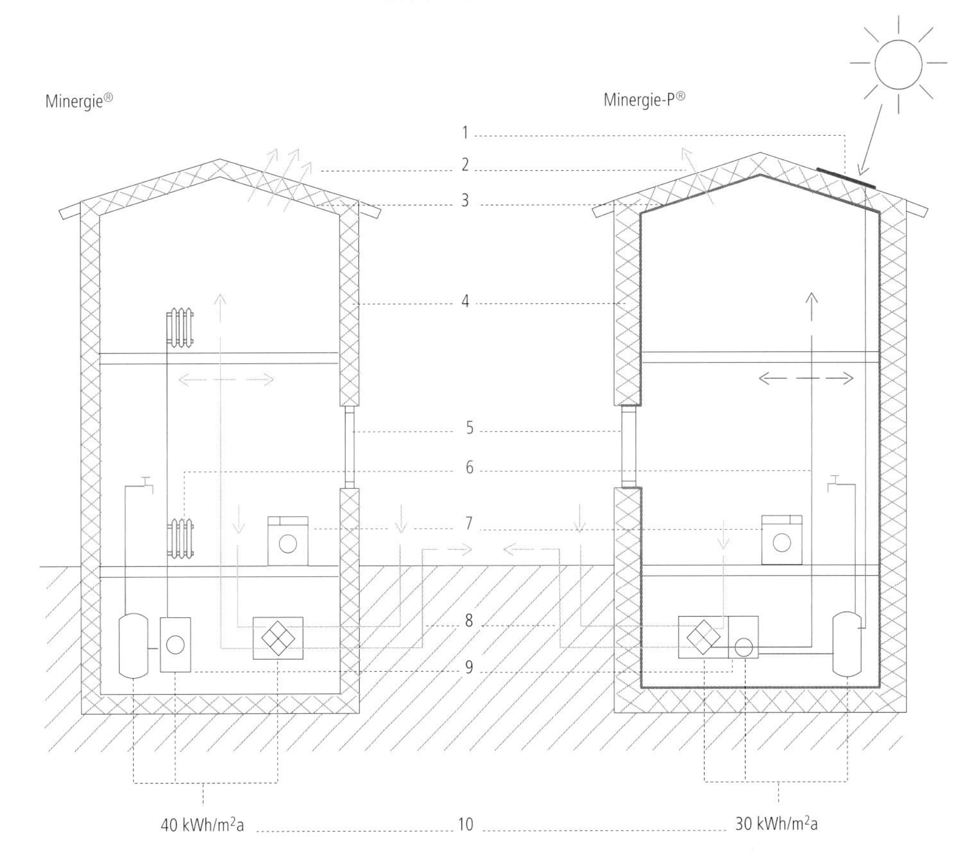

	Minergie®	Minergie-P®
1 Renewable energies	Recommended	Required
2 Heat demand	80% SIA Maximum	20% SIA Maximum
3 Air tightness	Good	Tested
4 Insulation	15-20 cm	20-35 cm
5 Thermal insulation glazing	double insulated	triple insulated
6 Heat distribution	typical	forced air possible
7 A*-Appliances	recommended	required
8 Controlled ventilation	required	required
9 Thermal power demand	no requirements	max. 10 W/m²
10 Energy power demand	40 kWh/m²a**	30 kWh/m²a

Figure 115
Differences between the Swiss energy standards Minergie® and Minergie-P®

* Low-energy-appliance rating
** The value of 40 kWh/m²a entails the energy demand for heating as well as the electrical demand. The primary energy demand for electricity x 0.5 is equal to the final energy demand for ventilation, cooling and water transport.

Source: Schweizer Baudokumentation, Switzerland

Bild 115
Der Unterschied zwischen Minergie® und Minergie-P®

Quelle: Schweizer Baudokumentation, Schweiz

Minergie-ECO®

This is a supplement standard to Minergie®. In addition to energy savings, it supports the use of sustainable building materials and healthy indoor environments. All other requirements with regard to energy efficiency and comfort of a Minergie® building need to be complied with as well.

Figure 116 explains the additional requirements for a Minergie-ECO® building compared to the regular Minergie® standards.

Minergie-ECO®

Der Standard Minergie-ECO® ist eine Ergänzung zum Minergiestandard. Während die Merkmale wie Komfort und Energieeffizienz in einem Minergiegebäude zu berücksichtigen sind, erfüllen zertifizierte Bauten nach Minergie-ECO® zudem Anforderungen an ein gesundes und ökologisch richtiges Bauen. Voraussetzung für eine Zertifizierung nach Minergie-ECO® ist eine konsequente Bauweise nach Minergie®- bzw. Minergie-P®-Standards.

Bild 116 zeigt die ergänzenden Ansprüche eines Minergie-ECO®-Hauses zum Minergiestandard.

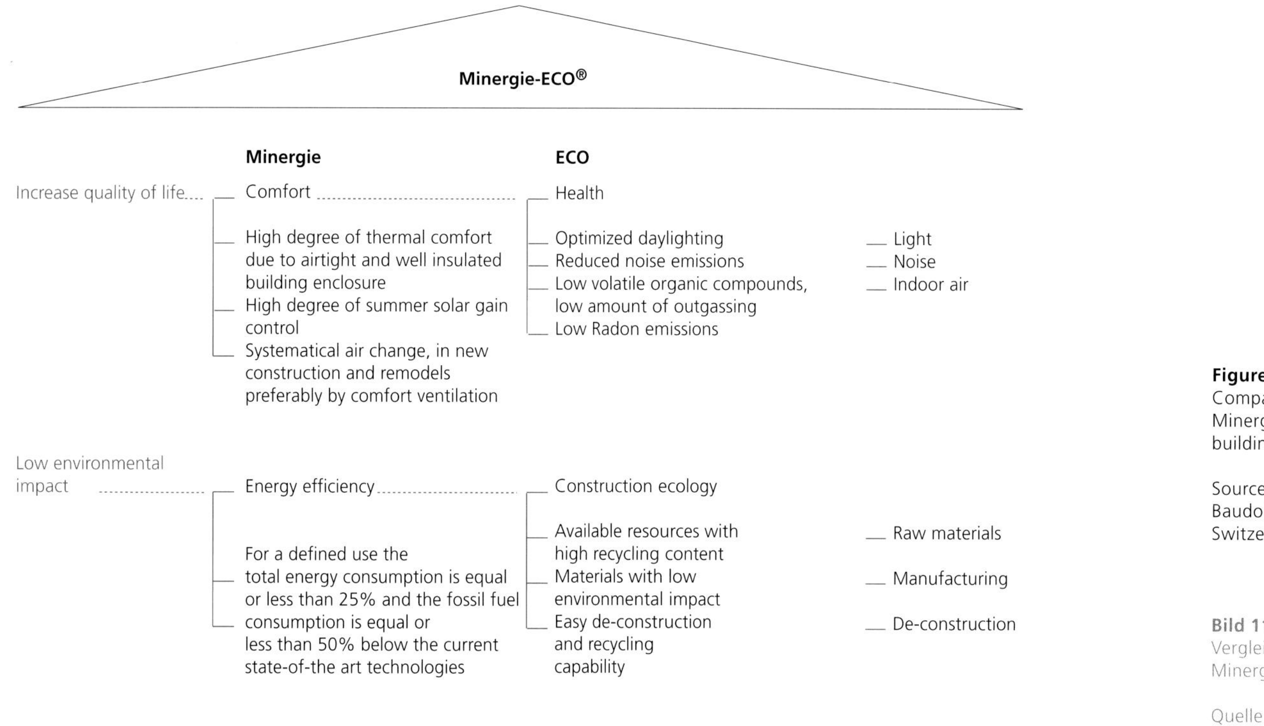

Figure 116
Comparison between Swiss Minergie® and Minergie-P® buildings

Source: Schweizer Baudokumentation, Switzerland

Bild 116
Vergleich Minergie®- zu Minergie ECO®-Häusern

Quelle: Schweizer Baudokumentation

Passivhaus, Germany

The basis for the development of the Passivhaus was laid in 1976 by Professors Waine Schick of the University of Illinois' Small Homes Council and S. "Bud" Konzo, a professor of mechanical engineering at the same university, who first published research findings and suggestions for super-insulated residential units that they called Lo-Cal houses. A first such highly insulated home was designed and built by the architect Michael McCulley in Urbana, IL.

The term Passivhaus (passive house) refers to the rigorous, voluntary, Passivhaus standard for energy use in buildings. It results in ultra-low-energy buildings that require little energy for space heating, similar to the above-described Minergie-P® of Switzerland. The first Passivhaus buildings were built in Darmstadt, Germany in 1990, and in September 1996 the Passivhaus Institute was founded in Darmstadt to promote and control the standard. Since then, more than 6,000 Passivhaus buildings have been constructed in Europe, mostly in Germany and Austria, with others being built in various countries worldwide. In North Illinois in 2003, [5], and the first to be certified was built at Waldsee, Minnesota in 2006.

Table 7 lists the differences between the Minergie®, Minergie-P®, and Passivhaus standards.

In order to successfully approach the design of a building that complies with Passivhaus standards, a so-called Passivhaus-Project-Package (PHPP 99) needs to be complied with. This standard establishes maximum values for heating energy and additional limits for primary energy use of all other building systems. As described in **Table 8**, various primary energy factors of the energy sources used in a Passivhaus need to be used to calculate the primary energy demand of the building. In addition to the primary energy factors, the table also lists carbon dioxide equivalent (CDE), the measure that describes how much global warming a given type and amount of greenhouse gas may cause, using the functionally equivalent amount or concentration of carbon dioxide (CO_2) as the reference.

Passivhaus, Deutschland

Die Grundlagen zur Entwicklung von „Passivhäusern" wurden 1976 an der Universität von Illinois gelegt (Prof. Waine Schick und S. „Bud" Konzo, „super-insulated Lo Cal") und, ein erstes Objekt wurde von Prof. Michael McCulley in Urbana, Illinois, gebaut.

Der Passivhausstandard setzt einen Grenzwert für den Energiekennwert Heizwärme und einen zusätzlichen Grenzwert für den Primärenergiebedarf sämtlicher Anwendungen fest. **Tabelle 7** beschreibt den Unterschied zwischen Minergie®-standard, Minergie-P® und Passivhausstandard.

Bei der Planung eines Hauses mit Passivhausstandard muss das Passivhaus-Projektierungs-Paket, PHPP 99, angewendet werden. Zusätzlich setzt der Passivhausstandard einen Grenzwert für den Energiekennwert Heizwärme und einen zusätzlichen Grenzwert für den Primärenergiebedarf sämtlicher Anwendungen fest. Zur Berechnung des Primärenergiebedarfs werden die Primärenergiefaktoren der verschiedenen Energieträger im Passivhaus-Projektierungs-Paket gemäß **Tabelle 8** angegeben. Diese zeigt neben den Primärenergiefaktoren auch die CO_2-Äquivalent-separat-Emissionsfaktoren.

	Minergie® Standard neue SIA 380/1	Minergie P®	Passivhaus-Standard
Reference category			
Reference area	EBF (Energy Reference area, height adjusted, (Gross, outside dimensions)	EBF (Energy Reference area, height adjusted, (Gross, outside dimensions)	Net square footage (NSF)
Volume	EBF_0 x average floor height x 0.8		Outside dimensions
Enclosure surface	Outside dimensions	Outside dimensions	Outside dimensions
Aussenhülle A/EBF			–
Window surface area	Clear opening	Clear opening	Shell opening
Energy reference values			
Heating demand, Qh Heating capacity	≤ 60% Hg	≤ 20% Hg	≤ 15 kWh/m² a NF
Heat Flow	No prerequisites	≤ 10 W/m² EBF	≤ 10 W/m² NF
Energy reference value, Heat weighted, (EW) (Final Energy)	New construction :	New construction and remodel :	
– Single family residential	≤ 42 W/m² a EBF	≤ 30 W/m² a EBF	–
– Multi-family residential	≤ 38 W/m² a EBF	≤ 30 W/m² a EBF	–
– Commercial buildings	≤ 40 W/m² a EBF	≤ 25 W/m² a EBF	–
Energy source	Weighting according to energy source	Weighting according to energy source	Not prescribed
Weighting	Final energy (Expected values)	Final energy (Expected values)	Primary energy ≤ 120 kWh/m² a NF
Electricity	2	2	2.97/2.72 [1]
Oil	1	1	1.09 [1]
Natural Gas	1	1	1.07 [1]
Wood	0.6 Ecological correction factor	0.6 Ecological correction factor	0.2 [1]
Mechanical systems			
Ventilation	Exhaust with controlled fresh air intake	Controlled ventilation with heat recovery	Controlled ventilation with heat recovery
Heat recovery			>75% (Counter flow heat exchanger)
Building envelope			
U value			
Opaque walls	Single building components ≤ 0.2 W/m²K including thermal bridging		≤ 0.15 W/m²K
Window			≤ 0.8 W/m²K (according EN 10077) Energy transmission value g ≤ 50%
Air tightness	Recommendation for airtight envelopes	$nL_{50} \leq 0.6\ h^{-1}$ air tightness measurement required	$nL_{50} \leq 0.6\ h^{-1}$ air tightness measurement required
Thermal bridging	Subject to calculation according SIA 180		No thermal bridges
Thermal bridging loss coefficient (ψ_a)	Thermal bridges need to be considered	Thermal bridges need to be considered	If $\psi_a \leq 0.01$ W/(mK), no requirement to proof
Evaluation			
Project evaluation	Calculation of project parameters		
Construction evaluation	Random samples	Air tightness measurement	air tightness measurement, quality control of entire mechanical system, control of thermal bridge elimination, control of insulation (elimination of air spaces)
Additional	Certified label, trade mark	Certified label, trade mark	Certificate, not protected

Table 7
Comparison between Swiss and German energy standards

1) Primary energy factor

	MINERGIE® Standard neue SIA 380/1	MINERGIE P®	Passivhaus Standard
Bezugsgrößen			
Bezugsfläche	EBF (Energiebezugsfläche) höhenkorrigiert (Brutto, Außenmaße)	EBF (Energiebezugsfläche) höhenkorrigiert (Brutto, Außenmaße)	Nettowohnfläche NF
Gebäudevolumen	EBF_0 x mittlere Geschosshöhe x 0.8		mit Außenmaßen
Hüllfläche	mit Aussenmaßen	mit Außenmaßen	mit Aussenmaßen
Aussenhülle A/EBF			–
Fensterflächen	lichte Maße	lichte Maße	Rohbaumaße
Energiekennzahlen			
Heizwärmebedarf (Qh) (Nutzenergie)	≤ 60% Hg	≤ 20% Hg	≤ 15 kWh/m² a NF
Heizwärmeleistung	keine Vorgaben	≤ 10 W/m² EBF	≤ 10 W/m² NF
Energiekennzahl Wärme gewichtet (EW) (Endenergie)	Neubau:	Neubau und Umbau:	
– EFH	≤ 42 W/m² a EBF	≤ 30 W/m² a EBF	–
– MFH	≤ 38 W/m² a EBF	≤ 30 W/m² a EBF	–
– Dienstleistungsbauten	≤ 40 W/m² a EBF	≤ 25 W/m² a EBF	–
Energieträger	Wertigkeit nach Energieträger (Gewichtung)	Wertigkeit nach Energieträger (Gewichtung)	nicht vorgegeben
Wertigkeit	Endenergie (voraussichtliche Werte)	Endenergie (voraussichtliche Werte)	Primärenergie ≤ 120 kWh/m² a NF
Elektrizität	2	2	2.97/2.72[1]
Öl	1	1	1.09[1]
Gas	1	1	1.07[1]
Holz	0.6 ökolog. Korrekturfaktor	0.6 ökolog. Korrekturfaktor	0.2[1]
Haustechnik			
Lüftung	Abluft mit kontrollierter Zuluft	kontrollierte Lüftung mit WRG	kontrollierte Lüftung mit WRG
Wärmerückgewinnung			> 75% (Gegenstrom-Wärmetauscher)
Gebäudehülle			
U-Wert			
Opake Hülle	Einzelbauteile ≤ 0.2 W/m²K inkl. Wärmebrücken		≤ 0.15 W/m²K
Fenster			≤ 0.8 W/m²K (nach EN 10077) Energiedurchlassgrad g ≤ 50%
Luftdichtigkeit	Empfehlung für möglichst luftdichte Gebäudehüllen	nL_{50} ≤ 0.6 h⁻¹ Luftdichtigkeitsmessung erforderlich	nL_{50} ≤ 0.6 h⁻¹ Luftdichtigkeitsmessung erforderlich
Wärmebrücken	Berücksichtigung in Berechnung gem. SIA 180		Wärmebrückenfreie Ausführung
Wärmebrückenvelustkoeffizient (ψ_a)	Wärmebrücken müssen berücksichtigt werden	Wärmebrücken müssen berücksichtigt werden	wenn $\psi_a \leq 0.01$ W/(mK), dann kein Nachweis
Kontrollen			
Projektierungskontrolle	Rechnerische Projektprüfung		
Ausführungskontrolle	Stichprobenprüfung	Luftdichtigkeitsmessung	Luftdichtigkeitsmessung Qualitätskontrolle über die gesamte Haustechnik Kontrolle Wärmebrückenvermeidung. Kontrolle Dämmung (Lufträume vermeiden).
Diverses	Bezeichnung unter Namensschutz: Label	Bezeichnung unter Namensschutz: Label	Zertifikat; Name nicht geschützt

Tabelle 7
Zusammenfassender Vergleich der verschiedenen Standards

1) Primärenergiefaktor

Energy source		Primary energy factor in kWh_{prim}/kWh_{End}	Carbon equivalent emissions factor in g/kWh_{End}
Fuels	Oil	1,08	293
	Natural gas	1,07	229
	Coal	1,07	396
	Wood	1,01	55
Electricity	Electricity mix	2,97	689
	Heat electricity	2,72	1018
District heat	Coal plant 70% CHP*	0,71	214
	Coal plant 35% CHP*	1,10	306
	Coal plant 0% CHP*	1,49	398
Cogeneration Power Plant, Natural Gas	Gas-fired 70% CHP	0,62	-84
	Gas-fired 35% CHP	1,03	113
	Gas-fired 0% CHP	1,43	311
Cogeneration Power Plant, Oil	Oil-fired 70% CHP	0,65	75
	Oil-fired 35% CHP	1,06	238
	Oil-fired 0% CHP	1,44	401

*Combined heat-power plant

Table 8
Primary energy factors and carbon equivalent emissions factors according to the Passivhaus Project Guideline 99

Source: EnEV 2007/9

Energieträger		Primärenergie-faktor in kWh_{prim}/kWh_{End}	CO_2-Äquivalent Emissions-faktor in g/kWh_{End}
Brennstoffe	Heizöl	1.08	293
	Erdgas	1.07	229
	Steinkohle	1.07	396
	Brennholz	1.01	55
Strom	Strom-Mix	2.97	689
	Heizstrom	2.72	1018
Fernwärme	StK HKW 70% KWK	0.71	214
	StK HKW 35% KWK	1.10	306
	StK HW 0% KWK	1.49	398
Blockheizkraftwerk Gas	Gas-BHKW 70% KWK	0.62	-84
	Gas-BHKW 35% KWK	1.03	113
	Gas-BHW 0% KWK	1.43	311
Blockheizkraftwerk Heizöl	Gas-BHKW 70% KWK	0.65	75
	Gas-BHKW 35% KWK	1.06	238
	Gas-BHW 0% KWK	1.44	401

Tabelle 8
Primärenergiefaktoren und CO_2-Äquivalent Emissionsfaktoren nach Passivhaus Projektierungspaket ´99 (Stk HKW = Steinkohleheizkraftwerk, KWK = Kraft-Wärme-Kopplung)

Quelle: EnEV (2007/9)

The term Passivhaus is not protected by a trademark, but the buildings designed and constructed according to the standard may be certified by the Passivhaus Institutein Darmstadt. In **Figures 117** and **118**, an example is used to illustrate the energy demand for heating under the different standards and the various building types. As shown in **Figure 118**, the heat energy demand is so similar that advantages of the Passivhaus in comparison with Minergie®-compliant structures cannot be detected. The differences between the two standards are in the way they calculate the reference surface area of the house. The Swiss Minergie® Standard uses the Swiss SIA 380/1 calculation to determine gross square footage or net usable space. But other parameters, such as internal heat loads, the climate, losses due to external shading, and the possible usage of all energy gains result in some differences between the two standards. The case study buildings in **Figure 117** were defined in such a way that they comply with the maximum Passivhaus energy consumption requirements of $Q_H < 15\ kWh/m^2$ a per net usable floor area.

Der Begriff des Passivhauses ist rechtlich nicht geschützt, jedoch können Passivhäuser durch das Passivhaus-Institut in Darmstadt zertifiziert werden. Um anhand eines Beispiels die gesamte Thematik nochmals zu verdeutlichen und die Unterschiede im Heizwärmebedarf bei verschiedenen Standards aufzuzeigen, sind in **Bild 117** und **118** verschiedene Gebäudetypen untersucht. Wie der ausgewiesene Heizwärmebedarf in **Bild 118** zeigt, liegen die Verbräuche eng beieinander, so dass ein wesentlicher Vorteil zugunsten des Passivhauses gegenüber dem Minergie®-standard nicht festzustellen ist. Die Unterschiede bei den Berechnungsverfahren des Passivhauses bzw. den Berechnungen gemäß SIA 380/1 betreffen zum einen die Flächen, auf welche die Werte bezogen werden (Bruttogeschossfläche oder Nettowohnfläche), aber auch die Randbedingungen wie z.B. die Größe der internen Wärmegewinne, das Klima, Verluste durch Beschattungen und die Ausnutzbarkeit der gesamten Wärmegewinne. Die Referenzgebäude gemäß **Bild 117** wurden so definiert, dass sie die Bedingungen für Passivhäuser, d.h. maximaler Heizwärmebedarf $Q_H < 15\ kWh/m^2a$, bezogen auf die Nettowohnfläche, erfüllen.

		Type I	Type II	Type III
		Multi-unit residential	Multi-unit residential	Four single-family residential units
Energy reference area	[m²]	943	943	1,045
Surface-to-volume ratio	[-]	1,21	1,70	1,98
Net usable space	[m²]	800	800	800
Window-to-wall ratio	[-]	20%	20%	20%
U-value wall	[W/(m²K)]	0,15	0,12	0,09
U-value window	[W/(m²K)]	0,8	0,8	0,8
g-value window	[W/(m²K)]	0,5	0,5	0,5
Ventilation	[m³/h]	700	700	630
Infiltration	[m³/h]	81	81	81
eta WRG*	[-]	0,75	0,75	0,75
Heat demand according Passivhaus Project Guideline '99	[kWh/(m²a)] (NWF)	13,4	15,6	14,9

Figure 117
Specification of the reference case building
* Eta: Coefficient of performance of heat recovery system

Source: Schweizer Baudokumentation, Switzerland

Bild 117
Definition der Referenzgebäude
* (Wirkungsgrad der Wärmerückgewinnung)

Quelle: Schweizer Baudokumentation

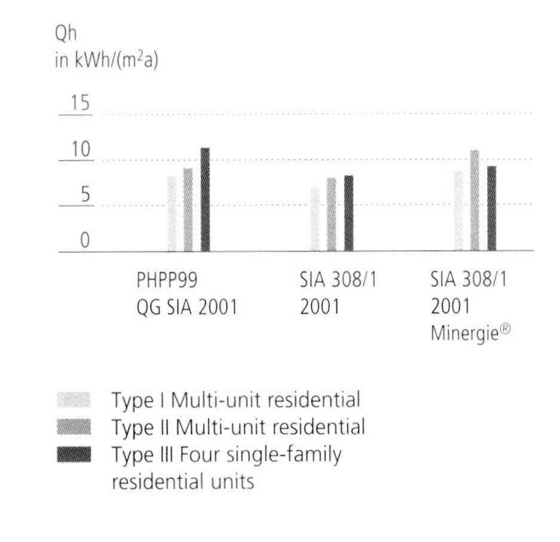

Figure 118
Heat demand in in kWh/(m²a) for various building types and German and Swiss energy standards.
SIA 380/1 "new" depicts values with mechanical ventilation and heat-recovery system. Differences are a result of varying parameters.

Explanation:
PHPP99: Passivhaus Project Guideline 99, Germany
Qg SIA: calculated including internal gains according SIA.

Bild 118
Heizwärmebedarf [in kWh/(m²a)] für verschiedene Gebäudetypen. SIA 380/1 neu stellen die Werte mit Lüftungsanlage und WRG dar. Die Unterschiede resultieren aus den unterschiedlichen Bezugsflächen und den unterschiedlichen Randbedingungen.

Erläuterungen:
PHPP99 = Passivhausprojektierungspaket 99,
Qg SIA 2001 = mit internen Gewinnen nach SIA gerechnet

9.2 The 2,000-Watt society

On average, humans consume approximately 17,500 kW/h annually and globally. This is the equivalent of an hourly power consumption of 2,000 W. Certainly, consumption in different countries varies enormously. In Western Europe, the hourly consumption is around 6,000 W per capita, and in the U.S. it is twice that amount. In many developing countries, and currently even in such rapidly emerging economies as China, the consumption is below 2,000 W, although it is increasing strongly.

The vision of a 2,000-watt society, originally developed by scientists at the Eidgenössische Technische Hochschule ETH (Federal Swiss Technology Institute, Zurich) to lower global energy consumption per person and day to 2,000 W can, according to the inventors, be achieved while maintaining comfort and standards of living, but it will require a rigorous increase in efficiency of all technical equipment and systems and a new understanding of energy service. Without greater material and energy efficiencies and a selective use of resources, the 2,000-watt society concept remains a mere declaration of intent.

It is important to remember that today the lion's share of the immense amount of primary energy consumed on Earth is wasted and has no direct use. This share, in excess of two-thirds of the entire energy used, is forever lost in energy conversion and energy transmission. The provision of final energy is intrinsically connected to high losses. Similarly wasteful is the operation of buildings, technically systems, and means of transportation, such as cars, buses, etc. As a result, only a third of the original primary energy ever reaches the end consumer.

The development of a 2,000-watt society also clearly involves "sustainable urban planning". Greater material and energy efficiency, gradual replacement of fossil by renewable primary energy sources, and, finally, an entrepreneurial adjustment that includes new professional processes will result in a post-fossil-fuel urban space worth living in.

The building projects presented in **Chapter 10** demonstrate how energy consumption in the built environment can be significantly reduced. The chapter will also introduce buildings in which internal heating loads due to occupants and equipment can be used for cooling.

9.2 2.000-Watt-Gesellschaft

Im globalen Mittel verbraucht der Mensch pro Jahr 17.500 kWh. Dies entspricht einer stündlichen Leistungsabnahme von 2.000 W. Dabei fällt der durchschnittliche Energieverbrauch nach einzelnen Ländern außerordentlich unterschiedlich aus. In Westeuropa beträgt der Energiebedarf ca. 6.000 W, in den USA sogar 12.000 W. In vielen Entwicklungsländern, selbst zurzeit noch in China, liegt der Energieverbrauch unter 2.000 W, jedoch mit deutlichen Steigerungen.

Die Vision der 2.000-Watt-Gesellschaft – ursprünglich im Bereich der ETH Zürich (Eidgenössische Technische Hochschule) – ermöglicht einen Ausgleich zwischen Industrie- und Entwicklungsländern und einen guten Lebensstandard für alle Menschen. Voraussetzung zur Erreichung einer 2.000-Watt-Gesellschaft ist die rigorose Anpassung von Bauten und Anlagen, von Fahrzeugen und Einrichtungen sowie ein neues Verständnis für Energiedienstleistungen. Ohne höhere Material- und Energieeffizienz und ohne selektiven Einsatz von Ressourcen bleibt die 2.000-Watt-Gesellschaft eine bloße Absichtserklärung.

Bezüglich des Handlungsbedarfs ist es nicht entscheidend, ob die Ölförderung in 20, 30 oder 40 Jahren ihrem Ende entgegengeht, sondern entscheidend ist, dass der Energiebedarf kontinuierlich auf 2.000 W gesenkt wird, was jedoch einen rigorosen Anpassungsprozess im Bereich der Infrastruktur und eine intelligente Lebensweise voraussetzt.

Bewusst machen muss man sich immer wieder, dass der Löwenanteil der gigantischen Mengen von Primärenergie zurzeit keinen direkten Nutzen bringt. Zwei Drittel der geförderten Primärenergien gehen durch Energieumwandlung und Energietransport verloren. Die Bereitstellung von Endenergie ist mit hohen Verlusten verbunden. Ebenso verlustreich ist der Betrieb von Bauten, Apparaten und Fahrzeugen. Hieraus resultiert, dass nur ca. ein Drittel der ursprünglichen Primärenergie als Energiedienstleistung tatsächlich beim Nutzer ankommt.

Die Entwicklung hin zur 2.000-Watt-Gesellschaft beginnt mit einer „nachhaltigen Stadtentwicklung". Durch eine höhere Material- und Energieeffizienz, durch Substitution von fossilen durch erneuerbare Energieträger und letztlich durch unternehmerische Anpassungen und Professionalisierungen von Prozessen entstehen neue lebensfreundliche Räume, postfossile Städte und Gebäude.

Residential buildings in Europe today consume approximately 1,400 W per capita (without so-called "grey energy". In contrast, Minergie®-conforming buildings or Passivhaus buildings use only 35 – 550 W of energy per person, which is equal to a technological reduction factor of 3 – 4 in the appropriate building typology.

The slow rhythm by which the building stock is gradually renewed over time is another factor to consider when making assumptions with regard to energy savings. Existing building stock with longevity will not be replaced or renovated within a short time frame. In central Europe, the percentage of new construction versus existing building stock is a mere 1 % annually. Therefore, a strong focus will be on the renovation of existing buildings to achieve greater energy efficiency. In the realm of transportation, greater efficiency can only be achieved as well-developed public transportation systems are coordinated with future individual light-vehicle technology to meet the demand for mobility. It is also important in this respect that urban sprawl be controlled and that the places of work and living move closer together.

The necessary reduction of energy demand between today and a possible 2,000-watt society is shown in **Figure 119**.

Nachfolgende Projekte (**Kapitel 10**) zeigen Gebäude, bei denen bauliche Maßnahmen den Energiebedarf so stark reduzieren, dass sich dieser durch ohnehin anfallende Wärme innerhalb des Gebäudes (Kühllasten) decken lässt.

Wohnhäuser in Mitteleuropa verbrauchen heute zurzeit ca. 1.400 W/Person (ohne graue Energie). Im Gegensatz hierzu benötigen Minergie-P-Häuser oder analoge Passivhäuser lediglich ca. 350 – 550 W/Person, was einem technologischen Reduktionspotenzial um den Faktor 3 – 4 bei entsprechenden Objekten mit sich bringt.

Der langsame Erneuerungsrhythmus in der Bausubstanz definiert einen weiten Zeithorizont der Veränderung, da selbstverständlich nicht sämtliche Altbaustrukturen in kurzer Zeit durch analoge Neubauten ersetzt werden können. In Mitteleuropa beträgt der jährliche Neubauanteil lediglich ca. 1 Prozent des bestehenden Bauvolumens. Insofern spielt gerade auch die Sanierung von Altbauten eine große Rolle. Im Bereich des Verkehrs kann nur dadurch ein hoher Effekt bei der Energieeinsparung entstehen, dass gut ausgebaute Nahverkehrsmittel mit zukünftigen Leichtfahrzeugen aller Art das gewünschte Mobilitätsangebot bieten. Hierbei jedoch auch von großer Bedeutung ist, dass die Zersiedelung städtischer Strukturen beendet wird und Wohnen und Arbeiten auf kürzestem Wege verbunden werden.

Der notwendige Umfang der Reduzierung des Energiebedarfs von der Jetztzeit zur 2.000-Watt-Gesellschaft ist in **Bild 119** bedingt dargestellt.

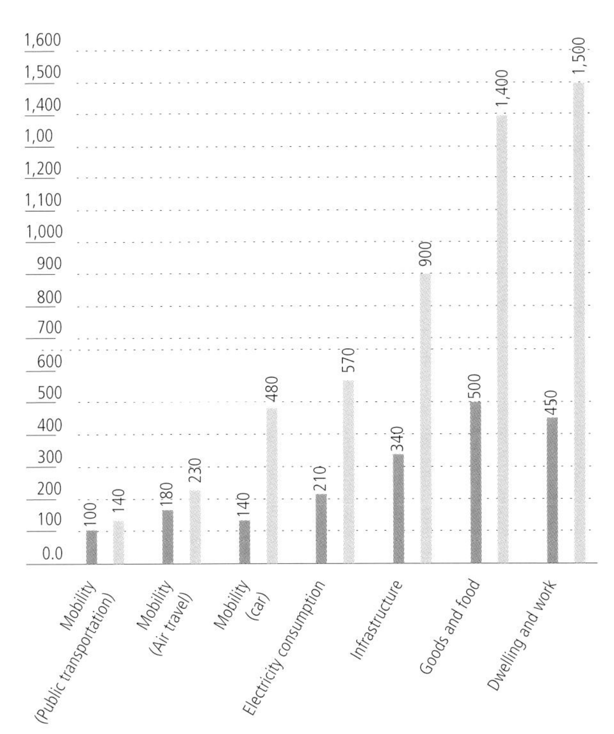

According to: 2,000-watt society
Swiss four-person household today

ÖV OV- Public transportation

Figure 119
Average energy demand of various sectors today compared to 2,000 Watt Society model

Source: Schweizer Baudokumentation, Switzerland

Bild 119
Durchschnittlicher Energiebedarf nach Sektoren in Watt – heute und in der 2.000-Watt-Gesellschaft

Quelle: Schweizer Baudokumentation

10 Real estate of the future – post-fossil fuel buildings

10 Die Immobilie der Zukunft – postfossile Gebäude

In the design of modern buildings that will have a future, the awareness of the finite nature of resources plays a major role. The requirement to be sustainable when it comes to the built environment is extensive and exacting; the term sustainability points both toward the past and the future because if we intend to look far ahead into the future we also have to look back to the past.

Vernacular buildings of past centuries were able to adapt, with simple means, to changing conditions of use and occupation. In contrast, newer, functionalistic or constructivist architecture of the modern movement have become obsolete after only historically brief periods, not only formally but also technically and functionally. In the area of "optimization of building", something seems to have gone wrong, especially in the past few decades. The same is true for the level of utilization of buildings – i.e., the degree to which they are occupied – which is often astonishingly limited. School buildings, for example, are only used for approximately 1,000 – 1,200 hours annually. In the private sector, new methods of work organization, such as "telecommuting" from home, are increasingly of interest, as are new functionalities related to the work itself. Prior to an investment decision with regard to real estate, it is worthwhile for businesses to consider "non-building solutions" – i.e., it could be more advantageous to reconfigure and renovate existing structures than to invest in new construction.

Lifecycle analysis is a tool that can also be used in the decision-making process. Even if demolition of an unfit existing building and construction from scratch is the chosen solution at first sight, consideration of the environmental and social consequences might very well lead to a renovation of existing building stock. A truly comprehensive, sustainable approach that includes social, cultural, and economic factors very often has to give preference to the preservation and re-adaptation of existing structures.

Other question worth contemplating are whether, in an urban environment, energy- buildings make sense at all, and whether the various supply grids could serve as storage vessels as well. If we consider that today's typically centralized and mono-functional grids will evolve over time into more decentralized, multi-functional and especially bi-directional grids, autonomy of a structure in an urban environment needs to be re-evaluated. On the other hand, if we feel that we need to be prepared for more temporary collapses of all infrastructures, then autonomy may still make at least some sense.

Bei der Entwicklung zukunftsfähiger Gebäude spielt das neue Bewusstsein der Endlichkeit der Ressourcen die wesentliche Rolle. Die Anforderungen an Nachhaltigkeit erweisen sich als wesentlich umfassender und anspruchsvoller, wobei der Begriff Nachhaltigkeit sowohl in die Vergangenheit als auch in die Zukunft zeigt: „wer weit nach vorne schauen will, muss weit zurückschauen."

Viele alte Gebäude aus den vergangenen Jahrhunderten konnten sich auf einfache Art immer wieder neuen Anforderungen anpassen, während die funktionalistischen und konstruktivistischen Gebäude der Moderne heute häufig schon nach kurzer Zeit als funktional, technisch und formal obsolet auszumachen sind. Insofern läuft seit Jahren in Sachen „Optimierung von Gebäuden" etwas falsch. Gleiches gilt für die Nutzungsintensität von baulichen Strukturen, die oft außerordentlich gering ist (z.B. Schulgebäude werden oft nur 1.000 – 1.200 Stunden pro Jahr genutzt). Durch neue Arbeitsformen (Telearbeit) und -funktionen können heute schon häufig Nicht-Bau-Lösungen eine Option sein, d.h. es kann unter Umständen von höherem Interesse sein, aufzuwendende Investitionen in die Ertüchtigung von Altbaustrukturen zu stecken, als Neubauten zu errichten.

Lebenszyklusanalysen können unter Umständen im Zweifelsfall zu Antworten beitragen. Der Abriss/die Ersatzoption kann als kurzfristige Maßnahme unter Umständen vorteilhaft erscheinen, will man eine Nutzungsstruktur den neuen Ansprüchen anpassen. Bezieht man jedoch die Umwelteinwirkungen und soziale Kriterien in die Betrachtung mit ein, so erscheint die Erhaltung oder Verbesserung oft langfristig genutzter Objekte sinnvoller. Im Sinne einer ganzheitlichen Betrachtung der Nachhaltigkeit sind unter Umständen aus sozialen, kulturellen und ökonomischen Gründen bestanderhaltende Lösungen vorzuziehen.

Eine weitere, interessante Frage ist, ob in einem städtischen Umfeld energetisch vollkommen autonome Gebäude sinnvoll sind oder ob sich nicht in vielen Fällen die Verwendung des Netzes als Speicher anbietet. Geht man davon aus, dass sich die heutigen zentralen und monofunktionalen Versorgungsnetze in Zukunft in dezentrale, mehrfunktionale und vor allem in zwei Richtungen nutzbare Netze verwandeln werden, macht Autonomie nicht unbedingt Sinn. Geht man allerdings vom Risiko eines zeitweisen Zusammenbrechens der Infrastrukturen aus, so wird die Autonomie zu einem wichtigen Kriterium von Zukunftsfähigkeit.

The sustainability of buildings is mainly determined by their durability and adaptability. The preservation of financial resources in the form of buildings is a measure of their capability to adapt to functional change, their controllability, their ability to be repaired, and the successful replacement of obsolete technology over the long run. This may very well result in the concept that lean and low-tech buildings are more sustainable and future-enabled than their high-tech counterparts. In the book Low-Tech, Light-Tech, High-Tech, the author introduced and promoted, as early as a decade ago, buildings that are adaptable and that gain the means for their operation mainly from the environment.

The tone of the current discourse in building design is being set by a few internationally recognized, signature architects. Many of their solutions involve highly transparent glass-steel or glass-aluminum designs that essentially require the massive inclusion of active building systems, such as air-conditioning, to provide the minimal comfort conditions for the building users. If, at the same time, building materials with a short life span and volatile and unhealthy characteristics are being used, it becomes difficult to speak of true progress in building design; such progress would be notable if buildings could be designed that require only a minimal amount of additional building system support.

The long-term concept for the operation of buildings combines goals in the material realm, such as facility maintenance; social goals, such as stable composition and reduction of changes; and economical goals, such as low operating cost, secured financial reserves for repairs, and replacement (**Figure 120**).

Die Zukunftsfähigkeit von Gebäuden wird im Wesentlichen durch ihre Dauerhaftigkeit und ihre Anpassbarkeit bedingt. In diesem Sinne ist die langfristige Erhaltung von Kapital sowohl an eine notwendige Nutzungsflexibilität als auch an ihre Kontrollierbarkeit, ihre Reparierbarkeit und ihre Ersetzbarkeit gebunden. Ob technisch sehr komplexe Lösungen tatsächlich zukunftsfähig sind, ist eine noch nicht beantwortete Frage. Insofern erweisen sich häufig schlanke Lösungen (leantech/lowtech) zukunftsfähiger als High-tech-Lösungen. Bereits vor annähernd zehn Jahren hat sich der Autor dieser Schrift im Buch „Low-Tech – Light-Tech – High-Tech" mit dieser Frage auseinandergesetzt und für umnutzbare Immobilien plädiert, die sich vorrangig aus den natürlichen Ressourcen des Umfeldes bedienen sollten.

Die derzeitige Architekturentwicklung, geprägt durch weltweit bekannte „Star-Architekten", hat zu hochtransparenten Glas-Stahl-Bauten geführt, bei denen häufig zur Einhaltung von minimalen Komfortanforderungen aufwändige Haustechnikkonzepte notwendig sind. Die gleichzeitige Verwendung von umweltbelastenden Baustoffen mit kurzer Lebensdauer, ungewissen Gesundheitsauswirkungen und ungünstigem Reziklierverhalten kann man nicht als Fortschritt betrachten. Ein Fortschritt ist es, Gebäude so zu bauen, dass sie nur noch eine sehr beschränkte Gebäudetechnische Ausrüstung benötigen.

Eine langfristige Strategie der Bewirtschaftung von Bauten verbindet materielle Ziele (bauliche Instandhaltung) mit sozialen Zielen (stabile Zusammensetzung, wenig Änderungen) und ökonomischen Zielen (niedrige laufende Kosten, Rückstellungen für Erneuerungen), **Bild 120**.

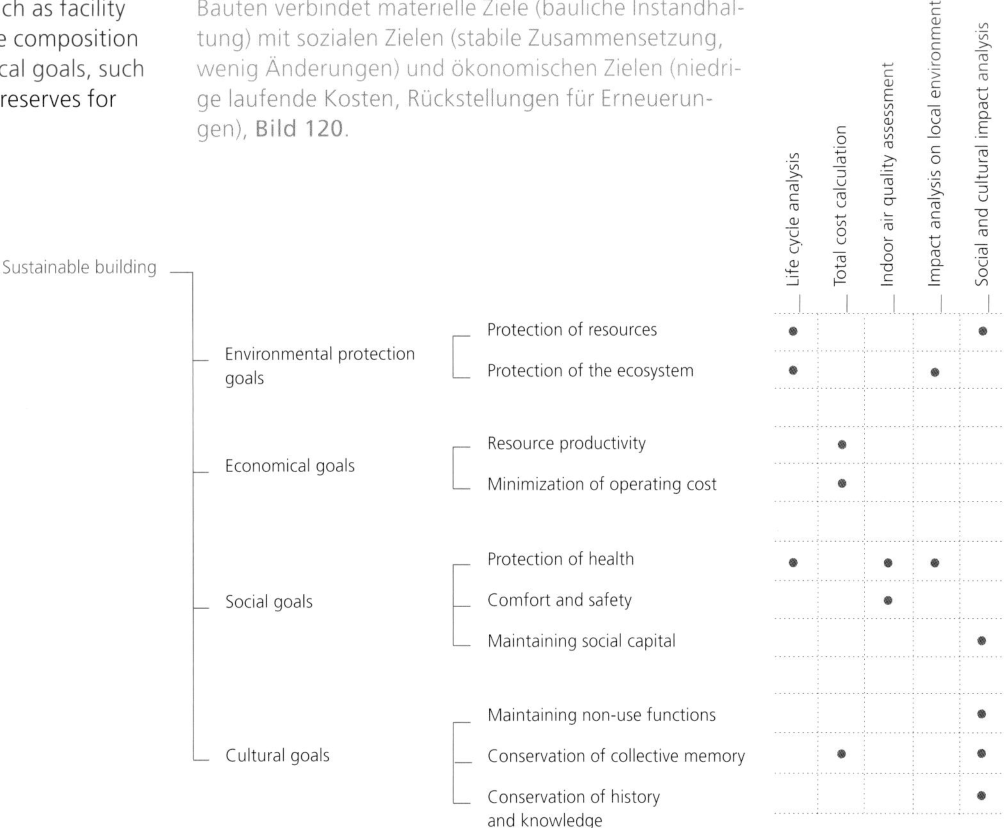

Figure 120
Goals for action towards sustainable building: Proof of sustainability can only be achieved with a combination of various assessment methods.

Source: archplus, Issue 184, October 2007 Niklaus Kohler

Bild 120
Handlungsziele für nachhaltiges Bauen: Der Nachweis der Nachhaltigkeit kann nur über die Kombination von mehreren Verfahren geleistet werden.

Quelle: archplus, Heft 184, Oktober 2007 Niklaus Kohler

If new construction is considered, we must include, as a matter of principle, strategic evaluation of the future of the new buildings. The question is: Which measures implemented today will increase the value of the facility in the long run? The art of design lies in the capability to establish or enhance the usability of a structure with many small corrections and interventions to secure its use without destroying its basic substance.

The future change in paradigms will create models such as these:

- Sustainable, long-term use is better than consumption.
- The capacity of buildings to be repaired and refurbished needs to be included into the initial design.
- The control of sustained efficiency and energy use of all involved systems is necessary.
- Increasing percentages of refurbishment of existing structures is interesting with respect to the labor market.
- Buildings need to become simplified through the use of better and simpler building materials.
- Construction "aids" that are problematic with regard to the environment and recycling need to be avoided.
- Recycling strategies need to be devolved during initial design.
- Building materials that require an unnecessarily great amount of energy for their production are not "modern."

Figures 120 and **121** illustrate the conditions that are mentioned in the above section.

Anlässlich der baulichen Erneuerung kann und muss grundsätzlich und strategisch über die Zukunft des Gebäudes nachgedacht werden. Mit welchen Maßnahmen kann der Wert eines Gebäudes langfristig erhöht werden? Die Kunst der Planer liegt in der Kombination von vielen kleinen Eingriffen, die die Gebrauchsfähigkeit des Gebäudes massiv erhöhen, ohne dessen Charakter vollständig zu verändern.

Im Zuge eines notwendigen Paradigmenwechsels werden Leitbilder sichtbar, die sich wie folgt kurz umschreiben lassen:

- Weiter nutzen ist besser als weiter verwenden.
- Die Reparaturfähigkeit neuer Konstruktionen und Sanierungsmaßnahmen sind zu planen.
- Effizienzkontrollen hinsichtlich des Energieverbrauchs vorhandener und neuer Anlagen sind notwendig.
- Ein steigender Anteil an Erneuerungsleistungen ist arbeitsmarktpolitisch interessant.
- Einfacher Bauen durch einfachere Baustoffe.
- Bauhilfsstoffe und Problemstoffe vermeiden.
- Trennbarkeit von Baustoffen muss bei der Planung und dem Bau bedacht werden.
- Die Notwendigkeit der Bereitstellung großer Energiemengen zur Herstellung von Baustoffen ist nicht zeitgemäß.

Die Bilder 120 und **121** dokumentierten das vorher Gesagte.

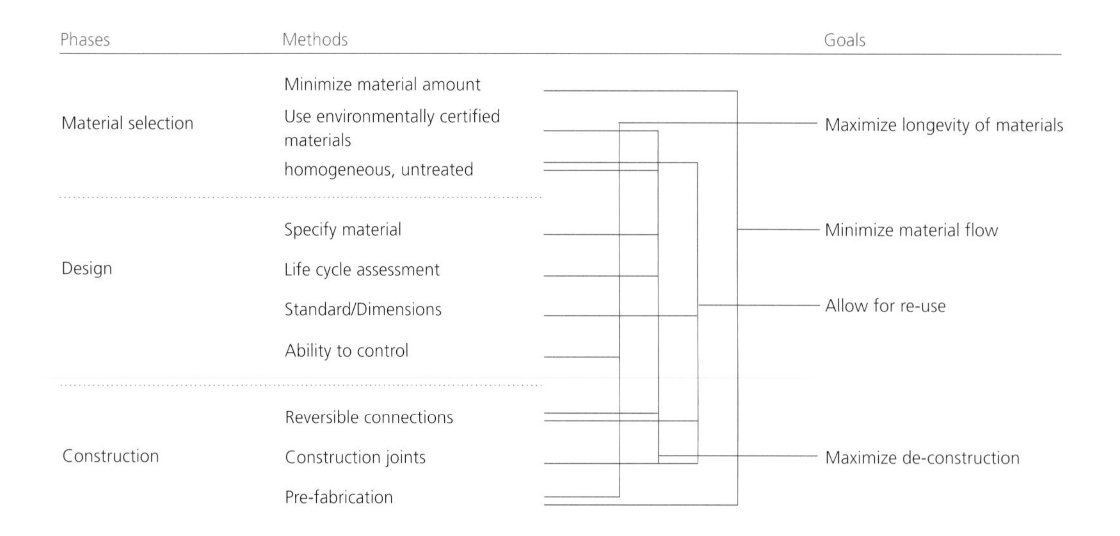

Figure 121
Principles of increased-lifespan design and construction: Reverse-engineering is incorporated into the initial design phase. All building components are evaluated in terms of their potential to be repaired,exchanged and recycled.

Source: archplus, Issue 184, October 2007 Niklaus Kohler

Bild 121
Prinzipien dauerhaften Konstruierens: Beim "Reverse Engineering" wird während des Entwurfsprozesses für jedes Bauteil überprüft, wie es zu reparieren, auszuwechseln und zu entsorgen ist.

Quelle: archplus, Heft 184 Niklaus Kohler

10.1 Stop the waste of resources

Werner Sobek strongly and persuasively argues for the necessity of future recycling-appropriate construction techniques in his essay "Bau schaffen – auch im Sinne der Nachhaltigkeit."

As discussed earlier, most of today's building materials cannot be recycled completely or easily – they are either characterized by inclusion of dangerous and/or unhealthy additives or highly valuable raw materials. The emissions not only as a result of the buildings' operation but also, and particularly, from the outgassing of materials are of great concern today. Most modern buildings insufficiently support a true circle-of-use and re-use of materials or a sustainable economy at large. It is therefore essential to require that materials and components used in a building's construction are capable of being almost entirely recycled and re-used. The mix and composition of various substances into composite materials that cannot be properly re-engineered and recycled has to come to an end if we do not want to get caught in another sustainability trap.

According to Sobek, ecologically conscious building can be extremely attractive and exciting, especially because it asks the designer to face new challenges.

Also, the need for complete recyclability does not end at the building itself, of course, but has to encompass all elements, substances, tools, and objects of our daily life. We have seen earlier how the automobile industry is now obliged to take back its products that have reached the end of their useable life and completely recycle them for future use. This concept needs to be the case for all other items we use as well. Today, we are already noticing the beginning of "wastetourism"; for example, Chinese recycling plants re-capture the valuable base resources out of large quantities of global computer and other electronic rubbish. Such recycling facilities can also be initiated and operated in the countries in the West in order to limit – or better yet, stop – the waste of valuable primary resources.

10.1 Stopp der Ressourcenverschwendung

In einem interessanten Beitrag „Bau schaffen – auch im Sinne der Nachhaltigkeit" von Werner Sobek weist dieser sehr eindringlich darauf hin, dass recyclinggerechtes Konstruieren in Zukunft dringend vonnöten ist.

Wie bereits vorher beschrieben, weisen heute häufig Baustoffe infolge der unzureichenden Trennung entweder problematische oder wertvolle Bestandteile auf. Das Emissionsaufkommen nicht nur aus dem Betreiben von Gebäuden (gebäudetechnische Anlagen), sondern auch und insbesondere von Baustoffen (freigesetzte Emissionen) stellt zurzeit ein großes Problem dar und erfüllt in keiner Weise die Ansprüche an eine Kreislaufwirtschaft, die die Ressourcen schont. Insofern sollen alle Baustoffe, die zusammengefügt werden, um ein Gebäude zu errichten, zu annähernd 100 Prozent dergestalt rückgebaut werden können, dass sie wieder verwendungsfähig sind. Die Durchmischung und Vermischung unterschiedlicher Baustoffe in nicht trennbarer Weise muss schnellstens beendet werden, will man nicht in Bälde „in die nächste Falle tappen".

Gemäß Sobek kann die Ökologie im Baubereich atemberaubend attraktiv und aufregend sein – insbesondere, wenn man sich neuen Ansprüchen stellen muss.

Die vollkommene Rezyklierbarkeit gilt jedoch nicht nur für Gebäude, sondern vielmehr auch für alle Gebrauchsgegenstände unseres täglichen Lebens. Wie bereits angesprochen, wird die Autoindustrie der Zukunft gezwungen, ihre Altfahrzeuge zurück zu nehmen und zu recyceln – vor allem Materialien und Grundstoffe wiederzuverwenden. Gleiches sollte auch für alle anderen Geräte des täglichen Bedarfs gelten. Bereits in unserer heutigen Zeit ist ein „Mülltourismus" festzustellen, der darauf aufbaut, dass zum Beispiel in China wertvolle Grundstoffe aus elektronischen Geräten rückgewonnen werden. Auch in unseren Ländern könnten unter vernünftigen Voraussetzungen entsprechende Recycelanstalten entstehen, die mithelfen, der Ressourcenverschwendung Einhalt zu bieten.

10.2 Design principles

In the book "Energieatlas, Nachhaltige Architektur", Manfred Hegger describes a systematic way to diagnose sustainable qualities in buildings. Such a comprehensive diagnostic approach replaces the current analysis of partial and isolated aspects – instead, all parameters of a building are combined in one evaluation matrix.

The hierarchical structure and the topics that are discussed in this diagnostic system are oriented in accordance with SIA-Swiss Society of Engineers and Architects Recommendation 112/1, which refers to sustainable building techniques.

10.2 Planungskriterien

Im Buch „Energieatlas, Nachhaltige Architektur" von Manfred Hegger beschreibt er ein Diagnosesystem für nachhaltige Gebäudequalität. Das Diagnosesystem ersetzt die bislang üblichen Beschreibungen einzelner Teilaspekte. Eine nachvollziehbare und alle wesentlichen Parameter umfassende Gliederung tritt an seine Stelle.

Das Diagnosesystem orientiert sich in seiner hierarchischen Struktur, seinen Themenfeldern und Erläuterungen an Vorgaben der SIA-Empfehlungen 112/1 (SIA = schweizerischer Ingenieur- und Architektenverband).

The topics are grouped in an analogue way, and additional explanations allow for the inclusion of all parameters and criteria. The system analyzes sustainable building quality according to such design categories as site, building object, and process quality, which are documented in **Table 9**. The catalogue is devised in such a way that designers have easy access to the criteria, and there is clarity in the various evaluation criteria and their use. The system is designed to allow for a comprehensive, holistic building evaluation, and it encourages the development of goals related to the individual criteria that can be used as design guidelines in the initial, schematic design phase.

Themen werden analog geordnet, erweiterte Erläuterungen beziehen sämtliche Kriterien und Indikatoren ein. Das Diagnosesystem ordnet die „nachhaltige Gebäudequalität" nach planungsbezogenen Kategorien wie Standort, Objekt- und Prozess-Qualität, **Tabelle 9**. Der Kriterienkatalog soll den Zugang für Planer erleichtern und zum anderen vermeiden, dass Zuordnungsprobleme von Kriterien, die alle Nach haltigkeitssäulen berühren, entstehen. Das Diagnosesystem soll zudem helfen, eine ganzheitliche Gebäudebeurteilung durchzuführen. Die mit den Kriterien verbundenen Zieldefinitionen und Erläuterungen dienen dabei auch als Planungsinstrument und zur vergleichenden Beurteilung von Planungen.

Area	Topic	Criteria
Site quality		Energy availability Services/functional mix Integration Social justice Utilization Mobility Noise/Vibration Radiation
Building quality	Access and communication	Traffic Social contacts Accessibility and ease of use
	Site	Site
	Design	Landscaping
	Well being and health	Safety Sound Light Air Comfort
	Structure	Existing building stock Structure and interior works
	Cost of construction	First cost Financing
	Cost of operation and maintenance	Operating and maintenance cost Repair
	Building materials	Raw materials/Availability Environmental impact Pollutants Restoration
	Energy sources	Building heating Building cooling Hot water generation Air supply Lighting Other electrical consumers Supply of energy needs
	Infrastructure	Operation waste water
Process quality		Sustainable building Building tradition Participation Integrated design process Analysis Monitoring Facility management

Table 9
Quality of the site, object and process of designing and construction

Source: Energieatlas, Nachhaltige Architektur, Hegger, Fuchs, Stark, Zeumer, Birkhäuser Verlag, 2007

Bereich	Thema	Kriterium
Standortqualität		Energieangebot Grundverorgung /Nutzungsmischung Integration /Durchmischung Solidarität /Gerechtigkeit Nutzung Mobilität Lärm /Erschütterungen Strahlung
Objektqualität	Erschließung / Kommunikation	Verkehr soziale Kontakte Zugänglichkeit und Nutzbarkeit
	Grundstück	Grundstücksfläche Freiflächen
	Gestaltung	Baukultur Personalisierung
	Wohlbefinden / Gesundheit	Sicherheit Schall Licht Raumluft Raumklima
	Gebäudesubstanz	Bausubstanz Gebäudestruktur /Ausbau
	Baukosten	Investitionskosten Finanzierung
	Betriebs- und Unterhaltskosten	Betrieb und Instandhaltung Instandsetzung
	Baustoffe	Rohstoffe /Verfügbarkeit Umweltbelastung Schadstoffe Rückbau
	Betriebsenergie	Gebäudeheizung Gebäudekühlung Warmwasserbereitung Luftförderung Beleuchtung sonstige elektrische Verbraucher Energiebedarfsdeckung
	Infrastruktur	Abfälle aus Betrieb und Nutzung Wasser
Prozessqualität		nachhaltiges Bauen Bautradition Partizipation integrale Planung Analysen Monitoring Facility Management

Tabelle 9
Standort-, Objekt- und Prozessqualität

Quelle: Energieatlas, Nachhaltige Architektur
Hegger, Fuchs, Stark, Zeumer
Birkhäuser Verlag, 2007

In the same publication, Manfred Hegger describes the use of energy concepts as part of the analysis of peripheral parameters (see **Table 10**).

Early development of an energy concept is an essential building block for future buildings. Similar to the form-finding aspects of the design work, the development of a sustainable building-systems concept is a creative process that defies standardization. The engineers of future building systems will need to be capable of both envisioning design-related concepts and mastering the engineering of energetic solutions.

In order to arrive at an energy concept, a thorough analysis of the determining local site parameters, organized in the groups of topics shown in **Table 10**, is mandatory.

Im gleichen Buch beschreibt Manfred Hegger zusätzlich eine Analyse der Randbedingungen bezüglich einzusetzender Energiekonzepte, **Tabelle 10**.

Die frühe Entwicklung eines Energiekonzeptes ist ein zentraler Baustein für eine zukunftsfähige Entwurfsplanung von Gebäuden. Analog zur architektonischen Formfindung ist das Erarbeiten eines technischen Konzepts ein kreativer Prozess, der nicht standardisiert werden kann. Hieraus entsteht der Anspruch, sich sowohl mit architektonischen Fragen als auch mit Fragen der Energiewirtschaft auseinandersetzen zu können.

Die Grundlage für die Entwicklung eines Energiekonzeptes ist die Ermittlung von Randbedingungen nach Themengruppen, wie sie in **Tabelle 10** dargestellt sind.

Parameter	Information	Action
Climate	Temperature maxima	Thermal performance of building envelope
	Diurnal swings	Potential for night flushing
	Average annual temperature	Potential for earth tube registers
	Relative humidity summer/winter	Potential for evaporative cooling
	Average wind speeds	Wind power generation
	Wind direction distribution	Natural cross ventilation
	Rainfall, amount and annual distribution	Evaporative cooling capacity
	Soil conditions	Geothermal probes
	Ground and surface water	Use as heat source and for passive cooling
	Energy gains solar radiation	Passive and active solar thermal and solar electric power generation
	Sun path	Optimizing summer thermal protection
Use	Requirements for heated spaces	Temperature minima and maxima
	Goals for thermal protection in summer	Room temperature and temperature difference i.e 22 °C ± 2 °C; 21 – 28 °C
	Air quality requirements	AGW-values; CO_2-maximum values
	Humidity requirements	Relative Humidity and humidity difference 50 % ± 10 %
	Illumination requirements	Sun shading and glare protection
Legal	Zoning	Optimizing site use up to maximum allowed density
	EnEV	Maximum primary energy consumption
	DIN 18599	Heat sources and heat sinks
	Use of infrastructure by law	Use of infrastructure, increase of capacity
	Water protection regulations	Use of soil and groundwater as energy source
	Regulations regarding functional use	i.e. Heat recovery in case of mechanical ventilation
	Historic landmark protection	Landmark preservation, i.e. by internal insulation
Design	Urban fabric and micro-climatic conditions	Use of renewable energy sources as design catalysts Use of primary and secondary solar energy
	Floor-area-ratio	–
	Ratio between usable space and solar gain surface	Definition of transparent wall surfaces according to building orientation

Table 10
Rules for the development of energy concepts

Source: Energieatlas, Nachhaltige Architektur, Hegger, Fuchs, Stark, Zeumer, Birkhäuser Verlag, 2007

Randbedingung	Information	Handlungsfeld
Klima	Temperatur Extremwerte	thermische Qualität der Gebäudehülle
	Temperaturdifferenz Tag / Nacht	Potenzial für freie Kühlung durch Nachtluft
	Jahresmitteltemperatur	Leistungspotenzial für Luft-Erdregister
	relative Luftfeuchtigkeit Sommer / Winter	Möglichkeit für direkte adiabate Kühlung
	mittlere Windgeschwindigkeiten	Stromerzeugung durch Windenergie
	Verteilung der Windrichtungen	natürliche Be- und Entlüftung durch Windbewegung
	Niederschlagsmenge und -verteilung	technischer Einsatz von Verdunstungskühlung
	geologische Erdschichten	Erschließung des Erdreichs über Erdsonden
	Grund- und Oberflächenwasser	Nutzung als Wärmequelle und für passive Kühlung
	Energiemenge Solarstrahlung	passive und aktive solare Wärme- und Stromerzeugung
	Sonnenbahnverlauf	Optimierung des sommerlichen Wärmeschutzes
Nutzung	Anforderung an beheizte Flächen	minimale und maximale Temperatur
	Zielvorgaben für sommerlichen Wärmeschutz	Raumtemperatur und Temperaturspreizung (z.B. 22 °C ± 2 °C; 21 – 28 °C)
	Anforderung an Luftqualität	AGW-Werte; CO_2-max.-Werte
	Anforderung an Luftfeuchtigkeit	relative Luftfeuchte und Spreizung (z.B. 50 % ± 10 %)
	Anforderung an Beleuchtung – Luxwerte	Sonnen- und Blendschutzsystem
Recht	B-Plan	Optimierung der Flächennutzung bis maximal zulässiger Bebauungsdichte
	EnEV	maximaler Primärenergieverbrauch
	DIN 18599	Wärmequellen und -senken
	Anschlusszwang	Infrastrukturnutzung und Erhöhung der Auslastung
	wasserrechtliche Vorgaben	Nutzung des Erdreichs und des Grundwassers als Energieträger
	rechtliche Vorgaben aus Nutzung	z.B. Wärmerückgewinnung bei erforderlicher maschineller Lüftung
	Denkmalschutz	Erhaltung der Raumwirkung z.B durch Innendämmung
Gestaltung	umgebende Bebauung und mikroklimatische Randbedingung	architektonische Gestaltung in Verbindung mit der Nutzung von Umweltenergien
	Verhältnis von Grundstücksgröße zu Bauvolumen	Nutzbarkeit primärer und sekundärer Solarenergie –
	Verhältnis Nutzfläche zu pot. Solarfläche	Anteil transparenter Wandflächen nach Himmelsrichtungen

Tabelle 10
Randbedingungen bei der Entwicklung von Energiekonzepten

Quelle: Energieatlas, Nachhaltige Architektur Hegger, Fuchs, Stark, Zeumer Birkhäuser Verlag, 2007

The base for all concept development can be found in the analysis of the characteristics of the climate at the future building site. Such an investigation will also influence the thermal characteristics of the building envelope. Identified temperature differences at the location, for instance (not only over a year, but also diurnal differences), will result in potential concepts for free or nighttime cooling.

The analysis of annual average temperatures at a site will help to establish the near-surface soil temperature and may result in concepts such as shallow geothermal applications, thermal labyrinths, earth tubes, slinky collectors, and water-source heat pumps.

Depending upon the conditions at the site, the relative and absolute humidities of the ambient air may be used as important criteria for developing humidifying or de-humidifying methods or adiabatic cooling strategies.

Prevailing wind directions and the speed of the wind may be used not only to determine the potential for natural ventilation of interior space but also for an evaluation of the degree to which wind power generation may be feasible.

Annual amounts of rainfall and its distribution over the course of a year will be the basis for the analysis of the potential for grey water usage and evaporative cooling techniques.

The annual solar radiation of a particular location as a result of the sun's path and altitude in the sky is the starting point for analysis of the possibilities of passive solar systems, the need for summer solar shading, and active solar system implementation, such as photovoltaic and/or solar thermal systems.

The use of a facility will be determined by comfort factors such as internal temperatures, relative humidity in the enclosed space, and hygienic quality, but also by factors providing visual comfort, such as visual connections from the inside to the outside and vice versa.

With their various regulatory frameworks such as building codes and zoning regulations individual countries will obviously influence the design process in various ways.

Locally available environmental energy potential will also have significant influence both on the architectural design and on the concepts used for building support systems. The same is true for form determinants, which are important to the projected design.

Figure 122 provides guidelines on how important the earliest possible analysis of local climatic conditions is for the design process and how architects and designers need to include them in the schematic design.

Klimatische Randbedingungen, d.h. das Klima des jeweiligen Standorts des Gebäudes, bildet die Voraussetzung für die Definition der thermischen Qualität der Gebäudehülle. Temperaturdifferenzen, nicht nur während des Jahres, sondern auch zwischen Tag und Nacht, geben Auskunft über das Potenzial einer passiven, freien Kühlung (Nachtauskühlung).

Die Jahresdurchschnittstemperaturen beeinflussen das Temperaturniveau des oberflächennahen Erdreichs und damit die mögliche Nutzung der oberflächennahen Geothermie (Thermolabyrinthe, Erdrohre, Flachsondensysteme mit Wärmepumpen).

Je nach Standort ist auch die relative und absolute Luftfeuchte ein wesentliches Kriterium zur Festlegung von Be- und Entfeuchtungseinrichtungen bzw. adiabater Kühlung.

Vorherrschende Windgeschwindigkeiten und Windrichtungen geben nicht nur Hinweise zur natürlichen Be- und Entlüftung einer Baustruktur, sondern auch zur Erzeugung von elektrischer Energie aus Windkraft.

Niederschlagsmengen und ihre Verteilung stellen das Potenzial einer teilweisen Grauwasserversorgung und unter Umständen einer Verdunstungskühlung dar.

Solare Strahlungsleistungen auf Gebäude im Zusammenhang mit den Sonnenumlaufbahnen bilden die Grundlage sowohl für die passive Solarnutzung, den sommerlichen Wärmeschutz wie auch die Entwicklung aktiver solarthermischer Systeme oder photovoltaischer Anlagen.

Die gewünschte Nutzung einer baulichen Struktur ergibt sich aus den Forderungen nach Temperaturen, Raumluftfeuchten, Ansprüchen an die hygienische Behaglichkeit sowie Ansprüchen an die visuelle Behaglichkeit mit Sichtbeziehungen von innen nach außen bzw. von außen nach innen.

Je nach Land können Vorgaben des Bauplanungs- und Bauordnungsrechts den Planungsprozess beeinflussen (Bebauungsdichte, Kubatur, Dachformen, Materialvorgaben usw.)

Die gestalterischen Randbedingungen können von Fall zu Fall bei der Entwicklung von Energiekonzepten einen wesentlichen Einfluss nehmen. Lokal verfügbare Umweltenergiepotenziale greifen sowohl in den Bereich der Architektur als auch in den Bereich der Technik ein.

Bereits im jüngsten Planungsstadium sollten Architekten und Planer ihre Bautypologien danach prüfen, welche Kriterien in der entsprechenden Klimaregion eine Rolle spielen. **Bild 122** zeigt hierzu erste Hinweise.

Figure 123 describes the changes in percentages of thermal energy demand in Central Europe for various building types with identical building mass. The great importance of the surface-to-volume ratio (S/V ratio) can be clearly seen – it is at the base of many governmental energy-savings regulations. **Figure 124** shows the solar gains through openings as a function of various building orientations for locations in Central Europe.

In order to evaluate solar energy gains due to insolation at various orientations and solar altitude angles, **Figure 125** is presented. This illustration is most important when integration of solar energy-generating systems in façades or roofs is considered and for determining what the approximate resulting positive energy gain will be.

Bild 123 beschreibt prozentuale Veränderungen des Wärmebedarfs in Mitteleuropa bei verschiedenen Gebäudetypen sowie Auflösung einer insgesamt gleichen Baumasse. Hier zeigt sich der Einfluss des A/V-Verhältnisses, das einigen Energieeinsparverordnungen und Regelwerken zugrunde liegt. In **Bild 124** ist für Mitteleuropa ausgewiesen, bei welcher Gebäudeausrichtung ein analoger Solareintrag während des Tages durch Fenster entsteht.

Bei der Beurteilung, welche Energieerträge durch aktive Solarmaßnahmen bei welchen Himmelsausrichtungen und Neigungen entstehen, zeigt **Bild 125**. Diese Darstellung ist insbesondere dann von Bedeutung, wenn es darum geht, in Fassadenbereichen oder Dachbereichen solartechnische Anlagen einzusetzen, um hieraus entsprechende Energieerträge zu generieren.

	Placement of auxiliary core areas	**Placement of areas for solar gains**	**Optimized length/width ratio**	**Placement of opaque building elements**	**Use of atria** a) as solar traps b) for ventilation and cooling	**Energy usage value of atria**
Cold			1 : 1		a	a
Moderate			1 : 1,6		a	
Dry			1 : 2		b	b
Tropical			1 : 3		b	

Figure 122
Building typologies according to conditions in various climate regions
Source: Ken Yeang

Bild 122
Bautypologien verschiedener Klimaregionen
Quelle: Ken Yeang

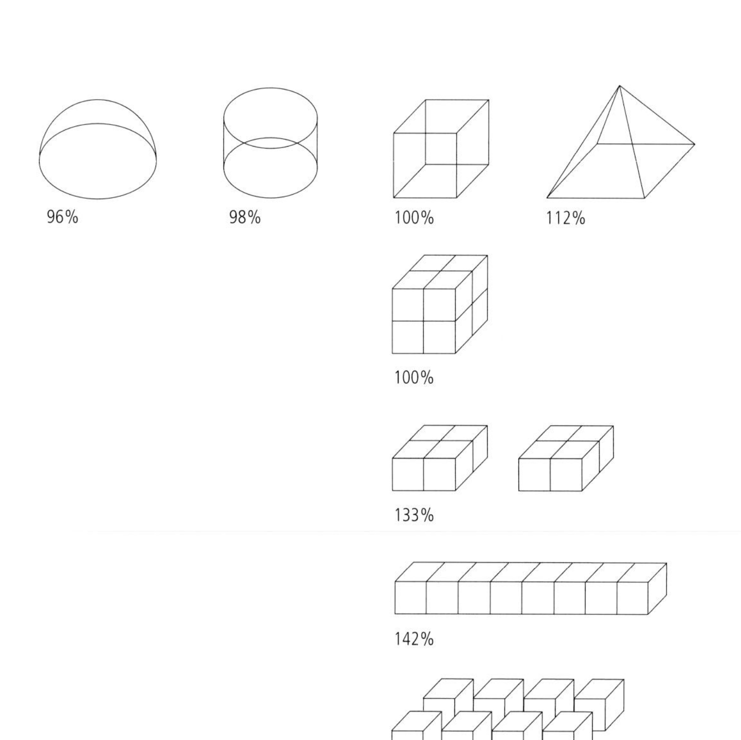

Figure 123
Change of heat energy demand for various building configurations with equal volume, in percent.

Source: Daniels/author

Bild 123
Prozentuale Veränderungen des Wärmebedarfs bei verschiedenen Gebäudetypen bei Auflösung einer insgesamt gleichen Baumasse

Quelle: Daniels/Autor

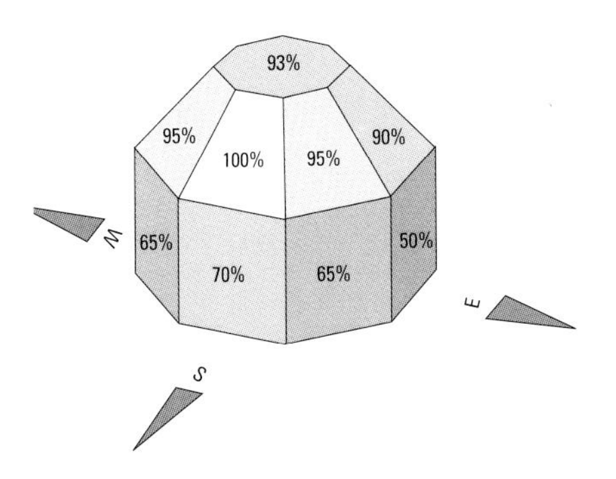

Figure 125
Approximate solar gains in % as a function of orientation

In Central Europe optimal energy gains (100 %) are achieved on surfaces facing south and sloped 30 degrees (versus a horizontal plane.) Orientations and angles different from the described values result in reduced energy gains.

Bild 125
Ca. solare Erträge (%) verschiedener Flächen und Ausrichtungen

Optimale Energieerträge in Mitteleuropa werden bei einer Anlagenausrichtung nach Süden und einem Winkel von 30 ° zur Horizontalen erzielt (100 %). Davon abweichende Ausrichtungen haben geringere Ertragswerte.)

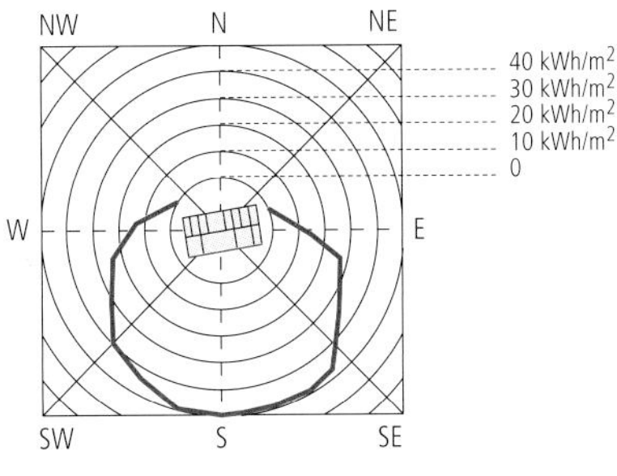

Solar radiation over the year
Orientation of fenestration:
170 degrees (South).
The yearly solar gain is
255.9 kWh/m² of
window surface area.

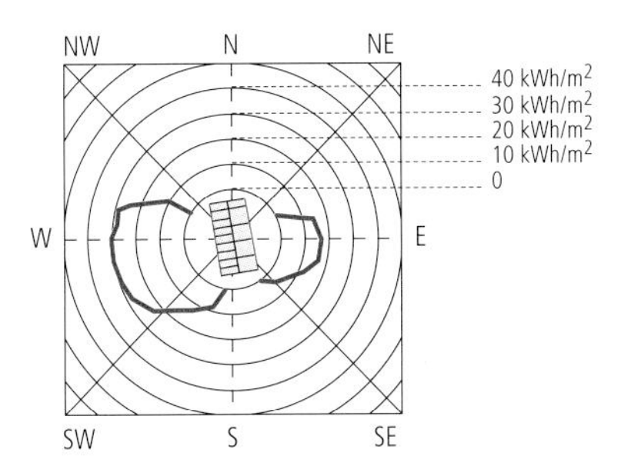

Solar radiation over the year
Orientation of fenestration:
260° degrees (West).
The yearly solar gain is
88.9 kWh/m² of
window surface area.

Solar radiation over the year
80 degrees (East).
The yearly solar gain
is 42.9 kWh/m² of
window surface area.

Figure 124
Solar radiation as a function of surface orientation, Central Europe

Source: Daniels/Technologie des ökologischen Bauens, Birkhäuser, 1999

Bild 124
Sonneneinstrahlung in Abhängigkeit der Oberflächenorientierung (Mitteleuropa)

Quelle: Daniels/Technologie des ökologischen Bauens, Birkhäuser, 1999

10.3 Building materials and grey energy

A building is constructed of matter, and it needs to protect the interior from too-great energy gains or losses and at the same time reduce the energy flow from the exterior to the inside – e.g., from hotter outside conditions to the interior space.

Energy is also required to transform raw matter into a specific building material. To a large degree, the production of building materials is responsible for the total amount of energy used in buildings. As a result, it becomes urgently important to select materials for buildings with consideration of the energy required for the manufacture and resulting consequences of the materials on the environment. The goal of a design process needs therefore to be to limit the number of matter flows and, additionally, to avoid their mixing. Entropy – that is, the intersection between resource and energy demand – should remain as little mixed as possible.

The production of the common building material aluminum may serve as an example. Besides a large amount of energy – smelters tend to be situated where electric power is both plentiful and inexpensive – aluminum production requires very large amounts of water for the processing and enrichment of bauxite. The waste water of the aluminum manufacturing process is also rich in toxic heavy metals. However, the advantage of the material aluminum lies in its durability and ease of recycling. To lower the entropy of the initial aluminum production process, aluminum needs to be recycled and re-used locally, and the manufacturing and derivation of raw materials also has to be local.

10.3 Baustoffe und graue Energie

Ein Gebäude besteht aus Materie. Diese soll den inneren Betrieb vor zu hohem Energieverbrauch schützen und gleichzeitig den Wärmefluss von innen nach außen oder von außen nach innen regulieren.

Unter Verwendung von Energie wird Materie in einen Baustoff oder Bauteil umgewandelt. Die Baustoffherstellung und der Bau bilden einen erheblichen Anteil am Gesamtenergiebedarf eines Gebäudes. Insofern ist die Materialauswahl auch unter dem Gesichtspunkt des Energieverbrauchs und der Umwelteinwirkungen zu treffen. Gleichzeitig sollte eine deutliche Verringerung der Materialmengen und der Massenströme für das Bauen erfolgen. Ziel der Planung muss es daher sein, wenige Stoffströme zu generieren und sie im Sinne einer geringen Entropie (Schnittstelle zwischen Ressourcen- und Energieverbrauch) möglichst wenig miteinander zu vermischen.

Bei zum Beispiel der Herstellung von Aluminium ist neben einem hohen Energie- auch ein hoher Wasserverbrauch zur Aufbereitung und Anreicherung des Bauxits notwendig. Die Abwässer aus dem Produktionsbereich enthalten stark toxische Schwermetalle. Der Vorteil des Aluminiums liegt in seiner hohen Dauerhaftigkeit und in seiner gleichermaßen guten Recycelfähigkeit. Um die Entropie bei der Aluminium-Produktion zu verringern, sind eine lokale Gewinnung und Verarbeitung sowie vor allem eine Wiederverwertung des Altmaterials sinnvoll und notwendig.

In stark contrast to the pure material aluminum, today's compound insulation materials consist of multiple carrier and fill materials, formed together by a variety of adhesives. This makes separation and recycling according to type nearly impossible.

Energy required for the production of a product is called embodied energy, or grey energy. It is defined as the sum total of the energy from the raw material extraction to the transport, manufacturing, assembly, and installation, as well as interest and other costs, required to produce a service or product, and ultimately for its disassembly, deconstruction and/or decomposition. The grey or embodied energy of building materials used in Germany amounts to approximately 20 times the energy necessary to operate the entire building stock on a yearly basis. With rising, improved energy efficiencies of building envelopes and the components of building systems, the grey energy embedded in the material only increases slightly, yet its proportional share of the total energy consumption over the entire lifespan of buildings increases significantly.
The example of a long-lasting Passivhaus may be used as an illustration. In the materials used in such a structure, up to 50 percent of the embedded energy remain captured over a long period of time – for example, 50 years. As a consequence, an increase of efficiency in the operation of buildings needs to include the increase of efficiency in material usage.

The primary energy content (PEI) defines the embedded grey energy content of a material, and this value is related to various reference units such as kg or m^3. Further distinctions are also made as to whether renewable or non-renewable energy was used in the material's manufacturing.

Energy consumed during the production process of a building material in relation to its mass can differ by a factor of > 2,000. Energy-intensive materials include metals, glass, and plastics, while energy-lean materials include clay and gypsum.

Designers need to be responsible and sensitive in using materials with a small content of grey energy based on available specifications concerning primary energy content. Having surveyed and compared the alternatives, the designer should select that material that uses the least amount of embedded energy while still meeting performance specifications.

Table 11 lists the primary energy content (grey energy) of various building materials, and **Table 12** shows the CO_2 equivalent of various materials and the use of renewable and non-renewable energy used in their production.

Table 13 allows a comparison between the variable yet significant percentage of primary energy content due to the transportation of the product.

Im Gegensatz zu Aluminium wird bei heutigen Wärmedämm-Verbundsystemen im Regelfall eine dämmende Schicht mit einem Untergrund verklebt, zusätzlich ergänzt durch eine vollflächige, kraftschlüssige Oberflächenbeschichtung. Eine sortenreine Trennung der einzelnen Materialschichten ist dadurch nicht möglich.

Die innerhalb der Herstellung verwendete und damit im Material gebundene Energie bezeichnet man als graue Energie. Sie definiert die Energiemenge, die für Herstellung, Transport, Lagerung sowie Entsorgung eines Produkts benötigt wird. Die im Gebäudebestand in Deutschland gebundene graue Energie beträgt ca. das Zwanzigfache des für den Gebäudebetrieb notwendigen jährlichen Energieeinsatzes. Mit fortschreitender energetischer Verbesserung der Gebäudehülle und Gebäudetechnik steigt die Menge der im Material vorhandenen grauen Energie nur gering an, ihr prozentualer Anteil am Gesamtenergieverbrauch über die Lebensdauer des Gebäudes steigt jedoch erheblich. Zum Beispiel verbleiben bei einem Passivhaus in den Materialien – über 50 Jahre gesehen – bis zu 50 Prozent des gesamten Energieverbrauchs gebunden. Insofern ergibt sich folgerichtig, dass einer Effizienzsteigerung im Gebäudebetrieb auch eine Effizienzsteigerung im Materialeinsatz folgen muss.

Der Primärenergieinhalt (PEI) weist die graue Energie eines Baustoffes aus. Die Kenngröße wird, bezogen auf verschiedene Bezugsgrößen (z.B. kg oder m^3), unterschieden. Eine weitere Unterscheidung kann sich dadurch ergeben, ob erneuerbare oder nicht erneuerbare Energien bei der Herstellung der Materialien zum Einsatz kommen.

Der für die Herstellung eines Baustoffes notwendige Energieverbrauch kann sich, bezogen auf das Gewicht des Materials, um mehr als um den Faktor 2.000 unterscheiden. Energieaufwändige Materialien sind z.B. dabei Metalle, Glas oder Kunststoffe. Energiearme Materialien sind Lehm oder Gips.

Dem Architekten und Planer obliegt es, auf der Basis der Informationen zum Primärenergieinhalt, Baustoffalternativen zu prüfen und bei gleicher Leistungsfähigkeit bevorzugt solche mit einem geringen Anteil an grauer Energie einzusetzen.

In **Tabelle 11** ist der Primärenergieinhalt (graue Energie) verschiedener Baustoffe dargestellt. In **Tabelle 12** ist wiederum der Primärenergieinhalt von Baustoffen ausgewiesen, wobei eine zusätzliche Information über den Einsatz erneuerbarer oder nicht erneuerbarer Energien sowie das CO_2-Äquivalent dargestellt ist.

Tabelle 13 zeigt den zwar variablen, aber nicht zu unterschätzenden Beitrag zum Primärenergieinhalt infolge des Produkttransportes.

When comparing the primary energy content of building materials, additional issues, such as the impact of the materials on the environment (e.g., could renewable resources be substituted?) need to be evaluated (Environmental Product Declaration, EPD).

It can be said that primary energy content is greatest in a building's primary structure (its frame and main support structure – approximately 56 %), followed by the interior architectural and technical components (approx. 20 %) and enclosing façades (14 %). Therefore, a structural optimization of the primary support structure will result in reduced grey energy content. Over the span of a building's useful life, the weighting of its materials according to their embodied energy may shift because they remain "locked in" for varying lengths of time. The goal is to select the materials and a primary structural system that are most appropriate according to the expected length of service, the type of usage, and the expected processes involved.

Bei der Betrachtung des Primärenergieinhaltes von Baustoffen sind zusätzlich die Fragen nach Funktionsäquivalenz, nachwachsenden Rohstoffen und Fragen nach relevanten Umweltwirkungen eines Baustoffes (Environmental Product Declaration, EPD) zu klären.

Bei nahezu allen Bauten ist der Primärenergieinhalt für die Tragkonstruktion am größten (ca. 56 Prozent der grauen Energie), gefolgt vom Innenausbau (ca. 20 Prozent) und der Erstellung von Fassaden (ca. 14 Prozent).

Die Optimierung eines Konstruktionsprinzips wirkt sich somit besonders positiv auf die Reduzierung der grauen Energie aus. Innerhalb des Lebenszyklus eines Gebäudes verändert sich die Gewichtung der Bauteile hinsichtlich des Energieaufwands, da sie unterschiedlich lange im Gebäude verbleiben. Die notwendigen Austauschprozesse ziehen während der Nutzungsphase erneut energetische Aufwendungen nach sich. Je häufiger der Austausch eines Bauteils notwendig wird, desto signifikanter ist sein Beitrag zur grauen Energie des gesamten Gebäudes. Ziel muss es sein, die Materialwahl und das konstruktive Gefüge auf die geplante Lebensdauer, Nutzungsart und die zu erwartenden Nutzungsprozesse abzustimmen.

Table 11
Primary energy content (grey energy) of building materials

Source: Energieatlas, Nachhaltige Architektur, Hegger, Fuchs, Stark, Zeumer, Birkhäuser Verlag, 2007

Tabelle 11
Primärenergieinhalt (graue Energie) von Baustoffen

Quelle: Energieatlas, Nachhaltige Architektur Hegger, Fuchs, Stark, Zeumer Birkhäuser Verlag, 2007

Material	PEI [MJ/m³]	PEI / E-Modul [J/kNm]	[%]
Concrete			
C35/40 Concrete	1,764	0.05	76 %
Reinforced concrete (2% steel content)	4,098	0.07	100 %
Brick, Stone materials			
Limestone brick	2,030	–	–
Brick	1,663	–	–
Wood			
Construction wood, pine	609	0.06	80 %
Wood sheet and board materials	3,578	0.33	469 %
Metals			
Steel (FE 360B)	188,400	0.89	1,281%
Weatherproof steel (WT St 27 2)	204,100	0.96	1,388%
Stainless steel (V2A)	411,840	1.96	2,827%
Aluminum (EN AW 7022)	753,380	10.76	15,513%
Float glass	35,000	0.50	721 %

Werkstoff	PEI [MJ/m³]	PEI / E-Modul [J/kNm]	[%]
Beton			
C 35/40 Beton	1.764	0,05	76 %
Stahlbeton (2 % Stahlanteil)	4.098	0,07	100 %
Ziegel, Werksteine			
Kalksandstein	2.030	–	–
Mauerziegel	1.663	–	–
Holz			
Konstruktionsholz, Kiefer	609	0,06	80 %
Brettschichtholz	3.578	0,33	469 %
Metalle			
Stahl (FE 360 B)	188.400	0,89	1.281%
wetterfester Stahl (WT St 27 2)	204.100	0,96	1.388%
Edelstahl (V2A)	411.840	1,96	2.827%
Aluminium (EN AW 7022)	753.380	10,76	15.513%
Floatglas	35.000	0,50	721 %

Table 13
Primary energy content of the product transportation (3.6 MJ=1 kWh)

Mode of transportation	Primary energy content, (non-renewable) [MJ/tkm]	Primary energy content, (renewable) [MJ/tkm]
Truck, 22 tons Gross vehicle weight rating (GVWR), 14.5 tons net, 85% utilization	1.50	0.00031
Inland river vessel, 1,250 tons, no currents	0.47	0.001
Freight train	0.40	0.053
Container ship, 27,500 tons, open sea	0.17	0.00004

Tabelle 13
Primärenergieinhalt des Produkttransportes (3,6 MJ=1 kWh)

Transportart	PEI nicht ern. [MJ/tkm]	PEI ern. [MJ/tkm]
LKW, 22 t zul. GGW, 14,5 t Nutzlast, 85 % Auslastung	1,50	0,00031
Binnenschiff, ca. 1.250 dwt, ohne Strömung	0,47	0,001
Güterzug	0,40	0,053
Containerschiff, ca. 27.500 dwt, Hochsee	0,17	0,00004

Material	Reference value	Calorific value [MJ]	Primary energy content (non-renewable) [MJ]	Primary energy content (renewable) [MJ]	Global warming potential (GWP) [kg CO_2eq]
Natural stone					
Granite (large transport distance) polished, $\rho = 2{,}750$ kg/m³	1 m³		9,837	332	626
Marble (medium transport distance), $\rho = 2{,}700$ kg/m³	1 m³		6,749	249	422
Slate (local), $\rho = 2{,}700$ kg/m³	1 m³		4,608	165	286
Sand Stone (local), $\rho = 2{,}500$ kg/m³	1 m³		4,099	153	253
Clay materials					
Rammed earth, $\rho = 2{,}200$ kg/m³	1 m³		158	1	9.7
Adobe brick, $\rho = 1{,}200$ kg/m³	1 m³		1,257	4	74.0
Building materials with mineral content					
Mortar and concrete					
Anhydrite mortar/concrete, compressive strength, 2,350 kg/m³	1 m³		655	11.0	43
Magnesite mortar/concrete, compressive strength, 2,000 kg/m³	1 m³		2,439	9.9	348
Cement mortar/concrete, 2,250 kg/m³	1 m³		2,161	27.0	389
Gipsum mortar, PIVa, $\rho = 1{,}300$ kg/m³	1 m³		1,477	9.6	177
Lime-cement mortar, PIIa, $\rho = 1{,}500$ kg/m³	1 m³		2,675	28.0	448
Masonry units					
Lime-sand stone, $\rho = 1{,}800$ kg/m³	1 m³		2,030	117	247
Concrete pavers, $\rho = 2{,}500$ kg/m³	1 m³		1,990	46	310
Concrete masonry unit, $\rho = 400$ kg/m³	1 m³		1,484	81	186
Light weight concrete masonry unit, $\rho = 600$ kg/m³	1 m³		787	35	97
Concrete					
Cast-in-place (C 25/30), $\rho = 2{,}340$ kg/m³	1 m³		1,549	17	251
Cast-in-place (C 35/45), $\rho = 2{,}360$ kg/m³	1 m³		1,764	23	320
Prefab 2% steel content, (FE 360 B, C 35/45), $\rho = 2{,}500$ kg/m³	1 m³		4,098	86	455
Boards					
Fiber cement boards, $\rho = 1{,}750$ kg/m³	1 m³		26,839	116	2,200
Drywall, $\rho = 850$ kg/m³	1 m³		2,655	251	150
Ceramic materials					
Hollow brick masonry, $\rho = 670$ kg/m³	1 m³		1,485	638	95
Brick, Interior hollow, $\rho = 750$ kg/m³	1 m³		1,663	715	107
Brick, high strength, solid, $\rho = 1{,}600$ kg/m³	1 m³		4,776	39	301
Ceramic tiles, glazed, $\rho = 2{,}000$ kg/m³	1 m³		6,322	0.060	393
Ceramic tiles, non-glazed, $\rho = 2{,}000$ kg/m³	1 m³		7,160	0.070	445
Bituminous materials					
Pure distillations bituminous (B 100-B 70)	1 kg		45.6	0.010	0.37
Polymer modified bituminous (PmB 65 A)	1 kg		35.3	0.020	0.50
Wood and wood-based materials					
Cut wood					
Pine wood, local, 12% wood moisture content, density at 0% humidity 450 kg/m³	1 m³	8,775	609	9,512	- 792[1]
Western red cedar, U.S., 12% wood moisture content, density at 0% Humidity 630 kg/m³	1 m³	12,285	4,485	14,359	- 907[1]
Teak wood, Brazil, 12% wood moisture content, density at 0% humidity 660 kg/m³	1 m³	12,870	3,217	13,435	- 1,013[1]
Plywood, 12% wood moisture content, density at 0% humidity 465 kg/m³					
Triple-layer plywood, 12% wood moisture content, density at 0% humidity 430 kg/m³	1 m³	9,300	3,578	13,870	- 662[1]
Veneer plywood, 5% wood moisture content, density at 0% humidity 490 kg/m³	1 m³	8,618	2,617	9,387	- 648[1]
Chipboard, 8.5% wood moisture content, density at 0% humidity 690 kg/m³	1 m³	10,175	4,729	15,041	- 636[1]
Oriented strand board (OSB), 4% wood moisture content, density at 0% humidity 620 kg/m³	1 m³	13,998	5,818	12,614	- 821[1]
Medium density fiber board (MDF), 7.5% wood moisture content,	1 m³	12,555	4,593	16,479	- 839[1]
Density at 0% humidity 725 kg/m³	1 m³	15,843	9,767	12,495	- 515[1]
Metals					
Iron metals					
Cast iron GG 20, secondary GJL	1 kg		10	0.49	0.97
Steel, rolled (FE 360 B)	1 kg		24	0.54	1.70
Steel mesh, secondary	1 kg		13	0.24	0.83
Weather resistant steel, Cold band (WT St 37-2), 2 mm	1 kg		26	0.56	2.00
Stainless steel (V 2 A, X 5 CrNi 18-10), 2 mm	1 kg		54	6.30	4.80
Non-ferrous metals					
Aluminum, (EN AW-7022 [AlZn5Mg3Cu]), sheets, 2 mm	1 kg		271	38	22.0
Lead, sheets, 2 mm	1 kg		34	1.9	2.3
Titanium zink, (Pure zink Z1, 0.003 % Titan), sheets, 2 mm	1 kg		45	3.8	2.6
Copper, sheets, 2 mm	1 kg		37	4.6	2.5
Glass					
Float glass, $\rho = 2{,}500$ kg/m³	1 kg		14	0.08	0.88
Plastics					
Polyethylene (PE-HD), foil	1 kg	41	75	0.09	1.82
Poly vinyl chloride (PVC-P), compound, as roof membrane	1 kg	17	61	2.10	2.28
Polymethyl methacrylate (PMMA acrylic), plate	1 kg	24	87	0.29	3.39
Polytetrafluorethylene (PTFE Teflon), coating	1 kg	8.3	295	2.50	16.20
Polystyrene resin (UP)	1 kg	32	115	0.45	4.68
Epoxy (EP)	1 kg	ca. 30	137	0.78	6.47
Chloride butadiene (CR Neoprene), support bearings	1 kg	ca. 25	96	0.96	3.65
Silicone (Si), caulking	1 kg	ca. 25	91	30.0	4.07

Table 12
Primary energy content of building materials (3.6 MJ=1 kWh)

Source: Energieatlas, Nachhaltige Architektur, Hegger, Fuchs, Stark, Zeumer, Birkhäuser Verlag, 2007

Material	Bezugs-wert	Heiz-wert [MJ]	PEI nicht ern. [MJ]	PEI ern. [MJ]	GWP Treibhaus-effekt [kg CO_2eq]
Naturstein					
Granit (große Transportentfernung), poliert, ρ = 2.750 kg/m³	1 m³		9.837	332	626
Marmor (mittlere Transportentfernung), poliert, ρ = 2.700 kg/m³	1 m³		6.749	249	422
Schiefer (ortsnah), ρ = 2.700 kg/m³	1 m³		4.608	165	286
Sandstein (ortsnah), gesägt, ρ = 2.500 kg/m³	1 m³		4.099	153	253
Lehmbaustoffe					
Stampflehm, ρ = 2.200 kg/m³	1 m³		158	1	9,7
Lehmsteine (Grünlinge), ρ = 1.200 kg/m³	1 m³		1.257	4	74,0
Baustoffe mit mineralischen Bindemitteln					
Mörtel und Estriche					
Anhydritmörtel/-estrich, Druckfestigkeitsklasse 20, 2.350 kg/m³	1 m³		655	11,0	43
Magnesiamörtel /-estrich, Druckfestigkeitsklasse 20, 2.000 kg/m³	1 m³		2.439	9,9	348
Zementmörtel /-estrich, Druckfestigkeitsklasse 20, 2.250 kg/m³	1 m³		2.161	27,0	389
Gipsmörtel, Putzmörtelklasse P IV a, ρ = 1.300 kg/m³	1 m³		1.477	9,6	177
Kalk-Zementmörtel, Putzmörtelklasse P II a, ρ = 1.500 kg/m³	1 m³		2.675	28,0	448
Werksteine					
Kalksandstein, ρ = 1.800 kg/m³	1 m³		2.030	117	247
Betonstein (Pflaster), ρ = 2.500 kg/m³	1 m³		1.990	46	310
Porenbetonstein, ρ = 400 kg/m³	1 m³		1.484	81	186
Leichtbetonstein, ρ = 600 kg/m³	1 m³		787	35	97
Beton					
Ortbeton (C 25/30), ρ = 2.340 kg/m³	1 m³		1.549	17	251
Ortbeton (C 35/45), ρ = 2.360 kg/m³	1 m³		1.764	23	320
Betonfertigteil, 2 % Stahl (FE 360 B, C 35/45), ρ = 2.500 kg/m³	1 m³		4.098	86	455
Platten					
Faserzementplatte, ρ = 1.750 kg/m³	1 m³		26.839	116	2.200
Gipsplatte (Typ A), ρ = 850 kg/m³	1 m³		2.655	251	150
keramische Baustoffe					
Hochlochziegel, Außenwand, ρ = 670 kg/m³	1 m³		1.485	638	95
Mauerziegel, Innenwand, ρ = 750 kg/m³	1 m³		1.663	715	107
Vollklinker (KMz), ρ = 1.600 kg/m³	1 m³		4.776	39	301
Steinzeug glasiert, ρ = 2.000 kg/m³	1 m³		6.322	0,060	393
Steinzeug unglasiert, ρ = 2.000 kg/m³	1 m³		7.160	0,070	445
bitumenhaltige Baustoffe					
reines Destillationsbitumen (B 100-B 70)	1 kg		45,6	0,010	0,37
polymermodifiziertes Bitumen (PmB 65 A)	1 kg		35,3	0,020	0,50
Holz und Holzwerkstoffe					
Schnittholz					
Kiefer, 12 % Holzfeuchte (HF) (ortsnah), Darrdichte 450 kg/m³	1 m³	8.775	609	9.512	-792[1]
Western Red Cedar, 12 % HF (Nordamerika), Darrdichte 630 kg/m³	1 m³	12.285	4.485	14.359	-907[1]
Teak, 12 % HF (Brasilien), Darrdichte 660 kg/m³	1 m³	12.870	3.217	13.435	-1.013[1]
Holzwerkstoffe					
Brettschichtholz (BSH), 12 % HF, Darrdichte 465 kg/m³	1 m³	9.300	3.578	13.870	-662[1]
Dreischichtplatte, 12 % HF, Darrdichte 430 kg/m³	1 m³	8.618	2.617	9.387	-648[1]
Bau-Furniersperrholz (BFU), 5 % HF, Darrdichte 490 kg/m³	1 m³	10.175	4.729	15.041	-636[1]
Spanplatte (P 5, V 100), 8,5 % HF, Darrdichte 690 kg/m³	1 m³	13.998	5.818	12.614	-821[1]
Oriented Strand Board (OSB), 4 % HF, Darrdichte 620 kg/m³	1 m³	12.555	4.593	16.479	-839[1]
mitteldichte Faserplatte (MDF), 7,5 % HF, Darrdichte 725 kg/m³	1 m³	15.843	9.767	12.495	-515[1]
Metall					
Eisenmetalle					
Gusseisen, Guss (GG 20; sekundär), GJL	1 kg		10	0,49	0,97
Baustahl, Warmwalzprofil (FE 360 B)	1 kg		24	0,54	1,70
Betonstahlmatten (sekundär)	1 kg		13	0,24	0,83
wetterfester Stahl, Kaltband (WT St 37-2), 2 mm	1 kg		26	0,56	2,00
Edelstahl (V 2 A, X 5 CrNi 18-10), 2 mm	1 kg		54	6,30	4,80
Nichteisenmetalle					
Aluminiumlegierung (EN AW-7022 [AlZn5Mg3Cu]), Blech, 2 mm	1 kg		271	38	22,0
Blei, Blech, 2 mm	1 kg		34	1,9	2,3
Titanzink (Reinzink Z1, 0,003 % Titan), Blech, 2 mm	1 kg		45	3,8	2,6
Kupfer, Blech, 2 mm	1 kg		37	4,6	2,5
Glas					
Floatglas, ρ = 2.500 kg/m³	1 kg		14	0,08	0,88
Kunststoff					
Polyethylen (PE-HD), Folie	1 kg	41	75	0,09	1,82
Polyvinylchlorid (PVC-P), Compound für Dachbahn	1 kg	17	61	2,10	2,28
Polymethylmethacrylat (PMMA "Plexiglas"), Platte	1 kg	24	87	0,29	3,39
Polytetrafluorethylen (PTFE "Teflon"), Beschichtung	1 kg	8.3	295	2,50	16,20
Polyesterharz (UP)	1 kg	32	115	0,45	4,68
Epoxidharz (EP)	1 kg	ca. 30	137	0,78	6,47
Chlor-Butadien-Kautschuk (CR "Neopren"), Lager	1 kg	ca. 25	96	0,96	3,65
Silikon (SI), Dichtungsmasse	1 kg	ca. 25	91	30,0	4,07

Tabelle 12
Primärenergieinhalt von Baustoffen (3,6 MJ=1 kWh)

Quelle: Energieatlas, Nachhaltige Architektur Hegger, Fuchs, Stark, Zeumer Birkhäuser Verlag, 2007

10.4 Insulation materials

A central role in the lowering of energy used in buildings is played by insulation materials—the better the insulation, the lower are the energy losses (or gains) between outside and inside, and temperatures of the building's surfaces and the ambient air temperatures become more and more the similar. Insulation materials, therefore, serve these functions:

- insulation against cold outside air temperatures in winter,
- insulation against high outside air temperatures in summer,
- protection of the building assembly against condensate formation, humidity, and freezing,
- limited acoustical protection against noise.

At the center of energetic optimization of materials is the value of thermal conductivity λ (W/m K), which is the property of a material to conduct heat from an area of an energy high to a low. This thermal energy may be transported by transmission, radiation, or convection, and all three processes are part of the transport of thermal energy at different magnitudes. The lower the thermal conductivity of a material, the lower it's energy loss or gain will be. Today's insulation materials are precisely categorized according to thermal conductivity specifications, and **Table 14** lists the data for insulation materials that are typically available and in use today, in Central Europe.

Thermal energy transport through a material is also dependent upon its density and internal structure: the less dense a material is, the lower will be its thermal conductivity because of a higher content of encapsulated air molecules per volume unit. Air serves as a good insulator when its convection remains small, which is achieved when it is entrapped in a material in small cavities. Inert gases, such as Xenon, Krypton or Argon, exhibit even smaller amounts of thermal conductivity than air, which makes them prime media for insulation glass fills. It is also important to remember that humidity in building assemblies increases thermal energy transmission because enclosed water molecules have a greater thermal conductivity than air. Therefore, it is essential to protect insulation materials from condensate or any other humidity saturation.

Independent of its insulation qualities, air becomes the energy carrier in convective processes. The thermal insulation capacity of a layer of air does not behave linearly but instead reaches its maximum at a thickness of around 60 mm. Consequently, thicknesses smaller or greater than this have less insulation capability. In order to increase the thermal insulation of materials or to reduce their convection capacity, fillings with inert gases are used. Also, convection does not occur in a vacuum, and this principle is utilized when vacuum insulating panels (VIPs) or evacuated glazing systems are employed.

10.4 Dämmstoffe

Von zentraler Bedeutung für die Reduktion des Energiebedarfs eines Gebäudes ist der Wärmeschutz. Je besser der Wärmeschutz, desto geringer sind die Energieverluste bei Temperaturdifferenzen zwischen innen und außen und umso mehr nähern sich die jeweiligen Oberflächen- und Lufttemperaturen einander an. Dämmstoffe dienen somit dem:

- Wärmeschutz im Winter (kalte Außentemperaturen),
- Wärmeschutz im Sommer (warme Außentemperaturen),
- Schutz der Baukonstruktion vor Kondensat, Feuchte und Frost,
- unter Umständen Schallschutz.

Im Mittelpunkt der energetischen Materialoptimierung steht die Wärmeleitfähigkeit λ (W/m K). Sie beschreibt den Wärmefluss, der als Bewegung von Atomen oder als Welle von einem Wärmeüberschuss zu einer Wärmesenke strömt. Die Wärmeenergie bewegt sich mittels Transmission, Strahlung und Konvektion. Alle drei Prozesse, wenn auch in unterschiedlichem Umfang, sind am Wärmetransport beteiligt. Je niedriger die Wärmeleitfähigkeit eines Materials ist, umso geringer ist der durch das Bauteil entstehende Energieverlust oder Energiegewinn. Die Isolierstoffe werden heute je nach Wärmeleitfähigkeit genau klassifiziert. In **Tabelle 14** sind Kenndaten von Dämmstoffen, wie sich üblicherweise in Mitteleuropa eingesetzt werden, dargestellt.

Die Wärmetransmission durch ein Material ist abhängig von seiner Masse und seiner inneren Struktur, d.h. je geringer die Dichte eines Baustoffes, desto geringer ist auch die Wärmeleitfähigkeit. Dies ergibt sich aus dem erhöhten volumenbezogenen Luftanteil im Material. Luft wirkt allerdings nur dann als Dämmung, wenn die Konvektion gering bleibt. Insofern wird dieser Anspruch dadurch erreicht, dass kleine Hohlräume gebildet werden. Edelgase besitzen eine noch geringere Wärmeleitfähigkeit als Luft. Diesen Umstand macht man sich zunutze bei der Isolierverglasung, bei der zwischen Scheiben von Wärmeschutzgläsern Argon, Krypton oder Xenon eingelagert wird.

Beachtenswert bei der Dämmung von Gebäuden ist, dass Feuchtigkeiten in Bauteilen die Transmission verringern, da das eindringende Wasser über eine höhere Wärmeleitfähigkeit verfügt. Insofern ist immer darauf zu achten, dass Wasserdampf nicht in Isoliermaterialien kondensiert und die Dämmung durchfeuchtet.

Unabhängig von der Dämmfähigkeit ist das Gasgemisch Luft bei Konvektionsprozessen der Energieträger. Die Wärmedämmleistung der Luftschicht ist nicht linear, sondern erreicht ihr Maximum bei einer Dicke von ca. 60 mm.

Table 14
Specifications of insulation materials

Source: Energieatlas, Nachhaltige Architektur, Hegger, Fuchs, Stark, Zeumer, Birkhäuser Verlag, 2007

Tabelle 14
Kenndaten von Dämmstoffen

Quelle: Energieatlas, Nachhaltige Architektur Hegger, Fuchs, Stark, Zeumer Birkhäuser Verlag, 2007

Insulation materials	Density [kg/m³]	Thermal transmission factor λ_B [W/mK]	Vapor diffusion resistance μ [-]	Class, flammables [1] [-]	Product norm	Product form
Anorganic						
Calcium silicate	115–290	0.045–0.070	2/20	A1–A2/bis A1	[2]	Board
Glass wool	12–250	0.035–0.050	1/2	A1–B1/bis A1	DIN EN 13162	Board, fleece, wool
Foam glass	100–150	0.040–0.060	Vapor tight	A1/A1	DIN EN 13167	Board, fill
Expanded polystyrene (EPS)	60–300	0.050–0.065	2/5	A1–B2/bis A1	DIN EN 13169	Boar, fill
Aerated clay	260–500	0.100–0.160	2	A1/A1	DIN EN 14063	Fill
Vermiculite	60–180	0.065–0.070	2/3	A1/A1	[2]	Fill
Organic						
Polystrole fiber	15–45	0.035–0.045	1	B1–2/bis B	[2]	Fleece
Polystyrene hard foam	15–30	0.035–0.040	20/100	B1/bis B	DIN EN 13163	Board
Extruded polystyrene	25–45	0.030–0.040	80/250	B1/bis B	DIN EN 13164	Board
Polyurethane foam	≥ 30	0.020–0.035	30/100	B1–2/bis B	DIN EN 13165	Board, Foam
Cotton	20–60	0.040–0.045	1/2	B1/bis B	[2]	Fleece, wool, blow-in
Flax	25	0.040–0.045	1/2	B1/bis B	[2]	Board, fleece, wool
Hemp fibers	20–70	0.040–0.045	1/2	B2/bis D	[2]	Board
Wood fiber insulation	45–450	0.040–0.070	1/5	B2/bis D	DIN EN 13171	Board
Wood wool board	360–570	0.065–0.090	2/5	B1/bis B	DIN EN 13168	Board
Coco fiber board	50–140	0.045–0.050	1/2	B1–B2/bis B	DIN 18165-1/-2	Fleece, wool
Expanded cork	80–500	0.040–0.055	5/10	B1–B2/bis B	DIN EN 13170	Fill, board
Sheep wool	20–80	0.035–0.040	1/2	B1–B2/bis B	[2]	Fleece, wool
Cellulose	30–100	0.035–0.040	1/2	B1–B2/bis B	[2]	Blow-in, board
New materials						
IR-Absorber modified extruded polystyrene	15–30	0.032	20/100	B1/bis B	DIN EN 13163	Board
Transparent insulation (TWD)	[4]	0.02–0.1[3]	Vapor tight	[4]	[2]	Panel
Vaccum insulation panels (VIP)	150–300	0.004–0.008	Vapor tight	B2 [2]	[2]	Panel

[1] The listed flammable classes are general values and need to be verified by manufacturer specification.
[2] Code compliant
[3] The material uses static insulation values and solar gains. The listed values include solar gains. Results may vary due to geographical location
[4] Varying due to manufacturer specification

Dämmmaterial	Rohdichte [kg/m³]	Bemessungswert der Wärmeleitfähigkeit λ_B [W/mK]	Dampfdiffusionswiderstandszahl μ [-]	Baustoffklasse / Brennbarkeitsklasse [1] [-]	Produktnorm	Produktform
anorganisch						
Kalziumsilikat	115–290	0,045–0,070	2/20	A1–A2/bis A1	[2]	Platte
Glaswolle/Steinwolle	12–250	0,035–0,050	1/2	A1–B1/bis A1	DIN EN 13162	Platte, Vlies, Stopfwolle
Schaumglas (CG)	100–150	0,040–0,060	prakt. dampfdicht	A1/A1	DIN EN 13167	Platte, Schüttung
expandierte Perlite (EPB)	60–300	0,050–0,065	2/5	A1–B2/bis A1	DIN EN 13169	Platte, Schüttung
Blähton	260–500	0,100–0,160	2	A1/A1	DIN EN 14063	Schüttung
Vermikulite (Blähglimmer)	60–180	0,065–0,070	2/3	A1/A1	[2]	Schüttung
organisch						
Polyesterfaser	15–45	0,035–0,045	1	B1–2/bis B	[2]	Vlies
Polystyrol-Hartschaum (EPS)	15–30	0,035–0,040	20/100	B1/bis B	DIN EN 13163	Platte
Polystyrol-Extruderschaum (XPS)	25–45	0,030–0,040	80/250	B1/bis B	DIN EN 13164	Platte
Polyurethan-Hartschaum (PUR)	? 30	0,020–0,035	30/100	B1–2/bis B	DIN EN 13165	Platte, Ortschaum
Baumwolle	20–60	0,040–0,045	1/2	B1/bis B	[2]	Matte, Filz, Stopfwolle, Einblasware
Flachs	25	0,040–0,045	1/2	B1/bis B	[2]	Platte, Matte, Filz, Stopfwolle
Hanffasern	20–70	0,040–0,045	1/2	B2/bis D	[2]	Platte
Holzfaserdämmplatte (WF)	45–450	0,040–0,070	1/5	B2/bis D	DIN EN 13171	Platte
Holzwollplatte (WW)	360–570	0,065–0,090	2/5	B1/bis B	DIN EN 13168	Platte
Kokosfaser	50–140	0,045–0,050	1/2	B1–B2/bis B	DIN 18165-1/-2	Matte, Filz, Stopfwolle
expandierter Kork (ICB)	80–500	0,040–0,055	5/10	B1–B2/bis B	DIN EN 13170	Schüttung, Platte
Schafwolle	20–80	0,035–0,040	1/2	B1–B2/bis B	[2]	Matte, Filz, Stopfwolle
Zellulosefaser	30–100	0,035–0,040	1/2	B1–B2/bis B	[2]	Einblasware, Platte
"innovative" Dämmstoffe						
IR-Absorber-modifiziertes EPS	15–30	0,032	20/100	B1/bis B	DIN EN 13163	Platte
transparente Wärmedämmung (TWD)	[4]	0,02–0,1[3]	prakt. dampfdicht	[4]	[2]	Paneel
Vakuumisolationspaneel (VIP)	150–300	0,004–0,008	prakt. dampfdicht	B2 [2]	[2]	Paneel

[1] Die angegebenen Brennbarkeitsklassen stellen Richtwerte dar. Sie sind mit den tatsächlichen Produktdaten abzugleichen.
[2] bauaufsichtlich zugelassen
[3] Das Dämmmaterial nutzt die statische Dämmwirkung sowie solare Gewinne. Die hier dargestellten Werte sind inklusive solaren Gewinnen über eine Heizperiode in Deutschland gemittelt. Es kann je nach Klima und Ausrichtung der Dämmung zu deutlichen Unterschieden kommen.
[4] stark produktabhängig

Thermal radiation through insulated glazing or panels is lowered by reduced transmission of infrared radiation or emissivity. Selective coatings on glazing lites helps to reflect long-wave infrared radiation, while the transmission of visible daylight radiation is not reduced. The advances in selective reflecting coatings for glazing lites have decreased thermal conductivity by 25 % or more.

In addition to today's advanced vacuum insulation systems, new methods of improving thermal insulation are being developed. One such system includes phase-change materials (PCMs), which are substances with a high heat of fusion that, through melting and solidifying at certain temperatures, are capable of storing or releasing large amounts of energy. A 3-ply insulating glass construction provides excellent heat insulation with a U-value of < 0.5 W/m²K, and a prismatic glass implemented in the space between the panes reflects the sun's rays with an angle of incidence of more than 40 ° (in summer, when the sun is high in the sky). On the other hand, the low winter sun passes through the sun protection at full intensity and falls on a PCM element such as that found in the product GLASSXcrystal, **Figure 126**. It is a heat storage module that receives and stores the solar energy and, after time, releases it again as radiant heat, **Figures 127.1 – 127.3**. PCMs use a salt hydrate as the storage material, and the heat is stored by melting the PCM; the stored heat is released again when the PCM cools. The salt hydrate is hermetically sealed in polycarbonate containers that are painted gray to improve the absorption efficiency, and on the interior side the element is sealed by 6-mm tempered safety glass that can also be printed with any ceramic silk-screen fritting.

Demnach haben dickere oder dünnere Luftschichten geringere Dämmwirkungen. Zur Verbesserung der Dämmwirkung, d.h. Reduktion der Konvektion, können dämmende Volumina mit Edelgasen gefüllt werden.

Im Vakuum erfolgt keine Konvektion. Diesen Umstand nutzt man durch die Verwendung von Vakuum-Isolationspanelen oder unter Umständen in Zukunft durch Vakuum-Scheiben.

Die Wärmestrahlung bei Dämmscheiben oder Dämmmaterialien wird durch die Verringerung des Infrarot-Strahlungsdurchgangs (Emissivität) beeinflusst. Selektiv reflektierende Schichten auf Materialien verringern oder verhindern den Infrarot-Strahlungsdurchgang. Selektiv reflektierende Schichten auf Scheiben helfen somit, den Durchgang langwelliger Strahlen (Infrarotstrahlung) zu verringern oder zu vermeiden. Der Strahlungsdurchgang sichtbaren Lichts wird dadurch nicht geschmälert. Die Aufbringung selektiver, reflektierender Schichten auf Gläser hat dazu geführt, dass die Wärmedurchgangskoeffizienten um ca. 25 Prozent und mehr verringert werden konnten.

Neben der Vakuum-Isolation als eins der innovativen Dämmmaterialien kommen heute zusätzlich PCM-gefüllte Fassadenelemente zum Einsatz (phase change material). Ein Dreifach-Isolierglasaufbau sorgt für eine exzellente Wärmedämmung bei einem u-Wert von < 0,5 W/m²K, ein in den Scheibenzwischenraum implementiertes Prismenglas reflektiert die hochstehende Sommersonne mit Einfallswinkeln > 40 ° nach außen. Tiefe Sonneneinfallswinkel (z.B. Wintersonne) führen dazu, dass ein verlustfreier Durchgang der Strahlung durch eine äußere Scheibe und das Prismenelement erfolgt und auf ein Wärmespeichermodul (z.B. GLASSXcrystal) auftrifft, **Bild 126**. Dieses Wärmespeichermodul nimmt solare Wärmeenergie auf, speichert sie und gibt sie zeitverzögert als angenehme Strahlungswärme während der Abendstunden oder Nacht ab, **Bilder 127.1 – 3**. Das Speichermaterial (PCM) in Form eines Salzhydrats speichert die Wärme durch Aufschmelzen des Wärmespeichermoduls. Beim Abkühlen (keine oder geringe Sonneneinstrahlung) wird die gespeicherte Wärme wieder abgegeben. Das Salzhydrat ist in Polycarbonatbehältern hermetisch eingeschweißt, die zur Verbesserung der Absorptionswirkung grau eingefärbt sein können. Raumseitig wird das Element durch eine 6 mm Einscheiben-Sicherheitsglasscheibe abgeschlossen und kann mit einem keramischen Siebdruck bedruckt werden.

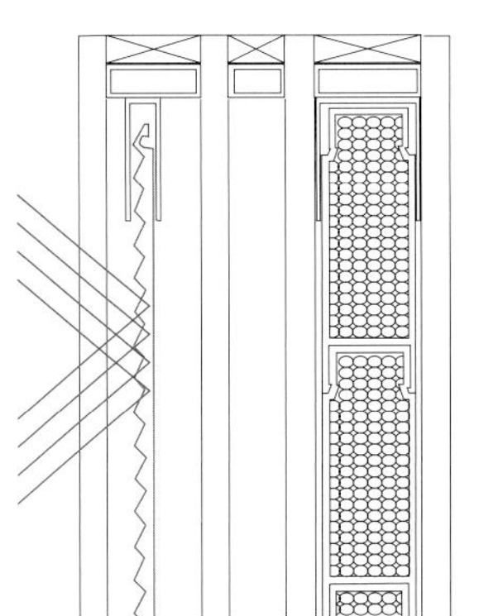

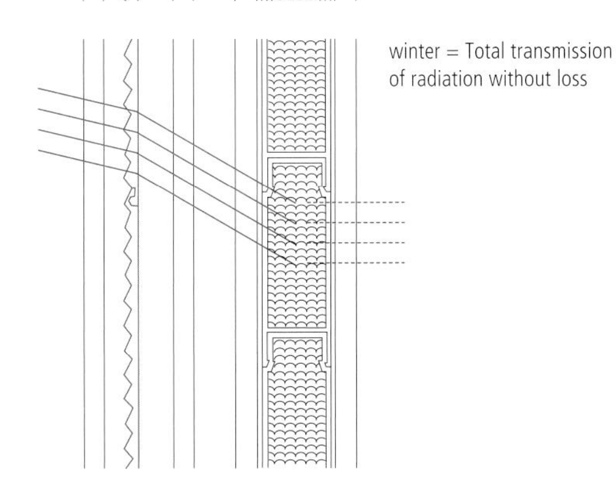

Figure 126
Functional principle of GLASSX®crystal function

Source: GLASSX®crystal, Zurich, Switzerland

Bild 126
Die Funktionsweise von GLASSX®crystal

Quelle: GLASSX®crystal, Zürich

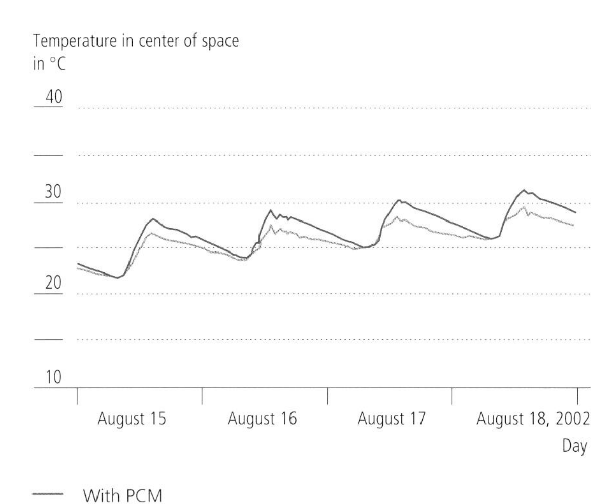

Figure 127.1
Functioning of phase-change-material (PCM).
Measurements on clear days.

Source: Knauf, ZAE Bayern

Bild 127.1
PCM-Wirkungsweise im System (Phase-Change Material)
Messungen an klaren Tagen

Quelle: Knauf, ZAE Bayern

Surface temperature
in °C

Global radiation, vertical
in W/m²

60 / 600
40 / 400
20 / 200
0 / 0

10 12 14 16 18 20 22 24 2 4 6
Time

- Surface temperature, exterior light, outsideTemperatur
- Surface temperature on PCM, outside
- Temperature in PCM center
- Surface temperature glass, inner light, inside
- Temperature on prismatic plate
- Air temperature, outside
- Surface temperature on PCM, inner
- Global radiation in W/m²

Figure 127.2
Summer temperatures on various surfaces of the GLASSX®crystal phase-change module under phase-change conditions.

Source: GLASSX®crystal, Zürich, Switzerland

Bild 127.2
Temperaturen im Sommer an verschiedenen Elementen bei Phasen- und Oberflächen-veränderungen mit PCM-Element „GLASSX®crystal"

Quelle: GLASSX®crystal, Zürich

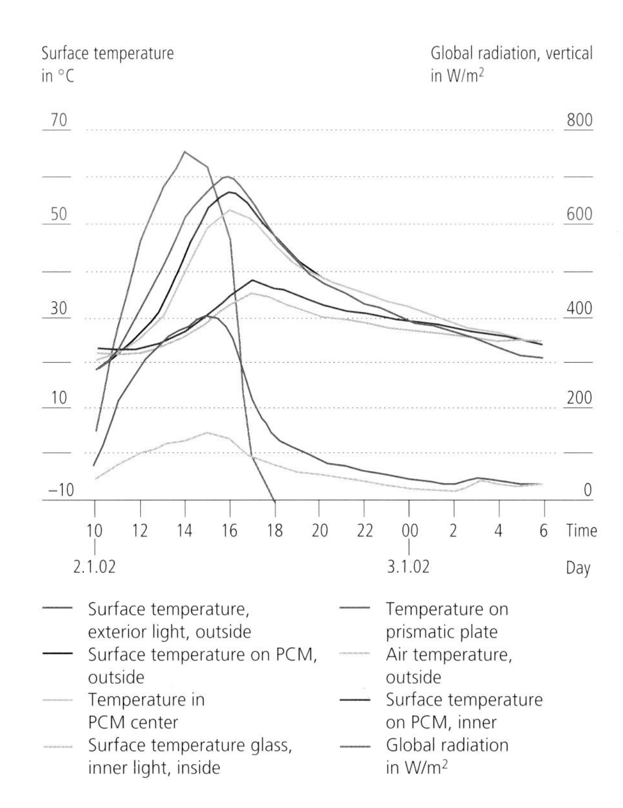

Figure 127.3
Winter temperatures on various surfaces of the GLASSX®crystal phase-change module under phase-change conditions.

Source: GLASSX®crystal, Zürich, Switzerland

Bild 127.3
Temperaturen im Winter an verschiedenen Elementen und Oberflächen „GLASSX®crystal"

Quelle: GLASSX®crystal, Zürich

Figure 128 shows an example of a building with a high-quality building envelope containing PCM modules. The **Figures 128.1 – 3** explain its function during summer months. The advantage of a PCM-equipped façade, like that of transparent insulation materials (TIM), is its capability to let daylight pass through the insulating layer and enter a space. Light transmission for crystalline PCM in winter is 0 – 20 %, and for liquid PCM in winter it is ~ 4 – 40 %. The maximum storage capacity of a glazing system equipped with PCMs is around 1,180 Wh/m² at temperatures of 26 – 28 °C.

Bild 128 zeigt ein Gebäude mit einer hochwertigen Fassadenstruktur mit dem PCM-Speicher, **Bilder 128.1 – 3** zeigen die Wirkungsweise im sommerlichen Betrieb. Der Vorteil einer entsprechenden Kombination, in seiner Wirkungsweise ähnlich einer TWD-Fassade (transparente Wärmedämmung), liegt darin, dass durch das Dämmmaterial gleichzeitig Licht dringt (Lichttransmission bei kristallinem PCM 0 – 20 Prozent, bei flüssigem PCM 4 – 40 Prozent. Die maximale Speicherfähigkeit einer entsprechenden Scheibe beträgt ca. 1.180 Wh/m² bei Speichertemperaturen von 26 – 28 °C.

Figure 128
Senior residence Domat/Ems with grey GlassX®crystal cavities, interior glazing fritted, white

Source: GlassX architecture, Zürich 2004, Switzerland

Bild 128
Altenwohnen Domat/Ems Referenz GlassX®crystal anthrazit eingefärbte PCM-Behälter, raumseitiges Glas mit Siebdruck weiß

Quelle: GLASSX architecture, Zürich
Speichern Wärmen Kühlen

Figure 128.1
GLASSX®crystal in crystalline state.

Source: GLASSX®crystal, Zürich, Switzerland

Bild 128.1
GLASSXcrystal im kristallinen Zustand

Quelle: GlassX®crystal

Figure 128.2
GLASSX®crystal under phase-change conditions.

Source: GLASSX®crystal, Zürich, Switzerland

Bild 128.2
GLASSXcrystal im Phasenwechsel

Quelle: GlassX®crystal

Figure 128.3
GLASSX®crystal in liquid state.

Source: GLASSX®crystal, Zürich, Switzerland

Bild 128.3
GLASSXcrystal im flüssigen Zustand

Quelle: GlassX®crystal

10.5 Façades

Façades are the separators between inside and outside conditions, and therefore they are the most important element of a building assembly for achieving reductions in the cost of primary energy, operating cost, and construction cost as well.

Building layers should be as capable of reacting to changing outside climatic conditions as humans themselves – e.g., during cold seasons, the façade should protect a building by providing excellent insulation. Alternatively, in summer, when outside temperatures are high, the building's envelope should have an equally high insulation capacity to prevent thermal gains from outside to inside during the day, yet during night the degree of insulation should be able to be decreased (for example, to a U-value of $< 5.0\,W/m^2K$) to allow the interior space to be cooled down by convection and radiation to the cooler outside environment.

During periods of cold outside temperatures, it is desirable to allow radiation to enter the building, while during the summer this effect must be avoided. Nevertheless, both during summer and winter operation a building's envelope should be capable of allowing as much daylight to enter as possible to illuminate the interior and mitigate the use of artificial lighting.

Surfaces of building façades should also be designed in such a way that long-wave radiation does not cause excessive heating of the envelope, and during cold seasons the loss of thermal energy due to radiation from the inside to the outside needs to be limited.

To achieve satisfactory thermal comfort for occupants, surface temperatures on façade interiors should be no more than 3 K below the air temperature of that space, and in summer the value should not exceed 3 – 4 K above ambient air temperature.

In addition, façades are of course the important visual connection element between inside and outside, and vice versa. Building envelopes, without question, need to respond to a multitude of requirements at once. **Figure 129** presents a matrix of active and passive technologies that can be implemented in order to adjust annual thermal loads, but it must be recognized that a building façade will not be capable of addressing all conditions to a user's full satisfaction, such as thermal, hygienic, and visual comfort. Active technical systems will need to be implemented as well to provide completely satisfactory interior conditions.

10.5 Fassaden

Fassaden, der Grenzbereich zwischen außen und innen, sind das wesentlichste Bauelement, geht es darum, Energie- und Betriebskosten und auch Investitionskosten zu senken.

Die die Lebensräume umgebenden Hüllen sollten sich im Jahresbetrieb so verhalten können, wie es der Mensch tut. Bei kalten Außenbedingungen sollte die Fassade den Innenraum (Körper) extrem gut dämmen. Bei hohen Außentemperaturen sollte die Dämmfähigkeit einer Fassade während des Tages gleichermaßen gegeben sein, um den Wärmefluss (Konvektion) von außen nach innen zu reduzieren. Während der Nacht soll die Dämmfähigkeit deutlich geringer sein (z.B. u-Wert $< 5{,}0\,W/m^2K$), um den Raum gegen den kühlen Außenraum zu entwärmen (Konvektion und Strahlung).

Während es bei kalten Außenbedingungen angenehm und wünschenswert ist, dass Strahlungsenergie in den Innenbereich von Gebäuden gelangt, muss dies im Regelfall im Sommer weitestgehend vermieden werden. Ob Sommer- oder Winterbetrieb – es ist wünschenswert, gleichzeitig und jederzeit einen möglichst hohen Tageslichteintrag zu erreichen, um Räume ohne Kunstbelichtung nutzen zu können.

Die Oberflächenstrukturen der Fassaden sollen zudem so ausgebildet werden, dass infolge der Aufheizung der Fassadenflächen kein zu hoher langwelliger Wärmeeintrag erfolgt – andererseits jedoch sollte die Wärmeabstrahlung im langwelligen Bereich (Infrarotstrahlung) im Winter von innen nach außen möglichst unterbunden werden.

Um eine hohe thermische Behaglichkeit zu erreichen, sollten die inneren Oberflächentemperaturen der Fassaden im Winter nicht mehr als 3 K unter Raumtemperatur und im Sommer möglichst nicht mehr als 3 – 4 K über Raumtemperatur liegen.

Zudem ist die Fassade ein wichtiges Verbindungselement zwischen Innen- und Außenraum, d.h. visuelle Bezüge von innen nach außen und unter Umständen von außen nach innen müssen gegeben sein. Wie unschwer zu erkennen, muss eine Fassade viel leisten. In **Bild 129** ist ein Übersichts- und Ablaufplan möglicher aktiver und passiver Maßnahmen zur Lastanpassung während des Jahres dargestellt. Eine Fassade wird nicht alle Ansprüche bezüglich thermischer Behaglichkeit, hygienischer Behaglichkeit und visueller Behaglichkeit erfüllen können, und insofern greifen unterstützend technische Maßnahmen ein, um alle gewünschten Raumzustände zu erreichen.

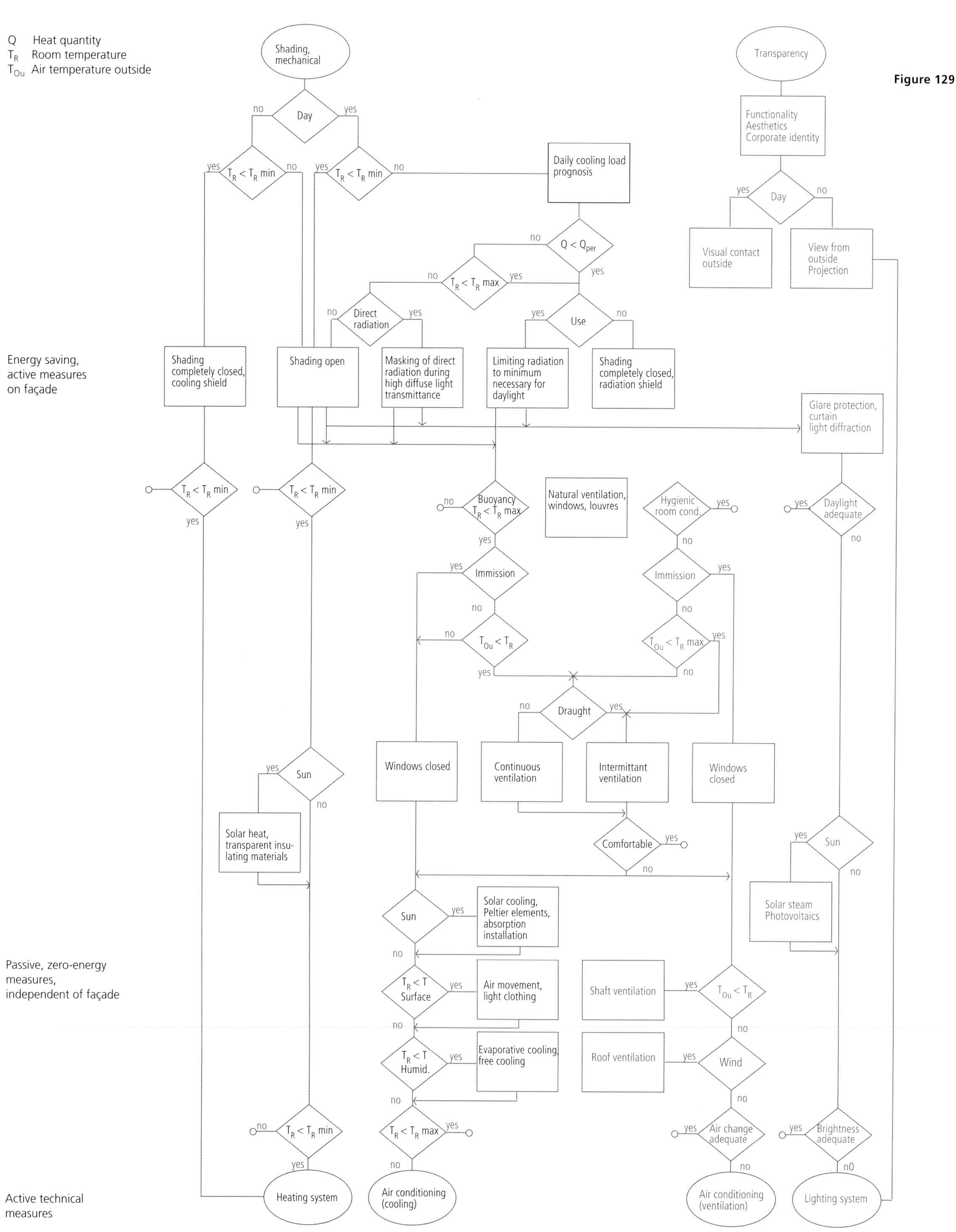
Thermal comfort
Air hygiene
Visual comfort
Q Heat quantity
T_R Room temperature
T_{Ou} Air temperature outside
Shading, mechanical
Day
no
yes
$T_R < T_R$ min
Daily cooling load prognosis
$Q < Q_{per}$
$T_R < T_R$ max
Direct radiation
Use
Transparency
Functionality
Aesthetics
Corporate identity
Visual contact outside
View from outside
Projection
Energy saving, active measures on façade
Shading completely closed, cooling shield
Shading open
Masking of direct radiation during high diffuse light transmittance
Limiting radiation to minimum necessary for daylight
Shading completely closed, radiation shield
Glare protection, curtain light diffraction
Buoyancy $T_R < T_R$ max
Natural ventilation, windows, louvres
Hygienic room cond.
Daylight adequate
Immission
$T_{Ou} < T_R$
$T_{Ou} < T_R$ max
Draught
Sun
Windows closed
Continuous ventilation
Intermittant ventilation
Windows closed
Solar heat, transparent insulating materials
Comfortable
Solar steam
Photovoltaics
Solar cooling, Peltier elements, absorption installation
$T_R < T$ Surface
Air movement, light clothing
Shaft ventilation
$T_R < T$ Humid.
Evaporative cooling, free cooling
Roof ventilation
Wind
Passive, zero-energy measures, independent of façade
Air change adequate
Brightness adequate
nO
Active technical measures
Heating system
Air conditioning (cooling)
Air conditioning (ventilation)
Lighting system

Figure 129

Figure 129
Overview and flow chart of possible active/passive measures for load compensation

Bild 129
Übersichts- und Ablaufplan möglicher aktiver/passiver Maßnahmen zur Lastanpassung

The following presents an overview of essential influential categories that are important to consider when designing façades:

Visible transmittance τ

Primary visible light transmittance is the percentage of visible light (wavelengths of 380–780 nm) striking the glass that penetrates to the interior, expressed as a number between 0 and 1.

U-value

The U-value is a measure of the thermal conductivity of a window – the lower the U-value, the better a window is at limiting heat losses or gains. A single-glazed window has a U-value of about 6 W/m^2K, while triple-glazed windows have U-values between 1 and 2. U-value is a function of the space between the lites of a window assembly and the gas with which this space is filled (i.e., air or inert gas). Improved U-values can also be achieved with low-emissivity (low-E) coatings, which are microscopically thin, virtually invisible, metal or metallic oxide layers deposited on a window or skylight glazing surface primarily to reduce the U-value by suppressing radiant heat flow.

Shading coefficient (b-value)

The shading coefficient, also called the b-factor of a window, is used to calculate the cooling load (VDI 2078, 1996). It is the ratio between the g-value of the respective glazing system and the g-value of a regular two-pane window. The g-value (see definition below) of this element is set at a constant of 80 %, but in the case of a single-pane window the constant is 87 %. Therefore, the average b-value is b = g ÷ 0.8.

Solar heat gain coefficient (SHGC, U.S.; UK/Europe: g-value)

The g-value is the sum of the primary transmittance (the T-value) and the secondary transmittance, which is the ratio of the inwardly flowing fraction of the solar energy absorbed by the window (or shading device) to the solar insolation for wavelengths of 320 – 2,500 nm.

Reduction factor F_c for shading systems

The reduction factor F_c (formerly Z) is a value between 0 and 1. The smaller the value, the better a shading device is at reducing solar transmittance.

Table 15 allows an overview of technical specifications of various insulated glazing types. Also given are additional categories for evaluating glazing, such as light reflection and color rendering index (CRI). CRI should be, in the case of a high-quality façade, > 90 % to assure that all colors of the spectrum are allowed to enter a space and that colors within the building are rendered correctly and an interior coloration scheme is not compromised.

Table 16 shows the values discussed above for currently available transparent building materials.

Bei der Konzipierung von Fassaden spielt eine Reihe von Faktoren eine Rolle, die es zu beachten und zu händeln gilt:

Lichtdurchlässigkeitsfaktor τ

Der Lichtdurchlässigkeitsfaktor t gibt an, wie viel Prozent des Tageslichtes von außen durch eine Scheibe hindurchtritt. Dabei erfolgt die Angabe der Lichtdurchlässigkeit im Bereich der Wellenlänge des sichtbaren Lichtes von 380 – 780 nm (Nanometer), bezogen auf die Helligkeitsempfindung des menschlichen Auges

Wärmedurchgangskoeffizient u

Der Wärmedurchgangskoeffizient u einer Isolierglasscheibe gibt an, wie viel Energie durch durch Temperaturdifferenzen (außen/innen) durch die Scheibenfläche verloren geht bzw. durch sie gewonnen wird. Je niedriger dieser Wert ist, desto weniger Wärme passiert die Scheibe. Der u-Wert von konventionellen Isolierglasscheiben ist im Wesentlichen abhängig vom Scheibenzwischenraum und dem darin enthaltenen Medium (Luft oder Edelgas). Eine Verbesserung des u-Wertes kann durch aufgedampfte Edelmetallschichten erreicht werden (low-e = low emission coating). **Bild 130** beschreibt die verschiedenen Vorgänge im Scheibenbereich einer hochwertigen Wärmeschutz-Isolierverglasung.

Energiedurchlassfaktor g/b

Die Angabe der Gesamtenergiedurchlässigkeit g erfolgt im Bereich der Wellenlänge von 320 – 2.500 nm. Sie ist die Summe aus der direkten Energiedurchlässigkeit und der Sekundärwärmeabgabe nach innen (Abstrahlung und Konvektion). Der mittlere Durchlassfaktor b (shading coefficient) beschreibt den prozentualen Energiedurchgang einer Doppelscheibe. Somit ist der mittlere Durchlassfaktor b = g÷0,8.

Die Tabelle 15 gibt eine Übersicht technischer Werte verschiedener Isolierglasscheiben an, wobei hier zusätzlich die Lichtreflexion und Farbwiedergabe (Farbwiedergabe-Index) dargestellt sind. Der Farbwiedergabe-Index sollte bei höherwertigen Gebäuden zumindest > 90 Prozent sein, damit sichergestellt wird, dass alle Farben des Farbspektrums des Tageslichts in den Raum gelangen und somit im Raum die Farbgestaltung nicht behindert wird.

Tabelle 16 zeigt die vorbesprochenen Werte für transparente Bauteile, die zurzeit am Markt verfügbar sind.

Figure 130
Solar radiation passing a thermal insulated glazing element

Source: Energieatlas, Nachhaltige Architektur, Hegger, Fuchs, Stark, Zeumer, Birkhäuser Verlag, 2007

Bild 130
Schematische Darstellung des Strahlungsdurchganges durch eine Verglasung

Quelle: Energieatlas, Nachhaltige Architektur Hegger, Fuchs, Stark, Zeumer Birkhäuser Verlag, 2007

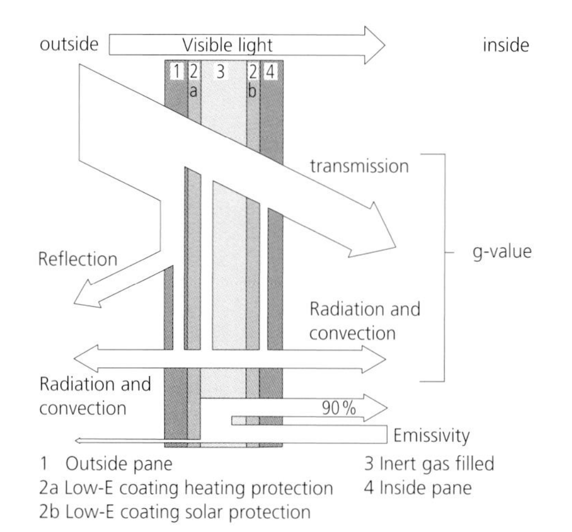

	Thermal insulated glazing Double pane, coating on one surface			Triple insulated glazing Coating on two surfaces.		Quadruple insulated glazing Coating on two surfaces	**Solar protection glazing** Double pane Coating on one surface		
Dimensions (Lite/air (gas) space/Lite) [mm]	4-15-4			4-12-4-12-4		6-12-6-12-6-12-6	6-16-4	6-16-4	6-16-4
	ε ≤ 0,05			ε ≤ 0,05		ε = 0,02	neutral [1]	blue [1]	green [1]
Media in intermediate space (Gas concentration >90%)	Air	Argon	Krypton	Argon	Krypton	Krypton	Argon	Argon	Argon
U_g value according EN ISO 10077-1 U_g [W/m²K]	1.5	1.2	1.1	0.8	0.5	0.3[1]	1.1	1.1	1.1
Total energy transmission coefficient [1] g [%]	64	64	64	52	52	38[1]	37	24	28
Visible light transmission [1] T_L [%]	81	81	81	72	72	59[1]	67	40	55
Light reflection [1] R_L [%]	12	12	9	14	14	18[1]	11/12[2]	10/33[2]	9/12[2]
Color rendition [1] R_a [%]	98	98	98	96	96	no inf.	96/94[2]	95 70[2]	86/88[2]

[1] Examples of manufacturer specification
[2] Values for inside and outside

Table 15

Source: Energieatlas, Nachhaltige Architektur, Hegger, Fuchs, Stark, Zeumer, Birkhäuser Verlag, 2007**Table**

	Wärmeschutzverglasung Zweischeiben-Isoliervergl. eine Scheibe beschichtet			Dreischeiben-Isoliervergl. zwei Scheiben beschichtet		Vierscheiben-Isoliervergl. vierfach beschichtet	**Sonnenschutzverglasung** Zweischeibenverglasung eine Scheibe beschichtet		
Maße (Scheibe / Zwischenraum / Scheibe) [mm]	4-15-4			4-12-4-12-4		6-12-6-12-6-12-6	6-16-4	6-16-4	6-16-4
	ε ≤ 0,05			ε ≤ 0,05		ε = 0,02	farbneutral [1]	blau [1]	grün [1]
Art des Glaszwischenraums (Gaskonzentration ≥ 90 %)	Luft	Argon	Krypton	Argon	Krypton	Krypton	Argon	Argon	Argon
U_g-Wert nach EN ISO 10077-1 U_g [W/m²K]	1,5	1,2	1,1	0,8	0,5	0,3[1]	1,1	1,1	1,1
Gesamtenergiedurchlasskoeffizient [1] g [%]	64	64	64	52	52	38[1]	37	24	28
Lichtdurchlässigkeit [1] T_L [%]	81	81	81	72	72	59[1]	67	40	55
Lichtreflexion [1] R_L [%]	12	12	9	14	14	18[1]	11/12[2]	10/33[2]	9/12[2]
Farbwiedergabe [1] R_a [%]	98	98	98	96	96	k. A.	96/94[2]	95 70[2]	86/88[2]

[1] exemplarische Herstellerangaben
[2] Werte gelten für innen / außen

Tabelle 15
Technische Werte verschiedener Isolierglasscheiben

Quelle: Energieatlas, Nachhaltige Architektur Hegger, Fuchs, Stark, Zeumer Birkhäuser Verlag, 2007

Transparent enclosure elements		U_g Value according DIN EN 673	Daylight transmission factor	g-Value according DIN EN 410
Single pane glazing	Float glass	5.8	0.90	0.85
	Float glass with low-E coating	3.8	0.67	0.62
	White glass	5.8	0.92	0.92
	Laminated single glazing with embedded solar protection film	5.8	0.75	0.52
Double insulated glazing	Thermal protection glazing, air filled	1.4	0.80	0.63
	Thermal protection glazing, argon filled	1.1	0.80	0.63
	Neutral solar protection glazing, air filled	1.1	0.70	0.41
	Neutral solar protection glazing, argon filled	1.1	0.62	0.34
	Neutral solar protection glazing, argon filled	1.1	0.51	0.28
	Neutral solar protection glazing, argon filled	1.1	0.40	0.24
	Neutral solar protection glazing, argon filled	1.1	0.30	0.19
Triple insulated glazing	Thermal protection glazing with coatings on two surfaces, argon filled	0.7 0.5	0.72 0.72	0.50 0.50
Quadruple insulated glazing	Thermal protection glazing with coatings on two surfaces, krypton filled	0.3*	0.59*	0.38*
Extruded hollow-core Polycarbonat Plate	poly carbonate	1.5*	0.70	0.60*

*Data provided by manufacturer

Table 16

Source: Energieatlas, Nachhaltige Architektur, Hegger, Fuchs, Stark, Zeumer, Birkhäuser Verlag, 2007

Transparente Bauteile		U_g-Wert nach DIN EN 673	Tageslicht-transmission	g-Wert nach DIN EN 410
Einfachglas	Floatglas	5,8	0,90	0,85
	Floatglas mit Low-E-Beschichtung	3,8	0,67	0,62
	Weißglas	5,8	0,92	0,92
	Verbund-Einfachglas mit einlaminierter Sonnenschutzfolie	5,8	0,75	0,52
Zweifach-Isolierglas	Wärmeschutzglas, luftgefüllt	1,4	0,80	0,63
	Wärmeschutzglas, argongefüllt	1,1	0,80	0,63
	neutrales Sonnenschutzglas, argongefüllt	1,1	0,70	0,41
	neutrales Sonnenschutzglas, argongefüllt	1,1	0,62	0,34
	neutrales Sonnenschutzglas, argongefüllt	1,1	0,51	0,28
	neutrales Sonnenschutzglas, argongefüllt	1,1	0,40	0,24
	neutrales Sonnenschutzglas, argongefüllt	1,1	0,30	0,19
Dreifach-Isolierglas	Wärmeschutzglas mit 2 Beschichtungen, argongefüllt	0,7	0,72	0,50
	Wärmeschutzglas mit 2 Beschichtungen, kryptongefüllt	0,5	0,72	0,50
Vierfach-Isolierglas	Wärmeschutzglas mit 4 Beschichtungen, kryptongefüllt	0,3*	0,59*	0,38*
Stegplatte	Polycarbonat	1,5*	0,70	0,60*

*Herstellerangabe

Tabelle 16
Technische Werte verschiedener Isolierglasscheiben

Quelle: Energieatlas, Nachhaltige Architektur Hegger, Fuchs, Stark, Zeumer Birkhäuser Verlag, 2007

A weak area of transparent building materials in an energy sense is the frames in which they typically are mounted. Under normal circumstances, window glass has to be inserted in a frame, which sometimes is operable or is the connecting element to the fixed materials of a wall. Typically, U-values are reduced in the area of a frame by around 0.2 W/m²K, which means that a double-insulated glazing unit (or insulating glass unit, IGU) with a U-value of 1.1 reaches, in practical terms, only 1.3 W/m²K when its frame properties are included in the calculation. For a more precise analysis of U-values, two different values are typically available: for the U-value of the glass itself it is u_g, while u_w, on the other hand, stands for the entire assembly of frame and glass.

Eine im energetischen Sinne hervorzuhebende Schwachstelle bei transparenten Bauteilen sind die Rahmen. Gläser müssen im Regelfall in Rahmen eingestellt werden, die sich unter Umständen bewegen lassen oder die Verbindung zu weiteren festen Bauteilen bilden. Üblicherweise verschlechtern sich die Wärmedurchgangskoeffizienten, u-Werte, durch ihren Rahmen um ca. 0,2 W/m²K, d.h. zum Beispiel eine Zweifach-Isolierverglasung mit einem u-Wert von ca. 1,1 erreicht letztendlich inklusive Rahmen einen solchen von 1,3 W/m²K. Bei genauer Betrachtung und Berechnungen von Wärmedurchgangswerten werden üblicherweise zwei getrennte u-Werte zum Ansatz gebracht – u-Wert Glas (u_g), u-Wert der gesamten Scheibe inklusive Rahmen (u_w).

During the past few years, research and development in the area of vacuum glazing systems has resulted in some promising products. Vacuum glazing consists of a pair of glass sheets with an evacuated spacing gap containing small support pillars. The space between the clear lites of the assembly is filled with air under a negative pressure of around 10^{-3} mbar. The U-value of such a system can be assumed to be < 0.5 W/m²K, and for a further reduced U-value the assembly can be enhanced by adding low-E coatings. The problematic aspects of vacuum glazing systems are currently still, on the one hand, the air-tightness necessary to prevent leaks, and on the other the numerous visible support pillars between the glass panes. They are structurally necessary to prevent the glass assembly from imploding. It is estimated that the cost for such a glazing system is comparable to triple-insulating glass.

The total energy transmission value g is typically too high during summer. G-values range normally around 20 %, which means that without additional solar shading devices up to 20 % of the solar gains reaching the glass will be transmitted into the interior space. This value is too high not only for warm and sunny regions on Earth but also for regions in Western Europe and parts of the U.S. Additional solar shading devices in the form of interior blinds or, even better, outside louvers need to be combined with the glazing system. **Figure 131** shows examples of interior and exterior shading systems with their g-values. A g-value of 0.14 is still too high for today's expectation for high-quality buildings. Effective systems can, for instance, consist of an external, back-ventilated, low-E coated glass pane, an interior solar shading louver behind it, and finally an internal double-insulated glass, again with a low-E coating on one of its glass surfaces. With such an assembly, a total g-value of 0.06 can be achieved – in other words, just 6 % of the thermal energy will enter the interior space. The importance of the reduction factor cannot be over-estimated in the design of proper glazing systems. The evaluation of the effectiveness of a shading system is done by the reduction factors f_c given in the table, and the specifications of various manufacturers need to be studied closely because results can vary significantly.

Um die Wärmeleitung und Wärmestrahlung von Scheiben weiter zu optimieren, wird seit Jahren an der Entwicklung einer Vakuum-Scheibe gearbeitet. Diese besitzt zwischen zwei klaren Gläsern als Füllgas Luft, jedoch mit einem Unterdruck von 10^{-3} mbar. Der u-Wert dieser Scheibe dürfte < 0,5 W/m²K betragen und kann wiederum mit low-e-Beschichtungen (low emission) arbeiten, um den g-Wert zu verringern. Das Problem von Vakuum-Scheiben liegt zurzeit noch im Bereich der Dichtigkeit des Randverbundes und vor allem in der Sichtbarkeit der Glasabstandshalter, die notwendig werden, damit entsprechende Scheiben nicht implodieren. Der angestrebte Preis entsprechender Vakuum-Gläser soll im Bereich einer Dreifach-Isolierverglasung liegen.

Die Gesamtenergiedurchlassgrade (g-Wert) sind im Regelfall bei Gläsern für den Sommer zu hoch. Sie liegen normalerweise über 20 Prozent, d.h. ohne zusätzlichen Sonnenschutz würden 20 Prozent der aufgestrahlten Energie in den Raum eintreten. Dieser Wert ist nicht nur für Westeuropa oder Großteile der USA, sondern vor allem für warme und sonnenreiche Gegenden der Welt zu hoch. Aus diesem Grund werden zusätzliche Sonnenschutzmaßnahmen in Form von Innenjalousien und vor allem Außenbeschattungselementen mit den Gläsern kombiniert. **Bild 131** zeigt beispielhaft einen innen liegenden wie auch einen außen liegenden Sonnenschutz mit den jeweiligen g-Werten. Ein g-Wert von 0,14 ist für die hohen Ansprüche der zukünftigen Gebäude jedoch noch deutlich zu groß, so dass üblicherweise Kombinationen entwickelt werden, bei denen eine äußere Glasscheibe (hinterlüftet) mit low-e-Beschichtung einen quasi dahinter liegenden, äußeren Sonnenschutz unterstützt und, wiederum von außen nach innen gesehen, dahinter eine Isolierverglasung mit low-e-Beschichtungen zum Einsatz kommt, so dass ein Gesamt-g-Wert von ca. 0,06 = ca. 6 Prozent erreicht wird. Die Ermittlung der Effektivität eines Sonnenschutzsystems erfolgt über die dargestellten Sonneneintragskennwerte. Der Minderungsfaktor des zusätzlichen Sonnenschutzsystems, beschrieben durch den Faktor f_c ist von hoher Bedeutung und kann außerordentlich unterschiedlich ausfallen. Insofern ist es ratsam, genaue Produktangabenzu benutzen.

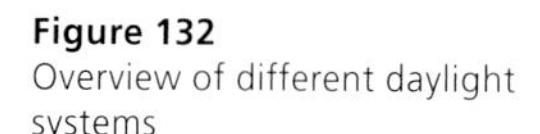

Figure 132
Overview of different daylight systems

Source: Energieatlas, Nachhaltige Architektur, Hegger, Fuchs, Stark, Zeumer, Birkhäuser Verlag, 2007

Bild 132
Übersicht über z.Zt. gebräuchliche Tageslichtsysteme

Quelle: Energieatlas, Nachhaltige Architektur Hegger, Fuchs, Stark, Zeumer Birkhäuser Verlag, 2007

Daylight systems

	Static (Fixed) systems	Tracking systems	Relocating systems	Electrochromatic (switchable) glazing
Light diffusion	Light diffusing holographic optical elements (HOE)			Photochromic Thermochromic Gasochromic
	Prism plate Fixed lamellae Light shelves Light directing glazing HOE	Rotating light redirection lamellae	Retractable light redirect lamellae	Photochromic Thermochromic Gasochromic
Light transport	Light pipes glassfiber	Heliostats		

With greatly improved solar shading, the amount of daylight entering a space is typically also reduced. In a case-by-case comparison, it is important therefore to analyze whether g-values can be reduced or whether special prismatic light redirection systems have to be used in order to increase day lighting.

Figure 132 shows the principal composition of various day lighting systems. Of great interest should also be the application of building-integrated, electrical energy-generating, thin-film photovoltaic systems (BIPV), which can be used as solar shading devices as well. Both moveable and fixed solar modules can be used, and g-values as low as 0.1 can be achieved. It is important to remember that photovoltaic elements heat up more than selective coatings, so an increase in long-wave radiation is the consequence.

It is apparent that the development of optimally functioning modern façades is a highly complex and challenging task. A great degree of imagination and technical expertise are necessary, in combination with necessary simulations to achieve maximum effectiveness with regard to a building's specific façade and as a function of its location characteristics.

Mit höherwertigem Sonnenschutz tritt im Regelfall das Problem auf, dass der Tageslichteinfall immer geringer wird. Insofern muss von Fall zu Fall abgewogen werden, ob zugunsten des Tageslichteintrags der g-Wert zurückgenommen wird oder aber ob spezielle lichtumlenkende Systeme zum Einsatz kommen sollen, die die Tageslichtsituation verbessern.

Bild 132 zeigt den prinzipiellen Aufbau unterschiedlicher Tageslichtsysteme, die zum Einsatz kommen können. Dabei von besonderem Interesse kann sein, dass Gebäudehüllen als Beschattungselemente Photovoltaiksysteme einsetzen, d.h. Dünnschicht-Photovoltaikelemente, die gleichzeitig elektrische Energie erzeugen. Sowohl starre wie auch bewegliche Elemente können Solarmodule aufnehmen, wobei im Regelfall g-Werte um minimal 0,1 erreicht werden können. Hierbei berücksichtigt werden muss jedoch, dass sich Photovoltaikelemente höher aufheizen als zum Beispiel selektive Beschichtungen, so dass mit einer erhöhten Strahlung im langwelligen Bereich zu rechnen ist.

Aus allem vorher Gesagten lässt sich leicht ablesen, dass die Entwicklung von Fassaden hin zum optimalen Betrieb ein hochkomplexes Thema ist. Viel Phantasie und Erfahrungen sind nötig – gegebenenfalls planungsbegleitende Simulationen, um das tatsächliche Optimum für ein Gebäude an einem Standort zu finden.

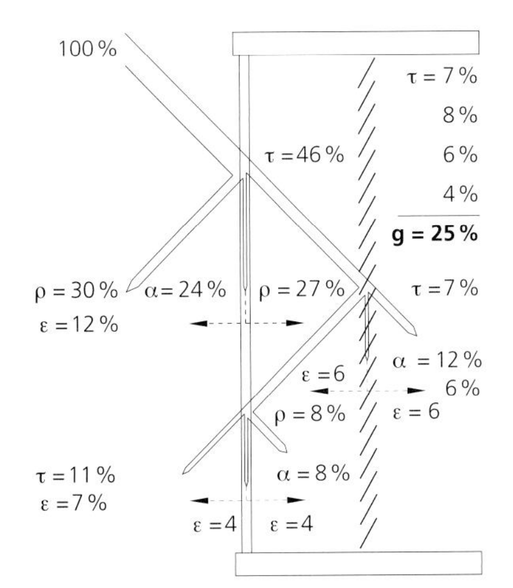

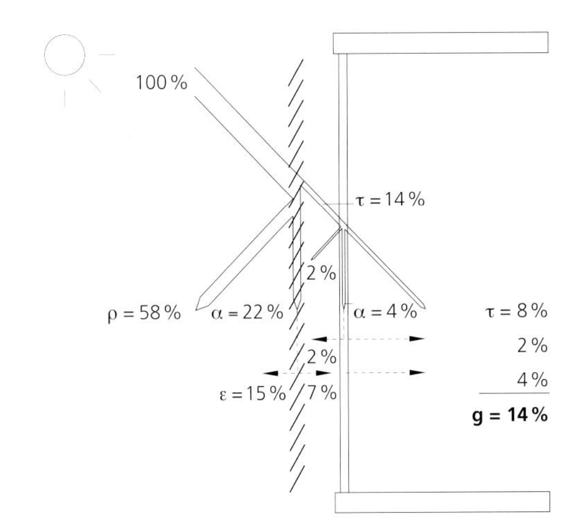

ρ Reflectivity
α Absorption
τ Transmission
ε Emissivity

Figure 131
Solar shading reduction factors

Source: Energieatlas, Nachhaltige Architektur, Hegger, Fuchs, Stark, Zeumer, Birkhäuser Verlag, 2007

Bild 131
Sonnenschutzmaßnahmen, Minderungsfaktoren bei Bestrahlung

Quelle: Energieatlas, Nachhaltige Architektur Hegger, Fuchs, Stark, Zeumer Birkhäuser Verlag, 2007

10.6 The concept of energy supply

The development of an energy supply concept for a building should be started only when the actual energy demand based on the building component assemblies can be quantified. It is evident that the energy demand target distributed across an entire year should be kept as small as possible. Ideally, a building, if properly designed with the most appropriate constructive measures, should allow for long intervals of non-active systems operation and still provide comfortable interior conditions throughout the year.

For the conception of a design for energy systems, the complete chain from primary energy source to final consumption needs to be evaluated, its efficiency needs to be optimized, and its use in the future has to be promising. **Figure 133** shows the path from concept development, with consideration of external factors and various energy sources, to final demand. The most important question for today's energy systems designer is which energy-consuming environmental systems and services are avoidable without compromising the quality and comfort of a building. Using highly creative yet simple means, even zero energy solutions are achievable. It is particularly important to avoid the creation of high operating cost through short-sighted savings on initial equipment cost.

10.6 Konzept der Energieversorgung

Die Entwicklung des Energiekonzepts sollte erst dann beginnen, wenn klar ist, welcher Energiebedarf durch geeignete bauliche Maßnahmen tatsächlich entsteht. Selbstverständlich soll dabei der Energiebedarf jahreszeitlich möglichst gering sein. Idealerweise erreicht man durch geeignete bauliche Maßnahmen, dass das Gebäude in langen Zeiten ohne technische Unterstützung ein behagliches Raumklima bereitstellen kann.

Bei der konzeptionellen Entwicklung der Energieversorgung muss die Kette von den Energiequellen bis zu gewünschten Energiedienstleistungen nachvollzogen und auf eine möglichst hohe Effizienz und Zukunftsfähigkeit hin untersucht werden. **Bild 133** zeigt den Weg der Konzeptentwicklung vom Angebot des Außenraums und verschiedene Energieträger bis hin zum Bedarf. Wesentlich ist bei allen Überlegungen zum Energiekonzept, sich vorrangig mit der Frage auseinanderzusetzen, in welchem Umfang spezifische Energiedienstleistungen ohne Qualitätsverlust für den Nutzer vermeidbar sind. Eine Nulloption kann unter Umständen infolge sehr einfacher und innovativer Lösungen liegen. Vermieden werden sollte in jedem Fall, dass zugunsten geringster Investitionskosten später höhe Betriebskosten ausgelöst werden.

Supply	Building systems					Demand and consumption
	Energy supply means	Energy conversion	Energy storage	Energy distribution	Energy delivery	
Crude oil	natural gas grid	Gas fired boiler	Water storage	Supply and return air system	Radiators	Space heating
Natural gas	Heating oil tank	Oil fired boiler	Buffer tank	Hot water loop	Radiant floor heating	Space cooling
Electricity grid	District heat transfer station	Electric boiler	Solar storage	Chilled water loop	Wall heating	Potable water heating
Long-distance heating	Wood pellet storage	Heat pump	Combi storage	warm air	facade heating	Humidification
District heating	Wood chips storage	Solar thermal plant	Long-time heat storage	Cold air	Building element activation	Dehumidification
Chilled water supply grid	Vegetable oil storage tank	Photovoltaic plant	Geothermal storage	Electrical conduit	Displacement ventilation register	Outside air
Solar radiation	Geothermal probes	Wood pellet stove	Latent heat storage	Natural gas line	High velocity supply nozzle	Lighting
Wood pellets	Groundwater well	Wood chips stove	Absorption storage	Heating circuit	Source	Electricity
Wood chips	Heat exchanger	Cogeneration power plant	Batteries	Cooling circuit		Process heat
Vegetable oil	Heat recovery	Compression chiller	Compressed air storage			Process cold
Geothermal energy		Absorption chiller	Fly wheel storage			
Ground water		Evaporative cooling	Hydrogen			
Surface water		Hydrogen fuel cell				
Outside air						
Wind						

Figure 133
Available options for energy balance and supply configuration

Source: Energieatlas, Nachhaltige Architektur, Hegger, Fuchs, Stark, Zeumer, Birkhäuser Verlag, 2007

Bild 133
Konzeptionelle Entwicklungsschritte zu Energieangebot und -Versorgung

Quelle: Energieatlas, Nachhaltige Architektur Hegger, Fuchs, Stark, Zeumer Birkhäuser Verlag, 2007

However, the opposite should be avoided as well – when a concentration on low operating cost creates unnecessarily high first-time investments, mistakes can be made. A truly holistic view of the climate design concept as a cooperative effort between architects and engineers will result in an optimized building. Again, a thorough evaluation of the conditions of the specific building site needs to be undertaken at a very early stage of the design process because only such an analysis can open up opportunities for the use of natural resources such as prevailing wind patterns, solar radiation, ambient temperatures, rain water, ground water, and soil temperatures. **Figure 134** explains the concepts and possibilities (Haus der Nachhaltigkeit).

Gleichermaßen muss vermieden werden, dass bei einer einseitigne Sicht auf möglichst niedrige Energiekosten, ein zu hoher Aufwand bei der Erstinstallation entsteht. In jedem Fall ist es notwendig, einen gesamthaften „Klima-Design Ansatz" gemeinsam, zwischen Architekt und Ingenieur zu entwickeln, der zum optimalen Ergebnis führt. Von besonderer Wichtigkeit ist dabei, dass eine sehr detaillierte Betrachtung des Standortes des Objektes und eine genaue Bedarfsanalyse erfolgt, um insbesondere die natürlichen Ressourcen des jeweiligen Standortes direkt am Objekt zu nutzen (Wind, Solarstrahlung, Temperaturen, Regenwasser, Grundwasser, Erdreichtemperaturen usw.). **Bild 134** verdeutlicht nochmals die denkbaren Ansätze und Möglichkeiten (Haus der Nachhaltigkeit).

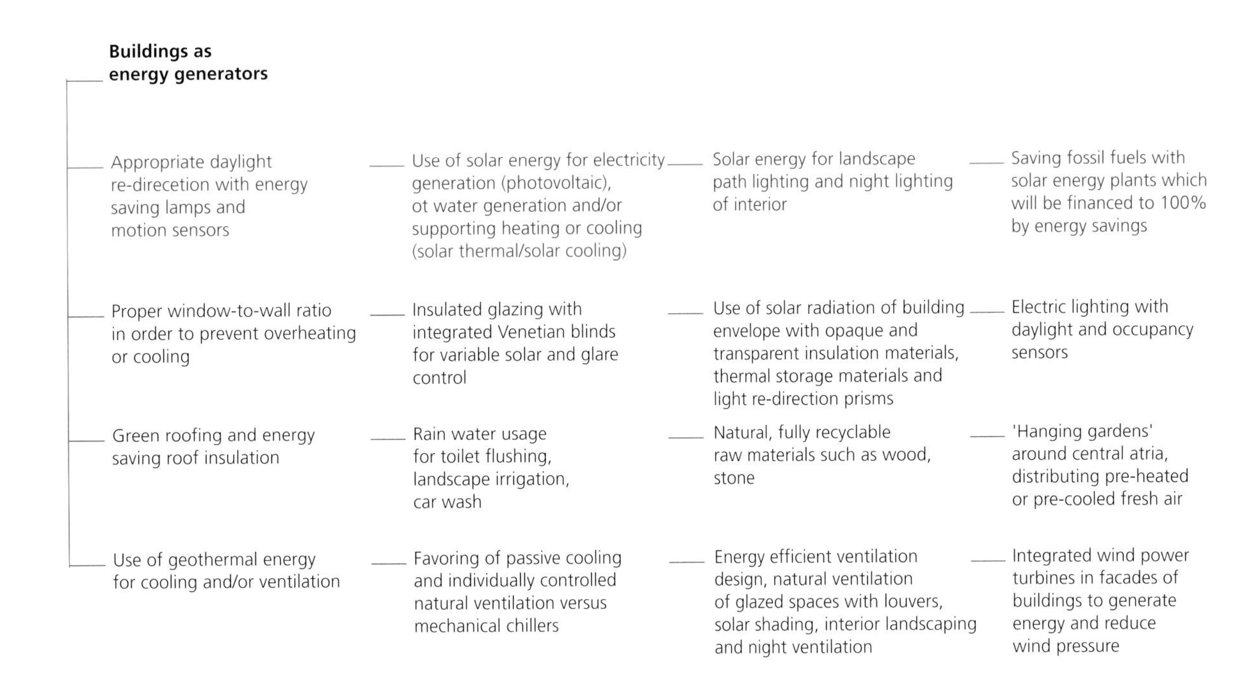

Figure 134
DEGI-House of sustainability

Source: DEGI RESEARCH, 2007

Bild 134
DEGI – Die Immobilie als Energieerzeuger

Quelle: DEGI RESEARCH, 2007

11 Small is beautiful! Exemplary small-scale projects

11 Klein, aber fein, beispielhafte Bauten

Before focusing attention on the large-scale and highly complex building typologies of the high-rise and super-tall tower, more "ordinary" examples of the built environment may serve as case studies of the issues discussed in earlier chapters. It goes without saying that this will be not a complete overview of all sustainable buildings "with a built-in future" or the architects who created them. Many more excellent projects by architects who for many years have dedicated their professional lives to the creation of high-quality, sustainable designs have been presented to the public recently. Such a complete presentation of the well-designed and environmentally progressive buildings that are currently being built around the world would certainly reach beyond the scope of this publication.

11.1 Office building Marché Restaurants Schweiz AG, Kempfthal, Switzerland

The building design by architects Beat Kämpfen, Zurich, Switzerland, in collaboration with the engineering firm Naef Engineers, also in Zurich, is conceived according to the Swiss standard of Minergie-P-ECO (ECO stands for ecologically correct, or the most healthy type of construction). The design is very straightforward: an elongated building with three floors and a means-of-egress staircase at either end describe the basic concept. The long façade of the building is precisely oriented toward the south and is fully glazed and characterized by protruding balconies and a terrace at ground level. On this side of the building, open-plan offices are located. The design brief asked for a direct access to the outside for employees in the offices, and this condition also coincides energetically very well with the shading provided by the required balconies during high solar altitude angles in summer. Spaces on the northern side of the building between the staircases include meeting rooms, single offices, and auxiliary spaces. Located on the ground floor are the entrance lobby, a café, a recycling station, and three studios for overnight guests. **Figures 135** and **136** show the project, and **Figure 137** displays an elevation. Because the building is designed according to the Minergie-P-ECO standard, it was necessary to use of building materials with the least possible content of volatile compounds, in both the material itself and its production, and even with regard to its future recycling. The design also needed to provide an optimum of day lighting, minimal noise emission, and excellent indoor air quality. The overarching goal for the client was to provide a working environment in which employees would "feel at home".

Bevor man sich den „Elefanten" – hochkomplexen Großbauten und Hochhäusern zuwendet, soll an eher „üblichen" Bauobjekten gezeigt werden, welche Architekten sich u.a. mit der zuvor beschriebenen Thematik bereits intensiv auseinandergesetzt haben. Selbstverständlich sind bei den Objektdarstellungen nicht alle Architekten mit Objekten vertreten, die sich ebenfalls seit Jahren und zum Teil Jahrzehnten der Planung zukunftsfähiger Gebäude verschrieben haben. Es würde den Umfang dieses Buches sprengen, alle die hervorragenden Beispiele zu zeigen, die es inzwischen auf der Welt gibt und bereits seit Jahren in richtigem Sinn arbeiten.

11.1 Bürogebäude Marché Restaurants Schweiz AG, Kempfthal, Schweiz

Das vom Architekten Beat Kämpfen, Zürich (mit Naef Ingenieurtechnik, Zürich), geplante Gebäude ist unter dem Standard Minergie-P-ECO (ECO = gesunde und ökologische Bauweise) konzipiert worden. Die Anlage des Gebäudes ist überaus einfach. Ein langgestreckter, dreigeschossiger Baukörper besitzt je ein Treppenhaus an den Enden des Objektes. An der exakt nach Süden geöffneten, vollverglasten Längsseite befinden sich Großraumbüros mit davor liegenden Balkonen bzw. einer Terrasse im Erdgeschoss. Der Anspruch eines direkten Ausgangs ins Freie für jeden Arbeitsplatz deckte sich bei diesem Objekt mit der energetischen und raumklimatisch sinnvollen Beschattung der hochstehenden Sommersonne. Die zwischen den Treppenhäusern liegenden Räume auf der Nordseite nehmen Sitzungszimmer und Nebenräume sowie Einzelbüros auf. Im Eingangsbereich befindet sich ein Café, eine Entsorgungsstation sowie 3 Studios für übernachtende Besucher. Die **Bilder 135** und **136** beschreiben das Projekt, **Bild 137** zeigt eine Ansicht. Das Gebäude wurde unter dem Energielabe Minergie-P-ECO und seinen Ansprüchen konzipiert. Aufgrund des planerischen Ansatzes musste darauf geachtet werden, geringe Umweltbelastungen bei der Herstellung, Verarbeitung und späterem Rückbau der Baustoffe entstehen zu lassen. Optimale Tageslichtverhältnisse sowie geringe Lärmimmissionen und eine geringe Schadstoffbelastung der Raumluft wurde nachgewiesen. Das Ziel des Bauherrn war, eine Arbeitsumgebung zu erhalten, in der sich die Mitarbeiter wohlfühlen sollten.

Figure 137
Office building Marché Restaurants Schweiz AG, View

Source: Beat Kämpfen Büro für Architektur, Zürich, Switzerland

Bild 137
Marche Restaurants Schweiz AG, Ansicht des Gebäudes

Quelle: Beat Kämpfen Büro für Architektur, Zürich, Schweiz

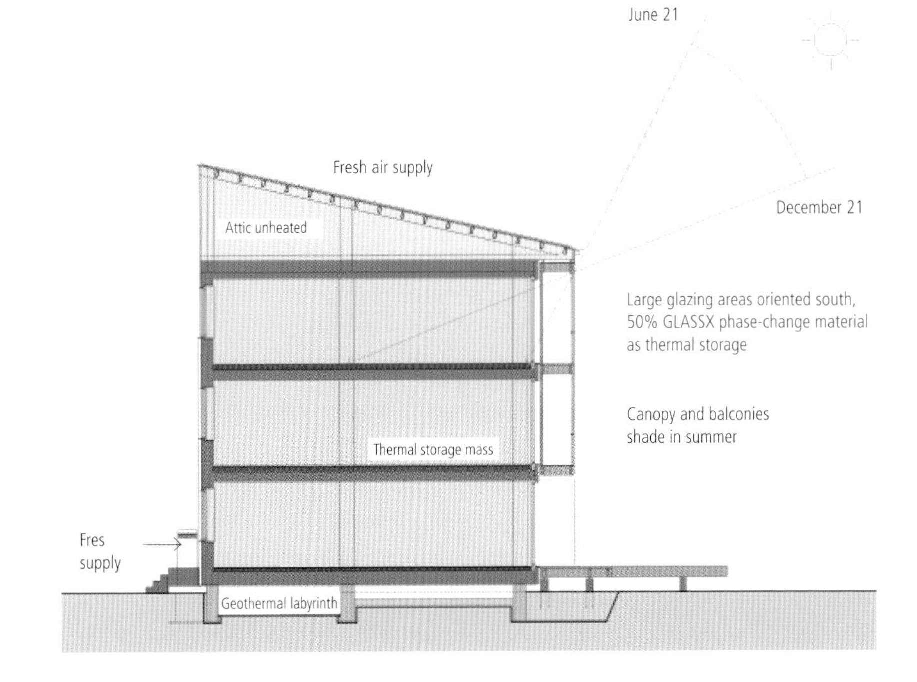

Figure 136
Marche Restaurants Schweiz AG, Switzerland, cross section
Due to budget considerations a basement was omitted. Mechanical equipment rooms (MER) are located in the attic.

Source: Architect Beat Kämpfen

Bild 136
Marche Restaurants Schweiz AG, Querschnitt: Aus Kostengründen wurde auf einen Keller verzichtet.
Die Haustechnik befindet sich im Dachraum.

Quelle: Architekt Beat Kämpfen

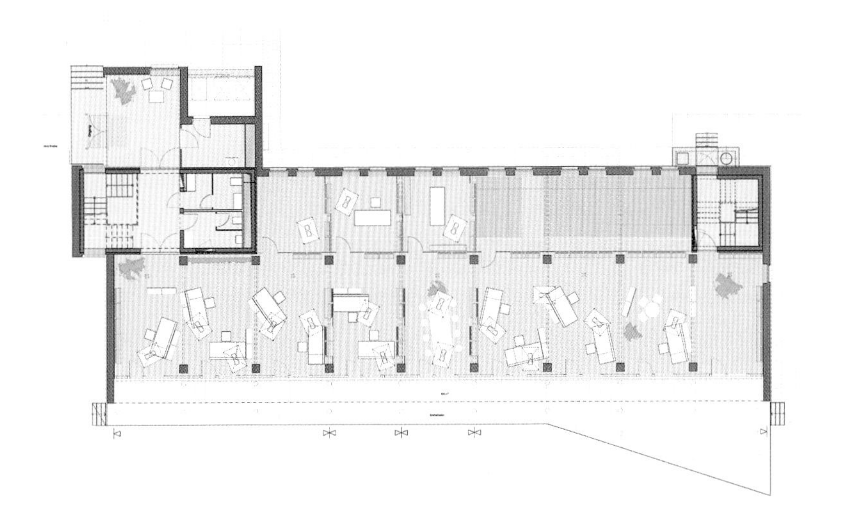

Figure 135
Marche Restaurants Schweiz AG Headquarters, Floor plan level 2: The structural grid allows free organization of office floor

Source: Architect Beat Kämpfen

Bild 135
Marche Restaurants Schweiz AG, Grundriss 1. OG:
Das Stützenraster erlaubt eine freie Einteilung der Bürogeschosse

Quelle: Architekt Beat Kämpfen

A crucial strategy during the design of the building was to develop the least complex concept for all technical systems, including the load-bearing structure, the building assembly, the building systems, and the interiors. On the other hand, great attention was paid to the detail and craft of execution. In the interior, planted walls provide a comfortable environment, whereas the outside envelope of the building uses solar glazing units with phase-change materials (PCM), as shown in **Figure 138**.

The mechanical equipment for heating, hot water generation, and ventilation, as well as the solar electrical generating system, is located inside the attic of the northwesterly core area. The horizontal distribution systems for air, heat, and electricity are placed inside a 30-cm-deep roof cavity that was back-filled with blow-in cellulose-flake insulation to achieve maximum thermal insulation. Media distribution happens along wall cavities of the northern façade, and in some instances in vertical slots placed in the structural columns (**Figure 138**). The primary load-bearing construction consists of timber, and the main load-bearing elements are located in the northeast and west-façades. The floor slabs are designed as wooden hollow-box elements with added stone aggregate to provide improved sound insulation.

The planted interior walls, with surface areas of 12 m^2 each, are located on every floor. They evaporate around 30 liters of water daily, which creates, especially in winter, very comfortable relative humidity of the interior air. The interior wood furnishings for the open-plan offices, which were also designed by the architects, contain sound-absorbing materials for necessary balanced sound attenuation inside the large office spaces.

The energy concept features a photovoltaic system covering the entire roof surface of the building, which is expected to meet the building's operation energy requirements. Typically, the energy produced by the photovoltaic roof is fed into the electricity grid and counted against electrical usage from the grid. The amorphous silicon-based solar cells used in the photovoltaic roof application are embedded in a double-pane glazing unit, and the formal result is a large, attractive, fish-scale expression. For the photovoltaic system, a so-called contracting model was used, in which the local electricity utility, the cantonal Elektrizitätswerke Zürich, is in charge of the operation and the initial funding of the photovoltaic roof. During winter operation, the building is minimally ventilated by the mechanical system, achieving 0.6 air changes per hour (ac/h), and heat energy is provided by the heat exchangers in the building and the water-based heat pumps connected to geothermal piles.

Eine wesentliche Strategie bei der Entwicklung des Gebäudes war es, auf allen Ebenen ein einfaches Konzept zu entwickeln (Statik, Konstruktion, Installation, Ausbau). Im Gegenzug dazu wurde auf eine sorgfältige Gestaltung Wert gelegt. Pflanzenwände innerhalb des Gebäudes unterstützen die Erreichung eines angenehmen Klimas, Solarglas-Fassadenelemente mit phase change material(PCM) kamen zum Einsatz, **Bild 138**.

Die Zentrale für Heizung, Warmwasser und Lüftung sowie die Solarstromsteuerung sind im Dachraum des nordwestlich gelegenen Kernbereichs untergebracht. Die zentrale Verteilung (horizontal) von Luft, Wärme und Strom liegt in einem Dachhohlraum (30 cm hoch), der mit Zelluloseflocken zugeschüttet wurde, um ein Höchstmaß an Dämmung zu erreichen. Die Verteilung der Medien erfolgt über Wandaussparungen (Nordwand) und zum Teil in Schlitzen der Stützen, **Bild 138**. Die primäre Konstruktion des Gebäudes ist eine Holzkonstruktion, wobei die wesentlichen Lastabtragungen über die Nord-Ost- und Westfassade erfolgen. Die Geschossdecken sind als Hohlkastenelemente ausgebildet, die zum Schallschutz mit Split bewehrt sind.

Pflanzwände, d.h. begrünte Wände mit einer Fläche von ca. 12 m^2, bestehen in jedem Geschoss und verdunsten pro Tag ca. 30 l Wasser, was im Winter zu einer angenehmen Erhöhung der relativen Raumluftfeuchte führt. Die von den Architekten entworfenen Holzmöbel nehmen auf ihren Rückseiten jeweils schallabsorbierende Elemente auf, um einen ausgewogenen Schallleistungspegel in den großflächigen Büroräumen zu erreichen.

Das Energiekonzept umfasst eine Photovotaikanlage, die das gesamte Dach überzieht. Durch diese soll soviel Strom produziert werden, wie die technischen Einrichtungen des Gebäudes verbrauchen.

Die elektrische Energie des Photovoltaikdaches wird im Allgemeinen ins Netz eingespeist und gegen die Verbräuche aus dem Netz gegengerechnet. Die amorphen Solarzellen auf dem Dach sind zwischen 2 Glasscheiben eingebettet und zeigen optisch eine großflächig geschuppte Dachhaut. Die Finanzierung und das Betreiben der Photovoltaikanlage liegt bei den Elektrizitätswerken des Kantons Zürich (Contracting-Modell). Im Winterbetrieb wird das Gebäude minimal mechanisch belüftet (ca. 0,6-facher Luftwechsel), wobei die Wärmeenergie durch Wärmerückgewinnungssysteme und eine Erdsonden-Wärmepumpe bereitgestellt wird.

Zur Verringerung der Wärmeverluste wurden die opaken Gebäudehüllen mit einem u-Wert von 0,084 – 0,104 W/m^{2}K ausgestattet. Die Dreifach-Isolierverglasung wurde kombiniert mit Holzrahmenelementen.

To minimize heat loss in winter, the opaque wall segments of the outside envelope achieve an excellent U-value of 0.084 – 0.104 W/m²K. The triple-pane insulated glazing of the transparent façade portions is framed by wooden elements, and half of the south-facing façade uses embedded PCM-elements by GLASSX to support the desired energy gains through the façade when appropriate (see **Chapter 10.4**).

A radiant floor embedded in an 80-mm concrete floor top can be used equally for cooling purposes during the summer months. Supply, with cold water, comes from the geothermal system, which reduces the building's cooling load capacity to 25 W/m². Completed in 2007, the building was constructed very economically, at cost of CHF 565.00/m³ gross volume (not including the photovoltaic roof surface). The annual electrical energy consumption is approximately 40,000 kWh/a, and the yearly heating demand is 7.8 kWh/m²a.

Die Hälfte der Südfassade ist mit PCM-Elementen (GLASSX-Elementen) ausgerüstet, um den Wärmeeintrag in das Gebäude zur richtigen Zeit zu unterstützen (vergleiche **Kapitel 10.4**).

Eine Fußbodenheizung in einem 80 mm hohen Estrichboden kann im Sommer über die Erdsonden mit kühlem Wasser gespeist werden (Kühllastaustrag ca. 25 W/m²). Das Objekt, fertiggestellt 2007, weist außerordentlich günstige Baukosten (ca. CHF 565,-/m³ umbauter Raum) auf, wobei die Kosten für die Photovoltaikanlage hier nicht enthalten sind. Die Energieverbräuche werden bei elektrischer Energie mit ca. 40.000 kWh/a angegeben, der Heizwärmebedarf mit ca. 7,8 kWh/m²a.

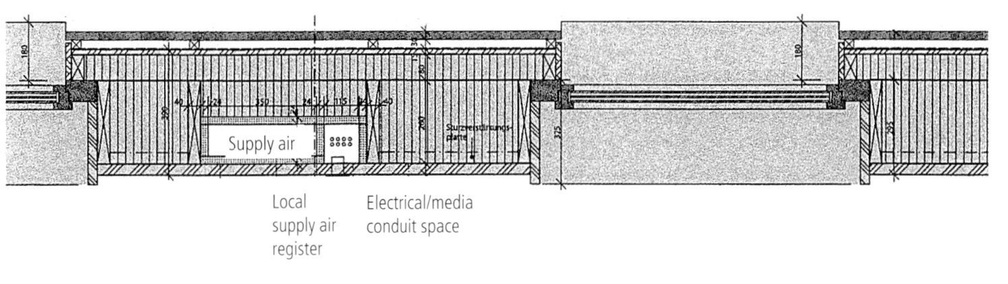

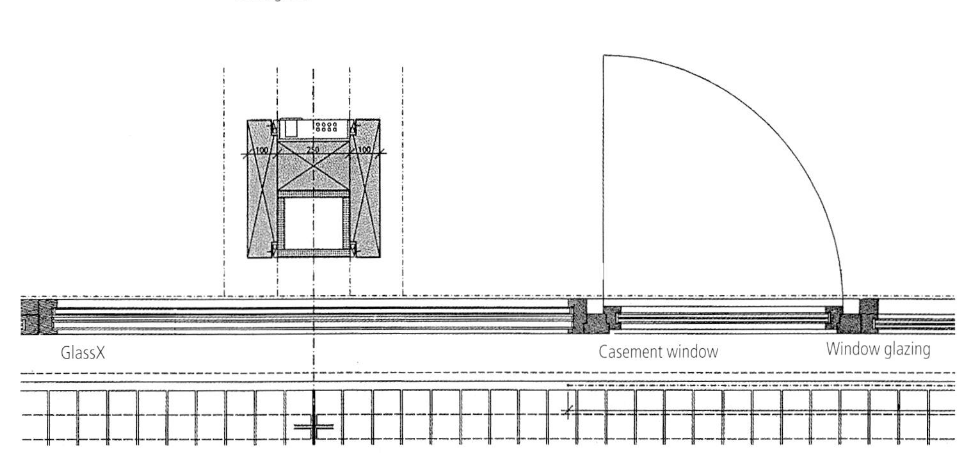

Figure 138
Horizontal section through exterior facade: Distribution of media such as air, electricity, and data is integrated in the structural columns and the niches of the north facing wall.

Source: Architect Beat Kämpfen

Bild 138
Detailschnitt: Die Verteilung der Medien (Luft, Strom, Telefon, EDV) passiert ausschließlich in den Stützen sowie in Nischen der Nordwand

Quelle: Architekt Beat Kämpfen

11.2 Wohnen am Lohbach

Wohnen am Lohbach, Innsbruck, Austria
Architects Baumschlager–Eberle und Grassmann

The projects presented in the following sections (**11.2 – 11.6**) were designed by the architects Baumschlager-Eberle, and they are all excellent examples of sustainable building design and operation. It is therefore prudent to have the architects explain their concepts themselves:

"The philosophy of Baumschlager-Eberle is that architecture needs to be understood as a holistic task, where the fulfillment of the complex topics at hand is only reached when all cardinal aspects of a project are brought to a conclusive balance; these are commonly the structural intelligence, ecology, economy, and social acceptance. It should be self-evident that responding to such principles should be the goal for any building design, whether it is a single-family residence, a multi-unit residential complex, or a factory."

Introduction

An important part of resource optimization is also the esteem in which the structure is held by its users and the public. Buildings that are liked will enjoy long lifespans, and longevity and durability are key aspects of resource preservation. In this respect it is important to precisely understand the relationship between the consumption of re-sources for the construction and operation of a building and the rising comfort expectations of the users. Evaluating this relationship will result in a process of comprehensive optimization.

For example, such an optimization process may address the heating of buildings. In Central Europe, approximately 40 % of the total primary energy is converted to heating energy, whereas in tropical or subtropical regions the same percentage of primary energy is spent on cooling. A reduction of this primary energy demand while also maintaining comfort conditions will certainly have major consequences for the resource demand of a society. The same is true for artificial lighting and the consumption of the natural resource "land." An optimization of the energy efficiency of buildings is accomplished by utilizing appropriate tools of design and planning, and the resulting energy savings will in most cases have significant positive repercussions for a country's economy. Thus, optimization is a part of a wider move toward greater sustainability.

In order to comply with such claims, the office of Baumschlager-Eberle uses project-related specific fact-finding methods together with their thorough and detailed understanding of the significant aspects of a proposed building. These aspects are discussed below.

11.2 Wohnen am Lohbach

Wohnwn am Lohbach, Innsbruck, Österreich
Architekten Baumschlager–Eberle und Grassmann

Da die Bauobjekte gemäß **Kapitel 11.2** bis **11.6** alle aus der Feder des Büros Baumschlager-Eberle stammen, soll hier der Architekt selbst für sich und sein Arbeiten sprechen. Die Ausführungen beinhalten alle wesentlichen Aspekte des nachhaltigen Bauens und nachhaltigen Betreibens.

"Kern des Selbstverständnisses von Baumschlager-Eberle ist es, Architektur als eine ganzheitliche Aufgabe zu betrachten, deren Komplexität erst dann erfüllt ist, wenn ein Gebäude allen Anforderungen wie konstruktive Intelligenz, Ökologie, Wirtschaftlichkeit und gesellschaftliche Akzeptanz entspricht. Ein solcher Anspruch an Nachhaltigkeit lässt sich grundsätzlich bei jedem Gebäude verwirklichen – gleichgültig, ob es sich um ein Einfamilienhaus, eine Wohnanlage oder einen Industriebau handelt."

Einführung

Die wichtigste Voraussetzung für die Ressourcenoptimierung ist die Wertschätzung des Gebäuden durch die Nutzer, vor allem aber durch die Akzeptanz der Passanten: Nur Gebäude, die geliebt werden, erzielen eine lange Lebensdauer. Diese Langlebigkeit ist die Basis für einen optimierten Ressourcenverbrauch. Eine zentrale Bedeutung kommt dem präzisen Wissen um das Verhältnis zwischen den Ressourcenverbräuchen für Erstellung und Betrieb und den steigenden Komfortansprüchen der Nutzer zu. Daraus kann eine weitreichende Optimierung abgeleitet werden.

Am Beispiel der Beheizung von Gebäuden werden die Auswirkungen von Optimierungen ersichtlich. In Mitteleuropa werden ca. 40 Prozent des gesamten Primärenergiebedarfs eines Landes in Heizenergie umgewandelt. In tropischen und subtropischen Regionen wird der Primärenergieeinsatz zum Heizen durch einen Primärenergieeinsatz zum Kühlen ersetzt. Eine Verringerung des Energiebedarfs zur Einhaltung der thermischen Behaglichkeit hat demnach direkte Auswirkungen auf den Ressourcenbedarf einer Gesellschaft. Vergleichbare Betrachtungen sind für Themen wie Beleuchtung und Landverbrauch legitim. Die Optimierungen hinsichtlich der energetischen Leistungsfähigkeit von Gebäuden lassen sich über Werkzeuge des Entwurfs und der Planung erreichen. Im Resultat haben die energetischen Einsparungen meist positive und signifikante Auswirkungen auf die Ökonomie des jeweiligen Landes. Sie können als Teil einer umfassenden Entwicklung zu mehr Nachhaltigkeit gesehen werden.

11.2 Wohnen am Lohbach

Land usage

The character of cities and settlements varies greatly in character across the globe. Tools to describe these characteristics are either relatively vague, such as general building dimensions, presence of green spaces, and average age of structures, or more precise, such as the units of building density (**Figure 139**). To use the tool of building density is relatively simple: it is the ratio between permanently used/usable space and the size of a building site.

The character of an area is therefore also determined by this density. For example, greater densities will have consequences for proximity of human population but also for economic processes. Higher densities are seen today as a locational advantage, and so are multi-functional urban spaces that provide greater appeal because of the experiential environment and improved quality of life. Resource optimization with regard to the parameter "density" means making decisions about the consumption of the resource called "land."

Water usage

Buildings require not only energy for heating, ventilation, lighting, and overall operation but also substantial amounts of water for consumption by their users. Because water is such a valuable – and, in many parts of the world, scarce – resource, the stringent optimization of water seems to make sense. Even in Europe, the availability of water varies widely. Therefore, the water consumption of buildings needs to be adjusted to the local conditions.

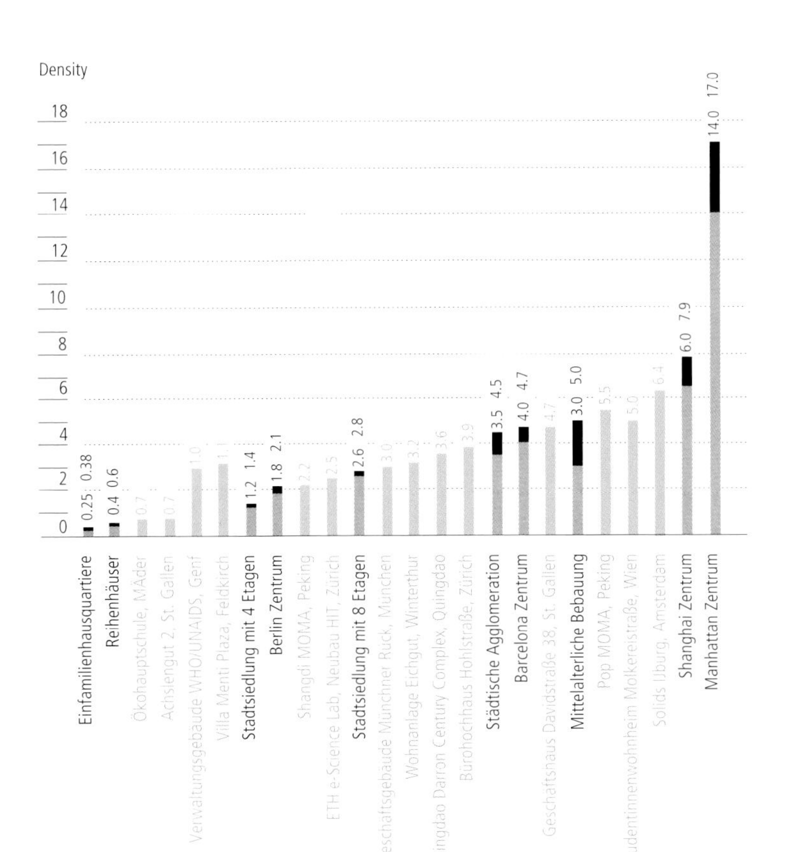

Figure 139
Densities of typical residential building compounds in comparison to buildings by Baumschlager–Eberle

Source: T. Keller, Professur für Geschichte des Städtebaus, gta, ETH Zürich/Architectural Devices AG, St. Gallen

Zur Umsetzung dieses Anspruchs an Nachhaltigkeit bedienen sich Baumschlage–Eberle der projektbezogenen Faktenlage sowie eines tiefen und detaillierten Verständnisses der Aspekte des Bauens. Eine systematische Bearbeitung des ganzheitlichen Ansatzes erleichtert die Umsetzung. Nachfolgend werden die Aspekte in Kurzform erläutert.

Landverbrauch

Städte und Siedlungen zeigen je nach Land und Region stark differenzierte Charaktere. Beschreiben lassen sich die Charaktere über vergleichsweise unscharfe Angaben zu generellen Gebäudedimensionen, zur Präsenz von Grünflächen, dem durchschnittlichen Alter der Gebäude, aber auch durch die präzisen Werte der baulichen Dichte, **Bild 139**. Die Bestimmung der baulichen Dichte ist ein vergleichsweise einfaches Werkzeug. Es beschreibt das Verhältnis von permanent genutzter Nutz- und Wohnfläche zur Grundstücksfläche.

Der Charakter eines Gebiets wird auch von der baulichen Dichte bestimmt. Größere Dichte hat ebenso Auswirkungen auf die grundsätzliche Nähe von Menschen wie auf wirtschaftliche Möglichkeiten. Multifunktionalität, wie sie verdichtete urbane Räume vorweisen können, wird heute als Standortvorteil, als Erlebnisdichte und als Lebensqualität wahrgenommen. Hinsichtlich eines ressourcenoptimierten Arbeitens beschreibt die bauliche Dichte den Umgang mit der Ressource „Land".

Wasserverbrauch

Bei der Nutzung von Gebäuden werden nicht nur Energie für Heizung, Lüftung, Licht und den allgemeinen Betrieb verbraucht, die Nutzer selbst konsumieren auch erhebliche Mengen von Wasser. Eine optimierte Verwendung von Trinkwasser scheint plausibel, wenn man sich bewusst wird, dass dies für weite Teile der Weltbevölkerung ein rares und kostbares Lebensmittel ist. Trinkwasser ist eine Ressource, die auch in Mitteleuropa lokal in sehr unterschiedlichen Mengen vorhanden ist. Je nach Standort eines Gebäudes müssen die Aspekte des Frischwasservorkommens und -verbrauchs gewichtet werden.

Bild 139
Bebauungsdichten bekannter Siedlungstypologien im Vergleich mit Bauten von Baumschlager–Eberle

Quelle: T. Keller, Professur für Geschichte des Städtebaus, gta, ETH Zürich/Architectural Devices AG, St. Gallen

In most cases, a reduction of water consumption can be achieved without sacrificing human comfort at all. Only 10 % of the water consumed in Central Europe is used as potable water; the remainder can be reduced by technical measures such as optimized fittings and different consumer behavior. For example, it is not necessary to use the foodstuff "water" for toilet flushing, irrigation, building maintenance, and laundry. The current potable water consumption of approximately 100 liters per capita and day in Europe could be lowered to around 40 liters per capita and day without any problem (see **Figures 70** and **71**).

Heat energy

Energy for heating is necessary in order to achieve comfort conditions if a building's internal temperature is lower than the thermostat setting (**Figure 140**). The heat energy demand is the energy input that is introduced by active measures into a building (energy gains by insolation do not fall into this category). Heat energy can be generated by various primary energy sources, either renewable or fossil-fuel based, including oil, gas, wood, electricity, and solar collectors, among others. The base used for a comparison is the so-called Primary Energy Equivalent. Heat loads ought to be reduced in order to lower heat energy consumption. Reducing loads first is also the method applicable to limiting consumption of other resources.

Generell kann der Verbrauch der Ressource Trinkwasser im Gebäudebetrieb durch geeignete Maßnahmen reduziert werden, ohne Komfort einzubüßen. Nur ca. 10 Prozent des Wasserverbrauchs in Mitteleuropa wären als reines Trinkwasser notwendig, Jeder weitere Verbrauch von Wasser in einem Gebäude lässt sich über verändertes Benutzerverhalten, optimierende und optimierte Armaturen und Geräte reduzieren. Zur Toilettenspülung, Gartenbewässerung, Gebäudereinigungen und zum Wäschewaschen ist es nicht zwingend notwendig, das Lebensmittel Trinkwasser einzusetzen. Insofern lässt sich feststellen, dass sich der zurzeit durchschnittliche Frischwasserbedarf um ca. 100 Liter/Person und Tag ohne weiteres auf 40 Liter/Person und Tag senken lässt (vergleiche **Bilder 70/71**)

Heizenergie

Die Heizenergie, **Bild 140**, ist jene Energie, die benötigt wird, um im Gebäude eine thermische Behaglichkeit zu erreichen, sollte die Temperatur im Inneren niedriger sein als angestrebt. Als Heizenergiebedarf wird die aktiv eingebrachte Heizenergiemenge betrachtet, also nicht die durch direkte Sonneneinstrahlung generierte Wärme im Gebäude. Heizenergie kann aus verschiedenen Energieträgern erzeugt werden, wie zum Beispiel Erdöl, Erdgas, Holz, Strom, Wärmekollektoren etc. Basis einer vergleichenden Betrachtung ist das Primärenergieäquivalent. Das Vermeiden von Heizlasten wird angestrebt, um geringe Heizenergieverbräuche zu erreichen. Dies ist dieselbe generelle Methodik, welche zur Reduzierung der Verbräuche anderer Ressourcen angewendet wird.

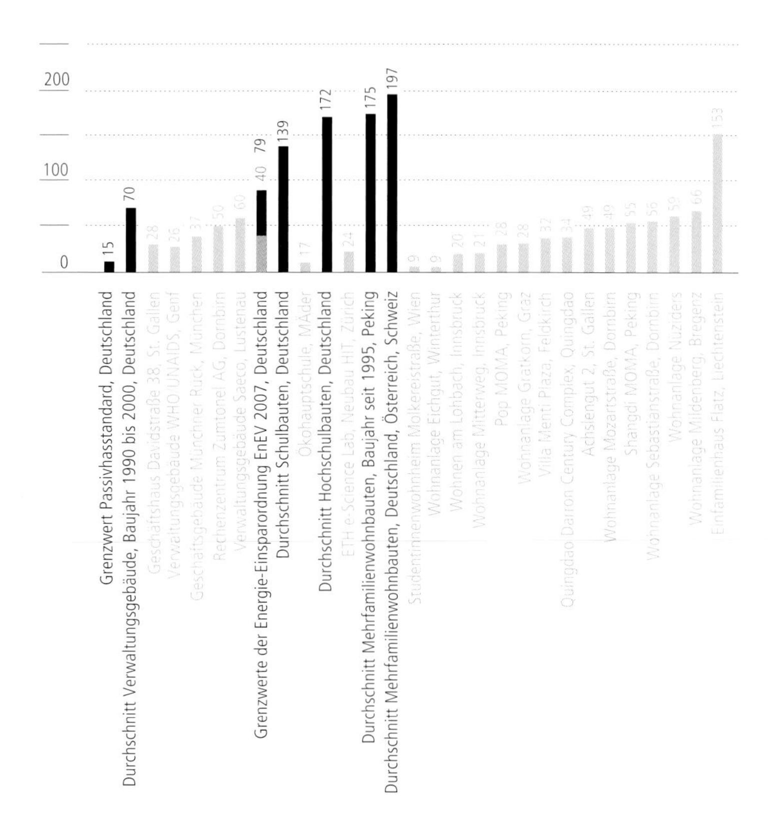

Figure 140
Heat energy demand of typical building types in comparison with projects by Baumschlager–Eberle

Source: VDI 3807/Daniels, author/Architectural Devices AG, St. Gallen

Bild 140
Heizenergieverbräuche bekannter Bautypen im Vergleich mit Bauten von Baumschlager–Eberle

Quelle: VDI 3807/Daniels, Autor/Architectural Devices AG, St. Gallen

11.2 Wohnen am Lohbach

Operating energy

Operating energy is that percentage of the total energy that is required for a building's cooling, lighting, warm water generation, and operation of equipment (**Figure 141**). It is important to provide for generation and transmission of that energy in the building with the least amount of production and transmission losses. In additional, cooling energy should remain in the building as long as possible without additional supply independent of the presence of internal loads such as people and equipment.To achieve reduced operating energy consumption, a weighting of aspects of building use, building systems, heat loads, and the internal distribution of energy is essential.

Warm water generation

The generation of warm water combines aspects of the resource consumption of water and heat energy. Therefore, this topic must be evaluated quite carefully. The distribution of warm water in the building with the least amount of losses is very important, as mentioned above. Since approximately 57 % of all water consumed in a building was previously heated, the optimization of this form of energy consumption needs to be done thoroughly.

Betriebsenergie

Die Betriebsenergie, **Bild 141**, gibt jene Energiemenge an, die für das Beheizen und Kühlen eines Gebäudes, das Aufbereiten des Warmwassers sowie der Elektrizität für Licht und Geräte notwendig ist. Das verlustarme Erzeugen und Verteilen der Energie im Gebäude ist relevant. Zudem sollte z.B. ein gekühltes Gebäude möglichst lange kühl bleiben, ohne dass Kühlenergie eingebracht werden muss, obwohl durch Personen und Geräte Wärmeenergie zugeführt wird.Im Resultat ist ein Abwägen von Aspekten der Nutzungsart, Gebäudetechnik, Wärmelasten und Art der Energieverteilung wichtig, um geringe Betriebsenergieverbräuche zu erreichen. Besonders zu beachten ist, dass die Aufwendungen für die Energie einen wesentlichen Teil der gesamten Betriebskosten eines Gebäudes darstellen.

Warmwasser

Warmwasser verbindet Aspekte des Wasserverbrauchs und der Heizenergie. Daraus entsteht die Notwendigkeit, den Energieaufwand für die Warmwassererzeugung möglichst effizient zu steuern. Zudem ist die verlustarme Verteilung des Warmwassers im Gebäude wichtig. Etwa 57 Prozent des verbrauchten Wassers eines Haushalts werden zuvor erwärmt und stellen damit einen bemerkenswerten Anteil des Betriebsenergieverbrauchs. Diesen Energieverbrauch gilt es zu optimieren.

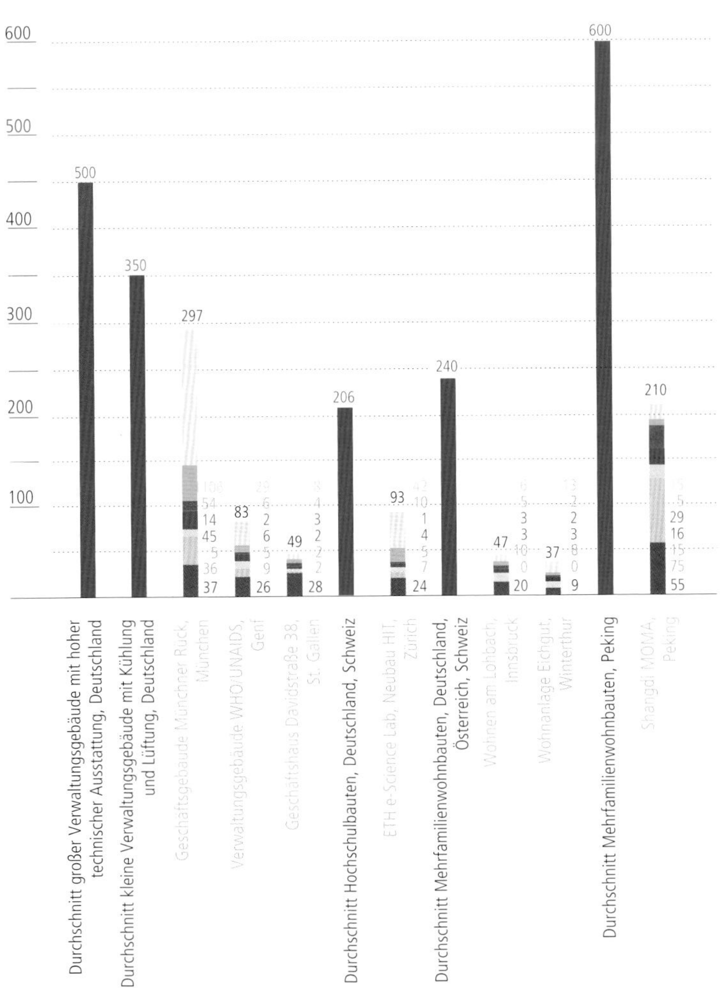

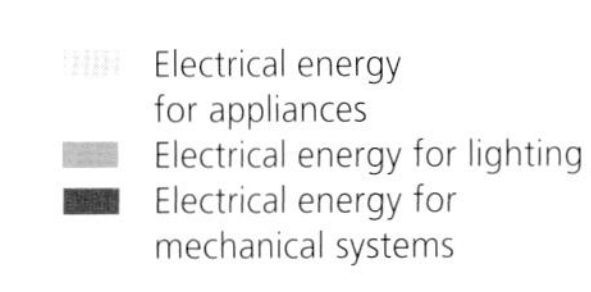
Electrical energy for appliances
Electrical energy for lighting
Electrical energy for mechanical systems

Electrical energy for ventilation
Hot water generation
Space cooling
Space heating

Figure 141
Operating energy consumption for typical building types compared to projects by Baumschlager–Eberle

Source: VDI 3807/Daniels, author/Architectural Devices AG, St. Gallen, Switzerland

Bild 141
Betriebsenergieverbräuche bekannter Bautypen im Vergleich mit Bauten von Baumschlager–Eberle

Quelle: VDI 3807/ Daniels, Autor/Architectural Devices AG, St. Gallen

Grey energy

For the production of building materials, the raw base materials provided by nature need to be transformed, which requires widely varying amounts of energy. Grey energy is that amount of energy which is required for the manufacturing of such building materials as concrete or building components such as windows or ventilation equipment. The calculation of grey energy starts with the mining of the raw material and includes all energy required for its transportation, manufacturing, and conversion into a finished product. Grey energy is therefore a clear indicator of a material's total energy consumption. The balance of grey energy allows for the optimization of structures and their components.

In Central Europe, the grey energy content of a building's materials is roughly equivalent to the amount of energy that would be required for the supply of heat energy to the finished structure for a period of about 30 years (this is the case only for a well-insulated structure). Because the grey energy content represents such an astonishingly large percentage of total energy requirements, it is necessary to include it in any sustainability balance analysis.

Grey energy is calculated as primary energy, and universally applicable, international standards are now being developed, in additional to the previously mentioned methods. According to current research, the actual building function, such as residential, commercial, or use as an educational or administrative building, is of no significant importance when it comes to the magnitude of grey energy use. Therefore, in Switzerland the so-called SIA Effizienzpfad Energie (Efficiency Path Energy) suggests that designs for all of the above-mentioned types of buildings use a target value of 830 kWh/m^2. However, the building's absolute size and its compactness will, of course, have a large influence on the percentage of grey energy embodied.

Method of application

The office of Baumschleger–Eberle has great awareness of the need to use resources sensibly during design, building, and operation of a structure. The results are buildings with minimized primary energy use, and the following are design tools used by that office to achieve these goals.

Graue Energie

Zur Herstellung von Baustoffen werden Rohstoffe aus der Natur sowie Energie benötigt. Als Graue Energie wird der gesamte kumulierte Energieaufwand bezeichnet, um einen bestimmten Baustoff wie beispielsweise Beton oder Bauteile wie Fenster oder Lüftungsanlagen herzustellen. Die Berechnung beginnt mit dem Rohstoffabbau und bilanziert sämtliche Transport-, Herstellungs- und Verarbeitungsprozesse bis zum verkaufsfertigen Produkt. Die Graue Energie ist damit ein einfacher, aussagekräftiger und praxistauglicher Indikator für den Verzehr von Energieressourcen. Die Bilanzierung der Grauen Energie erlaubt eine materialseitige Optimierung von Konstruktionen und Gebäuden.

In Mitteleuropa entspricht die für die Herstellung aller Baustoffe notwendige Graue Energie eines Gebäudes in ihrer Größenordnung etwa jener Energie, die zur Beheizung eines gut wärmegedämmten Gebäudes während 30 Jahre notwendig ist. Ihre Einbeziehung in Energiebilanzen von Gebäuden ist deshalb unumgänglich.

Graue Energie wird in Primärenergie gerechnet. Einheitliche und international vergleichbare Berechnungsmethoden werden zurzeit weiterentwickelt (und sind zum Teil zuvor dargestellt). Nach derzeitigem Stand des Wissens darf davon ausgegangen werden, dass die Nutzung der Gebäude wie beispielsweise für Wohnen, Unterricht oder Verwaltung auf die Größenordnung der Grauen Energie keinen entscheidenden Einfluss hat. In der Schweiz wird deshalb im so genannten „SIA Effizienzpfad Energie" für alle drei oben beschriebenen Nutzungen ein Zielwert für Graue Energie von 830 kWh/m^2 vorgeschlagen. Allerdings haben die absolute Gebäudegröße sowie vor allem die Kompaktheit wesentlichen Einfluss auf die Kennwerte für Graue Energie.

Methode zur Umsetzung

Baumschlager–Eberle ist sich der Notwendigkeit eines sensiblen und verhältnismäßigen Umgangs mit Ressourcen während der Planung, des Bauens und des Betriebes eines Gebäudes bewusst. Als Ergebnis dieser Überlegungen entstehen Gebäude, die minimale Primärenergieverbräuche erreichen. Baumschlager–Eberle hat Werkzeuge entwickelt, die eine präzise Integration der Aspekte des Themas Ressourcen in Entwurf und Planung ermöglichen.

11.2 Wohnen am Lohbach

Definition of comfort by clear target description

Because the major element in a building that controls resource consumption and defines sustainability standards is the building's envelope, the Baumschlager–Eberle project handbook lists in Chapter D, under the topics of "building enclosure" and "building systems", targets that are the result of a weighting of client and designer golas. For a majority of projects they can be described as follows:

1. Thermal comfort
   This is represented by values for air temperatures, air humidity, air velocity, and radiation that are seen as comfortable by the client.

2. Hygienic comfort
   This depends upon the quality of the room air. It will be determined by the quality of the supply air, space and use-dependent pollution, and the potential for treatment. Integrated into this category are aspects of air changes per hour, air delivery systems design, and material surface characteristics.

3. Visual comfort
   This category describes the optimization of the visual environment so that the human brain can process sensations without negative disturbances. On the user side, this needs to include the individual capacity of eyesight, the perceptive faculty, and speed and effectivity of discrimination. For instance, the perceptive faculty of interior space is compromised by strong luminous intensity differences, direct and indirect glare, and problematic color rendering characteristics.

4. Acoustical comfort
   This is not a scientifically precise term. For its definition, more information needs to be included than just the physical values of reverberation time, resonant frequency, white noise, attenuation, and others. Terms such as local cultural characteristics, subjective sensing, and the overall function of a space need to be included as well

Principles put into practice

The challenge is certainly now to convert the above-listed comfort criteria into a working design without compromising too much of their essence. To achieve this goal, the office of Baumschlager-Eberle uses, besides its own in-house experience and the expertise of external consultants, a specially developed, proprietary flow-chart software program (**Figure 142**).

Präzise Definition der Behaglichkeit durch Zielvorgaben

Eines der Schlüsselelemente für Nachhaltigkeit und Ressourcenverbräuche ist die Gebäudehülle. Das Projektbuch von Baumschlager–Eberle integriert deshalb im „Kapitel D – Gebäudehülle und Haustechnik" systematisch die Zielvorgaben aus dem Konsens zwischen Bauherrn und Planer. Diese Zielvorgaben lassen sich für die Mehrzahlt der Projekte in vier Arten von Behaglichkeit gliedern:

1. Thermische Behaglichkeit:
   Die Werte für Lufttemperatur, Luftfeuchte, Luftbewegung und Wärmestrahlung, die vom Nutzer als behaglich empfunden werden.

2. Hygienische Behaglichkeit:
   Sie ist stark abhängig von der Qualität der Raumluft. Diese wird bestimmt durch die Beschaffenheit der Zuluft, durch nutzungs- und raumbedingte Verunreinigungen und die Möglichkeiten zur Aufbereitung. Zusätzlich werden Aspekte der Luftführung, des Luftwechsels, der Oberflächen etc. in die Zielvorgabe integriert.

3. Visuelle Behaglichkeit:
   Sie beschreibt eine optimierte visuelle Umgebung, so dass der ungestörte Wahrnehmungsablauf im menschlichen Gehirn möglich ist. Dies berücksichtigt auf Nutzerseite die Grundempfindungen des Auges wie Sehleistung, die Wahrnehmungsgeschwindigkeit und Unterschiedsempfindlichkeit. Beispielsweise schränken falsche Leuchtdichteverteilungen im Raum, Blendungen, unrichtige Farbwiedergaben und nicht angepasste Raumgestaltung diesen Wahrnehmungsablauf ein.

4. Akustische Behaglichkeit.
   Ein nicht eindeutig definierter Begriff, da zur Definition mehr Informationen als rein physikalische Größen notwendig sind, wie z.B. Nutzung, subjektives Empfinden und kultureller Kontext. Es werden aber entsprechend den in Gebäuden stattfindenden Aktivitäten die zu erreichenden akustischen Anforderungen an Nachhallzeit, Grundgeräusche, Schallleistungen, Luftschalldämmungen etc. in Zahl und Art festgeschrieben.

Umsetzung in Entwurf und Planung

Die Herausforderung für den Entwurf ist es, die vorangehend beschriebenen Komfortstandards möglichst in ihrer Gesamtheit zu erreichen. Baumschlager-Eberle nutzen zu diesem Zweck eigene Erfahrungen, das Fachwissen von Ingenieuren und eine speziell entwickelte Software (Flussdiagramm, **Bild 142**).

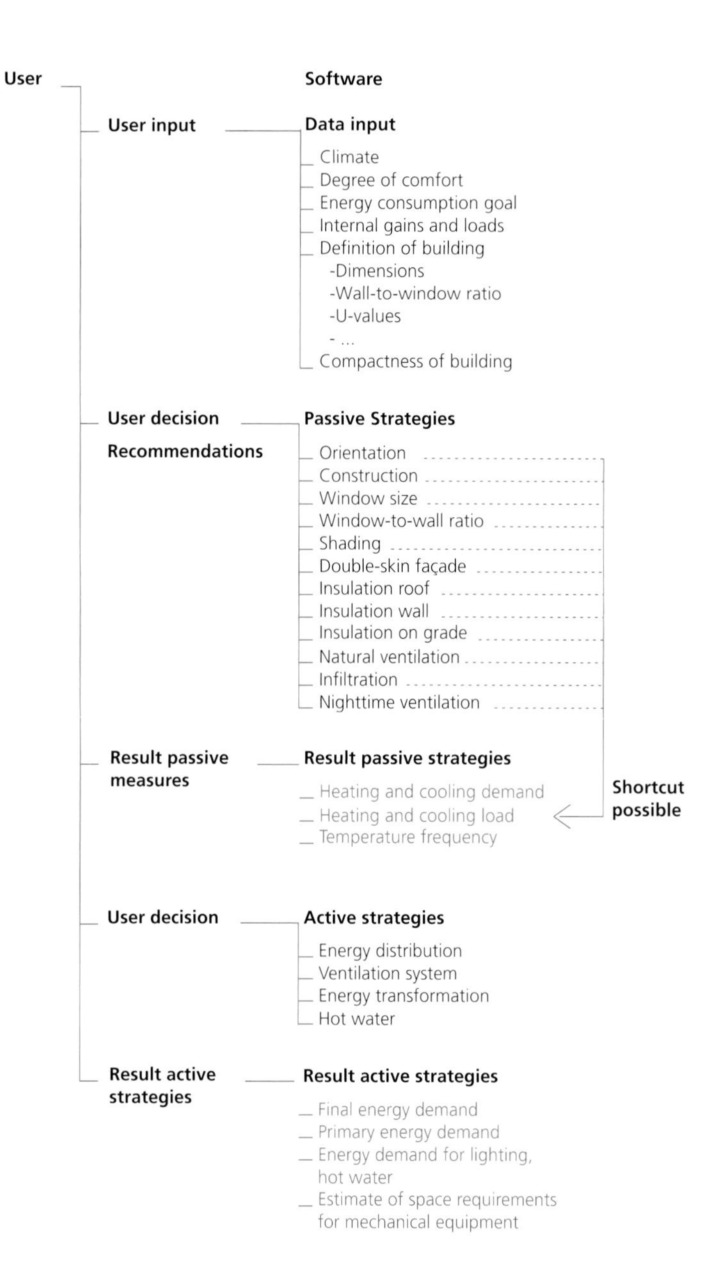

Figure 142
Flow diagram of Baumschlager–Eberle software to evaluate energy performance of building designs.

(The calculation is dynamic and based on yearly climate data of the particular location of the building.)

Source: Dr. Lars Junghans, Baumschlager–Eberle Architekten, Lochau, Austria.

Bild 142
Flussdiagramm der Baumschlager– Eberle Software zur Berechnung der energetischen Leistungsfähigkeit von Gebäuden.

(Die Berechnung erfolgt dynamisch auf Basis der während eines Jahres variierenden klimatischen Bedingungen am Gebäudestandort.)

Quelle: Dr. Lars Junghans, Baumschlager–Eberle Architekten, Lochau, Österreich

The consideration of various energy schemes is supported by the software tool as well, with parameters such as site-specific climate data, orientation, and existing building structures being incorporated into the decision process. As a first approximation, and with consideration of the desired energy savings goals, the initial behavior of the first design can be evaluated. This will include the achievable comfort degree, the materials to be used, expected solar energy gains, building dimensions and internal loads, and a first assumption concerning energy behavior. As described earlier, the surface-to-volume ratio of the building is of critical importance because about 30 % of primary energy savings can be achieved by an optimized building geometry.

Für die Abwägung von energetischen Varianten wird – unterstützt durch die Software – ein Prozess durchlaufen, der als Anfangswerte standortbezogene Daten wie Klima, Orientierung und Bestand integriert. Gemeinsam mit Energieeinsparzielen, zu erreichendem Komfortgrad, zu erwartenden internen Materialitäten, Energiegewinnen, Energielasten und Gebäudedimensionen lässt sich in erster Annäherung das grundsätzliche energetische Verhalten erkennen. Dabei ist der Einfluss der Kompaktheit eines Gebäudes auf den generellen Energieverbrauch signifikant. Bis zu 30 Prozent Einsparung beim Primärenergieverbrauch lassen sich bereits über die Wahl der Gebäudegeometrie erreichen.

Die Software generiert für den entwerfenden Architekten in einem ersten Zwischenergebnis Varianten von passiven Methoden des Energiesparens. Basierend auf diesen Vorschlägen kann dem digitalen Werkzeug die Wichtigkeit und Art der geplanten passiven Elemente vermittelt werden. Diese sind beispielsweise Angaben über Speichermasse, Wandaufbau,Nachtauskühlung sowie Merkmale der geplanten Gebäudehülle wie Schichtung, Verschattungsart und Luftdichtigkeit.

Als nächstes Zwischenergebnis erstellt die Software eine Abschätzung der energetischen Leistungsfähigkeit des Gebäudes. Dieses Ergebnis ist ebenso in Varianten aufgebaut, welche vom Nutzer zu werten sind,

11.2 Wohnen am Lohbach

The software mentioned above allows the design architect to make a first evaluation of the different options for energy savings. In additional, it allows the input of wall assembly composition, the thermal storage mass of the structure, nighttime cooling capability, and additional characteristics of the building envelope such as shading devices and air infiltration systems. The next intermediate result of the analysis is the energy generation capacity of the building. Alternatives are presented to the architect, which become the basis for further necessary decisions concerning active strategies such as the various options for energy generation, energy distribution, air treatment, energy conversion, and warm water generation.

The results of the process are a few precisely defined key target values for energy consumption, building systems, and the building envelope. The number of potential technical concepts is drastically reduced in the process, yet the overall precision is maintained. The designer retains an overview of the increasingly complex energy-related variations in and around the building design, but with the clear information provided by the process concerning the structure, the envelope, and the functional layout of the building. The designer is able to work in a goal-oriented manner and with a greatly reduced margin of error.

In subsequent design steps, and in close dialogue with the consulting engineers, the design will progress towards its physical manifestation. For each individual building, the described process will provide varying design alternatives because the base data will change from design to design, if only due to a change of site location.

Result

The design solution has to be able to harmonize the various recommendations, beyond the purely energetic targets and the aspects of comfort. With its software tool, Baumschlager-Eberle architects are capable of providing a decision-enabling process to the design architect to reduce the numerous available concepts down to the few that are most appropriate, without being biased by preconceived design notions.

Outlook

The comfort requirements in buildings and the careful use of resources can very well complement each other. The process further allows the integration of renewable energy resources into the design. Both a knowledge of available systems and a sensitivity to how they best need to be integrated in a comprehensive way into a design will be required to turn buildings into cooperative systems.

um die nachfolgende Auswahl der aktiven Energiestrategien zu definieren. Aktive Strategien entscheiden über die Arten der Energieerzeugung, Energieverteilung, Luftbehandlung, Energieumwandlung, Warmwassererzeugung etc.

Das Ergebnis dieses Prozesses sind wenige, aber präzise Kennwerte für Energiebedarf, Gebäudetechnik und Gebäudehülle, Die Anzahl der grundsätzlich möglichen technischen und konzeptionellen Varianten werden in der Menge drastisch reduziert, bei gleichbleibend hoher Korrektheit. Diese Methode ermöglicht es, das immense und permanent größer werdende Wissen über die energetisch relevanten Zusammenhänge in und um die Gebäude verständlich zu halten. Die Architekten erhalten für Entwurf und Planung klare Informationen bezüglich Struktur, Gebäudehülle, Layout etc. Daher können Baumschlager-Eberle zielorientiert mit geringer Fehlerquote arbeiten.

In weitern Schritten wird im intensiven Dialog mit Fachplanern eine Lösung entwickelt, die im Projekt physisch realisiert wird. Dieser Prozess ermittelt für jedes Gebäude und für jede Entwurfsvariante folglich verschiedene Ergebnisse, da Ausgangswerte immer variieren – und sei es nur der Standort.

Ergebnis

Der Entwurf hat die Vielzahl der Zielvorgaben miteinander zu verbinden, weit außerhalb der rein energetischen Vorgaben oder der Aspekte der Behaglichkeit. All diese Parameter haben in unterschiedlicher Stärke Einfluss auf die Energieverbräuche des Gebäudes. Baumschlager-Eberle schaffen es, Architekten Werkzeuge und Informationen in die Hand zu geben, welche die Vielzahl der Möglichkeiten innerhalb des Entscheidungsbaumes auf eine überschaubare Anzahl von sinnvollen Möglichkeiten reduzieren – ohne dabei der Starrheit vorgefertigter Konzepte zu verfallen.

Ausblick

Die Anforderungen an Komfortsteigerung und Ressourcenoptimierung können sich gegenseitig ergänzen und unterstützen. Die beschriebene Herangehensweise schafft jene Voraussetzungen, die den Einsatz von alternativen Technologien gesellschaftlich sinnvoll werden lassen.

Es ist eine Frage des Wissens und Verständnisses, mit welchen Systemen die höchsten Einsparungen erzielbar sind, Das wichtigste Handlungsprinzip wird in Zukunft sein, die für die jeweilige Bauaufgabe spezifischen Ressourcen – wie Wissen, Sonnenenergie, Biomasse, Luft, Geothermie etc. – auszunutzen. Das Gebäude selbst, Nutzerwünsche und die dafür vorgesehene Technik müssen als kooperative Systeme verstanden werden. Der wichtigste Beitrag, auf dem die weitere Entwicklung im Bereich Ressourcen basiert, ist das Verbreiten und Anwenden von vorhandenem Wissen.

Figure 143 shows the typical consumption data for the region of Central Europe and the countries Germany, Austria, and Switzerland. The energy consumption for existing building stock and new construction is juxtaposed with the Energiesparverordnung (Energy Savings Regulation). Similar to the Swiss Minergie standard, **Figure 143** displays the data for the residential projects Wohnen am Lohbach and another condominium complex in Vienna, Austria. They are all well below regular energy consumption schemes for comparable buildings.

Figure 144 shows the residences of Wohnen am Lohbach in Innsbruck, with their signature insulation elements at front of the building's balconies, which can be closed at night and opened during the day. If these insulation elements are used, the specific heat energy requirement for the complex is about 20 kWh/m²a. **Figure 145** shows a detail of the façade with its insulating window shutters, and **Figure 146** presents the floor plan. Another characteristic is the compactness of the building, which is similar to other designs of this architecture office, and which results in a favorable surface-to-volume ratio. A large, open-well staircase allows daylight to enter and penetrate deeply into the interior of the structure, **Figure 147**.

The installation of solar thermal collectors on the building's roof for warm water generation and rainwater collection complements the concept of the building's sustainability.

Bild 143 zeigt Heizenergie-Verbrauchskennwerte, wie sie üblicherweise in der Region Deutschland/Österreich/Schweiz anzutreffen sind. Die aus Benchmarkstudien entnommenen Energieverbräuche für Alt- und Neubauten sowie Neubauten sind ins Verhältnis gesetzt zur Energieeinsparverordnung, analog zum Minergiestandard der Schweiz. Im **Bild 143** sind die Wohnobjekte Wohnen am Lohbach sowie ein Wohnheim in Wien ausgewiesen und liegen deutlich unter den bekannten Zielwerten.

Bild 144 zeigt die Wohnanlage am Lohbach, Innsbruck, mit ihren charakteristischen Wärmedämmelementen vor den Balkonen, die während des Tages geöffnet und während der Nacht geschlossen werden. Beim konsequenten Betrieb und konsequenter Nutzung wurde bei dem Objekt ein spezifischer Wärmeenergiebedarf von 20 kWh/m²a erreicht. **Bild 145** zeigt ein Fassadendetail mit Wärmedämmläden, **Bild 146** den Grundriss des Gebäudes. Charakteristisch für viele der entworfenen Bauten ist, dass sie sehr kompakt sind und damit ein sehr günstiges A/V-Verhältnis erreichen. Auffällig, wie auch in **Bild 147** ausgewiesen, ist ein innerer großzügiger Treppenraum mit einem Treppenauge, durch das Tageslicht bis in den unteren Bereich des Gebäudes einfällt.

Auf den Dächern installierte Solarkollektoren zur Warmwassererzeugung sowie eine Regenwassernutzung ergänzen das ökologische Konzept.

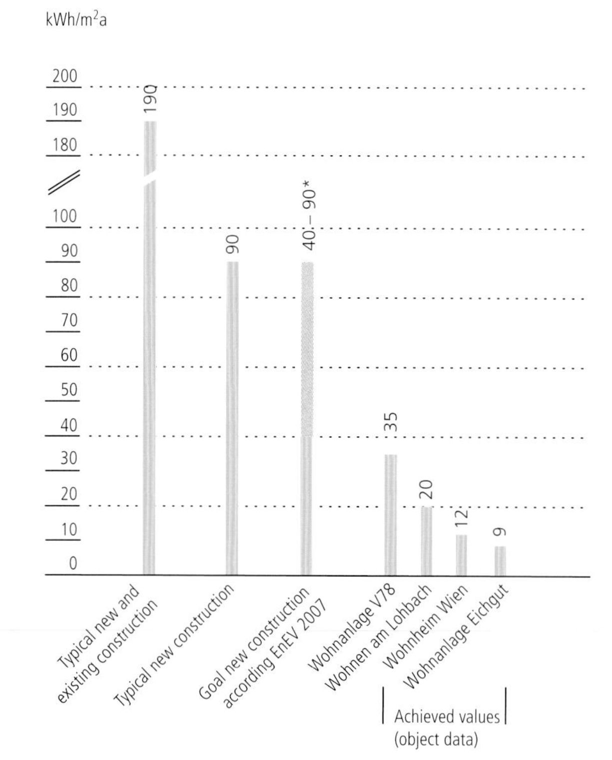

Figure 143
Heat energy demand for residential complexes

Bild 143
Wärmeenergie-Bedarf, bei Wohnanlagen

Figure 144
Wohnen am Lohbach,
Innsbruck, Austria

Architect:
Baumschlager–Eberle,
Lochau, Austria

Bild 144
Projekt Wohnanlage Lohbach,
Österreich

Architekt:
Baumschlager–Eberle,
Lochau, Österreich

Figure 145
Wohnen am Lohbach,
Innsbruck, Austria
Façade closeup

Bild 145
Wohnen am Lohbach,
Innsbruck, Österreich
Fassadendetail

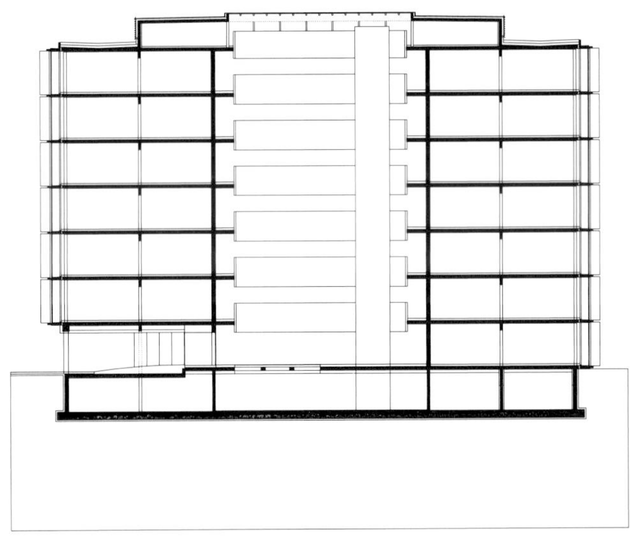

Figure 147
Wohnen am Lohbach,
Innsbruck, Austria
Cross section

Bild 147
Wohnen am Lohbach,
Innsbruck, Österreich
Schnitt-Darstellung

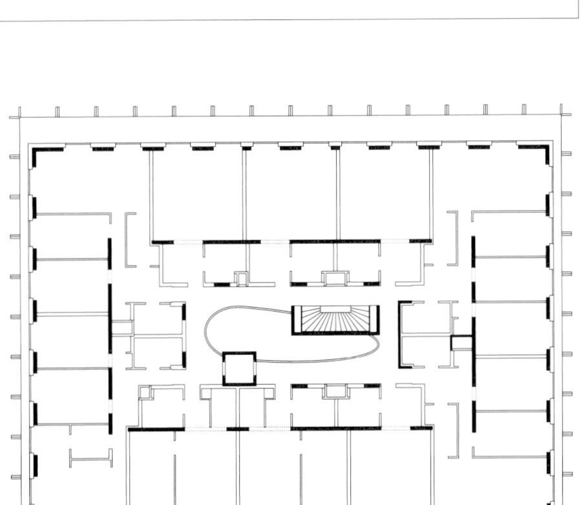

Figure 146
Wohnen am Lohbach,
Innsbruck, Austria
Plan, typical floor

Bild 146
Wohnen am Lohbach,
Innsbruck, Österreich
Grundriss Normalgeschoss

11.3 Student dormitories hall Molkereistraße, Vienna, Austria

Figure 148 shows a detail of the project's façade, which was designed after the previously described project. To further reduce the heat energy consumption, attractive copper-clad external insulation shutters were used to provide increased thermal insulation at night. **Figure 149** shows a floor plan of the project, and **Figure 150** shows the corresponding cross section. Due to site specifics and the function of the building, the design is not as compact as the projects described earlier, but the improved thermal insulation of both the opaque and transparent building envelope elements results in a greatly reduced overall energy demand.

11.3 Studentenwohnheim Molkereistraße, Wien, Österreich

Bild 148 zeigt einen Fassadenausschnitt des Objektes, das im zeitlichen Ablauf nach der Wohnanlage am Lohbach entstanden ist. Um den Wärmeenergiebedarf weiter zu senken, wurde eine sehr ansprechende Fassadenkonzeption entwickelt, bei der kupferbeschlagenen Wärmedämmläden die zusätzliche nächtliche Wärmedämmung übernehmen. **Bild 149** zeigt einen Grundriss, **Bild 150** den zugehörigen Schnitt. Obwohl das Studentenwohnheim infolge seiner Nutzungsansprüche und infolge seiner städtebaulichen Randbedingungen nicht mehr so kompakt entwickelt werden konnte wie das Objekt am Lohbach, wurde gleichwohl durch verbesserte Wärmeschutzmaßnahmen sowohl in den opaken wie auch in den transparenten Fassadenteilen geringere Wärmeenergieverbräuche erreicht.

Figure 148
Student dormitories Molkereistraße, Vienna, Façade

Architects: Baumschlager–Eberle, Lochau, Austria

Bild 148
Studentenwohnheim Molkereistraße, Wien Fassadenansicht

Architekten: Baumschlager–Eberle, Lochau, Österreich

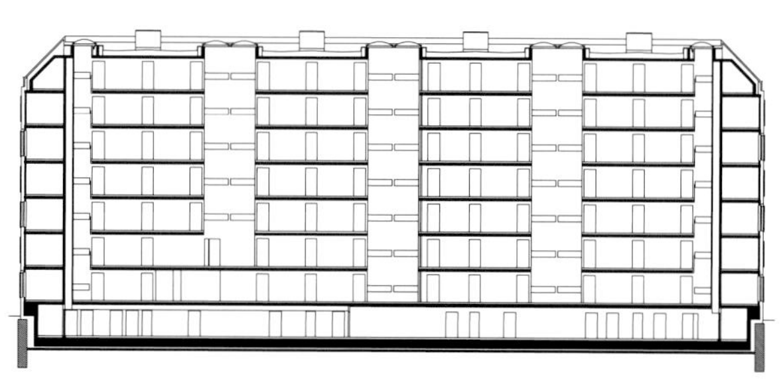

Figure 150
Student dormitories Molkereistraße, Vienna, Cross section

Bild 150
Studentenwohnheim Molkereistraße, Wien Schnitt

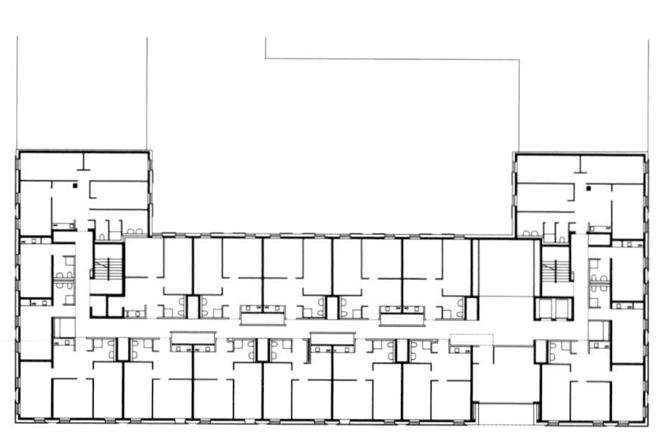

Figure 149
Student dormitories Molkereistraße, Vienna, Floor plan view

Bild 149
Studentenwohnheim Molkereistraße, Wien Grundriss

11.4 Ökohauptschule Mäder, Austria

Figure 151 allows, again, for the comparison of data on school buildings in Europe, broken down into existing building stock and new construction. The target data derived from EnEV or Minergie standards are shown for comparative evaluations of the thermal energy demand as well as primary electricity consumption.

Figures 152 and **153** show the so-called ecological or "green" secondary school in Mäder, Austria; **Figures 154** and **155** show the building's floor plan and section. Noteworthy is the amount of daylight that is introduced into the classrooms by a clap-board-like glazing system. The glazing units can be opened in summer and closed in winter to allow for proper ventilation of the building and the back ventilation of the façade. An interior triple-pane insulated glazing system achieves a U-value of 0.7 W/m²K and limits the convective transfer from outside to inside throughout the year. The target values for electricity consumption according to SIA 380/4 result in low operating cost for the building.

A large internal lobby is lit by a large light well that helps to reduce artificial lighting use.

11.4 Ökohauptschule Mäder, Österreich

In **Bild 151** sind wiederum Vergleichswerte aus einer Wertung von Schulbauten, geteilt nach Alt- und Neubauten sowie Neubauten, gegenübergestellt. Die Zielwerte für Neubauten analog EnEV oder Minergiestandard stehen zum Vergleich der erreichten Objektwerte, sowohl in Bezug auf den Wärmeenergiebedarf wie auch den Strom-Primärenergiebedarf.

Die **Bilder 152** und **153** zeigen die Schule, **154** und **155** Grundriss und Schnitt. Bei dem Schulgebäude ist die großzügige Verglasung auffällig, die für einen hohen Tageslichteinfall in die Klassenräume sorgt. Eine äußere, geschuppte Glasstruktur kann während des Winters geschlossen und im Sommer geöffnet werden, um die natürliche Belüftung und Hinterlüftung im Fassadenbereich zu unterstützen. Eine innere Dreifach-Isolierverglasung mit u-Werten um ca. 0,7 W/m²K sorgen dafür, dass der konvektive Wärmeübergang während des gesamten Jahres gleichbleibend gering ist. Die elektrischen Energieverbräuche liegen im Zielwertbereich (SIA 380/4) und führen somit zu geringen Betriebskosten.

Eine innere großzügige Erschließungshalle wird über einen eingestellten Lichtschacht mit Tageslicht versorgt, so dass der Kunstlichtbetrieb in Grenzen gehalten werden kann.

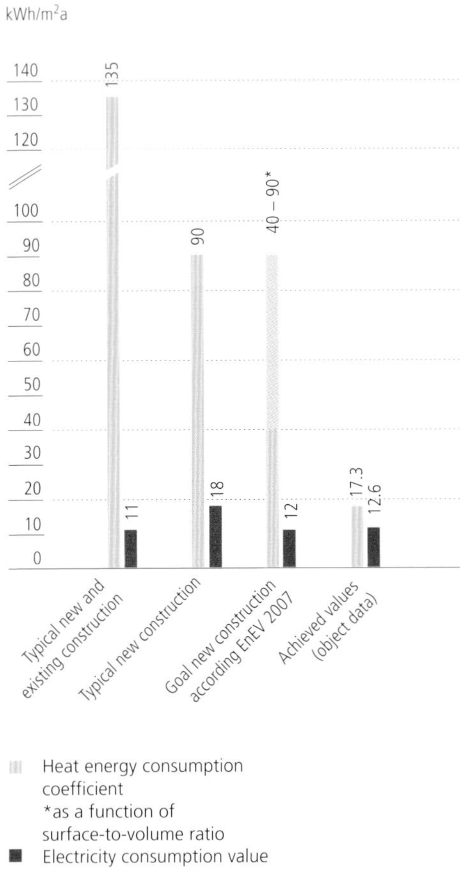

kWh/m²a
140 130 120 100 90 80 70 60 50 40 30 20 10 0
135 11
90 18
40 – 90* 12
17.3 12.6
Typical new and existing construction
Typical new construction
Goal new construction according EnEV 2007
Achieved values (object data)

Heat energy consumption coefficient
*as a function of surface-to-volume ratio
Electricity consumption value

Figure 151
Heat and electricity demand, Ökohauptschule Mäder, Austria

Bild 151
Wärme- und Stromenergie-Bedarf, Ökohauptschule Mäder, Österreich

Figure 152
Ökohauptschule,
Mäder, Austria

Architects:
Baumschlager–Eberle,
Lochau, Austria

Bild 152
Ökohauptschule Mäder,
Österreich

Architekten:
Baumschlager–Eberle,
Lochau, Österreich

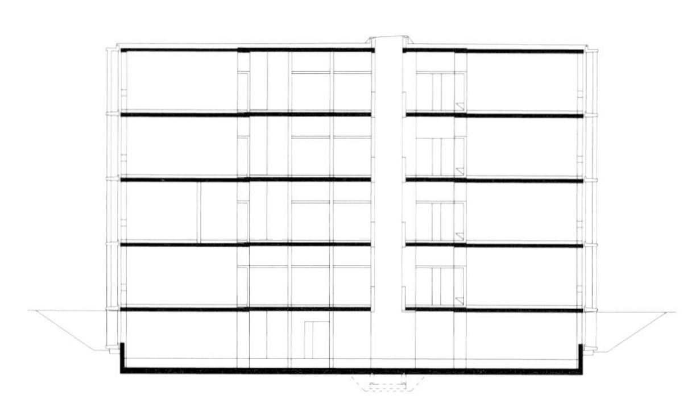

Figure 155
Ökohauptschule,
Mäder, Austria
Cross section

Bild 155
Ökohauptschule Mäder,
Österreich,
Schnitt

Figure 153
Ökohauptschule,
Mäder, Austria
Façade

Bild 153
Ökohauptschule Mäder,
Österreich,
Fassadenansicht

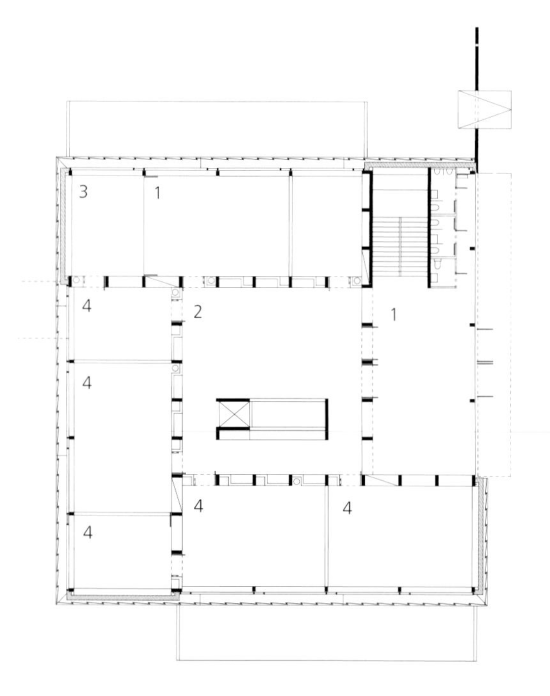

1 Foyer
2 Break
3 Kitchen
4 Work shop
5 Class room

Figure 154
Ökohauptschule,
Mäder, Austria
Floor plan

Bild 154
Ökohauptschule Mäder,
Österreich,
Grundriss

11.5 University building ETH-Z, E-science Lab, Zurich, Switzerland

According to **Figure 156**, educational facilities (existing and new construction), and especially university buildings, use approximately 166 kWh/m²a thermal energy and 34 kWh/m²a of electrical energy. This is caused mainly by long evening and nighttime operating hours. The values for new construction are almost identical to those of older buildings. Target values for electricity consumption, divided into that which is used for artificial lighting, cooling, and ventilation, are based on the Swiss Minergie standard. The building was, during the research for this publication, 2007 still under construction, and so model simulation was used to arrive at firm energy consumption predictions.

11.5 Lehrgebäude ETH-Z, E-science Lab, Zürich, Schweiz

Gemäß **Bild 156** verbrauchen Schulbauten (Alt- und Neubauten) und insbesondere Hochschulbauten ca. 166 kWh/m²a Wärmeenergie und ca. 34 kWh/m²a elektrische Energie, im Wesentlichen ausgelöst durch lange Beleuchtungszeiten während der Abendstunden und der Nacht. Die Durchschnittswerte von Neubauten liegen praktisch in der gleichen Größenordnung wie die Benchmarkwerte von Alt- und Neubauten. Die Zielwerte in Anlehnung an den Minergiestandard, Schweiz, zeigen Zielwerte des Stromenergiebedarfs, getrennt nach Beleuchtung, Lüftung und Kühlung. Da sich das Objekt zum Zeitraum Ende 2007 noch im Ausbaustadium befand, wurden umfangreiche Hochrechnungen gemacht, um die zu erreichenden Werte zu ermitteln.

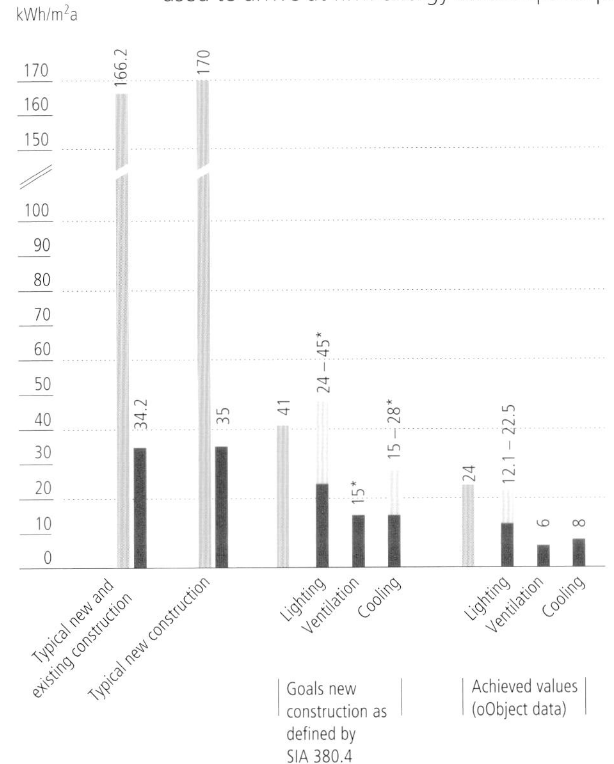

Figure 156
Heat and electricity demand, ETH E-science Lab, Zurich, Switzerland

Bild 156
Wärme-, und Stromenergie-Bedarf, ETH E-science Lab, Zürich, Schweiz

Heat energy consumption coefficient
Electricity consumption value
* as a function of surface-to-volume ratio

The compact design of the university building results in a favorable surface-to-volume ratio, **Figure 157** to **161**. In addition, high-quality insulating glazing units with U-values of < 0.7 W/m²K are used here. As a consequence, the heat energy necessary for the building will only be around 24 kWh/m²a. It is evident that part of reason for the low energy consumption is the supporting ventilation system in combination with a high-efficiency heat recovery system. Ventilation is provided by an air-water system with floor-integrated induction units. They are operated during the night as heat convectors without ventilation, and they are only operated when internal cooling loads are lower than heat losses. In highly insulated buildings, and in the computer simulation of this building, it can be observed that real heating scenarios only occur when outside temperatures fall significantly below an ambient air temperature of -5 °C.

Infolge der kompakten Bauweise, vergleiche **Bild 157** bis **161**, und des daraus resultierenden sehr günstigen A/V-Verhältnisses und weiterhin infolge der hochwertigen Wärmedämmung der großzügigen Verglasung (u-Werte < 0,7 W/m²K) können Verbrauchswerte der Heizenergie mit ca. 24 kWh/m²a festgestellt werden. Hierbei spielt selbstverständlich eine Rolle, dass das Gebäude unterstützend belüftet wird, ausgerüstet mit hochwertigen Wärmerückgewinnungssystemen. Die Belüftung über ein Luft-Wasser-System (fußbodenintegrierte Induktionsgeräte), die während der Nacht als Heizkonvektoren ohne Luftbetrieb arbeiten, soll nicht ganzjährig in Betrieb gehalten werden, sondern nur dann, wenn die inneren Kühllasten geringer sind als die Wärmeverluste. Bei hochgedämmten Gebäuden lässt sich insbesondere bei umfänglichen, computergestützten Arbeiten ermitteln, dass ein echter Heizfall erst dann eintritt, wenn die Außentemperaturen deutlich unter -5 °C liegen.

If outside temperatures reach 5 – 10 °C, the mechanical ventilation can be shut down in favor of natural ventilation, and at temperatures between 23 – 24 °C in summer the building will be cooled mechanically. Here, again, the induction units in the individual rooms are used, which are supplied with cold water from a centralized chiller plant. This building does not contain additional solar technologies to further reduce energy consumption. The supply of hot and cold water for the building is provided by the university's central combined-heat-power plant (CHP) on campus.

Architecturally, the building is characterized by shading elements whose design allows the use of desired insolation for passive solar energy gains during the winter, spring, and fall, and during the summer the façade is completely shaded. As shown in **Figure 158**, shaded skylights additionally allow for Zenith lighting. The images in **Figure 159** show detailed simulation of various floors, and **Figures 160** and **161** explain the building in floor plan and section.

Ab Außentemperaturen um 5 – 10 °C kann die mechanische Lüftung zugunsten einer natürlichen Belüftung abgeschaltet werden. Bei hochsommerlichen Temperaturen, d.h. Temperaturen über +23 – 24 °C muss das Gebäude sinnvollerweise gekühlt werden, um die anfallende Wärme ausreichend auszutragen. Hierzu dient wiederum die Induktionsanlage, die mit Kaltwasser einer Fernkälteanlage beschickt wird. Zur Reduzierung der Energieverbräuche wurden bei diesem Objekt keine aktiven solartechnischen Anlagen eingesetzt. Die zentrale Versorgung mit Wärme- und Kälteenergie er-folgt aus dem hochschuleigenen Nahversorgungsnetz auf Basis einer wärmegeführten Blockheizkraftwerk-Anlage.

Um einerseits in die Objekte ausreichend viel Tageslicht eintragen zu können und andererseits einen ausreichenden Sonnenschutz im Sommer zu erreichen, besitzt die Fassade signifikante Beschattungselemente, die so konzipiert sind, dass in der Übergangszeit und im Winter die Solarenergie passiv genutzt werden kann (Einstrahlung), während im Sommerbetrieb alle wesentlichen Fassadenelemente beschattet sind. Ergänzend zur inneren Belichtung sind, wie dem **Bild 158** zu entnehmen ist, beschattete gläserne Dachstrukturen vorgesehen, über die Zenitlicht einfallen kann. Ergänzend zeigen die **Bilder 159** eine Detailsimulation verschiedener Geschosse, **160** und **161** einen Grundriss sowie Schnitt.

Figure 157
E-sience Lab, Zurich, Switzerland

Architects:
Baumschlager–Eberle, Lochau, Austria

Bild 157
E-sience Lab, Zürich, Schweiz

Architekten:
Baumschlager–Eberle, Lochau, Österreich

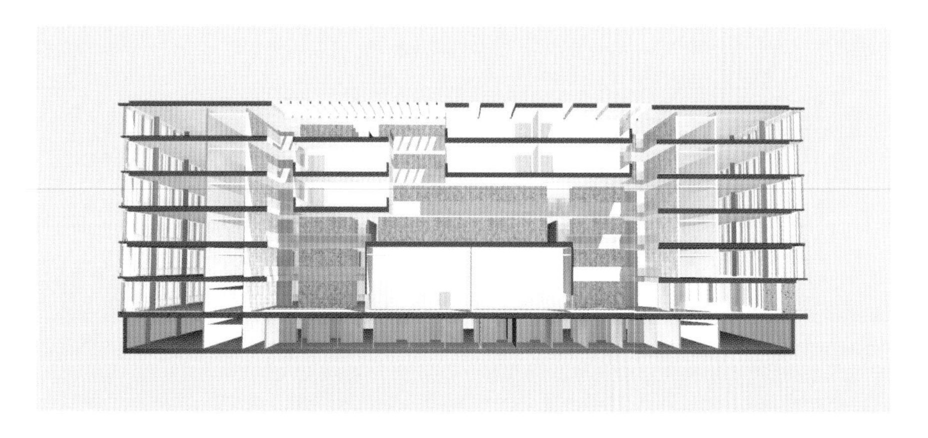

Figure 158
E-sience Lab, Zurich, Switzerland
Cross section

Bild 158
E-sience Lab, Zürich, Schweiz
Symmetrischer Schnitt durch dasGebäude

Figure 159
E-sience Lab, Zurich,
Switzerland
Detail simulation of interior hall

Bild 159
E-sience Lab, Zürich, Schweiz
Detail-Simulation der Innenhalle

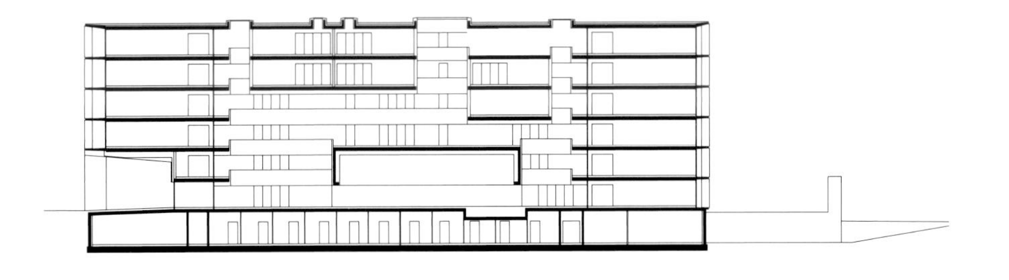

Figure 161
E-sience Lab, Zurich,
Switzerland
Cross section

Bild 161

E-sience Lab, Zürich, Schweiz
Schnitt

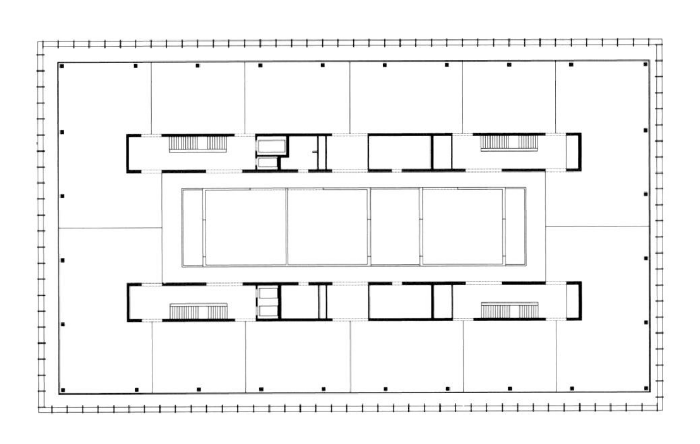

Figure 160
E-sience lLab, Zurich,
Switzerland
Floor plan

Bild 160
E-sience Lab, Zürich, Schweiz
Grundriss

11.6 Office building Münchner Rückversicherungs AG, Munich, Germany

Figure 162 provides general benchmark data for energy consumption in office buildings in Germany. They show that heat energy consumption values are typically around 125 kWh/m²a for existing, older buildings, and 90 kWh/m²a for new construction. The matching electrical energy consumption data were determined in accordance with SIA 380/4.

Compared with the above-mentioned targets, the values for this office building in Munich show significantly reduced energy consumption for artificial lighting, ventilation, and cooling.

Designers have, in most cases, a controlling influence on the consumption of electrical energy used for lighting, yet their influence diminishes when it comes to the consumption of electricity used for cooling and ventilation in cases where large-scale computer centers are part of the building program. As **Figure 162** demonstrates, the result of such functions in a building is always high consumption of energy for cooling and ventilation.

11.6 Bürogebäude Münchner Rückversicherungs AG, München, Deutschland

In **Bild 162** wiederum dargestellt sind Benchmarkzahlen für Bürobauten in Deutschland.

Alt- und Neubauten weisen Heizenergie-Verbrauchskennwerte von ca. 125 kWh/m²a aus, Neubauten solche von 90 kWh/m²a.

Zugehörige elektrische Energiebedarfswerte wurden analog zur SIA 380/4 ermittelt.

Die Zielwerte zeigen einen deutlich geringeren Wärmeenergiebedarf, gleichzeitig einen geringeren elektrischen Energiebedarf für Beleuchtung, Lüftung und Kühlung.

Während Architekten und Planer noch einen Einfluss auf den elektrischen Energiebedarf für Beleuchtung haben, wird der Einfluss im Bereich Lüftung und Kühlung dann obsolet, wenn im Gebäude umfangreiche EDV-Einrichtungen installiert werden. Hieraus resultieren im Regelfall hohe Energieverbräuche für Lüftung und Kühlung, wie in **Bild 162** gezeigt.

Figure 163
Münchner Rück
Munich, Germany
View of office building

Architects:
Baumschlager–Eberle,
Lochau, Austria

Bild 163
Münchner Rück
München,
Ansicht des Bürogebäudes

Architekten: Baumschlager–Eberle,
Lochau, Österreich

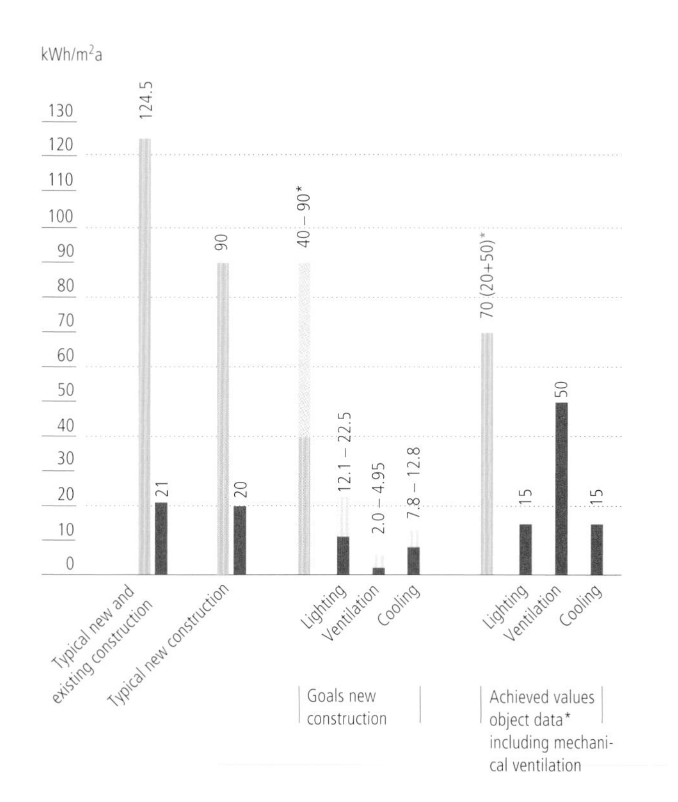

Heat energy consumption coefficient
*) as a function of Surface-to-volume ratio
■ Electricity consumption value

Figure 162
Heat and electricity demand of office buildings, Münchner Rück, Munich, Germany

Bild 162
Wärme-, und Stromenergie-Bedarf, bei Verwaltungsbauten, Münchner Rück, München

Figures 163 and **164** show the façade and a façade detail of this building, while **Figure 165** presents a floor plan and **Figure 166** presents the appropriate cross section through the structure.

The building features a wrapping glass structure, similar to the school building in Mäder, Austria, discussed previously. Although this element provides additional insulation during winter to defray heat energy losses, unfortunately it also causes increases in energy consumption used for necessary ventilation that exceed the target values. The building was conceived in accordance with the German heat energy regulation (Wärmeschutzverordnung) at the time and can be considered to be optimized in relation to those requirements.

The urban condition of the site and the functional requirements did not allow the building to be as compact as the previously introduced projects. The resulting surface-to-volume ratio cannot be considered to be very good, which is mostly a consequence of the building's partial single-loaded corridor layout.

Die **Bilder 163** und **164** zeigen eine Fassade und ein Fassadendetail, **Bild 165** einen Grundriss, **Bild 166** einen Schnitt durch das Gebäude.

Eine umlaufende, gläserne Struktur, ähnlich dem Schulhaus Mäder, dämmt das Gebäude zwar hinsichtlich der Wärmeenergieverluste in ausreichender Form, jedoch tritt durch den hohen Lüftungsbedarf ein zusätzlicher Wärmebedarf ein, der die Zielwerte überschreitet.

Da zum Zeitpunkt der Erstellung des Gebäudes die Zielwerte nach der Wärmeschutzverordnung in Deutschland zu definieren waren, ist das Gebäude unter diesen Maßgaben als optimal entwickelt anzusehen.

Aufgrund der städtebaulichen Situation sowie der räumlichen Konfigurationen konnte das Gebäude nicht so kompakt entwickelt werden wie zum Teil die zuvor dargestellten Objekte. Das A/V-Verhältnis entspricht keinem Optimum, die zum Teil einhüftigen Gebäudestrukturen verbieten dies.

Figure 164
Münchner Rück
Munich, Germany
Façade detail

Bild 164
Münchner Rück
Detailansicht Fassade

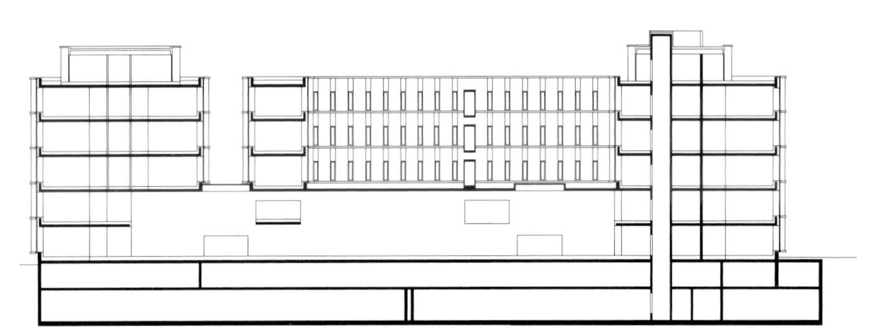

Figure 166
Munich Re-Insurance AG,
Munich, Germany
Cross section

Bild 166
Münchner Rück Versicherungs AG,
Schnitt-Darstellung

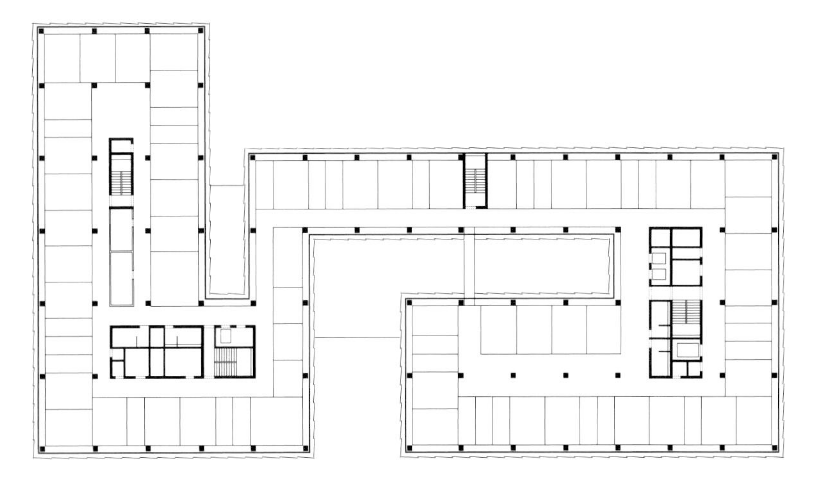

Figure165
Münchner Rück
Munich, Germany
Floor plan

Bild 165
Münchner Rück, München
Grundriss

11.7 Solar Decathlon Competition 2007, 1st prize Darmstadt, Germany; Washington D.C.; Phoenix

As in previous years the U.S. Department of Energy in 2007 solicited designs for the Solar International Decathlon Competition for the development of the vision of a marketable and energy-autonomous residential building prototype. Twenty solutions where presented on the Mall in Washington D.C. The focus of the 2007 competition was on building-integrated photovoltaics (BIPV), and the building systems of this "Year 2015 Prototype Home" needed to be fueled entirely by generation of energy from this renewable source. Another requirement for the participants was the need to provide – besides high design quality – a great degree of human comfort and design innovation of the living spaces.

The building design shown, which was the entry by a team of students and faculty of the Technical University of Darmstadt, Germany, has a size of approximately 100 m². As a main element of the goal for sustainability, the design team's philosophy was to provide a significant amount of flexibility of the floor plan and a certain neutrality of use. To satisfy the requirements, the following guiding principles for the Darmstadt entry were formulated:

- concept of layering of enclosure elements,
- concept of a platform design that provides a raised-access floor for furniture placement and technical systems integration,
- concept of a "whole-building plug-in" technical systems integration.

The concept of layering is explained in **Figure 167**, which shows (1) the greenhouse verandah of the house, (2) the living room space with large windows oriented toward the north and south and enclosed façades facing east and west, and (3) the internal utility core, including the bathroom.

11.7 Solar Decathlon Wettbewerb 2007, 1. Preis, Darmstadt/Washington/Phoenix

Im Jahr 2007 hat das US Energieministerium einen internationalen Wettbewerb ausgeschrieben „Solar Decathlon", um ein visionäres, energieautarkes Wohnhaus zu entwickeln. Im Herbst des gleichen Jahres wurden 20 konkurrierende Objekte in Washington D.C. vorgestellt. Schwerpunkt des Wettbewerbs 2007 war die Nutzung von gebäudeintegrierten Photovoltaikelementen. Die Gebäudetechnik des „Year 2015 Prototype Home" musste komplett auf der Nutzung von Solarenergie basieren. Eine weitere Forderung des Wettbewerbs war, dass neben einer hohen Architekturqualität ein behaglicher Wohnraum entstehen sollte, der innovative Wohnqualitäten ausweist.

Das ca. 100 m² große Haus sollte als entscheidendes Kriterium für Nachhaltigkeit im Wohnungsbau eine Nutzungsneutralität und hohe Flexibilität des Grundrisses ausweisen. Um alle Forderungen zu erfüllen, wurden drei wesentliche Grundprinzipien angewandt:

- Prinzip der Schichtung,
- Ausbildung einer Plattform für Möbel und Technik in Form eines doppelten Bodens,
- gebäudetechnisches Gesamtsystem nach einem einfachen Plug-in-Prinzip.

Das Prinzip der Schichtung zeigt **Bild 167** mit einer vorgelagerten Veranda (1, Wintergartenprinzip), dem Wohn- und Lebensraum (2) mit großen Fensterelementen nach Süd und Nord sowie geschlossenen Wandelementen nach Ost und West und dem inneren Versorgungskern mit Nasszelle (3).

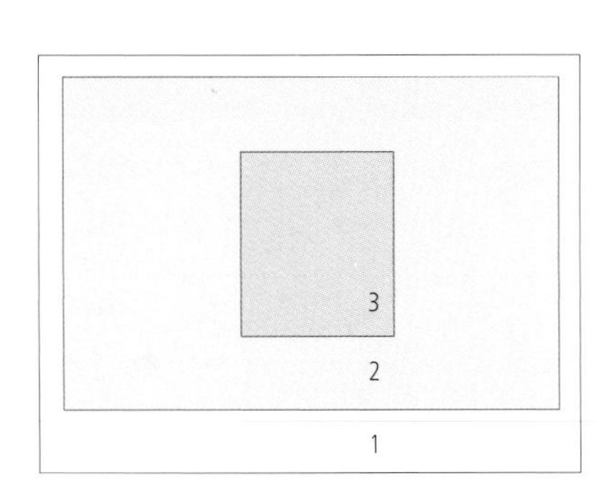

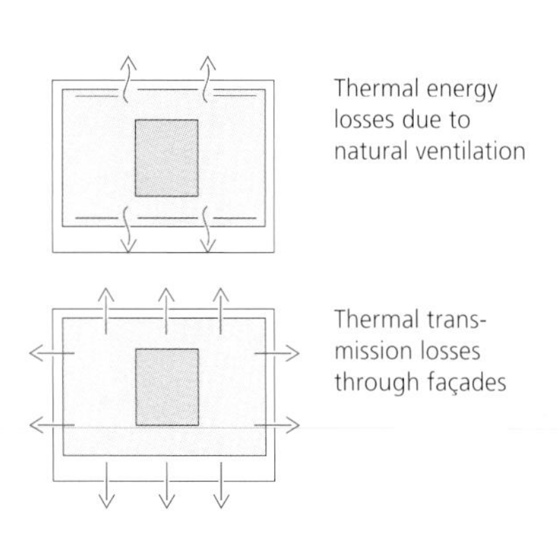

Figure 167
Solar Decathlon house 2007
Three layers define the space zoning

Source: Technical University Darmstadt, Germany

Bild 167
Solar-Deathlon-Haus, 2007
Drei Schichten beschreiben die Raumzonen

Quelle: TU Darmstadt

1 Greenhouse/Verandah
2 Living
3 Interior core with bathroom

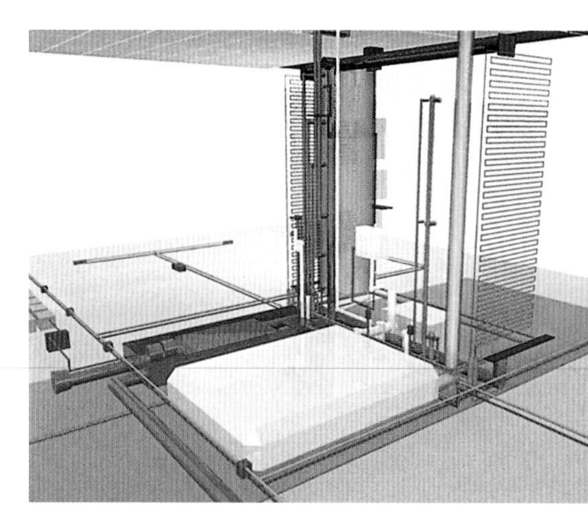

Figure 168
Detailed view of internal core with heat pump and thermal storage reservoir

Source: Solar Decathlon Competition 2007, Technical University Darmstadt, archplus 184, October 2007

Bild 168
Detailansicht der Installationen mit Wärmepumpe (WP) und Wärmespeicher (Tank)

Quelle: Wettbewerb Solar Decathlon TU Darmstadt, archplus 184, Oktober 2007

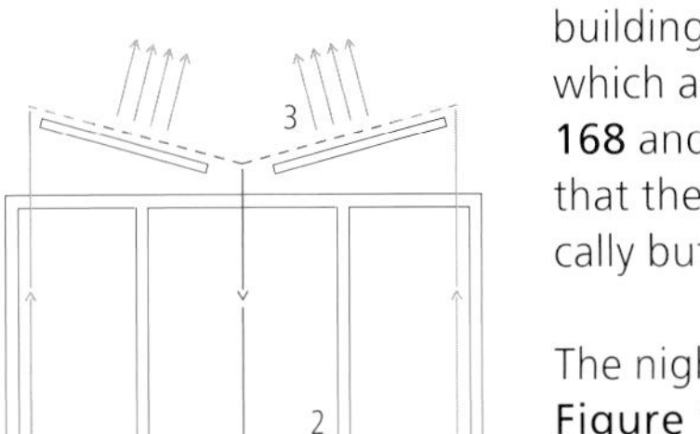

1 Reservoir
2 Distribution core
3 Back radiation

Figure 169
Diagram of thermal back-radiation of cooling water towards the night sky.

Source: Technical University Darmstadt, Germany

Bild 169
Prinzip der Aufbereitung von Kühlwasser (Tank) durch Wärmeabstrahlung gegen den nächtlichen Himmel

Quelle: TU Darmstadt

To provide an optimum of functional flexibility, the building is equipped with a raised floor system into which all technical systems are integrated (**Figures 168** and **169**). **Figures 170 – 173** allow the conclusion that the design is not only highly interesting energetically but also of superior esthetic and design quality.

The night view of the building unit, shown in **Figure 170**, communicates that the building provides an attractive, pleasurable living experience – the message seems to be: This is a wonderful place to live! **Figure 171** shows the verandah with an integrated photovoltaic roof in front of the south-facing façade and additional operable large-scale louvers. These are outfitted at the exterior with thin-film photovoltaic elements. **Figure 173** shows the living space with a sunken seating area; according to user requirements, daylight enters into this space and can be finely adjusted for either lighting or more direct solar gains.

Um ein hohes Maß an Nutzungsneutralität zu erreichen, wurde das Gebäude mit einem Doppelboden ausgerüstet, in dem alle wesentlichen Gebäudetechnischen Einrichtungen untergebracht waren. Die **Bilder 168** und **169** zeigen das wesentliche Prinzip. Dass das Gebäude nicht nur energetisch hochinteressant ist, sondern auch hohe Architekturqualität beweist, zeigen die **Bilder 170 – 173**.

Die Nachtaufnahme in **Bild 170** vermittelt einen wohligen Eindruck – hierin würde man gern leben. **Bild 171** zeigt die der Südfassade vorgelagerte Veranda mit einer Photovoltaikabdeckung und verfahrbaren Großlamellen, die an ihrer Außenseite mit Photovoltaik-Dünnschicht-Elementen bestückt ist. **Bild 173** vermittelt einen Eindruck im Wohnbereich mit einer Sitzgrube, in die geregelt Tageslicht und Direktstrahlung einfällt – je nach Wunsch des Nutzers.

Figure 170
Decathlon house night view

Bild 170
Decathlon Gebäude während der Nacht

Figure 172
Thin-film photovoltaic louvers integrated into façade

Bild 172
Lamellenstruktur mit PV-Dünnschicht-Elementen

Figure 171
Front loggia

Bild 171
vorgelagerter Wintergarten/Vorhof

Figure 173
Flexible living space with lounge

Bild 173
Flexibler Wohnraum mit Sitzgruppe

Figure 174 displays the floor plan with verandah space, kitchen, and core, and **Figure 175** presents a section through the building with descriptions of the various components of the assembly.

The passive energy concept of the prototype is an integral part of the design concept. The main elements are the excellent surface-to-volume ratio of the enclosure and the large, south-oriented windows that allow for solar gain. On the other hand, the active energy components of the photovoltaic shading devices protect the façade from overheating. The different layers of the south façade allow for excellent, fine-tuned response to changing conditions at different times of the day and under different seasonal conditions. Glazing oriented toward the north is of a quadruple-insulated type based on Passivhaus standards and provides a U-value of 0.3 W/m^{2}K. The south-oriented glazing is triple-insulated, also according to the recommendations of the Passivhaus standard, and has a U-value of 0.5 W/m^{2}K. The roof covering the verandah consists of a laminated glazing unit with integrated thin-film photovoltaic elements, which provide shading but also around 1.1 kW of electric power (Peak).

The roof of the building unit is composed of multiple layers, with the outer layer being laminated glass with integrated mono-crystalline photovoltaic elements. They are capable of providing 8.6 kW of electric energy (Peak). Above the bathroom unit, a solar collector area of roughly 2 m^2 provides the warm water used for the building's radiant heat floor and the warm water used in the bathroom and the kitchen. The system here is composed of evacuated solar tubes. Below the energy-generating roof, a vacuum insulating panel is installed that has an insulation capacity 10 times greater than that of a current typical modern roof.

One layer below the vacuum insulation of the multi-layered roof is a drywall ceiling attached to the wood joists that contains piping used for ceiling cooling during periods with higher outside temperatures.

The opaque wall sections facing north are highly insulated and show a U-value of 0.25 W/m^{2}K. An added outside louver allows the façade to be closed during the night for added privacy. The same is the case for the south façade, where the louvers not only protect against views from the outside but also generate 2 kW (Peak) of electrical energy with integrated thin-film photovoltaic elements.

Walls surfaces toward the east and west are equipped with an insulation layer of so-called phase-change material (PCM). By storing heat energy during the day and releasing it during the cooler evenings and at night, the PCM is effectively capable of reducing cooling loads, as shown in **Figure 127.1**.

Bild 174 zeigt den Grundriss mit seiner Veranda und den Kernbereich und Küche, **Bild 175** einen Schnitt mit Angaben zu den entscheidenden baulichen Strukturen.

Das passive Energiekonzept des Gebäudes ist integraler Bestandteil des Entwurfs. Dazu gehört neben einer hochgedämmten Hülle ein gutes A/V-Verhältnis zur Optimierung der Hüllflächen. Die großen Fensterflächen nach Süden ermöglichen passive solare Gewinne, wobei die energetisch aktivierten Verschattungssysteme die Veranda und die Wohnräume vor Überhitzung schützen. Durch die Schichtung der Gebäudehülle nach Süden kann auf unterschiedliche Tages- und Jahreszeiten hervorragend reagiert werden. Die nach Norden orientierten Fenster (Vierfachverglasung) weisen gemäß Passivhausstandard einen Wärmedurchgangskoeffizienten, u-Wert, von 0,3 W/m^{2}K auf. Die nach Süden orientierte innere Fensterfront besitzt eine Dreifachverglasung (Passivhausstandard) bei einem u-Wert von 0,5 W/m^{2}K. Die Dachfläche des Verandabereichs besitzt eine Verbund-Sicherheitsverglasung mit eingelegter PV-Dünnschichttechnik, die einerseits Schatten spendet und andererseits in der Lage ist, bis zu 1,1 kW Peak elektrische Energie zu produzieren.

Die Dachfläche des Gebäudes ist mehrschichtig aufgebaut, wobei die äußerste Haut wiederum eine Verbund-Sicherheitsglasabdeckung mit integrierten monokristallinen Photovoltaikzellen bildet, die in der Lage ist, bis zu 8,6 kW Peak elektrische Energie zu erzeugen. Oberhalb der Nasszelle ist ein ca. 2 m^2 großes Kollektorfeld (Vakuumröhren-System) eingelegt, um Warmwasser für eine Fußbodenheizung und Warmwasserbereitung zu erzeugen. Unterhalb des energieerzeugenden Daches befindet sich eine Vakuumisolierung, die eine in etwa 10 Mal höhere Isolierfähigkeit ausweist als übliche Dächer.

Unterhalb der Vakuumisolierung ist im Bereich der Holzträgerkonstruktion eine Gipsdecke installiert, die Kühlwasserleitungen in sich birgt, um als Kühldecke an warmen und heißen Tagen zu wirken.

Opake Wandflächen im Bereich der nördlichen Fassade sind hochgedämmt mit einem u-Wert von 0,25 W/m^{2}K. Zudem kann auch die Nordfassade durch Lamellenstrukturen geschlossen werden, um sich während der Nacht gegen Einblicke von außen zu schützen. Gleiches gilt selbstverständlich auch für die Lamellenstrukturen im Süden, die nicht nur Schatten spenden, sondern zudem auch durch die Photovoltaik-Dünnschichttechnik ca. 2 kW Peak elektrische Leistung erzeugen.

Die Wandflächen nach Ost und West sind durch eine Isolierschicht mit phase change material (PCM) bestückt, um zum Teil während des Tages Wärmeenergie zu speichern und in den kühleren Abendstunden wieder abzugeben. Hierdurch entsteht zudem eine Kühllastreduzierung, wie in **Bild 127.1** gezeigt.

The energetically active building systems concept entails – in addition to the various already-described photovoltaic systems and the solar thermal collector surface – an electric heat pump operating under air-air and partially air-water conditions. The orange element in **Figure 168** is the ventilation equipment fed by a cold-water storage tank, shown in the image as the lighter-colored container; both are integrated into the raised floor platform. The heat pump can switch between heating and cooling, with condenser heat that cannot be used for heating purposes being ejected to the outside by a fan. Power outlet boxes are integrated into the raised floor to allow for electrical supply at various areas of the space.

The materials used for the construction of the building were selected for their ability to be obtained from certified renewable sources, and they are recyclable.

Because the implementation of such advanced technology in buildings typically results in increased initial construction cost, it is important to study the potential for synergetic effects to lower first cost at the beginning of each design process.

Das aktive, energetische Konzept (Gebäudetechnik) umfasst neben den bereits beschriebenen Photovoltaikelementen und der Kollektorfläche eine elektrisch betriebene Wärmepumpe, die zum Teil im Luft-Luft- und zum Teil im Luft-Wasser-Betrieb arbeitet. Das Lüftungsgerät (in **Bild 168** orangefarben) liegt im Bodenbereich und bedient sich aus einem Kaltwasserspeicher, der ebenfalls im Boden untergebracht ist (heller Behälter). Die Wärmepumpe kann von Heizen auf Kühlen entsprechend umgestellt werden, wobei die Kondensatorwärme durch die Abluftströme der Lüftungsanlage ins Freie abgegeben wird, so die Wärmeenergie nicht genutzt werden kann. Die elektrische Energieversorgung innerhalb der Nutzflächen erfolgt über Bodentanks, so dass an verschiedensten Stellen elektrische Energie abgenommen werden kann.

Bei der Wahl der Materialien standen nachwachsende, naturnahe und vor allem recycelfähige Stoffe im Vordergrund.

Da die Verwendung innovativer Materialien und Technologien üblicherweise ein deutliches Ansteigen der Baukosten nach sich zieht, ist es wichtig, bereits in der Planung alle Elemente in richtiger Weise zu integrieren und durch Synergieeffekte die Baukosten zu senken.

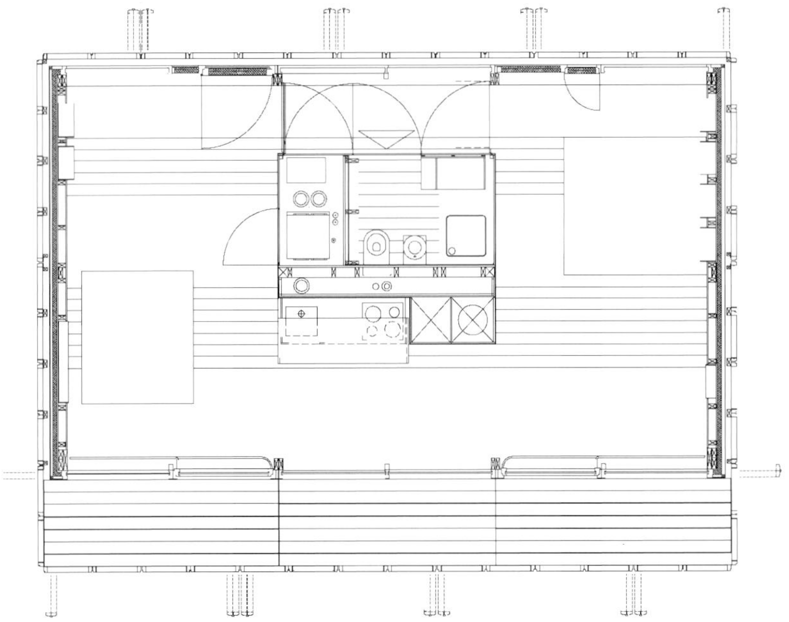

Figure 174
Solar Decathlon House
Floor plan

Source: Solar Decathlon Competition 2007, Technical University Darmstadt, archplus 184, October 2007

Bild 174
Grundriss des Solar Decathlon-Hauses

Quelle: Solar Decathlon Wettbewerb 2007, Deutschland, TU Darmstadt
Quelle: archplus 184, Oktober 2007

Figure 175
Solar Decathlon House
Cross section, details

Bild 175
Schnitt durch das Gebäude mit Detail-Angaben

1 Evaporative cooling system
2 Vacuum insulation (VIP) on roof
3 Cooling ceiling
4 PCM in east and west walls
5 Vacuum insulation (VIP) in east and west walls
6 Technical plattform in floor
7 Vacuum insulation (VIP) on floor
8 Heating floor radiant

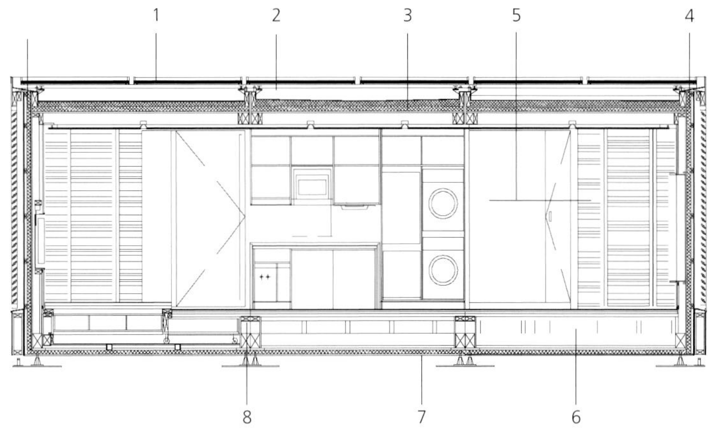

Ambient temperature for one year
(Meteonorm data set)
in °C

40
30
20
10
0
-10
-20

Time

Figure 176.1
Annual ambient temperature and relative humidity distribution for the location Darmstadt, Germany.

Bild 176.1
Jährliche Temperatur- und Feuchtigkeitsverteilung (rel. %) für den Standort Darmstadt/Deutschland

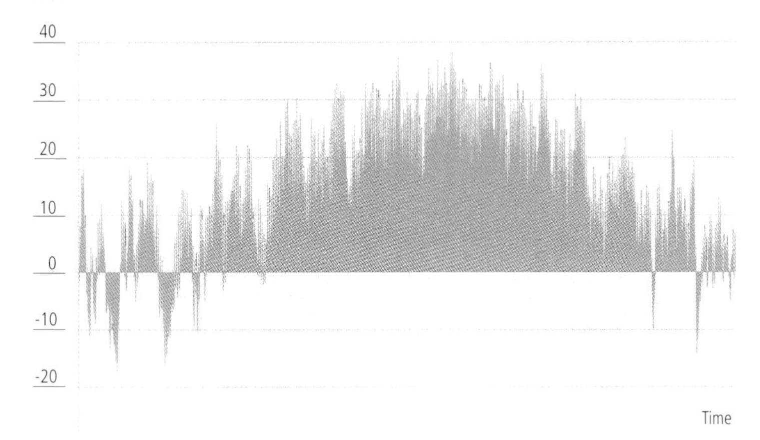

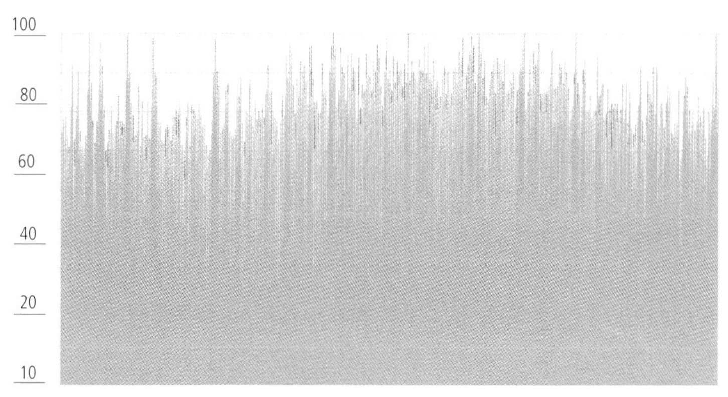

Figure 176.2
Annual ambient temperature and relative humidity distribution for the location Washington D.C.

Bild 176.2
Jährliche Temperatur- und Feuchtigkeitsverteilung (rel. %) für den Standort Washington DC.

Ambient temperature for one year
(Meteonorm data set)
in °C

50
40
30
20
10
0

Tim

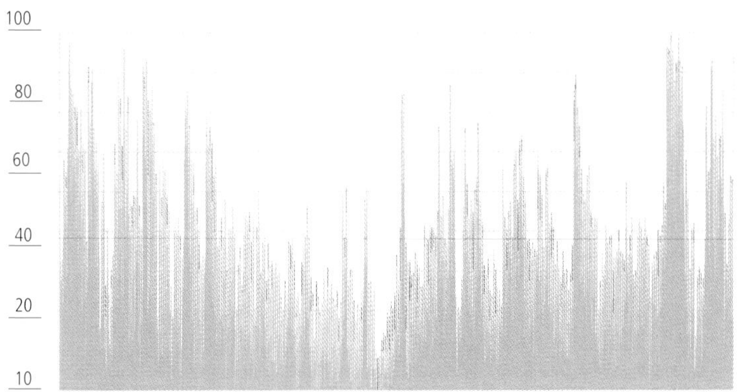

Source: Meteonorm data set, TU Darmstadt

Figure 176.3
Annual ambient temperature and relative humidity distribution for the location Phoenix, Arizona.

Bild 176.3
Jährliche Temperatur- und Feuchtigkeitsverteilung (rel. %) für den Standort Phoenix, Arizona.

Quelle: Meteonorm data set, TU Darmstadt

It is also of great interest that the building and technology concept be equally valid in different climate zones on Earth. In extensive modeling prior to construction, the functioning of the building in both hot and cold climates was analyzed, and, as **Figure 176** indicates, it performs under varying climatic conditions very well, indeed.

It was also important to analyze the annual temperature swings and changing relative humidity. The selected case study locations for the building were Germany, Washington, D.C., and Phoenix, AZ.

In the various simulation scenarios shown in **Figures 177**, the heat energy gains and losses of the structure at the different sites were studied to allow for the necessary and appropriate sizing of the building systems.

Interessant bei der Entwicklung des Gebäudes ist, dass es für unterschiedlichste Klimazonen einsetzbar wird. Ob heiß oder kalt – es kann überall seinen angestammten Platz finden. Die **Bilder 176** zeigen die im Zuge der Planung durchgeführten Simulationen.

Ein wesentlicher Aspekt für die Planung war die Darstellung und Analyse der jahreszeitlichen Außentemperaturen und relativen Feuchten. Diese wurden für die Standorte Darmstadt (Deutschland) sowie Washington und Phoenix Arizona, durchgeführt.

Unter anderem und neben vielen anderen Simulationen berechnet wurden gemäß den **Bildern 177** Wärmegewinne und Wärmeverluste für die einzelnen Standorte, um hieraus resultierend die notwendigen Dimensionierungen technischer Einrichtungen vornehmen zu können.

Gains and losses in
in kWh/m² per month
12.0
8.0
4.0
0.0
- 4.0
Jan
Feb
March
April
May
June
July
Aug
Sept
Oct
Nov
Dec
Month

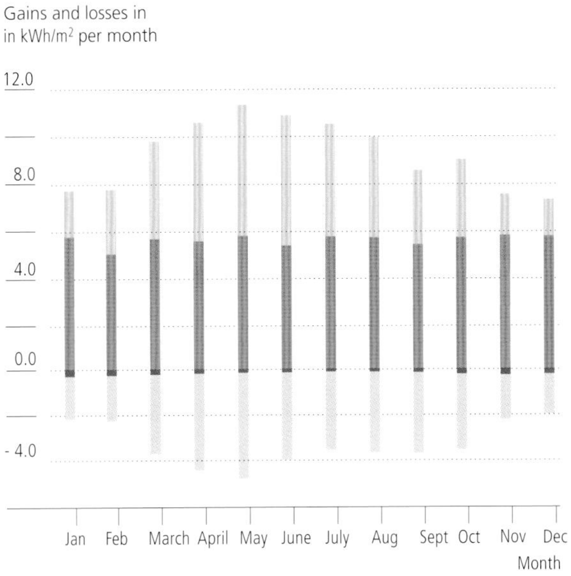

Figure 177.1
Analysis of heat gains and losses throughout the year for location Darmstadt, Germany

Bild 177.1
Analyse der Wärmegewinne und -verluste während des Jahres, für Darmstadt

Gains and losses in
in kWh/m² per month
12.0
8.0
4.0
0.0
- 4.0
Jan
Feb
March
April
May
June
July
Aug
Sept
Oct
Nov
Dec
Month

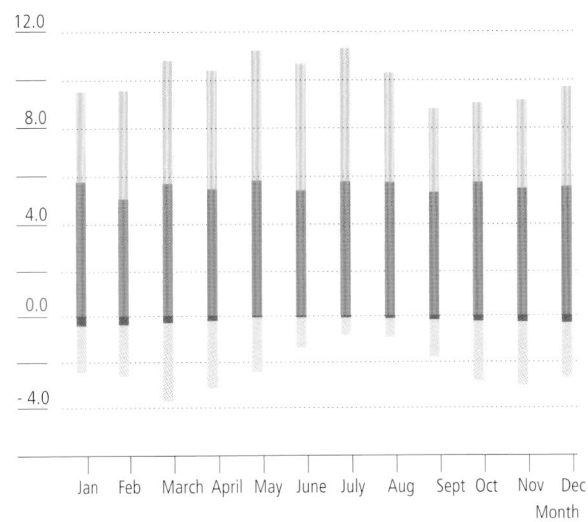

Figure 177.2
Analysis of heat gains and losses throughout the year for location Washington D.C.

Bild 177.2
Analyse der Wärmegewinne und -verluste während des Jahres, für Washington

Gains and losses in
in kWh/m² per month
12.0
8.0
4.0
0.0
- 4.0
Jan
Feb
March
April
May
June
July
Aug
Sept
Oct
Nov
Dec
Month

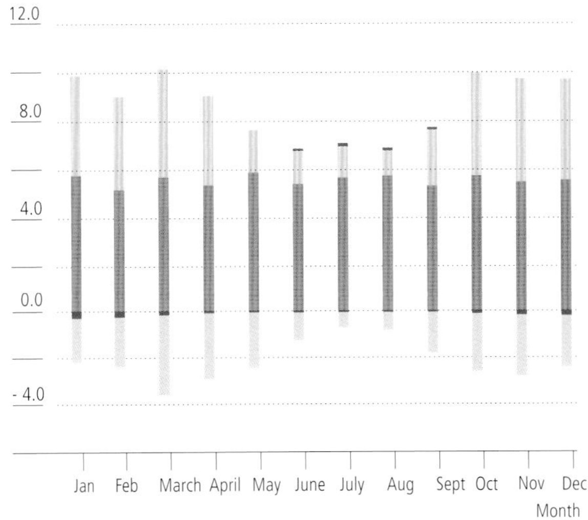

Figure 177.3
Analysis of heat gains and losses throughout the year for location Phoenix, Arizona

Bild 177.3
Analyse der Wärmegewinne und -verluste während des Jahres, für Phoenix

Source: Meteonorm data set, TU Darmstadt

Quelle: Meteonorm data set, TU Darmstadt

- Solar gains
- Internal heat loads
- Ventilation heat losses due to infiltration
- Ventilation heat losses through windows

Heat demand
Cooling demand

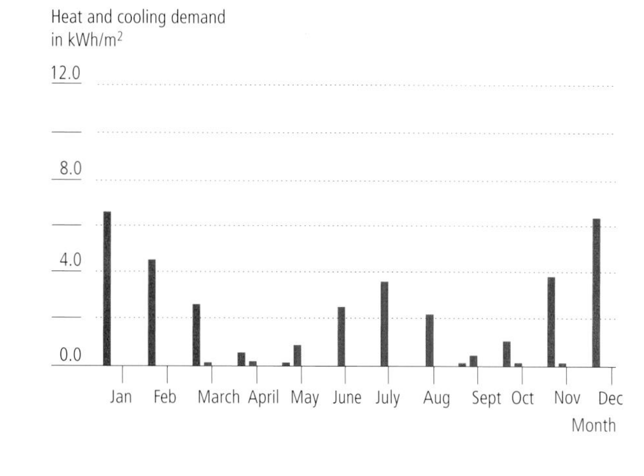

Figure 178.1
Analysis of heat gains and losses throughout the year for location Darmstadt, Germany

Source: Meteonorm data set, TU Darmstadt

Bild 178.1
Überblick über Wärme- und Kühlleistungsbedarf pro Jahr für Darmstadt

Quelle: Meteonorm data set, TU Darmstadt

Figure 178 displays the necessary heat energy for the locations in Germany and Washington D.C., and the energy required for cooling at a hot-arid location such as Phoenix.

Figure 179 shows a comparison of the annual heat energy demand and cooling energy for all three test locations. If the potentials of the Passivhaus standard (PHPP) had been exhausted in their totality, the listed amount of energy consumption could have been reduced even further.

Die **Bilder 178** zeigen die notwendigen bereitzustellenden Heizleistungen (Standort Darmstadt bzw. Washington) und andererseits vor allem die bereitzustellenden Kälteleistungen (Standort Phoenix).

Bild 179 zeigt einen Vergleich des jährlichen Wärmeenergiebedarfs bzw. der jährlichen Kälteleistung für die Standorte Washington, Phoenix und Darmstadt. Bei völliger Ausschöpfung des Passivhausstandards (PHPP) könnten die Energieverbräuche noch etwas geringer ausfallen.

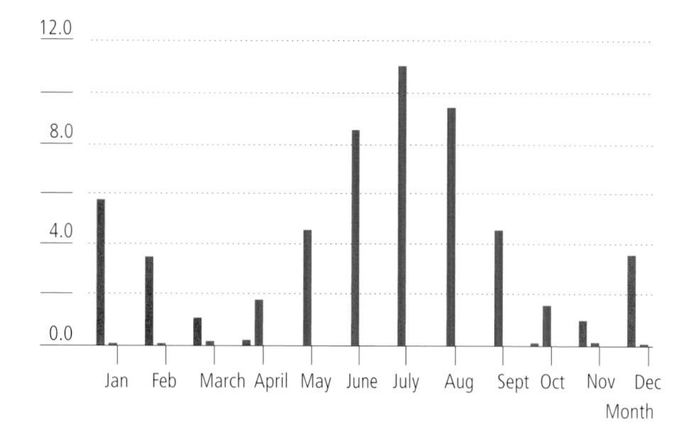

Figure 178.2
Overview of annual heat and cooling demand for Washington D.C.

Bild 178.2
Überblick über Wärme- und Kühlleistungsbedarf pro Jahr für Washington

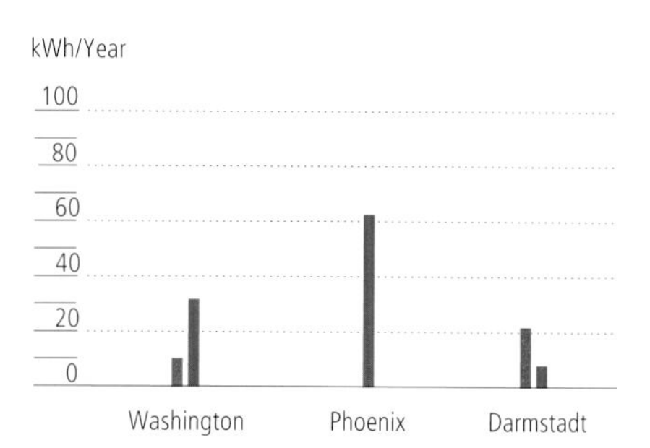

Figure 179
Overview of different energy demand for heat and cooling for the three locations

Bild 179
Übersicht über den unterschiedlichen Endenergiebedarf für Wärme und Kälte

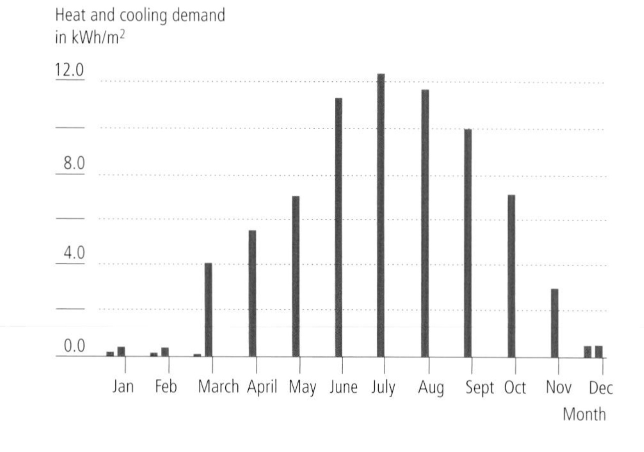

Figure 178.3
Overview of annual heat and cooling demand for Phoenix, Arizona

Bild 178.3
Überblick über Wärme- und Kühlleistungsbedarf pro Jahr für Phoenix

Such an energy-optimized building and its installed systems need to be controlled by a small building-management system that ensures best operating performance on a continuing basis. **Figure 180** displays a diagram of the controls. In this regard, it is not only important to control for optimal solar use but also to provide indoor temperature conditions according to user demand at all times. During night operating hours, the heat energy gained by the cooled ceiling piping can be radiated back into space, lowering the 19 °C warm return water from the cooled ceiling piping to approximately 16 °C. The result is a lower cooling load and smaller energy consumption by the heat pump.

By using energy efficient-lamps and light-emitting diodes (LED) for interior lighting, the energy used was lowered by approximately 30 % as compared with standard lighting.

Sinnvollerweise wird ein derartig energieoptimiertes Gebäude durch eine kleine Gebäudeleittechnik so gesteuert und geregelt, dass alle Elemente und technischen Einrichtungen ihre volle Leistung erbringen. **Bild 180** zeigt das Prinzipschaltbild der Steuerung. Hierbei ist es nicht nur wichtig, auf Solarstrahlungen (Nutzung) jederzeit richtig zu reagieren, sondern die Raumtemperaturen nach den Wünschen der Nutzer zu regeln. Darüber hinaus wird im Nachtbetrieb das Potenzial des schwarzen Weltalls auch dazu genutzt, Warmwasserströme aus den Kühldecken über den Speicher infolge Wärmestrahlung abzukühlen. Ca. 19-grädiges Warmwasser aus Kühldecken konnte so auf 16 °C abgekühlt werden, d.h. Kälteleistung gewonnen werden, die ausschließlich geringe Pumpenenergie erfordert.

Durch den Einsatz von energiesparenden Lampen und LED-Lichtsystemen konnte die Anschlussleistung gegenüber herkömmlichen Beleuchtungsanlagen um ca. 30 Prozent gesenkt werden.

1 Building type (all main informations about the construction of the house, the internal gains and losses).
2 This component computes the solar radiation on a vertical receiver shaded by an overhang and/or wingwall.
3 This shading component controls the strategy of the lamellae.
4 Input component for the weather data sets used in the simulation.
5 This component interpolates radiation data, calculates serval quanties related to the position of the sun, and estimates insolation on a number of surfaces of either fixed or variable orientation.
6 This component takes as input the dry bulb temperature and relative humidity of moist air and calls the TRNSYS Psychometrics routine, returning the following correspondenting moist air properties: dry bulb temperature, relative humidity, absolute humidity ratio and enthalpy.
7 This equation component converts the data from kj in kW.
8 This component simulates the heat exchanger which is modelled as a constant effectiveness device and is independent of the systems configuration.
9 This component controls the bypass function. By the bypass function the ambient temperature can directly be transported into the room, without going through the heat exchanger.
10 These types control the function about heating and cooling for the bypass component.
11 These components exports the simulation results to computer monitor and seval text files.

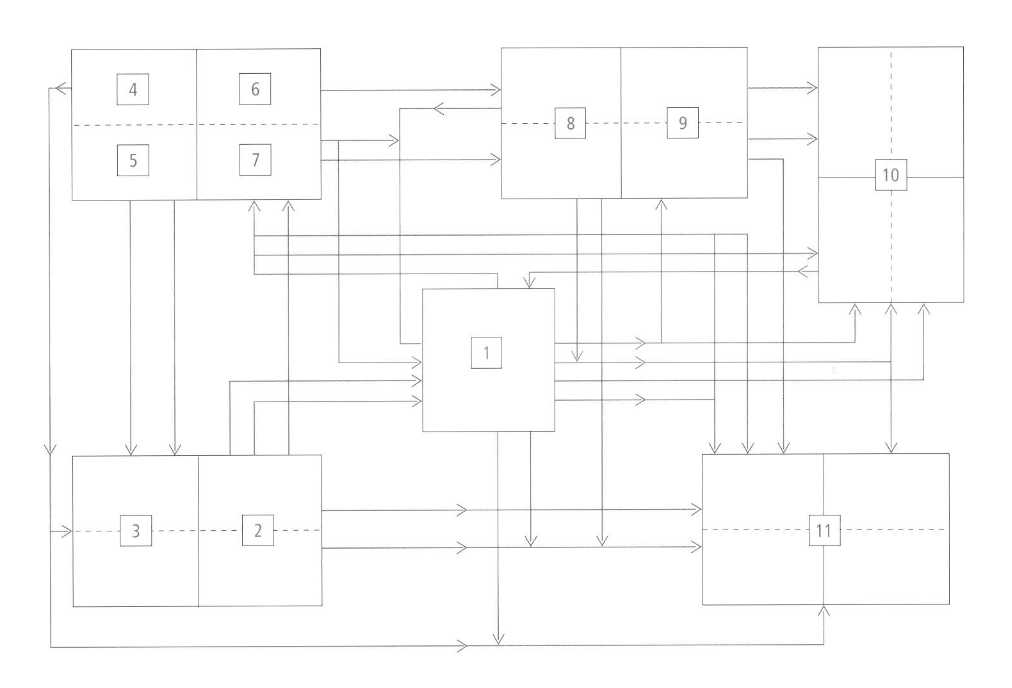

Figure 180
Systems diagram of energy control

Source: Solar Decathlon Competition 2007, TU Darmstadt

Bild 180
Gebäude Energie-Steuerung

Quelle: Wettbewerb Solar Decathlon 2007, TU Darmstadt

Figure 181 illustrates a potential air-to-air heat pump configuration with the necessary components to thermally condition the building. Note that these images are for demonstration purposes only and do not show the system used in this particular building.

The excellent contribution by the students under the direction of Manfred Hegger of the Technical University Darmstadt, Germany, deserves high praise. The U.S. Department of Energy selected their contribution out of twenty university-led teams from the U.S. and as far away as Puerto Rico, Spain, and Canada as the most attractive and energy-efficient solar-powered home. It serves as an excellent example of the methods that can be used to address energy issues facing us in the near future.

Bild 181 zeigt einen denkbaren (nicht original im Gebäude installierten) Aufbau einer Wärmepumpenanlage (Luft-Luft-System) mit allen Komponenten, die notwendig werden, um das Haus entsprechend betreiben zu können. Das von Manfred Hegger und seinen Studenten entworfene Haus, zu dem man nur gratulieren kann, zeigt alle Merkmale der notwendigen Architektur, um die auf uns zukommenden Problemen zu lösen.

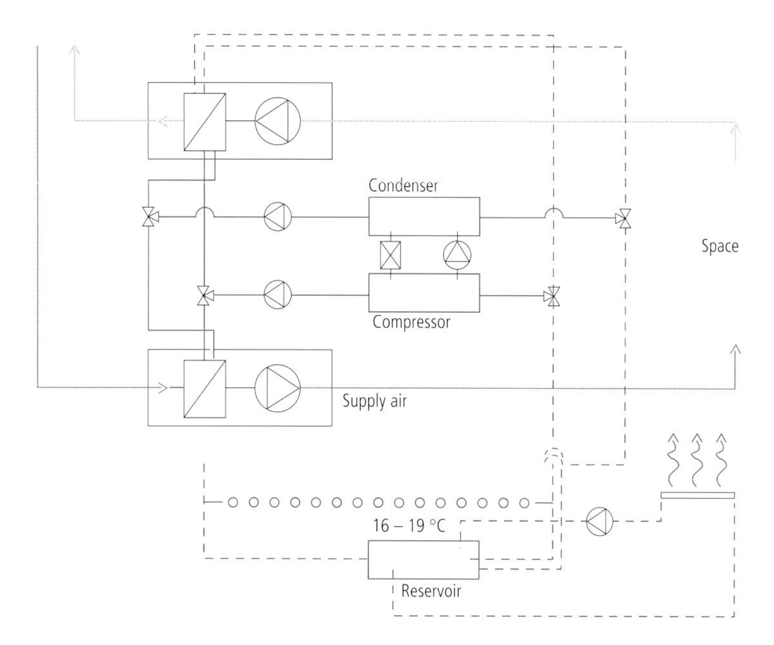

Figure 181
Diagram of heat pump, ventilation concept with chilled ceiling elements and "free cooling"

Source: Daniels/author

Bild 181
Denkbares Wärmepumpen-(WP-) Lüftungssystem mit Kühldeckenelementen und „freier Kühlung"

Quelle: Daniels/Autor

11.8 Micro-compact home

The architect Richard Horden is well known for developing design ideas for buildings that are minimized in their need for materials and the energy required for their operation. Together with his students at the Technische Universität München, Germany, he designed, built and tested the so-called "micro-compact home" by actually living in it. The cube unit has a dimension of 2.65 x 2.65 x 2.65 m and can be used under different dwelling configurations, especially when a quick need for temporary shelter is required. As a stimulating design primer, Horden and the students looked at the interior design of aircraft. Systems found there, such as indirect lighting, direct mini-ventilation systems, integrated flat screen TVs, integrated Internet connection, and the design of a kitchen module, were used in the design.

11.8 Micro-compact home, Mini-Wohnobjekt

Richard Horden ist bekannt für seine Arbeiten, Häuser zu entwickeln, die mit einem absoluten Minimalaufwand an Material und Energie auskommen. Im Zuge seiner vielen Studien zu verschiedensten Anwendungszwecken hat er das Micro-compact home zusammen mit Studenten der Technischen Universität München geplant, gebaut und probebewohnt. Das Objekt besitzt eine Kantenlänge von 2,65 x 2,65 x 2,65 m. Diese Wohnkuben sind für die verschiedensten und unterschiedlichsten Anwendungszwecke einsetzbar, insbesondere dann und dort, wo es darum geht, temporär und kurzfristig Wohnobjekte zu schaffen. Bei der architektonischen Entwicklung der Wohnkuben orientiert sich Richard Horden an der Innenarchitektur und dem Innenausbau von Flugzeugen. Indirekte Beleuchtung, direkte (Mini-) Belüftung, integrierte Flachbildschirme, Internet-anschlüsse werden wie auch eine Küchenzeile in den Ausbau einbezogen.

As shown in **Figures 182** and **183**, there is also a low-energy version of the micro-home that is equipped with 8 m^2 of photovoltaic elements, and, as an alternative, with wind power generation. The vertical wind turbine allows for energy generation also at night and stores it in a battery to be used during the day. **Figure 184** presents the energy balance for such a low-energy unit. By using multi-layered, high-tech building materials similar to the ones presented in the previous project, the micro-compact house can be operated completely autonomously. It therefore can be deployed to disaster regions to provide shelter for one or two people for a short – or even longer – time span, because provisional arrangements sometimes prove to have longevity. The **Figures 185** and **186** provide additional information concerning the design.

Ein Gebäude für eine Person und für eine temporäre Zeit kann bei allem Komfort extrem klein sein und selbstverständlich extrem wenig benötigen, **Bilder 182 – 183**. Eine low-e (Niedrigenergie)-Homeversion soll durch Photovoltaikelemente (8 m^2) und ein kleines Windrad (Alternativenergiebereitstellung) mit elektrischer Energie versorgt werden, **Bild 184**. Das Vertikalwindrad, das einen Generator treibt, kann auch während der Nacht elektrische Energie produzieren und in eine Batterie einladen, um diese Energie während des nächsten Tages zu nutzen. **Bild 184** zeigt eine Energiebilanz für ein entsprechendes Micro-compact low-e-home. Durch den Einsatz modernster Baumaterialien in Schichtbauweise, in ähnlicher Struktur wie das zuvor gezeigte Objekt, kann das Mini-Wohnhaus absolut energieautark betrieben werden und ist somit auch gerade in Katastrophenbereichen einsetzbar, geht es darum, für ein oder zwei Personen jeweils einen Wohncontainer bereitzustellen, der eine Unterbringung über einen längeren Zeitraum (Provisorien sind oft sehr langlebig) ermöglicht. **Bilder 185** und **186** ergänzen die Informationen zum Micro-compact home.

Figure 182
Micro-compact home
Munich

Architect: R. Horden,
Technical University Munich

Bild 182
Micro-compact home,
München

Architekt: R. Horden,
TU München

to Figure 182
Micro-compact home
Munich
Outside, inside view

Architect: R. Horden,
Technical University Munich

zu Bild 182
Micro-compact home,
München
Aussen- und Inneransicht

Architekt: R. Horden,
TU München

Figure 183
Micro-compact home
Munich
Sitting and sleeping area

Bild 183
Micro-compact home,
Sitz- und Schlafbereich

to Figure 183
Micro-compact home
Munich
Sitting and sleeping area

zu Bild 183
Micro-compact home,
Sitz- und Schlafbereich

Energy balance	Winter (Jan)	Summer (July)
For +19°C internal temperature	kWh/month	kWh/month
Direct natural solar energy gains:	+33.56	+88.26
Internal energy gains:	+71.83	+59.11
Transmission energy losses:	−235.67	−31.56
Infiltration energy losses:	−132.73	−70.04
Total passive energy losses:	−263.01	+43.77
Energy: heating + cooling	−266.34	−41.31
Electrical system:	+81.85	+81.85
Total consumption:	−348.19	−123.16
Photovoltaic panels 8 qm:	+303.50	+1,280.00
Wind energy:	+45.00	+25.00
Total loss/gain	**−0.31 kWh/m**	**+1,221.84**

Figure 184
Micro-compact home
Model and energy budget

Bild 184
Modell und Energiebilanz des
Micro-compact low-e home

Figure 186
Micro-compact home
Cross section

Bild 186
Micro-compact home,
Schnitt-Darstellung

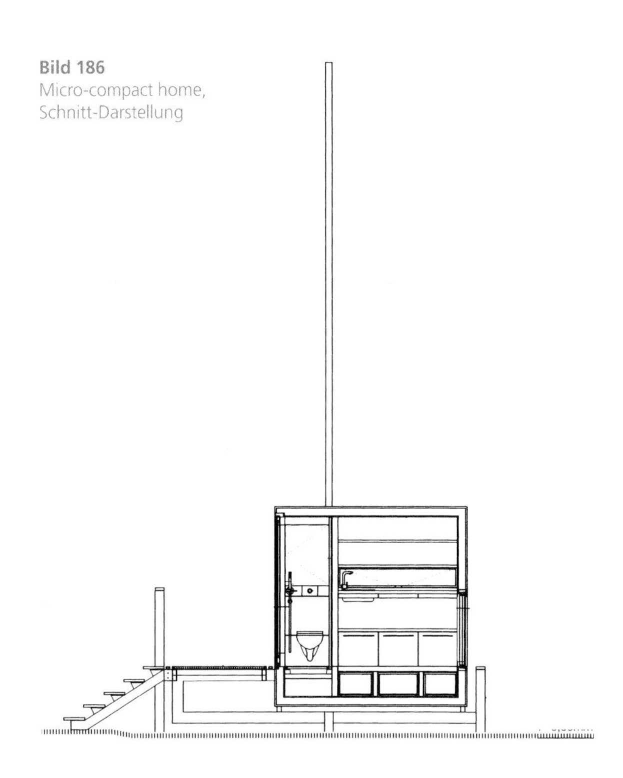

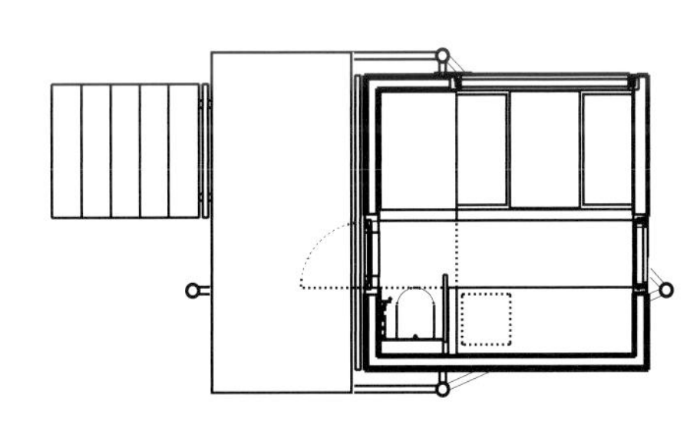

Figure 185
Micro-compact home
Floor plan

Bild 185
Micro-compact home
Grundriss

11.9 Main building Eawag Empa, Dübendorf, Switzerland

Eawag, the Swiss Federal Institute of Technology's Aquatic Science and Technology Department at the Forum Chriesbach is a design by the architects Bob Gysin and Partners, Zurich, the consulting engineers 3-Plan Haustechnik AG and Büchler and Partner AG, Zurich, the acoustical consultants Büro für Bauphysik, and Kopitsis Bauphysik AG in Wohlen, Switzerland.

Figure 187
Forum Chriesbach, Eawag Europe, Zurich, Switzerland
Front elevation

Architect: BGP,
Bob Gysin+Partner AG,
Architekten ETH SIA BSA,

Source: archplus 184, October 2007

Bild 187
Forum Chriesbach,
Eawag Europa, Zürich
Frontansicht

Architekt: BGP,
Bob Gysin+Partner AG,
Architekten ETH SIA BSA,

Quelle: archplus 184, Oktober 2007

11.9 Hauptgebäude Eawag Empa, Dübendorf, Schweiz

Das Hauptgebäude der Eawag Empa (Forum Chriesbach) wurde vom Architekturbüro Bob Gysin und Partner, Zürich, und den Ingenieurbüros 3-Plan Haustechnik AG und Büchler & Partner AG, Zürich, zusammen mit dem Büro für Bauphysik, Akustik und Simulation Kopitsis, Bauphysik AG in Wohlen entworfen.

Das als annähernd „Nullenergiehaus" realisierte Hauptgebäude der Eawag ist einer der vier Foschungsanstalten der ETH, die an langzeitorientierten Lösungen für Problemstellungen wie den Umgang mit Wasser, Energie, Boden, Mobilität und Materialien arbeiten, **Bilder 187** bis **190**.

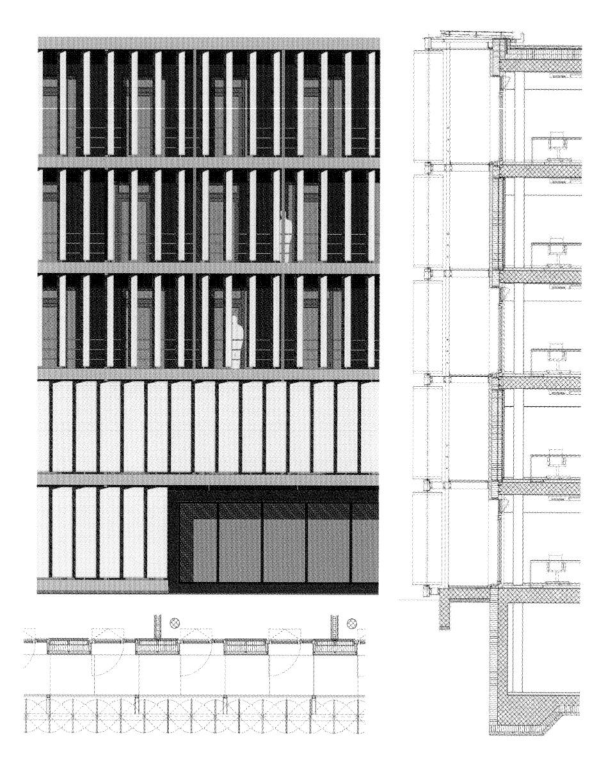

Figure 188
Elevation and façade section

Bild 188
Fassadenansicht und -schnitt

to Figure 188
Façade detail: rotating, screen-printed outer glass louvers

zu Bild 188
Datail-Ansicht, bedruckte, drehbare äußere Lamellen

This building almost achieves zero-energy status. Here, the results of the institute's own in-house research, such as concerning water resources, energy, soil, and materials,is being tested (**Figures 187 – 190**).

Since about a third of all energy is consumed in work and residential environments, it is important to study the development of efficient buildings. Despite the highly advanced standards such as Minergie and Minergie-P in Switzerland, the Forum Chriesbach demonstrates with its own building that it is indeed possible to be more efficient than required by the Swiss energy regulation, and this by a factor 4 (**Figure 191**).

In ihrem Immobilienprojekt werden auch Erkenntnisse aus der eigenen Forschung getestet. In der Schweiz wird rund ein Drittel des Gesamtenergiebedarfs durch „Wohnen und Arbeiten" verbraucht, was die Bedeutung aufzeigt, energieeffiziente Gebäude zu realisieren. Trotz etablierter Standards wie Minergie und Minergie-P wird dieses große Energieeinsparpotenzial nicht ausgeschöpft. Das Forum Chriesbach geht einen Schritt weiter, indem sein Gesamtenergieverbrauch um den Faktor 4 besser ist als ein heute nach dem Schweizer Energiegesetz realisiertes Gebäude, **Bild 191**.

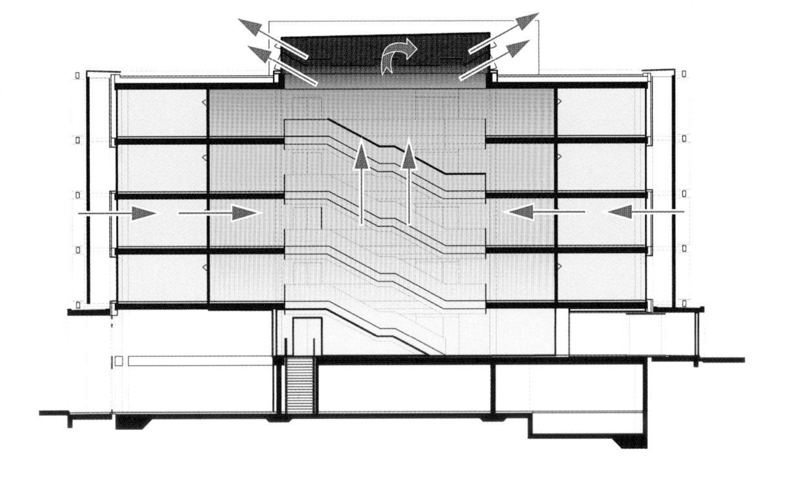

Figure 190
Cross section and natural ventilation concept of the building

Depending on season the outer glass louvers are either providing shade or allow for solar gains. Thermal gains are stored in opaque building components such as the concrete slabs, composite magnesite screed-wood floors, and clay walls. Additionally, the clay material regulates interior humidity. In summer the stored energy is released by night cooling of the building, enhanced by natural stack effect ventilation of the atrium.

Bild 190
Schnitt durch das Gebäude mit natürlichem Lüftungskonzept

Je nach Jahreszeit werden die Glaslamellen vor der hochgedämmten Gebäudehülle auf Sonnenschutz oder Sonnendurchlass gestellt. Die Wärme wird in den Massivbauteilen (Betondecke, Hartsteinholzbelag, Lehmwände) gespeichert. Der Lehm reguliert zusätzlich den Feuchtehaushalt. Im Sommer wird die Wärme mittels Nachtauskühlung abgeführt, begünstigt durch die Kaminwirkung des Atriums.

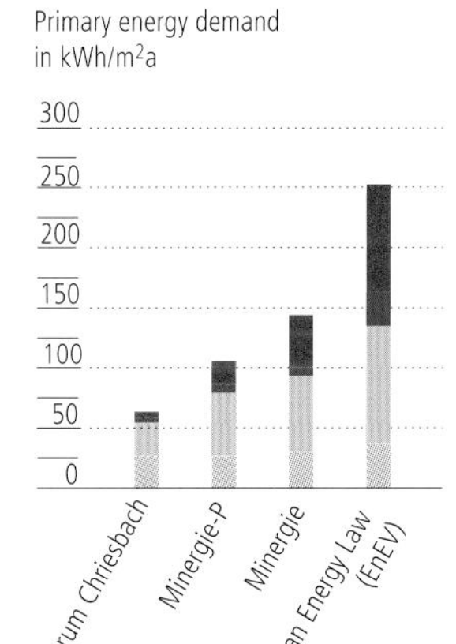

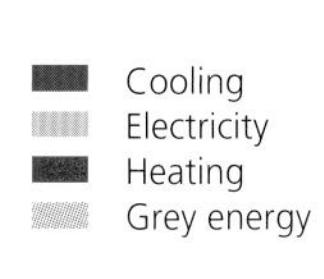

Figure 191
Comparison between primary energy demand

Primary energy "grey energy": 28.6 kWh/m²a as compared to primary energy for heating: 2.7 kWh/m²a.Heating and cooling energy demand was almost reduced to zero, however the building requires primary energy for electricity 32.4 kWh/m²a. The primary energy consumption including grey energy is 64.9 kWh/m²a, which is equal to a reduction by a factor 4 compared to the Swiss Energy Standard.

Bild 191
Vergleichende Betrachtung des Primärenergiebedarfs

Primärenergie Graue Energie: 28,6 kWh/m²a. Demgegenüber beträgt die Primärenergie Wärme 2,7 kWh/m²a. Während Wärme- und Kälteenergie fast auf Null reduziert werden konnten, benötigt das Gebäude Primärenergie Elektrizität 32,4 kWh/m²a. Der Primärenergieverbrauch inkl. Grauer Energie beträgt 64,9 kWh/m²a – eine Verringerung um den Faktor 4 gegenüber dem üblichen Schweizer Energiestandard.

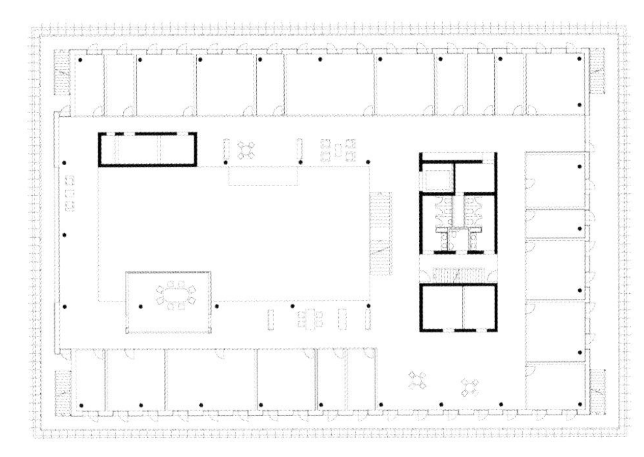

Figure 189
Floor plan

Bild 189
Grundriss

The building is a signature part of the Eawag Empa campus. The fritted glass lamellae of its façade are capable of tracking the position of the sun: they serve as solar shading devices during summer and allow for passive energy gains during winter. As a weather screen, they are capable of providing protection all year long. The thermal layer behind the outer layer consists of glass elements alternating with highly insulated wooden panels, clad with blue-colored fiber-cement boards. The functional areas of the building are arranged around a U-shaped, 5 storey atrium lobby that is characterized by floating meeting rooms and an internal staircase. This arrangement is meant to optimize the building's functional processes and provide a high degree of internal transparency. The building provides flexible work arrangements for around 150 researchers and staff. Communication areas, exhibition space, seminar rooms, a lecture hall, multi-media rooms, a library, and a restaurant complete the building's program.

The primary structure of the building is a reinforced-concrete steel skeleton with circulation cores and load-bearing façade columns. The building's envelope is prefabricated in order to allow for a minimized construction phase and a high degree of user flexibility. All technical building systems remain exposed to allow for future simplified access, easier operation, lower maintenance cost, and facilitation of later recycling with clear separation of materials. In addition to the target values for embodied grey energy and overall sustainability, other criteria such as aesthetics, usage conformity, fire safety, and economy were essential. Also, the coloration scheme and the selection of a material palette were of great importance. With the exception of the space reserved for more public and representation functions, the colors selected are subtle and represent in most cases the natural material coloration.

The building orientation, its volume, the thermal storage capacity, the envelope design, and the solar shading system are conceived with the goal of operating the building entirely without an additional conventional fossil-fuel heating system. The energy required for heating is provided by internal loads such as people, equipment, and the lighting. Also, excess heat, geothermal energy, and solar thermal energy gains are used to compensate for peak loads (to recharge excess heat, the building is part of an existing district heating utility grid). The necessary cooling in summer for such spaces as the conference, seminar, and meeting rooms is provided by chilled ceilings that are connected to the district chilled-water loop of the grid. The building façade, with its shiplap structure, the nighttime flushing of the interior spaces, and the ventilation system render conventional heating and cooling to a large degree superfluous (**Figure 190**). The roof-mounted photovoltaic system generates one third of the required electric energy, and other parts of the roof consist of extensively vegetated areas that serve for water retention and time-delayed release of rain water. Part of the research conducted within the

Der Baukörper setzt einen neuen städtebaulichen Akzent innerhalb des Eawag Empa-Areals. Seine siebbedruckten Glaslamellen können dem Sonnenstand nachgeführt werden und haben die Funktion des Sonnenschutzes im Sommer, der passiven Sonnenenergienutzung im Winter und des ganzjährigen Wetterschutzes. Die dahinter liegende thermische Schicht besteht aus Glas im Wechsel mit vorfabrizierten, hochwärmegedämmten Holzelementen, die mit blau eingefärbten Eternitplatten verkleidet sind. Die unterschiedlichen Nutzungszonen liegen U-förmig um ein 5-geschossiges Atrium, das durch schwebende Sitzungsboxen und den Treppenaufgang strukturiert wird. Diese Anordnung soll die Funktionsabläufe optimieren und den Innenraum transparent gestalten. Das Forschungs- und Verwaltungsgebäude bietet frei vernetzbare Arbeitsplätze für 150 Personen. Kommunikationszonen, Ausstellungsflächen, Seminarräume, ein Vortragssaal, ein Multimediaraum, eine Bibliothek sowie ein Personalrestaurant komplettieren das Raumangebot.

Das Gebäude ist in Stahlbeton-Skelettbauweise mit aussteifenden Erschließungskernen und tragenden Fassadenstützen errichtet. Die Fassadenhülle ist vorfabriziert. Diese Konstruktion ermöglichte eine kurze Bauzeit und große Nutzungsflexibilität. Alle haustechnischen Anlagen sind offen geführt, so dass die Installationen jederzeit zugänglich sind, was den Betrieb vereinfacht, den Unterhalt begünstigt und bei einem späteren Rückbau eine einfache Materialtrennung gewährleistet. Neben den Kennwerten zur Grauen Energie und Umweltverträglichkeit waren Kriterien wie Ästhetik, Nutzungskonformität, Wärmespeicherung, Brandverhalten und Wirtschaftlichkeit auch ausschlaggebend für das Farb- und Materialkonzept. Die Farbgebung ist eher zurückhaltend und entspricht den jeweiligen Materialfarben – mit Ausnahme der repräsentativen, öffentlichen Räume.

Das Gebäude ist von seiner Volumetrie, Speichermasse, Gebäudehülle, Ausrichtung und dem Sonnenschutz so konzipiert, dass es ohne eine herkömmliche Heizungsanlage auskommt und seinen Heizwärmebedarf über die im Gebäude anfallende Wärme (Menschen, Computer, Beleuchtung) und durch die Nutzung von Abwärmequellen, Erdwärme und solarer Energiegewinne decken kann. Für die Spitzenabdeckung und die Rückspeisung überschüssiger Energie ist es an ein bestehendes Arealwärmenetz angeschlossen. Die benötigte Kühlung in Seminar-, Vortrags- und Sitzungsräumen erfolgt über Kühldecken, die an das zentrale Kältenetz des Areals angeschlossen sind. Die optimierte Gebäudehülle samt Lamellenschicht macht in Kombination mit dem Lüftungssystem und der Nachtauskühlung eine Heizung und aktive Kühlung zum Teil überflüssig, **Bild 190**. Eine Photovoltaikanlage auf dem Dach erzeugt ein Drittel des Strombedarfs. Das extensiv begrünte Dach dient zur Rückhaltung und Sammlung von Regenwasser. Getestet wird am Gebäude außerdem der Einsatz regenwassergespülter NoMix-Toiletten zur separaten Sammlung von Urin.

building concerns so-called NoMix toilets, which are capable of separating urine from flush water and saving 80 % of the water in the process.

The compact design of the building, the mixed composition of the envelope with wooden elements, and the simple interior outfitted with hardwood floors and other interior materials such as clay boards used as railings for staircases were able to lower the grey energy content of the building significantly. Also, only recycled concrete and recycled plastics were used. Completely omitted from the building and its construction process were all construction chemicals based on volatile solvents. The images **192 – 194** illustrate the most important aspects of the building systems.

Durch die kompakte Gebäudeform, Mischbauweise mit Holzelementen in der Außenhülle und dem einfachen Innenausbau mit Hartsteinholzbelag und einer mit Lehmbauplatten beplankten Holzständerkonstruktion konnte die Graue Energie maßgeblich reduziert werden. Zum Einsatz kamen zudem Recyclingbeton und Kunststoffrecyclate. Auf Bauchemikalien auf Lösungsmittelbasis wurde verzichtet. Die **Bilder 192** bis **194** zeigen die wesentlichen Elemente des technischen Ausbaus.

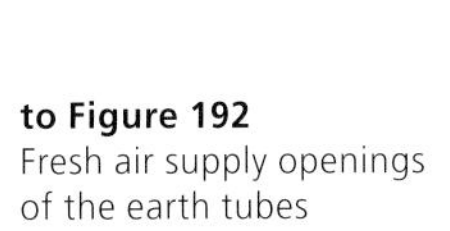

to Figure 192
Fresh air supply openings of the earth tubes

zu Bild 192
Frischluftaustrittsöffnungen des Erdregisters

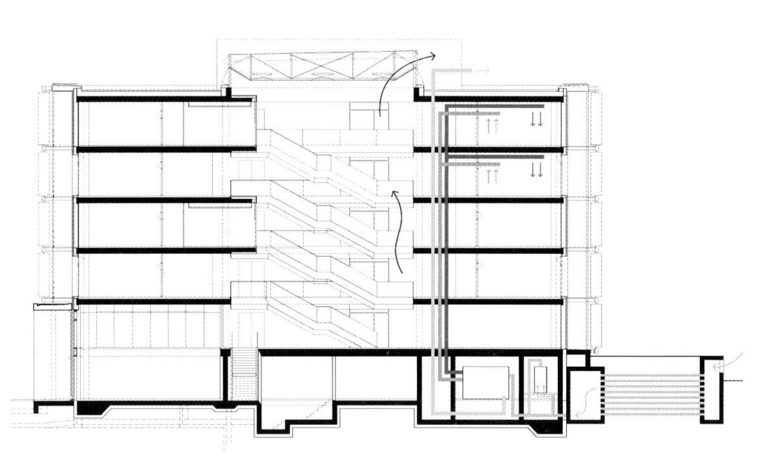

Figure 192
Diagram of ventilation with earth tubes

Bild 192
Prinzipbild der Lüftungstechnik mit Erdregister

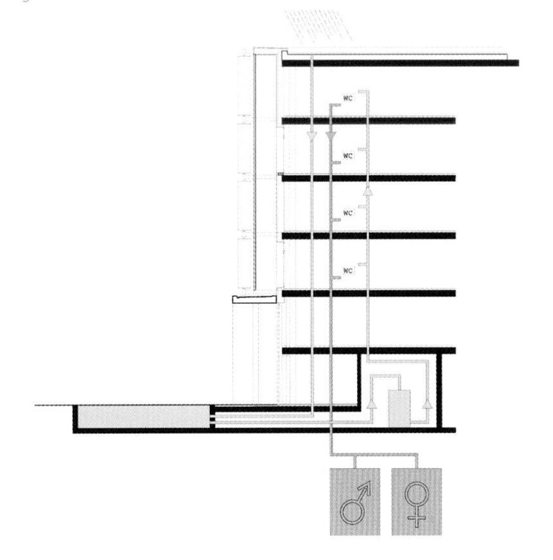

Figure 193
Rain water usage and black water processing

The system saves potable water yet is also used for research purposes. Water-less urinals and toilets in the building separate urine which contains common fertilizers but also problematic trace elements which need to be eliminated prior to further use in agricultural applications.

Bild 193
Regenwassernutzung/ Abwasseraufbereitung

Die Sanitärtechnologie dient dem sparsamen Umgang mit Trinkwasser und außerdem Forschungszwecken.
Urin enthält neben problematischen Spurenstoffen auch einen Großteil der Düngestoffe des häuslichen Abwassers. Mit wasserlosen Toiletten und Urinalen wird Urin getrennt gesammelt, um zu erforschen, wie Schadstoffe daraus eliminiert, Dünger extrahiert und in den landwirtschaftlichen Kreislauf rückgeführt werden könnte.

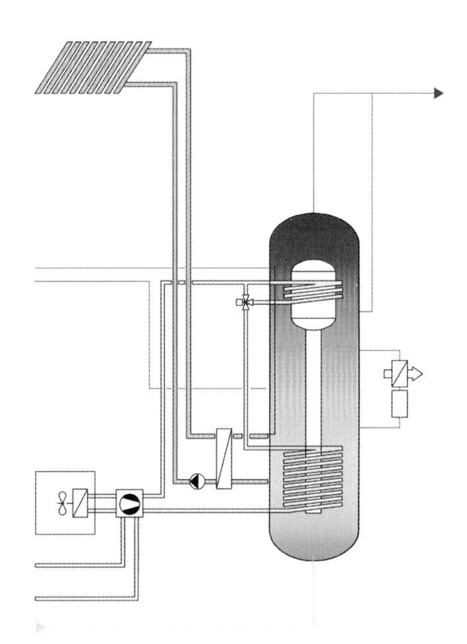

Figure 194
Heating energy supply

A central hot water storage tank with a capacity of 12 m³ provides 55 °C hot water and 40 °C heating energy. The reservoir uses the excess heat of the chiller plant of the building. Additionally, the water is heated by evacuated solar tubing. The remainder of the necessary energy is provided by the local heat energy grid of the campus. Excess heat, not needed in the building, is reintroduced into this grid and can be used in other university buildings.

Bild 194
Wärmeenergiebereitstellung

Der 12 m³ große zentrale Wärmespeicher stellt Warmwasser (55 °C) und Heizwärme (40 °C) bereit. Er nutzt die Abwärme aus der gewerblichen Kälteanlage. Das Wasser wird mittels Vakuumröhrenkollektoren zusätzlich erhitzt. Den Restwärmebedarf sichert das Arealwärmenetz. Überschüssige Wärme wird in das Netz eingespeist und steht anderen Gebäuden zur Verfügung.

11.10 Training academy Herne

Architects: Jourda et Perraudin, Lyon/Jourda Architectes, Paris
Hegger Hegger Schleif Architekten, Kassel
Building systems: HL-Technik, Munich/Frankfurt
Structural engineers: Schlaich Bergermann und Partner, Stuttgart

The continuing training academy of the Ministry of the Interior of the State of North Rhine-Westphalia in Germany, **Figure 195** is part of an ecological and social science campus on the site of the former coal mine of Mont Cenis in Herne, Germany. The master plan of the area consists of the academy as a part of the International Building Exhibition (IBA), a public park, residential units, and a market square. The building's main concept is a large, micro-climatic, glazed space in which the individual buildings of the academy, such as the administration, hotel, restaurant, public library, and large auditorium and recreational facilities are inserted, **Figure 196** to **198.**

As early as the 1980 s, the architects Jourda and Perraudin began researching designs that would provide various internal climate zones. The building for the Herne academy is 72 meters wide and 180 meters long, and it has a 16-meter-high glass hull. This superstructure serves as a solar electric power plant and provides a temperate, "Mediterranean" climate even under the sometimes adverse weather conditions of Central Germany (**Figures 196,197**).The effect of this climatic hull is that energy consumption is greatly reduced. The inside of the glass hull is accessible to the public and provides a lively, highly attractive urban space that is fitted with terraces, a reflecting pool, and plants otherwise found only in warm climates, such as tall palm trees. The glazing of the outside envelope consists of laminated, non-insulated glass, which is able to allow for solar gains. Another significant advantage of the concept of the large glass superstructure is the fact that the inserted buildings have minimized circulation area requirements: circulation space typically necessary inside a building can here be located on the inside of the glass hull, external to the inserted building itself.

11.10 Fortbildungsakademie Herne

Architekten: Jourda et Perraudin, Lyon/Jourda Architectes, Paris
Hegger Hegger Schleif Architekten, Kassel
Gebäudetechnik: HL-Technik, München/Frankfurt
Tragwerk: Schlaich Bergermann und Partner, Stuttgart

Die Fortbildungsakademie Herne (Deutschland) ist Teil eines ökologischen und sozialen Gesamtkonzeptes auf der Zechenbrache Mont Cenis, auf der im Rahmen einer Neuentwicklung des IBA Emscher Park neben einem Park durch Wohngebiete, Einkaufszentrum und Marktplatz ein neues Zentrum entstehen sollte, **Bild 195**. Geplant war eine „mikroklimatische Hülle". Bereits in den 80er Jahren hatten Jourda & Perraudin zu einer Architektur unterschiedlicher Klimazonen innerhalb eines Gebäudes geforscht. Die 72 m breite, 180 m lange und 16 m hohe gläserne Hülle der Akademie fungiert einerseits als Solarkraftwerk und bildet andererseits eine klimatische Pufferzone, in der unter Regen-, Wind- und Kälteschutz milde Temperaturen herrschen, **Bilder 196/197**. Dadurch verringert sich der Energiebedarf der in die Hülle eingestellten so genannten Innenhäuser. Sie beherbergen Funktionen wie Verwaltung, Restaurant, Hotel und Freizeiteinrichtungen der Akademie, eine öffentliche Bibliothek, Stadtverwaltung und einen Bürgersaal, **Bild 198**.

Figure 195
Academy Mont-Cenis Herne-Sodingen, Germany

Ministry of the Interior of the State of North Rhine-Westphalia

Architects: Jourda & Perraudin Paris, France
Hegger Hegger Schleif · HHS Planer +Architekten BDA, Kassel

Bild 195
Bildungszentrum Herne „Mont Cenis", Ansicht des Projektes

Architekten: Jourda & Perraudin Architectes, Paris,
Hegger Hegger Schleif · HHS Planer +Architekten BDA, Kassel

Durch diese Funktionsmischung konnte sich das Innere der einfachverglasten Glashaut (Verbundsicherheitsglas, VSG) zu einem öffentlich zugänglichen und belebten Raum entwickeln. Befördert wurde dies durch die attraktive Gestaltung mit Terrassen, Wasserbecken und Bepflanzung. Ein weiterer Vorteil der Klimahülle sind Einsparungen von beheizten Flächen, da ein Großteil der Erschließungsflächen außerhalb der Gebäude in der Halle untergebracht werden konnte.

Figure 197
Interior, with reflecting pool

Bild 197
Ansicht des Innenraumes mit Wasserbecken

Figure 196
Interior showing primary structure

56 spruce trunks and other rectangular wooden sections form the framework for the micro-climatic hull

Bild 196
Innenansicht mit Primärkonstruktion

56 Fichtenstämme und andere nachwachsende Holzelemente bilden das Traggerüst der Primärkonstruktion der Mikroklima-Hülle

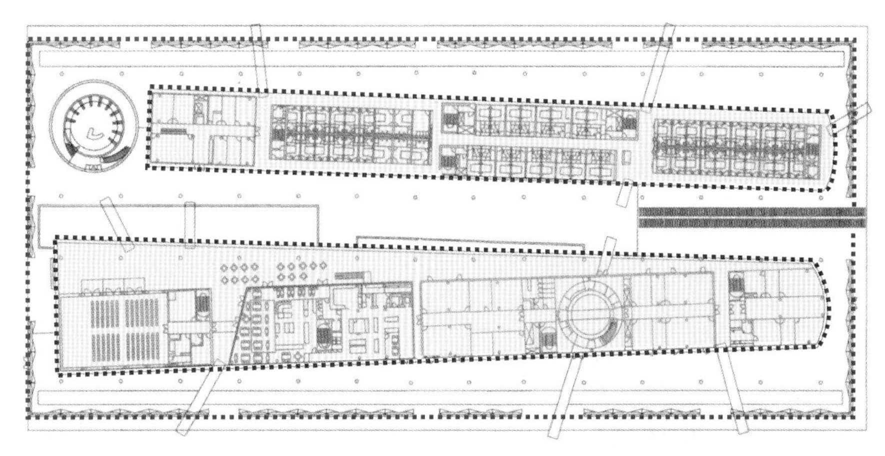

Figure 198
Floor plan showing climate zones

The single-glazed envelope of the building serves as a climatic buffer. It reduces the energy consumption of the internal buildings and provides additional usable and comfortable space within during times of inclement weather.

Bild 198
Grundriss Erdgeschoss mit klimatischer Zonierung.

Die einfachverglaste Hülle wirkt als Klimapuffer. Sie reduziert nicht nur den Energiebedarf der eingestellten Funktionsgebäude, sondern eröffnet einen zusätzlichen öffentlich nutzbaren Raum, der mit seinen gemäßigten Temperaturen fast das ganze Jahr über von verschiedenen Nutzern belegt wird.

Zone 1 Outside
Zone 2 Inner outside
Zone 3 Inside

When the building was completed in 1998, it represented the world's largest over 10,000 sqm BIPV system, generating power up to 1 MW and approximately 750,000 kWh/a (**Figure 199**). In addition to electricity generation, the integrated photovoltaic elements also serve as shading devices for the glass skin. They are arranged in varying densities according to requirements for shading, which creates an attractive cloud-like pattern on the roof when seen from the inside of the glass space. Shading was omitted along the center promenade, but here ventilation openings in the glass hull can be operated to prevent overheating of the inside space. The building contains additional holographic-optical elements (HOE) that redirect sunlight to the interior buildings and spaces. The interior is daylit by light shelves and cooled by earth registers. The interior concrete slab-on-grade, with its large thermal mass, serves as a temperature-balancing element for the entire building. All materials used in the construction were selected according ecological criteria, and compound materials were omitted from the structure to ensure easy maintenance and full recycling capability.

Mit dem Gebäude wurde 1998 die weltgrößte gebäudeintegrierte Photovoltaikanlage mit einer Modulfläche von über 10.000 m^2 auf dem Dach und an der Süd-Westfassade der Hülle mit einer Leistung von insgesamt bis zu 1 MW und ca. 750.000 kWh/a in Betrieb genommen, **Bild 199**. Neben der Stromgewinnung übernehmen die Module auch die Verschattung der gläsernen Klimahülle. Dazu wurden sie je nach Beschattungsbedarf in unterschiedlicher Dichte auf dem Dach angebracht, woraus sich eine wolkenartige Struktur ergab. Direkt über den eingestellten Gebäuderiegeln verdichten sich die Module, um einer Überhitzung der Innenhäuser vorzubeugen. Entlang der Promenade wurde dagegen gänzlich auf die Verschattung verzichtet. Hier befinden sich Lüftungsklappen, die verhindern sollen, dass sich die Klimahülle zu stark aufheizt. Zusätzlich ist die Glashülle mit lichtlenkenden holographisch-optischen Elementen ausgerüstet, um in den in eingestellten Häusern liegende Bereiche wie die Bibliothek mit Tageslicht zu versorgen. Die Innenhäuser werden über „light shelves" mit zusätzlichem Tageslicht versorgt und außerdem über Erdregister luftgekühlt. Der Betonsockel im Inneren trägt zur Gebäudetemperierung der gesamten Anlage bei. Die Auswahl der Materialien erfolgte unter ökologischen Aspekten. Auf Verbundwerkstoffe wurde zugunsten einer Recyclingfähigkeit und einfacher Instandhaltung weitgehend verzichtet.

Figure 199
Glass roof with ventilation opening, building integrated photovoltaic modules.

Bild 199
Glasdach mit Öffnungselementen zur natürlichen Durchlüftung sowie Photovoltaik-Elementen

The primary structural system of the glass hull uses renewable timber such as 130 old-fir tree trunks from the Sauerland region. As part of an overall energy concept, the building is served by a combined heat-power plant (CHP) that provides additional necessary thermal and electric energy. Due to the special condition of the former coal mining site, the building can also be operated by mine gas from the abandoned mine shafts. Especially during overcast, low-pressure weather conditions, an increased amount of mine gas is exhausted from the ground, which compensates for the reduced solar energy generated by the photovoltaic array under such conditions (**Figure 200**).

Originally, the "micro-climatic" glass container was simply meant as a weather barrier for the inside buildings and an electric power-generating surface to lower fossil fuel consumption. Yet, during operation it became evident that this buffer zone actually could be used very well as an interior action space during almost the entire year. All energy consumed in the building remained well below the original prognosis, and more electric energy is being produced by the building's glass surfaces than is consumed inside. The annual thermal energy consumption of the internal buildings is below 32 kWh/m^{2}a. Although the interior provides the mild climate of Southern Europe, with average temperatures of 20 °C in summer and winter, some users of the building still complained about too-low temperatures during very cold seasons. Nevertheless, since its construction the climatic hull has never experienced temperatures below freezing levels despite the fact that the simulation predicted minimal temperatures of around –5 °C. During the summer months, the originally modeled temperatures are being achieved, with interior glass hull temperature not exceeding 2 K above outside ambient temperature.

Bei der Tragkonstruktion kamen nachwachsende Rohstoffe wie 130 Jahre alte Fichtenstämme aus dem Sauerland zum Einsatz. Eingebettet in ein energetisches Gesamtsystem, wird die Akademie über ein Blockheizkraftwerk mit Wärme und mit Strom zur Spitzenauslastung versorgt. Das Kraftwerk kann aufgrund der besonderen Situation vor Ort mit Grubengas aus den ehemaligen Bergwerkschächten betrieben werden. Gerade bei bewölkten Tiefdrucklagen strömt besonders viel Gas aus der Tiefe. Dann können witterungsbedingte Ertragseinbußen der Photovoltaikanlage ausgeglichen werden, **Bild 200**.

Ursprünglich sollte die „mikroklimatische Hülle" lediglich Witterungsschutz für die eingestellten Gebäude bieten und solare Energieeinträge einfahren, um den Energieverbrauch der in ihr untergebrachten Gebäude zu senken. Im Betrieb hat sich gezeigt, dass diese Pufferzone zu drei Vierteln des Jahres tatsächlich nutzbar ist und als öffentlicher Raum wider Erwarten in Beschlag genommen wird. Die prognostizierten Energiekennwerte wurden bisher unterschritten und durch die Solarstromanlage mehr Strom produziert, als im Jahresmittel verbraucht wurde. Der Jahresheizwärmebedarf der eingestellten Häuser liegt unter 32 kWh/m^{2}a. In der subjektiven Wahrnehmung der Nutzer, die sich unter mediterranen Temperaturverhältnissen milde 20 °C winters wie sommers vorstellen, stießen die kühlen Temperaturen im Winter zwar auf Kritik, jedoch ist die Klimahülle seit Eröffnung des Komplexes trotz vorausberechneter Minimaltemperaturen von –5 °C frostfrei geblieben. Auch in den Sommermonaten konnten die im Vorfeld errechneten Temperaturen eingehalten werden und liegen nur um ca. 2 K über den Außentemperaturen.

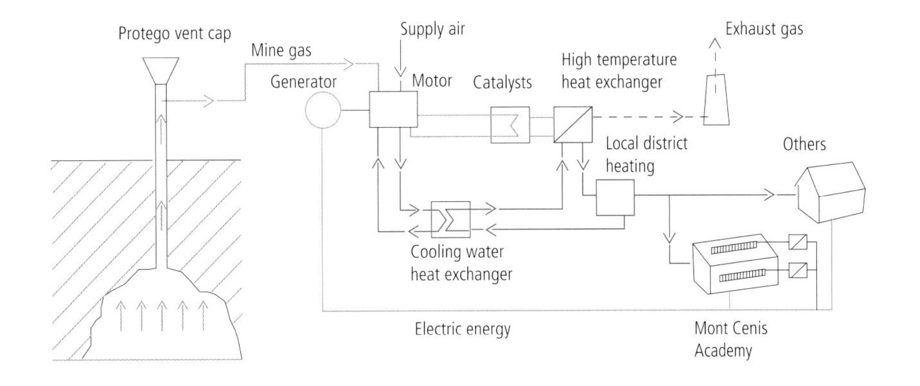

Figure 200
Integrated energy system grid with cogeneration power plant, mining gas-fired

Source: archplus 184, October 2007

Bild 200
Energieverbund mit BHKW-System, mit Grubengas betrieben

Quelle: archplus 184, Oktober 2007

11.11 Exhibition Hall 26, International Hanover Trade Fair, Hanover, Germany

Architects: Thomas Herzog & Partners, Munich
Building Systems: HL-Technik, München
Structural Engineers: Schlaich Bergermann and Partners, Stuttgart

In a most effective integral design process between architects, structural engineers, and building systems consultants, the iconic exhibition hall 26 of the International Trade Fair in Hannover was conceived in 1994. The single-story building is 225 meters long with a width of 130 meters and a height of 26 meters. **Figures 201 – 203** show the building floor plan, the development of the structural system, and the completed building.

11.11 Messehalle 26, Deutsche Messe AG, Hannover, Deutschland

Architekt: Thomas Herzog & Partner, München
Gebäudetechnik: HL-Technik, München
Tragwerk: Schlaich Bergermann und Partner, Stuttgart

In einem integralen Planungsansatz zwischen Architekt, Tragwerksplaner und Gebäudetechniker wurde im Sommer 1994 die Messehalle 26 in Hannover geplant. Ihre Abmessungen (eingeschossig) betragen ca. 225 x 130 m, ihre maximale Höhe ca. 26 m.

Die **Bilder 201 – 203** zeigen neben dem Grundriss die Entwicklung der Hauptkonstruktion und die fertige Halle.

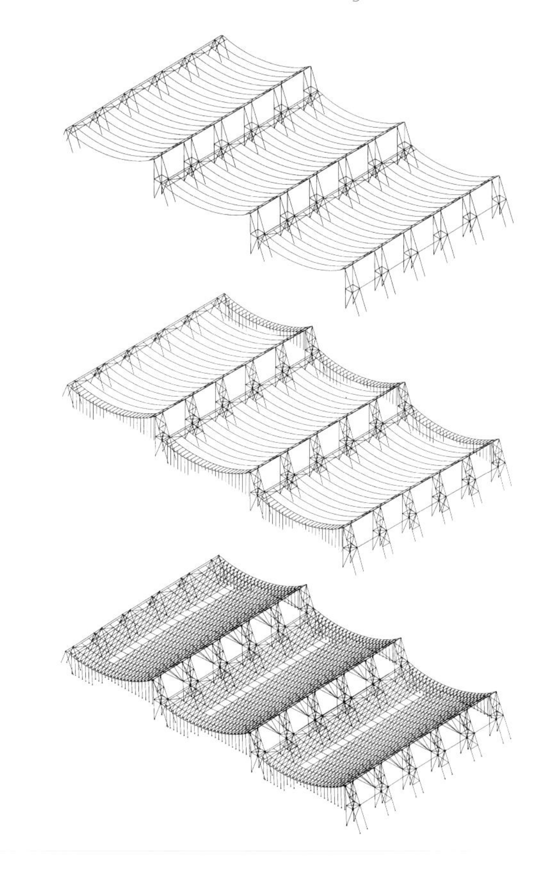

Figure 202
Main load-bearing structure of exhibition hall

Source: Sbp, Schlaich Bergermann + Partner

Bild 202
Entwicklung der Hauptkonstruktion

Quelle: Sbp, Schlaich Bergermann + Partner

Figure 203
Night view of exhibition hall

Bild 203
Nachtansicht der Halle

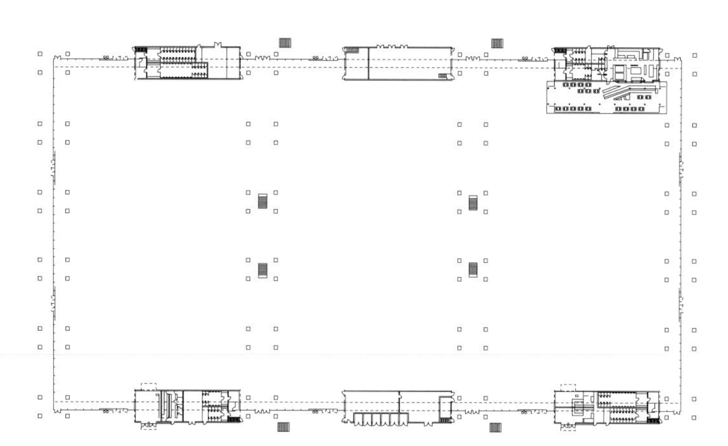

Figure 201
Floor plan view

Source:
plusminus 20 °/40 ° latitude
Hindrichs · Daniels
Edition Axel Menges,
Prof. Dr. Herzog + Partner,
Munich

Bild 201
Grundriss

Quelle:
plusminus 20 °/40 ° latitude
Hindrichs · Daniels
Edition Axel Menges,
Prof. Dr. Herzog + Partner,
München

In addition to the requirement to provide 30,000 m² of exhibition space, the design brief required observing the following parameters:

- Development of a structural system providing maximum uninterrupted spans. This condition led to the design of a hanging roof structure.
- Functionally required clear height of the interior space of 10 – 11 meters. This was used to develop high points in the design to support natural ventilation by stack effect, buoyancy, and wind effects with outside negative pressures.
- Large expanses of glass to provide daylight for the interior and simultaneous reduction of solar energy gains.

The hall that was developed uses, to a high degree, renewable environmental energy and contains two different interior zones:

- A large, wide-span, column-less exhibition space that can be freely configured as needed is provided under the hanging tensile-steel roof, which is covered by wood panels.
- Narrow bands between the exhibition spaces and along its perimeter are used for circulation and recreation.

Additional program spaces, which include three restaurants, bathroom areas, and the mechanical space necessary for the thermal conditioning of the hall and for its service requirements, are located in rectangular side buildings. They are clad on the exterior with wooden louvers and so are able to harmonize the integration of the necessary visitor doors, sliding gates, supply and exhaust air openings, and other elements inserted into the façade.

In most cases of exhibition hall design, the distribution of air is arranged from the top of the space down, according to the principle of air mixing. In contrast to this conventional air distribution strategy, Hall 26 is supplied by cooled air from below, which heats up subsequently and is exhausted at the top of the space. This technology allowed the cooling load to be lowered to approximately 160 W/m², which is a reduction by half compared with standard system designs. Because the demand for cooling is limited to just that area of around 2.5 meters in height in which people circulate, it was possible to reduce the energy required for cooling substantially. Air handling in this large space is provided by six individual units, located in the perimeter buildings, from where the supply air is moved into the 3-meter-tall, glass supply air channel. The perforated bottom of the glass channel allows the air to flow in a downward direction to the exhibition hall floor from a height of 4.7 meters, and the air is then distributed along that surface in a manner similar to "a lake of cool, stratified air".

Für die Halle 26 bestand die spezielle Anforderung darin, bei einer Größe von ca. 30.000 m² einen Typus mit einer Querschnittsgeometrie zu entwickeln, der folgende Merkmale integriert:

- Form eines Tragwerks, das für große Spannweiten prädestiniert ist, was zur Wahl eines Hängedachs führte,
- einerseits Festlegung in weiten Bereichen funktional nötige Raumhöhe (ca. 10 – 12 m), andererseits Ausbildung von Hochpunkten zur natürlichen Entlüftung unter Nutzung des thermischen Auftriebs- und von Windeffekten (Sogeffekte/Unterdruck, ca. 2 m),
- Ausbildung großer Bereiche zum Einlass von Tageslicht bei Reduzierung von direkter solarer Einstrahlung.

Der entwickelte Hallentypus, bei dem mit hoher Wirkung Umweltenergien zum Betreiben eingesetzt werden, enthält zwei unterschiedliche Grundrisszonen:

- weiträumige, stützenlose, frei disponierbare Ausstellungsbereiche unter dem leichten zugbeanspruchten Hängedach aus Stahl, auf dem eine Holzdecke aufliegt,
- schmale Bereiche zwischen den Ausstellungsflächen und an ihrem Rand mit Stahlpylonen zur Aufnahme der Kräfte aus den Hängedächern. Diese Zonen dienen der Erschließung der Ausstellungsflächen und der Rekreation.

Weitere Einzelfunktionen, wie drei gastronomische Betriebe, Toilettenanlage und alle Räume für die technische Ver- und Entsorgung sowie die Raumkonditionierung sind seitlich in sechs selbstständigen, kubischen Bauwerken untergebracht. Sie besitzen eine äußere Bekleidung mit kräftigen Holzlamellen, deren Konstruktion die Integration von Zu- und Abluftöffnungen, Türen und Schiebetoren erlaubt.

Während in den meisten Fällen Messehallen mit einer Luftführung von oben nach oben und damit durch Mischlüftung betrieben werden, erfolgt bei der Halle 26 eine Luftführung im Kühlbetrieb von unten nach oben. Dadurch war es möglich, die vorgegebene Kühllast mit ca. 160 W/m² auf die Hälfte zu reduzieren, da im Regelfall im Aufenthaltsbereich (bis ca. 2,5 m Höhe) nur ca. 50 Prozent der Kühllasten anfallen und zu kompensieren sind. Die Luftaufbereitung für die Halle erfolgt in sechs Einzelgeräten, die jeweils im 1. OG der seitlich angestellten Boxen untergebracht sind. Von hier aus wird die Luft durch ca. 3 m hohe, gläserne Kanäle in den Hallenraum eingeführt und über die Unterseite der Luftkanäle nach unten (auf den Boden) ausgetragen. Somit erfolgt die Belüftung durch großformatige Auslässe (Lochdeckenfelder mit Stützstrahlauslässen), die als Baldachin ausgebildet sind.

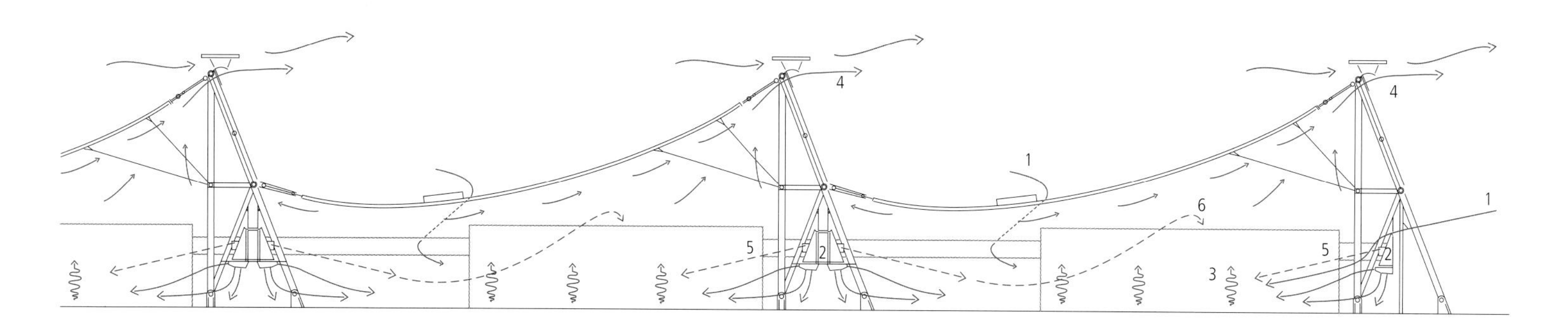

Cooling:
1 Fresh air intake through façade louvers
2 Glass air tunnel with displacement vents at bottom for cooled air
3 Stack effect ventilation, internal heat loads
4 Natural ventilation through roof openings

Heating:
5 Mechanical distribution of heated intake air through wide-angle nozzles
6 Return air ventilated through lateral ventilation

Figure 204.1
Diagram of natural and mechanical ventilation of the exhibition hall

Bild 204.1
Prinzipdarstellung der natürlichen und mechanischen Be- und Entlüftung der Ausstellungshalle. Schnitt.

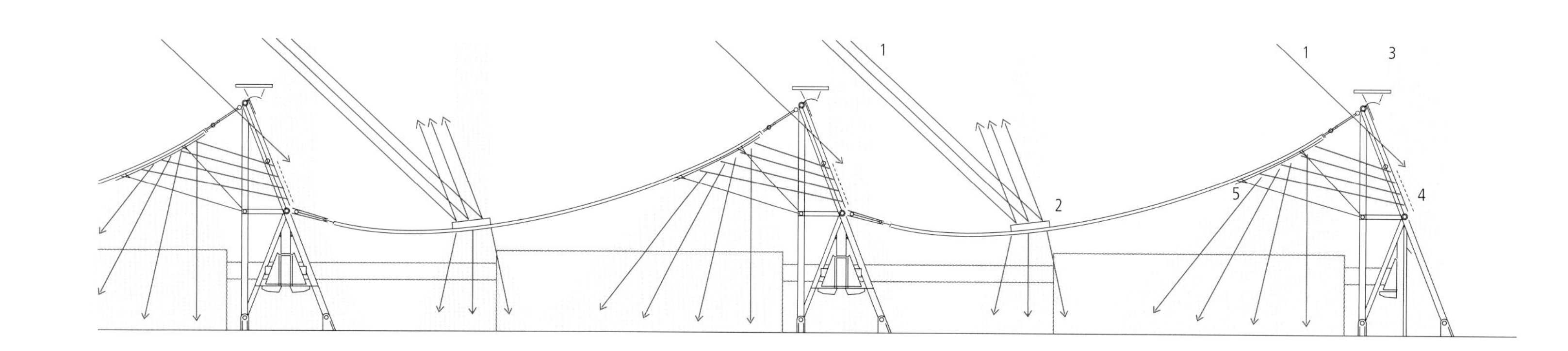

Figure 204.2
Principle of daylighting

Bild 204.2
Prinzip der Tagesbelichtung

Figure 205
Roof structure (view from below), showing light redirection deflection elements and glass supply air duct.

Bild 205
Innenansicht mit gläsernem Zuluftkanal und Tageslicht-Umlenkelementen

Figure 206
Inside view of glazed supply air duct

Bild 206
Gläserner Zuluftkanal mit zusätzlichen Heiz-Luft-Auslässen

Wherever the supply air encounters heat sources in the exhibition hall, such as people or equipment, it rises slowly as a "plume", similar to displacement ventilation. The air rises up, eventually reaches the high point of the space, and is exhausted by operable louvers in the roof membrane of the hall's gable surfaces. These are operated and controlled according to local prevailing wind conditions such as speed and direction to maximize the utilization of negative pressures for exhaust.

The undeniable advantage of such a system lies in its increased interior air quality, based on the important condition that 100 % fresh air is used for space conditioning – conventional mixing of a small amount of fresh air with a majority of recirculating air is thus avoided. The used air moves to the top of the space by thermal buoyancy forces. During heating seasons, the air-handling units operate in recirculation mode, ejecting the supply air by high pressure, horizontally placed "eye-ball" nozzles deep into the space. During the design phase, the solution was tested extensively in computer simulations and physical wind tunnel studies at a scale of 1:5, with special attention being given to the design and effectiveness of the upper gable exhaust flaps to determine whether sufficient negative pressure for effective exhaust would exist. Figure **204.1** shows, with blue arrows, the cooling case that is typical for large exhibition halls, as well as the heating scenario (red dashed lines). **Figures 205** and **206** show the primary structure, with its triangulated steel columns and the suspended large glass supply air duct.

Aus einer Höhe von ca. 4,7 m strömt die Zuluft nach unten und verteilt sich dort über den gesamten Hallenboden gleichmäßig in die gesamte Raumtiefe. Analog einer Quellbelüftung steigt die Luft an den im Raum befindlichen Wärmequellen (Menschen, Maschinen, Geräte, Computer, Leuchten) langsam wieder auf und erreicht endlich den Hochpunkt der jeweiligen Hallendächer. Im Firstbereich befinden sich durchgehende Öffnungen, die durch einzeln steuerbare Klappen unterschiedlich – je nach Windanströmungsrichtung – individuell schließbar oder öffenbar sind, so dass Sogkräfte wirksam werden können.

Die eindeutigen Vorteile liegen bei diesem System in der Verbesserung der Luftqualität und der Verbesserung des Komforts im Aufenthaltsbereich, da diesem ohne Vermischung mit weiterer Raumluft reine Außenluft zugeführt wird. Die erwärmte und belastete Zuluft strömt infolge des thermischen Auftriebs nach oben in Richtung Dachflächen. Im Heizfall (im Regelfall vor dem Messebetrieb im Winter) arbeiten die Raumlufttechnischen Anlagen ausschließlich im Umluftbetrieb, wobei nunmehr die warme Zuluft über horizontal ausblasende, verstellbare Weitwurfdüsen ausgebracht wird. Das System der Luftführung wurde im Modellversuch (1:5) weiterentwickelt und auf seine Wirksamkeit hin überprüft. Gleichermaßen wurde im Rahmen einer Windkanalstudie die Wirksamkeit der öffenbaren Firstklappen optimiert, um ausreichend hohe Sogwirkungen zur Entlüftung zu erzielen. **Bild 204.1** zeigt sowohl den bei Messen üblichen Kühllastfall (blaue Pfeile) wie auch den Heizbetrieb (gestrichelte rote Pfeile). **Bilder 205** und **206** zeigen einen Teil der Dachstruktur mit den Stahlböcken (Hauptkonstruktion) sowie den eingehängten gläsernen Zuluftkanal.

Figures 207.1 and **2** show the annual operation of the air-handling system for heating and cooling scenarios. Yellow areas show time intervals of so-called "free-cooling", during which the thermal buoyancy inside the hall is sufficient to ventilate the space by natural means; no mechanical system operation is required in such periods. These times coincide with the recurrent installation and disassembly times required for exhibitions. From April until October, and in some instances until November, cooling is necessary during times of exhibition events, and between November and March heating is normally required. Such heating typically does not take place during the opening hours of the exhibition itself, since internal loads of the lounges in a magnitude of approximately 80 W/m^2 are guided into the hall itself. An additional demand of around 80 W/m^2 is due to thermal losses via the hall's roof structure.

Figure 207.2 shows a comparison of systems with regard to annual energy consumption between conventional air supply (mixing) and the described solution of displacement ventilation. As can be seen here, significant energy savings can be gained in the process.

An additional important component of a strategy for sustainability can be recognized in the fact that the hall uses extensive day lighting. **Figure 204.2** shows the large north-facing glazed surfaces located at the primary structural columns and the clerestory windows at the low points of the hanging roof. It is important to realize that the interior roof surface is designed as a "super-reflector" to distribute the daylight entering the hall, via light re-directing elements, deep into the space.

The interior of the roof surface serves the function of a diffuse light reflector also for cases in which artificial lighting is required.

Die **Bilder 207.1/2** zeigen einerseits den jährlichen Anlagenbetrieb der Raumlufttechnik sowie die Heiz- und Kühlfälle. Die gelben Felder, freie Lüftung, deuten den Zeitraum an, in dem in der Halle Auf- und Abbaubetriebe ablaufen, bei denen die Halle mechanisch nicht belüftet wird. Infolge der jährlichen Witterungsbedingungen in Hannover ergeben sich größere Zeiträume, bei denen eine natürliche Lüftung bzw. eine Lüftung ohne zusätzliche Kühlung und Heizung möglich ist. Im Zeitraum April bis Oktober/zum Teil November wird zum Messezeiten zum Teil eine Kühlung notwendig, und im Zeitraum November bis März im Wesentlichen eine Beheizung. Die Beheizung während des Messebetriebes ist im Regelfall nicht notwendig, da die inneren Kühllasten mit ca. 80 W/m^2, wirksam in der Aufenthaltszone, ausreichend hohe Wärmemengen in die Hallenräume eintragen. Weitere 80 W/m^2 decken die Wärmeverluste der Dächer.

In **Bild 207.2** ist ein Systemvergleich bezüglich des Jahresenergieverbrauchs zwischen einem konventionellen Anlagenbetrieb (Mischlüftung) bzw. dem hier eingesetzten Betrieb (Quelllüftung) dargestellt. Wie dem Bild zu entnehmen, ist infolge der Anlagen- und Belüftungsstruktur ein hohes Maß an Energie-einsparung in den jeweiligen Monaten gegeben.

Ein weiteres wesentliches Element der Energieeinsparung betrifft die Tagesbelichtung der Messehalle. **Bild 204.2** zeigt die natürliche Belichtung über große Nordverglasungen im Bereich der Stahlhauptkonstruktionen bzw. in den Feldern dazwischen über mit Lichtrastern bestückte Oberlichter im Bereich des Tiefpunktes der Hängekonstruktion des Daches. Durch lichtlenkende Elemente wird das Tageslicht über den „Großreflektor" Hallendach in den Aufenthaltsbereich gelenkt.

Nach dem gleichen Prinzip erfolgt die Tageslicht-Ergänzungsbeleuchtung durch Kunstlicht, bei dem die gekrümmten großen Dachflächen zur Lichtstreuung genutzt werden.

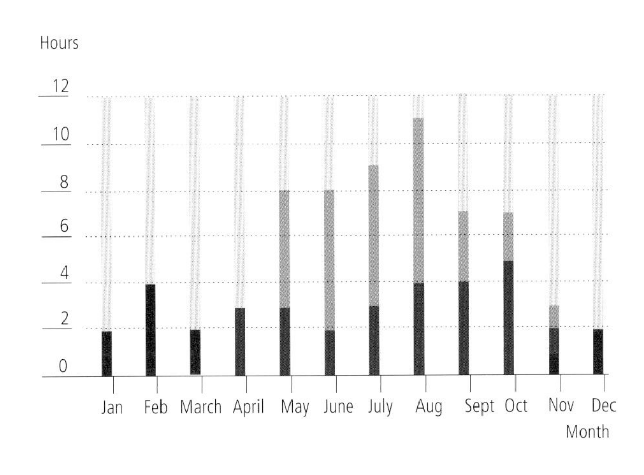

Figure 207.1
Ventilation strategy including cooling for Hall 26 during exhibitions and off-hours. Operating hours per exhibition day

Source: Hall 26, Hanover Fair, Germany
Prof. Dr. T. Herzog + Partner
Concept for space conditioning

Bild 207.1
Belüftungsstrategie mit Kühlung und Heizung zu Messe- und Nichtmessezeiten (Betriebsstunden je Messetag)

Quelle: Halle 26,
Prof. Dr. T. Herzog + Partner,
(Konzept für das Raumklima)

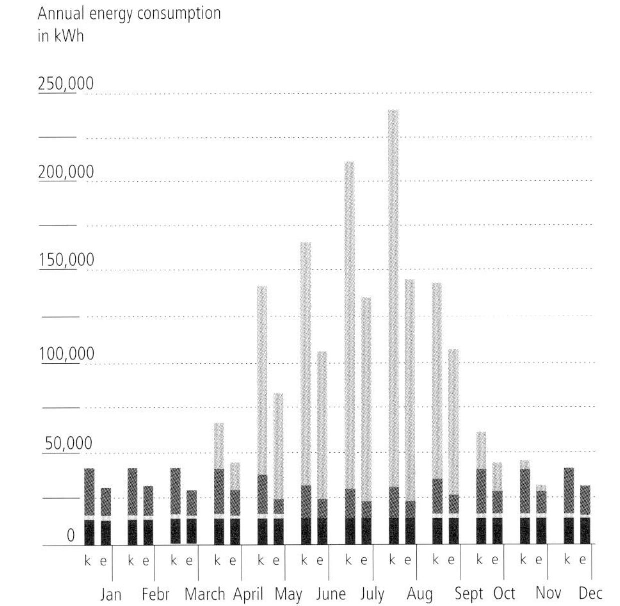

Figure 207.2
Comparison between systems and their annual electric energy consumption

Bild 207.2
Systemvergleich/Jahres-Energieverbrauch Elt-Energie, in kWh

12 Zero-energy super-tall buildings: vision or illusion?

12 Zero-energy super-tall buildings – Vision oder Illusion

The vision of constructing tall or super-tall buildings that are capable of generating more energy than is required for their operation is fascinating. If it indeed would be possible to build such structures, the ratio of their energy-generating surfaces to the floor surface area that such systems would have to supply would be between 1:1.5 or 1:3. If this could be accomplished, it might become possible to build buildings in the future that no longer require any electricity for their operation.

In reality, the vision of an energy-autonomous building is related to the consumption of energy for heating and cooling, but it remains primarily a problem of energy demand for very "electricity-hungry" machines and services.

Depending upon solar position at the location of the building site and the correct surface ratio of energy-generating to energy-consuming areas, the sun is capable of providing sufficient solar energy for heating a building. If subterranean cold water streams in the soil are used as well, almost the entire energy demand for the cooling of a tall building can be satisfied – although this will only be the case for cooler climatic regions on Earth.

The real problem continues to be the generation of enough electrical energy with the help of renewable sources. Stefan Behling and the scientists at the Technical University of Stuttgart, with industry support from the German façade manufacturer Schüco International, are in the process of developing a new façade in which the necessary structural elements are combined with energy supply components. The result will be an energy-generating façade – in other words, the building façade will become a power plant.

The integrated Behling-Schüco façade concept corresponds to the following requirements:

1. Invisible integration of operable casement windows in post-and-bar façade systems that allow for natural ventilation. The automated opening and closing system, including all necessary mechanical hardware, is hidden in vertical façade posts with no more than 8.5 cm width, and the windows can be operated centrally, in a de-centralized manner, or individually by the user.

2. The outside shading device consists of micro-lamellae, which effectively shade the glazing units and can stand up to wind speeds of 100 km/h. The lamellae are lowered or raised in two guiding channels in the façade posts on either side.

Photovoltaic elements and/or thermal solar components can be integrated into the façade at the same time. The goal for future systems is to generate on the surface of the façade – in addition to electricity – high-temperature thermal energy that can subsequently be

Die Vision, Gebäude zu konzipieren, die mehr Energie erzeugen als sie verbrauchen, gilt für bauliche Strukturen, deren Energieerzeugungsflächen in etwa im Verhältnis von 1:1,5 bis 1:3 der energieverbrauchenden Flächen liegen, es sei denn, es gelänge tatsächlich, ein Gebäude zu konzipieren, in dem im Wesentlichen keine elektrische Energie mehr benötigt wird.

Es ist nicht unbedingt ein Problem des Heizens oder zum Teil Kühlens, will man ein energieautarkes Gebäude konzipieren, sondern es ist primär ein Problem des elektrischen Energiebedarfs für „stromfressende" Maschinen und Geräte.

Je nach Sonnenangebot des jeweiligen Standortes kann beim richtigen Flächenverhältnis energieerzeugender Flächen zu verbrauchenden Flächen ausreichend viel Wärme produziert werden. Bei Nutzung von Kühlwasserströmen aus dem Untergrund in kühlen Regionen der Welt kann annähernd der gesamte Kühlenergiebedarf aus dem Erdreich gedeckt werden.

Das eigentliche Problem liegt darin, ausreichend viel elektrische Energie durch die natürlichen Ressourcen zu erzeugen. An der Universität Stuttgart, Lehrstuhl Stefan Behling, wurden mit Unterstützung der Firma Schüco vor Jahren Entwicklungen aufgenommen, eine Fassade zu konzipieren, die funktional notwendige Baukomponenten mit Energieertragselementen kombiniert, so dass eine Energiefassade entsteht.

Das integrale Fassadenkonzept (Behling/Schüco) erfüllt dabei folgende Forderungen:

1. Die unsichtbare Integration von öffenbaren Fensterflügeln in die Pfosten-Riegelfassade sorgt für natürliche Lüftung. Die Automation aller Öffnungsarten mit verdeckt liegenden Systemantrieben sind in Pfostenbreiten von ca. 8,5 cm integriert und können zentral, dezentral oder individuell am Fassadenelement betrieben werden.

2. Der außen liegende Sonnenschutz besteht aus einer Mikrolamelle, die bei Windgeschwindigkeiten bis ca. 100 km/h zuverlässig verschattet. Die Mikrolamelle wird seitlich in den Pfosten geführt und ist im eingefahrenen Zustand nicht sichtbar.

Sowohl Photovoltaikelemente wie auch thermische Kollektorflächen können gleichwertig in die Fassade integriert werden. Vision in der weiteren Entwicklung ist, dass in Zukunft die Gebäudehülle nicht nur Strom erzeugt, sondern auch hochtemperierte Wärme, um diese mittels einer Absorptionskälteanlage zu Kühlenergie umzuwandeln. Hierzu wurde im ersten Ansatz ein optisch durchlässiger Flachkollektor entwickelt, der den Bezug nach außen zulässt und ein interessantes Licht- und Schattenspiel erzeugt.

Figure 208
Elevation,
façade-integrated grid collector

Source:
Schueco International KG

Bild 208
Fassadenansicht,
in Glasfassade integrierter Raster-Kollektor

Quelle:
Schüco International KG

Figure 209
Evacuated solar tube collectors as façade elements

Source: archplus 184, October 2007

Bild 209
Vakuum-Röhrenkollektoren als Fassadenelemente

Quelle: archplus 184, Oktober 2007

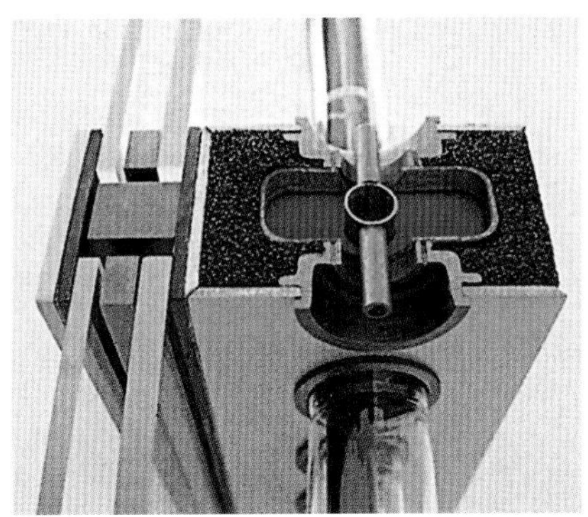

to Figure 209
Connection detail of façade integrated evacuated solar tube collector

Source: archplus 184, October 2007

zu Bild 209
Detailanschluss an Vakuum-Röhrenkollektor

Quelle: archplus 184, Oktober 2007

transformed into cooling energy by means of absorption chillers. The first step toward this goal is the development of an optically transparent flat-collector plate, which will allow for visual connection from the inside to the outside and create an interesting play of light and shadow.

Figure 208 shows the sample façade with the detail of the plumbing connections linking the solar collectors. **Figure 209** shows the detail of evacuated solar tubing, where the necessary plumbing is integrated into the façade post.

Currently, solutions exist that integrate flat-plate solar collectors into the envelope of a building. However, the complex issue of integration of evacuated solar tubing is still partly unsolved, although this would, without question, be of great aesthetic appeal when integrated into the skin of buildings. Vacuum solar tubing reaches a higher operating temperature than flat-plate collectors, and, consequently, it has a 35 % greater efficiency due to the fact that the absorber elements embedded into the tubes can be oriented to the most effective position.

Bild 208 zeigt eine Musterfassade mit einem Detail der Verrohrung der Solarkollektoren. **Bild 209** zeigt Röhrenkollektoren in Fassaden mit Anschlussdetail (Rohrsystem in Pfosten integriert).

Während für Flachkollektoren bereits Systeme bestehen, die eine Integration in die Gebäudehülle ermöglichen, ist der Einsatz von Vakuum-Röhrenkollektoren bisher nur bedingt integrativ gelöst, obwohl sie aufgrund ihrer ästhetischen Struktur ein großes Potenzial im Fassadenbereich anbieten würden. Dabei von besonderer Bedeutung ist, dass Vakuum-Röhrenkollektoren ein höheres Temperaturniveau erreichen als Flachkollektoren und somit einen um ca. 35 Prozent höheren Energieertrag, da die integrierten Absorberelemente in den Röhren ideal ausgerichtet werden können.

12.1 Building integrated wind power systems (BIPW)

The integration of wind turbines into buildings is a special topic. Researchers around Stefan Behling at the University of Stuttgart analyzed various building configurations and formal designs of buildings in which wind turbines for the generation of electric energy are integrated. **Figure 210** gives an overview of the various models, and **Figure 211** shows specifically a twin-tower arrangement with a boomerang-like footprint. Experiments have shown that if the rotors are placed in the jet-like building configuration, the air velocity at the rotor is increased by 1 m/s; the turbine therefore generates a slightly greater amount of energy than a free-standing system.

As documented by Behling's research, aerodynamic systems in buildings represent neither spatial nor energetic optimums. Near the rotors, negative acoustic, vibration, and electromagnetic disturbances develop that are undesirable in a building.

In summary, it can be said that evolving integrations of solar thermal electric systems into the façades of buildings will result in a much greater energy harvest than the inclusion of wind-power systems near or in buildings

12.1 Gebäudeintegrierte Windkraftanlagen (BIWP, building integrated wind power)

Die Integration von Windturbinen in bauliche Strukturen beinhaltet ein besonderes Thema. Die Forschergruppe um Stefan Behling hat mit weiteren Fachleuten verschiedenste Ausdrucksformen und Gebäudestrukturen untersucht, die über Windkraftwerke elektrische Energie erzeugen sollen. **Bild 210** zeigt verschiedenste Modelle, **Bild 211** einen Zwillingsturm mit „Bumerang-Grundriss". Wurden die Rotoren innerhalb der düsenförmigen Gebäudekonfiguration platziert, so erhöhte sich die Windgeschwindigkeit am Rotor um ca. 1 m/s und erzeugte hierdurch etwas mehr Energie als eine freistehende Anlage.

Wie die Forschergruppe herausfand, sind aerodynamische Konstruktionen sowohl räumlich als auch energetisch nicht optimal. In der Nähe der Rotoren treten akustische, schwingungsbedingte oder elektromagnetische Beeinträchtigungen auf.

Zusammenfassend lässt sich feststellen, dass die Entwicklung fassadenintegrierter, energieschöpfender Elemente zu deutlich höheren Erträgen führen dürfte als der Einsatz von Windkraftanlagen an und im Gebäude.

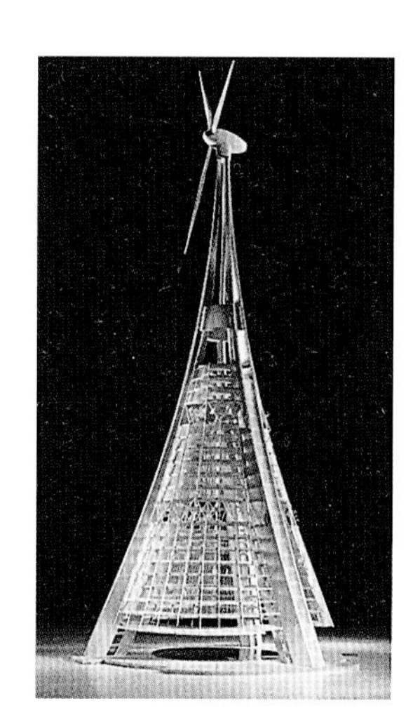

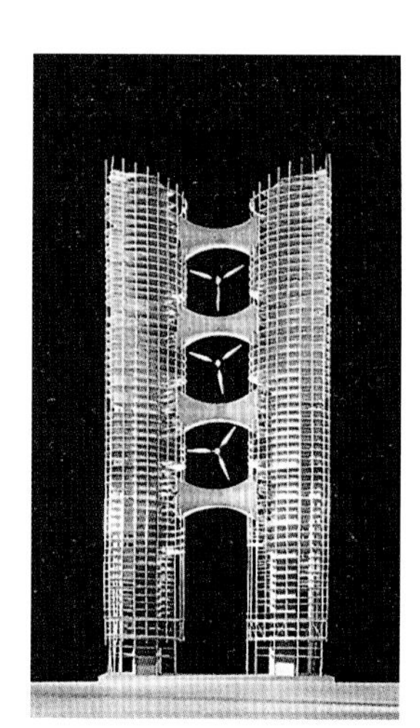

Figure 210
Building integrated wind power generation (BIWP)

The research project "Web-Integrated Wind Turbines in the Built Environment" was initiated by IBK2, Insitute of Stefan Behling, Universität Stuttgart. The project leader was Jörg Hieber in cooperation with BDSP Partnership Ltd. London (UK) and the Imperial College, Department of Aeronautics, London (UK). The research was funded by the European Union's JOULE III program.

Various building configurations were tested as scaled models and in CFD simulations in regards to their aerodynamic behavior. The most optimal flow velocities could be achieved by building footprints with smooth, aerodynamical "free forms", such as kidney or boomerang shapes.

Venturi wings or other connector elements were introduced to reduce turbulent airflow and the related loss of flow velocity.

The achievable output is equal to an increase in wind speed, other than the theoretical assumption of an increase in the power of 2. The highest achievable power amplification remains smaller than factor 2.

Source: Stefan Behling, Universität Stuttgart, Germany

Bild 210
Gebäude ntegrierte Windkraft-Anlagen

Das Forschungsprojekt "Web-Integration von Windturbinen in die gebaute Umwelt" entstand am IBK 2, Insitut Stefan Behling, der Universität Stuttgart (Projektleitung: Jörg Hieber) in Zusammenarbeit der Projektpartner BDSP Partnership Ltd, London (UK), MECAL Applied Mechanics BV, Enschede (NL), dem Imperial College, Dept. of Aeronautics, London (UK), und wurde durch die EU im JOULE III Programm unterstützt.

Verschiedene Gebäudeformen wurden auf ihre aerodynamischen Eigenschaften in Windkanalversuchen an maßstäblichen Modellen und in CFD-Simulationen untersucht. Optimale Durchströmgeschwindigkeiten wurden mit glatten, aerodynamischen Freiformen erzielt, insbesondere bei Türmen mit nieren- oder bumerangförmigen Grundrissen.

Venturiflügel oder andere Verbindungselemente verhindern zusätzlich Verwirbelungen und damit Verlust an Durchströmgeschwindigkeit. Die erzielbare Leistung ist proportional zur Vergrößerung der Windgeschwindigkeit (entgegen des theoretischen Wertes einer Steigerung in der 2. Potenz). Die maximale Kraftverstärkung, die direkt erreicht werden kann, bleibt kleiner als Faktor 2.

Stefan Behling
Universität Stuttgart

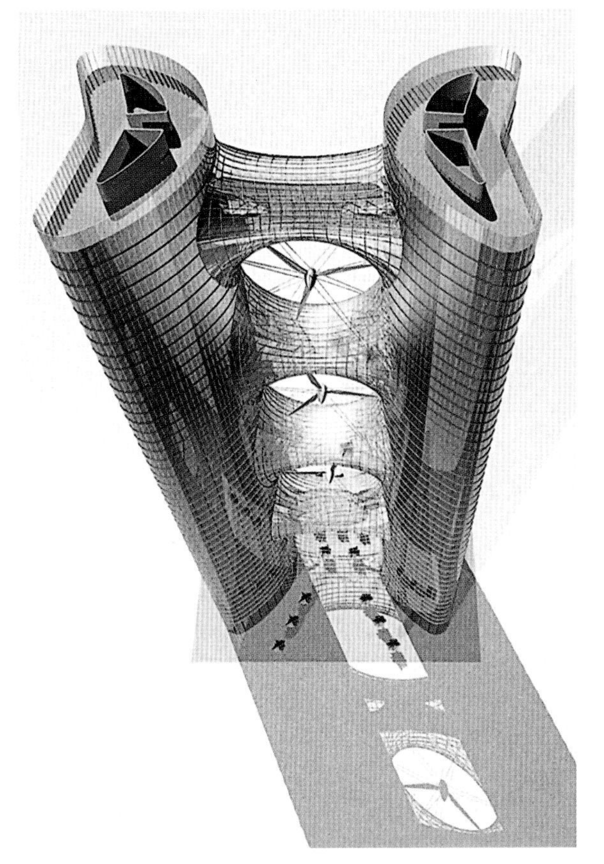

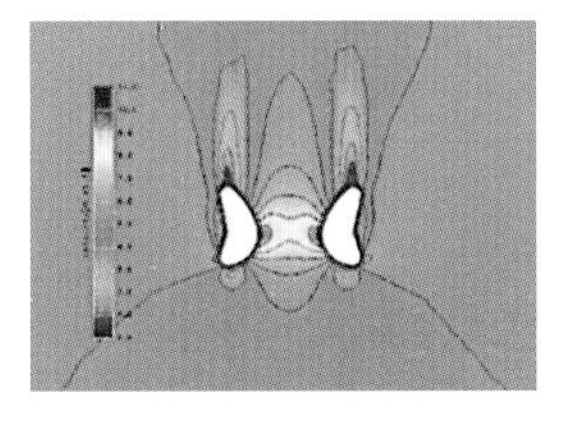

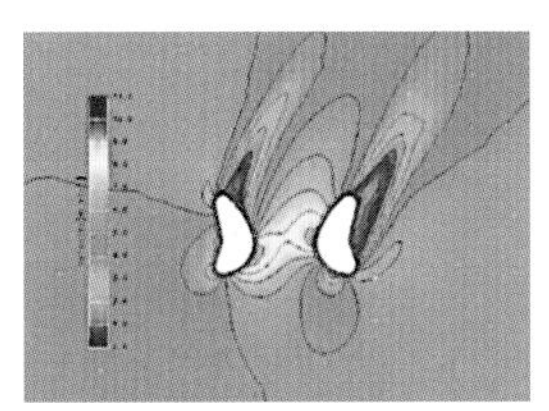

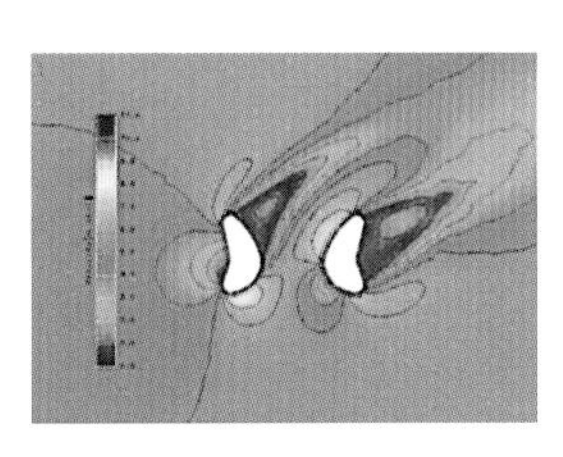

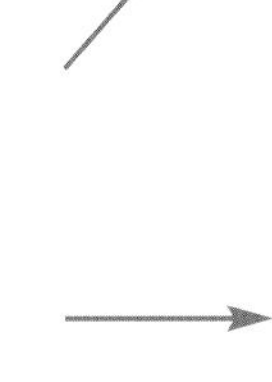

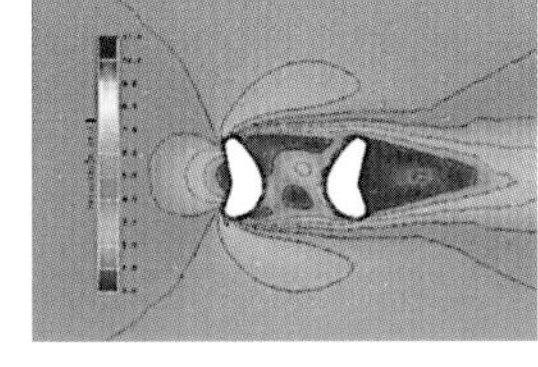

Red = Positive pressure
Blue = Negative pressure

Wind direction

Figure 211
Twin towers with boomerang footprint and building integrated wind power generation (BIPW)

Simulation of negative and positive pressure zones as a function of varying angles of flow

Source: Stefan Behling, Universität Stuttgart, Germany

Bild 211
Zwillingstürme mit "Bumerang" – Grundriss und integrierten Windrädern
Simulation von Über- und Unterdruckfedern bei Windanströmung aus verschiedenen Richtungen

Quelle: Stefan Behling
Universität Stuttgart

12.2 Pearl River Tower, Guangzhou, China

Figure 212 shows the 71-story high-rise of the Guangdong Tobacco Company, designed by Adrian D. Smith, a former partner at Skidmore, Owings Merrill (SOM). According to the firm, the tower will be equipped with the newest environmen-tal technologies, and it is being designed to be the "most energy-efficient tall building in the world".

The building, which will be completed in 2010, will have a total gross floor area of approximately 230,000 m² and a height of 310 m. The goal of the designers, who won an international design competition, is to generate partly more energy within the building than it actually consumes. At each third of its height, the tower shows air-intake openings behind which wind turbines are located (vertical axis-wind turbines), generating electric power. Large expanses of glass in the building result in effective daylight utilization, while the horizontal solar shading louvers placed inside the façade corridor of the building's double-skin envelope help to reduce the significant thermal gains. The double-skin façade reduces the solar gains with the help of an effective solar shading systems made of Venetian blinds; it is planned that rainwater that reaches the surface of the building envelope will be collected and treated for future use as grey water. Photovoltaic elements, on the south-, east- and west-facing façades contribute partly to electrical energy provisions for the generation of warm water.

It was planned that a large portion of the tower's energy would be provided micro-turbines supplied by natural gas biogas, kerosine etc. They generate thermal energy (also usable for cooling purposes) as well as electrical energy with a coefficient of performance of around 85 %. In order to reduce the energy required for cooling the office floors, the plan is to use displacement ventilation, which allows for small air changes per hour and small air volumes. The necessary energy to cool the office floors will be further lowered by the use of hang seilling panels, integrated into radiant, chilled, ceilling panels. The originally conceived "concrete core activation" (cooled concrete ceiling slabs) had to be abandoned due to budget concerns. The soil around the site was supposed to be used as a heat sink to accept the condenser heat of the building's chiller plants, but this concept had to be altered due to the high ground water temperatures of the site. Return air from the office floors in the building is conducted back to the small air handling units where some energy is recaptured in heat exchangers. During the exchange between the cooler, dryer air of the inside of the building and the warmer and moister air of the outside, the water vapor of the air stream condenses, and this condensate is used in the building.

12.2 Pearl River Tower, Guangzhou, China

Bild 212 zeigt ein 71-geschossiges Hochhaus, das vom Architekturbüro SOM, Chicago (Skidmore, Owings Merrill) geplant wurde bzw. zurzeit geplant wird. Der Turm soll das Hauptquartier der Guangdong Tobacco Company aufnehmen. Vorstellungen der Planer im Hause SOM, Chicago, ist es, den Turm mit neusten Technologien zur Nutzung erneuerbarer Energien auszurüsten. Hierdurch soll die „energieeffizienteste Hochhausstruktur der Welt" entstehen.

Das im Jahr 2010 fertigzustellende Gebäude besitzt eine Bruttogeschossfläche von ca. 230.000 m² und eine Gebäudehöhe von ca. 310 m. Die Zielvorstellungen von SOM auf der Basis eines gewonnenen internationalen Wettbewerbs sind, dass das Gebäude teilweise mehr Energie erzeugt als es verbraucht. In jeweils einem Drittel der Fassadenhöhe sind Lufteinlässe gestaltet, hinter denen sich Windkraftwerke (Vertikal-Rotoren) verbergen, die elektrische Energie erzeugen. Großflächige Verglasungen führen zu einer hohen Tagesbelichtung, wobei eine doppelschalige Glasstruktur die Solargewinne durch hochwertigen Sonnenschutz reduziert. Regenwasser, das auf die Fassade fällt, soll gesammelt und zu Grauwasser aufbereitet werden, um es nutzen zu können. Zur Erzeugung von elektrischer Energie werden partiell in die Süd, Ost- und Westfassade Photovoltaikelemente integriert, die einen Anteil der elektrischen Energie bereitstellen.

Ein wesentliches Element der Energiebereitstellung sollte durch Mikroturbinen erfolgen, die mit Ferngas, Biogas, Kerosin etc. betrieben werden. Diese können sowohl Wärmeenergie (nutzbar auch als Kälteenergie) und elektrische Energie, mit einem Wirkungsgrad von um mehr als 85 Prozent, erzeugen. Im Bereich der Gebäudetechnik ist vorgesehen, die wesentlichen Flächen durch eine Quelllüftung zu durchlüften, um mit möglichst kleinen Luftmengen und Luftwechselzahlen auszukommen. Zur Erzeugung ausreichender Kühlenergie in den Räumen sind Kühldecken geplant, und das Erdreich soll als Wärmesenke genutzt werden. Die Abluftströme des Hauses werden rückgeführt, um einen Teil der Kühlenergie rückgewinnen zu können. Durch den Wärmetausch zwischen kühler und trockner Abluft der Räume im Gegenzug zu feucht-warmer Luft von außen (kreislaufverbundene Systeme) fällt Schwitzwasser aus der Außenluft aus, das im Gebäude genutzt wird.

12.2 Pearl River Tower, Guangzhou, China

Figure 212
Pearl River Tower
Guangzhou, China

Architects: Skidmore, Owings & Merrill LLP (SOM), Adrian Smith, Consulting Design Partner, Chicago

Client: CNTC Guangdong Company, China

Bild 212
Pearl River Tower, Guangzhou, China

Architekten: Skidmore, Owings & Merrill LLP (SOM), Adrian Smith, Consulting Design Partner, Chicago

Bauherr: CNTC Fa. Guangdong, China

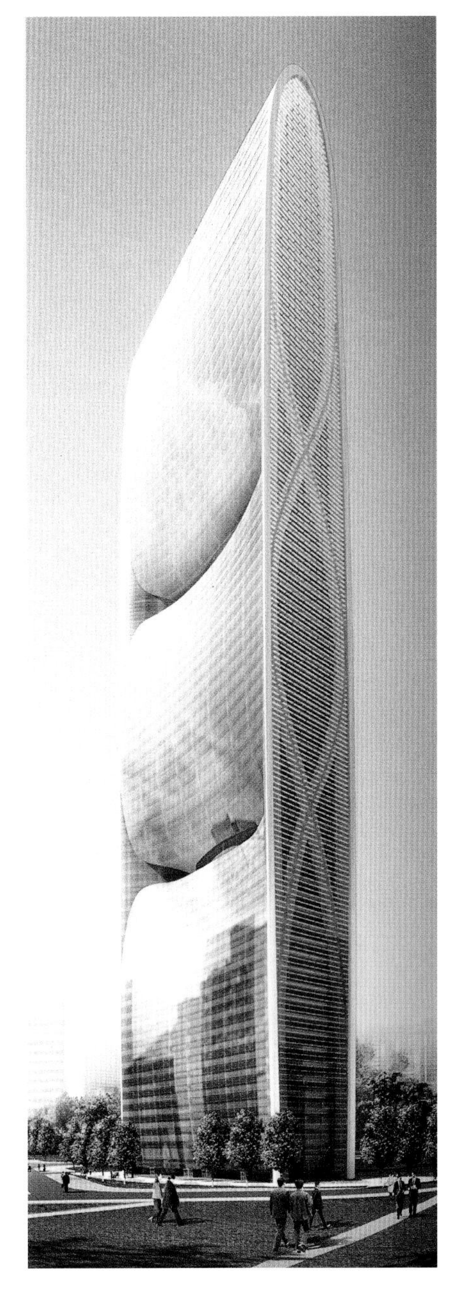

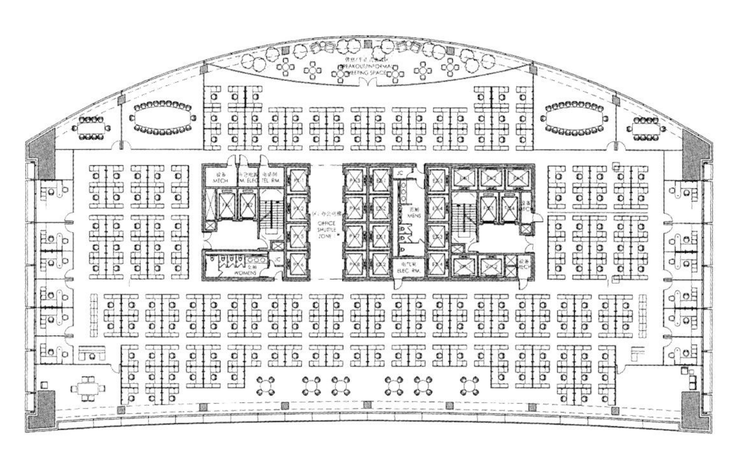

to Figure 212
Typical floor plan

zu Bild 212
Standard-Grundriss

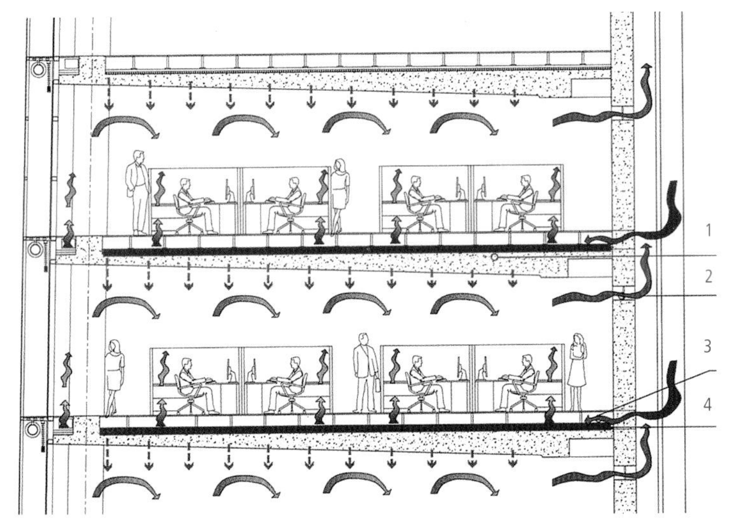

to Figure 212
Cross section and diagram of ventilation concept

1 Concrete cooling radiant slab
2 Air return
3 Air supply
4 Chilled water supply

zu Bild 212
Gebäudeschnitt und Lüftungssystem

to Figure 212
Section

zu Bild 212
Schnitt

Figures 213 and **214** show a section, the principle of air flow direction towards a wind funnel opening, and the principal working of the wind turbines. To conduct a preliminary evaluation of whether and to what degree the energy requirements of the tall building can be supplied by renewable sources, we had to conduct a detailed analysis of the local climatic conditions. **Figure 215** shows the temperature curve and enthalpy of the outside air at the location, Guangzhou. Throughout the year, a high relative humidity exists, as seen in **Figure 216**. In **Figure 217**, we can see that the annual ambient temperature range is clearly above the freezing level and that very often temperatures are in the range of 18 – 30 °C. Solar gains, as seen in **Figure 218**, reach a theoretical total amount of 1,391 kWh/m², and thus they are 30 % greater than in Central Europe. The degree to which these solar gains can be converted into energy may be approximately in the range of 70 – 80 kWh/m² of façade surface.

Die **Bilder 213** und **214** zeigen neben den Prinzipien der Luftzuführung zu den Windturbinen einen Schnitt des Gebäudes sowie den Einlauftrichter in der Fassade (Teilansicht). Um eine erste grobe Analyse vornehmen zu können, ob und in welchem Umfang die Energieverbräuche des Hauses durch erneuerbare Energien zu erzeugen sind, ist eine Studie der äußeren Randbedingungen am Standort notwendig. **Bild 215** zeigt den Verlauf von Temperaturen und den Verlauf der Enthalpie der Außenluft in Guangzhou. **Bild 216** weist aus, dass am Standort Guangzhou während des gesamten Jahres relativ hohe relative Feuchten bestehen. Gemäß **Bild 217** liegen die Außentemperaturen im Regelfall deutlich über dem Gefrierpunkt und sehr häufig zwischen 18 – 30 °C. Der Solareintrag, **Bild 218**, mit einem theoretischen Gesamtwert während des Jahres von 1.391 kWh/m² ist um ca. 30 Prozent höher als in Mitteleuropa und dürfte eine Größenordnung nutzbarer elektrischer Energie von ca. 70 – 80 kWh/m² Fassadenfläche erreichen..

Form

Site

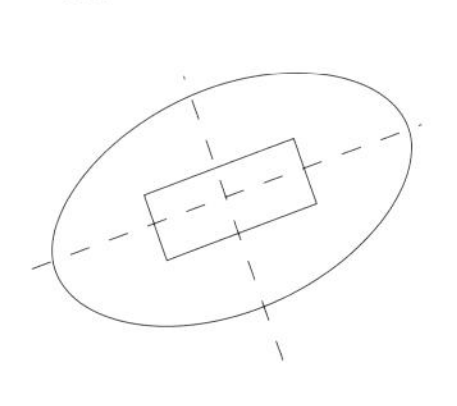

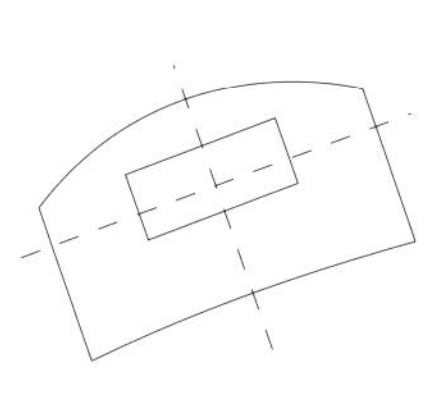

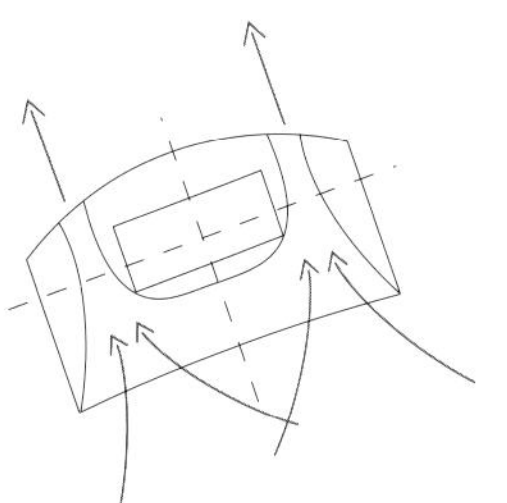

Figure 213
Concept of building-integrated wind power (BIWP) and its influence on building footprint developement

Pearl River Tower, Guangzhou, China

Source: SOM, Chicago

Bild 213
Konzeptentwicklung zu gebäudeintegrierten Windturbinen, Entwicklung der Grundrissstruktur zum Einbau von Windturbinen
Pearl River Tower, Guangzhou, China

Quelle: SOM, Chicago

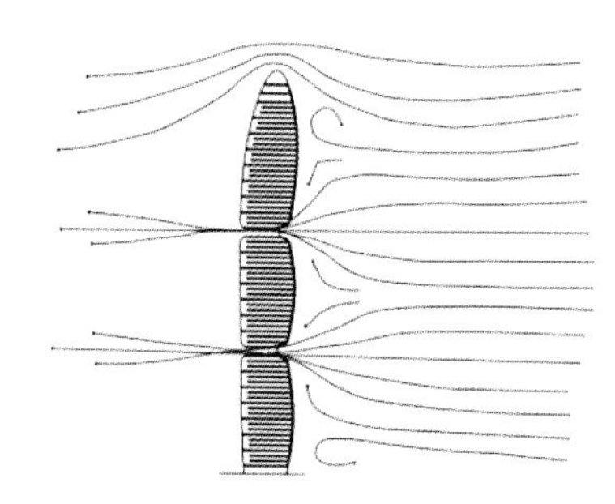

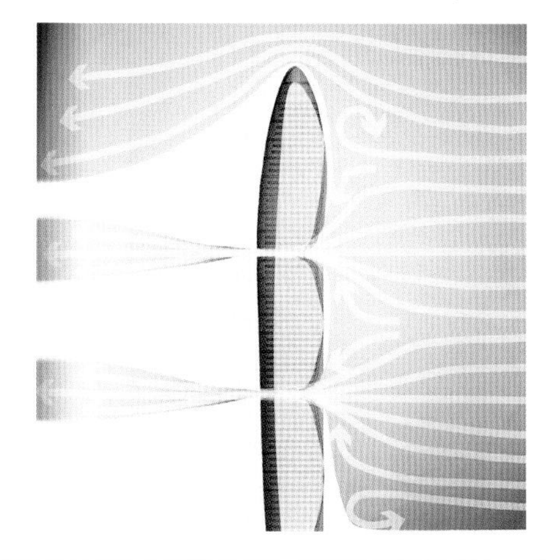

Figure 214
Idealized diagram of flow conditions through building integrated wind turbines

Source: SOM, Chicago

Bild 214
Idealisierte Darstellung der Winddurchströmung durch das Gebäude – durch die Windturbinen

Quelle: SOM, Chicago

12.2 Pearl River Tower, Guangzhou, China

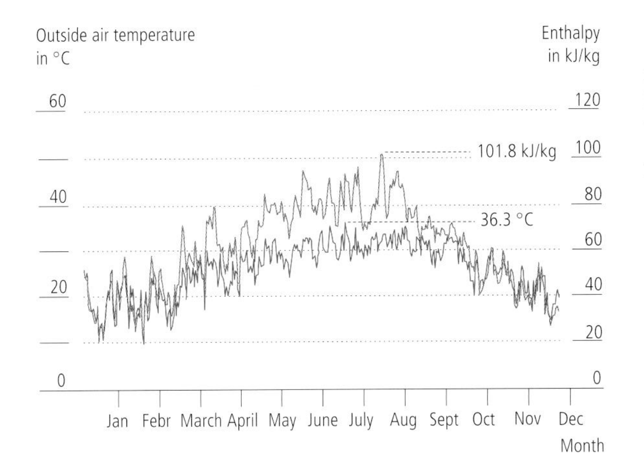

Figure 215
Daily outside air temperature and enthalpy maxima, Guangzhou, China

Source: Meteonorm 5.0 meterological data

Bild 215
Tägliche Maximalwerte der Aussenluft: Temperatur und Enthalpie, Guangzhou, China

Quelle: Meterological data – Meteonorm 5.0

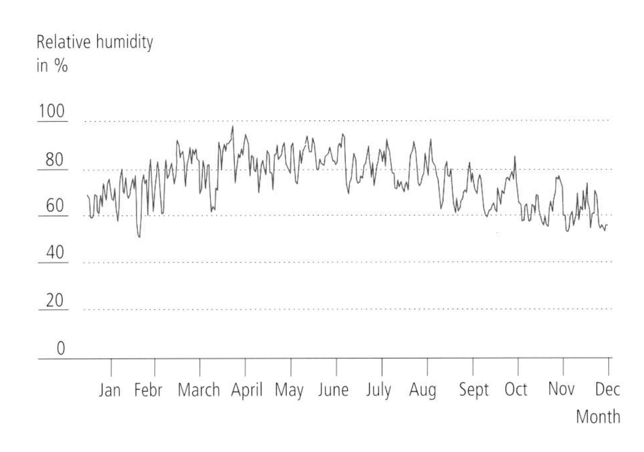

Figure 216
Daily relative humidity average Guangzhou, China

Source: Meteonorm 5.0 meterological data

Bild 216
Täglicher Durchschnittswert der relativen Feuchtigkeit, Guangzhou, China

Quelle: Meterological data – Meteonorm 5.0

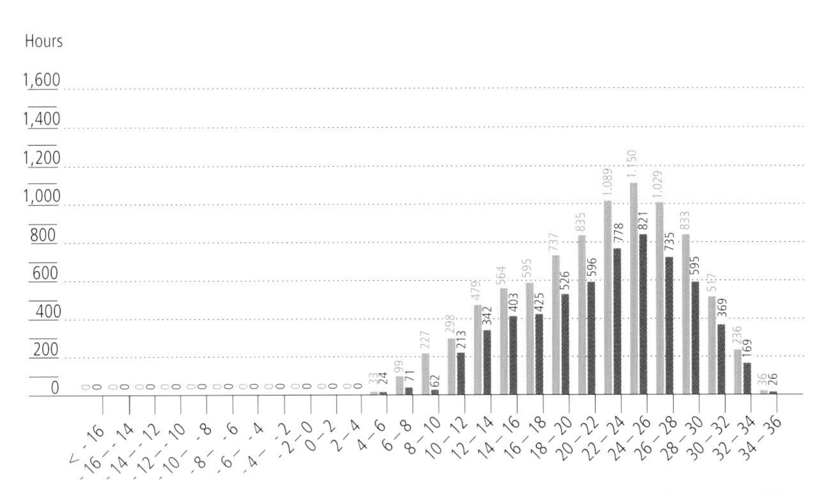

Figure 217
Outside temperature statistic for a 24 hour period January – December Guangzhou, China

Source: Meteonorm 5.0 meterological data

åBild 217
Aussentemperatur Statistik (24-Stunden-Messung, Januar – Dezember) Guangzhou, China

Quelle: Meterological data – Meteonorm 5.0

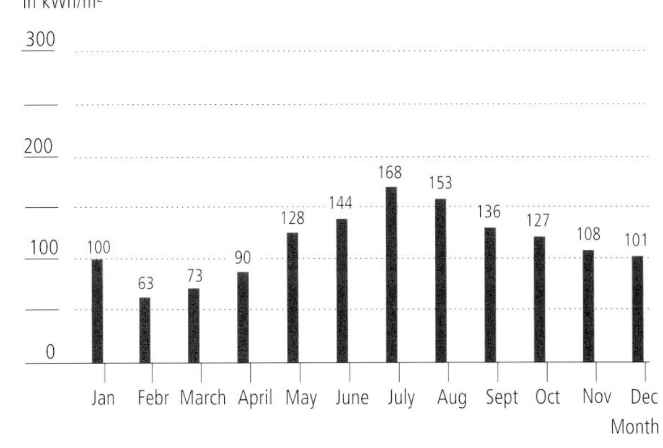

Figure 218
Global solar radiation, Guangzhou, China

Source: Meteonorm 5.0 meterological data

Bild 218
Globalstrahlung, Guangzhou, China

Quelle: Meterological data – Meteonorm 5.0

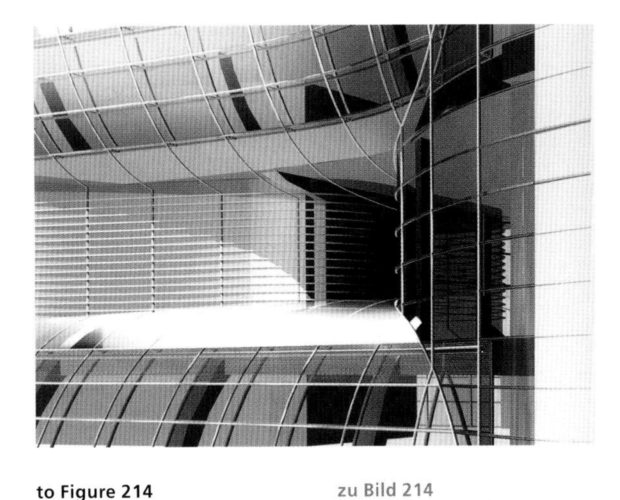

to Figure 214
Close-up of funnel-shaped facade cone leading to the wind turbine.

Source: SOM, Chicago

zu Bild 214
Detail eines Einlauftrichters in der Fassade

Quelle: SOM, Chicago

Figure 219 shows the amount of annual rainfall in Guangzhou that can be collected.

In order to evaluate the concept of wind turbine usage, the wind speeds and directions on the site need to be researched. Wind speeds shown for Guangzhou in **Figure 220** are based on a height above ground of 10 m, and wind speeds increase with increased height, as shown in **Figure 221**. The diagram shows conditions for a city-center location. As seen in the diagram, an average wind speed of approximately 8.9 m/s can be assumed at a height of 100 m, versus 3.5 m/s on the ground. At a height of 200 m, 10.9 m/s can be observed.

The diagram in **Figure 222** shows the frequency at which certain wind speeds can be observed on the ground plane. In this respect, it is important to realize that wind power plants generate noticeable power only above a level of 4 m/s and that typically their rated power output is only reached at wind speeds between 12 and 15 m/s.

Figure 223 shows the annual wind distribution diagram (wind rose) for Guangzhou. It is evident that prevailing wind directions are from north-east to south-south-west. The planned building therefore needs to be situated on the site on an east-west axis, in order to present large south- and north-facing façades to the wind.

Bild 219 weist die Regenmengen für den Standort Guangzhou aus, die zum Teil über die Fassaden nach unten fallen und hier gesammelt und aufbereitet werden können

Bezüglich der Erzeugung elektrischer Energie per Windkraftanlagen ist eine Studie der Windgeschwindigkeiten während des Jahres notwendig. Die in **Bild 220** angegebenen Windgeschwindigkeiten beziehen sich auf eine Höhe von ca. 10 m. Mit zunehmender Höhe (Lage der vorgesehenen Windräder) nimmt die Windgeschwindigkeit, gemäß **Bild 221**, zu (Stadtzentren). Insofern ist in einer Höhe von ca. 100 m anstatt einer mittleren Windgeschwindigkeit von ca. 3,5 m/s im Bodenbereich eine solche von 8,9 m/s feststellbar und in der Höhe von ca. 200 m eine solche von 10,9 m/s.

Bild 222 zeigt die Häufigkeit bestimmter Windgeschwindigkeiten, wiederum bezogen auf Bodennähe. Hierbei ist beachtenswert, dass Windkraftanlagen eine nennenswerte Leistung ab 4 m/s abgeben und im Regelfall ihre Nennleistung erst bei ca. 12 – 15 m/s.

Bild 223 zeigt die Windrose für den Standort Guangzhou während des gesamten Jahres. Die Windrose stellt dar, dass die Haupt-Windanströmrichtungen im Bereich von Nord-Ost bis Süd-Süd-West liegen. Das Gebäude muss sich demgemäß in seiner Längsachse von Ost nach West ausdehnen, so dass große Nord- und Südfassaden entstehen.

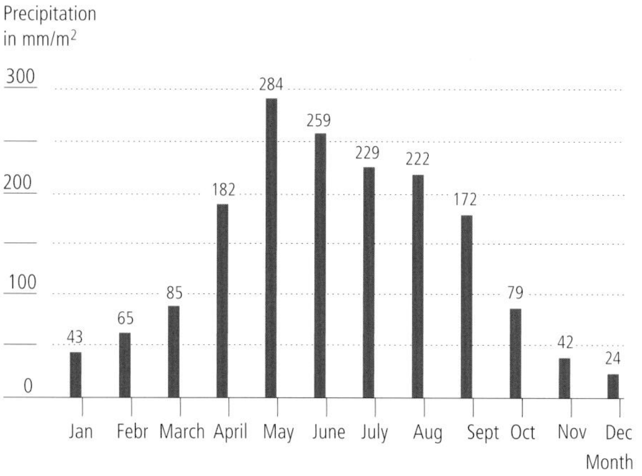

Precipitation
in mm/m²
300
200
100
0
43
65
85
182
284
259
229
222
172
79
42
24
Jan
Febr
March
April
May
June
July
Aug
Sept
Oct
Nov
Dec
Month

Figure 219
Precipitation
Guangzhou China

Source: Meteonorm 5.0
meterological data

Bild 219
Niederschlag
Guangzhou, China

Quelle: Meterological data – Meteonorm 5.0

Daily average wind speed
in m/s
10
8
6
4
2
0
Jan
Febr
March
April
May
June
July
Aug
Sept
Oct
Nov
Dec
Month

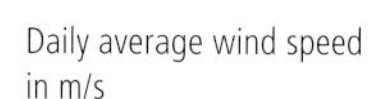

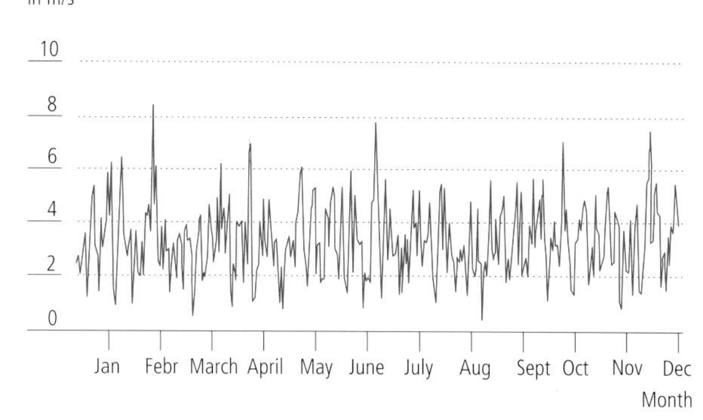

Figure 220
Daily average wind speed
at a height of 10 meters
Guangzhou, China

Source: Meteonorm 5.0
meterological data

Bild 220
Täglicher Durchschnittswert
der Windgeschwindigkeit
in ca. 10 m Höhe
Guangzhou, China

Quelle: Meterological data – Meteonorm 5.0

12.2 Pearl River Tower, Guangzhou, China

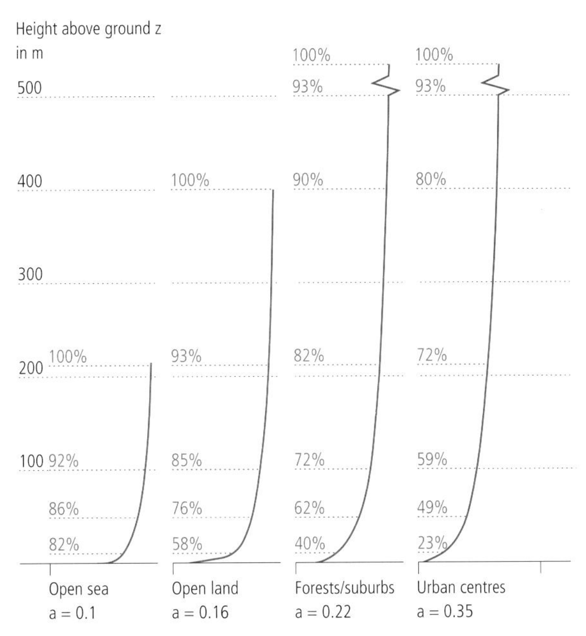

Figure 221
Wind velocity increases proportionately with height above water or flat ground. Above built environments, velocities decrease because the boundary layer begins at a greater height.

$w(z) = w(z_{ref})\ (z/z_{ref})^{\alpha}$

$w(z)$ wind velocity
$w(z_{ref})$ gradient velocity $= 100\ \%$
z_{ref} reference height
α profile exponent

Source:
plusminus 20°/40° latitude
Hindrichs · Daniels
Edition Axel Menges

Bild 221
Windgeschwndigkeitsprofile in Abhängigkeit der Bodenrauhigkeit und Höhe

$w(z) = w(z_{ref})\ (z/z_{ref})^{\alpha}$

$w(z)$ Windgeschwindigkeit
$w(z_{ref})$ Gradietengeschwindigkeit $= 100\ \%$
z_{ref} Referenzhöhe
α Profilexponent

Quelle:
plusminus 20°/40° latitude
Hindrichs · Daniels
Edition Axel Menges

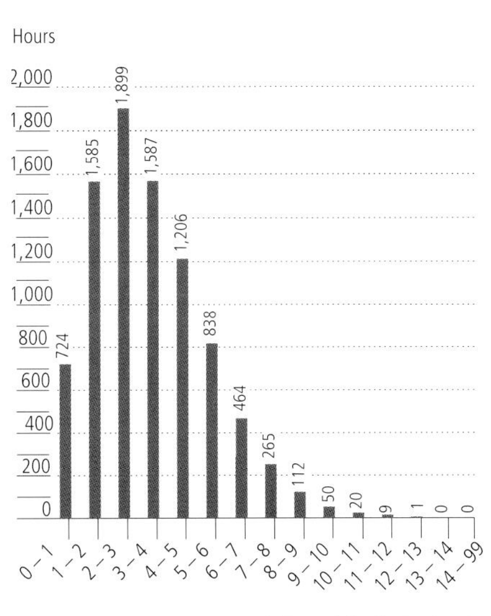

Figure 222
Wind speed distribution near ground
Guangzhou, China

Source: Meteorological data
Meteonorm 5.0

Bild 222
Verteilung der Windgeschwindigkeit in Bodennähe
Guangzhou, China

Quelle: Meteorological data – Meteonorm 5.0

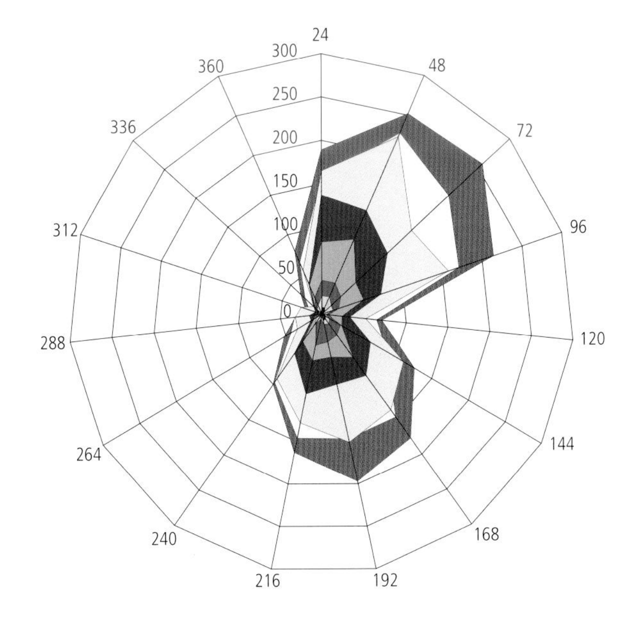

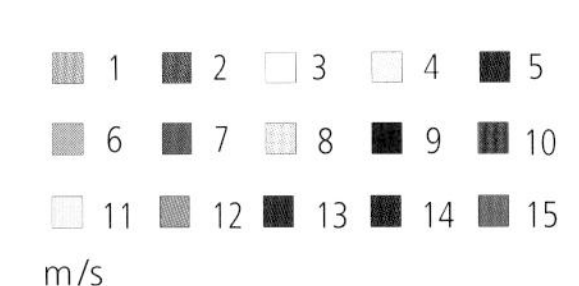

Figure 223
Wind rose diagram, year
Guangzhou, China

Wind speed and frequency distribution of wind direction in hours/annum

Source: Meteorological data
Meteonorm 5.0

Bild 223
Windrose über das gesamte Jahr, Goangzhou, China

Windgeschwindigkeit in m/s und Häufigkeit der Windrichtungsverteilung (h/a)

Quelle: Meteorological data – Meteonorm 5.0

To allow for a preliminary calculation of the nominal wind energy generation of various types of wind turbine designs for a wind speed between 12 and 15 m/s as a function of rotor diameter, the diagram in **Figure 224** is used. As shown, the resulting energy generated by the vertical rotors and the planned diameter of the façade aperture of 6 m will result in only a small amount of energy being generated. Furthermore, wind will always seek the "path of least resistance" – a handicap particularly for flat buildings – and flow off to the building's shorter sides (**Figure 225**). To counter-balance this aerodynamic effect, the designers have deformed the surface of the tower in such a way that a quasi "diffuser" is formed that guides the air to the rotors.

Zur ersten groben Kalkulation ist in **Bild 224** der Verlauf von Nennleistungen bei Windgeschwindigkeiten zwischen 12 – 15 m/s verschiedener Windkraftanlagen in Abhängigkeit der Rotordurchmesser dargestellt. Da voraussichtlich die Windräder lediglich eine Höhe (Vertikalrotoren) bzw. einen Durchmesser um ca. 6 – 8 m besitzen, ist selbst bei günstigen Windgeschwindigkeiten die Nennleistung je Rotor noch gering. Ein weiteres Handicap bei flachen Gebäuden ist, dass selbst bei strömungsbedingten Ausformungen der Fassaden die Luft üblicherweise den Weg des geringsten Widerstandes geht, d.h. seitlich über die Fassaden abfließt, **Bild 225**. Um diesem entgegenzuwirken, haben die Planer im Hause SOM die Fassaden mehrfach gewölbt, um hierdurch quasi einen Diffusoreinlass zu bilden, d.h. die von außen anströmende Luft gezielt auf die Windräder zu leiten.

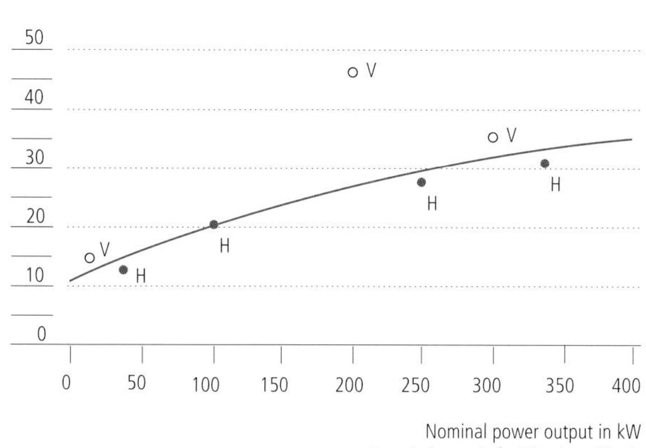

H Horizontal axis wind turbine
V Vertical axis wind turbine

Figure 224
Approximate nominal power output of wind turbine units as a function of rotor diameter at wind speeds of 12 – 15 m/s (air density = 1,2 kg/m³)

Source: Daniels/author

Bild 224
Ca. Nennleistungen von Windkraftanlagen in Abhängigkeit der Rotorendurchmesser bei Windgeschwindigkeiten von 12 – 15 m/s (Luftdichte r = 1,2 kg/m³)

Quelle: Daniels/Autor

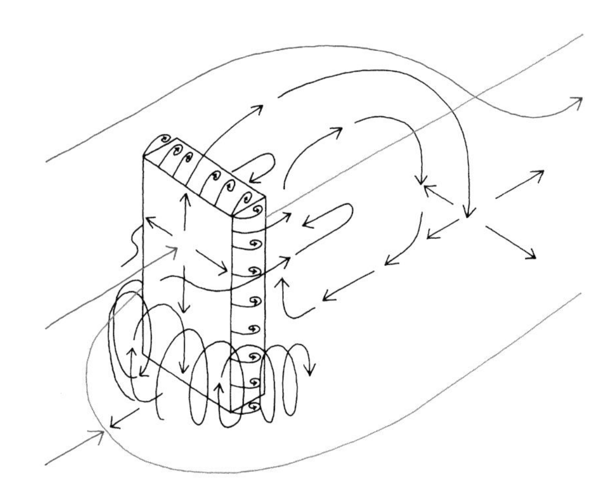

Figure 225
Turbulent flow near building (Turbulent air flow field near building has very complex structures which are difficult to calculate).

Source:
plusminus 20 °/40 ° latitude
Hindrichs/Daniels
Edition Axel Menges

Bild 225
Turbulentes Strömungsfeld an einem Gebäude (Gebäudenahe turbulente Strömungen haben sehr komplexe Strukturen und sind schwer zu berechnen).

Quelle:
plusminus 20 °/40 ° latitude
Hindrichs/Daniels
Edition Axel Menges

Based on observation of all of the relevant conditions documented above and the assumed typical energy usage, the resulting energy budget is as follows:

- Heat energy required
  warm/hot climate regions approx. 0 W/m² gross floor area (GFA)
  moderate climate regions approx. 20 W/m² GFA
  cold climate regions approx. 50 W/m² GFA
- cooling energy required, exterior floor plate zones
  cold climate regions approx. 0 W/m2 GFA
  moderate climate regions approx. 50 W/m² GFA
  warm/hot climate regions approx. 65 W/m² GFA
- cooling energy required, interior floor plate zones
  all climate regions approx. 35 W/m² GFA
- electrical power requirements
  low building standard approx. 15 W/m² GFA
  medium standard approx. 25 W/m² GFA
  high standard up to approx. 50 W/m² GFA

The power data shown are for the internal energy use of the building, and the following annual energy consumption data show the required energy.It can be assumed that the energy requirement for heating of the tower is no greater than 10 kWh/m²a. The absolute amount of heat energy required in the building, mainly for the months of December and January, is therefore approximately 2,300 kW_P, and the annual heat energy required is 920 MWh/a. The specific cooling loads in the exterior floor plate zones of the building, near the building envelope, are approximately 36 W/m², and those in the internal zone are around 24 W/m² GFA. It can be assumed that the weighted average of the cooling loads will be approximately 32 W/m² GFA, that the absolute cooling load will be around 7,360 kW_P for the period between May and October, and that the total annual cooling energy consumption will be 21,190 MWh/a.

If we assume that the cooling energy requirement is mostly provided by the excess heat of the micro-turbine systems (total 50 micro turbines/3 MW electrical power), which is used by absorption chillers, the resulting required electrical energy demand will be around 8,000 kW_{PE} and the annual electrical energy demand will be approximately 12,400 MWh/a.

The solar energy harvested by the photovoltaic panels on the east- and west-facing façades of the building is converted into electrical energy, resulting in a gain of 420 MWh/a if those PV panels cover 50 % of the building's façades, or twice as much if they fully cover the east and west façades.

Energy as a result of wind power can be calculated as 13 kW at an average wind speed of 9.9 m/s (averaged over the height of the entire building). If this is weighted by around 3,330 hours of available wind, the result will be an annual gain of approximately 173 MWh/a. The total amount of energy generation from solar and wind energy is therefore around 600 MWh/a. All data shown are related to the gross floor area.

Unter Berücksichtigung aller genannten Faktoren und der üblichen Energieverbräuche, wie nachfolgend dargestellt, ergibt sich eine erste grobe Bilanz:

- Wärmeendenergiebedarf
  warme Regionen ca. 0 W/m² BGF
  gemäßigte Regionen ca. 20 W/m² BGF
  kalte Regionen ca. 50 W/m² BGF
- Kühlenergiebedarf Außenzonen
  kalte Regionen ca. 0 W/m² BGF
  gemäßigte Regionen ca. 50 W/m² BGF
  heiße Regionen ca. 65 W/m² BGF
- Kühllast Innenzonen
  alle Regionen ca. 35 W/m² BGF
- Elektrischer Endenergiebedarf
  geringer Ausbaustandard ca. 15 W/m² BGF
  gehobener Ausbaustandard ca. 25 W/m² BGF
  hoher Ausbaustandard bis 50 W/m² BGF

Alle aufgeführten Leistungsdaten entsprechen dem Endenergieverbrauch innerhalb des Gebäudes, alle nachfolgenden Energiemengen per annum zeigen in etwa die Endenergieverbräuche.

Der Wärmeenergiebedarf des Turms dürfte in etwa 10 kWh/m²a betragen. Die absolute Heizleistung des Gebäudes beträgt somit in etwa 2.300 kW_P (Dezember/Januar) und der Jahres-Heizwärmebedarf ca. 920 MWh/a. Die spezifischen Kühllasten der Außenzone liegen bei ca. 36 W/m², die der Innenzone bei ca. 24 W/m². Ein gewichtetes Mittel über die Gesamtfläche dürfte bei ca. 32 W/m² BGF liegen, die absolute Kälteleistung beträgt ca. 7.360 kW_P (Mai bis Oktober) und der Jahres-Kälteenergiebedarf ca. 21.190 MWh/a.

Geht man davon aus, dass die Kälteenergie praktisch ausschließlich aus der Abwärme der Mikroturbinen (50 Turbinen/3 MW_{el}) über Absorptionskältemaschinen erzeugt würde, so ergäbe sich ein gleichzeitiger elektrischer Energiebedarf von grob 8.000 kW_{PE} und ein Jahres-Elektro-Energiebedarf von ca. 12.400 MWh/a.

Der Energieertrag des Hauses errechnet sich bei 50-prozentiger Belegung der Ost- und Westfassade mit Photovoltaikelementen mit rund 420 MWh/a und bei vollflächiger Belegung mit ca. dem doppelten Wert.

Die Windgeneratoren mit je einer Leistung von ca. 13 kW (Nennleistung bei ca. 9,9 m/s mittlerer Anströmgeschwindigkeit über alle Geschosse) beträgt bei einer gewichteten Stundenanzahl per annum von 3.330 ca. 173 MWh/a. Somit liegt der Gesamtenergieertrag aus Solarenergie und Windenergie bei ca. 600 MWh/a.

Even if we assume that the photovoltaic elements used in the building will have an efficiency of approximately 15 % and that they are placed in an optimal orientation toward the solar incidence, and even if we furthermore believe that the wind turbines are always in an optimal flow condition, the notion of a supply of energy for the building solely from renewable sources needs to be rejected. In this case, the percentage of renewable energy of the total energy demand budget may be no greater than 4 %.

Alle zuvor angegebenen Kenndaten sind jeweils bezogen auf den Quadratmeter Bruttogeschossfläche. Selbst wenn man davon ausgeht, dass sowohl die Photovoltaikelemente (Wirkungsgrad ca. 15 Prozent angenommen) im Mittel in der vertikalen Fassade optimal positioniert sind und die Windkraftanlagen optimal angeströmt werden, kann eine jährliche Versorgung ausschließlich durch erneuerbare Energien nicht festgestellt werden. Sie dürfte einen Beitrag von ca. 2,5 bis 4 Prozent erbringen.

Figure 226
La Tour Phare, La Defense, Paris, France (Competition entry)

Architect:
UNIBAIL_Jaques Ferrier, Architecte ADC & CE Ingenierie

Bild 226
"Phare" la Defense, Skyline

Architekt:
UNIBAIL_Jaques Ferrier, Architecte ADC & CE Ingenierie

12.3 The "Phare" (lighthouse) high-rise in Paris

The design of the Phare high-rise in Paris by architect Jacques Ferrier, working together with HL-Technik Engineers, in Munich, Germany and others, is the result of a competition entry for the business district La Défense in the French capital. The entire ensemble consists of a smaller, 100-meter-tall, narrow, plate-like building and a tower of 71 floors. The project's total gross area is 140,000 m^2.

Figures 226 – 228 show the most important aspects of the design. Because the building is part of a dense urban fabric, the possibilities of integration of renewable energy-generating sources were limited to the use of building-integrated photovoltaics (BIPV) in the façade and integrated wind turbines. This is a common limitation when it comes to the integration of renewable energy sources – the rest of the necessary infrastructure such as heat- and cooling-energy generation is derived from centralized local grids, and the same is the case for medium-voltage electricity.The meteorological conditions at the site in Paris are presented in **Figures 229 – 234**, and **Figure 235** shows the building ensemble in the context of the dense surrounding fabric of the La Défense business park in Paris.

12.3 Hochhaus „Phare", Paris

Das Hochhaus „Phare", entworfen von Architekt Jacques Ferrier, unter Mitwirkung u.a. von HL-Technik, München, u.a. ist ein Wettbewerbsentwurf in La Défense bei Paris. Das Gesamtensemble besteht aus einem kleineren, ca. 100 m hohen Scheibengebäude sowie dem Turm mit 71 Geschossen bei einer Gesamtfläche von ca. 140.000 m^2.

Die **Bilder 226 – 228** zeigen die wesentliche Struktur. Die Möglichkeiten des Einsatzes erneuerbarer Energien waren bei dem Turm infolge seiner städtebaulichen Einbindung eher bescheiden und mussten sich auf den Einsatz von Photovoltaik in den Fassadenflächen und die Nutzung von Windenergie beschränken. Somit stellt man eine häufig übliche Einschränkung bei der Nutzung erneuerbarer Energien fest. Die Wärme- und Kälteerzeugung erfolgte aus vorhandenen zentralen Netzen, gleichermaßen die Elektroversorgung auf Mittelspannungsebene.Die meteorologischen Bedingungen des Standortes Paris sind in den **Bildern 229 – 234** beschrieben. **Bild 235** zeigt die Lage des Hochhauses innerhalb einer dichten städtebaulichen Umgebung.

Figure 227
La Tour Phare, La Défense, Paris, France (competition entry), Rendering

Bild 227
Hochhaus "Phare", La Défense, Paris
Ansicht (Rendering) des Towers

Figure 228
La Tour Phare, La Défense
Rendering

Bild 228
Ansicht oberer Dachbereich (Rendering) des Towers

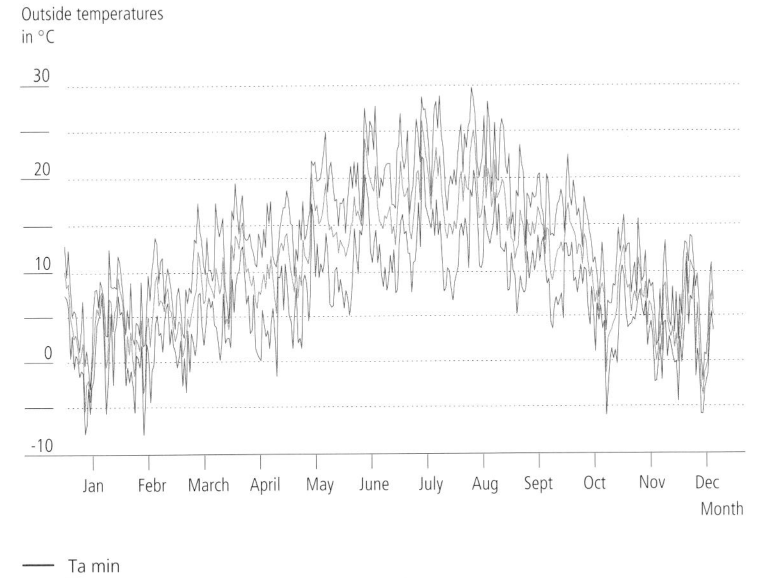

Figure 229
Daily outside temperatures
Paris, France

Bild 229
Tageswerte der Außentemperatur, Paris

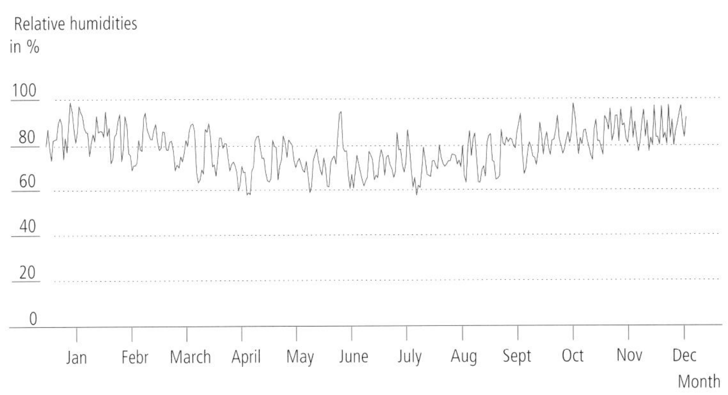

Figure 230
Daily average relative humidities
Paris, France

Bild 230
Täglicher Durchschnittswert der relativen Feuchtigkeit, Paris

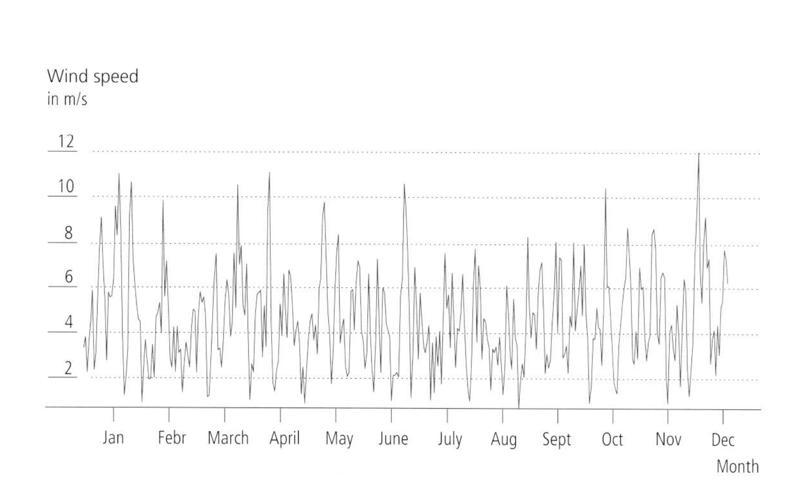

Figure 231
Average daily wind speed
Paris, France

Source: Meteorological data – Meteonorm 5.0

Bild 231
Täglicher Durchschnittswert der Windgeschwindigkeit, Paris

Quelle: Meteorological data – Meteonorm 5.0

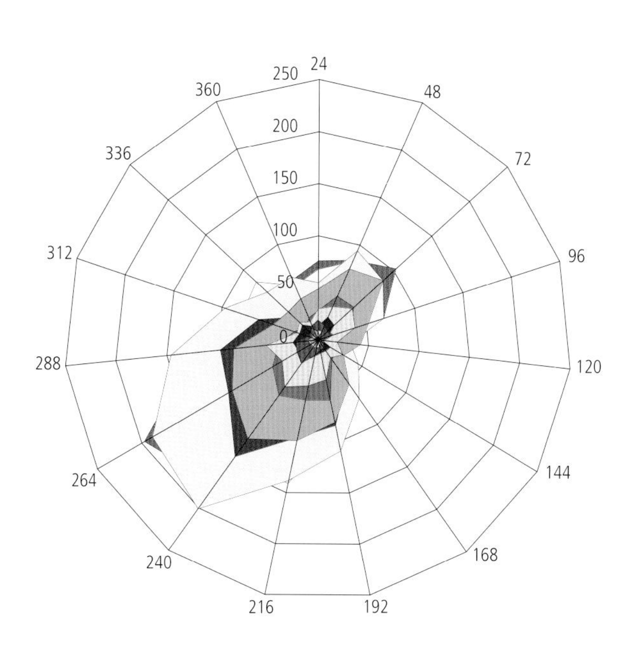

Figure 232
Wind rose diagram, year
Paris, France

Wind speed and frequency distribution of wind direction in hours/annum

Source: Meteorological data – Meteonorm 5.0

Bild 232
Windrose über das gesamte Jahr, Paris

Windgeschwindigkeit in m/s und Häufigkeit der Windrichtungsverteilung (h/a)

Quelle: Meteorological data – Meteonorm 5.0

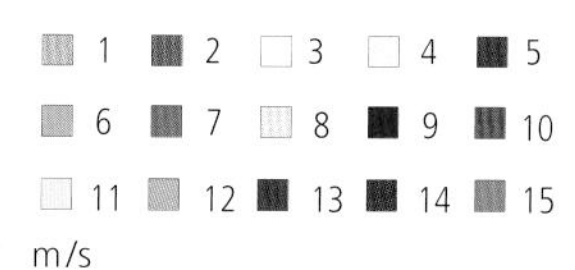

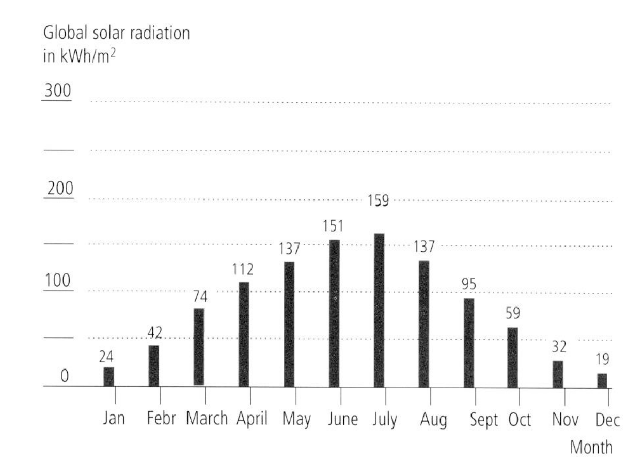

Figure 233
Global solar radiation
Paris, France

Source: Meteorological data – Meteonorm 5.0

Bild 233
Globalstrahlung, Paris

Quelle: Meteorological data – Meteonorm 5.0

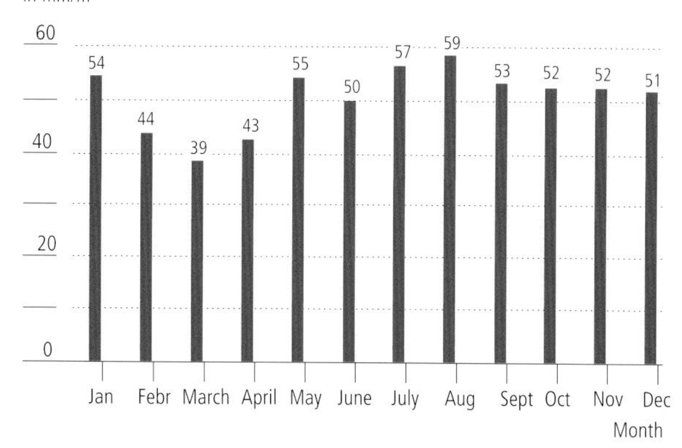

Figure 234
Yearly amount and distribution of precipitation,
Paris, France

Source: Meteorological data – Meteonorm 5.0

Bild 234
Niederschlagsmengen und -verteilung über das Jahr, Paris

Quelle: Meteorological data – Meteonorm 5.0

Prevailing wind conditions at the site indicate wind from the southwest, and based on this analysis there was discussion during the initial schematic design phase of the project of which building shape would support the greatest possible natural ventilation potential. According to **Figure 236**, initial tests showed that a flat and smooth building surface would result in large negative pressure components at the southeast and northwest sides of the building, and, as a result, would create high internal air velocities. To reduce this undesirable effect, the external façade surface was consciously differentiated to reduce outside surface air velocities and the resultant negative surface pressures. **Figure 237** shows additionally that inside the building various green spaces were designed to provide pleasant, high-quality break areas for the employees in the building.

Da die Hauptwindströmungsrichtung in Paris aus Südwest festzustellen ist, wurde am Anfang der konzeptionellen Arbeiten darüber diskutiert, in welcher Form die Fassaden auszubilden seien, um ein Höchstmaß an natürlicher Belüftbarkeit zu erreichen. Gemäß **Bild 236** würde bei einer sehr glatten Oberflächenstruktur am Gebäude bei Überströmung ein hoher Unterdruck auf sowohl der Südost- wie auch Nordwestfassade erreicht, wodurch es bei natürlicher Belüftung innerhalb des Gebäudes zu starken Luftbewegungen kommen kann. Um diesen Effekt zu vermindern, wurde die Fassadenstruktur bewusst sehr stark aufgegliedert, äußere konstruktive Elemente und außen liegende Fassadenstrukturen verringern im fensternahen Bereich die Außenluftgeschwindigkeiten und damit die Unterdrücke erheblich. **Bild 237** zeigt zudem, dass innerhalb des Gebäudes verschiedenste Grünzonen entstehen sollten, um den Nutzern des Gebäudes angenehme Aufenthaltsräume und Pausenzonen zu schaffen.

Figure 235
La Tour Phare, La Défense, Paris, France (Competition entry), urban context

Architect:
UNIBAIL_Jaques Ferrier, Architecte ADC & CE Ingenierie

Bild 235
"Phare" in La Défense Städtebauliche Situation des Gebäudes (Wettbewerbsentwurf), städtischer Kontext

Architekt:
UNIBAIL_Jaques Ferrier, Architecte ADC & CE Ingenierie

Figure 237
La Tour Phare, La Défense, Paris, France (project)

Integrated green skygardens and green roofs

Bild 237
Grünraum- und Atriumkonzept des "Phare", Paris (Projekt)

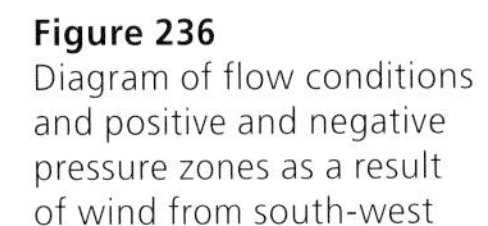

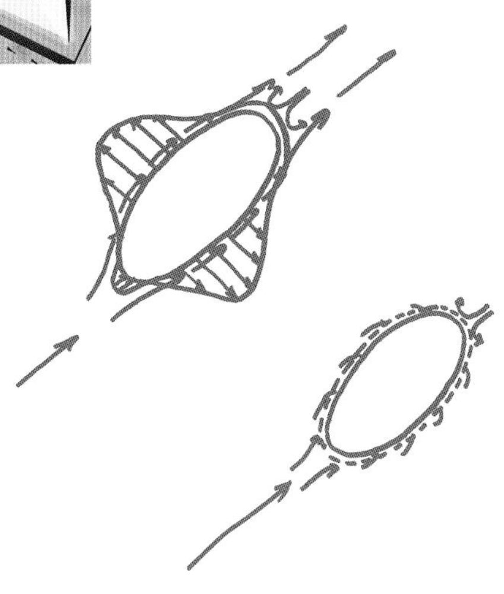

Figure 236
Diagram of flow conditions and positive and negative pressure zones as a result of wind from south-west

Source: Daniels/author

Bild 236
Skizzenhafte Darstellung der Druckfelder und Überströmungen des Baukörpers bei Wind aus SW

Quelle: Daniels/Autor

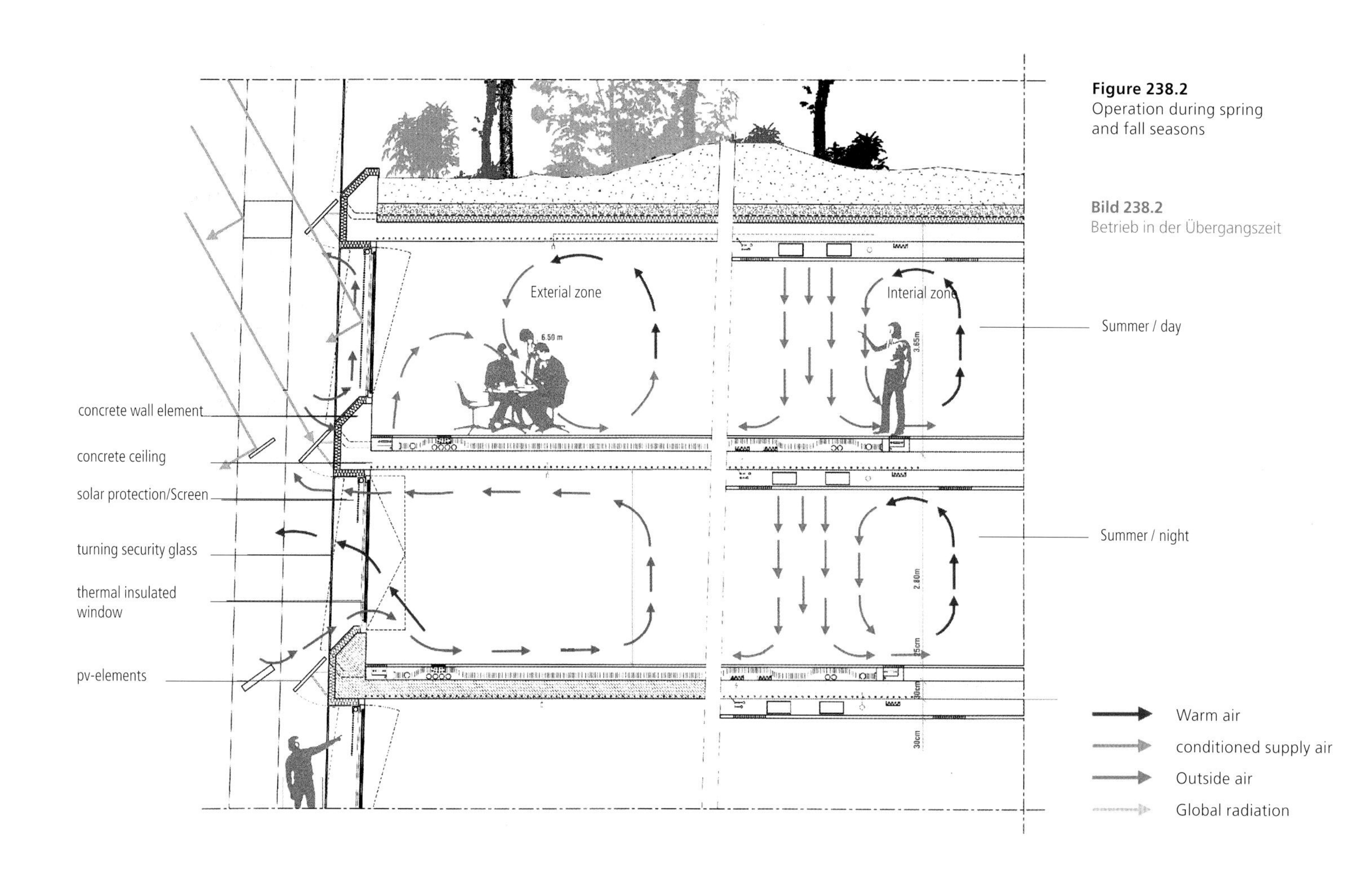

Figure 238.2
Operation during spring and fall seasons

Bild 238.2
Betrieb in der Übergangszeit

Figures 238.1 – 3 show the operational conditions for the building for day and night scenarios. During the summer, with temperatures sometimes reaching 30 °C or higher, it is advisable to operate the building with effective shading and closed operable windows – at these times, the cooling load needs to be lowered. During the night, operable windows can be opened to allow for nighttime ventilation. In the building mass, stored heat load will be given off.

At the location in Paris, during spring and autumn the building can be operated predominantly with natural ventilation, except for the internal zones that do not have sufficient access to external façades. In winter, with ambient air temperatures below +5 °C, the building's skin should be closed in order to use the least amount of thermal energy for comfort conditioning. It is important to realize in this regard that a large amount of the heat energy used in our buildings during winter periods is already provided by the excess heat of artificial lighting and office equipment.

Figure 239 provides a view into the office floor, and **Figure 240** shows a sketch by the architect with respect to renewable energy use and important visual connections inside the building.

Bilder 238.1 – 3 zeigen, in welcher Form im Sommer während des Tages und der Nacht das Gebäude betrieben werden soll. Im Hochsommer bei Außentemperaturen zum Teil über 30 °C ist es sinnvoller, das Gebäude hervorragend zu beschatten und mit geschlossenen Fenstern zu arbeiten, um den Kühllasteintrag zu verringern. Während der Nacht können die Fenster so aufgestellt werden, dass das Gebäude zwangsweise natürlich durchlüftet wird, um gespeicherte Wärmeenergie abzuführen.

In der Übergangszeit soll das Gebäude weitestgehend ausschließlich natürlich belüftet werden, ausgenommen die Innenzonen, die keine ausreichende Außenluftzufuhr erhalten. Bei Außentemperaturen unter +5 °C soll das Gebäude zunehmend geschlossen werden, um es mit geringstem Einsatz an Wärmeenergie zu temperieren. Hier kann davon ausgegangen werden, dass ein wesentlicher Teil der abfließenden Wärme durch die Abwärme der Beleuchtung und von Büromaschinen ersetzt wird.

Bild 239 zeigt einen Einblick in eine Bürofläche, **Bild 240** eine Skizze des Architekten hinsichtlich des Einsatzes erneuerbarer Energien sowie wesentlicher Sichtbeziehungen zum Außenraum.

Figure 238.1
La Tour Phare, La Défense, Paris, France (project)

Diagrammatic concept of natural ventilation in summer and cooling of office floors (Summer day = mechanical ventilation with cooling, summer night = natural ventilation, mass de-charging)

1 Warm space air
2 Cool fresh air
3 Ambient air
4 Solar radiation

Bild 238.1
Konzept der sommerlichen Durchlüftung und Kühlung der Büroflächen
(Sommer Tag = mechanische Belüftung mit Kühlung
Sommer Nacht = natürliche Belüftung zum Wärmeaustrag)

1 Erwärmte Raumluft
2 Gekühlte Zuluft
3 Außenluft
4 Solarstrahlung

Figure 238.3
Operation in winter

with natural ventilation, upper floor concrete-core cooling, mechanical ventilation and heating, lower floor.

Bild 238.3
Betrieb in der Winterzeit

mit natürlicher Belüftung und Bauteilkühlung (obere Etage), mit mechanischer Belüftung und Heizung (untere Etage)

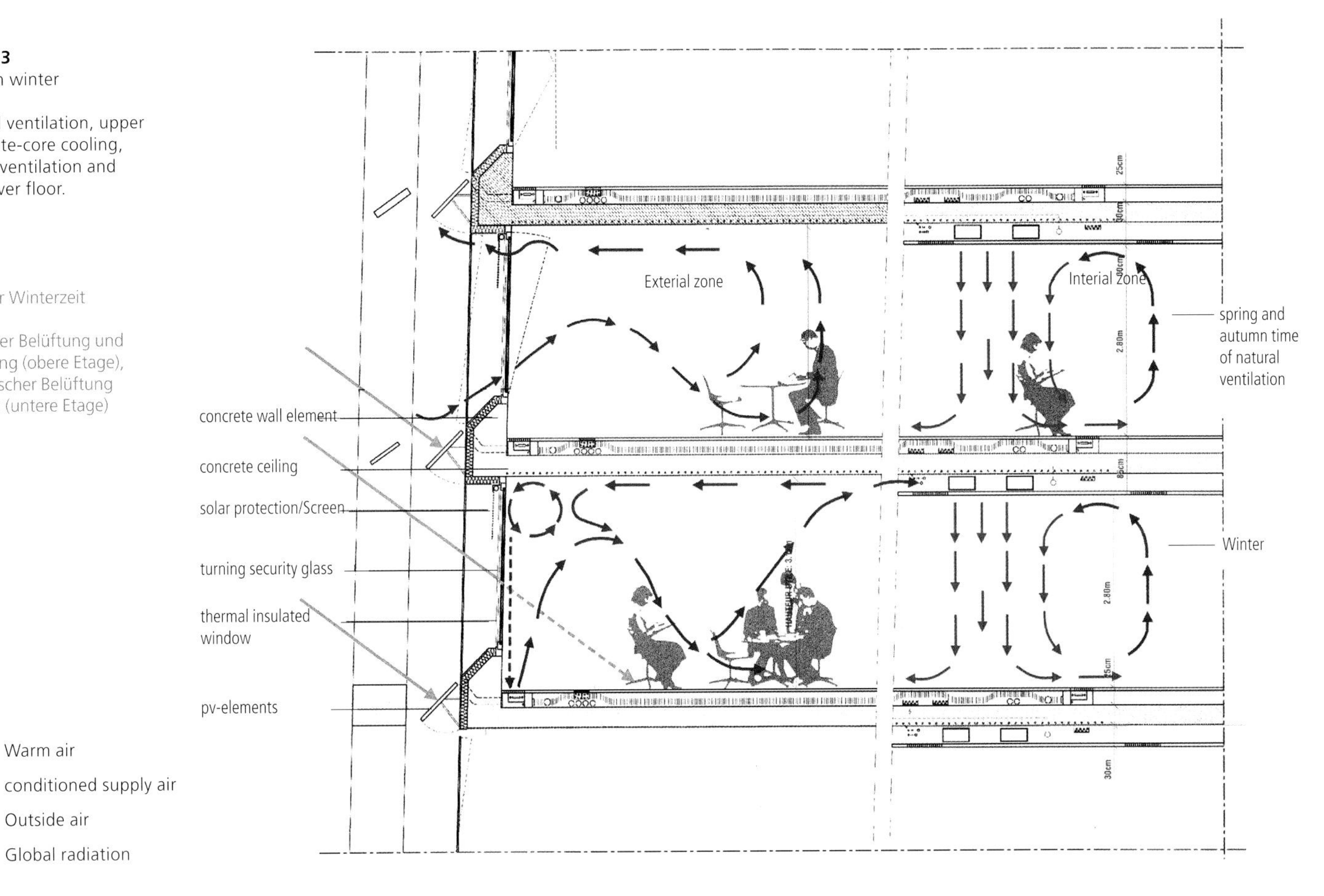

Figure 239
Interior view of office floor

Bild 239
Ansicht einer Büroebene

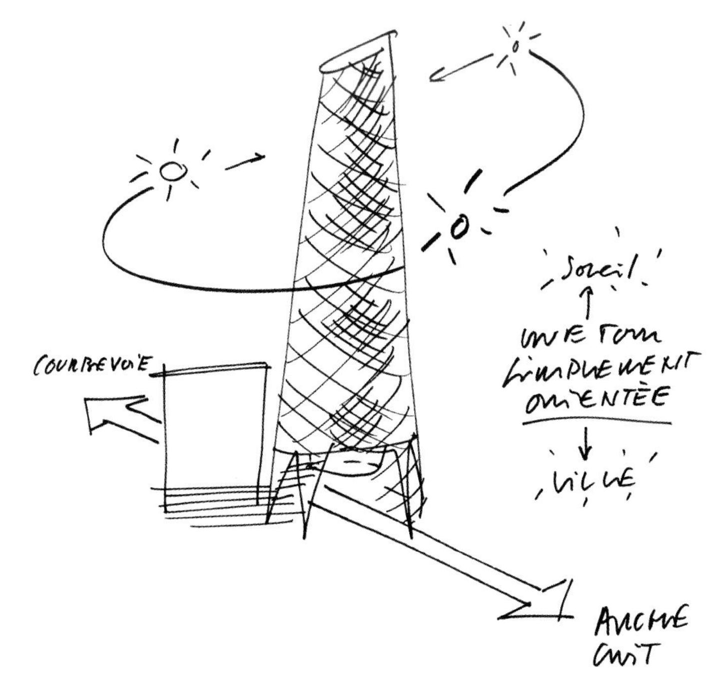

Figure 240
La Tour Phare, La Défense, Paris, France (project),

Conceptual sketch by the architect showing sun path and views towards sky and city.

Bild 240
Skizze des Architekten bezüglich der Sonnenumlaufbahn sowie der Sichtbeziehungen/Himmel – Stadtraum

The façade of the building is conceived as a box-window type envelope, as seen in **Figures 238**, with parts of the façade being opaque elements. An exterior laminated glass pane can be opened, but during night hours and winter seasons it will remain closed. Behind the laminated glass outer layer the wind-protected shading system is located. An internal insulated glazing system with a U-value of 1.2 W/m^{2}K is certainly not the optimum of thermal insulation for winter, but it was used to allow for partial flow of excess thermal energy to the outside during evening and summer operating conditions.

The box-window design of the façade is operable, mainly for nighttime flushing. The interior insulated glazing unit can be opened by turn-tilt hardware in such a way that only the outer single-laminated glass remains as a barrier against heat gain. Its U-value is approximately 5.8 W/m^{2}K.

The exterior elements of the façade are shading devices, limited by the structural conditions of the outside envelope; they also could have been conceived as photovoltaic energy-generating modules. **Figure 241** shows the various façade details.

Several different studies using various parameters were developed to analyze the potential for renewable energy usage. **Figure 242** shows the temperature hysteresis for the site of the building for the operating periods 12 a.m. – 12 p.m. and 7 a.m. – 7 p.m. Shown are those times in which the building needs additional heating and cooling with active mechanical systems to provide comfort. Between such periods of active building system use, long time intervals remain in which the building can be operated entirely by natural ventilation, especially during times of moderate outside wind velocities. At such times, the internal heat loads are compensated by natural ventilation, and odors and interior pollutants are expelled naturally to provide good interior air quality.

Wie bereits die **Bilder 238** ausweisen, ist die Fassade des Gebäudes als Kastenfensterstruktur mit zum Teil opaken Wandteilen konzipiert. Eine äußere Verbund-Sicherheitsglasscheibe, hinter der sich der Sonnenschutz verbirgt, kann langzeitig geöffnet bleiben und wird lediglich in den Nachtstunden bzw. im tiefen Winter geschlossen. Eine innere Wärmeschutz-Isolierverglasung mit einem u-Wert von 1,2 W/m^{2}K stellt zwar nicht das Optimum eines Wärmeschutzes für den Winterbetrieb dar, wurde jedoch eingesetzt, um in den Abendstunden und im Sommer die Wärmeenergie zum Teil abfließen lassen zu können.

Die Kastenfensterstruktur kann während der Nachtstunden geöffnet werden, d.h. die innere Isolierverglasung wird aufgedreht (Dreh-Kippflügel), so dass als Barriere nur noch die äußere VSG-Verglasung der Entwärmung entgegensteht (u-Wert ca. 5,8 W/m^{2}K).

In der äußeren, statisch bedingten Konstruktion vor den Fassaden sind Beschattungselemente eingesetzt, die in Gänze auch hätten Photovoltaikflächen sein können. Die **Bilder 241** zeigen verschiedene Fassadenausschnitte.

Zur groben Bilanzierung des Gebäudes wurden verschiedene Studien und überschlägige Berechnungen angestellt, um den Einsatz erneuerbarer Energien soweit wie möglich zu nutzen. **Bild 242** zeigt die Temperaturhysteresis des Standortes für die Betriebszeiträume 0 – 24 bzw. 7 – 19 Uhr. Hierin eingetragen sind die Zeiten, in denen das Gebäude zunehmend aktiv geheizt werden muss bzw. die Zeiten, in denen das Gebäude aktiv gekühlt werden sollte, um angenehme Raumbedingungen zu schaffen. Zwischen Heizen und Kühlen verbleibt ein großer Zeitraum, in dem bei moderaten Windgeschwindigkeiten und Winddrücken das Gebäude im Wesentlichen nur natürlich belüftet werden muss, da über die natürliche Belüftung auch in der Übergangszeit die im Inneren entstehende Wärme ausgetragen wird und gleichermaßen Schad- und Geruchsstoffe.

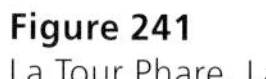

Figure 241
La Tour Phare, La Défense, Paris, France (competition entry), façade with sun shading and photovoltaic elements

Bild 241
Fassadenstrukturen mit Sonnenschutzelementen und Photovoltaik-Strukturen.

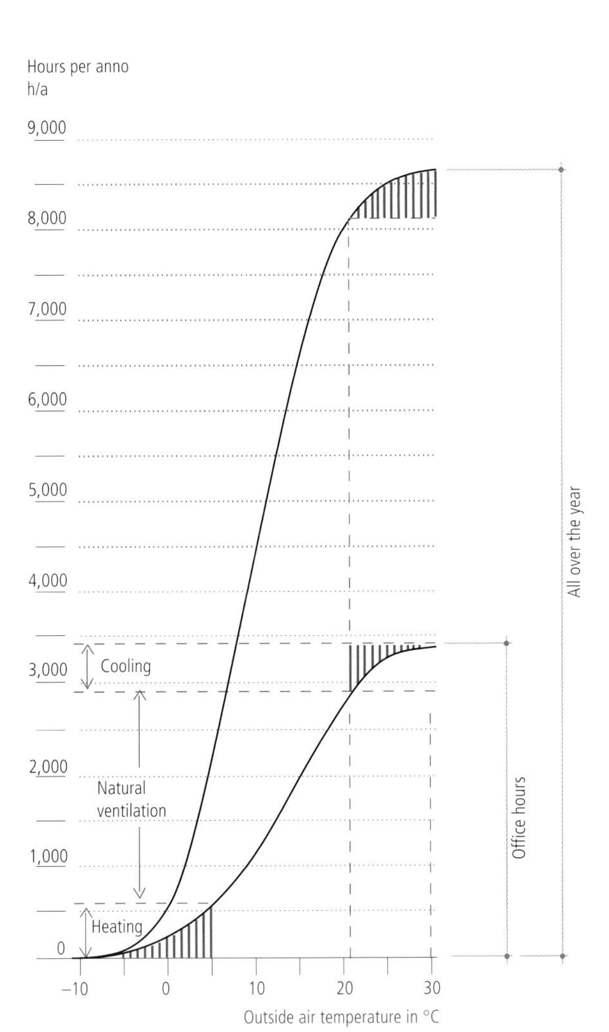

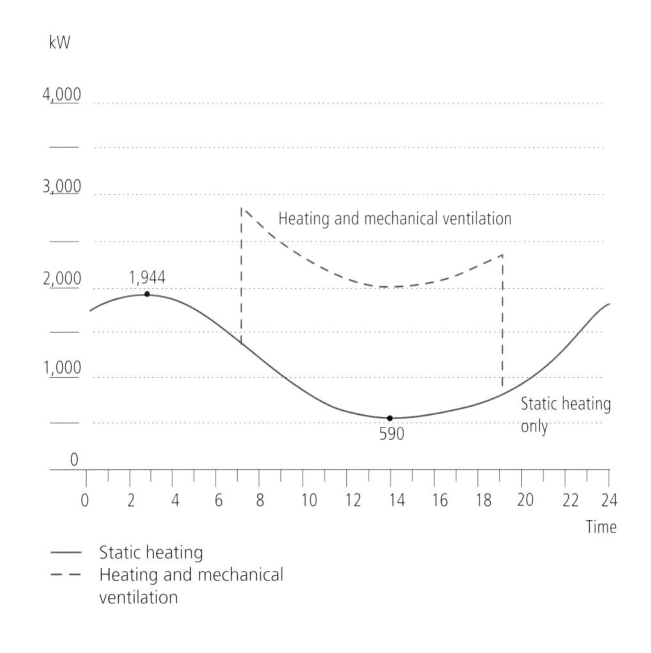

Figure 243.1
Heat energy requirement for a overcast and cold day in January
Source: Daniels/author

Bild 243.1
Heizung/Wärmebedarf Januar-Tag, bewölkt und kalt

Quelle: Daniels/Autor

Figure 242
Annual temperature hysteresis for office hours to determine operational mode of a building (natural ventilation, heating, cooling)

Source: Daniels/author

Bild 242
Temperatur-Hysteresis per annum bzw. zu den Büroarbeitszeiten zur Bestimmung der Heiz- und Kühlperiode sowie der Zeit, zu der das Gebäude ausschließlich mit natürlicher Lüftung betrieben werden kann

Quelle: Daniels/Autor

Figure 243.1 shows the curve of maximum heat required during a cold but sunny January day. **Figure 243.2** displays the cooling loads for a sunny and hot July day. **Figure 243.3** shows an estimate of expected electrical energy requirements for a summer day with maximum cooling loads and maximum need for mechanical ventilation.

Heating and cooling energy required for the building is supplied by city utility grids, yet a part of the cooling energy can be compensated by electrical energy generated on the building's surface itself. The percentage that would be contributed to the overall energy consumption of the building is only around 6.7 %, although this does not include the electrical energy generated by wind turbines mounted on top of the high-rise building.

Bild 243.1 zeigt den Verlauf des maximalen Wärmebedarfs an einem Januartag, sonnig, kalt. **Bild 243.2** stellt einen sonnigen, heißen Juli-Tag dar und die daraus resultierenden Kühlleistungen für das Gebäude. **Bild 243.3** zeigt den in etwa zu erwartenden elektrischen Energiebedarf eines warmen Sommertages mit maximal notwendiger Kälteleistung und mechanischer Belüftung.

Während der Wärmeenergie- bzw. Kälteenergiebedarf für das Gebäude ausschließlich aus Fernnetzen bedient wird, kann ein Teil der am Gebäude gewonnenen elektrischen Energie der Kälteerzeugung zugute kommen. Der Beitrag würde in etwa 6,7 Prozent ausmachen. Hierbei nicht berücksichtigt ist der elektrische Energiebeitrag durch auf der Turmspitze stehende Windräder. **Bild 243.3** zeigt den elektrischen Energiebedarf mit seinen Spitzenwerten und wiederum den Anteil, der durch die Photovoltaikelemente als auch durch Windräder beigesteuert wird.

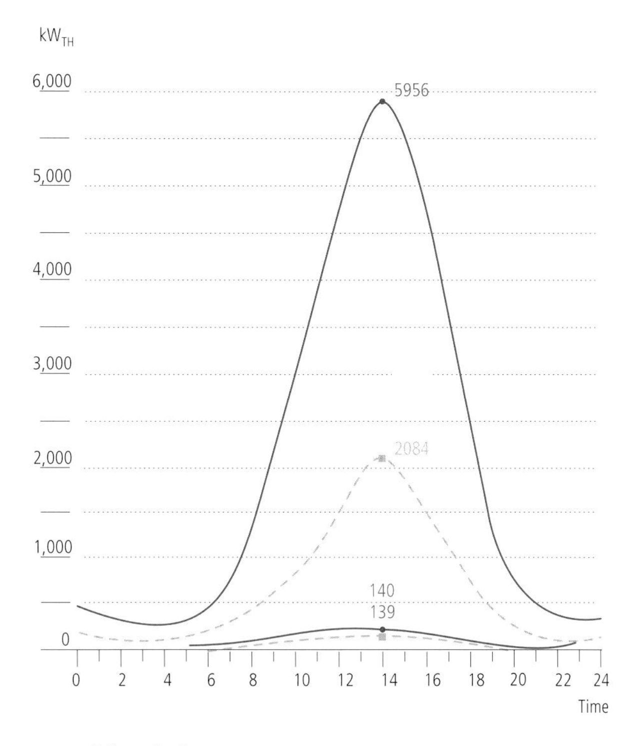

Figure 243.2
Cooling energy requirement for a sunny July day

Source: Daniels/author

Bild 243.2
Kühlung/Kältebedarf Juli-Tag, sonnig

Quelle: Daniels/Autor

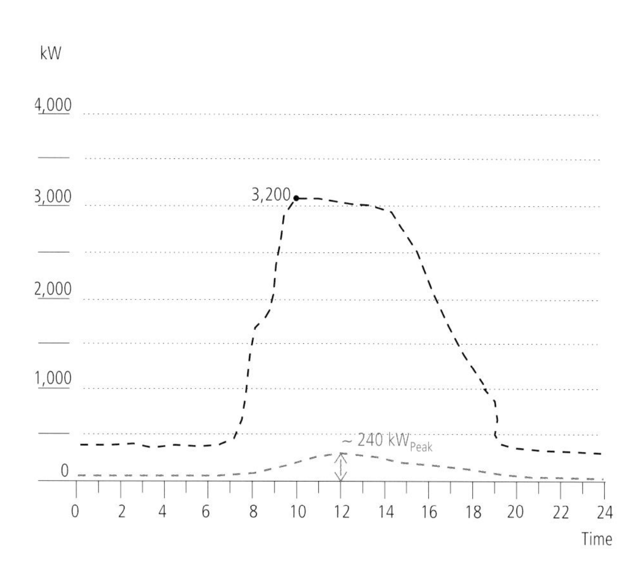

Figure 243.3
Electrical energy consumption and generation for a sunny, windy July day

Source: Daniels/author

Bild 243.3
Elektrischer Energiebedarf/-ertrag
Juli-Tag, sonnig und windig

Quelle: Daniels/Autor

To improve the energy generation of the south-façade-mounted photovoltaic panels, it would be necessary to equip them as sun-tracking panels. **Figure 244.1** shows the energy generation of the panels for the month of June and the location, Paris, and **Figure 244.2** allows conclusions concerning the inclination of the panel surfaces for maximum energy generation. A horizontally oriented photovoltaic panel surface generates a maximum of energy during the summer months, whereas winter gains are minimal. In contrast, a vertical photovoltaic surface reaches maximum electrical output during spring and fall, whereas summer gains are minimal.

On top of the building, wind power systems with vertical rotors are conceived with a nominal power output of 110 kW at approximately 13 – 15 m/s. The expected wind velocities for a building height of 280 meters are shown in **Figure 245**.

Details concerning wind rotors with a power output of 100 kW are listed in **Table 17**.

Um einen möglichst hohen Jahresertrag über die an den Südfassaden vorgelagerten Photovoltaikelemente zu erreichen, wäre es sinnvoll, diese nachzuführen. **Bild 244.1** zeigt den Energieertrag für den Monat Juni während eines Tages sowie angedeutet die Spitzenwerte für weitere Monate am Standort Paris. **Bild 244.2** gibt eine Information darüber, welche Einstellwinkel bei welchen Erträgen zu den verschiedensten Monaten optimal sind. Eine horizontale Photovoltaikfläche erreicht in den Winterzeiten sehr minimale Werte und einen Spitzenwert im Juni/Juli. Eine vertikale Fassade erreicht im Gegenteil zur horizontalen Fläche in den Übergangszeiten hohe Werte und im Sommer hingegen sehr niedrige Erträge.

Auf dem Gebäude sollen Windräder mit Vertikalrotoren zum Einsatz kommen, die elektrische Energie liefern, Nennleistung 100 kW bei ca. 13 – 15 m/s. Die zu erwartenden Windgeschwindigkeiten in der Höhe von 280 m sind in **Bild 245** ausgewiesen.

Der Lieferumfang, bezogen auf ein Windrad mit einer Nennleistung von ca. 100 kW ist in **Tabelle 17** dargestellt.

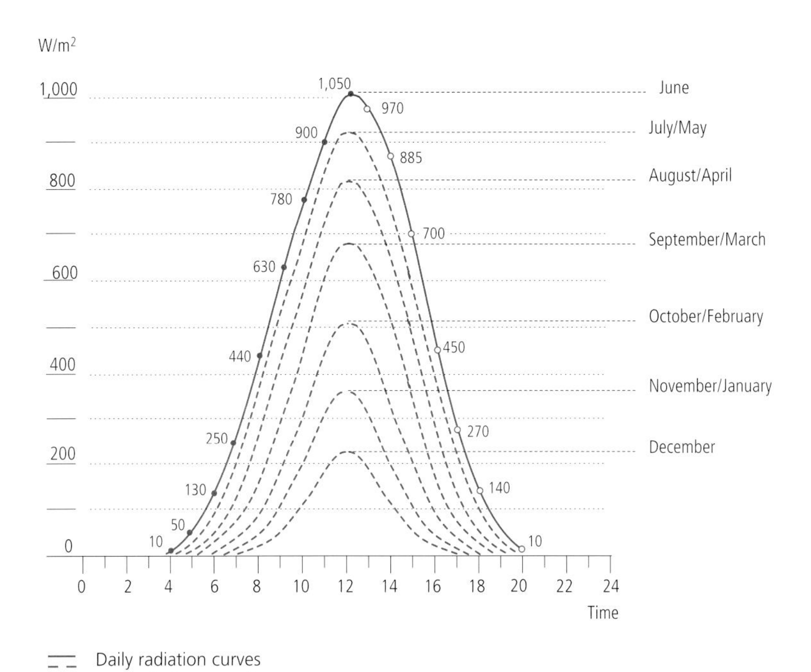

Figure 244.1
Monthly average global radiation on a south-facing wall, without consideration of turbidity.

Source: Daniels/author

Bild 244.1
Mittlere Monatswerte der Globalstrahlung an der Südfassade (ohne Berücksichtigung eines Trübungsfaktors),

Quelle: Daniels/Autor

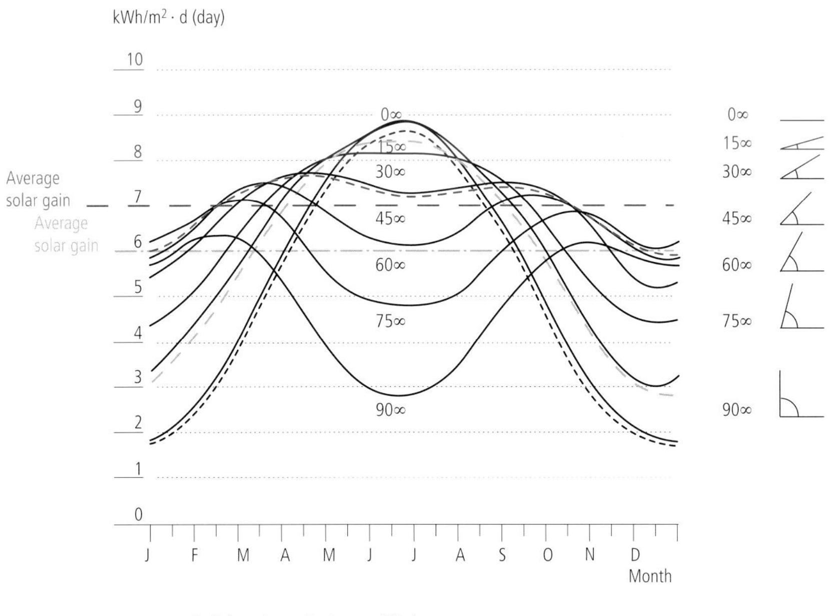

Figure 244.2
Overall global solar exposure level (G max) by month of the year, for different PV-cells inclinations and southerly orientation by a max. solar altitude of 63°.

Source: Daniels/author

Bild 244.2
Auftreffende Globalstrahlung auf südorientierte Flächen unterschiedlicher Neigung (ohne Berücksichtigung von PV-Wirkungsgraden)

Quelle: Daniels/Autor

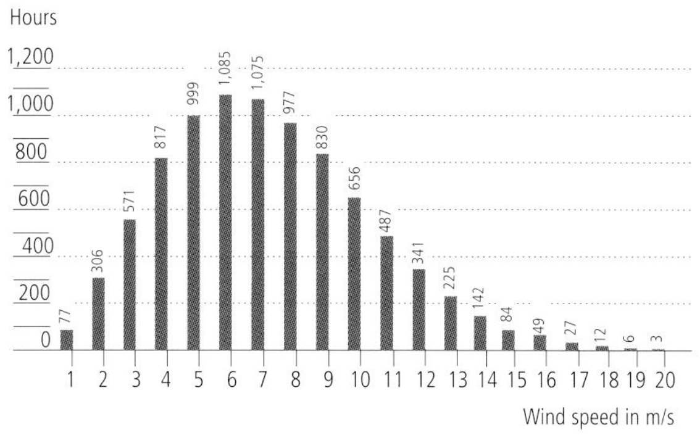

Hours
1,200
1,000
800
600
400
200
0
77 306 571 817 999 1,085 1,075 977 830 656 487 341 225 142 84 49 27 12 6 3
1 2 3 4 5 6 7 8 9 10 11 12 13 14 15 16 17 18 19 20
Wind speed in m/s

Figure 245
Wind speed distribution at a height above ground of 280 meters
Paris, France

Source: Daniels/author

Bild 245
Windgeschwindigkeitsverteilung in 280 m Höhe (Paris)

Quelle: Daniels/Autor

m/s	Wind velocity and -direction													$P_{nominal}$ = 100 kW	
	360°	30°	60°	90°	120°	150°	180°	210°	240°	270°	300°	330°	Hours	max. capacity 100 [kW]	max. capacity 100 [kW]
1	5.8	4.8	5.2	4.8	5.4	5.4	7.3	5.9	4.9	5.7	11.5	10.8	77.3	0	0
2	26.2	22.8	22.3	20.6	21.0	21.2	26.2	26.5	20.4	22.7	39.1	37.4	306.5	0	0
3	51.8	45.0	42.4	39.0	39.1	39.6	48.0	50.0	40.8	44.5	67.8	63.2	571.3	0	0
4	75.5	67.8	61.8	54.8	54.3	55.9	66.9	73.6	62.2	67.2	92.6	84.0	816.7	0	0
5	92.1	85.1	78.2	64.4	62.6	67.3	81.4	94.2	81.7	87.0	109.8	94.8	998.7	3	2,996
6	96.6	94.1	87.9	65.7	62.6	71.7	88.7	107.9	96.3	101.2	116.5	96.3	1085.4	13	14,110
7	89.5	93.4	90.8	59.1	55.2	69.5	88.0	112.8	105.0	108.8	113.6	89.1	1074.9	26	27,948
8	73.5	83.7	85.6	47.7	43.2	61.4	80.7	109.9	106.0	106.9	103.1	75.4	977.3	43	42,023
9	54.4	67.8	75.2	34.6	30.5	50.0	69.8	100.1	101.1	99.3	87.8	58.9	829.5	58	48,111
10	35.2	50.5	61.8	22.3	18.9	37.5	55.3	85.4	90.4	85.1	69.7	42.4	654.5	73	47,780
11	20.5	33.9	46.9	13.1	10.7	26.1	41.4	68.7	75.8	69.1	51.6	28.7	486.5	86	41,841
12	10.9	20.8	33.5	6.6	5.4	16.8	29.1	51.0	60.3	52.0	36.3	18.0	340.6	96	32,694
13	5.1	11.8	22.3	3.1	2.5	9.8	18.9	36.3	44.7	36.9	23.9	10.1	225.3	98	22,078
14	1.9	5.5	14.1	1.3	0.8	5.4	11.6	24.5	32.1	24.6	14.3	5.7	142.1	97	13,782
15	0.6	2.8	8.2	0.4	0.4	2.7	6,5	14.7	21.4	15.1	8.6	2.9	84.4	90	7,598
16	0.0	1.4	4.5	0.0	0.0	1.1	3.6	8.8	13.6	9.5	4.8	1.4	48.7	83	4,041
17	0.0	0.7	2.2	0.0	0.0	0.5	3.2	4.9	7.8	4.7	2.9	0.7	26.6	77	2,052
18	0.0	0.0	0.7	0.0	0.0	0.0	0.7	2.9	3.9	2.8	1.0	0.0	12.1	71	859
19	0.0	0.0	0.7	0.0	0.0	0.0	0.7	1.0	1.9	0.9	1.0	0.0	6.3	65	409
20	0.0	0.0	0.0	0.0	0.0	0.0	0.0	1.0	1.0	0.9	0.0	0.0	2.9	66	191
Total	639.48	692.04	744.6	437.56	412.54	542.3	727.08	980.14	971.39	945.13	955.79	719.76	8,767.6		308,515 kWh/a

Table 17
Approximate annual energy generation of a wind turbine in Paris, France.
(Wind velocity in 280 meters)

Source: Daniels/author

m/s	Windgeschwindigkeit und -richtung													P-Nenn = 100 kW	
	360°	30°	60°	90°	120°	150°	180°	210°	240°	270°	300°	330°	Anzahl Stunden	Förder-Nennleistung FL 100 [kW]	FL 100 [kWh]
1	5.8	4.8	5.2	4.8	5.4	5.4	7.3	5.9	4.9	5.7	11.5	10.8	77.3	0	0
2	26.2	22.8	22.3	20.6	21.0	21.2	26.2	26.5	20.4	22.7	39.1	37.4	306.5	0	0
3	51.8	45.0	42.4	39.0	39.1	39.6	48.0	50.0	40.8	44.5	67.8	63.2	571.3	0	0
4	75.5	67.8	61.8	54.8	54.3	55.9	66.9	73.6	62.2	67.2	92.6	84.0	816.7	0	0
5	92.1	85.1	78.2	64.4	62.6	67.3	81.4	94.2	81.7	87.0	109.8	94.8	998.7	3	2.996
6	96.6	94.1	87.9	65.7	62.6	71.7	88.7	107.9	96.3	101.2	116.5	96.3	1085.4	13	14.110
7	89.5	93.4	90.8	59.1	55.2	69.5	88.0	112.8	105.0	108.8	113.6	89.1	1074.9	26	27.948
8	73.5	83.7	85.6	47.7	43.2	61.4	80.7	109.9	106.0	106.9	103.1	75.4	977.3	43	42.023
9	54.4	67.8	75.2	34.6	30.5	50.0	69.8	100.1	101.1	99.3	87.8	58.9	829.5	58	48.111
10	35.2	50.5	61.8	22.3	18.9	37.5	55.3	85.4	90.4	85.1	69.7	42.4	654.5	73	47.780
11	20.5	33.9	46.9	13.1	10.7	26.1	41.4	68.7	75.8	69.1	51.6	28.7	486.5	86	41.841
12	10.9	20.8	33.5	6.6	5.4	16.8	29.1	51.0	60.3	52.0	36.3	18.0	340.6	96	32.694
13	5.1	11.8	22.3	3.1	2.5	9.8	18.9	36.3	44.7	36.9	23.9	10.1	225.3	98	22.078
14	1.9	5.5	14.1	1.3	0.8	5.4	11.6	24.5	32.1	24.6	14.3	5.7	142.1	97	13.782
15	0.6	2.8	8.2	0.4	0.4	2.7	6,5	14.7	21.4	15.1	8.6	2.9	84.4	90	7.598
16	0.0	1.4	4.5	0.0	0.0	1.1	3.6	8.8	13.6	9.5	4.8	1.4	48.7	83	4.041
17	0.0	0.7	2.2	0.0	0.0	0.5	3.2	4.9	7.8	4.7	2.9	0.7	26.6	77	2.052
18	0.0	0.0	0.7	0.0	0.0	0.0	0.7	2.9	3.9	2.8	1.0	0.0	12.1	71	859
19	0.0	0.0	0.7	0.0	0.0	0.0	0.7	1.0	1.9	0.9	1.0	0.0	6.3	65	409
20	0.0	0.0	0.0	0.0	0.0	0.0	0.0	1.0	1.0	0.9	0.0	0.0	2.9	66	191
Summe	639.48	692.04	744.6	437.56	412.54	542.03	727.08	980.14	971.39	945.13	955.79	719.76	8.767.6		308,515 kWh/a

Tabelle 17
Grobe Kalkulation des zu erwartenden Energieertrags einer Wind-Kraftanlage Standort Paris
(Windgeschwindigkeit in m/s in 280 m Höhe)

Quelle: Daniels/Autor

If we compare the peak power outputs for heat, cooling, and electricity, the annual consumption, or generation, is equal to:

- Heating load (without internal loads)
  approx. 3,360 MWh/a
- Heating power (without internal loads)
  approx. 660 MWh/a
  (offices)
- Cooling load
  approx. 2,980 MWh/a
- Electrical power (including cooling energy transmission)
  approx. 5,810 MWh/a

Total resulting final energy demand ca. 53.2 kWh/m^2 gross area x a.

The power generation of solar photovoltaic systems is approx. 154 MWh/a;
the gain from wind power generation is approx. 308 MWh/a.

In conclusion, it can be determined that 8 % of the building's total energy demand will be supplied by renewable sources such as solar and wind power.

Führt man anhand der jeweiligen Peakleistungen einen Vergleich für Wärmeenergie, Kälteenergie, elektrische Energie durch, so ergibt sich ein Jahresverbrauch bzw. ein Jahresertrag in etwa wie folgt:

- Heizleistung (ohne innere Lasten)
  ca. 3.360 MWh/a
- Heizleistung (abzüglich innere Wärmelasten)
  ca. 660 MWh/a
  (Bürobetrieb)
- Kälteleistung ca. 2.980 MWh/a
- Elektrischer Energiebedarf (inklusive Kältetransport)
  ca. 5.810 MWh/a

Hieraus resultiert ein Endenergiebedarf von ca. 53,2 kWh/m^2 BGFa.

Der Solarertrag der photovoltaischen Anlagen errechnet sich mit ca. 154 MWh/a,
der Windertrag liegt bei ca. 308 MWh/a.

Somit wird ca. 8 Prozent der im Gebäude benötigten Energie durch erneuerbare Energien bereitgestellt.

12.4 High-rise project "Phare" Middle East

Figure 246 shows the above-described Paris high-rise now being transplanted to Abu Dhabi, United Arab Emirates.

Figures 247 – 250 explain the necessary outside climatic conditions, such as temperatures, relative humidity, air velocity, rain water, and solar radiation.

As seen in **Figure 247.2**, even in such a region buildings can be operated naturally, at least at times. In fact – astonishingly – on 25 % of all days of the year the building can be operated with natural ventilation strategies without compromising comfort at all. The maximum amounts of energy based on external climatic conditions and the usual technologies, as in the case of a building in Paris, are as follows:

- Heating load max.
  - Paris approx. 2,800 kW
  - Abu Dhabi approx. 760 kW
- Cooling load
  - Paris approx. 5,960 kW
  - Abu Dhabi approx. 15,500 kW
- Electrical energy demand
  - Paris approx. 3,200 kW
  - Abu Dhabi approx. 4,460 kW

(Data based on the assumption that cooling energy is generated by absorption chillers. Conventional electric chillers would increase the demand by approx. 4,650 kW.)

The analog annual energy demands are approximately:

- Annual heat
  - Paris approx. 5,077 MWh/a
  - Abu Dhabi approx. 456 MWh/a
- Annual cooling requirement
  - Paris approx. 3,170 MWh/a
  - Abu Dhabi approx. 36,179 MWh/a
- Total annual energy demand
  - Paris approx. 7,335 MWh/a
  - Abu Dhabi approx. 9,846 MWh/a

(based on the assumption that cooling energy is generated mainly by heat energy)

12.4 Projekt „Phare" – Mittlerer Osten

Zum Vergleich mit einem feucht-heißen Standort wird gedanklich der Turm nunmehr nach Abu Dhabi versetzt (**Bild 246**).

Die **Bilder 247** bis **250** zeigen anfänglich wiederum die Konditionen des Außenraums mit seinen Temperaturen, relativen Feuchten, Luftgeschwindigkeiten, Regenwassermengen und solaren Strahlungsenergien.

Wie **Bild 247.2** zu entnehmen ist, können auch und selbstverständlich in dieser Region, zeitweise Gebäude mit natürlicher Belüftung betrieben werden. An ca. 25 Prozent aller Tage ist ein natürliches Belüften ohne Einbußen an Komfort möglich. Die maximal notwendigen Energiemengen aufgrund dimensionierender Außenzustände und unter Nutzung auch der in Europa üblichen Technologien ergeben sich nach überschlägiger Berechnung unter ansonsten gleichen Bedingungen folgende Daten:

- Heizleistung max.
  - Paris ca. 2.800 kW
  - Abu Dhabi ca. 760 kW
- Kälteleistung
  - Paris ca. 5.960 kW
  - Abu Dhabi ca. 15.500 kW
- Elektrischer Energiebedarf
  - Paris ca. 3.200 kW
  - Abu Dhabi ca. 4.460 kW

(unter Berücksichtigung, dass der Kälteenergiebedarf durch Absorptionskältemaschinen erbracht wird. Bei elektrisch betriebenen Kältemaschinen würde sich der elektrische Energiebedarf um ca. 4.650 kW vergrößern).Die analogen, jährlichen Energiebedarfswerte liegen in etwa bei:

- Jahreswärmeenergiebedarf
  - Paris ca. 5.077 MWh/a
  - Abu Dhabi ca. 456 MWh/a
- Jährlicher Kälteenergiebedarf
  - Paris ca. 3.170 MWh/a
  - Abu Dhabi ca. 36.179 MWh/a
- Jährlicher elektrischer Energiebedarf
  - Paris ca. 7.335 MWh/a
  - Abu Dhabi ca. 9.846 MWh/a

(hier wiederum unter der Bedingung, dass die Kälteenergie primär durch den Einsatz von Wärmeenergie erzeugt wird)

Figure 246
Le Phare Tower, for comparative research analysis placed in a different climate region. United Arab Emirates. (Rendering)

Source: UNIBAIL_Jaques Ferrier, Architecte ADC & CE Ingenierie

Bild 246
Vergleichsobjekt "Phare" an einem Standort in den Vereinigten Arabischen Emiraten

Quelle: UNIBAIL_Jaques Ferrier, Architecte ADC & CE Ingenierie

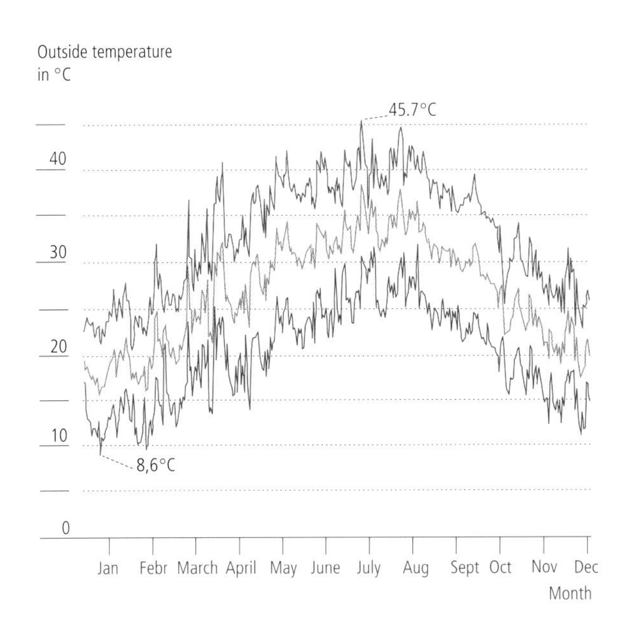

Figure 247.1
Daily temperatures Abu Dhabi, UAE

Source: Meteonorm 5.1 meterological data

Bild 247.1
Tageswerte der Außentemperatur, Abu Dhabi

Quelle: Meteorological data – Meteonorm 5.1

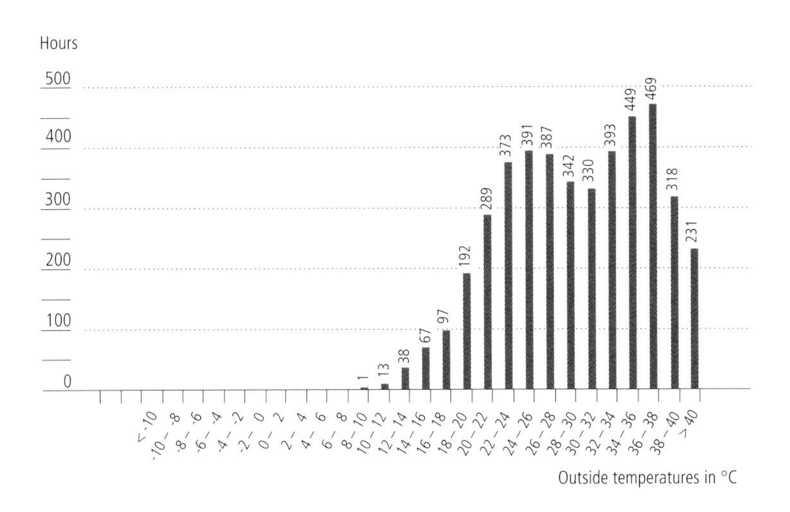

Figure 247.2
Outside temperature statistic between 7 p.m. and 7 a.m. Abu Dhabi, UAE

Source: Meteonorm 5.1 meterological data

Bild 247.2
Außentemperaturstatistik zwischen 7 Uhr und 19 Uhr, Abu Dhabi

Quelle: Meteorological data – Meteonorm 5.1

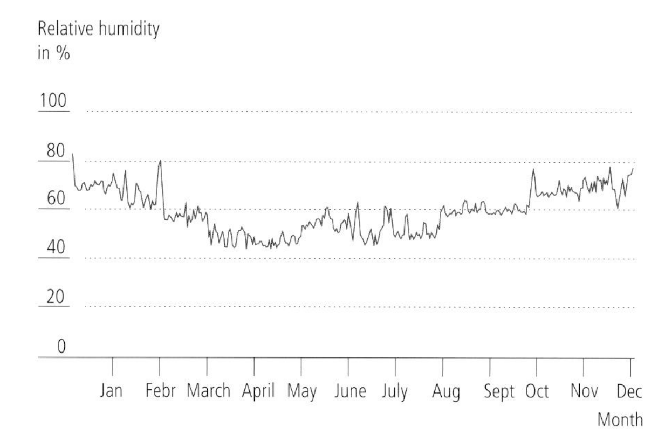

Figure 248.1
Daily average relative humidity
Abu Dhabi, UAE

Source: Meteorological data meteonorm 5.1

Bild 248.1
Täglicher Durchschnittswert der relativen Feuchtigkeit
Abu Dhabi, VAE

Quelle: Data source meteorological data – Meteonorm 5.1

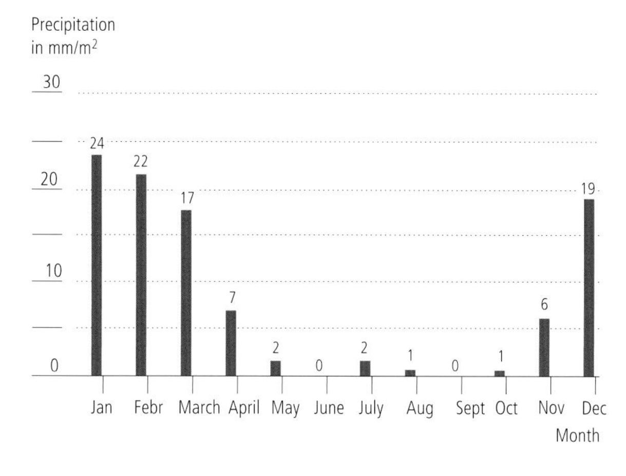

Figure 248.2
Annual precipitation
Abu Dhabi, UAE

Source: Meteorological data meteonorm 5.1

Bild 248.2
Niederschlag während des Jahres
Abu Dhabi, VAE

Quelle: Meteorological data – Meteonorm 5.1

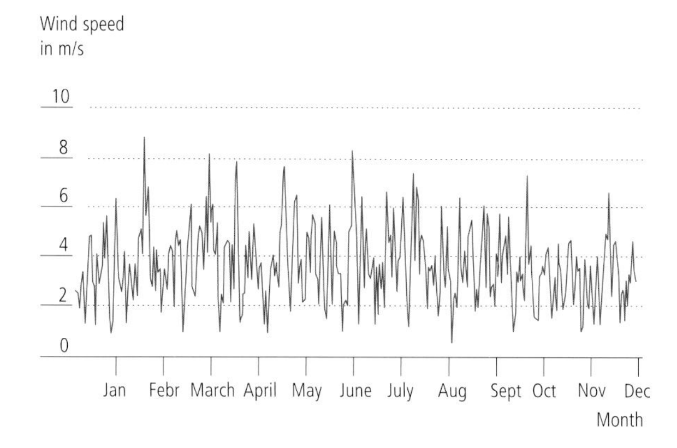

Figure 249.1
Daily average wind speed near ground surface
Abu Dhabi, UAE

Source: Meteorological data meteonorm 5.1

Bild 249.1
Täglicher Durchschnittswert der Windgeschwindigkeit in Bodennähe
Abu Dhabi

Quelle: Meteorological data – Meteonorm 5.1

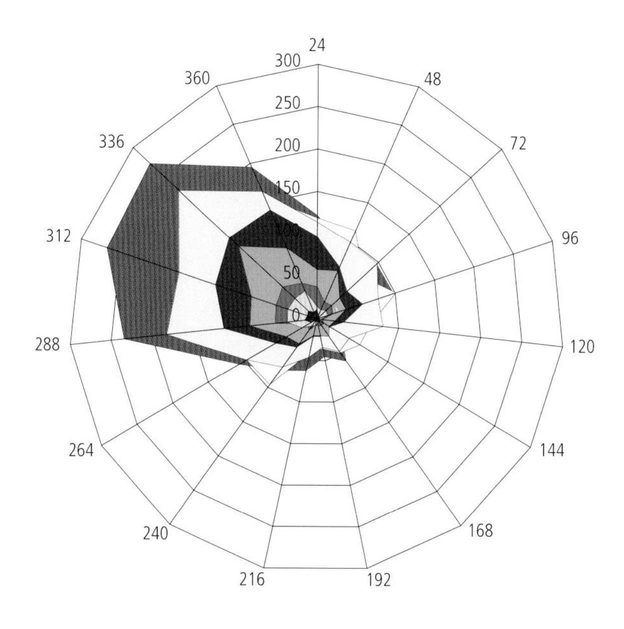

Figure 249.2
Wind rose diagram, year
Abu Dhabi, UAE

Wind speed and frequency distribution of wind direction in hours/annum

Source: Meteorological data meteonorm 5.0

Bild 249.2
Windrose über das gesamte Jahr, Abu Dhabi

Windgeschwindigkeit in m/s und Häufigkeit der Windrichtungsverteilung (h/a)

Quelle: Meteorological data – Meteonorm 5.0

1 2 3 4 5
6 7 8 9 10
11 12 13 14 15
m/s

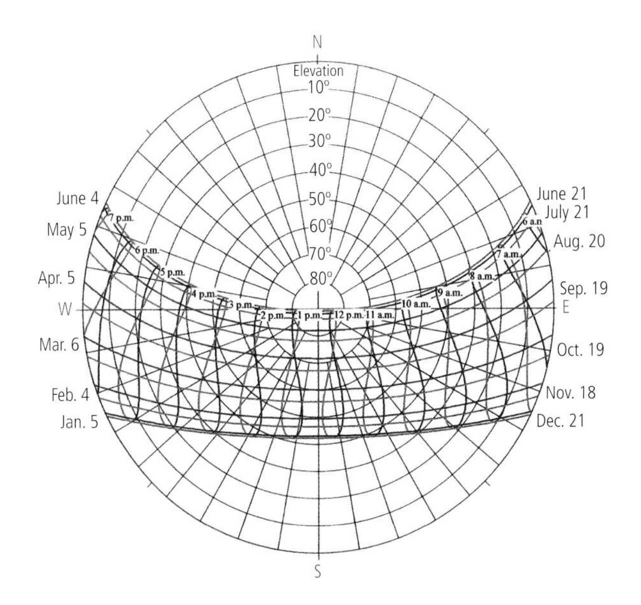

Figure 250.1
Sun orbit diagram
Abu Dhabi, UAE

Bild 250.1
Sonnenumlaufbahn-Diagramm
Abu Dhabi, VAE

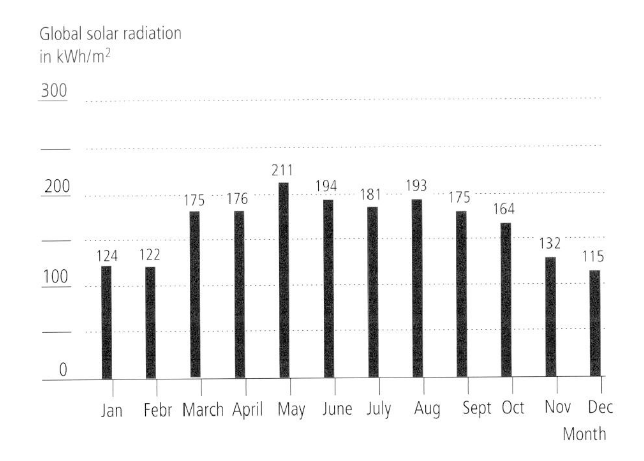

Figure 250.2
Distribution of global solar radiation
Abu Dhabi, UAE

Source: Meteorological data meteonorm 5.1

Bild 250.2
Verteilung der Globalstrahlung
Abu Dhabi, VAE

Quelle: Meteorological data – Meteonorm 5.1

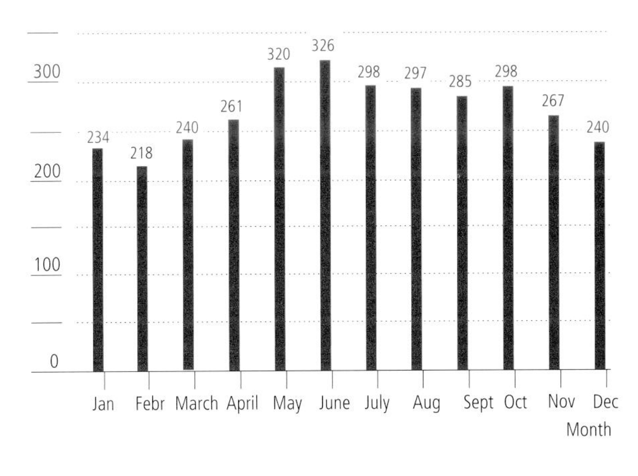

Figure 250.3
Distribution of hours of sunshine
Abu Dhabi, UAE

Source: Meteorological data meteonorm 5.1

Bild 250.3
Verteilung der Sonnenscheindauer
Abu Dhabi, VAE

Quelle: Meteorological data – Meteonorm 5.1

The building in Abu Dhabi will also be equipped with a heat-recovery system that is capable of recapturing around 60 % of the exhaust warm-air energy and using it to supplement the cooling system. If, on the other hand, evacuated solar-tube collectors are introduced to supply heat to operate the absorption chiller cooling system, under the same conditions as in Paris, around 1,000 MWh of cooling energy could be generated. This is equal to around 2.76 % of the annual energy demand.

If we want to supply the entire cooling energy demand with the help of solar systems (evacuated solar tubing), the surface area of a collector field of approximately 75,000 to 80,000 m^2 would be required. **Figure 251** shows the diagram of a combined solar-thermal absorption cooling system arrangement. The necessary heat energy in this case is being supplied mainly by evacuated solar tubes; however, heat from gas-fired chillers or hydrogen fuel cells can be used as an alternative.

If we consider the building in Abu Dhabi to be equipped with wind turbines with similar dimensions as those in Paris, the generated energy would be an annual amount of approximately 380 MWh/a, equal to about 3.8 % of the annual energy demand.

In summary, 6.5 % to a maximum of 7 % of the required total energy to operate the building could be generated by renewable resources directly on or in the building (not included is the heat energy required to generate cooling energy).

Therefore, it can be concluded that the absolute energy generation potential at the location in Abu Dhabi is greater than at the Central European location of Paris, although the resulting overall percentages are very similar because the absolute energy demand in the desert location of Abu Dhabi is significantly greater.

In conclusion, it is not yet possible to equip a high-rise with systems for renewable energy systems capable of supplying 100 % of a tall building's energy needs. Under the most optimistic assumptions and the most ideal site conditions, and with the implementation of the most efficient current technologies, the total amount of energy generated with renewable, building-integrated systems will not exceed 20 – 50 %, at the most.

Wie auch in Paris wird ein Wärmerückgewinnungssystem eingesetzt, das in der Lage ist, ca. 60 Prozent der Kälteenergie durch Wärmetausch mit Abluftströmen rückzugewinnen.

Setzt man zur Erzeugung eines Teils der Kälteenergie Vakuum-Röhrenkollektoren ein, so wären diese unter gleichen Bedingungen wie in Paris in der Lage, ca. 1.000 MWh Kühlenergie bereitzustellen. Dies entspricht in etwa 2,76 Prozent des Jahresenergiebedarfs.

Zur Erzeugung der gesamten Kälteenergieleistung auf solarer Basis wäre ein Kollektorfeld von ca. 75 – 80.000 m^2 notwendig. **Bild 251** zeigt den prinzipiellen Aufbau der kombinierten solar- und wärmebetriebenen Absorptionskühlung. Die Wärmeenergie wird vorrangig durch Vakuum-Röhrenkollektoren erzeugt – ergänzend kann eine direkte Gasbefeuerung der Absorptionskältemaschine erfolgen oder der Einsatz von Wärmeenergie aus einer Brennstoffzelle.

Geht man weiterhin davon aus, dass auf dem Gebäude ein ähnlich großes Windrad wie beim Objekt in La Défense installiert würde, so wäre der elektrische Energieertrag ca. 380 MWh/a, was in etwa 3,8 Prozent des gesamten jährlichen Energiebedarfs ausmacht.

Direkt am Gebäude würden somit insgesamt ca. 6,5 bis maximal 7 Prozent der notwendigen Energie (hier nicht berücksichtigt die Wärmeenergie zur Erzeugung von Kälteenergie) erreicht.

Es lässt sich somit feststellen, dass zwar einerseits der Energieertrag in absoluten Größenordnungen in Abu Dhabi höher ist als in Mitteleuropa, der prozentuale Anteil jedoch in ähnlicher Größenordnung liegt, da die absoluten Verbräuche in Abu Dhabi deutlich höher sind.

Zusammenfassend lässt sich feststellen, dass es nicht ohne weiteres möglich ist, ein Hochhaus mit am Gebäude installierten Energieertragsflächen bis zu 100 Prozent zu versorgen. Selbst bei Idealbedingungen und ausgefeiltesten Techniken dürfte der Ertrag maximal bis an 20 – 50 % heranreichen, je nachdem, in welchem Verhältnis Energieertragsflächen zueinander stehen und was der Standort bietet.

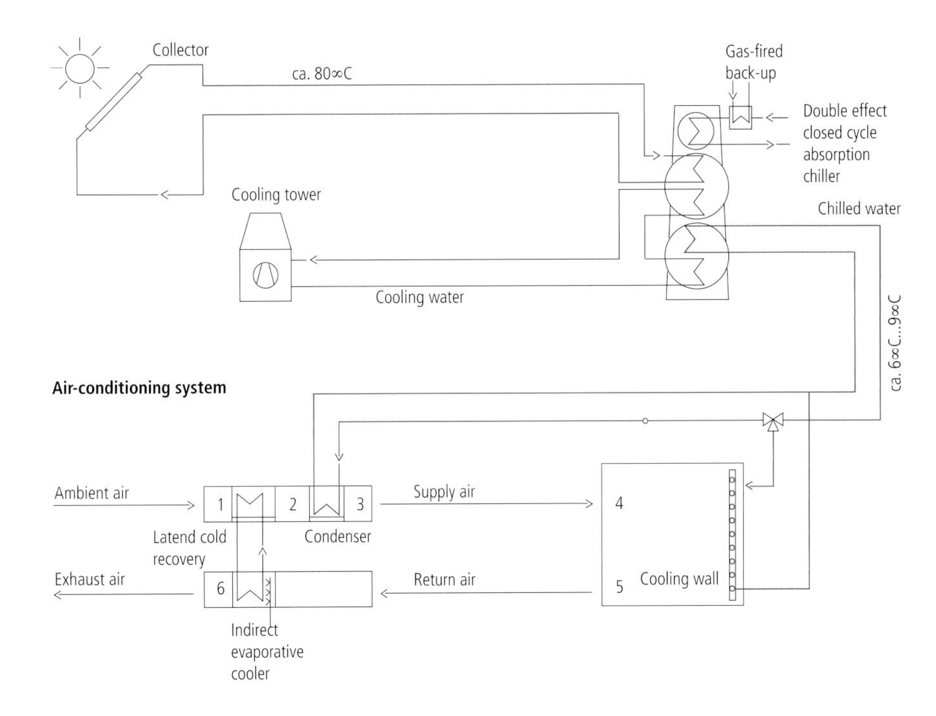

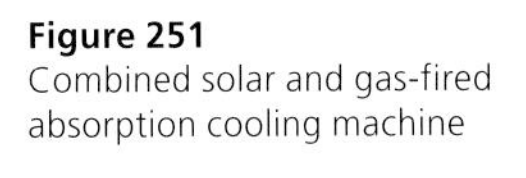

Figure 251
Combined solar and gas-fired absorption cooling machine

Source:
plusminus 20°/40° latitude
Hindrichs · Daniels
Edition Axel Menges

Bild 251
Kombinations-Absorptions-kältemaschine, Solar-thermischer und gasbefeuerter Hybrid

Quelle:
plusminus 20°/40° latitude
Hindrichs/Daniels
Edition Axel Menges

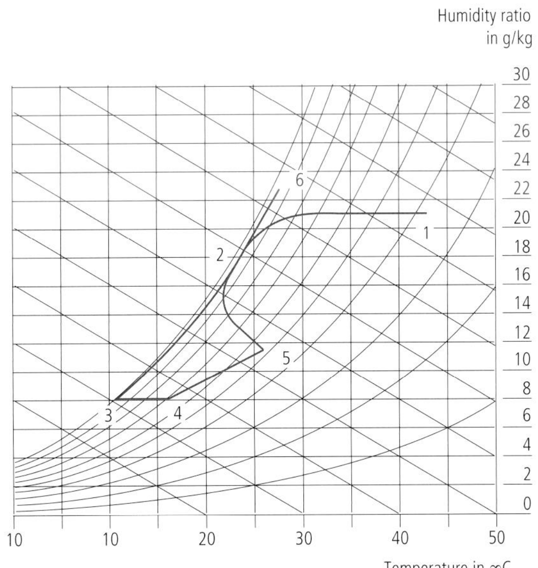
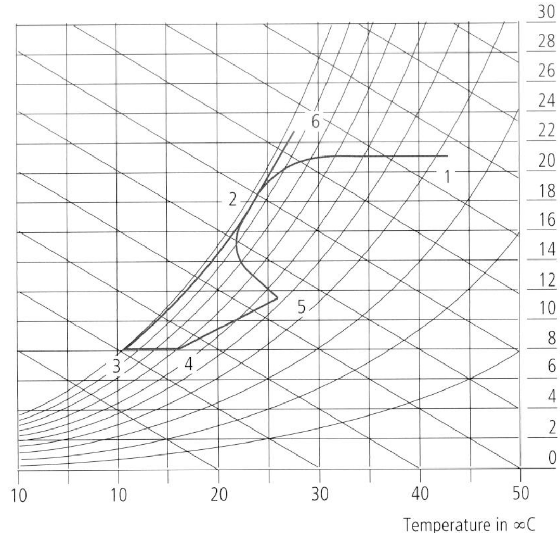

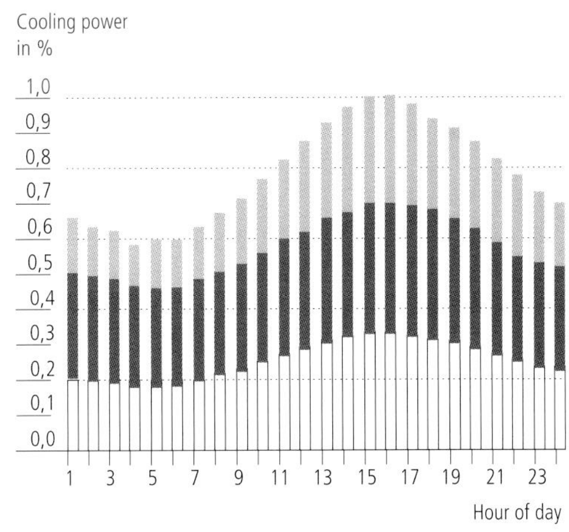

to Figure 251
Typical Middle East design conditions
Psychromatic (Mollier) chart with an air-conditioning process with latent cold recovery

Numbers in psychrometric chart refer to positions in Figure 251 systems diagram.

Source:
plusminus 20°/40° latitude
Hindrichs/Daniels
Edition Axel Menges

zu Bild 251
Charakteristische Design - Bedingungen in Nahost-Klimata. Das h/x (Mollier) Diagramm zeigt Klimatisierung mit Kühlenergie-Rückgewinnung.

Die Zahlen beziehen sich auf das Systemdiagramm in Bild 251

Quelle:
plusminus 20°/40° latitude
Hindrichs/Daniels
Edition Axel Menges

to Figure 251
Typical cooling load profile system with high efficient latent cold recovery

Source:
plusminus 20°/40° latitude
Hindrichs/Daniels
Edition Axel Menges

zu Bild 251
Typischer Verlauf der Kühllast. System mit hocheffizienter, latenter Kälterückgewinnung.

Quelle:
plusminus 20°/40° latitude
Hindrichs/Daniels
Edition Axel Menges

12.5 IcadeTower, Central Europe (continental climate region)

Figures 252 and **253** illustrate the concept for a variable-usage high-rise project with adjacent podium structure for a location in Central Europe. The mixed-use tower has a gross area of 22,010 m²; half of the building will be used for hotel apartments, and the other half will be dedicated to office functions. The podium functions contain conference areas, a restaurant and kitchen, the lobby and event spaces, and a small-scale shopping arcade. The goal was that the tower should be operated largely with the natural resources available at the site.

The high-rise façade surface areas facing south are approximately 5,500 m² large, and 4,500 m² of the building's envelopes are oriented to the northeast and northwest. Approximately 36 % of these enclosure surfaces are made of glass attached to a post-bar curtain wall frame, allowing the exchange between opaque and transparent wall surfaces depending upon internal use. **Figures 254.1** and **254.2** show the system and its design.

12.5 Icade Tower, Mitteleuropa (Kontinentalklimazone)

Für einen mitteleuropäischen Standort wurde ein Turm mit angegliedertem Flachbau – **Bilder 252** und **253** – entwickelt, der als umnutzbare Immobilie betrieben werden soll. Auf ca. 22.010 m² BGF soll jeweils zur Hälfte eine Apartment-Hotel- und eine Büronutzung entstehen. Im Flachbaubereich sind Konferenzzonen, Restaurant und Küche, Lobby- und Eventflächen sowie kleinteiliges Shopping untergebracht. Dabei sollte sich der Turm mit seinem Flachbau weitestgehend aus den natürlichen Ressourcen des Standorts zum Betreiben bedienen.

Die Fassadenfläche des Turms beträgt ca. 5.500 m² südorientiert und ca. 4.500 m² nordwest- und nordostorientiert. Ca. 36 Prozent der Fassadenfläche besteht aus Glasflächen, die in einer Pfosten-Riegel-Konstruktion eingestellt sind, um je nach Nutzung Fenster- und Wandflächen austauschen zu können. Die **Bilder 254.1** und **254.2** zeigen die angestrebten Lösungen und Ansätze zur Gestaltung.

to Figure 252
Icade Tower,
Perspective south

zu Bild 252
Icade Tower,
Süd-Perspektive

Figure 252
Icade Tower,
Perspective north

Source: Project Icade Tower,
van SANTEN & Associates
HL Technik, Klaus Daniels
PSP Architekten Ingenieure

Bild 252
Icade Tower,
Nord-Perspektive

Quelle: Projekt Icade Tower,
van SANTEN & Associés,
HL-Technik, Klaus Daniels
Architekten Ingenieure PSP

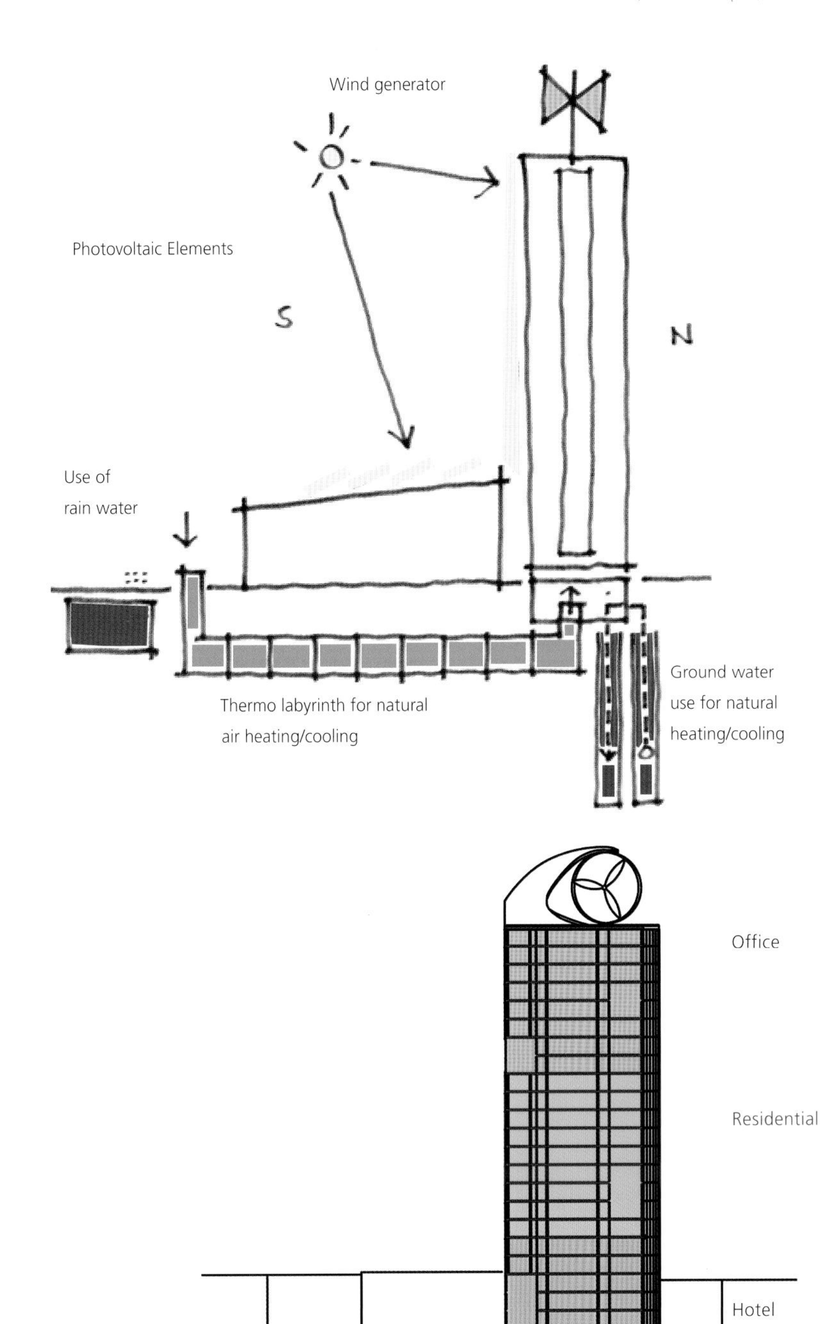

Figure 252.1
Energy concept:
minimization of energy loss
maximization of energy gains
with earth, wind and solar
potentials

Bild 252.1
Energiekonzept:
Verringerung des Energieverlustes, Maximierung der erwünschten Energiegewinne aus geothermischem, Wind-, und solarem Angebot

Figure 253
Urban & building concept:
A multi -use development with residential, office and hotel ensures a balanced disposition of utilisation, reduces traffic and optimizes synergy effects.

Bild 253
Städtebauliches und Gebäudekonzept: Der Funktionsmix aus Wohnen, Büro- und Hotelnutzung verringert innerstädtisches Verkehrsaufkommen und erzeugt positive Synergie-Effekte.

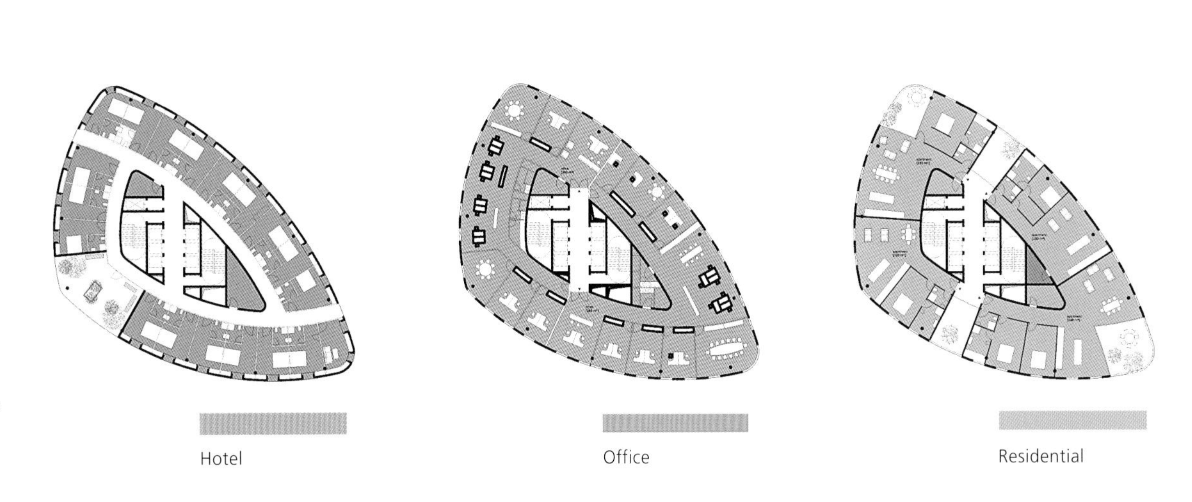

to Figure 253
Urban & building concept:
typical floors with different uses

zu Bild 253
Grundrisskonzepte zu den Funktionsbereichen (von links):
Hotel, Büro, Wohnen

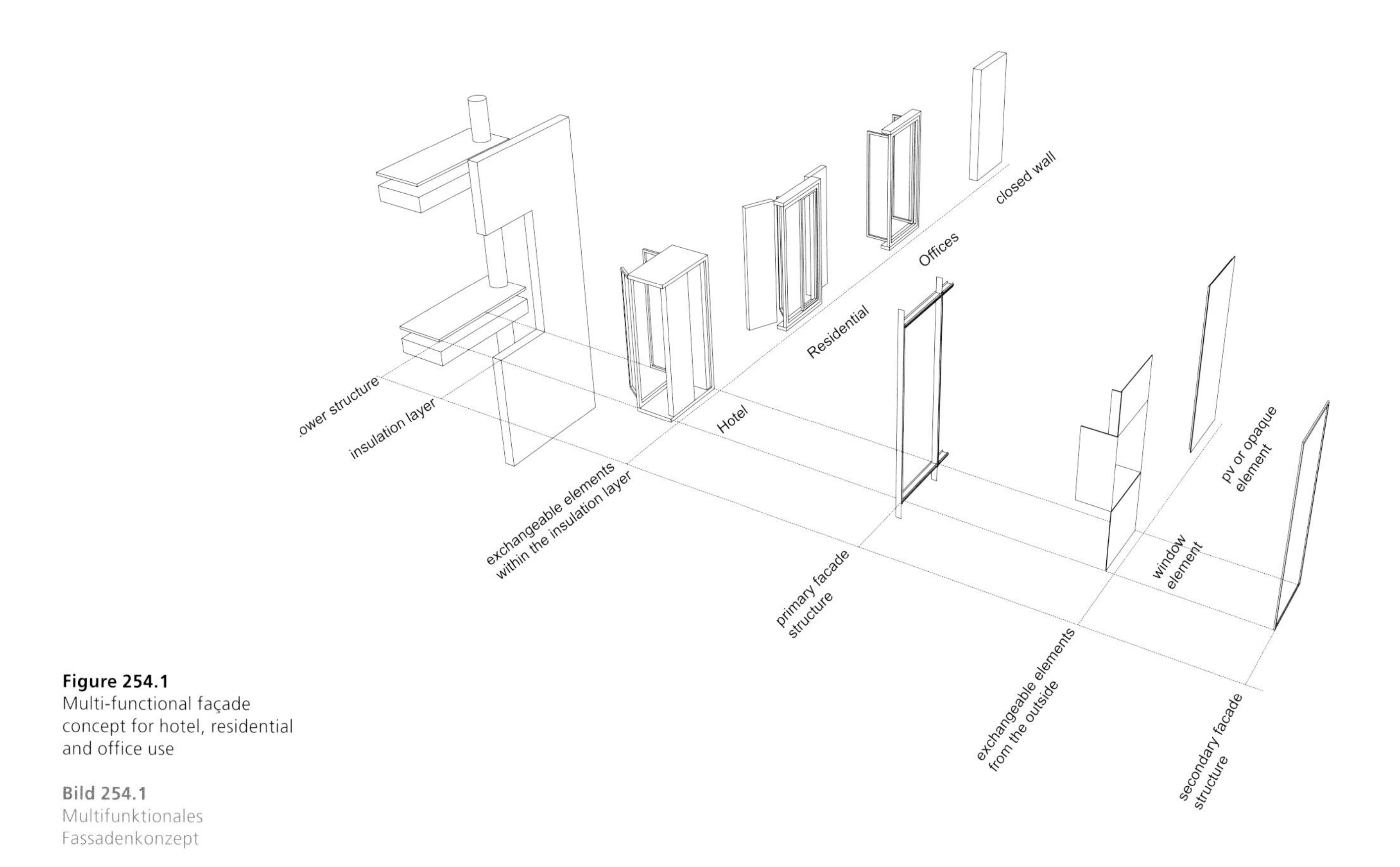

Figure 254.1
Multi-functional façade concept for hotel, residential and office use

Bild 254.1
Multifunktionales Fassadenkonzept

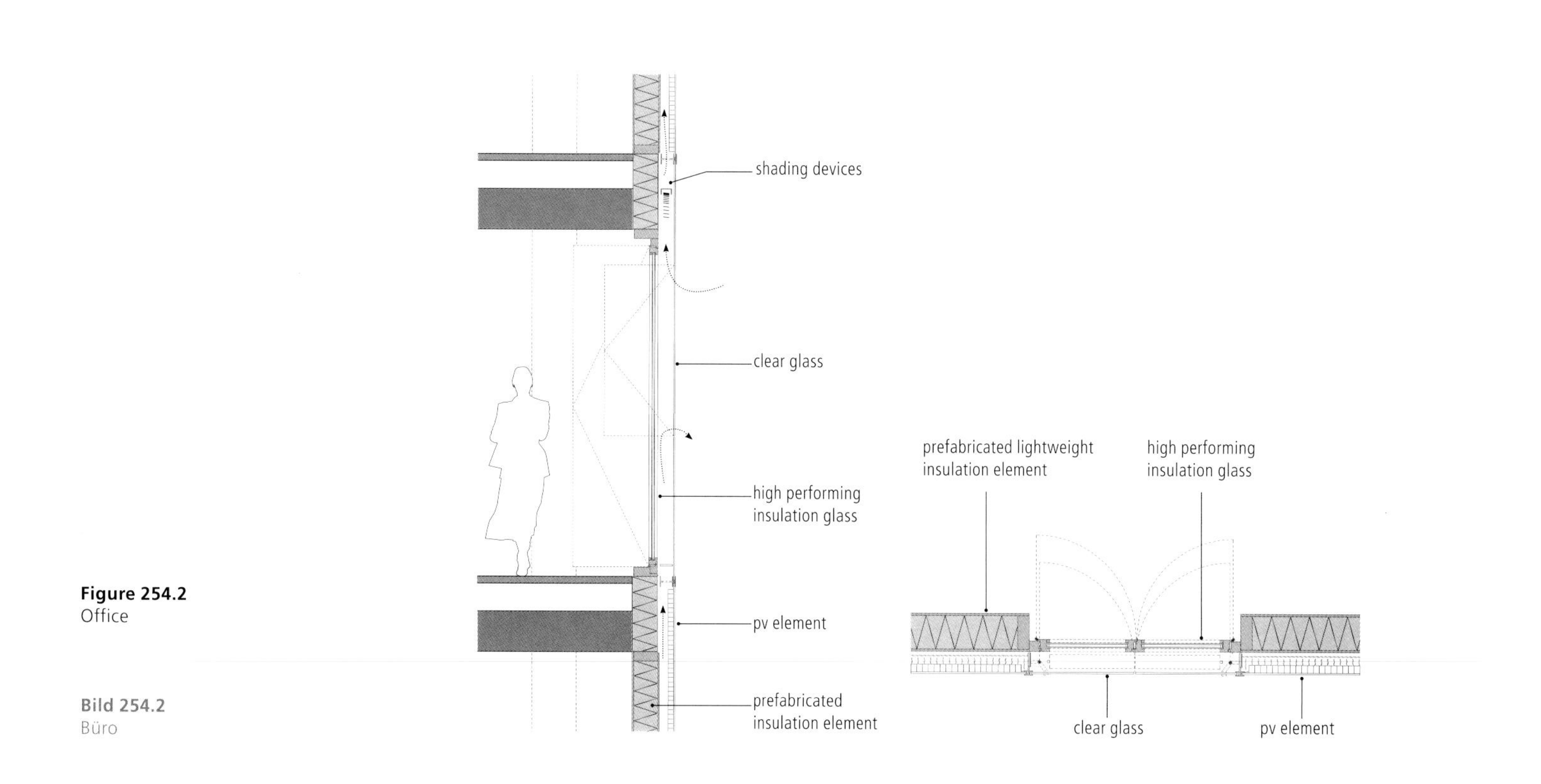

Figure 254.2
Office

Bild 254.2
Büro

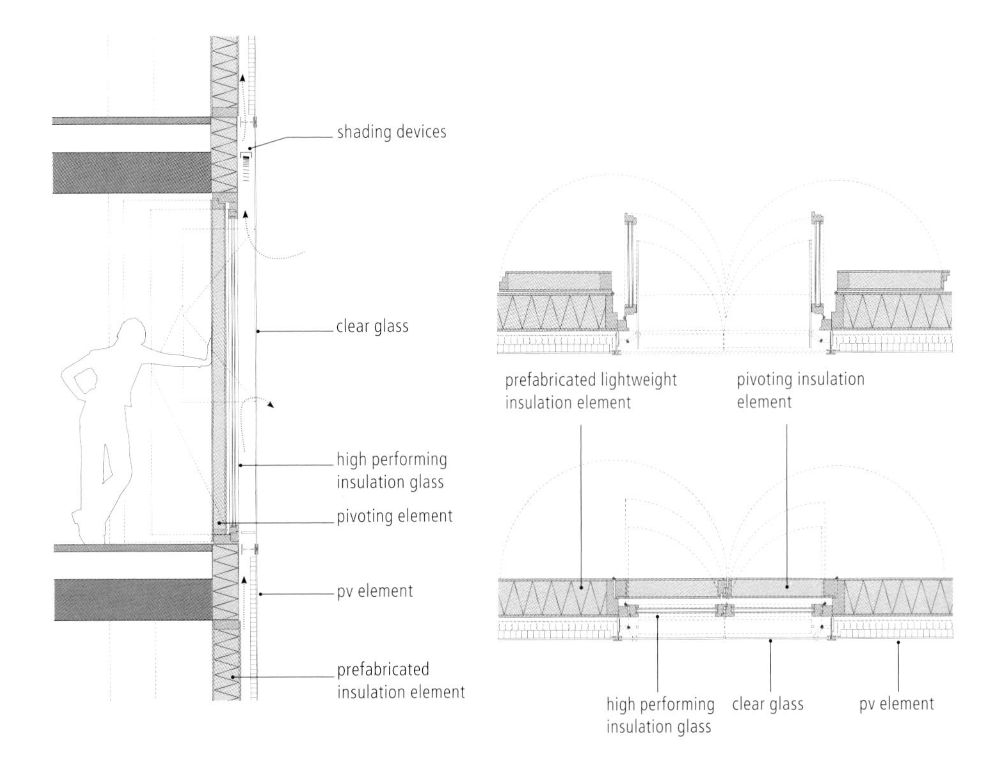

to Figure 254.2
Residential

zu Bild 254.2
Wohnung

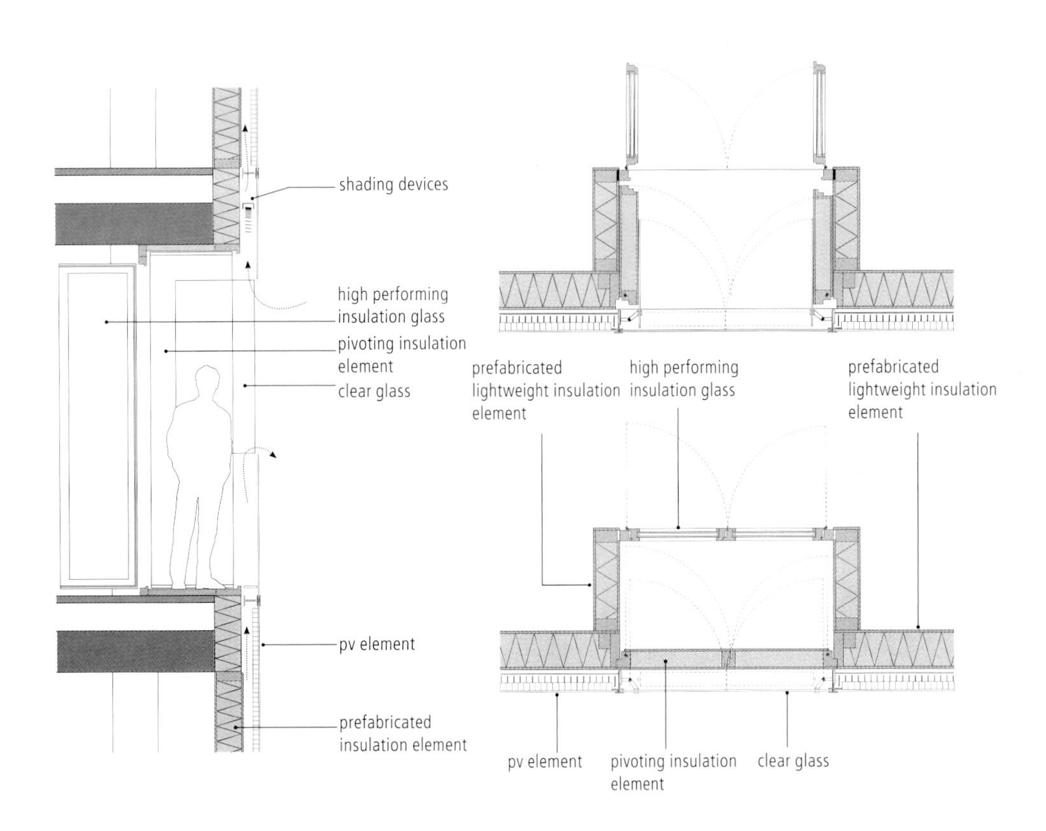

to Figure 254.2
Hotel

zu Bild 254.2
Hotel

In designing a sustainable and highly efficient building systems concept, a first step needs to entail the limitation and reduction of heat, cooling, and electrical energy demands as much as possible.

In order to achieve this goal, the following U-values for enclosure surfaces were selected:

- opaque enclosure surfaces 0.2 W/m²K
- window elements 0.7 W/m²K
- natural ventilation due to infiltration 0.3 ac/h

To provide an optimum of solar shading during the summer and a high degree of natural ventilation during spring and fall seasons, a box-window type was selected that allows for the variation of U-values for the window surfaces from 0.7 W/m²K to approximately 5.5 W/m²K.

The window system design consists of an exterior laminated glass pane, behind which a highly effective solar shading system is placed. In unison with the insulated glass facing the interior space (U-value: 1.0 W/m²K), the assembly's g-values result in the following:

- g-value without solar shading approx. 0.46
- g-value with solar shading approx. 0.06

Due to the fact that the rooms of the building are mostly daylit and oriented along the façade of the building, a zoned lighting concept was developed that limits power consumption for electric lighting to 7 W/m², providing 350 lux of illuminance.

Ventilation of the offices, apartments, and hotel rooms is provided at a rate of 0.94 ac/h, equal to an outside air ventilation rate of 35 m³/hP for hotel rooms and 28 m³/hP for offices. Similarly, the conference area, restaurant, lobby, and event spaces were designed within these values. The total outside air supply volume, without parking, is approximately 75,060 m³/h.

The following table shows the loads and power and the respective energy consumption and generation, achieved by various technologies of the building, based on the above-mentioned parameters.

Erster Schritt in der Planung musste das Bemühen sein, die Wärmeenergieverbräuche, Kühlenergieverbräuche und Verbräuche elektrischer Energie soweit als möglich zu reduzieren.

Aus diesem Grund wurden Wärmedurchgangskoeffizienten gewählt wie folgt:

- opake Wandelemente 0,2 W/m²K
- Fensterelemente 0,7 W/m²K
- natürlicher Luftwechsel infolge Leckage (Fugendichtigkeit) 0,3-fach

Zur optimalen Beschattung im Sommerbetrieb und vor allem zur weitestgehend natürlichen Lüftung in der Übergangszeit wurde eine Kastenfensterkonstruktion konzipiert, die die Veränderung der u-Werte im Fensterbereich von 0,7 W/m²K auf ca. 5,5 W/m²K zulässt.

Hinter einer öffenbaren äußeren Verbund-Sicherheitsglasscheibe befindet sich ein hochwertiger Sonnenschutz, der in Verbindung mit der dahinter liegenden Isolierverglasung (u-Wert 1,0 W/m²K) zu g-Werten führt wie:

- g-Wert ohne Sonnenschutz ca. 0,46
- g-Wert mit Sonnenschutz ca. 0,06

Infolge der im Wesentlichen nach außen orientierten, tagesbelichteten Räume wurde auch unter Berücksichtigung der Reduzierung der elektrischen Energieverbräuche eine zonale Beleuchtung geplant, die bei ca. 350 Lux lediglich eine Anschlussleistung von ca. 7 W/m² besitzt.

Die Belüftung der Büroflächen sowie Apartments und Hotelzimmer erfolgt über einen ca. 0,94-fachen Luftwechsel, was einer Außenluftrate von 35 m³/hP in den Hotelzimmern und einer solchen von 28 m³/hP in den Büros entspricht. Analoge Ansätze erfolgten in den Bereichen Konferenz und Restaurant sowie Küche, Lobby und Eventräume. Die Gesamtaußenluftmenge liegt (ohne Parking) bei ca. 75.060 m³/h.

Auf der Basis der vor aufgeführten Randbedingungen sind in der nachfolgenden tabellarischen Struktur sowohl die dimensionierenden Lasten und Leistungen sowie die Energieverbräuche und Energieerträge durch die verschiedensten Technologien dargestellt.

Peak/hour

Energy for	Loss/loads by transmiss. radiation	Loss/loads total (incl. all consumer)	Peak	Energy serving by		
				Boiler chiller	EP-system	Ground-water soil system Therm. lab
	W/m² GFA	W/m² GFA	kW	kW	kW	kW
Cooling	14.56	37.3	446	–	–	Groundw. 446 (76m³/h)
				–	–	Therm. lab 216
Heating	13.16	18.30	279	100	180	
				–	–	Therm. lab 339
				–	–	Heat recov. 351
				–	–	Heat loss EPS 43
Electr. Power	–	21.30	469	E.P.S. net 313	120	
			Photovoltaic 293_P	–	–	–
			Wind mill 96_P	–	–	–

max. Leistung/Stunde

Energie für	Wärmeverl., Kühllasten d. Baukörpers	spezifische Kühlleist./ Heizleistung	max. Leistung	Energieerzeugung durch		
				Kessel, Kälte-masch.	EPS-system	Grund-wasser und system Therm. Lab.
	W/m² GFA	W/m² GFA	kW	kW	kW	kW
Kühlung	14.56	37.3	446	–	–	Grundw. 446 (76m³/h)
				–	–	Therm. Lab. 216
Heizung	13.16	18.30	279	100	180	
				–	–	Therm. Lab. 339
				–	–	Wärmerückg. 351
				–	–	Wärmeangeb. 43
Strom	–	21.30	469	Netz 349	120	
			Photovoltaic 293_P	–	–	–
			Wind mill 96_P	–	–	–

Consumption/year

Consumption	Consumption	Production	Energy use	District systems	Reduction CO_2
MWh	MWh/m²	MWh	MWh	MWh	kW
Groundw. 254.2*	11.54	Groundw. 43,000 m³	–	–	100% *renewable energy
Therm. lab 88.4*	–	-	–	–	
737	33.48	EPS 900		163	ca. 14% *renewable energy (only using air cond.)
71.6*	–	Boiler 51	Gas for EPS 1,715	–	
74.2	–	–		–	
215	–	–		–	
799	36.3	(199)			
–		EPS 600		223	*52.8% renewable energy
	./. 9.35	Photovoltaic			
–		206*	total		
	./. 9.81	Wind mill	./. 422*		
–		216*			
			Gas for EPS [illegible]		

Verbrauch/Jahr

Verbrauch	Verbrauch	Produktion	Erzeugung	Energiever brauch Netz	CO_2 Verringerung
MWh	MWh/m²	MWh	MWh	MWh	kW
Grundwasser 254,2*	11,54	Grundwasser 43.000 m³	–	–	100% * erneuerbare Energie
Therm. Labyr. 88.4*	–	-	–	–	
737	33,48	EPS 900		163	ca. 14% * erneuerbare Energie (Nur Klimatisierung)
71,6*	–	Boiler 51	Gas für EPS 1.715	–	
74,2	–	–		–	
215	–	–		–	
799	36,3	(199)			
–		EPS 600		223	*52,8% * erneuerbare Energie
	./. 9,35	Photovoltaic			
–		206*	total		
	./. 9,81	Windturbine	./. 422*		
–		216*			
			Gas für EPS 1.293		

Project Icade Tower (22,010 m²GFA), Overview of consumption/production of end energy

* renewable energy
EPS emergency power supply

Entwurf des Icade Towers (22.010 m²GFA) Übersicht über Endenergie-verbrauch und -erzeugung

* erneuerbare Energie
EPS BHKW-System

For the geographical location of the building, a detailed analysis of the potential wind energy generation based on average wind velocities was conducted, which is shown in the diagrams of **Figures 255.1, 255.2**. Because the building's height of 100 m is fairly moderate, the amount of annual energy generated from wind will be moderate as well.

Electric power generated by photovoltaic technology is the result of the global radiation at the particular site and the south-oriented photovoltaic panels and their slope. **Figure 256.1** depicts the typical profile of global radiation for countries situated between 40 ° and 65 ° latitude. The building will have photovoltaic elements integrated in vertical façades at 90 ° with a net surface area of approximately 1,940 m², as well as 15 °-sloped roof-surface photovoltaic elements with a net surface area of 1,480 m².

Für den Standort wurde eine detaillierte Untersuchung der Windenergieerträge mit mittleren Windgeschwindigkeiten erarbeitet, die in Diagrammen, **Bilder 255.1, 255.2** dargestellt sind. Da das Gebäude lediglich eine Höhe von ca. 100 m besitzt, sind die jährlichen Windenergieerträge in einem moderaten Bereich.

Die elektrischen Energieerträge auf Basis Photovoltaik ergeben sich aus der Globalstrahlung des angenommenen Standortes sowie der Lage/Stellung der Photovoltaikanlage, südorientiert. Die Darstellung der Globalstrahlung, **Bild 256.1** zeigt das typische Profil von Ländern, die im Bereich von 40 – 65 ° (Längengrad) liegen. Beim Gebäude vorgesehen sind Photovoltaikelemente in vertikalen Fassaden (90 °, ca. 1.940 m² netto) sowie annähernd horizontalen PV-Dachflächen (15 °, 1.480 m² netto).

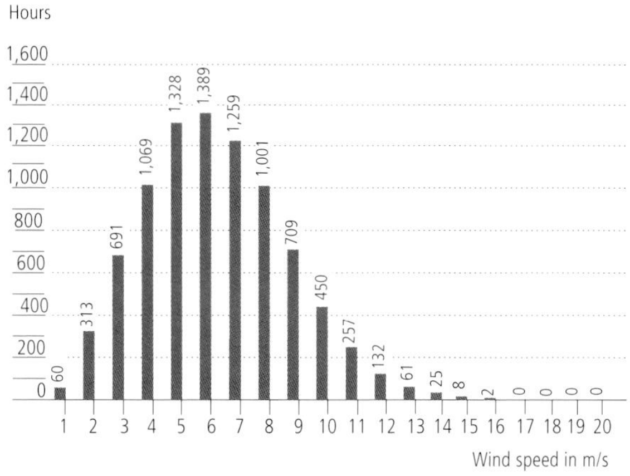

Hours
1,600
1,400
1,200
1,000
800
600
400
200
0
60 313 691 1,069 1,328 1,389 1,259 1,001 709 450 257 132 61 25 8 2 0 0 0 0
1 2 3 4 5 6 7 8 9 10 11 12 13 14 15 16 17 18 19 20
Wind speed in m/s

Figure 255.1
Typical distribution of wind speeds at heights of 100 meters,Central Europe

Source: TRY 14 Weather data

Bild 255.1
Windgeschwindigkeitsverteilungin 100 m Höhe (Beispiel mitteleuropäischer Standort)

Quelle: Wetterdaten TRY 14

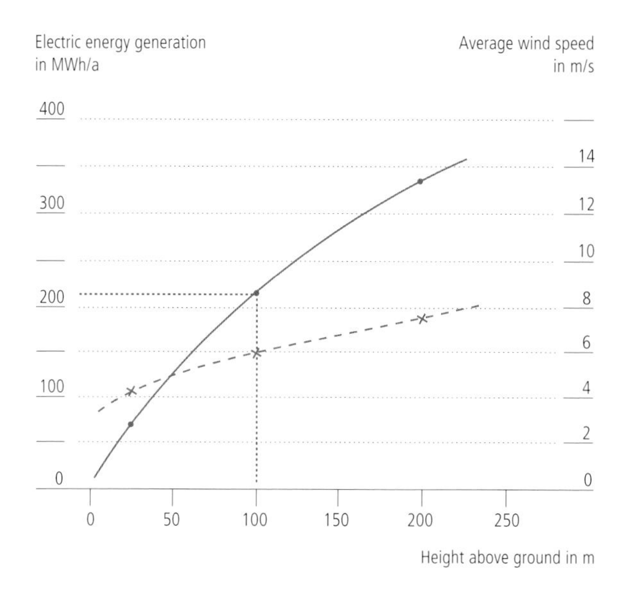

— Electric energy generation
-- Wind speed

Figure 255.2
Approximate annual electric energy generation of a wind turbine with a rotor diameter of 20 meters, a nominal power of 100 kW, at a wind speed of 15 m/s

Source: Daniels/author

Bild 255.2
Elektrischer Energieertrag (ca.) per annum einers Windrades, Rotorendurchmesser 20 m, Nennleistung 100 kW bei 15 m/s

Quelle: Daniels/Autor

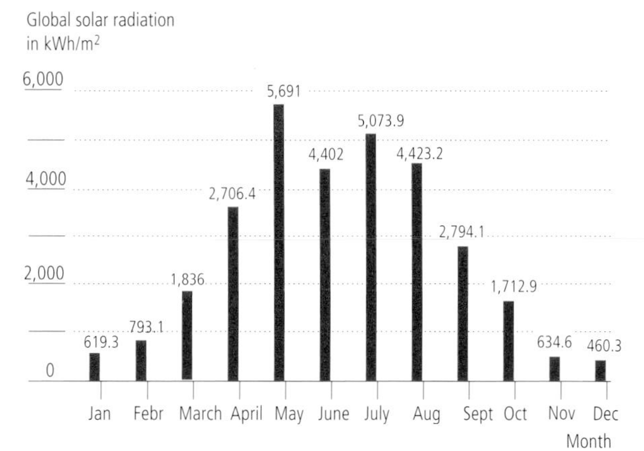

Figure 256.1
Global solar radiation for a Central European location

Source: TRY 14 weather data

Bild 256.1
Globalstrahlung, (Beispiel mitteleuropäischer Standort)

Quelle: Wetterdaten TRY 14

Figure 256.2
Voltage characteristics of mono-crystalline photovoltaic module (Siemens SM 55) for different module temperatures. Incident solar energy 1,000 W/m²

Source: The Technology of Ecological Building, K. Daniels Birkhäuser Verlag 1995

Bild 256.2
Strom-Spannungskennlinien eines monokristallinen PV-Moduls (Siemens SM 55) bei unterschiedlichen Modul-Temperaturen
Bestrahlungsstärke 1,000 W/m²

Quelle: Technologie des ökologischen Bauens, K. Daniels Birkhäuser Verlag 1995

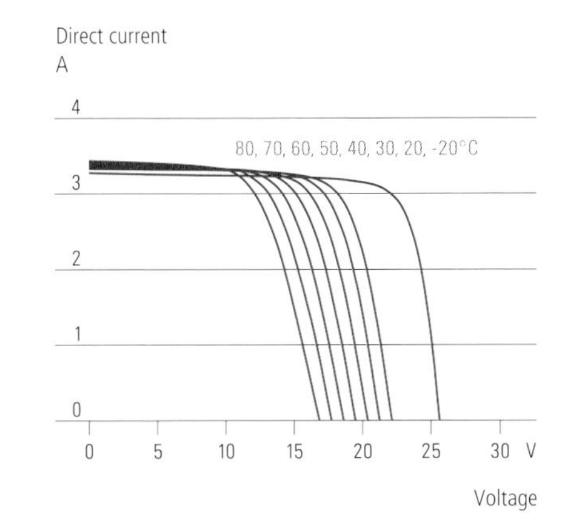

Figure 256.3
Voltage characteristics of mono-crystalline photovoltaic module (Siemens SM 55) for different incident solar energy in W/m².
Module temperature 25 °C

Source: The Technology of Ecological Building, K. Daniels Birkhäuser Verlag 1995

Bild 256.3
Strom-Spannungskennlinien eines PV-Moduls (Siemens SM 55) bei unterschiedlichen Bestrahlungsstärken in W/m²
Modultemperatur 25 °C

Quelle: Technologie des ökologischen Bauens, K. Daniels Birkhäuser Verlag 1995

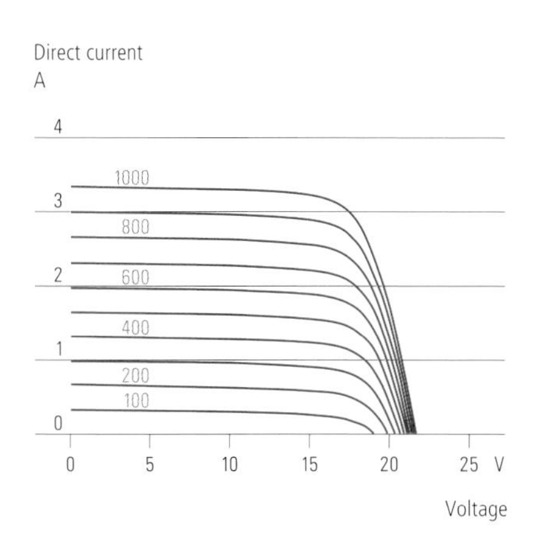

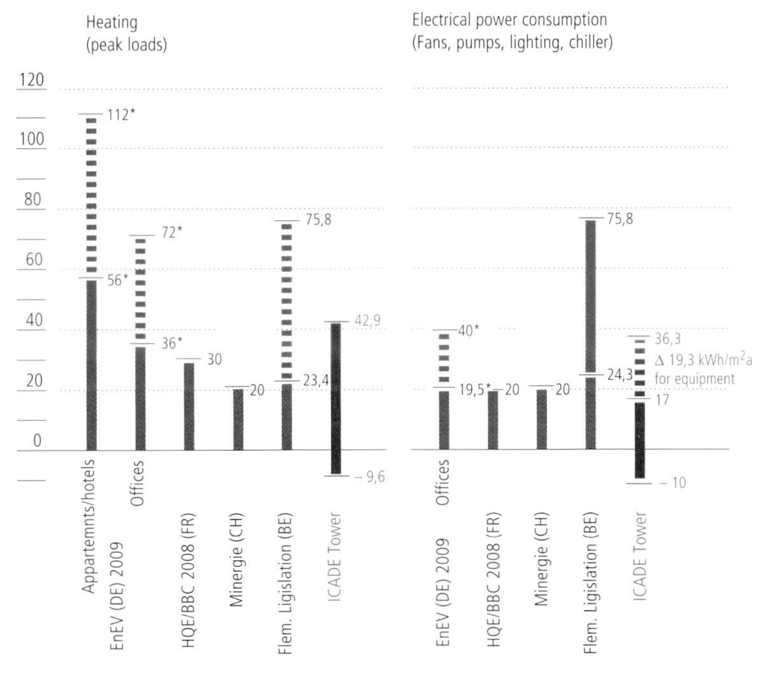

Figure 257.1
Primary energy consumption in kWh/m²a

Source: Project Icade Tower, van SANTEN & Associates
HL Technik, K. Daniels
PSP Architekten Ingenieure

Figure 257.1
Primärenergieverbrauch in kWh/m²a

Quelle: Projekt Icade Tower, van SANTEN & Associés
HL Technik, K. Daniels
PSP Architekten Ingenieure

Figures 257.1 – 257.3 introduce a comparison between primary and final energy consumption of the building project in relationship with the various energy standards such as EnEV , Germany; Low-Consumption-Building label (BBC-HQE), France; and Minergie, Switzerland.

Additionally, the annual energy consumption and generation as a result of the employed combined-heat power plant (CHP), the photovoltaic system, and the energy generated from wind are shown.

Energy generation in kWh/m² month are shown in **Figure 256.2** for various slopes of the photovoltaic panels. They complement themselves in a nearly ideal sense and result in a relatively constant annual energy generation pattern, although changes in voltage must be noticed (**Figure 256.3**).

Die Darstellungen auf den **Bildern 257.1** bis **257.3** zeigen einen Vergleich von Primär- und Endenergieverbräuchen des Objektes im Verhältnis zu verschiedenen Energiestandards wie EnEV (Deutschland), BBC (HQE) (Frankreich) und Minergie-Standard (Schweiz).

Eine weitere Darstellung zeigt den Verlauf der jährlichen Energieverbräuche und Energieerträge infolge des Einsatzes einer Blockheizkraftwerk-Anlage, Photovoltaik und Windenergie auf.

Wie die Darstellung, **Bild 256.2**, der Energieerträge in kWh/m² Monat für die verschiedenen Neigungen der Photovoltaikelemente zeigt, ergänzen sich diese in idealer Weise und führen zu einem relativ gleichmäßigen jährlichen Gesamtertrag, wobei jedoch die Veränderungen der Stromspannungskennlinien – **Bild 256.3** – zu berücksichtigen sind.

Figure 257.3 shows a comparison between (1) the designed dimensions and power of the building systems for heating, cooling, and electrical energy, and (2) annual consumption. The building is connected to the grid of a district heating plant and a medium-voltage electricity grid that serves as electric storage. As shown in the diagram, the building's 450-meter-long thermal labyrinth placed underneath the podium part of the building is an important element for defraying peaks in the cooling and heating systems, as well as for ventilation equipment; however, on a yearly basis the labyrinth's contribution is small because only outside air is treated. In summer, the entering outside air is pre-cooled in the thermal labyrinth, yet a subsequent cooling by ground water provides the majority of the required passive cooling. Electrical energy consumption for mechanical chiller power is therefore reduced to a minimum.

Electrical energy supply is composed of the connection to the electricity utility grid on one hand and the building-integrated cogeneration power plant on the other. With a designed annual operation of 5,000h/a, the cogeneration plant contributes significantly to electricity generation. Both the photovoltaic system and the installed wind turbines make noteworthy contributions to the energy generation within the building, yet surplus energy needs to be either sold to third party users or stored.

Bild 257.3 zeigt eine vergleichende Betrachtung im Bereich Heizung, Kälte und elektrische Energie, jeweils bezogen auf die dimensionierenden (zu installierenden) Systemtechniken und ihre Leistungen sowie die jährlichen Verbräuche und Erträge unter gleichzeitiger Nutzung eines angeschlossenen Fernwärmesystems bzw. eines Mittelspannungsnetzes der städtischen Versorgung, das als Elektrospeicher dient. Wie aus der Darstellung ersichtlich, ist das ca. 450 m lange und unter dem Flachbau liegende Thermolabyrinth ein wesentliches Element zur Reduzierung der Spitzenleistung (Wärmeenergie/Kälteenergie Raumlufttechnischer Anlagen), erzeugt jedoch geringere Erträge über das gesamte Jahr, da nur die zugeführte Außenluft behandelt wird. Im Sommer soll die Außenluft für das Gebäude anfänglich über ein Thermolabyrinth abgekühlt werden – die nachgeschaltete Grundwasserkühlung übernimmt den wesentlichen Teil der nötigen Kühlenergieerzeugung bei geringstem Einsatz elektrischer Energie (Pumpenleistungen).

Die elektrische Energieversorgung setzt sich aus einer Netzversorgung und der Versorgung über ein Blockheizkraftwerk zusammen, wobei dieses bei einer angenommenen Laufzeit von 5.000 h/a einen wesentlichen Beitrag zur Stromversorgung liefert. Sowohl die Photovoltaikelemente als auch das Windrad tragen mit nennenswerten Anteilen zur elektrischen Energieversorgung bei, jedoch ist es notwendig, jederzeit die überschüssigen Energieerträge entweder an Drittnutzer abgeben zu können oder anderweitig zu speichern.

Figure 257.3
Production of energy

Source: Project Icade Tower, van SANTEN & Associates
HL Technik, K. Daniels
PSP Architekten Ingenieure

Bild 257.3
Energieerzeugung

Quelle: Projekt Icade Tower, van SANTEN & Associés
HL Technik, K. Daniels
PSP Architekten Ingenieure

Electrical power production and consumption in MWh

Σ PV-Systeme/Windgenerator/E.P.-System

EPS — EPS — EPS

80, 60, 40, 2, 0

Jan Feb Mar Apr May Jun Jul Aug Sept Oct Nov Dec
Month

— Consumption = 800,000 kWh/a 66.6 MWh/month
— Energency power supply (EPS) 120 $kW_{E,\ peak}$ 600,000 kWh/a by 5,000 h/a operating hours/year electrical energy + 900.000 kWh/a heating energy
— Wind generator average generation of windpower = 216,000 kWh/a
- - Photovoltaic-system generation 206,410 kWh/a

Bild 257.2
Energy concept, electrical power production and consumption per month in MWh

Source: Project Icade Tower, van SANTEN & Associates
HL Technik, K. Daniels
PSP Architekten Ingenieure

Bild 257.2
Energiekonzept, elektrische Energieerzeugung und -Verbrauch pro Monat in MWh

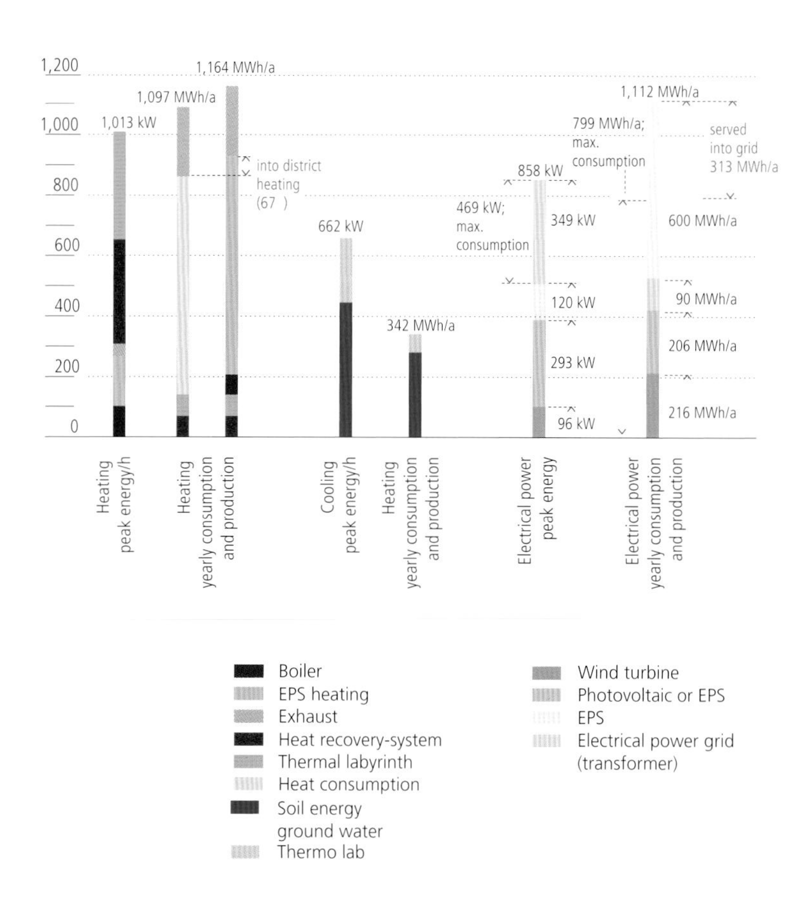

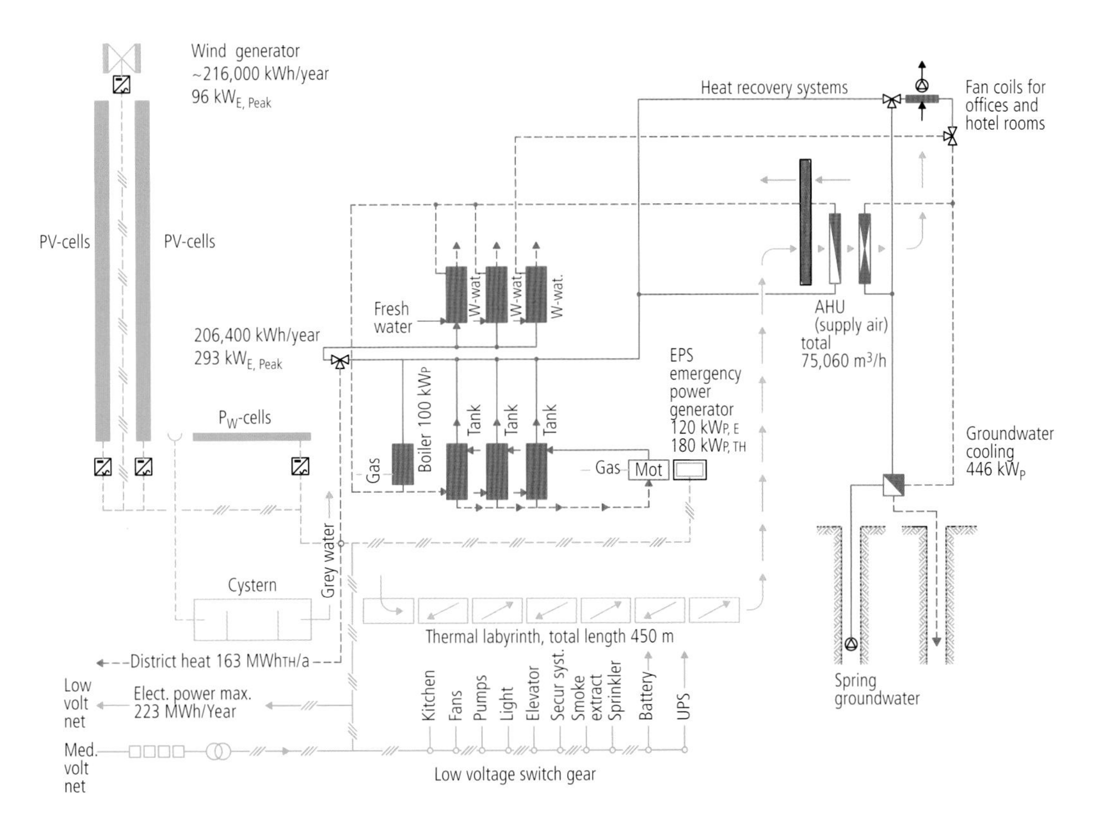

Figure 258
Energy concept:
technical MEP-scheme
Overview of energy serving systems

Source: Project Icade Tower, van SANTEN & Associates
HL Technik, K. Daniels
PSP Architekten Ingenieure

Bild 258
Energiekonzept mit Uebersicht der energieerzeugenden Systeme

Quelle: Projekt Icade Tower, van SANTEN & Associés
HL Technik, K. Daniels
PSP Architekten Ingenieure

Figure 258 shows diagrammatically the concerted interaction between the elements of air processing, heating, cooling, and electricity supply of the building. Cooling energy is derived completely from renewable resources, yet the percentage of renewable energy sources for heating only amounts to 14 %. The percentage of electricity generated by renewable technologies is 52.8 %, an astonishingly high amount, with wind power and photovoltaic power generating almost equal amounts.

To summarize all energy consumption sectors of the building such as cooling and heat energy, including the exhaust gas losses and the electric energy consumption of 2,010 MWh/a, and the related grid supply, including the supply by natural gas to building-generated power, the percentage of renewable energy contribution is 64 %. The project is an example of how clearly coordinated, integrated planning of all participants in the design process results in a very significant decrease in energy consumption out of fossil sources – and, hence, in CO_2 emissions.The remainder of the energy consumption of ca. 36 % for the particular building needs to be covered by energy from large power plants, situated in or around the urban centers – the virtual power plant.

In contrast to the Pearl River Tower high-rise, the wind turbine on top of the Icade tower is freely exposed to the oncoming air stream and orients itself always optimally toward the direction of the wind, thus maximizing wind power efficiency. The Icade turbine configuration prevents the buildup of higher pressure differentials that will cause the wind to flow off the edges of the building surface and, as a result, not reach the turbines with maximum impact.

Das Zusammenspiel der verschiedensten Elemente der Luftaufbereitung, Heizung, Kühlung und Stromversorgung, ergibt sich aus der schematischen Darstellung der gesamten Systemtechnik für das Gebäude, **Bild 258**. Während die Kälteenergie zu 100 Prozent aus erneuerbarer Energie bezogen wird, beträgt der Anteil erneuerbarer Energien bei der Wärmeerzeugung lediglich 14 Prozent. Der Anteil erneuerbarer Energie mit 52,8 Prozent bei der Erzeugung elektrischer Energie ist außerordentlich hoch, wobei sowohl die Photovoltaik als auch das Windrad jeweils einen annähernd hälftigen Beitrag liefern.

Fasst man sämtliche Energieverbräuche für Kühlenergie, Wärmeenergie (einschließlich Abgasverluste) und elektrische Energie mit 2.010 MWh/a zusammen und stellt sie ins Verhältnis zur bezogenen Energie sowohl aus dem Elektronetz wie auch des einzusetzenden Gases, so ergibt sich ein Gesamtertrag aus erneuerbarer Energie von ca. 64 Prozent. Somit zeigt sich, dass bei konsequenter Planung und richtigem Zusammenwirken aller Planungsbeteiligten ein erheblicher Beitrag zur Reduzierung des Energieverbrauchs fossiler Brennstoffe sowie des CO_2-Ausstoßes möglich ist. Der verbleibende Energieverbrauch von ca. 36 Prozent muss aus Versorgungsnetzen erfolgen, die aus Großanlagen im oder um den Stadtraum herum kommen – dem virtuellen Kraftwerk.

Im Gegensatz zum Gebäude Pearl River Tower wird das Windrad auf dem Icade-Tower frei angeströmt und dieses jeweils in die Hauptwindrichtung ausgerichtet, um die Windenergie optimal umzusetzen. Dabei wird vermieden, dass durch höhere Druckveränderungen die Windströmungen über die Gebäudeoberfläche abfließen – somit nicht auf die Windräder mit vollem Energieinhalt auftreffen.

Part 3

Utilizing renewable energies

Energy storage

Conclusion

Teil 3

Nutzung erneuerbarer Energien

Energiespeicherung

Fazit

13 Utilizing renewable energies

13 Nutzung erneuerbarer Energien

As presented in previous chapters, the use of renewable energy sources is at present still resulting in significantly greater technical complexities and cost than the use of fossil sources.

As a consequence, it will be necessary to reduce overall energy consumption as
a whole. Additionally, materials used in construction need to be recyclable to the greatest possible degree because natural resources are limited. Our knowledge about such de-construction, with full awareness of the true value of raw materials, is still at the very beginning, and its importance is being underestimated gravely. Effective use of renewable energy sources will provide us, in next 30 – 40 years, with the means for survival, but we need to realize that their implementation will not be a case of choice but of necessity.

The two areas of concern – energy and reuse – need continued innovation as an absolute imperative. The passive and direct use of the resources given to us by nature will play a major role, and only when the potential for renewable energy use is fully exhausted in any building application – depending upon the local climatic and other conditions – should additional service demand be met using active technologies.

13.1 Active technologies in and at buildings

Natural ventilation, day lighting, passive heating, and passive cooling should be in the foreground of any project development today.

As will be described in the following chapters, active building systems should only be considered to close a comfort gap when passive system provisions are insufficient, and this should be particularly the case for electricity.

13.1.1 Thermal storage and night cooling

The thermal storage capacity of a building's interior plays a major role in the passive reduction of internal cooling loads and the resulting room temperatures. The effects of the thermal storage capacity of a structure become apparent when heat gains exist as a result of radiation or in changing room temperatures. In such cases, the temperature changes over time are delayed by thermal storage inside the building in the sense that the load-time function changes to a cooling load-time function. A room unit accepts thermal energy in its space-defining surfaces due to solar gains, artificial lighting, occupants, equipment, or other sources of thermal energy. This energy penetrates the various enclosing building materials to various degrees and at different depths, and in doing so it causes a temperature increase due to radiation and/or convection. Materials of the enclosing surfaces and the

Die Nutzung erneuerbarer Energien ist, wie alle vorherigen Ausführungen dargestellt haben, zurzeit mit einem deutlich höheren Aufwand verbunden als der Einsatz fossiler Brennstoffe.

Insofern muss es vorrangiges Ziel sein, den Energiebedarf auf breiter Front deutlich zu senken. Weiterhin muss es unter Berücksichtigung der Endlichkeit der eingesetzten Materialien ein vorrangiges Ziel sein, Werkstoffe so zu verbauen, dass sie beim Abbruch eines Gebäudes (Rückbau) in höchstmöglichem Maß recycliert werden können. Wie bereits festgestellt, steht sein Jahren dieses Kapitel immer noch am Anfang und wird offensichtlich nach wie vor massiv unterschätzt.

Der Einsatz erneuerbarer Energien sichert uns in ca. 30 – 40 Jahren unser Überleben. Es ist in Zukunft nicht eine Frage des Wollens, sondern des Müssens. Unter der Erkenntnis, dass die eingesetzten Stoffe und technischen Geräte wiederum nur endlich sind, ist es eine zwingende Notwendigkeit, innovative Lösungen zu finden, die beide Anspruchsfelder decken. Dabei spielt der direkte Einsatz (passive Einsatz) der uns von der Natur geschenkten Ressourcen eine ganz wesentliche Rolle. Erst wenn alle Möglichkeiten des passiven Einsatzes des Außenraums je nach Standort ausgenutzt sind, sollten aktive Technologien an und im Gebäude die Rest-Dienstleistung der Versorgung übernehmen.

13.1 Aktive Technologien an und im Gebäude

Die natürliche Belüftung, natürliche Belichtung, passive Beheizung, passive Kühlung usw. stehen bei einer Projektentwicklung vorrangig im Vordergrund.

Die aktiven Systeme, wie nachfolgend aufgeführt, sollten lediglich eine Ergänzungsfunktion für den Fall darstellen, dass die passiven Systeme und Möglichkeiten nicht ausreichen, die gewünschten und erforderlichen „Energie-Dienstleistungen" zu decken. Dies gilt insbesondere für die elektrische Energie.

13.1.1 Wärmespeicherung/Nachtauskühlung

Bei der Reduzierung von Kühllasten und bei der Reduzierung der hieraus folgenden Raumtemperaturerhöhungen spielt die Wärmespeicherung im Gebäude eine wesentliche Rolle. Sie macht sich dann bemerkbar, wenn die Wärmebelastung entweder infolge von Strahlung entsteht oder wenn sich Raumtemperaturen verändern. In diesen Fällen werden durch speichernde Bauelemente die zeitlichen Belastungen so verändert, dass aus der Belastungszeitfunktion eine Kühllast-Zeitfunktion entsteht. Ein Raum als Einheit nimmt Wärmeenergie über seine Umschließungsflächen auf, die in diesem durch Sonneneinstrahlung, Beleuchtung,

enclosed air mass are then in a process of continuous and mutual radiation exchange. Depending upon the temperature of the surfaces and the air, the result is a convective and radiation thermal energy exchange-from the room's surfaces to the air and vice versa. Under varying room temperatures, increasing temperature means a reduction in cooling load because the energy is now stored in building elements, such as walls, floors, and ceilings.

On the other hand, reducing room temperatures means an increase in cooling load due to discharge of stored thermal energy.

The results of thermal storage processes in buildings can be expressed by cooling-load factors shown as a distribution over the course of a day. They are able to depict the dampening and the associated time lags.

The room types relevant to thermal storage capacity can be distinguished as follows:

- XL = very low thermal mass, total mass < 200 kg/m²
- X = low thermal mass, mass ~ 200 – 400 kg/m²
- M = medium thermal mass, mass ~ 400 – 600 kg/m²
- S = high thermal mass, mass > 600 kg/m².

Currently, this somewhat broad classification is being replaced by more precise model simulations that allow a close description of the ability of diverse building materials to store energy. They form the basis for the subsequent calculation of cooling loads that in some instances need to be compensated by such active mechanical means as air or water cooling.

Figure 259 shows as an example different room types with various enclosure surfaces and the resulting different thermal storage behavior.

Personen, Maschinen usw. auf ihn einwirken.
Die Wärmeenergie dringt infolge Wärmestrahlung und Konvektion je nach Materialaufbau und Fähigkeit des Wärmeübergangs mehr oder weniger tief in die speichernden Bauteile ein und führt dabei zu einer Erhöhung der Oberflächentemperaturen. Die Oberflächen Raumumschließender Bauteile stehen dabei miteinander im Strahlungsaustausch. Je nach Raumluft- und Oberflächentemperatur kommt es zu einem konvektiven Wärmeübergang und einem Strahlungsaustausch von den Oberflächen an den Raum und umgekehrt. Bei veränderlichen Raumtemperaturen bedeutet ein Temperaturanstieg eine Reduzierung des Anstiegs der Kühllast durch die Einspeicherung von Wärme in Bauteile

Eine Absenkung der Raumtemperaturen bedeutet eine Erhöhung der Kühllast durch Entspeicherung.

Die Ergebnisse der Wärmespeichervorgänge werden durch Tagesgänge der Kühllastfaktoren wiedergegeben, die den Kühllastverlauf in seiner Dämpfung und Zeitverzögerung beschreiben.

Prinzipiell unterscheidet man wärmespeichernde Räume wie folgt:

- XL= sehr leicht speichernd
  Gesamtspeichermassen < 200 kg/m² Bodenfläche
- X = leicht speichernd
  Gesamtspeichermassen ca. 200 – 400 kg/m² Bodenfläche
- M = mittelschwer speichernd
  Gesamtspeichermassen ca. 400 – 600 kg/m² Bodenfläche
- S = schwer speichernd,
  Gesamtspeichermassen > 600 kg/m² Bodenfläche.

Dieses sehr grobe Raster wurde in den letzten Jahren zugunsten genauerer Simulations-Berechnungsverfahren ersetzt, so dass zum heutigen Zeitpunkt bei entsprechenden Planungen von Gebäuden das Speicherverhalten und die zeitlichen Verzögerungen der im Raum wirksam werdenden Kühllasten sehr genau beschrieben werden können. Sie werden somit Grundlage der Bemessung der tatsächlich zeitlich eintretenden Lasten, die gegebenenfalls durch technische Einrichtungen (Luftkühlung, Wasserkühlung) kompensiert werden müssen.

Bild 259 zeigt beispielhaft Raumvarianten mit unterschiedlichen Umschließungsflächen und hieraus resultierend unterschiedlichem Speicherverhalten.

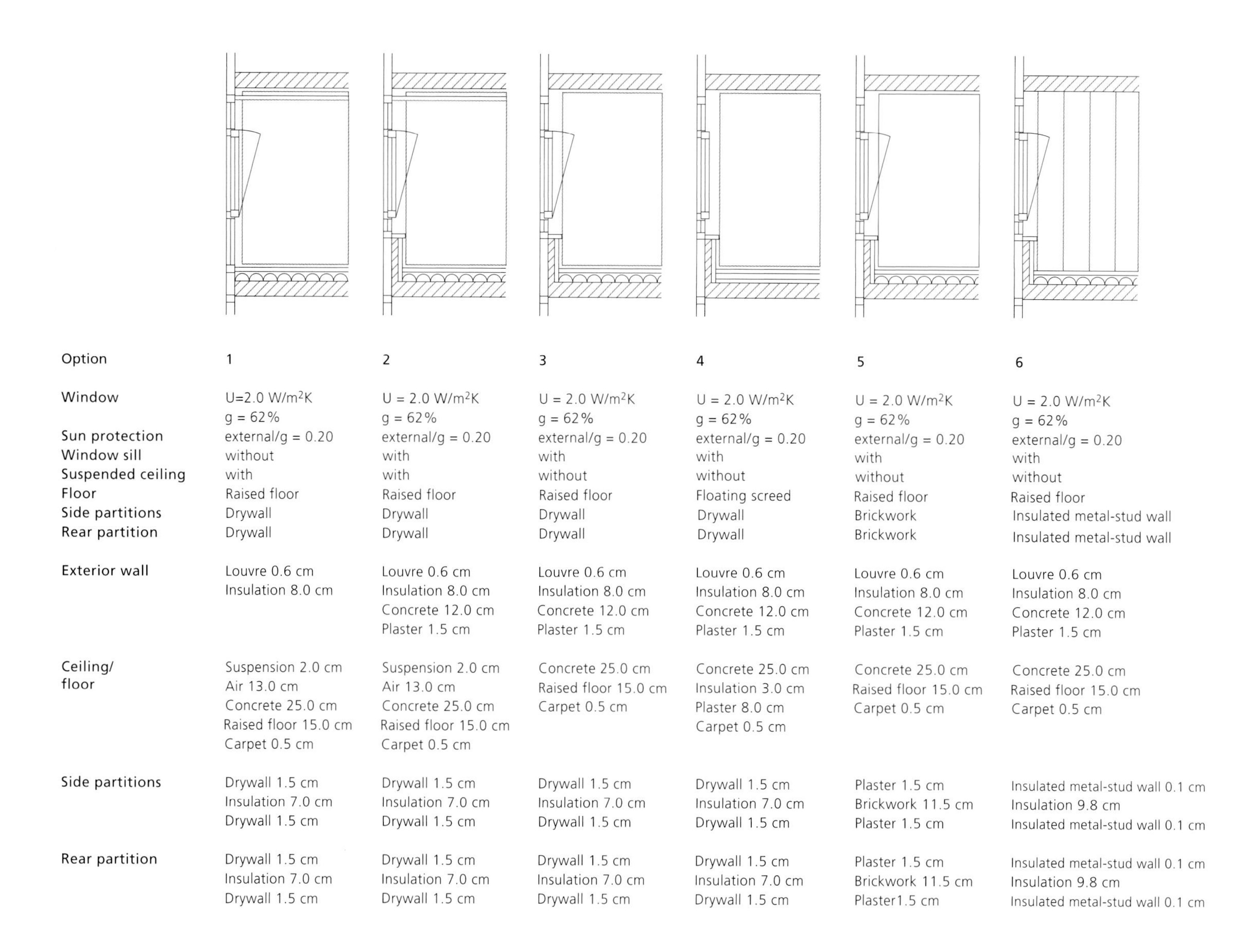

Option	1	2	3	4	5	6
Window	U=2.0 W/m²K g = 62%	U = 2.0 W/m²K g = 62%	U = 2.0 W/m²K g = 62%	U = 2.0 W/m²K g = 62%	U = 2.0 W/m²K g = 62%	U = 2.0 W/m²K g = 62%
Sun protection	external/g = 0.20	external/g = 0.20	external/g = 0.20	external/g = 0.20	external/g = 0.20	external/g = 0.20
Window sill	without	with	with	with	with	with
Suspended ceiling	with	with	without	without	without	without
Floor	Raised floor	Raised floor	Raised floor	Floating screed	Raised floor	Raised floor
Side partitions	Drywall	Drywall	Drywall	Drywall	Brickwork	Insulated metal-stud wall
Rear partition	Drywall	Drywall	Drywall	Drywall	Brickwork	Insulated metal-stud wall
Exterior wall	Louvre 0.6 cm Insulation 8.0 cm	Louvre 0.6 cm Insulation 8.0 cm Concrete 12.0 cm Plaster 1.5 cm	Louvre 0.6 cm Insulation 8.0 cm Concrete 12.0 cm Plaster 1.5 cm	Louvre 0.6 cm Insulation 8.0 cm Concrete 12.0 cm Plaster 1.5 cm	Louvre 0.6 cm Insulation 8.0 cm Concrete 12.0 cm Plaster 1.5 cm	Louvre 0.6 cm Insulation 8.0 cm Concrete 12.0 cm Plaster 1.5 cm
Ceiling/ floor	Suspension 2.0 cm Air 13.0 cm Concrete 25.0 cm Raised floor 15.0 cm Carpet 0.5 cm	Suspension 2.0 cm Air 13.0 cm Concrete 25.0 cm Raised floor 15.0 cm Carpet 0.5 cm	Concrete 25.0 cm Raised floor 15.0 cm Carpet 0.5 cm	Concrete 25.0 cm Insulation 3.0 cm Plaster 8.0 cm Carpet 0.5 cm	Concrete 25.0 cm Raised floor 15.0 cm Carpet 0.5 cm	Concrete 25.0 cm Raised floor 15.0 cm Carpet 0.5 cm
Side partitions	Drywall 1.5 cm Insulation 7.0 cm Drywall 1.5 cm	Drywall 1.5 cm Insulation 7.0 cm Drywall 1.5 cm	Drywall 1.5 cm Insulation 7.0 cm Drywall 1.5 cm	Drywall 1.5 cm Insulation 7.0 cm Drywall 1.5 cm	Plaster 1.5 cm Brickwork 11.5 cm Plaster 1.5 cm	Insulated metal-stud wall 0.1 cm Insulation 9.8 cm Insulated metal-stud wall 0.1 cm
Rear partition	Drywall 1.5 cm Insulation 7.0 cm Drywall 1.5 cm	Drywall 1.5 cm Insulation 7.0 cm Drywall 1.5 cm	Drywall 1.5 cm Insulation 7.0 cm Drywall 1.5 cm	Drywall 1.5 cm Insulation 7.0 cm Drywall 1.5 cm	Plaster 1.5 cm Brickwork 11.5 cm Plaster1.5 cm	Insulated metal-stud wall 0.1 cm Insulation 9.8 cm Insulated metal-stud wall 0.1 cm

Alternatives 1 and 2 are low-storage spaces because their thermal storage mass is concealed, and thus rendered ineffective, by suspended ceiling systems and raised access floors. Thermal storage capacity increases in alternatives 3 and 4 due to the omission of such shielding elements, which results in their being medium storage rooms. Alternatives 5 and 6 show rooms with increased storage capacity up to the classification of heavy thermal storage. The resulting feedback with regard to room temperature under equal internal and external heat load scenarios is shown in **Figure 260.1**.

Während die Varianten 1 und 2 als leicht speichernde Raumeinheiten zu sehen sind, da die Speichermassen jeweils durch Deckenabhängungen oder Doppelböden verdeckt sind, nimmt das Speicherverhalten bei den Varianten 3 und 4 infolge des Fortfalls entsprechender Verkleidungen zu, d.h. die Räume sind mittelschwer speichernd. Bei den Varianten 5 und 6 nimmt die Speicherfähigkeit zu – hin zu schwer speichernden Strukturen.

Figure 259
Room types with different enclosure conditions

Orientation:	South
Dimensions:	
Floor space:	21.0 m²
Occupants:	2
Room height:	2.90 m
Lighting:	158 W
Ceiling suspension:	0.15 m
Machines:	315 W
Window sill height:	0.70 m

Source:
Advanced Building Systems
Klaus Daniels
Birkhäuser Verlag, 2003

Bild 259
Raumtypen mit verschiedenen Speicherstrukturen

Orientierung:	Süden
Raumabmessungen:	
Fußbodenfläche:	21,0 m²
Personen:	2
Raumhöhe:	2,90 m
Beleuchtung:	158 W
Deckenabhängung:	0,15 m
Maschinen:	315 W
Brüstungshöhe:	0,70 m

Quelle:
Advanced Builing Systems
Klaus Daniels
Birkhäuser Verlag, 2003

Figure 260.1
Room temperatures following five days of good weather in summer.
Weather data from TRY (TRY region 8)

Room temperatures are indicated as perceived temperatures (from DIN 1946/part 2), under consideration of building component and surface temperatures.

Source:
Advanced Buidling Systems
Klaus Daniels
Birkhäuser Verlag, 2003

Bild 260.1
Raumtemperaturen nach 5-tägiger sommerlicher Schönwetterperiode.
Wetterdaten nach Testreferenzjahr (TRY-Region 8)

Die Raumtemperaturen werden als operative (empfundene) Temperaturen (nach DIN 1946/Teil 2) angegeben, also unter Berücksichtigubng der Bauteil-Oberflächentemperaturen.

Quelle:
Advanced Builing Systems
Klaus Daniels
Birkhäuser Verlag, 2003

If the goal is a significant improvement of a space's thermal storage capacity, it will be necessary to use deliberate night-flushing with operable windows that can let in cool outside air and operable windows. **Figure 260.2** shows the previously presented case-study spaces of alternatives 2 and 5 with and without nighttime cooling. It was here assumed that 3 – 4 air changes per hour can be achieved with slightly opened, tilt-turn windows and average outside air velocities. When we compare the graphs, it becomes clear that a low to medium thermal storage capacity of the enclosing surfaces results in a decrease in air temperature of approx. 2 – 3 K, that those with heavy storage mass reduce the temperature by 3 – 4 K, and that the temperature of the heavy-storage-mass room in alternative 5 is 4 K lower than that of the low-storage configuration.

If energy consumption and the related investment cost for mechanical cooling and ventilating equipment needs to be lowered, a building needs to be conceived with heavy storage mass, and such a building then needs to be "de-warmed" by nighttime cooling in order to be again pre-conditioned to accept energy during the following day. Such conditioning of internal storage mass can also be achieved by so-called "free cooling", in which chilled water for cooling purposes, generated by cooling towers and ground water streams, is used for so-called active concrete core conditioning – the cooling of concrete storage components of the building by circulating, cooled water. As a result, the nighttime storage capacity is increased significantly by such thermo-active building component cooling.

Die Auswirkungen bezüglich des Temperaturverhaltens im Raum bei gleichen äußeren und inneren Wärmebelastungen sind in **Bild 260.1** dargestellt.

Will man die Speicherfähigkeit eines Raumes deutlich verbessern, so ist es notwendig, diesen während der Nacht mit kalter Außenluft gezielt zu durchlüften. **Bild 260.2** zeigt für die zuvor beschriebenen Testräume (Variante 2/5) das Temperaturverhalten mit und ohne Nachtauskühlung. Im Fall der Nachtauskühlung wurde von einem ca. 3- bis 4-fachen Luftwechsel ausgegangen, wie er sich üblicherweise bei mittleren Außenluftgeschwindigkeiten und leicht gekippten Fenstern einstellen wird. Vergleicht man die Kurven mit und ohne Nachtlüftung untereinander, so lässt sich leicht ausmachen, dass bei einem leicht bis mittelschwer speichernden Raum eine Temperaturreduzierung um ca. 2 – 3 K einstellt. Bei einem schwer speichernden Raum erreicht die Temperaturreduzierung ca. 3 – 4 K, wobei das Temperaturniveau des schwer speichernden Raumes (Variante 5) um ca. 4 K niedriger liegt als das eines leicht speichernden Raumes.

Will man den Energieverbrauch und die daraus resultierenden Investitionskosten im Bereich der Lüftungs- und Kältetechnik deutlich reduzieren, so ist es in jedem Fall sinnvoll, einen Raum nicht nur schwer speichernd auszubilden, sondern auch während der Nacht soweit wie möglich zu entwärmen, um für den folgenden Tag eine erhöhte Speicherfähigkeit zu erreichen. Die Entwärmung von Baumassen kann unter Umständen auch sinnvoll dadurch erreicht werden, dass eine Freie Kühlung zum Einsatz kommt, d.h. Aufbereitung von Kühlwässern über Rückkühlwerke oder Grundwasserströme, die durch Betonmassen geleitet werden (thermoaktive Bauteilsysteme), um die nächtliche Auskühlung und somit Speicherfähigkeit zu erhöhen.

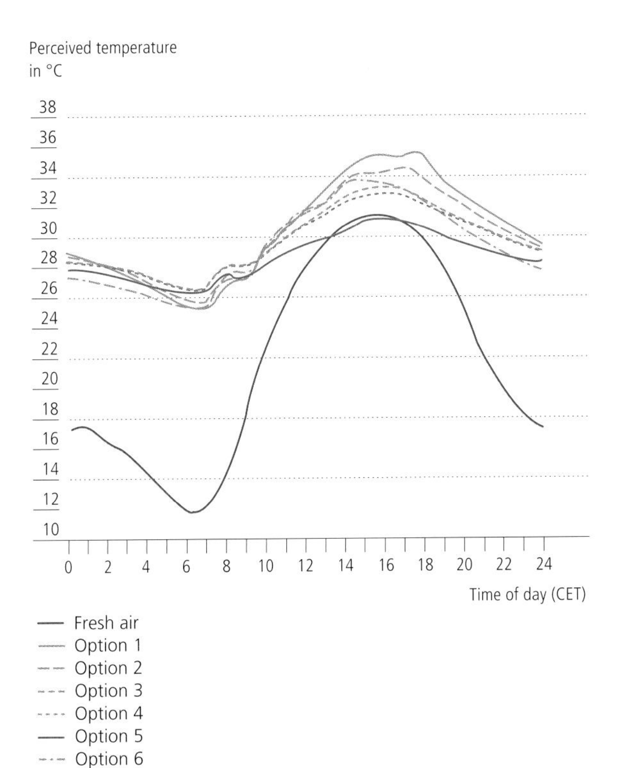

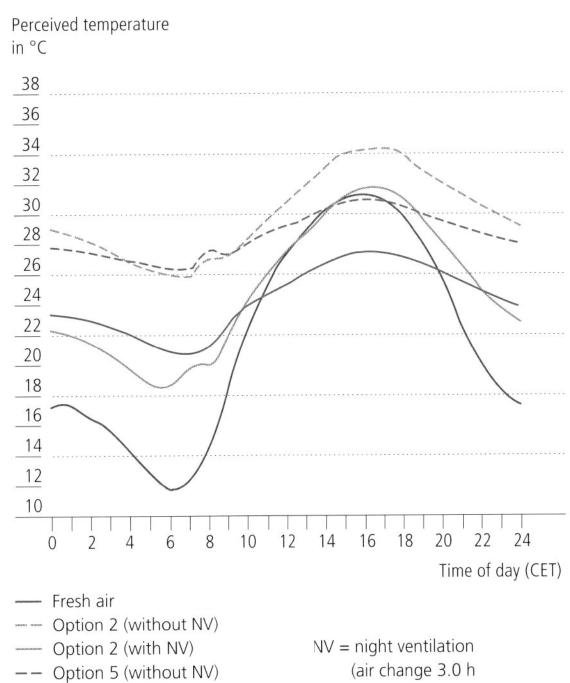

Figure 260.2
Room temperatures following five days of good weather in summer
Comparison: with/without night ventilation
Weather data from TRY (TRY region 8)

Source:
Advanced Building Systems
Klaus Daniels
Birkhäuser Verlag, 2003

Bild 260.2
Raumtemperaturen nach 5-tägiger sommerlicher Schönwetterperiode.
Vergleich mit/ohne Nachtlüftung
Wetterdaten nach Testreferenzjahr (TRY-Region 8)

Quelle:
Advanced Builing Systems
Klaus Daniels
Birkhäuser Verlag, 2003

13.1.2 Utilization of biomass

For many countries, the use of biomass will be an important pillar of their energy palette in the future. Biomass used for domestic heating and cooking purposes is as old as mankind. The modern use of biomass in more focused applications began around 20 years ago, starting a true renaissance of this old, renewable energy source. However, biomass is almost equally as costly as fossil fuels. Biogenic fuel availability is limited, fluctuates significantly from country to country, and will be capable of supplying only a small percentage of the energy a country's economy requires.

Plant-based biomass is capable of delivering carbon-neutral energy because during its burning only that amount of CO_2 is released to the atmosphere that was absorbed and stored by the plant during its growth period. However, one negative aspect of biomass fuel burning is important to recognize and alleviate. Fine particulate matter (PM) or, simply, fine dust, causes health-related risks, so its release into the air needs to be eliminated in the process.

The term biomass describes that spectrum of natural resources derived from organic matter, and, in a broader sense, also those materials that are the result of a transformation of organic matter or its final products, such as paper, organic household waste, vegetable oils, biogas, and others.

Primary biomass products are energetically processed energy sources such as wood chaff, wood pellets, vegetable oils, and substances such as liquid manure and sludge that are derived as a result of organic transformation processes in sewage treatment plants. The most important groups are:

- Leftover products from harvesting and forestry
- Organic byproducts as a result of production processes in woodworking, liquid manure from large cattle operations, and biogas
- Organic waste such as sludge or methane mine gas
- "Energy plants" such as the oil-rich species of sunflowers, canola, corn, and others
- Grasses with a high degree of cellulose content, and bush and tree species with low maintenance requirements.

In order to gain and use biogenic energy sources, people first have to cultivate and harvest them. Sources used in a solid state include those derived from wood or straw biomass. They need to be prepared for energy use by shredding (cut wood) or cutting or sawing (wood pellets).

Liquid-state biomass energy is delivered in the form of vegetable oils and alcohols, with soybean and canola oil being of significant importance because they are capable of trans-esterification of their short-chain alkyl (methyl or ethyl) esters and

13.1.2 Nutzung von Biomasse

Die Nutzung von Biomasse wird in etlichen Ländern der Welt eine der wichtigen Säulen zur Energiebereitstellung in der Zukunft sein. Den entsprechenden Einsatz zu Heiz- und Kochzwecken hat es schon so lange gegeben, wie die Menschheit existiert. Der gezielte Einsatz von Biomasse zur Energiebereitstellung hat, beginnend vor ca. 20 Jahren, eine Renaissance erlebt, wobei zum heutigen Zeitpunkt in Mitteleuropa der Einsatz von Biomasse praktisch gleich teuer ist wie der fossiler Brennstoffe. Aufgrund der endlichen Verfügbarkeit biogener Brennstoffe ist der Einsatz immer nur ein Teil der gesamten Energiewirtschaft und sehr unterschiedlich ausgeprägt.

Pflanzliche Biomasse birkt als nachwachsender Rohstoff das Potenzial, in der Energiebereitstellung einen CO_2-neutralen Kreislauf zu gewährleisten, da bei der Verbrennung von Biomasse nur die Menge an CO_2 entweicht, die während des Wachstums von den Pflanzen aufgenommen wurde (CO_2-neutraler Energieträger). Ein Negativum bei der Nutzung von Biomasse, das es zu beachten gilt, ist die Entwicklung von Feinstäuben, die es zu eliminieren gilt, um gesundheitliche Risiken zu vermeiden.

Biomasse beschreibt das Spektrum von Energieträgern organischer Herkunft und im weitesten Sinne auch alle die Stoffe, die durch eine Umwandlung bzw. stoffliche Nutzung daraus entstehen (Papier, organischer Hausmüll, Pflanzenöle, Biogas usw.).

Primärprodukte der Biomasse (entstanden durch direkt photosynthetische Umwandlung der Solarstrahlung) umfassen die Bereiche Holz, Stroh, Gräser usw..

Sekundärprodukte der Biomasse sind energetisch aufbereitete Energieträger wie z.B. Hackschnitzel, Holzpellets, Pflanzenöl usw. sowie durch Ab- oder Umbau organischer Substanzen in Organismen entstehende Stoffe (Gülle, Klärschlamm usw.). Die im Regelfall wichtigsten Bereiche der Nutzung von Biomasse umfassen:

- Ernterückstände und Rückstände aus der Waldwirtschaft
- Organische Nebenprodukte, die durch Verarbeitungsprozesse entstehen wie z.B. Abfall der Holzverarbeitung oder Gülle aus der Nutztierhaltung (u.U. auch Biogas)
- Organische Abfälle wie Klärschlamm und Deponiegas
- Energiepflanzen wie ölhaltige Pflanzen (Raps, Sonnenblumen, Soja, Mais usw.) sowie Gräser mit hoher Zellulosemasse und Hecken- und Baumgewächse mit geringem Pflegebedarf

Um biogene Energieträger zu gewinnen und zu nutzen, müssen diese im Regelfall angebaut und geerntet werden (Verfügbarmachung).

being converted into biodiesel. Alcohol from the sugar of starch-rich plants such as sugar cane, corn, sugar beet, and potatoes, can be used in fermentation processes and converted to energy. In the absence of air, fermentation processes are also capable of producing gases from biomass such as sludge, organic waste, and liquid manure in special large-scale plants. Jatropha, an inedible plant that can be cultivated on non-arable land and thus is not competing with food crops, has seeds that are capable of producing three times more oil than soybeans. If, for example, only 10 % of India's non-arable land surfaces of around 60 million hectares (600,000 km^2) were used to cultivate Jatropha, 10 % of the country's biodiesel demand would be satisfied. Countries such as South Africa, Ghana, Brazil, and Madagascar grow Jatropha to a significant extend in order to prevent land for food production being used for energy generating purposes.

New processes such as biomass-to-liquid (BML, or BMTL, a multi-stage process to convert biomass into biofuels), are currently being developed, and they are capable of transforming various biomass sources into universal oil, comparable to fossil oils. All such oils are being used in principle as fuels for motors in the transportation sector or for power-heat coupling and cogeneration processes.

As shown in **Figure 261**, several combustion technologies are available for the burning of biogenic energy sources. Biogas and bio oil, on the other hand, are already used in combustion motors and in gas-fired applications such as furnaces.

Als Festbrennstoff werden holzartige und halmgutartige Biomassen genutzt. Die Aufbereitung zur energetischen Nutzung erfolgt durch Zerkleinern (Stückholz), Zerhacken (Holzhackschnitzel) oder durch verdichtetes Pressen von Holzmehl (Holzpellets).

Biomasse in Form flüssiger Energieträger werden bereitgestellt durch Pflanzenöle und Alkohole. Soja und Raps sind hierbei von besonderer Bedeutung, da Rapsöl durch den Prozess der Umesterung zu Rapsmethylester veredelt werden kann (Biodiesel).

Alkohole aus zucker- oder stärkehaltigen Pflanzen werden aus z.B. Zuckerrohr, Mais, Zuckerrüben, Kartoffeln usw. durch Gärungsprozesse gewonnen und anschließend zur Energiegewinnung genutzt. Jatropha, eine nicht essbare Pflanze, die auf landwirtschaftlich nicht nutzbaren Flächen angebaut werden kann, produziert dreimal mehr Öl als Sojabohnen. Wenn z.B. nur 10 Prozent von Indiens nicht nutzbaren Flächen mit einer Größenordnung von 60 Millionen Hektar bepflanzt würden, könnten ca. 10 Prozent des Biodieselbedarfs Indiens hierdurch gedeckt werden.

Länder wie Südafrika, Ghana, Brasilien und Madagaskar bauen Jatropha-Pflanzen in größerem Umfang an, um zu vermeiden, dass zur Ernährung wichtige Flächen verloren gehen.

Neue Verfahren der „biomass-to-liquid-Verfahren" werden zurzeit entwickelt, um aus verschiedenen Biomassearten ein hochwertiges und universell einsetzbares Öl zu erreichen, das in seinen Eigenschaften den Erdölprodukten entspricht. Alle Öle werden im Wesentlichen als Treibstoff für Motoren (Kraft-Wärme-Kopplung) oder im Verkehr zum Einsatz gebracht.

Aus Gärungsprozessen unter Luftabschluss können aus Biomasse Gase gewonnen werden (Gülle, organische Abfälle, Biohausmüll, Klärschlamm), die eine großmaßstäbliche Nutzung zulassen. Zur Verbrennung biogener Energieträger stehen verschiedene Feuerungsprinzipien bereit, **Bild 261**. Biogase und Bioöl wird in den bekannten Formen genutzt, d.h. z.B. durch Einsatz von Gasfeuerungsstätten (Gaskessel) sowie Motoren.

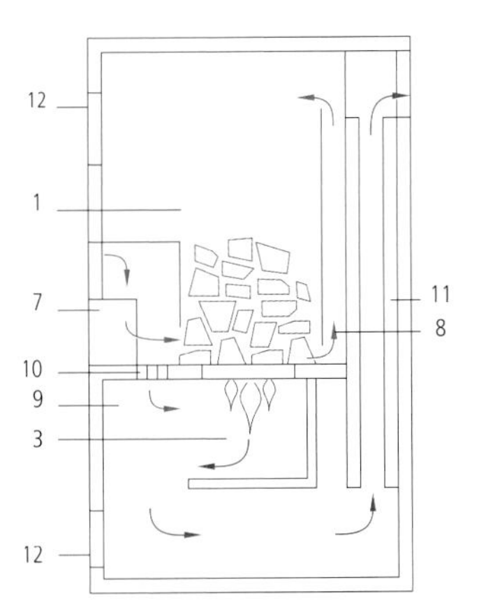

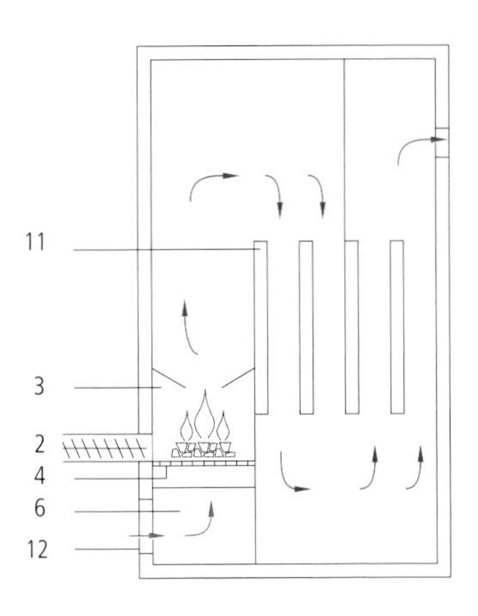

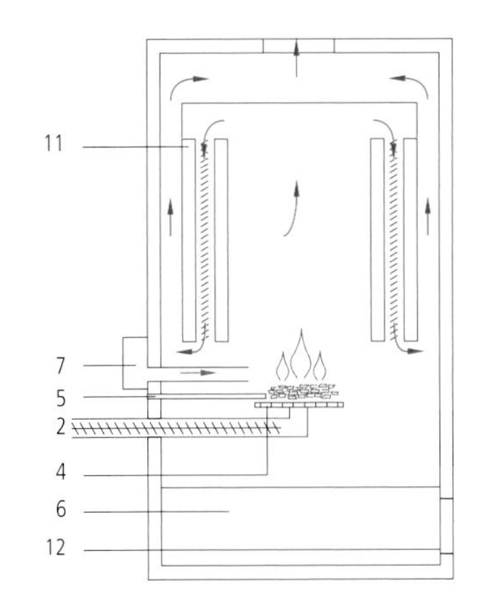

1 Fill chamber
2 Feeder scroll
3 Combustion chamber
4 Burner top plate
5 Igniter
6 Ash box
7 Fan
8 Primary air
9 Secondary air
10 Secondary air control
11 Heat exchanger
12 Service access

Figure 261
Biomass stove systems

Source: Energieatlas, Nachhaltige Architektur, Hegger, Fuchs, Stark, Zeumer, Birkhäuser Verlag, 2007

Bild 261
Feuerungssysteme zur Nutzung von Biomasse

Quelle: Energieatlas, Nachhaltige Architektur Hegger, Fuchs, Stark, Zeumer Birkhäuser Verlag, 2007

13.1.3 Shallow geothermal applications

It is common knowledge that beneath the Earth's surface incredible amounts of energy can be found, yet the awareness of its potential for energy generation is not yet fully developed. The energy stored in the upper 3 km of the Earth's crust would be sufficient to supply the total energy demand of mankind for the next one hundred thousand years.

The heat energy of the crust is the result of radioactive decay during the formation of planet Earth and the solar radiation affecting upper levels of the soil. That energy is available on a constant basis, and its utilization – in contrast to solar and wind energy – is not surface intensive because it can be extracted by pointed exploration.

Depending upon the depth from which we gain geothermal energy, the following distinctions are made: shallow, near-surface geothermal energy exploitation takes place up to a depth of 500 m below the Earth's surface, while deep geothermal applications reach beyond 3,000 m. **Figure 262** gives an overview of the technology.

13.1.3 Untiefe Geothermie

Uns allen bekannt, jedoch wenigen bewusst ist, dass unter unseren Füßen ungeheure Energien aus dem Erdinneren schlummern. Allein durch die Wärmevorräte der oberen 3 km der Erdkruste könnte der derzeitige Energiebedarf der Menschheit für die nächsten hunderttausend Jahre gedeckt werden.

Erdwärme als Restwärme aus der Zeit der Erdentstehung ist Wärmeenergie aus radioaktiven Zerfallsprozessen und Wärmeenergie aus Sonneneinstrahlung in oberflächennahe Schichten. Sie steht andauernd zur Verfügung und ist wenig flächenintensiv (entgegen Solar- und Windenergie), wenn sie punktuell dem Erdreich entzogen wird.

Je nach Tiefe der Wärmeentnahme unterscheidet man zwischen oberflächennaher (bis ca. 500 m) und tiefer Geothermie (unterhalb ca. 3.000 m). **Bild 262** gibt einen ersten Überblick.

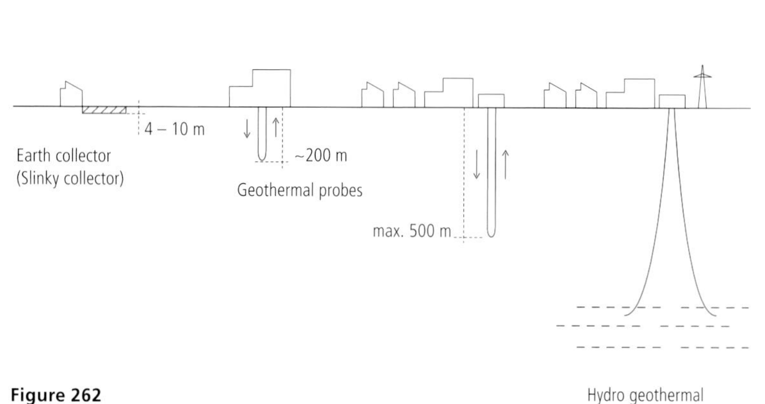

Figure 262
Geothermal probe systems for near-surface applications

Source: archplus 184, October 2007

Bild 262
Sondensysteme zur Erdwärmenutzung, Oberflächennahe Geothermie

Quelle: archplus 184, Oktober 2007

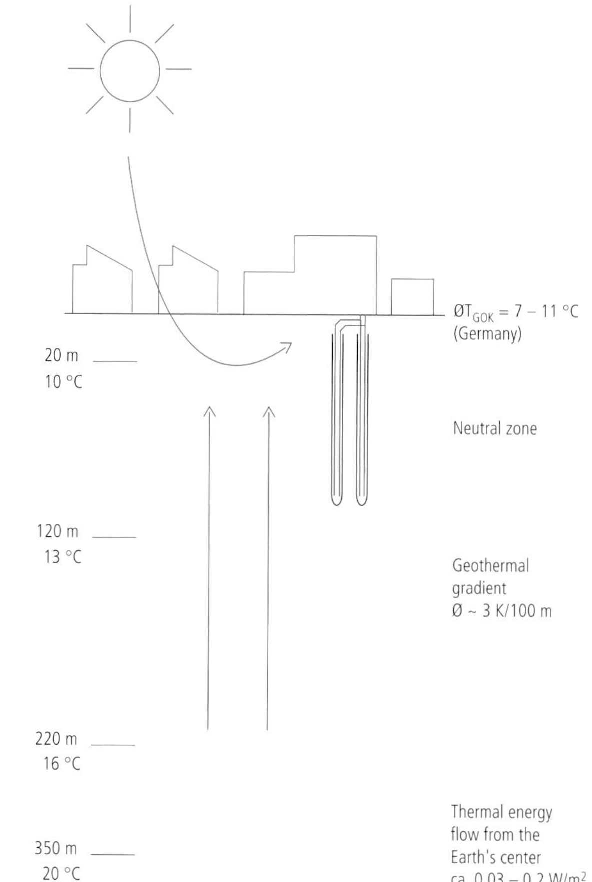

Detail to Figure 262
Geothermal energy basics. Near-surface geothermal up to a depth of 400 meters deep geothermal beyond 400 meters

Source: Dr. Claus Heske & Dr. Marec Wedewardt

Detail zu Bild 262
Geothermie Basisinformationen, Oberflächennahe (bis 400 m u. GOK) und tiefe Geothermie (ab 400 m u.GOK)

Quelle: Dr. Claus Heske & Dr. Marec Wedewardt

To make use of shallow geothermal energy, probes as seen in **Figure 263** are normally inserted into the ground. **Table 18** shows the typical energy gains achievable in Central Europe. Before larger-scale geothermal well fields are installed, test boreholes are drilled to evaluate the potential of the soil for energy extraction. **Figure 264** shows such a so-called enhanced geothermal response test.

Zur Nutzung der oberflächennahen Geothermie werden im Regelfall Erdwärmesonden im Erdreich niedergebracht, wie sie **Bild 263** zeigt. Typische Entzugsleistungen, wie sie in Mitteleuropa feststellbar sind, sind in **Tabelle 18** dargestellt. Vor dem endgültigen Ausbau größerer Erdsondenanlagen werden üblicherweise Testsonden niedergebracht, über die die Leistungsfähigkeit des Untergrunds festgestellt wird. **Bild 264** zeigt beispielhaft den Betrieb einer Testsonde (enhanced geothermal response test).

1 Soil heat extraction with RAUGEO geothermal probes
2 with RAUGEO earth collectors
3 with RAUGEO energy piles
4 with air-soil heat exchanger AWADUKT Thermo

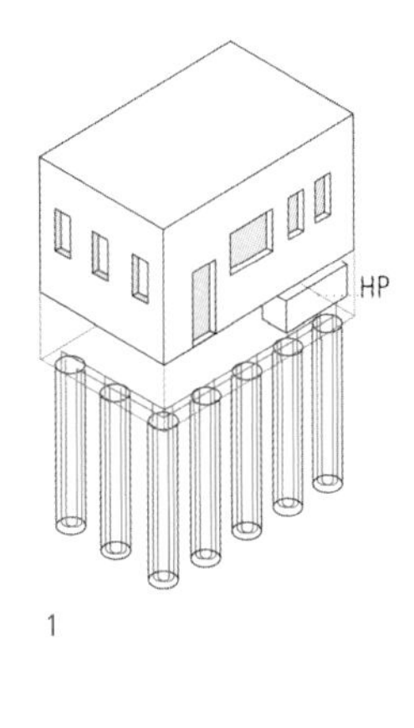

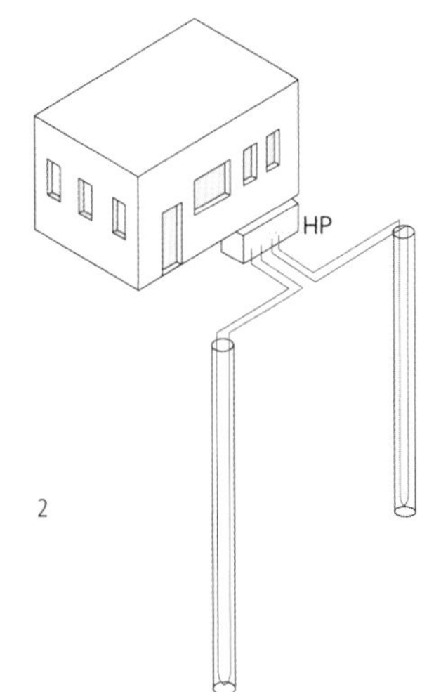

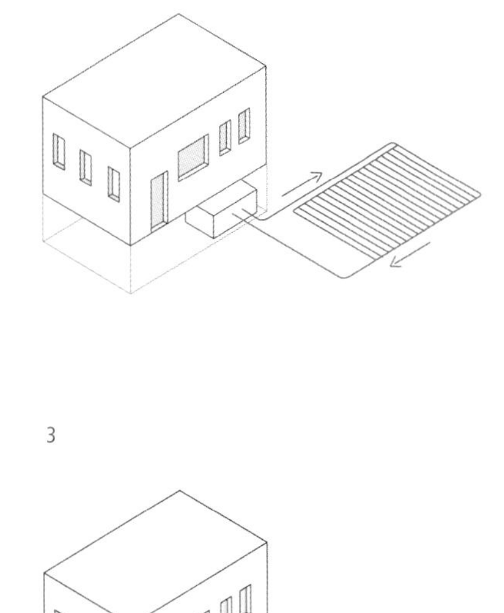

Figure 263
Near-surface utilization of geothermal energy

Source: REHAU, Germany

Bild 263
Nutzung der oberflächennahen Wärme des Erdreichs

Quelle: REHAU, Deutschland

Figure 264
Example of temperature development of a test probe (November 2007)

Source: Enhanced Geothermal Response Test CDM, Unterhaching, Germany
Dr. Claus Heske &
Dr. Marec Wedewardt

Bild 264
Beispiel des Temperaturverlaufs einer Testsonde (Nov. 2007)

Quelle: Enhanced Geothermal Response Test CDM, Unterhaching, Deutschland
Dr. Claus Heske &
Dr. Marec Wedewardt

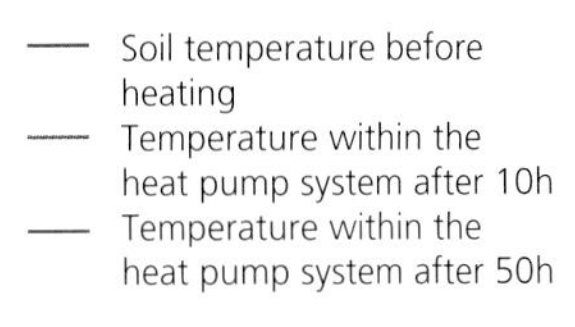

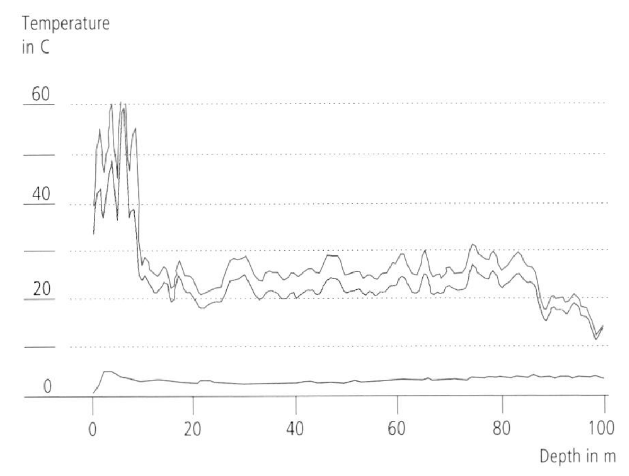

Soil type	Specific power production for 1,800 h/a	for 2.400 h/a
General values:		
Inferior soil (dry sediment) [λ<1,5 W/(m·K)]	25 W/m	20 W/m
Normal hard soil and saturated sediment [λ=1,5 – 3,0 W/(m·K)]	60 W/m	50 W/m
Rock with high thermal conductivity [λ>3,0 W/(m·L)]	84 W/m	70 W/m
Individual rock:		
Gravel, sand, dry	<25 W/m	<20 W/m ⟵
Gravel, sand, water carrying	65 – 80 W/m	55 – 65 W/m
Under strong ground water movement in gravel and sand	80 – 100 W/m	80 – 100 W/m ⟵
Clay, loam, moist	35 – 50 W/m	30 – 40 W/m
Lime stone (massive)	55 – 70 W/m	45 – 60 W/m
Sand stone	65 – 80 W/m	55 – 65 W/m
Igneous magmatic rocks (i.e. granite)	65 – 85 W/m	55 – 70 W/m
Alkaloid magmatic rocks (i.e. basalt)	40 – 65 W/m	35 – 55 W/m
Gneiss	70 – 85 W/m	60 – 70 W/m

The shown values can fluctuate significantly as a result of fissure veins, weathering and slating

Table 18
Potential thermal power extraction values for geothermal probes

- Heat extraction only (heating and hot water)
- Length of individual probes between 40-100 meters
- Closest spacing between probes:
  minimum 5 meters between probes with a length of 40-50 meters
  minimum 6 meters between probes with a length of >50 – 100 meters
- Probes: Double-U DN 20, DN25 or DN 32 mm or coaxi al probes with a minimum diameter of 60 mm
- Not applicable for larger arrays of small units on limit ed surface area

from VDI 4640

Source: Enhanced Geothermal Response Test CDM, Unterhaching, Germany
Dr. Claus Heske &
Dr. Marec Wedewardt

Untergrund	Spezifische Entzugsleistung für 1.800 h/a	für 2.400 h/a
Allgemeine Richtwerte:		
Schlechter Untergrund (trockenes Sediment) [λ<1,5 W/(m·K)]	25 W/m	20 W/m
Normaler Festgesteins- Untergrund und wassergesättigtes Sediment [λ=1,5 – 3,0 W/(m·K)]	60 W/m	50 W/m
Festgesteinmit hoher Wärmeleitfähigkeit [λ>3,0 W/(m·L)]	84 W/m	70 W/m
Einzelne Gesteine:		
Kies, Sand, trocken	<25 W/m	<20 W/m ⟵
Kies, Sand, wasserführend	65 – 80 W/m	55 – 65 W/m
Bei starkem Grundwasserfluss in Kies und Sand, für Einzelanlagen	80 – 100 W/m	80 – 100 W/m ⟵
Ton, Lehm, feucht	35 – 50 W/m	30 – 40 W/m
Kalkstein (massiv)	55 – 70 W/m	45 – 60 W/m
Sandstein	65 – 80 W/m	55 – 65 W/m
Saure Magmatite (z.B. Granit)	65 – 85 W/m	55 – 70 W/m
Basische Magmatite (z.B. Basalt)	40 – 65 W/m	35 – 55 W/m
Gneis	70 – 85 W/m	60 – 70 W/m

Die Werte können durch die Gesteinausbildung wie Klüftung, Schieferung, Verwitterung erheblich schwanken.

Tabelle 18
Mögliche spezifische Entzugsleistungen für Erdwärmesonden

- nur Wärmeentzug (Heizung einschließlich Warmwasser)
- Länge der einzelnen Erdwärmesonden zwischen 40 und 100 m
- kleinster Abstand zwischen zwei Erdwärmesonden:
  mindestens 5 m bei Erdwärmesondenlängen 40 – 50 m
  mindestens 6 m bei Erdwärmesondenlängen >50 – 100 m
- als Erdwärmesonden kommen Doppel-U-Sonden mit DN 20, DN 25 oder DN 32 mm oder Koaxialsonden mit mindestens 60 mm Durchmesser zum Einsatz
- nicht anwendbar bei einer größeren Anzahl kleiner Anlagen auf einem begrenzten Areal

aus VDI 4640

Quelle: Enhanced Geothermal Response Test CDM, Unterhaching, Deutschland
Dr. Claus Heske &
Dr. Marec Wedewardt

Figure 265
Temporal variation in earth temperatures at depths of 5 m, 50 m and 85 m, and at 50 m distance from bore hole in Elgg, Switzerland, over a period of five years of operation (Heating-degree-days HDD 20/12 according Eugster et al.)

Source:
Advanced Building Systems
Klaus Daniels
Birkhäuser Verlag, 2003

Bild 265
Der zeitliche Verlauf der gemessenen Erdreichtemperaturen in 5 m, 50 m und 85 m Tiefe und 50 cm Abstand von der Erdwärmesonde in Elgg, Schweiz, über fünf Betriebsjahre (Heiz-Gradtage 20/12) (nach W.J. Eugster et.al.)

Quelle:
Advanced Building Systems
Klaus Daniels
Birkhäuser Verlag, 2003

Soil temperatures fluctuate throughout the year and are a function of the depth of the application, as shown in **Figure 265**. This is an important consideration when using a heat pump to extract heat. Such heat pumps, which work according to the concept of a chiller, reject cooling energy (evaporator power) into the soil during winter operations, simultaneously providing heat energy. The cold water supplied to the ground then heats up by approximately 3 – 4 K. The gained heat energy is conducted through the compressor of the heat pump and raised to the temperature level – typically between 40 and a maximum of 55 °C – needed in the building for space heating. During the summer, the heat pump will work as a cooling machine, now supplying warm water to the ground. **Figure 266** shows a typical energy balance of a heat pump system. If the local conditions of the site do not provide a suitable aquifer for heat storage, under certain circumstances an artificial aquifer can be designed, as shown in **Figure 267**. **Figure 268** shows the connection of the aquifer to a heat-recovery system of a large heat pump, and **Figure 269** shows an analogue heat pump system.

Im jahreszeitlichen Verlauf schwanken die Erdreichtemperaturen je nach Tiefe, wie in **Bild 265** dargestellt. Dies gilt es zu berücksichtigen, wenn über eine Wärmepumpe Wärme entzogen werden soll. Die Wärmepumpe, im Prinzip eine Kältemaschine, gibt ihre Kühlenergie (Verdampferleistung) an das Erdreich ab (Winterbetrieb), wobei gleichzeitig Wärmeenergie erzeugt wird. Das in den Untergrund geförderte Kaltwasser erwärmt sich im Erdreich um ca. 3 – 4 K. Die hierbei gewonnene Wärmeenergie wird durch den Verdichter der Wärmepumpe auf das notwendige Temperaturniveau angehoben, das zur Beheizung von Gebäuden dient (im Regelfall 40 bis maximal 55 °C). Im Sommerbetrieb kann die Wärmepumpe als Kälteanlage arbeiten, wobei nunmehr Warmwasser in den Untergrund eingespeist wird (Heizung) und somit die Ergiebigkeit im nachfolgenden Heizbetrieb (Kühlung des Erdreichs) angehoben wird. **Bild 266** zeigt eine typische Energiebilanz einer Wärmepumpenanlage. Kann das Erdreich nicht als natürlicher Aquifer (Wärmespeicher) dienen, so werden unter Umständen künstliche Aquifere eingesetzt, wie in **Bild 267** ausgewiesen. **Bild 268** zeigt die Einbindung eines Aquiferspeichers in die Wärmerückgewinnungsanlage einer großen Wärmepumpe, **Bild 269** ein analoges Wärmepumpensystem.

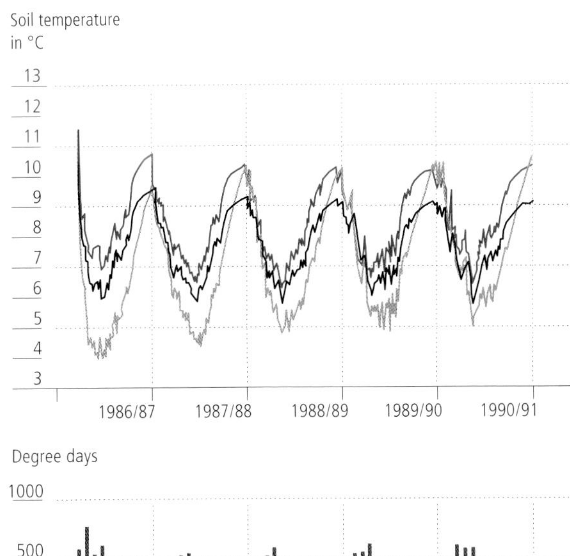

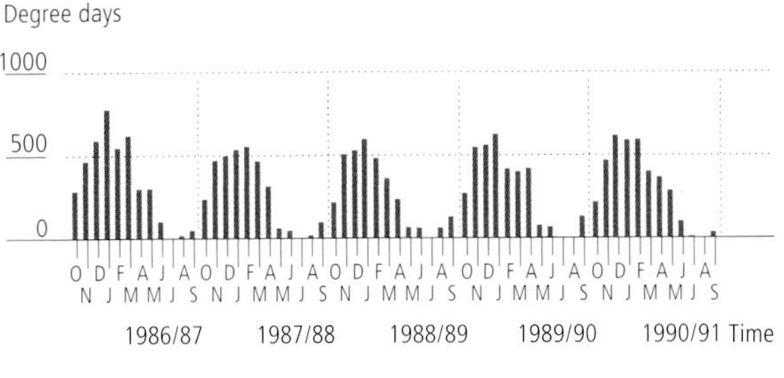

— 85 m
— 50 m
— 5 m

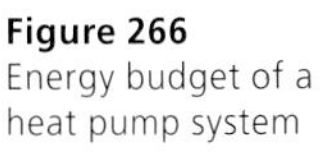

Figure 266
Energy budget of a heat pump system

Source: Enhanced Geothermal Response Test CDM, Unterhaching, Germany
Dr. Claus Heske &
Dr. Marec Wedewardt

Bild 266
Energiebilanz einer Wärmepumpenanlage

Quelle: Enhanced Geothermal Response Test CDM, Unterhaching, Deutschland
Dr. Claus Heske &
Dr. Marec Wedewardt

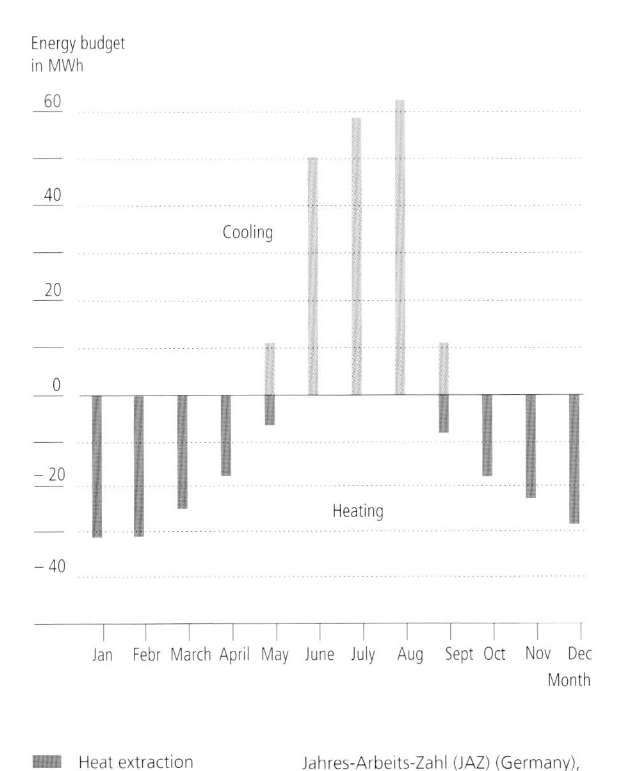

Heat extraction
JAZ: 99999,0
Heat injection
JAZ: 4,0

Jahres-Arbeits-Zahl (JAZ) (Germany),
US: Coefficient of Performance (COP)
JAZ: 9,9999.0

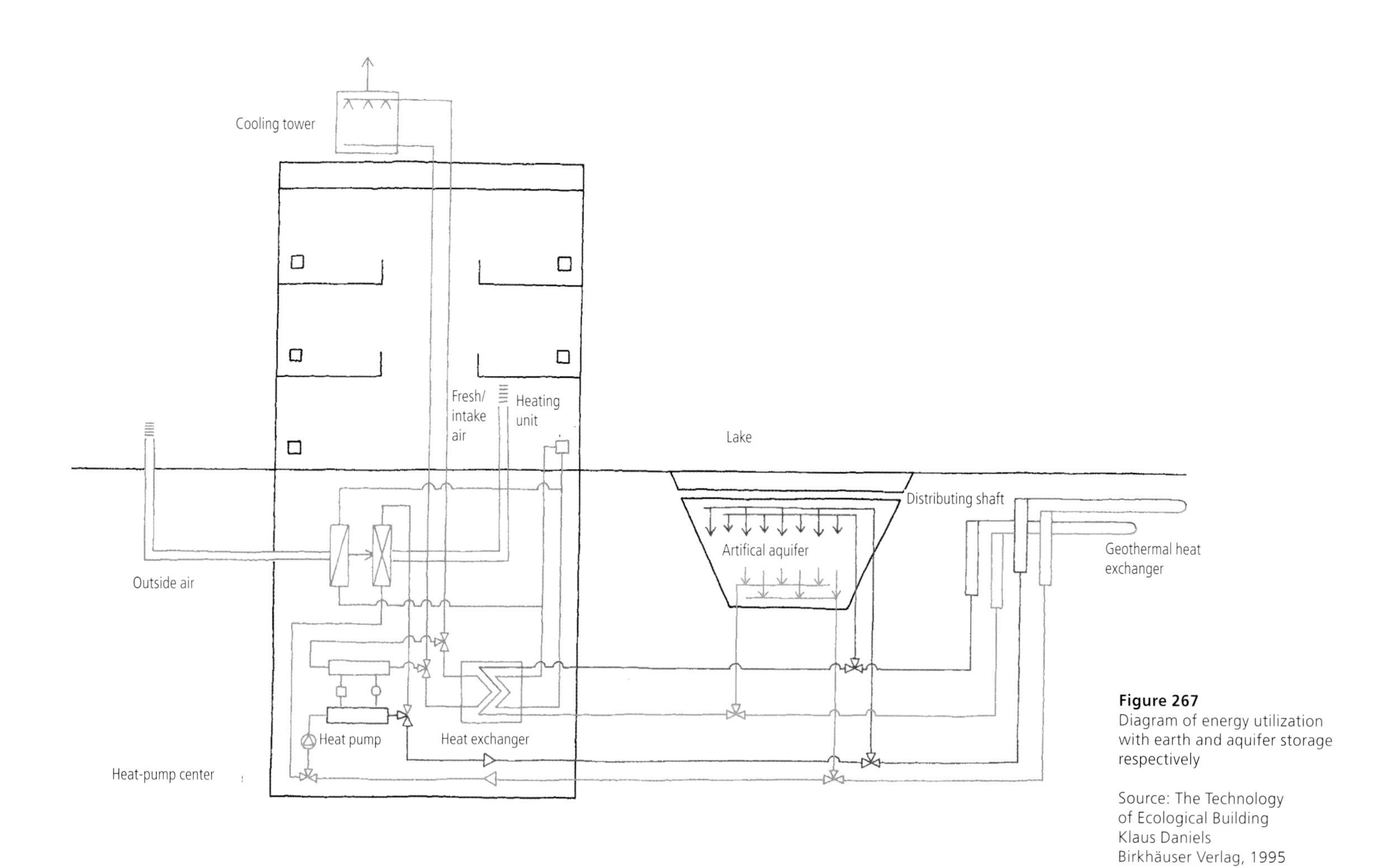

Figure 267
Diagram of energy utilization with earth and aquifer storage respectively

Source: The Technology of Ecological Building
Klaus Daniels
Birkhäuser Verlag, 1995

Bild 267
Prinzipielle Systemdarstellung der Energienutzung mit Erd- bzw. Aquiferspeicher

Quelle: Technologie des ökologischen Bauens
Klaus Daniels
Birkhäuser Verlag, 1995

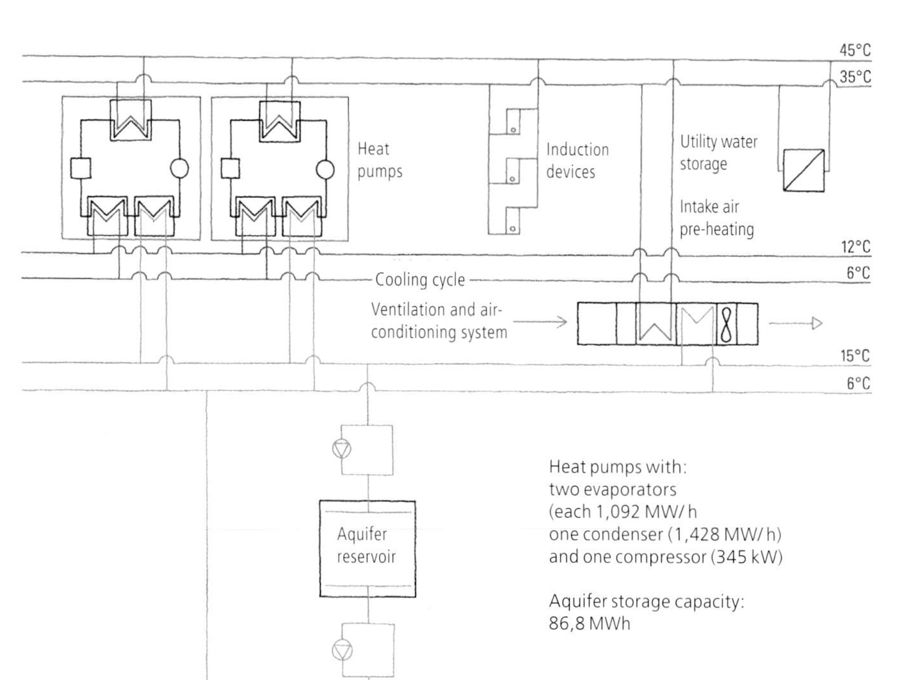

Heat pumps with:
two evaporators
(each 1,092 MW/h
one condenser (1,428 MW/h)
and one compressor (345 kW)

Aquifer storage capacity:
86,8 MWh

Figure 268
Integration of an aquifer reservoir into the heat-recovery installation of a large heat-pump system
(Imtech, Hamburg/München)

Source: The Technology of Ecological Building
Klaus Daniels
Birkhäuser Verlag, 1995

Bild 268
Einbindung eines Aquiferspeichers in die Wärmerückgewinnungsanlage einer großen Wärmepumpenanlage
(Imtech, Hamburg/München)

Quelle: Technologie des ökologischen Bauens
Klaus Daniels
Birkhäuser Verlag, 1995

Figure 269
Central heating installation with gas heat pumps

Source:
Advanced Building Systems
Klaus Daniels
Birkhäuser Verlag, 2003

Figure 269
Wärmezentrale mit Gaswärmepumpe

Quelle:
Advanced Building Systems
Klaus Daniels
Birkhäuser Verlag, 2003

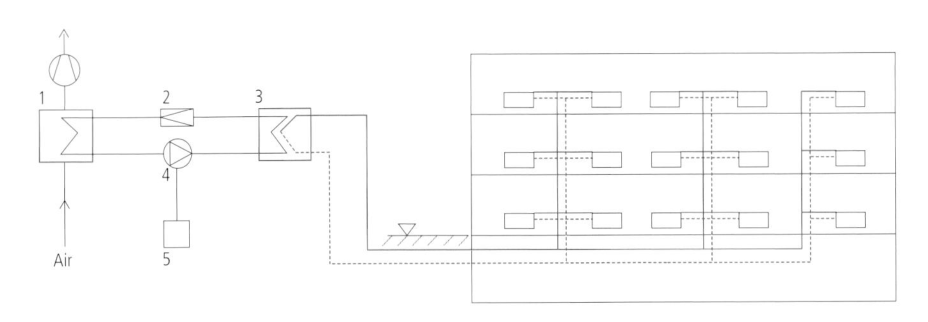

Figure 270
Diagram of gas heat pump

Source:
Advanced Building Systems
Klaus Daniels
Birkhäuser Verlag, 2003

1 Evaporator
2 Expansion valve
3 Condenser
4 Compressor
5 Gas motor
— Heating network

Bild 270
Prinzipschaltbild einer gasbetriebenen Wärmepumpe (Luft/Wasser)

Quelle:
Advanced Building Systems
Klaus Daniels
Birkhäuser Verlag, 2003

In addition to the soil as heat source, in some instances media such as air, warm sewage water, or ground water can be used as a substitute. **Figure 270** shows a diagram of a gas heat pump (air/water) that gives cooling energy off to warm air exhaust streams, essentially recapturing the heat energy of the exhaust air. **Figure 271** shows a cross section of a sewage channel with an inserted raw sewage heat utilization (Rabtherm®) system. The technology is designed to recapture thermal energy from large-scale sewer grids.

Neben dem Erdreich als Wärmequelle stehen unter Umständen auch Luft, warmes Abwasser oder Grundwasser zur Verfügung. **Bild 270** zeigt das Schema einer Gaswärmepumpenanlage (Luft/Wasser), bei der die Kühlenergie gegen warme Abluftströme abgegeben wird und somit die Wärmeenergie der Abluftströme im Wesentlichen rückgewonnen wird. **Bild 271** zeigt den Querschnitt eines Abwasserkanals mit einem Rabtherm-Wärmetauscher, bei dem es da-rum geht, die Wärmeenergie ausgedehnter Abwasserkanalsysteme zu nutzen.

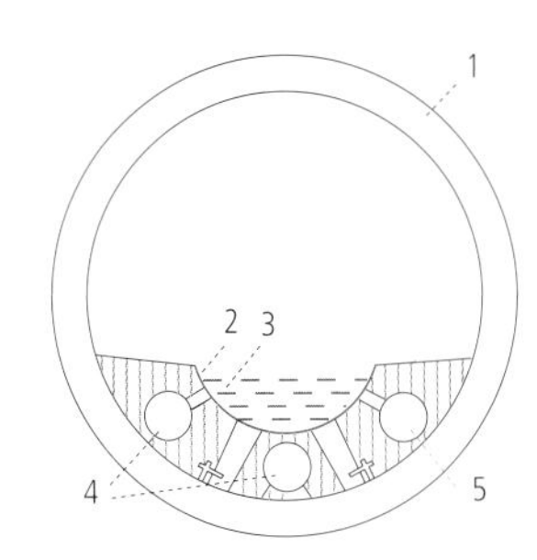

1 Sewer
2 Heat exchanger half shell
3 Sewage water
4 Intermediate medium, feed
5 Intermediate medium, return

Figure 271
Cross-section of sewer with Rabtherm heat exchanger
(Drawing: Studer + Partner)

Source:
Advanced Building Systems
Klaus Daniels
Birkhäuser Verlag, 2003

Bild 271
Querschnitt eines Abwasserkanals mit Rabtherm Wärmetauscher
(Werkbild Studer + Partner)

Quelle:
Advanced Building Systems
Klaus Daniels
Birkhäuser Verlag, 2003

Besides the typically used deep borehole probes, the water from wells can be used to capture energy as well. In this case, the water is typically cooled indirectly with a heat exchanger, and the cooler groundwater is then re-injected into the deep soil. **Figure 272** shows a diagram of a groundwater-source heat pump. In a case-by-case analysis, it is necessary to determine whether the groundwater can also be used for the cooling of chiller plants, where it would serve as a heat sink.

Neben den heute üblicherweise eingesetzten Tiefensonden kann auch Grundwasser als Wärmequelle aus Brunnen genutzt werden, das im Regelfall indirekt (über Wärmetauscher) entwärmt wird, um anschliessend das entwärmte (gekühlte) Grundwasser wieder in den Untergrund zurückzupumpen. **Bild 272** zeigt das Prinzipschema einer Grundwasser-Wärmepumpenanlage. Von Fall zu Fall kann das Grundwasser auch zur Rückkühlung von Kältemaschinen dienen, das Grundwasser dient nunmehr als Wärmesenke.

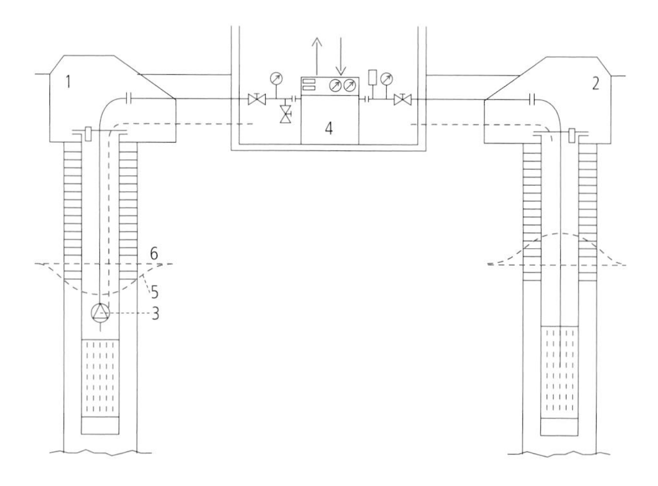

1 Heat source
2 Heat sink
3 Submerged pump
4 Water/water heat pump
5 Feed water level
6 Rest water level

Notation for heat pumps

X – Y – HP
(Source) (Sink)

Figure 272
Diagram of groundwater heat pump installation with heat source, heat sink, submerged or deep-well pump, water/water heat pump

Source:
Advanced Building Systems
Klaus Daniels
Birkhäuser Verlag, 2003

Bild 272
Prinzipschema einer Grundwasser-Wärmepumpenanlage, mit Förderbrunnen, Schluckbrunnen, Unterwasserpumpe und Wasser/Wasser-Wärmepumpe

Quelle:
Advanced Building Systems
Klaus Daniels
Birkhäuser Verlag, 2003

13.1.4 Earth tubes and geothermal labyrinths

To pre-cool air in summer prior to its use in a building, it can be guided through tubes placed into the ground or a subterranean concrete labyrinth. In addition to cooling, this air may be pre-warmed in the winter to provide heating. To some degree, dehumidification of outside air in such systems is possible as well.

Thermo channels or earth tubes are typically employed if the air volume used in a building is not too great, and concrete, PVC, and clay are the most common materials used for earth tubes.

Figure 273 shows the typical soil temperatures found in Central Europe – 8 °C in winter and about 15 °C in summer – and the temperature changes of the air after passing through the earth tubes. **Figure 274** gives data for temperatures of outside air after passing through an earth tube placed at a depth of 3 m.

13.1.4 Erdrohre/Thermolabyrinth

Um Außenluft im Sommer vor Einleitung in ein Gebäude abzukühlen, gegebenenfalls zu entfeuchten bzw. im Winter vorzuheizen, bietet es sich an, die Außenluft durch im Erdreich verlegte Rohre anzusaugen oder durch ein unter dem Gebäude liegendes, betoniertes Thermolabyrinth hindurchzuführen.

Thermokanäle (Erdrohre) werden im Regelfall dann eingesetzt, wenn die dem Gebäude zuzuführenden Außenluftströme nicht zu groß sind. Zum Einsatz kommen dabei Tonrohre, Betonrohre oder Kunststoffrohre, die im Erdreich verlegt werden.

In Mitteleuropa und den hier typischen Erdreichtemperaturen zwischen ca. 8 (Winter) bzw. 15 °C (Sommer) zeigt sich ein Temperaturverlauf beim Durchströmen der Erdrohre, wie in **Bild 273** dargestellt. **Bild 274** zeigt ein in ca. 3 m Tiefe verlegtes Erdrohr mit dem durch dieses Rohr entstehenden Temperaturfeld bei Durchströmung von Außenluft.

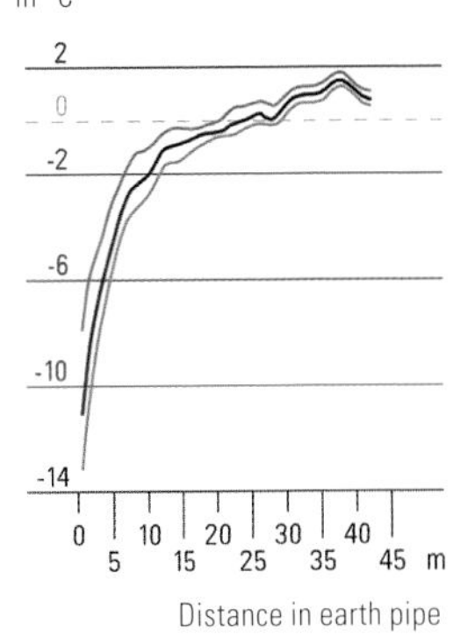

— Values during max. outside temperature
— Daily average values
— Values during min. outside temperature

Figure 273
Temperature in the direction of the earth pipe wall, winter

Source: The Technology of Ecological Building
Klaus Daniels
Birkhäuser Verlag, 1995

Bild 273
Temperaturverlauf entlang der Erdrohrwand, Winter

Quelle: Technologie des ökologischen Bauens
Klaus Daniels
Birkhäuser Verlag, 1995

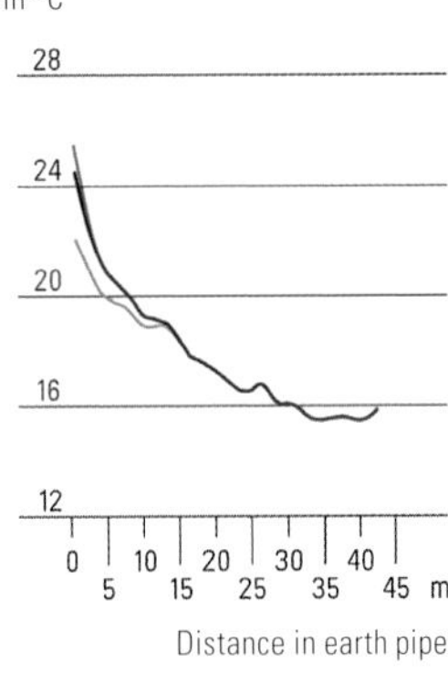

Temperature in the direction of the earth pipe wall, summer

Operation of earth pipe:
9 a.m to 9 p.m.

Temperaturverlauf entlang der Erdrohrwand, Sommer

Betriebszeit des Erdrohres: 9.00 – 21.00 Uhr

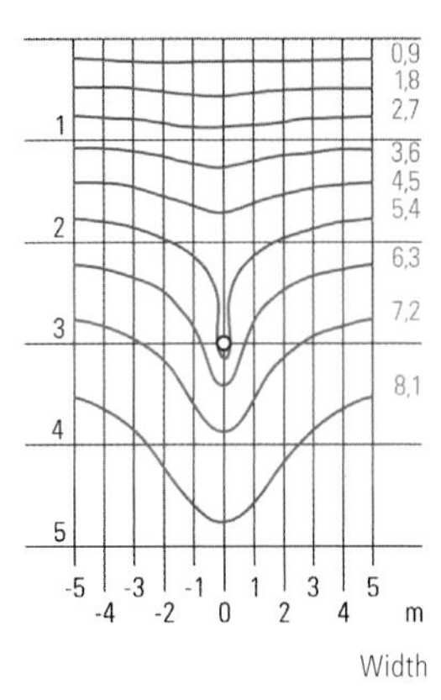

Figure 274
Cone of depression in ground surrounding earth pipe (Isothermal lines)

Earth surface: 0 °C
Pipe surface: 4,5 °C

Source: The Technology of Ecological Building
Klaus Daniels
Birkhäuser Verlag, 1995

Bild 274
Temperaturfeld des Erdreiches mit Erdrohr (Isothermen)

Erdoberfläche: 0 °C
Rohroberfläche: 4,5 °C

Quelle: Technologie des ökologischen Bauens
Klaus Daniels
Birkhäuser Verlag, 1995

With a spacing of the tubes of about 5 – 6 m, a clear increase in temperature of the outside air in winter and a distinct cooling in summer are noticeable. **Figure 275** depicts the temperature data for an earth tube at a depth of 3 m. The use of earth tubes should not be limited to cooler climate regions because they are very feasible for warmer regions as well, although the soil temperatures will be 20 °C higher than in more moderate climates. If, for example, in a hot climate region outside air with a temperature of 45 °C passes through earth tubes placed in the ground at a soil temperature of 28 °C, after approximately 400 m that outside air would have been cooled down to the surrounding soil temperature level. Thus, the passively achieved decrease of air temperature by 15 – 17 K can be considered an extraordinary energy saving.

If required outside fresh air volumes for ventilation are high (> 20,000 m³/h), earth tubes tend to become too voluminous and expensive, and in such cases thermal labyrinths may be used instead. They consist of concrete channels with a large heat transmission surface between outside air and the soil.

As early as two decades ago, a large thermal labyrinth was designed for use in an opera house by the European technical services provider Imtech Hamburg in Germany. Its channels are approximately 140 meters long. **Figure 276** shows the working principle of such a labyrinth, and **Figure 277** depicts the system during construction.

Bei einem Abstand der Rohre um ca. 5 – 6 m ergibt sich im Winter ein deutlicher Wärmegewinn, im Sommer eine deutliche Abkühlung entsprechender Außenluftströme. **Bild 275** zeigt beispielhaft die Lufttemperaturverläufe innerhalb eines Erdrohres, verlegt in 3 m Tiefe. Die Verlegung von Erdrohren ist nicht nur in kühlen Regionen von Interesse, sondern kann auch einen wesentlichen Beitrag in warmen und heißen Regionen liefern, wobei nunmehr jedoch das gesamte Temperaturniveau unter Umständen um bis zu 20 °C höher liegt. Dementsprechend wäre in heißen Regionen mit einer maximalen Außenlufttemperatur von 45 °C zu rechnen bei einer Bodentemperatur in etwa um 28 °C. Nach ca. 400 m durchströmter Erdrohrlänge würde sich die Außenluft annähernd auf die Erdreichtemperatur einpendeln. Somit würde eine Temperatursenkung um ca. 15 – 17 K erreicht, was einem nicht zu unterschätzenden Energiegewinn entspricht.

Bei großen Luftmengen, z.B. über 20.000 m³/h Außenlufteintrag in ein Gebäude, werden Erdrohre zu großvolumig und aufwändig. In diesen Fällen werden vornehmlich Thermolabyrinthe eingesetzt, d.h. Betonkanäle mit einer großen Wärmeübergangsfläche zwischen durchströmender Außenluft und Erdreich.

Bereits vor ca. 20 Jahren wurde durch die Firma Imtech, Hamburg, ein großes Thermolabyrinth für ein Opernhaus entwickelt, bei dem die Länge der Betonkanäle ca. 140 m betrug. **Bild 276** zeigt das Arbeitsprinzip des Thermolabyrinths, **Bild 277** ein Thermolabyrinth im Bauzustand.

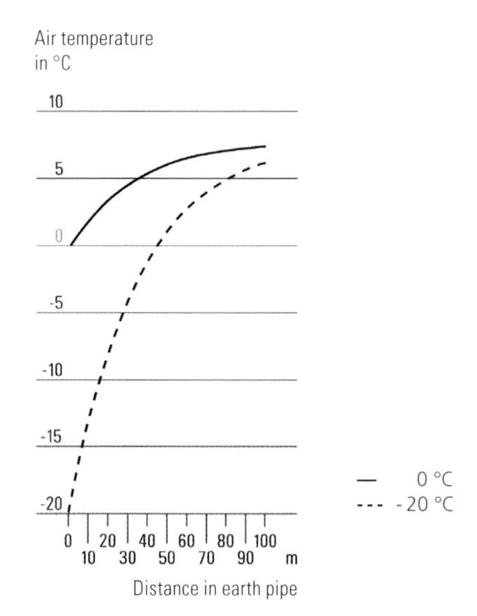

Figure 275
Air temperature in the direction of earth pipe (winter) at 3 m depth (ø 127 mm, $\dot{V}$ = 140 m³h) for various outside temperatures

Source: The Technology of Ecological Building
Klaus Daniels
Birkhäuser Verlag, 1995

Bild 275
Lufttemperaturverlauf entlang des Erdrohres (Winter) in 3 m Tiefe (ø 127 mm, $\dot{V}$ = 140 m³h) bei verschiedenen Außentemperaturen

Quelle: Technologie des ökologischen Bauens
Klaus Daniels
Birkhäuser Verlag, 1995

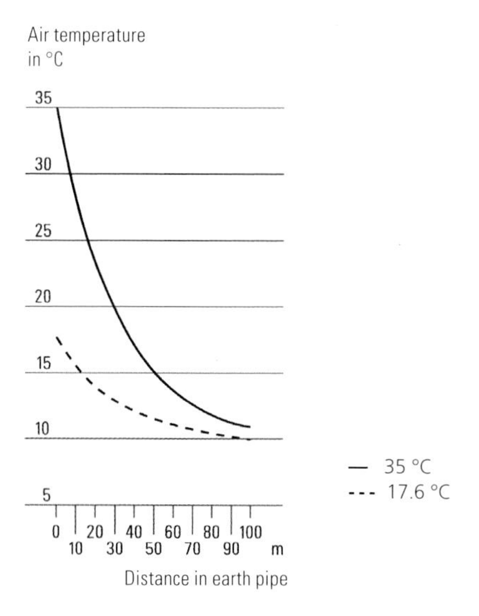

Air temperatures in direction of earth pipe (summer) at 3 m depth (ø 127 mm, $\dot{V}$ = 140 m³h) for various outside temperatures

Lufttemperaturverlauf entlang des Erdrohres (Sommer) in 3 m Tiefe (ø 127 mm, $\dot{V}$ = 140 m³h) bei verschiedenen Außentemperaturen

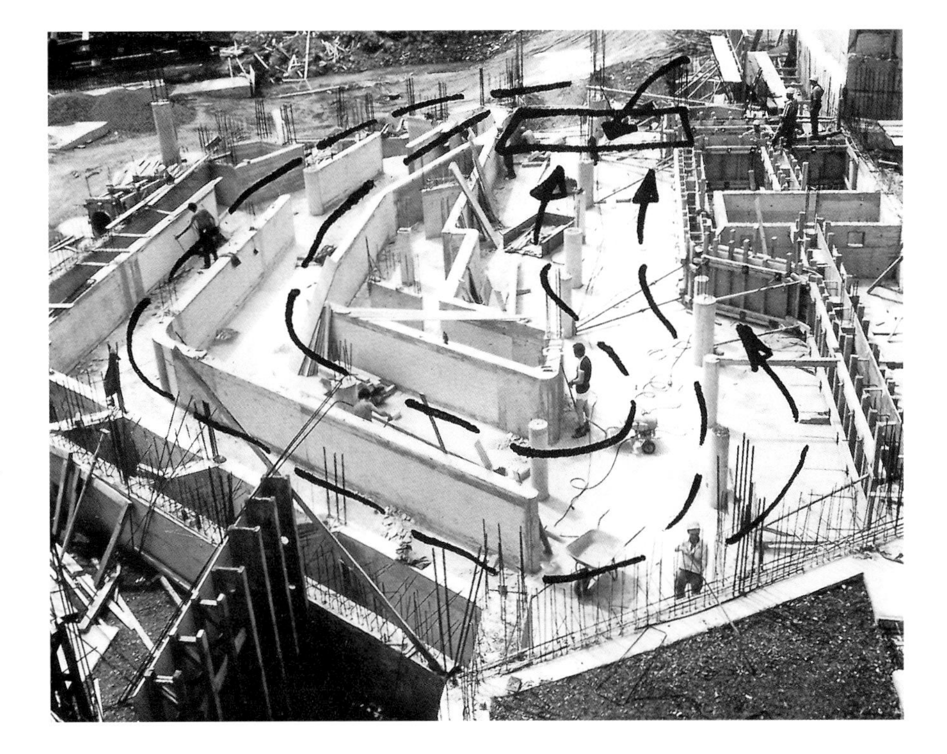

Figure 277
Thermal labyrinth in new municipal theatre in Heilbronn, Germany, during construction (with sketched-in air flow)

Source: The Technology of Ecological Building
Klaus Daniels
Birkhäuser Verlag, 1995

Bild 277
Das Thermolabyrinth im neuen Stadttheater Heilbronn im Bau mit skizzierter Luftzuführung

Quelle: Technologie des ökologischen Bauens
Klaus Daniels
Birkhäuser Verlag, 1995

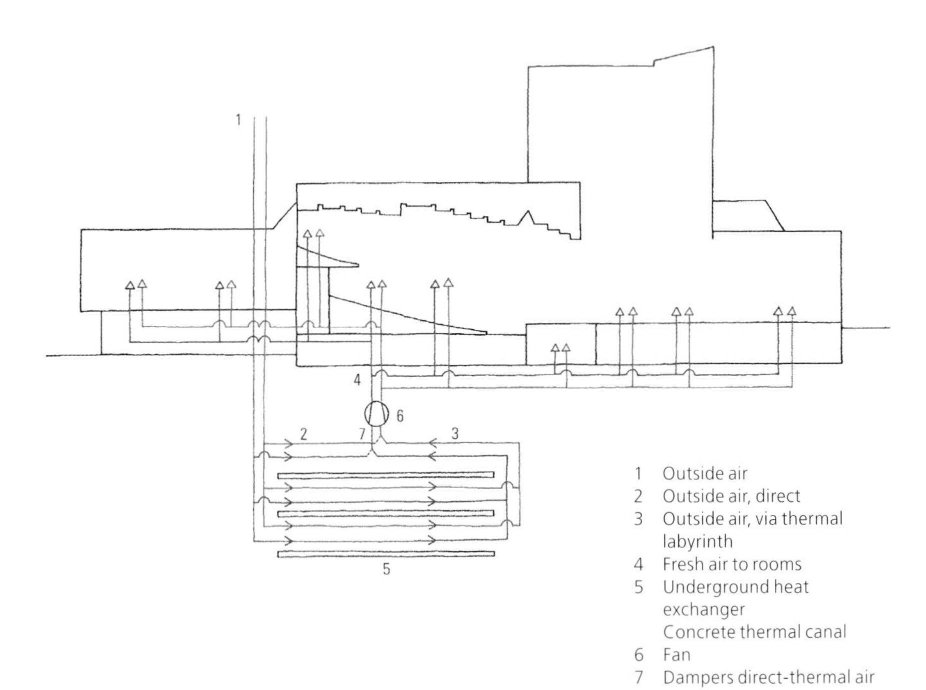

1 Outside air
2 Outside air, direct
3 Outside air, via thermal labyrinth
4 Fresh air to rooms
5 Underground heat exchanger
Concrete thermal canal
6 Fan
7 Dampers direct-thermal air

Figure 276
Diagram of outside air flow, reaching the air-conditioning system either directly or by way of the labyrinth.
(Imtech, Hamburg/München)

Source: The Technology of Ecological Building
Klaus Daniels
Birkhäuser Verlag, 1995

Bild 276
Schema Außenluftstrom und Luftführung in die Lüftungszentrale direkt oder über das „Thermolabyrinth"
(Imtech, Hamburg/München)

Quelle: Technologie des ökologischen Bauens
Klaus Daniels
Birkhäuser Verlag, 1995

Any labyrinth placed beneath a building needs to be well insulated against the structure above. **Figure 278** shows the achievable temperature differences (power). Since the building site is located in Germany, the typical soil temperatures are +8 °C in winter and +14 °C in summer.

Because the velocity of air passing through a labyrinth may change based on the air volumes required for ventilation, it is important to remember that temperature gains in winter and losses in summer become significantly smaller with reduced length of stay of the air in the channel. Therefore, labyrinths need to be designed in such a way that the velocity of air passing through is limited to a maximum of 1 m/s to ensure high efficiency of the system.

Figure 279 shows the achievable energy gain by using a thermal labyrinth. In order to calculate approximate transmission power, **Figure 280** serves as a tool.

Das unter dem Gebäude liegende Thermolabyrinth (gut isoliert gegen den aufgehenden Hochbaukörper) erzielte Leistungen (Temperaturveränderungen), wie sie in **Bild 278** dargestellt sind. Da das Gebäude in Deutschland gebaut wurde, ist mit den typischen Bodentemperaturen im Winter und im Sommer zu rechnen (Winter ca. +8 °C, Sommer ca. +14 °C).

Durch die sich verändernden Durchströmungsgeschwindigkeiten durch das Thermolabyrinth (analog zu den geförderten Volumenströmen) zeigt sich, dass infolge kürzerer Verweilzeiten der Luft im Bodenkanal die Temperaturgewinne bzw. Temperaturverluste deutlich kleiner werden. Insofern ist es wichtig zu beachten, dass entsprechende Thermolabyrinthe mit Durchströmungsgeschwindigkeiten um maximal 1 m/s ausgelegt werden, um einen höchstmöglichen Effekt zu erzielen.

Die durch das Thermolabyrinth gewonnenen Energiemengen sind in **Bild 279** ausgewiesen. Zur überschlägigen Berechnung und Feststellung der Übertragungsleistungen dient **Bild 280**.

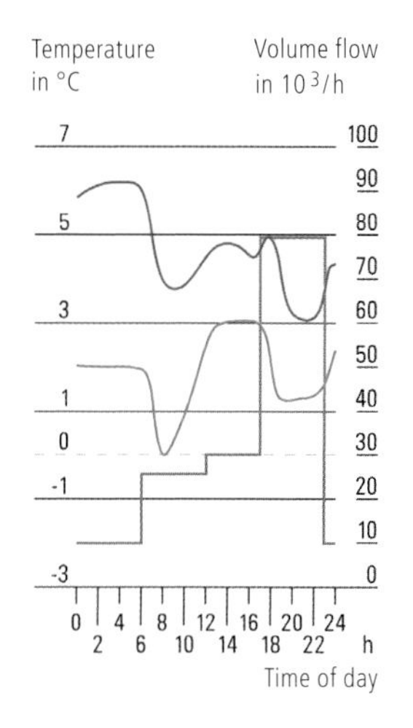

— Start of duct
— End of duct
— Volume flow
Length of duct = 140 m

Figure 278
Recorded temperatures and analysis, one day in winter 1984

Source: The Technology of Ecological Building
Klaus Daniels
Birkhäuser Verlag, 1995

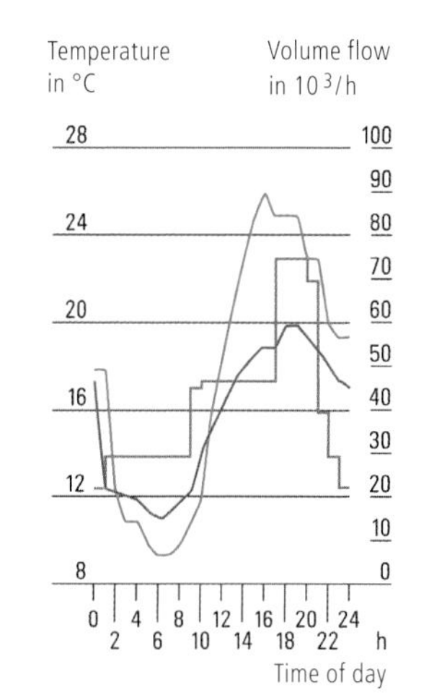

— Air shaft/entrance
— Air shaft/exit
— Volume flow

Recorded temperatures and analysis, temperature increase and drop, one day in May 1983

Temperaturaufzeichnung und Auswertung, Erwärmung und Kühlung in einem Tagesablauf im Mai 1983

Bild 278
Temperaturaufzeichnung und Auswertung, ein Tag im Winter 1984

Quelle: Technologie des ökologischen Bauens
Klaus Daniels
Birkhäuser Verlag, 1995

Required energies
■ Cold energy
■ Heat energy
■ Savings potential trough thermal labyrinth

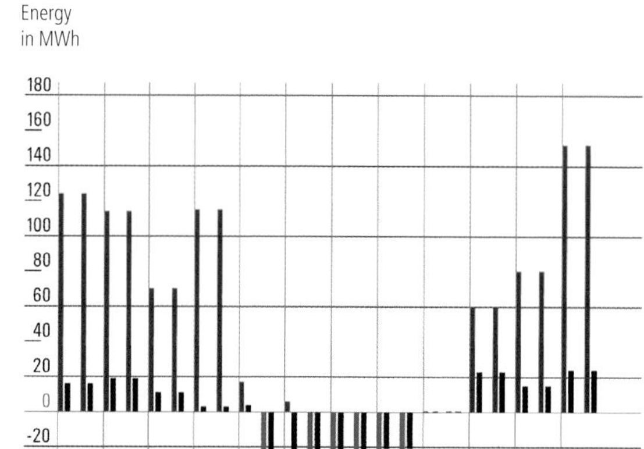

Figure 279
Energy balance (municipal theatre Heilbronn).
Averages over three years

Source: The Technology of Ecological Building
Klaus Daniels
Birkhäuser Verlag, 1995

Bild 279
Energiebilanz (Stadttheater Heilbronn).
Mittlere Ergebnisse aus drei Jahren

Quelle: Technologie des Ökologischen Bauens
Klaus Daniels
Birkhäuser Verlag, 1995

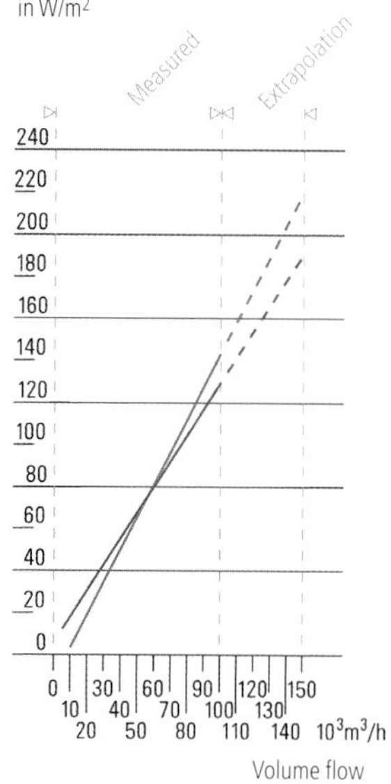

— Heat
— Cold
(August, July)

Figure 280
Specific maximum tunnel transmission rates (from representative values)

Source: The Technology of Ecological Building
Klaus Daniels
Birkhäuser Verlag, 1995

Bild 280
Spezifische maximale Kanalübertragungsleistungen (ermittelt aus representativen Werten)

Quelle: Technologie des ökologischen Bauens
Klaus Daniels
Birkhäuser Verlag, 1995

13.1.5 Cogeneration, or combined heat-power plants (CHP)

Cogeneration power plants use the heat of an engine or a power station to simultaneously generate both electricity and useable heat. In simple terms, it is a more efficient use of final energy than in singular power plants or singular thermal-energy plants. In such "pure" traditional power plants, the achievable performance efficiency is around 35 % – in other words, only 35 % of the power input is being converted to electricity, with the rest being lost in the form of heat energy.

To achieve more efficiency, many years ago a large number of power plants were converted to heat-power plants. In such power plants, the heat loss as a byproduct of electricity generation is used for thermal energy generation, with a resulting increase in efficiency of up to 90 % of the energy input. **Figure 281** allows for an overview of heat-power plants and CHPs useable in building applications. In principle, CHPs are small power plants with combustion engines driving a generator to generate electricity. The excess heat of the combustion engine's cooling water and its exhaust gas heat energy are used for heating applications. Alternatively, such excess heat may be transformed to cooling energy in absorption chiller plants. A plant producing electricity, heat, and cooling energy is sometimes called a trigeneration or, more generally, a polygeneration plant. The results are very high year-round performance coefficients for such cogeneration power plants.

13.1.5 Kraft-Wärme-Kopplung

Bei der Kraft-Wärme-Kopplung (KWK) geht es schlichtweg nur darum, die Endenergie deutlich besser auszunutzen als bei reinen Kraftwerken bzw. Heizwerken. Bei reinen Kraftwerken lässt sich lediglich ein Wirkungsgrad, bezogen auf die eingesetzte Energie, um ca. 35 Prozent feststellen, d.h. lediglich ca. 35 Prozent der Energie werden in Strom umgewandelt, 65 Prozent treten als „Verlustwärme" auf.

Um den Energieeinsatz deutlich besser nutzen zu können, wurde bereits vor vielen Jahren damit begonnen, Kraftwerke zu Heizkraftwerken umzugestalten. Hierbei wird die Verlustwärme neben der Stromproduktion genutzt, so dass ein Gesamtwirkungsgrad bis zu ca. 90 Prozent der eingesetzten Energie erreicht werden kann. **Bild 281** gibt einen Überblick über Heizkraftwerke bzw. im Gebäude zu installierende Blockheizkraftwerke.

Blockheizkraftwerksanlagen sind im Prinzip Kleinkraftwerke mit Verbrennungsmotoren, die einen Generator zur Stromerzeugung betreiben und die Abwärme der Motoren aus Kühlwasser und Abgas zur Heizung nutzen. Die Wärmeenergie, die in einem Blockheizkraftwerk freigesetzt wird, kann auch mittels Absorptionskälteanlagen zu Kühlenergie umgewandelt werden, so dass das Blockheizkraftwerk (BHKW) ganzjährig einen außerordentlich hohen Wirkungsgrad erreicht.

Figure 281
Power-heat coupling (PHC)

Source:
Advanced Builing Systems
Klaus Daniels
Birkhäuser Verlag, 2003

Bild 281
Systeme der Wärme-Kraft-kopplung (KWK)

Quelle:
Advanced Builing Systems
Klaus Daniels
Birkhäuser Verlag, 2003

System	**Heating power stations (HPS)**			**Combined heat and power station (CHP)**		
	Heating power station with steam turbines	Heating power station with gas turbines	Combined heating power station	Combined heat and power station, "to scale"	Standard combined heat and power station	Small combined heat and power station (TOTEM)
Drive system	Steam turbines	Gas turbines	Gas and steam turbines combined	Gas motor with three-way catalytic converter	Gas motor with three-way catalytic converter	Gas motor
Fuel	Coal, heavy oil (fluidized combustion), natural gas, heating oil (conventional steam boiler)	Natural gas, EL heating oil, gasified coal (in future)	Natural gas, EL heating oil, gasified coal (in future)	Natural gas, biogas (e.g. in sewage treatment plants)	Natural gas, biogas (e.g. in sewage treatment plants)	Natural gas, biogas (e.g. in sewage treatment plants)
Principal application (examples)	Integrated district heating	Process heat for industry, hospitals (steam, hot water)	Integrated district heating	Integrated local heating for larger single buildings	Integrated local heating for larger single buildings	SFH development, single buildings
Performance range	5 MW and up	0.5 MW and up	20 MW and up	50... 1'000 kW	150... 200 kW 2) 3)	15... 50 kW 3)
Current parameter1)	0.30... 0.60	0.40... 0.70	0.80... 1.20	0.45... 0.65	0.45... 0.65	0.35... 0.45
Efficiency	0.85	0.75... 0.85	0.75... 0.85	0.85... 1.00 4)	0.85... 1.00 4)	0.85... 1.00 4)

1) Current parameter = electricity production/heat production
2) Good performance range in terms of economic efficiency and potential for application
3) Several units can be coupled to increase performance

4) Efficiencies of 1.00 and above (in relation to the net calorific value) are possible in theory when relevant measures are taken (use of latent heat by means of waste gas condensation)

In relation to the energy input, CHPs achieve an efficiency of 30 – 40 % for electricity generation and approximately 55 % thermal yield, resulting in a total of around 85 – 95 % efficiency. **Figure 282.1** shows a CHP hall containing large units with an electrical power output of 3.2 MW and a thermal power of around 7 MW. In colder regions, cogeneration plants are in certain instances driven not by electricity but by heat, which means that they need to reach long, continuous times of operation to provide the necessary thermal energy. **Figure 282.2** shows the use of a CHP module based on the thermal power requirements of a large energy user.

Cogeneration plants should reach or exceed an annual minimum operation time of 4,500 h/a in order to be economical. If large building complexes require emergency power systems driven by diesel generators, it can be of interest to use a cogeneration power plant instead in order to provide continuous operation. Cogeneration power plants are today typically fueled by light fossil heating oil or natural gas, but the future fuels for such systems will be of the family of renewable fuels such as bio diesel and biogas to decouple the plants from fossil energy use.

Bezogen auf den Brennstoffeinsatz, erreichen BHKW's eine Stromausbeute von ca. 30 – 40 Prozent und eine Wärmeausbeute von etwa 55 Prozent, so dass sich ein Gesamtwirkungsgrad von ca. 85 – 95 Prozent feststellen lässt. **Bild 282.1** zeigt eine Maschinenhalle mit großen BHKW-Blöcken, wobei die elektrische Leistung ca. 3,2 MW und die thermische Leistung ca. 7 MW entspricht.

In kalten Regionen werden BHKW-Systeme unter Umständen nicht strom-, sondern wärmegeführt, d.h. es kommt im Wesentlichen darauf an, dass die BHKW's eine möglichst lange Betriebszeit zur Wärmebereitstellung erreichen. **Bild 282.2** zeigt den typischen betrieblichen Einsatz entsprechender BHKW-Module aufgrund des vorher berechneten Wärmebedarfs eines Großabnehmers.

BHKW-Systeme sollten zumindest eine Betriebszeit von ca. 4.500 h/a erreichen, um sie wirtschaftlich einsetzen zu können.Benötigen große Objekte eine Notstromversorgung, so ist es von Interesse, die Notstromversorgungsanlage (im Regelfall Notstrom-Diesel-anlage) als Blockheizkraftwerk einzusetzen, damit ein entsprechender Dauerbetrieb gegeben ist.

Blockheizkraftwerke werden zurzeit im Regelfall noch mit leichten Heizölen bzw. Ferngas betrieben. In Zukunft müssen entsprechende Systeme mit Biogas bzw. Bioöl gefahren werden, um sich von den fossilen Brennstoffen abkoppeln zu können.

Figure 282.1
Mechanical equipment room with CHP (Thermal output approx. 7 MW)

Source:
Advanced Builing Systems
Klaus Daniels
Birkhäuser Verlag, 2003

Bild 282.1
Zentrale einer Kraft-Wärme-Kopplungsanlage (CHP/BHKW)

Quelle:
Advanced Buidling Systems
Klaus Daniels
Birkhäuser Verlag, 2003

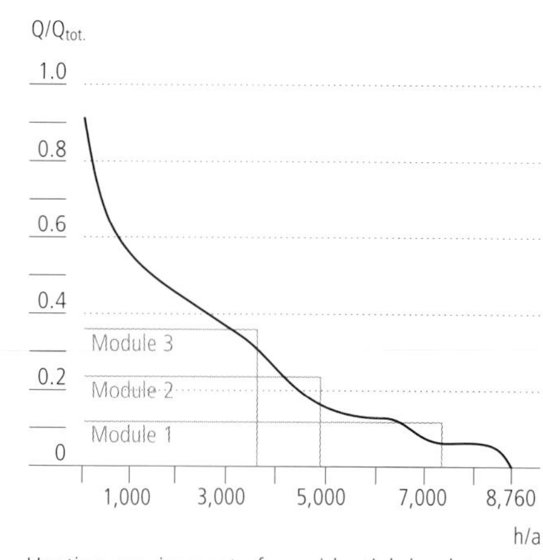

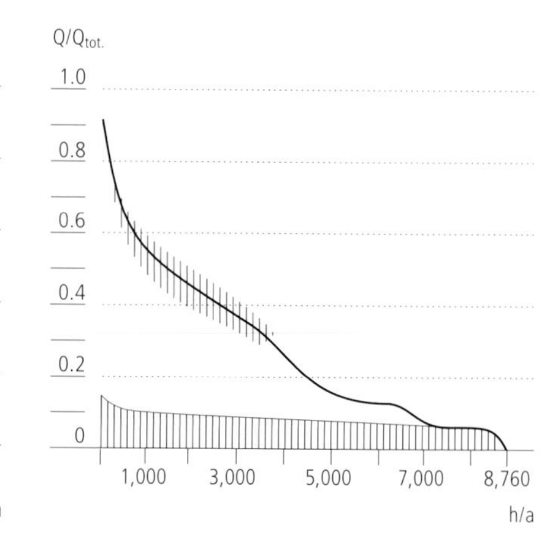

Heating requirement of a residential development

||||| Tolerance range determined by external temperature
||||| Warm water

Figure 282.2
Annual heating requirements of a residential development and interpretation of graph for CHP sizing

Source:
Advanced Building Systems
Klaus Daniels
Birkhäuser Verlag, 2003

Bild 282.2
Jahresdauerlinie des Wärmebedarfs einer Wohnsiedlung und Auswertung der Jahresdauerlinie für die BHKW-Dimensionierung

Quelle:
Advanced Buidling Systems
Klaus Daniels
Birkhäuser Verlag, 2003

13.1.6 Solar thermal energy

As the term suggests, solar thermal energy plants convert the energy of the Sun into thermal energy. Depending upon the specific temperature range, a variety of system solutions exist today (**Figure 283**). **Figure 284** shows the areas where such systems are used most appropriately and the related necessary operating temperatures.

13.1.6 Solarthermie

Solarthermische Anlagen dienen – wie der Name schon sagt – dazu, Solarenergie in Wärmeenergie umzuwandeln. Hierbei gibt es je nach gewünschtem Temperaturbereich eine Vielzahl von Systemlösungen, die in **Bild 283** überschlägig dargestellt sind.

Bild 284 zeigt die Einsatzbereiche mit zugehörigen Systemtemperaturen (Warmwassertemperaturen), die für die unterschiedlichsten Nutzungen benötigt werden.

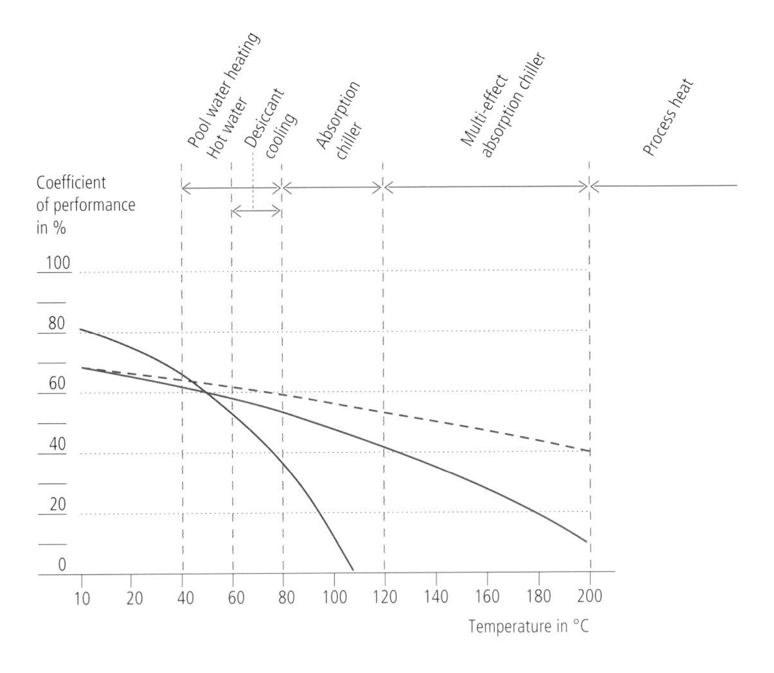

— Flat-plate collector
— Vacuum-tube collector
-- CPC-collector

Figure 284
Possible fields of application of solar heat (depending on the temperature)

Source:
Advanced Building Systems
Klaus Daniels
Birkhäuser Verlag, 2003

Bild 284
Einsatz verschiedenartiger solarthermischer Kollektoren zur Versorgung unterschiedlicher Verbraucher

Quelle:
Advanced Building Systems
Klaus Daniels
Birkhäuser Verlag, 2003

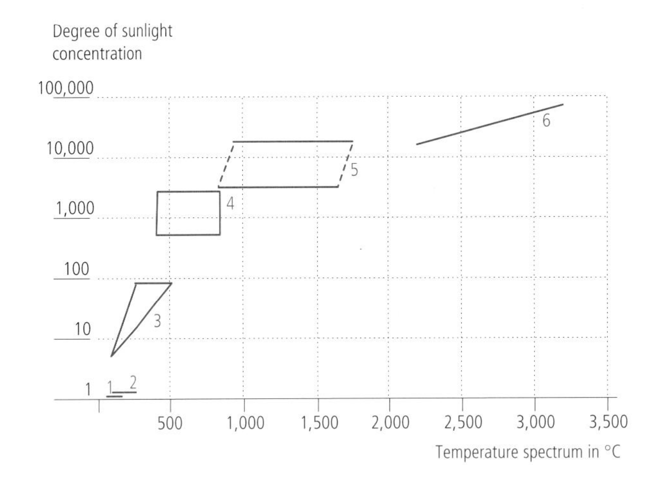

Figure 283
Areas of application of various solar systems

Source:
Advanced Building Systems
Klaus Daniels
Birkhäuser Verlag, 2003

Bild 283
Einsatzbereich solarthermischer Anlagen

Quelle:
Advanced Building Systems
Klaus Daniels
Birkhäuser Verlag, 2003

1 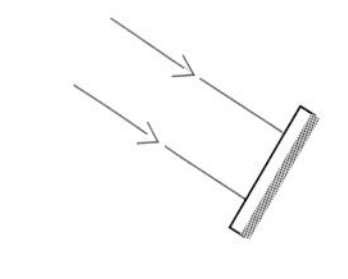

4 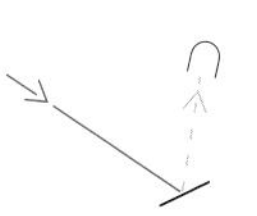

2 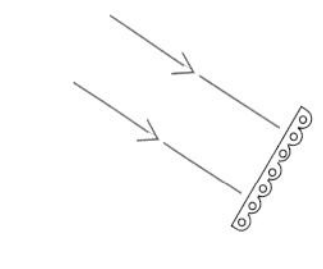

5 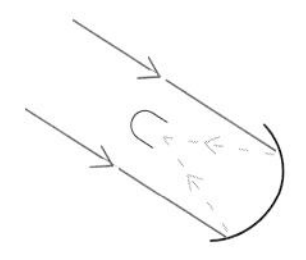

3 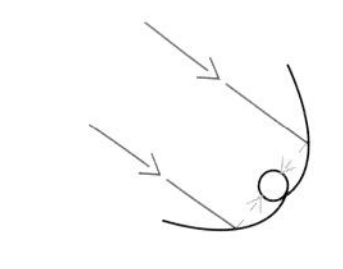

6 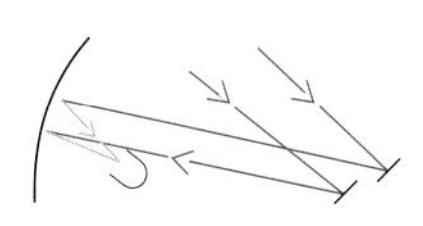

1 Flat-plate collectors
2 Non-focussed systems
3 Trough systems
4 Focal-point focused systems
5 Parabolic concentrators
6 Solar kilns

Solar absorbers

Solar absorbers are simple absorber mattes that are typically made of high-quality EPDM neoprene and that have vulcanized collector and distributor piping integrated into them. They need to be temperature resistant in a range between -50 to +60 °C and be capable of being cleaned out (**Figure 285**). Solar absorbers can be added to, or integrated into, roof surfaces, but they can also be placed "on grade." They reach water temperatures of up to 60 °C, and their main use is to heat the water of outdoor pools. The necessary absorber surface area for such an application is approximately 50 – 80 % of the water surface of the pool that needs to be heated.

Solarabsorber

Solarabsorber sind einfache Absorbermatten aus hochwertigem Kautschuk mit anvulkanisierten Verteiler- und Sammelrohren (reinigungsbeständig, temperaturbeständig von -50 bis +60 °C), **Bild 285**.

Solarabsorber lassen sich sowohl auf Dachflächen als auch auf dem Boden großflächig verlegen und erreichen Warmwassertemperaturen von bis zu 60 °C. Sie werden in der Regel vornehmlich für die Beheizung von Schwimmbädern genutzt. Ihre Fläche beträgt ca. 50 – 80 Prozent der zu beheizenden Wasseroberfläche.

Figure 285
Solar absorber for swimming pool, 300 m^2 absorber surface

Source:
Advanced Building Systems
Klaus Daniels
Birkhäuser Verlag, 2003

Bild 285
Solarabsorber für ein Schwimmbad 300 m^2 Absorberfläche

Quelle:
Advanced Building Systems
Klaus Daniels
Birkhäuser Verlag, 2003

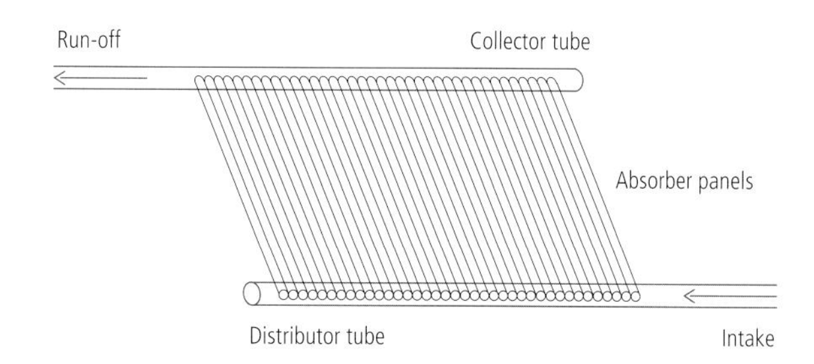

Structure of solar absorber

Aufbau eines Solarabsorbers

Flat-plate collectors

Flat-plate collectors consist of heat-absorbing piping and surfaces that are joined tightly together, and often they are protected by a translucent glass cover. The solar radiation is absorbed largely due to the special coatings on the collector, the emissions of which are kept to a minimum, and the captured heat energy is transported to storage tanks using water as the medium. Such tanks need to be designed to resist corrosion and freezing, depending upon the climate location. To achieve maximum energy gain, collectors need to be adjusted precisely to the position of the sun in the sky (**Figure 286**). The efficiency of solar collectors, which should be oriented due south (northern hemisphere) or due north (southern hemisphere) is compromised if the optimal azimuth angle of the sun (e.g., 180° in the northern hemisphere) cannot be achieved. Decreased efficiencies can also result when collector surfaces are not sloped appropriately to the angle of the sun's rays. Ideally, the angle of solar incidence in relation to the collector's surface is 90°, as seen in **Figure 287**.

Flachplattenkollektoren

Flachplattenkollektoren bestehen aus wärmeabsorbierenden Flächen und Rohren, die wärmeschlüssig miteinander verbunden sind. Sie werden im Regelfall durch hochlichtdurchlässige Glasabdeckungen gegen den Außenraum geschützt. Die Solarstrahlung absorbierende Beschichtung eines Kollektors weist eine hohe Absorption bei niedriger Emission auf. Die vom Kollektor gewonnene Wärmeenergie wird mittels Wasser zu einem Speicher transportiert, wobei das gesamte System je nach Standort frost- und korrosionssicher ausgebildet wird. Um einen maximalen Energieeintrag zu erreichen, müssen entsprechende Kollektoren optimal zur Sonne ausgerichtet werden, **Bild 286**. Die Ausrichtung der Kollektoren nach Süden (nördliche Halbkugel) bzw. nach Norden (südliche Halbkugel) verändert ihren Wirkungsgrad dann, wenn der optimale Azimutwinkel (180 °, nördliche Halbkugel) nicht erreicht wird. Gleichermaßen tritt eine Wirkungsgradverschlechterung dann ein, wenn der Kollektor nicht optimal zur Sonne hin geneigt wird. Idealerweise beträgt der Sonneneinfallswinkel zur Kollektorfläche 90 °, **Bild 287**.

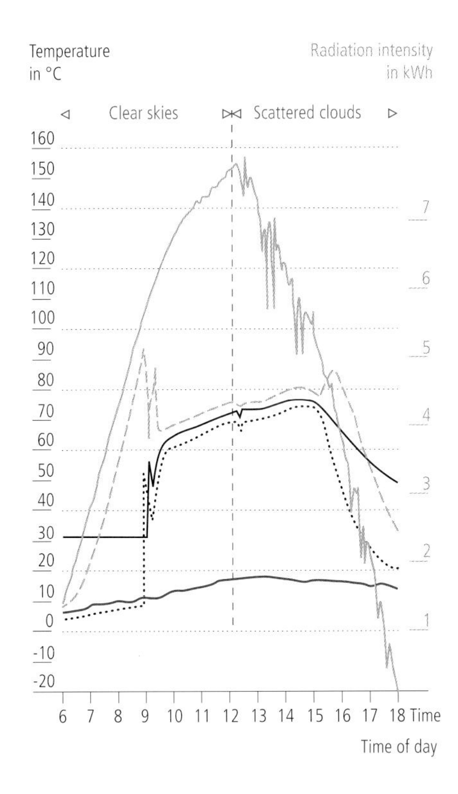

- Radiation intensity
- Surface temperature of collectors
- Water temperature (exiting from collector)
- Water temperature (entering collector)
- Outside temperature

Figure 286
Energy curve of flat plate collector over the course of one day
(measurements Bell + Gossett)

Source:
Advanced Building Systems
Klaus Daniels
Birkhäuser Verlag, 2003

Bild 286
Daten über einen Tagesablauf einer Flachplattenkollektor-anlage
(Messwerte Bell + Gossett)

Quelle:
Advanced Building Systems
Klaus Daniels
Birkhäuser Verlag, 2003

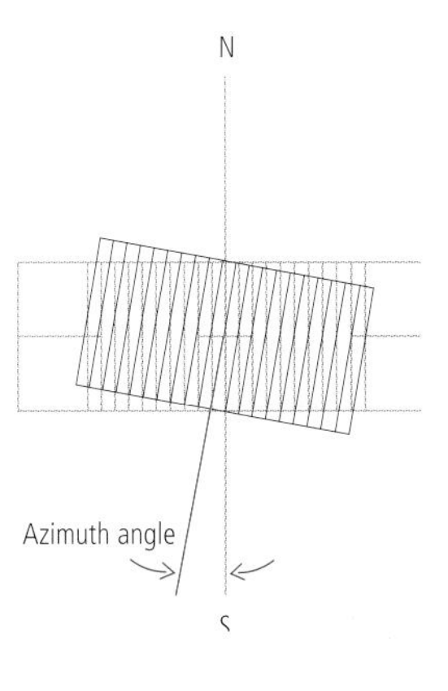

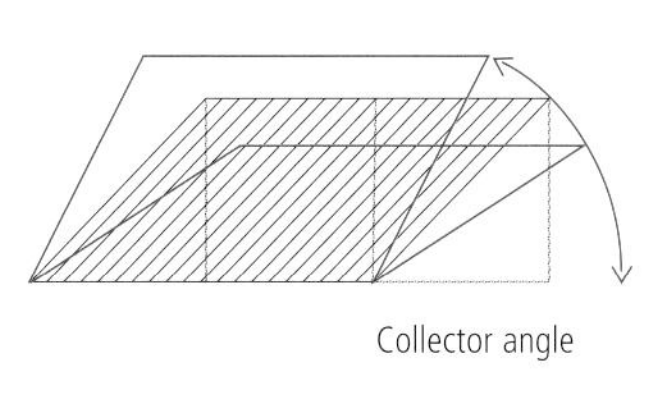

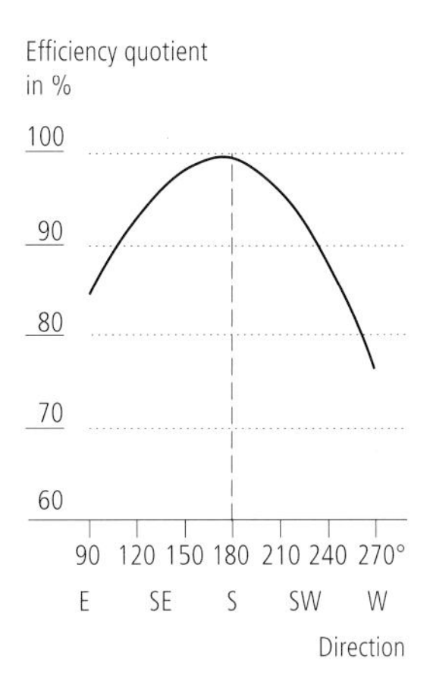

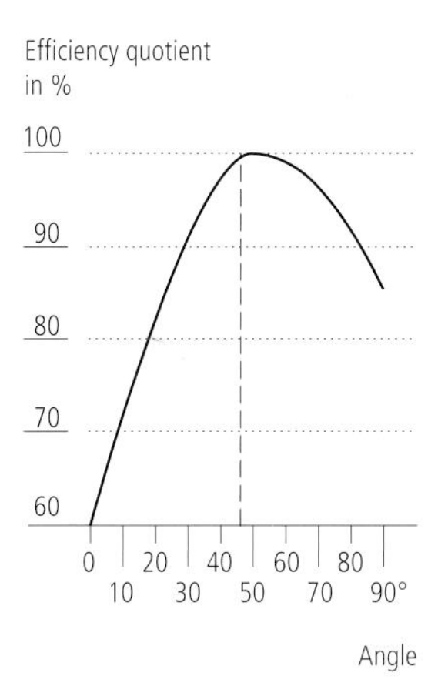

Period of active use	Optimum angle
Jan. – Dec. (year round)	30 – 50°
April – Sept. (seasonal)	25– 45°
May – August (seasonal)	20 – 40°
Sept. – April (heating period)	50 – 70°

Figure 287
Adjustment of collector installation in case of less than optimum placement

Source:
Advanced Building Systems
Klaus Daniels
Birkhäuser Verlag, 2003

Bild 287
Korrekturfaktoren von Kollektoranlagen bei nicht optimaler Aufstellung

Quelle:
Advanced Building Systems
Klaus Daniels
Birkhäuser Verlag, 2003

Figure 288 shows a newer type of flat-plate collectors that are called vacuum flat-plate collectors.

Evacuated solar-tube collectors are used today to achieve higher operating temperatures, and they also provide esthetically appealing coloration effects, which may add to a building's design appeal. In evacuated solar tubes, the solar radiation penetrates glass tubes and reaches absorber surfaces placed inside. Each evacuated tube consists of two glass tubes made from strong borosilicate glass. The outer tube is transparent, allowing light rays to pass through with minimal reflection, and the inner tube is coated with a special selective coating. The tops of the two tubes are fused together, and the air contained in the space between the two layers of glass is pumped out while exposing the tube to high temperatures. This "evacuation" of the gasses forms a vacuum and is the essential factor in the performance of the evacuated tubes. Liquid in the absorber tube evaporates and reaches as steam the condenser of the system. Here the energy is transferred to a stream of water; the resulting steam condenses and flows back to the absorber tube. **Figure 289** shows a diagram of the working principle of evacuated solar absorber tubes.

Bild 288 zeigt eine weiterentwickelte Form des Flachplattenkollektors zum Vakuum-Flachkollektor.

Um höhere Raumtemperaturen zu erreichen und unter Umständen eine interessante farbliche Gestaltung, werden Vakuum-Röhren-Kollektoren genutzt.

Bei Vakuum-Röhren-Kollektoren durchdringt die Sonnenstrahlung evakuierte Glasrohre und trifft auf die sich in den Rohren befindlichen Absorberflächen. Durch eine hochwertige, selektive Beschichtung der Absorberflächen und durch das Vakuum werden die Wärmeverluste an die Umgebung fast völlig unterbunden. Die auf der Absorberfläche gesammelte Wärmeenergie wird auf Wärmerohre übertragen, die im Regelfall auf der Unterseite des Absorbers angelegt sind. Durch die Wärmeübertragung verdampft die Flüssigkeit im Wärmerohr und gelangt als Dampf zu einem Kondensator. Im Kondensator wird die Wärmeenergie wiederum auf einen Wasserstrom übertragen, wodurch der Dampf kondensiert und im Wärmerohr zurückfließt. **Bild 289** zeigt das Arbeitsprinzip des Vakuum-Röhren-Kollektors.

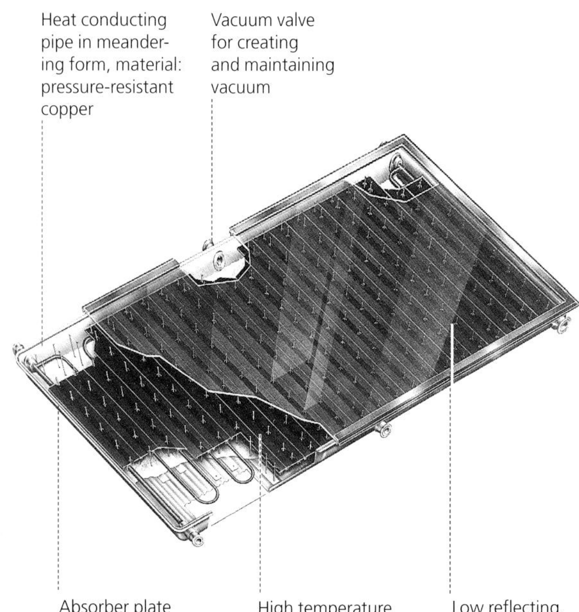

Figure 288
Vacuum flat plate collector (Thermosolar)

Source:
Advanced Building Systems
Klaus Daniels
Birkhäuser Verlag, 2003

Bild 288
Vakuum-Flachkollektor (Thermosolar)

Quelle:
Advanced Building Systems
Klaus Daniels
Birkhäuser Verlag, 2003

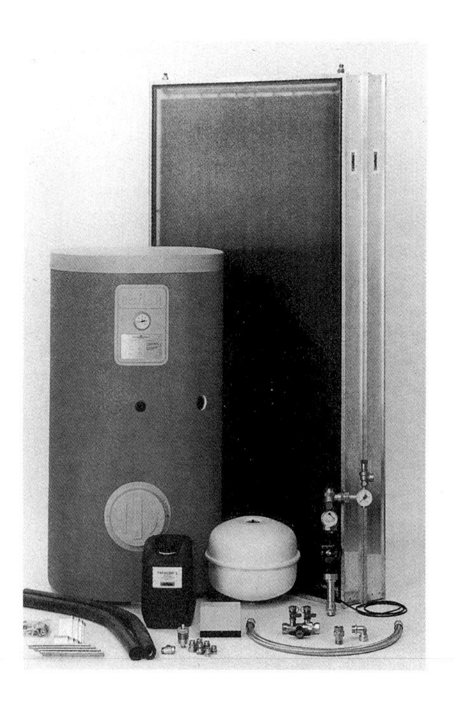

Installation-ready solar system with high-performance flat plate collector (TiNOx-coating), (Reflex Solar-Unit)

Montagefertige Solaranlage mit Hochleistungsflachkollektor (TiNOx-Beschichtung), (Reflex Solar-Unit)

1 Heat radiation
2 Glass tube
3 Conducting tube with filling
4 Absorber, selectively coated
5 Condenser
6 Heat insulation
7 Heat exchanger tube

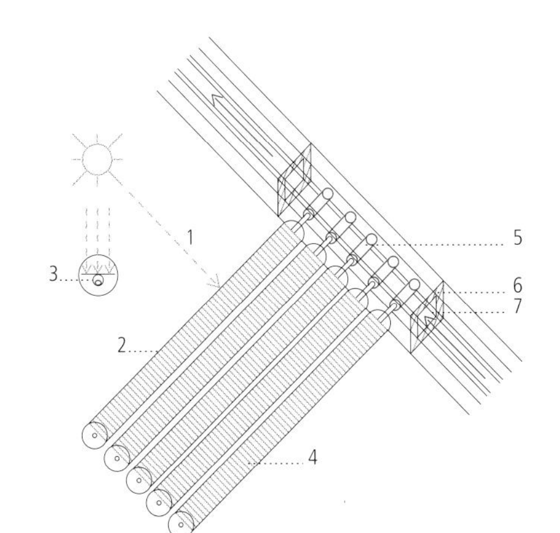

-- Flat plate collector
— Vacuum flat plate collector
— Evacuated tube collector

ΔT = Temperature difference between surroundings and average collector temperature (Flow/return)

Indoor measurement; wind velocity 4 m/s and radiation strength 730 and 750 W/m² respectively

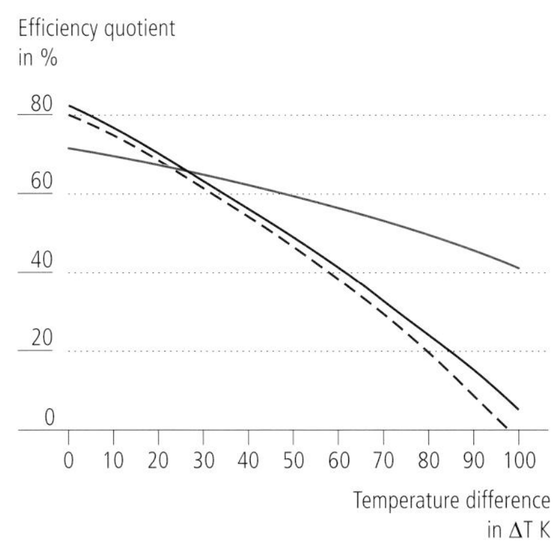

Figure 289
Principle of evacuated tube collector

Source:
Advanced Building Systems
Klaus Daniels
Birkhäuser Verlag, 2003

Figure 290
Efficiency quotient of flat plate collector / vacuum flat plate collector and evacuated tube collector (Stiebel-Eltron)

Source:
Advanced Building Systems
Klaus Daniels
Birkhäuser Verlag, 2003

Bild 289
Arbeitsprinzip eines Vakuum-Röhrenkollektors

Quelle:
Advanced Building Systems
Klaus Daniels
Birkhäuser Verlag, 2003

Bild 290
Wirkungsgrad von Flachkollektor/Vakuum-Flachkollektor und Vakuum-Röhrenkollektor im Vergleich (Stiebel-Eltron)

Quelle:
Advanced Building Systems
Klaus Daniels
Birkhäuser Verlag, 2003

Figure 291
Evacuated tube collectors on solar house Heliotrop in Freiburg, Germany as architectural components (Image: Viessmann)

Source:
Advanced Building Systems
Klaus Daniels
Birkhäuser Verlag, 2003

Bild 291
Vakuum-Röhrenkollektoren auf dem Freiburger Solarhaus Heliotrop als Elemente der Architektur
(Werkbild: Viessmann)

Quelle:
Advanced Building Systems
Klaus Daniels
Birkhäuser Verlag, 2003

In comparison, **Figure 290** shows coefficients of performance and operating temperatures of flat-plate collectors, vacuum flat-plate collectors, and evacuated solar tubes. The German architect Rolf Disch designed the solar house Heliotrop so that it rotates as a whole and tracks the sun for optimal use of solar energy. The house contains evacuated solar tubes for the generation of energy used for heating and for warm water generation, and large photovoltaic elements on the roof supply the majority of the electricity needed in the building (**Figure 291**).

Im **Bild 290** sind zur ersten Auslegung die Wirkungsgrade und Betriebstemperaturen von Flachkollektoren, Vakuum-Flachkollektoren und Vakuum-Röhren-Kollektoren vergleichend dargestellt. Zur optimalen Nutzung der Solarenergie hat der Architekt Disch ein Solarhaus (Heliotrop) entwickelt, das sich mit der Sonne dreht. Das Haus besitzt sowohl Vakuum-Röhren-Kollektoren zur Beheizung und Warmwasserbereitung als auch ein großes Photovoltaikelement auf dem Dach, das einen großen Teil der notwendigen elektrischen Energie liefert, **Bild 291**.

In both warm regions with great amounts of annual sunshine and climates with seasons of varying hot and cold temperatures, the use of evacuated solar tubes is of particular interest because they are capable of producing energy for both cooling and heating. **Figure 292** shows a heating and cooling system supplied with energy from a solar system. It needs to be pointed out, however, that thermal solar energy alone is typically not sufficient to supply the energy needed for large buildings.

In warmen und sonnenreichen Regionen oder in sonnenreichen Regionen mit Wechselklima (kalt und warm) ist der Einsatz von Vakuum-Röhren-Kollektoren zur Erzeugung der notwendigen Kälteenergie und Wärmeenergie von besonderem Interesse. **Bild 292** zeigt eine mit Solarenergie betriebene Heizungs- und Kälteanlage, wobei festgestellt werden muss, dass es im Regelfall bei größeren Objekten selten oder gar nicht möglich ist, ausschließlich ein Gebäude mit thermischer Solarenergie zu betreiben.

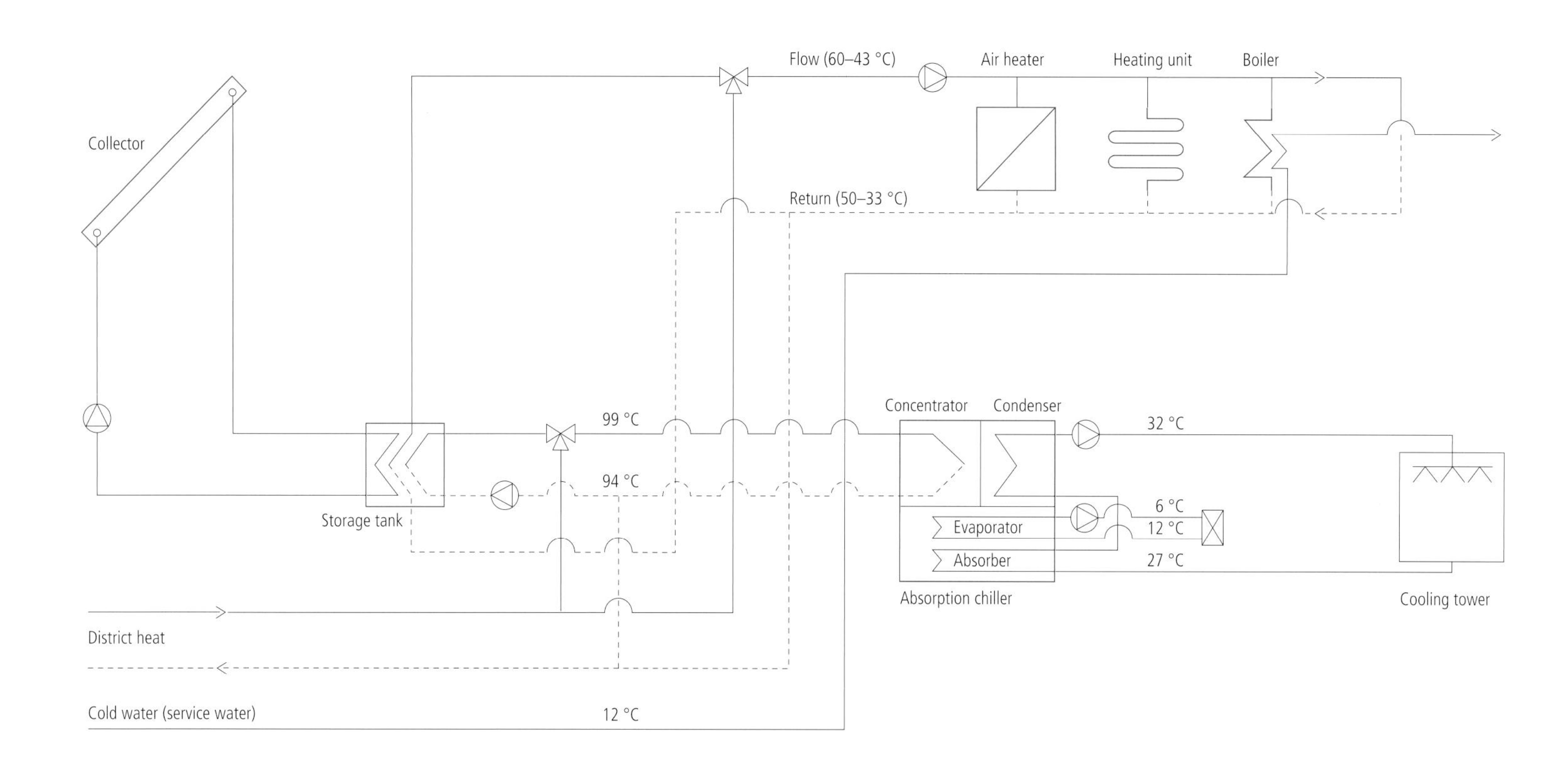

Figure 292
Solar energy heating and cooling installation

Source:
Advanced Building Systems
Klaus Daniels
Birkhäuser Verlag, 2003

Bild 292
Mit Solarenergie betriebene Heizungs- und Kälteanlage

Quelle:
Advanced Building Systems
Klaus Daniels
Birkhäuser Verlag, 2003

13.1.7 Photovoltaic systems

A solar cell or photovoltaic element is a component which converts energy contained in light directly into electric energy. The underlying principle of this process is called the photovoltaic effect, a subcategory of the internal photovoltaic effect. The future for photovoltaic systems seems to be very bright. Such systems will produce a large amount of required energy, and perhaps the largest percentage of all renewable sources.

The theoretically achievable thermal efficiency of systems using the Sun's rays to generate energy is 85 %. It is the result of the Sun's temperature of 5,800 – 6,000 K, the maximum absorber temperature, < 2,500 K, and the surrounding temperature, 300 K.

Because only a limited spectrum of the sunlight is being used, the theoretical value of 85 % is reduced to approximately 29 % for solar cells based on the material silicon. This condition clearly illuminates the disadvantage of photovoltaic technology when compared with solar thermal power plants because solar cells are only capable of using a limited amount of the sun's radiation (**Figure 293**).

13.1.7 Photovoltaik

Der Bereich der photovoltaischen Solarnutzung wird in Zukunft der Bereich sein, der zur Erzeugung erneuerbare Energien einen großen, wenn nicht den größten Anteil einnimmt.

Der theoretisch maximale thermodynamische Wirkungsgrad der Energiegewinnung aus Sonnenlicht beträgt 85 %. Er ergibt sich aus der Sonnentemperatur von 5.800 – 6.000 K, der maximalen Absorbertemperatur (< 2.500 K) und der Umgebungstemperatur (300 K).

Da nur ein begrenzter Wellenlängenbereich aus dem Sonnenspektrum genutzt wird, reduziert sich der aufgeführte theoretische Wert zum Beispiel bei Solarzellen auf Siliziumbasis auf 29 %. Hieraus entsteht der prinzipielle Nachteil von Solarzellen gegenüber von solarthermischen Kraftwerken, da die Solarzellen nur einen Teil des Spektrums nutzen können, **Bild 293**.

Eine Solarzelle oder photovoltaische Zelle ist ein Bauelement, das die im Licht enthaltene Strahlungsenergie direkt in elektrische Energie umwandelt. Die physikalische Grundlage der Umwandlung ist der photovoltaische Effekt – ein Sonderfall des inneren photoelektrischen Effekts.

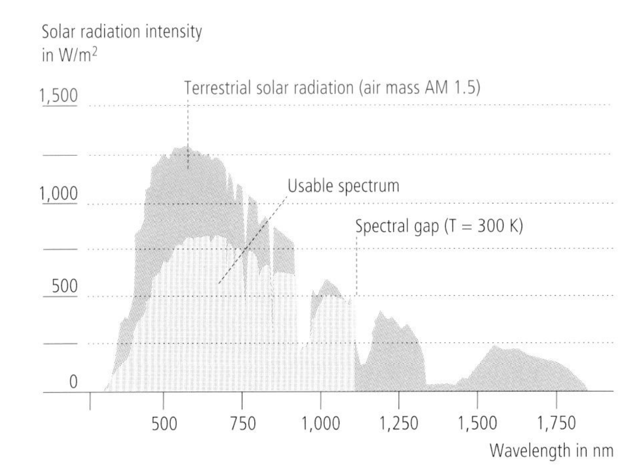

Figure 293
Utilization of solar radiation by mono or poly-crystalline silicon

Source:
Fachausschuss Photovoltaik der Deutschen Gesellschaft für Sonnenenergie (DGS), Germany, Ralf Haselhuhn

Bild 293
Ausnutzung der Sonnenstrahlung durch Silizium (mono- und polykristall)

Quelle:
Fachausschuss Photovoltaik der Deutschen Gesellschaft für Sonnenenergie (DGS), Deutschland, Ralf Haselhuhn

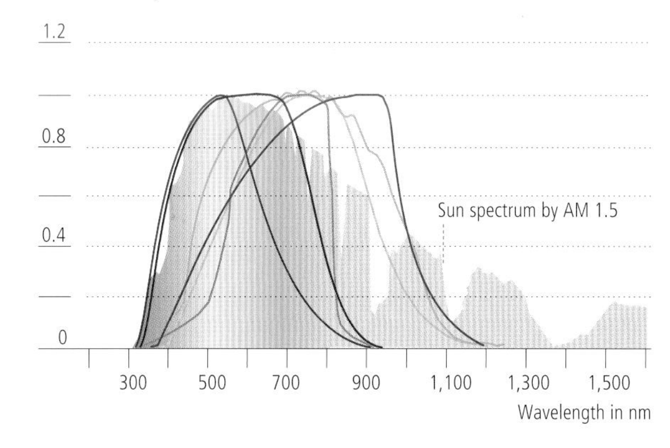

— Amorphous silicon (ASi)
— Micro-morphous silicon
— Crystalline silicon
— Crystalline silicon on glass
— Cadmium telluride
— CIS, Copper indium selenium

Figure 293.1
Spectral sensitivity of various photovoltaic cell materials as determined by the Institute of Solar Energy Supply Technology (ISET) at the University of Kassel, Germany, and manufacturers CSGSolar, mikromorph, and Mitsubishi Heavy.

Source:
Fachausschuss Photovoltaik der Deutschen Gesellschaft für Sonnenenergie (DGS), Germany, Ralf Haselhuhn

Bild 293.1
Spektrale Empfindlichkeit verschiedener Zellmaterialien nach Messung am ISET-Kassel und Herstellermessung (CSG: CSGSolar, mikromorph, Mitsubishi Heavy)

Quelle:
Fachausschuss Photovoltaik der Deutschen Gesellschaft für Sonnenenergie (DGS), Deutschland, Ralf Haselhuhn

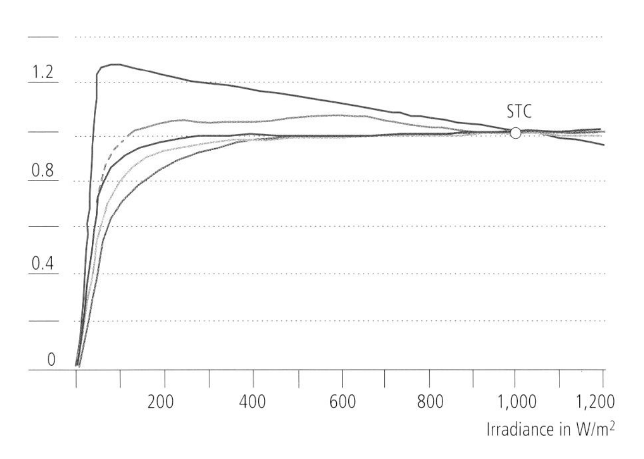

— Triple-amorphous (ECNOO)
— CdTe (TÜV04)
— CIS (ECNOO)
— Crystalline Wafer Standard module (ECNOO)
---- Crystalline Wafer Modul, improved (Grun06)

STC:
Standard Test Conditions (STC) of 1000 W/m² solar irradiance and 25°C PV module temperature

Figure 293.2
Relative coefficient of performance for various solar irradiances according to measurements by the Energy Research Center of the Netherlands (ECN) and TÜV Rheinland, Germany, 2004

Source:
Fachausschuss Photovoltaik der Deutschen Gesellschaft für Sonnenenergie (DGS), Germany,
Ralf Haselhuhn

Bild 293.2
Relativer Modulwirkungsgrad beiunterschiedlichen Einstrahlungen
(1) nach Messungen am niederländischen Forschungsinstitut ECN 2000 und dem TÜV Rheinland, 2004

Quelle:
Fachausschuss Photovoltaik der Deutschen Gesellschaft für Sonnenenergie (DGS), Deutschland,
Ralf Haselhuhn

Solar cells can be grouped into the categories listed in **Table 19** and are distinguished by:

- thickness (thick photovoltaic cells or thin-film cells)
- material
- structure of material (mono or polycrystalline; amorphous)
- materials of semiconductors used (organic solar cells/pentacene, dye-sensitized solar cells)

Solar cells are typically made of silicon in wafer technology. In this process, molten silicon is cut into thin slices, the so-called wafers. Efficiencies are typically between 10 and 18 %, although in some special instances they are around 30 %.

In an alternative method of production, the silicon is turned into thin, uninterrupted ribbons, which then are cut. This method results in less material waste and reduced energy used for the production. Module efficiency for such cells is around 13 %.

A new technology, called thin-film, uses a chemical vapor deposition method to deposit a thin semiconductor layer onto a surface, either coated glass or stainless steel sheets designed in the final shape of the panel. The 3-μm coatings are 100 times thinner than a typical silicon wafer which result in amorphous structures with a red color, which unfortunately only have an efficiency of around 6 – 8 %.

Solarzellen lassen sich nach verschiedenen Kriterien einordnen, die in **Tabelle 19** ausgewiesen sind:

- Materialdicke (Dickschicht-, Dünnschichtzellen)
- Material
- Kristallstruktur (mono-, polykristallin, amorph)
- Halbleitermaterialien (organische Solarzellen, Farbstoff-Solarzellen)

Solarzellen werden überwiegend aus Silizium in Wafertechnologie hergestellt. Dabei wird das Silizium geschmolzen, und aus den kristallin erstarrten Blöcken werden dünne Scheiben (Wafer) gesägt. Die Wirkungsgrade der Solarzellen liegen bei ca. 10 – 20 Prozent, die Modulwirkungsgrade bei 10 – 18 Prozent, in Sonderfällen bis 30 Prozent.

In einem alternativen Verfahren zur Wafertechnologie werden die Silizium-Solarzellen in dünnen Bändern (ribbons) kontinuierlich produziert und abgeschnitten (geringere Produktionsschritte, geringere Materialverbräuche, geringere Herstellungsenergie). Die Modulwirkungsgrade liegen um ca. 13 Prozent.

Eine neue Technologie – Dünnschichttechnologie – nutzt ein Verfahren, bei dem ein Halbleitermaterial auf einen Träger (meistens Glasscheibe) in der Größe des endgültigen Solarmoduls aufgedampft wird. Die Schichten mit 3 Mikrometer Stärke sind ca. 100 mal dünner als herkömmliche Wafer. Dabei entstehen amorphe Strukturen (rötliche Farbe), die jedoch nur einen Wirkungsgrad von ca. 6 – 8 Prozent erreichen.

Another production method is called crystalline silicon on glass (CSG), which deposits a < 2 μm thin semiconductor directly on a glass pane and heats it up to the stage of crystallization. Efficiency for cells manufactured in this way is around 8 %.

In addition to glass, a variety of other carrier materials are conceivable, including stainless steel foils, polymers, and various ceramic materials, but they all need to be capable of sustaining vapor application temperatures of up to 500 °C.

Ein weiteres Herstellungsverfahren im Bereich der Dünnschichttechnik ist das CSG-Verfahren (Crystalline Silicon on Glass), bei dem eine äußerst dünne Siliziumschicht (< 2 μm) direkt auf eine Glasscheibe aufgebracht und durch Erhitzung kristalliert wird. Die Zellwirkungsgrade erreichen hierbei etwa 8 Prozent.

Als Trägermaterial für Solarzellen in Dünnschichttechnologie kommen neben Glas alle Materialien in Frage, die eine Aufdampfungstemperatur bis 500 °C vertragen (Edelstahlfolien, Polymerfolien, Keramikplatten usw.).

Table 19
Typical photovoltaic cell conversion efficiencies for various technologies

Due to the natural aging process of the modules their efficiency decreases slightly per year.
Manufacturers guarantee 80-90 % efficiency after a 20 year operating period. The table assumes a reduction factor of 0.8 % which results in an efficiency of 85 % after 20 years of operation compared to the original efficiency of the module.

Source: SunTechnics

Solar cell material	Cell conversion efficiency ηz Laboratory result	Cell conversion efficiency Production ηz	Module performance Mass production ηm
Monocrystalline silicon	24.7 %	18.0 %	14.0 %
Polycrystalline silicon	19.8 %	15.0 %	13.0 %
Hand-drawn silicon	19.7 %	14.0 %	13.0 %
Crystalline thin-film silicon	19.2 %	9.5 %	7.9 %
Amorphous silicon*	13.0 %	10.5 %	7.5 %
Micro amorphous silicon*	12.0 %	10.7 %	9.1 %
Hybrid HIT (Heterojunction with Intrinsic Thin layer) solar cell	20.1 %	17.3 %	15.2 %
Copper indium diselenide (CIS) Copper indium gallium selenide (CIGS)	18.8 %	14.0 %	10.0 %
Cadmium telluride CdTe	16.4 %	10.0 %	9.0 %
III-V Semiconductors	35.8 %**	24.7 %	27.0 %
Dye-sensitized solar cells	12.0 %	7.0 %	5.0 %***

* stable state
** measured at concentrated insolation
*** small scale manufacturing

Tabelle 19
Übliche Wirkungsgrade unterschiedlicher PV-Module

Aufgrund der natürlichen Alterung der Module sinkt der Anlagenertrag von Jahr zu Jahr geringfügig. Dabei werden nach 20 Betriebsjahren in der Regel noch 80 – 90 % der Nennleistung von den Herstellern garantiert. Entsprechend wurde für die Wirtschaftlichkeitsberechnungmit einer jährlichen Minderleistung von 0,8 % gerechnet, so dass nach 20 Jahren noch 85 % des Anfangsertrags erzielt werden.

Quelle: SunTechnics

Solarzellenmaterial	Zellwirkungsgrad ηz (Labor)	Zellwirkungsgrad ηz (Produktion)	Modulwirkungsgrad ηm (Serienproduktion)
monokristallines Silizium	24,7 %	18,0 %	14,0 %
polykristallines Silizium	19,8 %	15,0 %	13,0 %
handgezogenes Silizium	19,7 %	14,0 %	13,0 %
kristallines Dünnschicht-Silizium	19,2 %	9,5 %	7,9 %
amorphes Silizium*	13,0 %	10,5 %	7,5 %
mikromorphoses Silizium*	12,0 %	10,7 %	9,1 %
hybride HIT-Solarzelle Silizium	20,1 %	17,3 %	15,2 %
CIS, CIGS	18,8 %	14,0 %	10,0 %
Cadmium-Tellurid	16,4 %	10,0 %	9,0 %
III-V-Halbleiter	35,8 %**	24,7 %	27,0 %
Farbstoffzelle	12,0 %	7,0 %	5,0 %***

* in stabilem Zustand
** gemessen bei konzentrierter Einstrahlung
*** Kleinproduktion

Thick silicon cells

In this category, monocrystalline cells (c-Si) are being used that have efficiencies up to 20 %. However, their manufacturing requires a large amount of energy, which results in a high degree of embedded, or grey energy content. In addition to monocrystalline cells, polycrystalline cells (mc-Si) are being used, which contain significantly less grey energy and reach an efficiency of 16 % (**Figure 294**).

Fresnel lens photovoltaic concentrator

In order to reduce the use of silicon, some processes employ a concentrator technology in which Fresnel lenses focus the daylight onto a smaller surface area of the solar cell, thus reducing the necessary energy-generating surface (**Figure 294.1**). One disadvantageous thing in such a design is the fact that the solar cell needs to be always oriented perfectly toward the sun in order to achieve maximum efficiency.

Monocrystalline solar cell

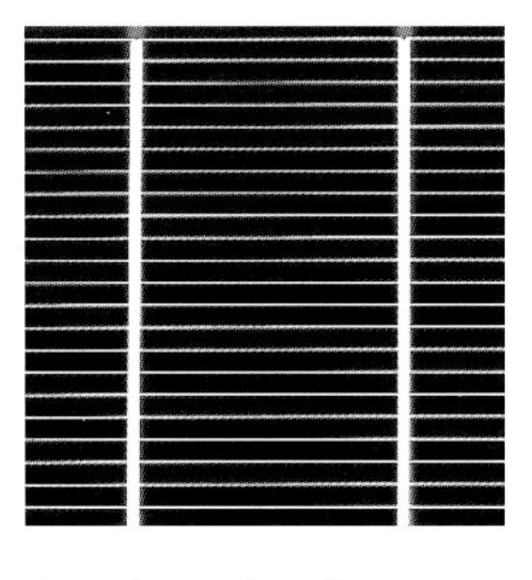

Polycrystalline solar cell

Amorphous solar cell, semi-transparent

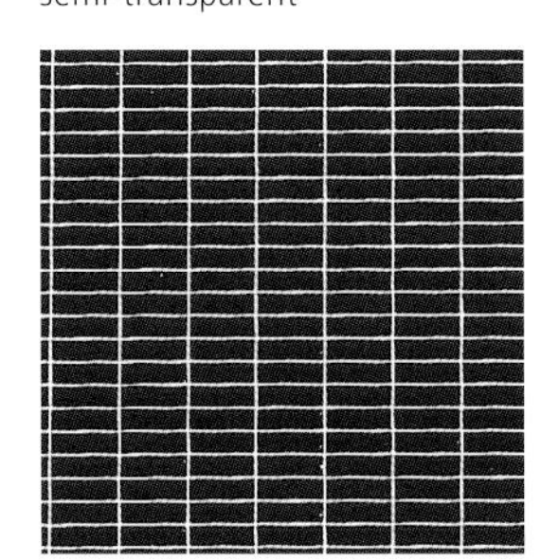

Amorphous solar cell, opaque

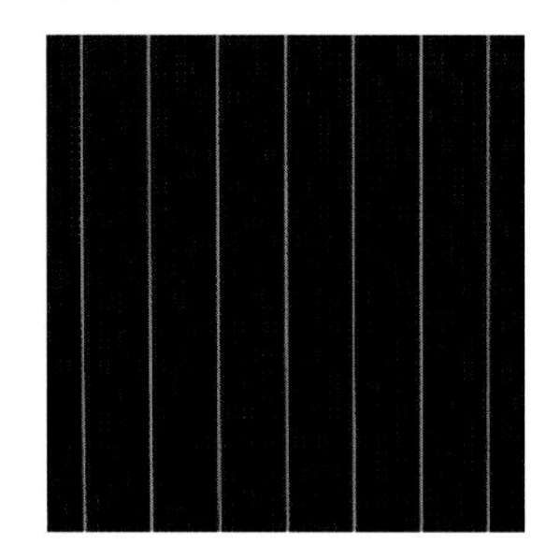

Figure 294
Photovoltaic modules, thick-film technology

Source:
Advanced Building Systems
Klaus Daniels
Birkhäuser Verlag, 2003

Bild 294
Photovoltaische Module, Dickschichttechnik

Quelle:
Advanced Building Systems
Klaus Daniels
Birkhäuser Verlag, 2003

Siliziumzellen, Dickschicht

In diesem Bereich eingesetzt werden monokristalline Zellen (c-Si), die bis zu 20 Prozent Wirkungsgrad erzielen. Die Herstellung erfordert jedoch einen hohen Energieeinsatz (gaue Energie). Neben den monokristallinen Zellen sind auch die polykristallinen Zellen (mc-Si) bekannt, die Wirkungsgrade von mehr als 16 Prozent erreichen und deutlich weniger Graue Energie verbrauchen, **Bild 294**.

Zur Verringerung des Verbrauchs von Silizium wird von Fall zu Fall eine Konzentrator-Technologie eingesetzt, bei der durch Fresnel-Linsen das Tageslicht gebündelt wird, um somit die Energie erzeugende Fläche (Solarzelle) deutlich zu verkleinern, **Bilder 294.1**. Nachteil dieser Systemtechnik ist, dass das gesamte System ständig exakt zur Sonnenlaufbahn nachgeführt werden muss, um wirksam zu sein.

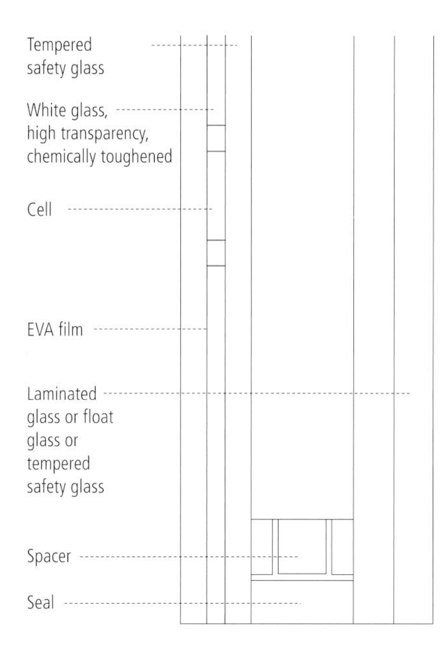

to Figure 294
Insulating glass in module, inclosing photovoltaic element in thick-film technology

zu Bild 294
Modul Isolierglasaufbau, mit photovoltaischem Element (Dickschichttechnik)

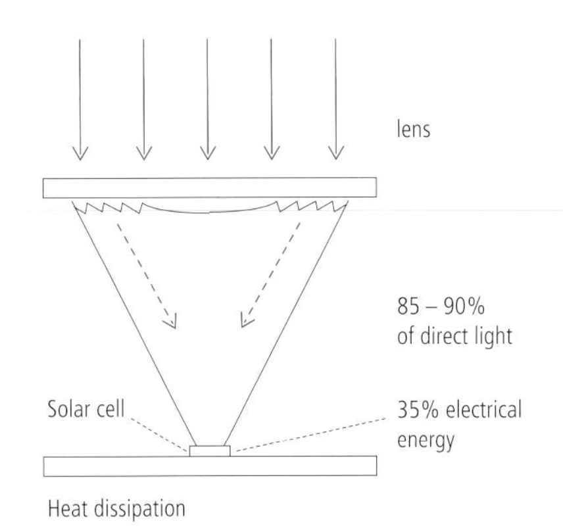

Figure 294.1
The Sol* Con™-concentrator-technology focuses the sunlight through fresnel lenses on a solar cell of 2 by 2 mm size.

Source: Solar Tec AG Munich, Germany

Bild 294.1
Solarzelle mit einem Konzentrator (Fresnel-Linse) zur Verringerung der Solarzellenfläche

Quelle: Solar Tec AG München, Deutschland

to Figure 294.1

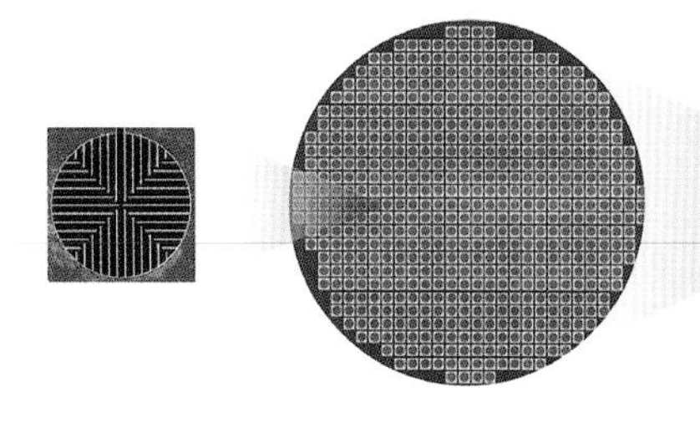

Each wafer delivers up to 2,500 Sol* Con-high effiency solar cells, Sol* Con™-cells have a size of only 2 x 2 mm²

Impressive material savings: 0.01m² Sol* Con™-cells generate the same amount of electricity as a 12 m² solar module of conventional design.

Thin-film silicon cells

Thin-film silicon cells made of amorphous silicon (a-Si) are now able to produce efficiencies of 5 – 10 % (**Figure 295**).They currently have a significant market share, and material shortages for the materials involved in their production do not exist.

Crystalline silicon – for example monocrystalline silicon cells (μc-Si) in combination with amorphous silicon – has efficiencies of 10 % and also a significant share of the photovoltaic market

Siliziumzellen, Dünnschicht

Dünnschichtzellen aus amorphem Silizium (a-Si) erreichen zurzeit Modulwirkungsgrade von ca. 5 – 10 Prozent und besitzen den höchsten Marktanteil, da keine Materialengpässe bestehen, **Bild 295**.

Kristallines Silizium (z.B monokristallines Silizium, μc-Si) in Kombination mit amorphem Silizium besitzt Wirkungsgrade bis 10 Prozent und findet gleichermaßen einen hohen Marktanteil

Figure 295
Thin-film solar cells

1 Flexible thin-film solar cell made of crystalline silicon. Efficiency 20 %
Solare Energiesysteme Freiburg (ISE)

2 ETFE laminate with imbedded solar cells made of amorphous thin-film silicon
Solar Next AG

3 Solar cell made of 0.2 mm diameter glass beads with copper-iridium coating, current efficiency: 5 % Laboratory sample of Scheuten Solar company

4 Copper-iridium sulfide band for further use as a laminate for applications such as electricity-producing bags, membranes.
Odersun

5 Semi-transparent module made of crystalline silicon with laser-cut perforation
Sunways AG

6 CIS cells with colored cover glass.
Zentrum für Sonnenenergie und Wasserstoff-Forschung (ZSW)

7 Colored multi-crystalline silicon cells
Sunways AG

Bild 295
Dünnschicht-Solarzellen

1 Flexible Dünnschichtsolarzelle aus kristallinem Silizium, Wirkungsgrad 20 %.
Fraunhofer Institut für solare Energiesysteme Freiburg (ISE),

2 Laminat aus 2 ETFE-Folien mit dazwischenliegenden Solarzellen aus amorphem Dünnschicht-Silizium.
Solar Next AG

3 Solarzelle aus 0,2 mm kleinen Glaskügelchen mit Kupfer-Indium-Disulfid-Beschichtung. Bisheriger Wirkungsgrad: 5 %, Labormuster der Fa. Scheuten Solar

4 Kupfer-Indium-Disulfid auf Kupferband zur Weiterverarbeitung z.B. im Kunststofflaminat für stromproduzierende Taschen, Membrankonstruktionen o.ä.
Odersun

5 Halbtransparente Module aus kristallinem Silizium mit gelaserter Lochstruktur.
Sunways AG

6 CIS-Zellen mit farbigen Deckgläsern.
Zentrum für Sonnenenergie- und Wasserstoff-Forschung (ZSW)

7 Farbige multikristalline Siliziumzellen.
Sunways AG

Semiconductor solar cells

III-V-semiconductor solar cells have high performance values and are resistant to temperature and UV radiation. Because of the multitude of advantages they offer, they are often used in space applications, in which they are applied as multilayer cells with efficiencies of up to 30 % (50 W/kg).

II-IV-semiconductor solar cells (CdTe) are at the beginning of their market availability, but they can be produced economically in large-scale production, and their efficiency is around 16 %.

CIS, CISGS cells

Copper indium gallium selenide (CIGS) is a new semiconductor material comprising copper, indium, gallium, and selenium. It is currently being developed for market introduction and will reach efficiencies of 12 – 15 %.

Dye-sensitized solar cells (DSC)

DSG cells are a relatively new group of low-cost solar cells. They are based on a semiconductor formed between a photo-sensitized anode and an electrolyte to create a photo-electrochemical system similar to photosynthesis in nature. These cells were invented by Michael Grätzel and Brian O'Regan at the École Polytechnique Fédérale de Lausanne in Switzerland and are also known as Grätzel cells. Their operative lifespan is limited, and they reach efficiencies of slightly more than 10 %.

So-called **organic solar cells** currently undergoing development also have reduced efficiency and a shorter life span.

Semiconductor solar cells for energy generation are generally joined electrically to large "modules". The individual cell has on its front and back thin electric conductor material. If the cells are connected in series, they achieve a desired output voltage, and if they are connected in parallel they provide a desired amount of current source capability. Diodes – so-called bypass diodes – are included to avoid overheating of cells in case of partial shading. The principle of electric connectivity is shown in the diagram in **Figure 296**.

Silicon, as the base material for the manufacturing of photovoltaic cells, is available in almost unlimited abundance. However, for necessary additional material for the manufacturing of the compound cell, such as indium, gallium, tellurium, and selenium, the current global demand exceeds annual production several times over. For iridium, it can be stated that the total global availability of this resource will already be exhausted during the current decade. Selenium and tellurium (Te) are semi-metallic chemical elements that can be found in the anode sludge of the process of copper electrolysis, but this substance is not

Halbleiter-Solarzellen

III-V-Halbleiter-Solarzellen besitzen hohe Wirkungsgrade und sind temperaturbeständig sowie robust gegen UV-Strahlung. Aufgrund der vielen Vorteile werden sie häufig in der Raumfahrt eingesetzt und erreichen hier als Multischichtzellen Wirkungsgrade bis 30 Prozent (50 W/kg).

II-IV-Halbleiter-Solarzellen (CdTe) sind großtechnisch günstig herstellbar, wobei bis zu 16 Prozent Modulwirkungsgrade erreicht werden konnten. CdTe-Solarzellen stehen erst vor der Markteinführung.

CIS-, CIGS-Zellen

(CIS = Kupfer-Indium-Diselenid / Kupfer-Indium-Disulfid) wird zurzeit für den Markt vorbereitet, wobei mit diesen Systemen Wirkungsgrade von 12 – 15 % erreicht werden.

Farbstoffzellen

(auch Grätzelzellen) nutzen organische Farbstoffe zur Umwandlung von Licht in elektrische Energie, angelehnt an die Photosynthese. Die maximalen Wirkungsgrade liegen bei etwas mehr als 10 %, haben sie jedoch eine begrenzte Lebensdauer.

Eine ebenfalls kurze Lebensdauer mit schlechterem Wirkungsgrad sind **organische Solarzellen**, die zurzeit noch in der Entwicklung sind.

Halbleiter-Solarzellen werden zur Energiegewinnung im Regelfall zu großen Solarmodulen verschaltet. Die Zellen werden dafür mit Leiterbahnen an Vorder- und Rückseite in Reihe geschaltet, wodurch sich die Spannung der Einzelzellen addiert und dünnere Drähte für die Verschaltung verwendet werden können als bei einer Parallelschaltung. Zum Schutz vor einem „Lawinendurchbruch" in einzelnen Zellen, zum Beispiel Teilabschattung, müssen zusätzliche Schutzdioden (Bypass-Dioden) parallel zu den Zellen eingebaut werden, um die von Abschattung betroffenen Zellen überbrücken zu können. Das Prinzip der Verschaltung zeigt **Bild 296**.

Während Silizium als Grundstoff für die Solarzellenproduktion in fast unbegrenzter Menge zur Verfügung steht, ist bei Zusatz-Solarzellenmaterialien genau das Gegenteil festzustellen. Bei Indium, Gallium, Tellur und Selen überschreitet zurzeit der weltweite Verbrauch die jährliche Produktionsmenge um ein Mehrfaches. Insofern wird Indium noch in diesem Jahrzehnt als Ressource aufgebraucht sein. Selen und Tellur als Halbmetalle liegen in geringer Konzentration im Anodenschlamm der Kupferelektrolyse vor und werden zurzeit in nicht ausreichendem Maße gewonnen. Die ökonomisch erschließbaren Selen- und Tellurreserven (ca. 82.000 t / ca. 43.000 t) stehen nur begrenzt zur Verfügung – zum Vergleich Buntmetall-Kupferreserven 550 Millionen Tonnen.

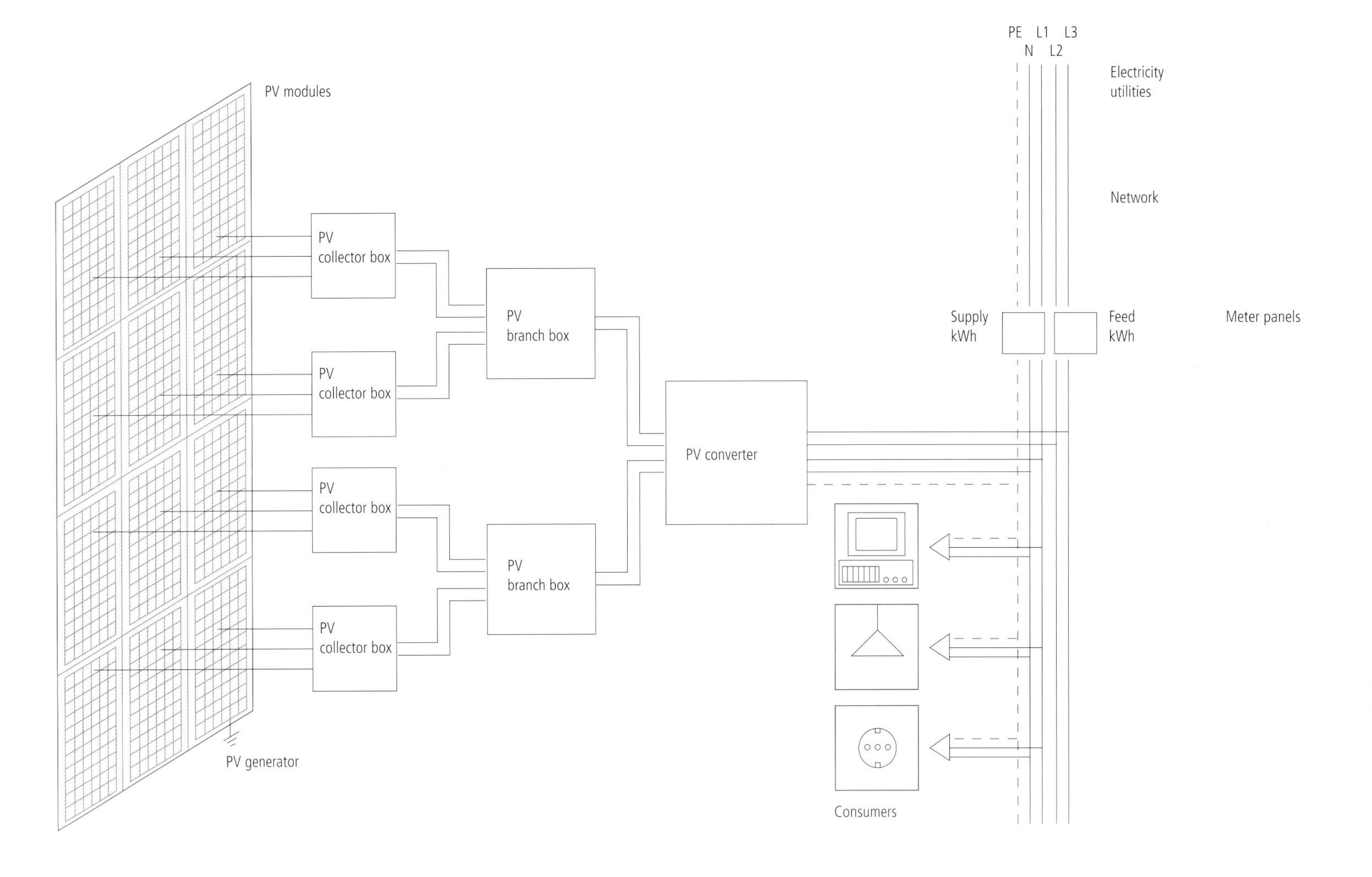

Figure 296
Wiring diagram of a photovoltaic installation

Source:
Advanced Building Systems
Klaus Daniels
Birkhäuser Verlag, 2003

Bild 296
Schaltschema einer Photovoltaik-Anlage

Quelle:
Advanced Building Systems
Klaus Daniels
Birkhäuser Verlag, 2003

sufficiently available. The deposits of selenium and tellurium that can be accessed economically – approximately 82,000 tons of selenium and around 43,000 tons of Te – represent an extremely limited supply, indeed, when compared with the still-available known resources of copper (550 million tons).

Recycling is unfortunately not a feasible option because such materials are typically combined to create complex, multi-layer, and multi-material compounds and are so finely distributed that recycling is unachievable.

In terms of energetic amortization, which is the time the solar cell exceeds the energy required for its production by its own energy generation, frameless thin-film photovoltaic cells have the advantage. Their energy amortization time is between 2 to 3 years. In the case of polycrystalline solar cells, that time span increases to 3 – 5 years, and for mono crystalline cells it increases to 4 – 6 years. For all systems, that time span increases by approximately one year when physical mounting systems, collectors, and inverter accessories necessary for the conversion from DC to AC power are added.

Recyclingansätze wie zum Beispiel beim Kupfer greifen bei Gallium, Indium, Selen und Tellur nicht, da die Materialien meist in komplexe Vielstoff-Schichtstrukturen eingebunden und so fein verteilt sind, dass sie nicht rückgewonnen werden können.

Bei der energetischen Amortisation (Zeitpunkt, zu dem die Energie, die für die Herstellung einer Photovoltaikzelle aufgewandt wurde, durch selbige wieder erzeugt wird) schneiden am besten die Dünnschichtzellen (ohne Rahmen) mit ca. 2 – 3 Jahren ab. Bei polykristallinen Solarzellen beträgt die energetische Amortisation etwa 3 – 5, bei monokristallinen Solarzellen ca. 4 – 6 Jahre. Die jeweiligen Zeitdauern erhöhen sich um ca. ein Jahr infolge des Energieeinsatzes zur Herstellung von Montagesystemen, Sammelsystemen und Wechselrichtern.

Die Umweltbelastungen bei der Herstellung photovoltaischer Solarzellen sind nicht unproblematisch, da im Herstellungsprozess chemische Prozesse angewandt werden, bei denen gasförmige, flüssige und feste

Environmental impacts as a result of the manufacturing process of photovoltaic cells are problematic because they require the extensive use of toxic and/or carcinogenic substances in either gaseous, solid, or liquid state, such as carcinogenic cadmium, arsenic, and copper gallium diselenide. Health risks may occur when old photovoltaic cells are not properly disposed of, and the negative impact on the environment is one characteristic of photovoltaic energy generation which distinguishes this technology from all other renewable energy-generation methods.

The first comprehensive study concerning the environmental impact of lead, cadmium, photovoltaic cell manufacturing, and mercury as they are related to the manufacturing process of photovoltaic cells was conducted by the renowned Brookhaven National Laboratory, which conducts research for the U.S. Department of Energy. The results were first published in 2008 in the Journal of Environmental Science and Technology. The study, which analyzes the manufacturing practices of thirteen international PV producers, concludes that using photovoltaic technology results in 90 percent less emission of environmentally hazardous materials than the use of fossil fuel sources if all aspects of exploration, processing, and energy production are taken into account.

13.1.8 Fuel cell technology

Fuel cells for the generation of electric and thermal energy are in use in many test applications and have an electric power generation capability of around 200 kW. Additional systems, with power outputs of around 5 kW, are being offered for applications in residential units. For applications in buildings, high-temperature fuel cells using gas – and in the future, hydrogen – are being utilized. They provide an efficiency of electricity generation of 50 % and produce exhaust gas temperatures that are between 80 and 180 °C. Fuel cell technology is seen as an environmentally friendly alternative to cogeneration power plants because no carbon oxide is released into the atmosphere.

The working principle of the fuel cell is a "flameless" combustion of natural gas or hydrogen in the presence of an electrolyte. They are typically either distinguished by the employed electrolyte or according to their operating temperature (high- and low-temperature fuel cells).

Fuel cells for use in cogeneration processes are:

- phosphoric acid fuel cells (PAFCs)
- molten carbonate fuel cells (MCFCs)
- solid oxide fuel cells (SOFCs)

Chemikalien zum Einsatz kommen, die grundsätzlich umweltschädlich sind (z.B. toxisches und/oder karzinogenes Cadmium, Arsen, Kupfergalliumdiselenid usw.). Langzeitschäden werden auch durch die unsachgemäße Entsorgung von Altzellen verursacht, und insofern unterscheidet sich die Photovoltaik im negativen Sinne von den anderen Technologien zur Nutzung regenerativer Energie.

Die erste umfassende Studie zum Thema Umweltbelastung bei der Herstellung von Photovoltaikzellen durch Blei, Quecksilber und Cadmium, durchgeführt vom renommierten Brookhaven National Laboratory, das für das US Energieministerium Forschung betreibt, wird Mitte 2008 im Journal „Environmental Science & Technology" erscheinen. Die Studie, die die Herstellungsverfahren von dreizehn internationalen PV-Herstellern untersucht hat, kommt zu dem Schluss, dass die Photovoltaiktechnologie 90 Prozent weniger umweltbelastende Substanzen erzeugt als die der fossilen Brennstoffe, wenn alle Faktoren der Gewinnung, Verarbeitung und Energieerzeugung umfassend einbezogen werden.

13.1.8 Brennstoffzelle

Die Brennstoffzellentechnologie zur Erzeugung von sowohl elektrischer wie auch Wärmeenergie ist mit einer Reihe von Versuchsanlagen bis ca. 200 kW elektrischer Leistung im Einsatz. Darüber hinaus werden Brennstoffzellen für die Versorgung von Wohnobjekten mit einer elektrischen Leistung bis ca. 5 kW zurzeit angeboten. Im Bereich der Gebäudetechnik bieten sich Hochtemperatur-Brennstoffzellen an, die Erdgas – später Wasserstoff – mit einem Wirkungsgrad von mehr als 50 % in elektrischen Strom umwandeln und dabei Abgastemperaturen liefern, die im Bereich zwischen 80 – 180 °C und höher liegen. Brennstoffzellen verstehen sich im Wesentlichen als Alternative zu Blockheizkraftwerken, wobei bei diesen keine Stickoxide im Betrieb freigesetzt werden.

Das Funktionsprinzip der Brennstoffzelle ist die flammlose Verbrennung von Erdgas oder Wasserstoff in einem Elektrolyten. Die Brennstoffzellen werden üblicherweise nach dem verwendeten Elektrolyten unterschieden bzw. nach der Betriebstemperatur (Hoch-, Niedertemperatur-Brennstoffzellen).

Die für die Nutzung in der Kraft-Wärme-Kopplung geeigneten Brennstoffzellentypen sind:

- phosphorsaure Brennstoffzellen (PAFC)
- karbonatschmelze-Brennstoffstelle (MCFC)
- oxidkeramische Brennstoffzelle (SOFC)

PAFCs are technically the most developed; however, with 40 % electric performance, they have the lowest efficiency.

MCFCs are currently in the early stages of use in test and demonstration facilities, whereas SOFC technology has matured and is at the end of its development stage.

In contrast to PAFC and MCFC fuel cells, whose liquid electrolyte is suspended in a solid substance matrix, the electrolyte of the SOFC is a ceramic – or, as the name implies, a solid oxide.

An SOFC fuel cell consists of a 0.2 mm ion zirconium oxide ceramic electrolyte sandwiched between two layers of porous, 0.03 mm-thick electrodes made of an electronically conducting metallic ceramic material (**Figure 297**).

Die PAFC-Brennstoffzellen sind technisch am weitesten entwickelt, weisen jedoch den geringsten elektrischen Wirkungsgrad mit ca. 40 % auf.

Die MCFC-Brennstoffzellen werden zurzeit primär als Demonstrationsanlagen in Betrieb gehalten, während die SOFC-Brennstoffzellen zurzeit am Ende der Entwicklung stehen.

Im Gegensatz zu den PAFC- und MCFC-Brennstoffzellen, bei denen der flüssige Elektrolyt in einer Festkörpermatrix gehalten wird, ist bei den SOFC-Brennstoffzellen die Keramik der Elektrolyt.

Eine SOFC-Zelle besteht aus einem keramischen Elektrolyt (0,2 mm Ion leitendes Zirkon-Oxid) und darauf beidseitig aufgebrachten porösen Elektroden (0,03 mm) aus einer elektrisch leitenden, metallischen Keramik, **Bild 297**.

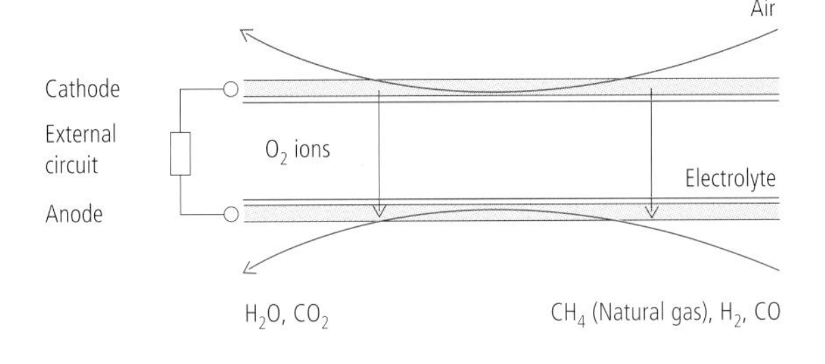

Figure 297
Principle of fuel cells

Oxygen ions from the air travel from the cathode via the electrolyte to the anode and react with the conditioned natural gas. This creates electricity and heat.

Source:
Advanced Building Systems
Klaus Daniels
Birkhäuser Verlag, 2003

Bild 297
Das Prinzip der Brennstoffzelle

Sauerstoffionen aus der Luft wandern von der Kathode durch den Elektrolyten zur Anode und reagieren dort mit aufbereitetem Erdgas. Dabei entsteht elektrischer Strom und Wärme.

Quelle:
Advanced Builing Systems
Klaus Daniels
Birkhäuser Verlag, 2003

In the cathode, the oxygen of the entering air is being ionized at temperatures of around 800 °C – in other words the oxygen accepts electrons. The oxygen ions travel through the electrolyte to the anode as a result of the concentration differential, and then the oxygen reacts with a mixture of natural gas (CH_4), Hydrogen (H_2), and carbon monoxide (CO), forming CO_2, water, and heat. The surplus electrons of the oxygen are released and return via a separate circuit back to the cathode. The direct current between both electrodes is below 1 V, which requires many cells to be layered on top of each other. The resulting so-called stack is required in order to generate the required voltages. An optimal flow design allows the developing heat to be used simultaneously for pre-heating of the air, conditioning of the natural gas (reforming of natural gas to water and CH_4, H_2, and CO), heating purposes, and post-combustion of the unused natural gas (**Figure 298**). The complete system diagram of a fuel cell is shown in **Figure 299**. Because the current investment costs are about three times greater than in the case of a cogeneration/combined heat and power plant, this technology still has marginal market penetration. Even under the assumption that fuel cells are being mass produced, cost very likely will remain high, so it remains to be seen what the future of fuel cell implementation will be.

In der Kathode wird Sauerstoff der einströmenden Luft bei Temperaturen um 800 °C ionisiert, d.h. der Sauerstoff nimmt Elektronen auf. Die Sauerstoff-Ionen diffundieren aufgrund des entstehenden Konzentrationsgefälles durch den Elektrolyten zur Anode. Dort reagiert der Sauerstoff mit dem Gemisch aus Erdgas (CH_4), Wasserstoff (H_2) und Kohlenmonoxid (CO) unter Bildung von CO_2 und Wasser bei Freisetzung von Wärme. Die überzähligen Elektronen aus dem Sauerstoff werden dabei ebenfalls frei und fließen über einen externen Stromkreis wieder zur Kathode zurück. Die Gleichspannung, die zwischen den beiden Elektroden entsteht, liegt unter 1 V, so dass viele Zellen in Serie geschaltet werden müssen, um ausreichende Ausgangsspannungen zu erreichen. Hierzu werden die Brennstoffzellen übereinander gestapelt. Durch eine optimierte Strömungsführung in einem Modul kann die freiwerdende Wärmeenergie gleichzeitig zur Vorerwärmung der Zuluft, zur Aufbereitung des Erdgases (Reformierung von Erdgas und Wasser zu CH_4, H_2 und CO) zu Heizzwecken und zur Nachverbrennung des unverbrauchten Erdgases genutzt werden, **Bild 298**. Den Gesamtaufbau einer Brennstoffzelle zeigt **Bild 299**. Da voraussichtlich die zurzeit sehr hohen Investitionskosten (ca. 3 Mal höher als BHKW-Systeme) selbst bei großer Serienfertigung nicht wesentlich zu senken sind, ist es offen, welche Rolle die Brennstoffzellen in Zukunft spielen werden.

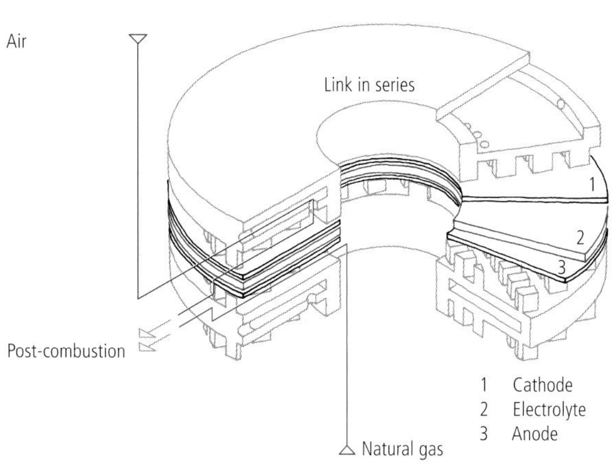

Figure 298
Section of cell stack

Only when several cells are stacked are units created which meet the required performance in the kW range.

Source:
Advanced Building Systems
Klaus Daniels
Birkhäuser Verlag, 2003

Bild 298
Schnitt durch einen Zellenstapel

Erst durch die Serienschaltung mehrerer Zellen entstehen Einheiten mit der geforderten Leistung im kW-Bereich

Quelle:
Advanced Building Systems
Klaus Daniels
Birkhäuser Verlag, 2003

Figure 299
Diagram of fuel cell installation

Source:
Advanced Building Systems
Klaus Daniels
Birkhäuser Verlag, 2003

1 Fuel cell stack
2 Air preheating chamber
3 Startup and auxiliary burner
4 Natural gas treatment
5 Flue gas heat exchanger
6 Controls, d.c./a.c. inverter

Bild 299
Aufbau einer Brennstoffzellenanlage

Quelle:
Advanced Building Systems
Klaus Daniels
Birkhäuser Verlag, 2003

Nevertheless, the use of fuel cells is very interesting in cases in which great and long-duration solar energy gains exist. In such cases, the production of hydrogen with electricity gained from photovoltaic systems may be economical. It can be assumed that large-scale fuel cell applications, which for the most part still need to be developed, will take place in sunny regions of the Earth in the future. **Figure 300** shows the systems integration of the components for photovoltaic electricity generation, the catalyst, the hydrogen tank, and the fuel cell.

Interessant ist der Einsatz von Brennstoffzellen allemal dort, wo hohe und langzeitige Potenziale von Solarenergie bestehen. Durch die Nutzung von photovoltaischem Strom kann die Herstellung von Wasserstoff durch Elektrolyseure kostengünstig betrieben werden, und insofern ist damit zu rechnen, dass zumindest in sonnenreichen Gegenden großformatige Brennstoffzellen – die zum Teil erst entwickelt werden müssen – zum Einsatz kommen. **Bild 300** zeigt den Verbund aus photovoltaisch erzeugtem Strom, Elektrolyseur, Wasserstoffspeicher und Brennstoffzelle.

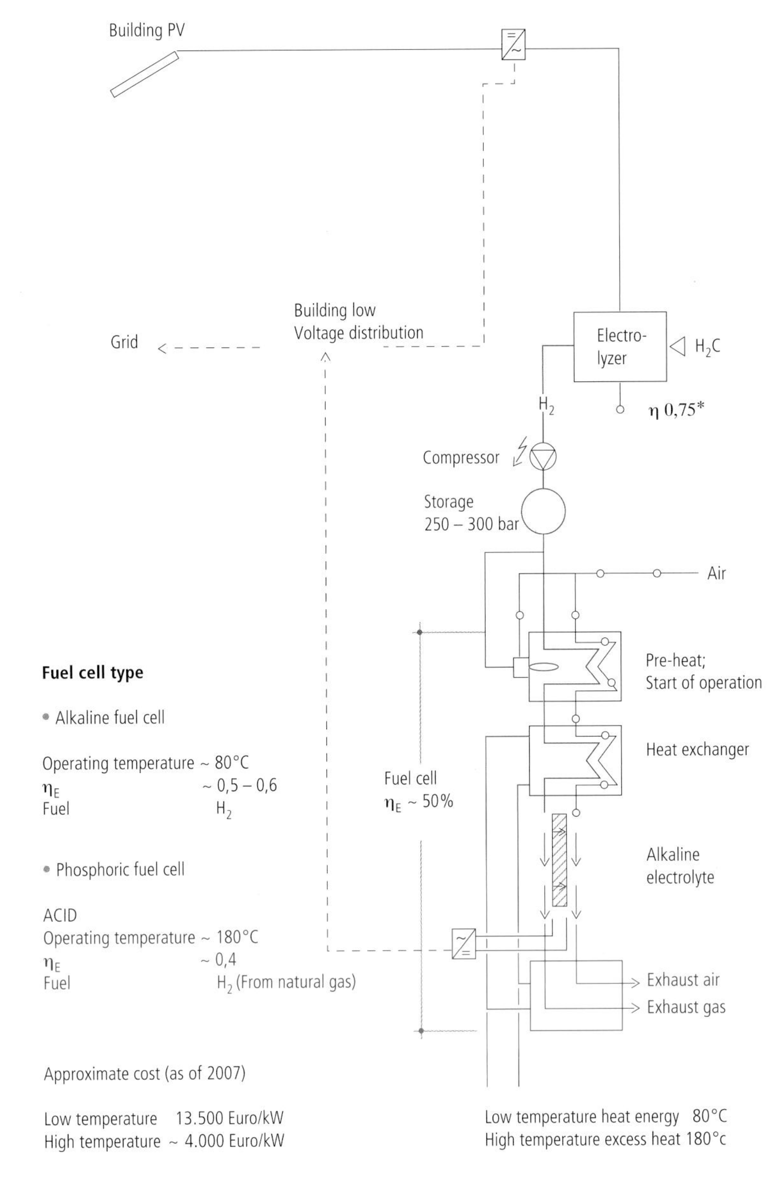

Figure 300
Diagram of solar operated fuel cell with added hydrogen generation and storage

Source: Daniels/author

Bild 300
Solarbetriebene Brennstoffzellen-Anlage mit vorgeschalteter Wasserstoffproduktion und -speicherung

Quelle: Daniels/Autor

13.1.9 Wind power

The generation of electricity by wind turbines near, on, or integrated into buildings seems to be problematic and most likely will not experience the great breakthrough once hoped for. The technology will not constitute a promising research area for the future. Vertical-axis rotors, as shown in **Chapter 12**, mounted on the roof of buildings are capable of supplying only a small amount of the buildings' energy needs – in other words, they are far from making a large energy-supply contribution. This technology is therefore destined for such small applications as were presented in **Chapter 12.1**; nothing needs to be added here.

13.1.9 Windkraft

Der Einsatz von Windkraftwerken in Gebäuden oder direkt an Gebäuden erscheint eher problematisch und wird daher nicht den großen Durchbruch erleben, den man sich vorstellen könnte. Windenergetische Anlagen innerhalb von Gebäuden werden im Regelfall nicht das Gebiet der Zukunft sein. Auf Gebäude aufgesetzte Windräder (Vertikalrotoren) können, wie bei den Beispielen in **Kapitel 12** gezeigt, einen kleinen Anteil der notwendigen Energien erzeugen – die Masse wird es nicht sein. Insofern ist den Darstellungen gemäß **Kapitel 12.1** nichts Wesentliches hinzuzufügen.

13.2 Large-scale technology (virtual power plants)

Energy production of the future – as is already the case today – will most likely take place outside the perimeters of our buildings. Virtual power plants, wind farms, large solar fields, some underwater or other maritime systems, or systems inserted into the depth of the Earth's crust will be the major contributors to future renewable power generation. **Figure 301** shows a variety of technologies that are currently in different stages of research, development, and pilot demonstration, or that have already been introduced to the marketplace. As shown, only a small percentage of such technologies are directly connected to or part of the structures of buildings. A brief overview will be presented of system solutions placed near large energy-consuming centers in order to reduce the transmission losses. Today, we speak of virtual power plants, which describe clusters of distributed individual generation installations collectively run by a central control entity. Their concerted operational mode will provide the renewable energy we will use in the future. Because the power required for the operation of large buildings will be generated external to their structures, we will be dependent upon delivery by outside utilities. **Figure 302** shows today's potential for renewable energy generation, and **Figure 303** shows the per capita energy use in 2006. Based on the shown consumption, which most certainly will increase in regions such as China and India, we need to develop comprehensive concepts of renewable energy generation that are most appropriate for a specific global region.

13.2 Großformatige Technologien (virtuelle Kraftwerke)

Die eigentliche Energiewirtschaft wird sich – wie schon heute – außerhalb unserer Häuser abspielen – in Heizkraftwerken, Windfarmen, großen Solarfeldern, zum Teil auf und unter dem Wasser bzw. durch das Wasser und in der tiefen Erdrinde. **Bild 301** zeigt eine Vielzahl von Technologien im Bereich Forschung, Entwicklung, Demonstration und Markteinführung zur jetzigen Zeit. Nur ein kleinerer Teil der in der Entwicklung befindlichen energieerzeugenden Systemtechnologien betrifft Strukturen, die am und im Gebäude zur Anwendung kommen. Insofern soll zum Abschluss ein kurzer Überblick gegeben werden über Systemtechniken, die sich möglichst nahe an die großen Verbrauchszentren anschließen, um die hohen Übertragungsverluste zu vermeiden. Insofern wird heute bereits von den virtuellen Kraftwerken gesprochen, technischen Einzellösungen, die in ihrer Gesamtheit je nach Anfall der erneuerbaren Energien in der Lage sind, die Energien bereitzustellen, die wir benötigen. Da im Regelfall die in größeren Bauten notwendigen Energien nicht am und im Haus produziert werden können, werden wir immer auf die Zulieferung von „außen" angewiesen sein. **Bild 302** zeigt einen Überblick über das Potenzial an regenerativer Energiegewinnung heute, **Bild 303** den Energieverbrauch pro Kopf im Jahr 2006. Basierend auf dem Pro-Kopf-Verbrauch, der sich allerdings in einigen Regionen der Welt wie Indien und China erheblich vergrößern wird, muss man sich von Fall zu Fall damit auseinandersetzen, in welcher Region welches Potenzial erneuerbarer Energien genutzt werden kann, und vor allem, was die jeweiligen Standorte an Ressourcen bereithalten.

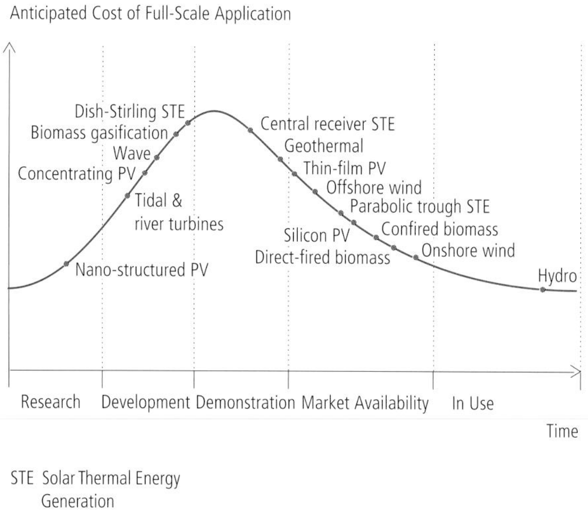

Anticipated Cost of Full-Scale Application
Dish-Stirling STE
Biomass gasification
Wave
Concentrating PV
Tidal & river turbines
Nano-structured PV
Central receiver STE
Geothermal
Thin-film PV
Offshore wind
Parabolic trough STE
Silicon PV
Confired biomass
Direct-fired biomass
Onshore wind
Hydro
Research Development Demonstration Market Availability In Use
Time

STE Solar Thermal Energy Generation

Figure 301
Various stages of development of renewable energy technologies

Source: John Douglas, science and technology writer in report 1012730, EPRI 2007

Bild 301
Entwicklung Erneuerbarer Energien

Quelle: John Douglas, science and technology writer in report 1012730, EPRI 2007

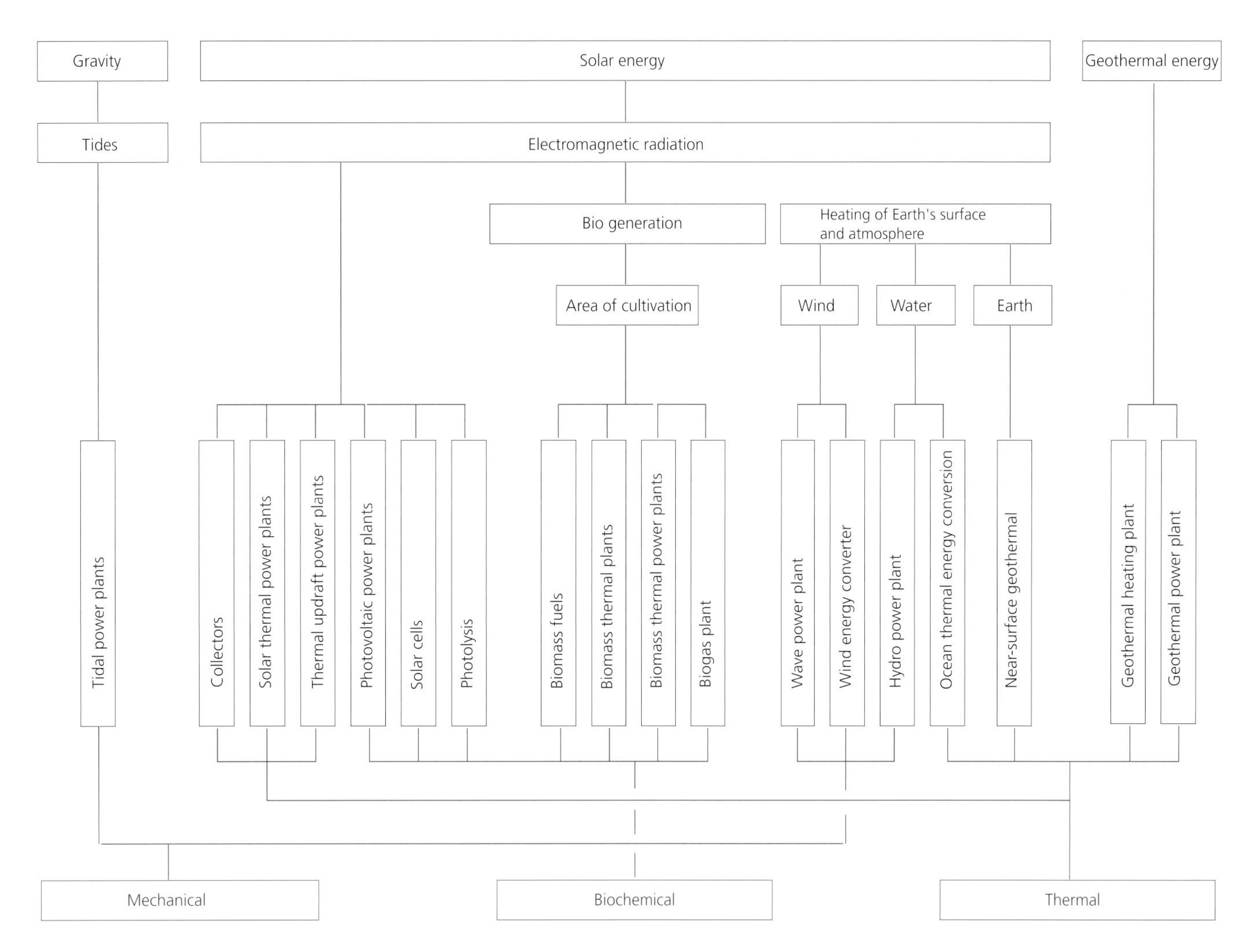

Figure 302
Today's potential for renewable power generation

Source: archplus 184, October 2007

Bild 302
Potenzial an regenerativer Energiegewinnung heute

Quelle: archplus 184, Oktober 2007

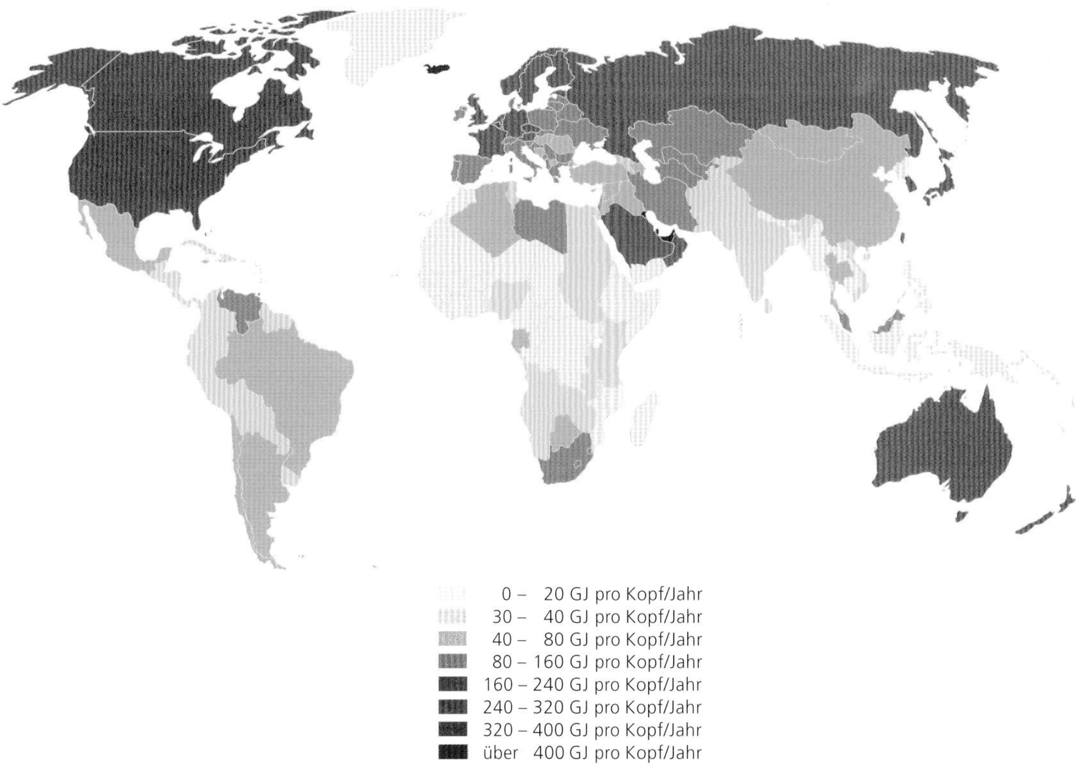

Figure 303
The world conmsumption of energy per person is 74.1 GJ

Source: International Energy Agency, Key World Energy Statistics 2006

Bild 303
Im Weltmaßstab beträgt der durchschnittliche Pro-Kopf-Verbrauch 74,1 GJ

Quelle: International Energy Agency, Key World Energy Statistics 2006

13.2.1 Solar thermal power plants

Solar thermal power plants use the energy of the sun to produce heat with the help of turbines, and then they convert the thermal energy into electricity. Of all solar technologies, parabolic trough collectors are the most mature technology with the highest coefficient of performance and the least amount of investment cost. This is the technology with the best future prospects (**Figure 304**). It is fully developed – what it needs now is the support of legislatures and the confidence of investors.

Although photovoltaic technology represents a viable de-centralized technology to generate energy, the strength of energy generated by solar energy generating systems (SEGSs) lies in centralization. The unpopulated deserts of Northern Africa would be sufficient to supply a major portion of the power required in all of Europe. Solar thermal technology therefore represents a significant energy option for the future, and the application of this technology is also directly linked to the European Union's carbon dioxide reduction strategy. From today to the year 2025, according to a study by Greenpeace, SEGS power plants could be capable of eliminating approximately 362 million tons of carbon dioxide emissions into the atmosphere globally.

SEGSs are particularly efficient if solar intensity is high. Therefore, they are of great interest not only for the Sunny regions of southern Europe but also for the economically deprived regions in the global "sun belt." Such areas would certainly benefit economically from expanded renewable energy generation by SEGSs.

The Club of Rome study demonstrates that the potential solar thermal power generation would exceed global demand many times. In addition, cost-effective transmission technologies would be capable of providing clean electricity from the deserts to more than 90 % of the world's population.

Solar thermal power plants can be combined with fossil-fuel steam power plants, and parabolic trough collector installations can be linked to conventional power supplies. Especially in the global "sun belt", such combinations of SEGSs and fossil plants would represent a logical addition. SEGSs with a power of 100 – 200 MW_{el} could replace conventional plants that would be operated as medium-load plants – and this without qualitative changes of the grid structure.

13.2.1 Solarthermische Kraftwerke

Solarthermische Kraftwerke nutzen die Sonnenenergie zur Erzeugung von Wärme, die durch Turbinen in Strom umgewandelt wird. Mit einem hohen Wirkungsgrad und den niedrigsten Stromgestehungskosten unter den Solartechniken haben insbesondere die technisch ausgereiften Parabolrinnen-Kraftwerke eine hervorragende Zukunftsperspektive, **Bild 304**. Die Technik ist marktreif, sie braucht jetzt die Unterstützung der Politik und das Vertrauen potenzieller Investoren.

Während die Photovoltaik die richtige Technik zur dezentralen Nutzung der Sonnenenergie ist, liegt die Stärke der solarthermischen Kraftwerkstechnik (STKT) in der zentralen Energieerzeugung. Allein in den unbesiedelten Wüsten Nordafrikas ließe sich ein Vielfaches des europäischen Strombedarfs erzeugen. Damit ist die solarthermische Kraftwerkstechnik eine bedeutende Technologieoption für einen nachhaltigen Energiemix der Zukunft. Sie wird unmittelbar zur CO_2-Minderungsstrategie der Europäischen Union beitragen. Nach einer Greenpeace-Studie kann durch den Einsatz der solarthermischen Kraftwerkstechnik (STKT) weltweit bis 2025 die Emission von 362 Millionen Tonnen CO_2 verhindert werden.

Die STKT ist besonders effizient bei einer hohen Intensität der Sonneneinstrahlung. Sie bietet deshalb nicht nur den südlichen Mitgliedsstaaten der EU, sondern auch vielen wirtschaftlich benachteiligten Regionen im Sonnengürtel der Erde sehr gute Entwicklungschancen. Angesichts des drohenden Klimawandels wird durch die STKT eine Möglichkeit eröffnet, den Anteil der erneuerbaren Energien an der Stromerzeugung gerade in diesen Ländern erheblich zu erhöhen.

Der Club of Rome betont in einer Studie, dass das Potenzial solarthermischer Kraftwerke den weltweiten Energiebedarf um ein Vielfaches übersteigt. Kostengünstige und verlustarme Übertragungsleitungen könnten sauberen Strom aus den Wüsten zu mehr als 90 Prozent der Weltbevölkerung bringen.

Als Dampfkraftwerke sind solarthermische Kraftwerke mit fossilen Energiequellen kombinierbar: Parabolrinnen-Solarfelder lassen sich an konventionelle Kraftwerke ankoppeln. Gerade im Sonnengürtel der Erde können solarthermische Kraftwerke damit in einem ersten Schritt zu einer sinnvollen Ergänzung des fossil befeuerten Kraftwerkparks werden. STKT-Kraftwerke in einer Größe von 100 bis 200 MW_{el} können konventionelle Kraftwerke ersetzen, die in Mittellast betrieben werden – und zwar ohne qualitative Änderungen der Netzstruktur.

Figure 304
Power plants in the Mojave Desert

Proven technology: the solar trough power plants in the Mojave Desert have been an important source of power for California for 15 years.

Source: Schott AG, Deutschland

Bild 304
Power plants in der Mojave Wüste

Erprobte Technologie:
Die Solar-Kraftwerke in der kalifornischen Mojave-Wüste leisten seit 15 Jahren einen erheblichen Beitrag zur Energieversorgung mittels erneuerbarer Energie.

Quelle: Schott AG, Deutschland

to Figure 304
Solar trough receiver of a solar thermal power plant

Source: Schott AG, Germany

zu Bild 304
Receiver in solarthermischen Kraftwerken

Quelle: Schott AG, Deutschland

to Figure 304
Large-scale solar thermal power plant:
pipeline configuration connecting solar receivers to heat exchanger

Source: Schott AG, Germany

zu Bild 304
Pipeline-System einer großen solarthermischen Anlage (Verbindung zwischen Solar-Receivern und Wärmetauscher)

Quelle: Schott AG, Deutschland

to Figure 304

zu Bild 304

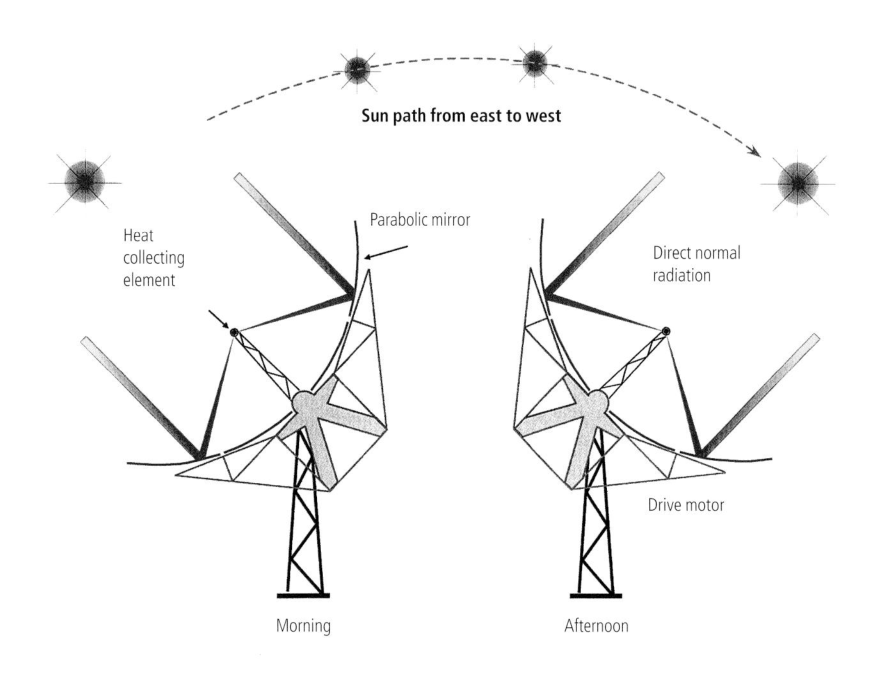

to Figure 304
Parabolic mirror system

zu Bild 304
Parabol-Rinnen-System, nachgeführt

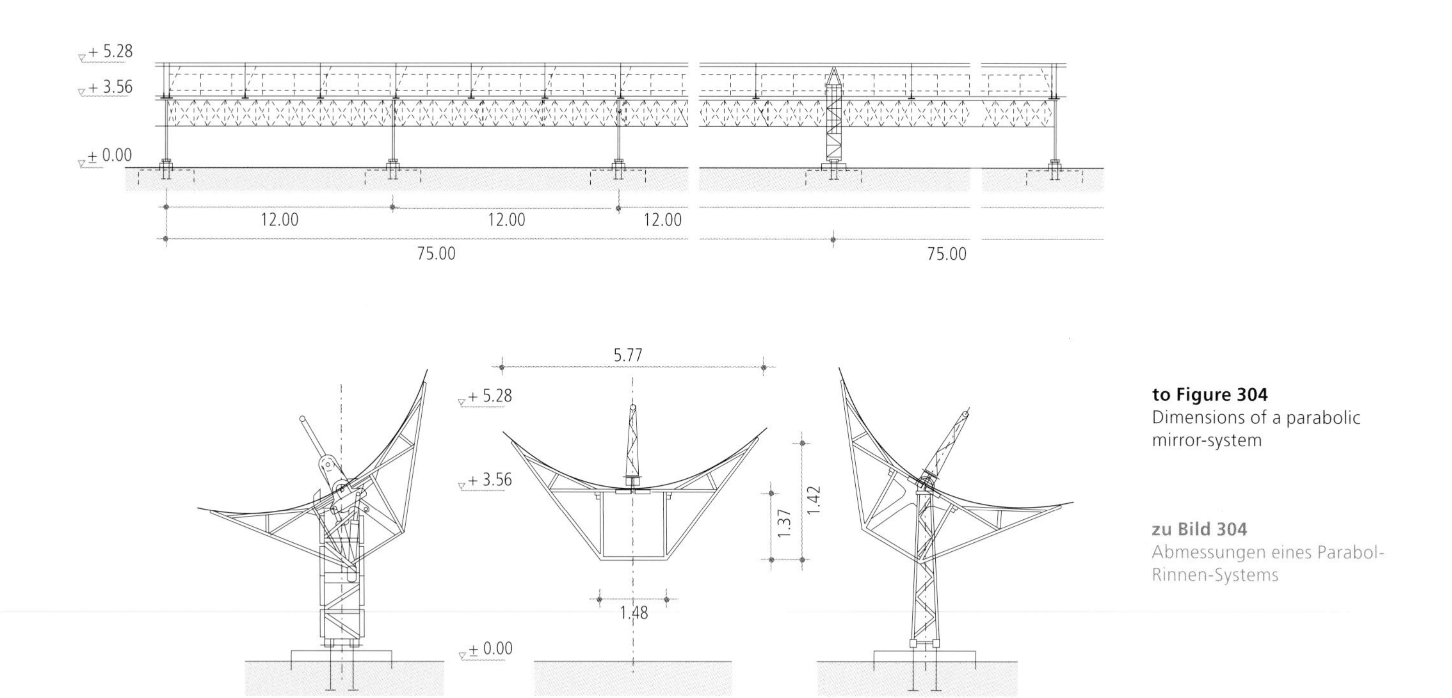

to Figure 304
Dimensions of a parabolic mirror-system

zu Bild 304
Abmessungen eines Parabol-Rinnen-Systems

Working principle of parabolic trough collectors

Solar thermal power plants are, in principle, steam power plants that generate electricity from a high-temperature heat source (**Figure 305.1**). In the solar fields of the power plant, the sun's rays are concentrated by a factor of 80 onto an absorber tube. The concentration takes place by long rows of parabolic mirrors. A high-temperature thermal oil circulates in the absorber tube located at the focal point of the mirror, where it is heated up to approximately 400 °C. In a heat exchanger positioned in a centrally located power plant block, the super-heated steam powers conventional steam turbo generators.

Advanced thermal storage technology makes the solar electricity available for use at night and during overcast weather conditions. Already, the volume of millions of liters of circulating thermal oil represent a large portion of storage capacity, capable of bridging shorter periods of cloud cover occurrences. Molten salt storage technology provides an additional reliable storage of energy for uninterrupted power supply.
If salt storage is used, the turbo generators are also be capable of being operated under constant peak loads, which results in the highest performance efficiency of the power plant. Molten salt storage has proven to be reliable and is approved by utility grid operators, and the Spanish national power utility has confirmed that the technology is as reliable as conventional fossil power generation. Additionally, no technical conflicts of integration of such plants into the existing regular grid structure seem to exist.

Arbeitsweise der Parabolrinnen-Kraftwerke

Solarthermische Kraftwerke sind nichts anderes als Dampfkraftwerke, die aus Hochtemperaturwärme Strom erzeugen, **Bild 305.1**. Im Solarfeld des Kraftwerks konzentrieren parabolisch geformte und in langen Reihen angeordnete Spiegel die einfallende Sonnenstrahlung 80-fach auf ein Absorberrohr, in den ein Spezialöl als Wärmeträger auf rund 400 Grad erhitzt wird. Im zentralen Kraftwerksblock wird dann in einem Wärmetauscher der Dampf erzeugt, der konventionelle Dampfturbinen antreibt.

Moderne Speichertechnik macht den Solarstrom bei ungünstigem Wetter und in der Nacht verfügbar: Die im Solarfeld umlaufenden Millionen Liter Wärmeträgeröl stellen bereits eine beträchtliche Speicherkapazität dar, die kurzfristige Bewölkungsphasen überbrücken kann. Salzspeicher gewährleisten zusätzlich eine verlässliche Stromversorgung rund um die Uhr. Durch die Anwendung der Speichertechnik kann die Turbine außerdem immer unter Volllast und dadurch mit einem optimalen Wirkungsgrad laufen. Das führt zu einer hohen Wirtschaftlichkeit des Kraftwerks. Die Salzspeicher-Technik ist erprobt und wird von Netzbetreibern als verlässlich eingestuft. Der spanische nationale Netzbetreiber hat zum Beispiel der hier beschriebenen Kraftwerksanordnung denselben Zuverlässigkeitsstatus zuerkannt wie fossil beheizten Kraftwerken. Er sieht keine Probleme bei der Integration solcher Kraftwerke in bestehende Netze.

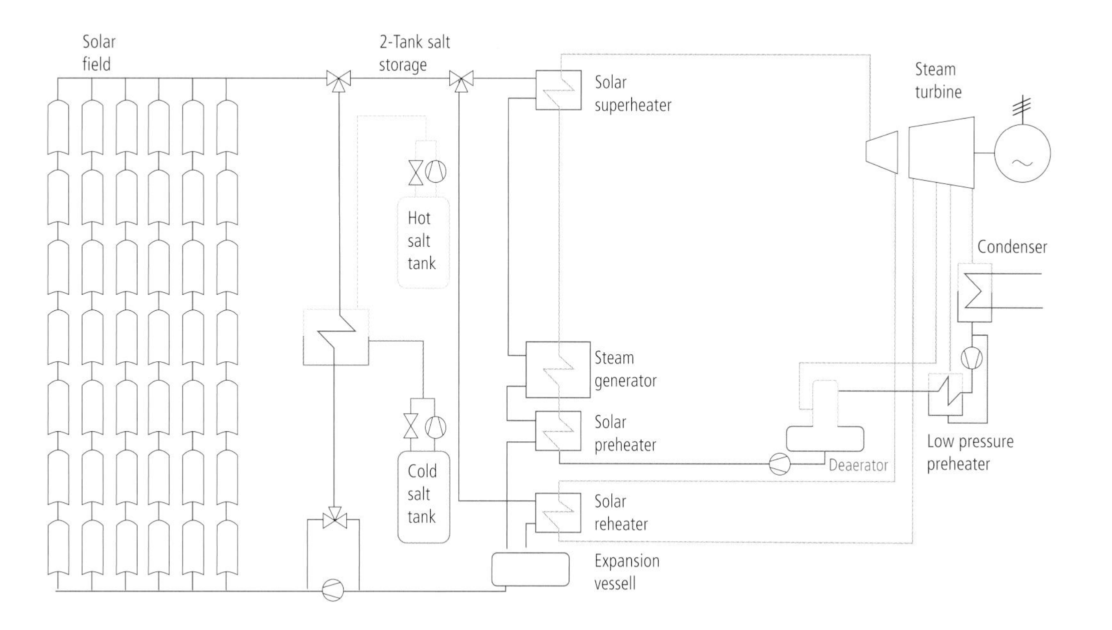

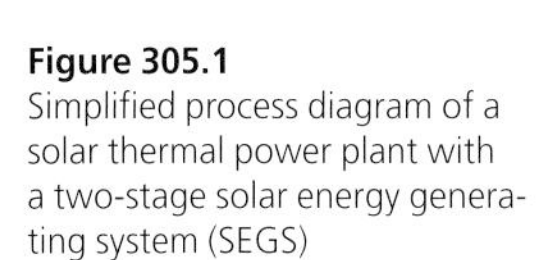

Figure 305.1
Simplified process diagram of a solar thermal power plant with a two-stage solar energy generating system (SEGS)

Bild 305.1
Vereinfachte Darstellung einer solarthermischen Anlage mit 2-stufiger Turbinenanlage, SEGS- Kraftwerk (Solar Energy Generating System)

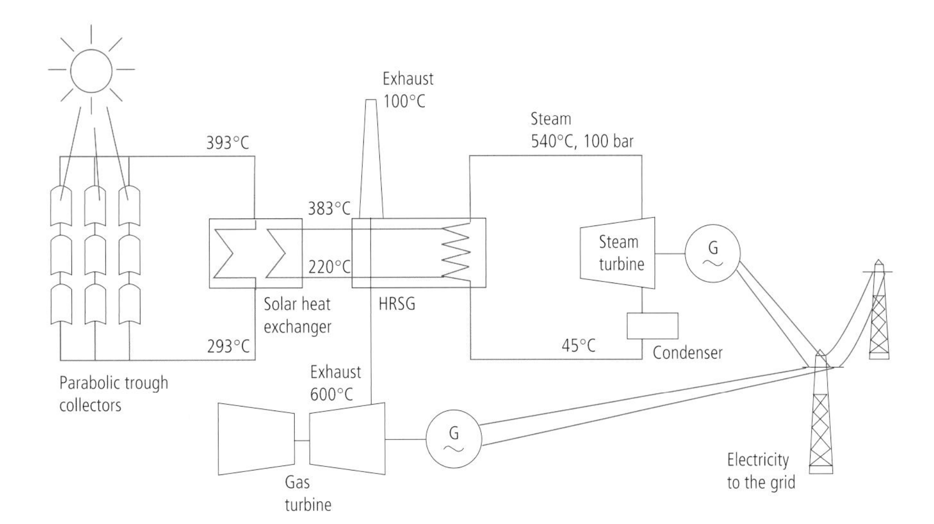

Figure 305.2
Flow of an ISCC process (150 MW)
(ISCCS –Integrated Solar Combined Cycle System)

Source: Lahmeyer international

Bild 305.2
Flussdiagramm eines Integrierten Solar-Kreislauf-Systems (ISCCS)

Quelle: Lahmeyer international

The concept of hybrid power plants is feasible. Because solar fields feed their energy into conventional steam turbines, they can be quite well combined with new generation of clean, natural gas-fired power to hybrid installations. Also possible is the addition of parabolic trough collector fields to conventional steam power plants for added solar steam generation. This hybrid concepts means:

- maximum-capacity use of the turbines optimizes operation of the joint power plants' blocks,
- economical electricity cost by compensatory pricing,
- when compared with molten salt storage, more economical buffering of the solar radiation fluctuations by temporary, supplemental fossil energy supply,
- smooth transition from fossil fuel power plant technology,
- potential for sustainable upgrade of existing fossil-fuel power plant.

Hybrid technology (**Figure 305.2**) will improve the competitive position versus conventional fossil plants significantly. Substantial cost advantages already develop when SEGSs are supplemented by additional fossil-fuel generation, as seen in the case of the solar thermal power plants in California. Electricity costs are cut in half by using Integrated solar combined cycle systems (ISCCS), in which solar heat for steam generation and the gas turbine waste heat can be used for both preheating and superheating of the steam, approximately doubling steam turbine capacity. Electricity generation cost in this case – even without additional public incentives – is only slightly higher than in the case of conventional power plants. Solar thermal power plants therefore represent a most important link between today's fossil-fuel power generation and the future solar-based energy supply.

Der Bau von Hybrid-Kraftwerken ist möglich: Da Solarfelder ihre Wärmeenergie in einen konventionellen Kraftwerksteil mit Dampfturbine einspeisen, können sie problemlos zum Beispiel in relativ saubere erdgasbefeuerte Kombi-Kraftwerke der neuen Generation integriert werden. Es ist auch möglich, bereits bestehende konventionelle Dampfkraftwerke mit Parabolrinnen-Solarfeldern als zusätzliche solare Dampferzeuger nachzurüsten. Die Hybridtechnik bedeutet:

- verbesserte Auslastung der Turbinen und damit optimaler Betrieb des gemeinsamen Kraftwerksblocks,
- günstige Strompreise durch Mischkalkulation,
- im Vergleich zur Salzspeichertechnik kostengünstigere Pufferung des gelegentlich schwankenden Strahlungsangebots durch fossile Zusatzfeuerung,
- sanfter Einstieg in den überwiegend aus fossilen Kraftwerken bestehenden Kraftwerkspark,
- die Möglichkeit einer ökologischen Aufwertung fossiler Kraftwerke.

Durch die Hybridtechnik, **Bild 305.2**, wird die Konkurrenzfähigkeit gegenüber konventionellen Kraftwerken stark verbessert. Erhebliche Kostenvorteile ergeben sich schon bei der fossilen Zusatzbefeuerung im Rahmen von SEGS-Kraftwerken (Solar Energy Generating System), wie sie in den kalifornischen Standorten betrieben werden. Die Stromgestehungskosten halbieren sich bei ISCCS-Kraftwerken (Integrated Solar Combined Cycle System). Bei diesem Typ werden Parabolrinnen-Kraftwerke mit modernen Gas- und Dampfkraftwerken kombiniert. Die Stromgestehungskosten liegen hier auch ohne Fördermittel nur geringfügig über denen konventioneller Kraftwerke. Solarthermische Hybrid-Kraftwerke sind damit ein wichtiges Bindeglied zwischen der heutigen fossilen und der künftigen solaren Energieversorgung.

The above-listed advantages of hybrid power plant technology have resulted in the concentration of funding in the magnitude of US$ 200 million by the Global Environmental Facility (GEF). The GEF is an independent financial organization that provides grants for projects that benefit the global environment and promote sustainable livelihoods in local communities. In particular, hybrid steam-solar power plants benefit from the organization's funding. The capability of combining renewable with fossil energy generation increases market opportunities greatly, especially in the U.S., where large-scale investment in new-generation gas-fired power plants is underway. In Algeria, large-scale plans already exist to refine gas-generated electricity by solar thermal supplements.

Parabolic-trough collector plants can be operated in cogeneration configurations. This, for instance, may result in the possibility of using the decoupled heat energy to generate power for the production of potable water in desalination plants. In the case of co- generating combined heat-power plants, efficiencies of around 55 % are conceivable.

The advantages of solar thermal plants are evident:

- Parabolic-trough power plants are suitable for large-scale applications in the range of 10 – 200 MW of electrical power, and their modular character is capable of creating a range of power supply magnitudes. Currently, the optimal size lies between 150 and 200 MWel. Parabolic plants can replace conventional fossil plants without qualitative changes to the grid structure.
- Because storage concepts for thermal energy exist, the turbo generators of thermal plants are capable of producing energy even during the night or at other times of reduced radiation intensity. They are capable of providing safe, grid-stable, and predictable energy.
- Solar thermal electricity generation can be integrated into conventional thermal power plant configurations which means that the difficulty of use in developing countries is eased.
- Parabolic plants are capable today of providing electricity at costs of 10 – 20 cents/kWh, depending upon the solar intensity of the specific geographical place of implementation. In comparison with the high initial cost of construction of such plants, the operating costs are low, currently around just 3 cents/kWh. It is expected that electricity cost of such plants will be priced competitively with medium-load fossil-fuel generating by the year 2015.
- Use of solar energy provides supply reliability because it is independent of raw energy cost fluctuations, and the unlimited supply in turn allows for sound cost calculation over the entire period of investment.

Die aufgeführten Vorteile haben die Global Environmental Facility (GEF) bewogen, ihre Fördermittel in Höhe von 200 Millionen US-Dollar ausschließlich für Hybrid-Kraftwerke des Typs Gas- und Dampf/Solar zur Verfügung zu stellen. Die Kombinierbarkeit mit herkömmlichen Kraftwerken erhöht auch die Marktchancen in den USA erheblich. Dort wird zurzeit in großem Stil in Gaskraftwerke der neuen Generation investiert. Auch in Algerien plant man, den aus Gas erzeugten Strom durch solare Zusatzleistung zu veredeln.

Parabolrinnen-Kraftwerke können in Kraft-Wärme-Kopplung betrieben werden, indem neben elektrischer Energie Wärme ausgekoppelt wird, um zum Beispiel Trinkwasser aus einer Meerwasserentsalzungsanlage zu gewinnen. In Kraft-Wärme-Kopplung sind solare Wirkungsgrade von bis zu 55 Prozent denkbar.

Die Vorteile der solarthermischen Kraftwerkstechnik liegen auf der Hand und sind:

- Parabolrinnen-Kraftwerke eignen sich für den großtechnischen Einsatz im Bereich von 10 bis 200 MW elektrischer Leistung. Der Modulcharakter des Solarfeldes erlaubt den Einstieg in beliebiger Leistungsgröße. Zurzeit liegt die optimale Größe bei 150 – 200 MWel. Parabolrinnen-Kraftwerke können konventionelle thermische Kraftwerke ersetzen – und zwar ohne qualitative Änderungen der Netzstruktur.
- Durch die Möglichkeiten thermischer Speicherung können die Turbinen solarthermischer Kraftwerke auch in einstrahlungsarmen Zeiten und nachts Strom produzieren. Solarthermische Kraftwerke können ihre Leistung sicher, planbar und netzstabil bereitstellen.
- Die solarthermische Stromerzeugung ist in konventionelle thermische Kraftwerke integrierbar. Die kombinierte Nutzung führt zu einer erheblichen Kostensenkung und erleichtert damit gerade in Schwellenländern den Einstieg in die Nutzung erneuerbarer Energien.
- Parabolrinnen-Kraftwerke können heute – je nach standortbedingter Einstrahlungsintensität – zu Preisen zwischen 10 und 20 Cent/kWh kostengünstigen Solarstrom produzieren. Den hohen Kosten in der Investitionsphase stehen niedrige Betriebskosten von zurzeit nur 3 Cent/kWh gegenüber. Bis 2015 werden Stromerzeugungskosten mit denen fossil befeuerter Mittellast-Kraftwerke vergleichbar sein.
- Die Nutzung der Sonnenenergie sorgt für Planungssicherheit: Die Unabhängigkeit der Betriebskosten von schwankenden Brennstoffpreisen und die uneingeschränkte Verfügbarkeit erlauben eine sichere Kalkulation über den gesamten Investitionszeitraum.

- In times when demand for cooling energy in hot climates is highest, the solar thermal plant is most efficient. Such demand peaks are already today supplied by new solar thermal plants in places such as California.
- High-voltage direct-current (HVDC) is the technical standard to transmit large amounts of energy over long distances, such as from Northern Africa to Europe. The cost of such transmission lines is approximately 2 cents/kWh.
- Parabolic thermal plants are tried and tested, and the first generation of such plants, totaling 354 MW_{el}, are operating in the U.S. with great reliability. The technology has proved its potential impressively, with almost 12TW of solar electricity, which in economic terms is the equivalent of U.S. $1.6 billion. No negative impact on the typically fragile environments of their application have been noted during the 20 years of their operation.
- Excess heat from solar thermal power plants can be used in absorption chiller plants to provide cooling energy.
- Solar power plants use globally available and fully recyclable materials such as steel, glass, and concrete. In addition, the majority of the construction tasks can be awarded to local contractors because it does not require specialized expertise. The modular character of the technology allows for mass production and great economic potential.
- Solar thermal plants are sustainable. The energy recapturing period is short, approximately 5 months, which compares well with other renewable energy generation systems. Among SEGS technologies, parabolic-trough collectors have the least amount of construction material needs.
- Of all the renewable energy supply technologies, such as biomass, wind energy, and hydro power (except for high alpine water reservoirs), solar thermal plants have the smallest space requirements, and because they are best located in arid regions and deserts they do not compete negatively with arable land use. Typical placements are between 35° northern and southern latitude.
- For countries in Northern Africa and the Middle East particularly, the aspect of potable water generation by desalination with the excess heat of solar thermal plants is of great importance. Here, the excellent efficiency of such locations for thermal energy generation can be combined favorably with the provision of water.

- Gerade dann, wenn im Sonnengürtel am meisten Strom zur Kühlung nachgefragt wird, produziert die STKT am effektivsten. Diese Stromspitzen werden heute schon konkurrenzfähig durch die neuen solarthermischen Kraftwerke in Kalifornien abgedeckt.
- Hochspannungs-Gleichstrom-Übertragungsleitungen, die heute Stand der Technik sind, können den Strom über große Distanzen transportieren – zum Beispiel von Nordafrika nach Mitteleuropa. Die Kosten belaufen sich auf rund 2 Cent/kWh.
- Parabolrinnen-Kraftwerke sind erprobt. In den USA laufen die Kraftwerke der ersten Generation mit einer Gesamtkapazität von 354 M zuverlässig. Mit fast 12 Terrawattstunden produziertem Solarstrom im Wert von 1,6 Milliarden Dollar hat die Parabolrinnen-Technologie ihr Potenzial eindrucksvoll bewiesen. In der nahezu 20jährigen Betriebspraxis sind keinerlei nachteilige Auswirkungen auf das soziale oder das fragile natürliche Umfeld bekannt geworden.
- Die Abwärme aus solarthermischen Kraftwerken kann im Regelfall dazu genutzt werden, um über Absorptionskälteanlagen Kälteenergie aufzubereiten.
- Solarthermische Kraftwerke verwenden weltweit verbreitete, recycelbare und preiswerte Baustoffe: Stahl, Glas und Beton. Ein großer Teil der Bauarbeiten kann an Unternehmen vor Ort vergeben werden. Der modulartige Aufbau des Solarfeldes erlaubt den Einstieg in die Massenproduktion mit erheblichem Rationalisierungspotenzial.
- Solarthermische Kraftwerke weisen eine sehr gute Ökobilanz auf. Die Energierückgewinnungszeit ist mit fünf Monaten – auch im Vergleich zu anderen regenerativen Energien – gering. Die Parabolrinnen-Technologie hat unter den STKT-Technologien den geringsten Materialbedarf.
- Solarthermische Kraftwerke haben einen erheblich geringeren Flächenbedarf als Biomasse, Windenergie und Wasserkraft – von Staudämmen im Hochgebirge abgesehen. Da sie zudem allein in den Trockenzonen der Erde errichtet werden, entsteht kaum Landnutzungskonkurrenz. Solarthermische Kraftwerke können im Sonnengürtel der Erde zwischen 35 ° nördlicher und südlicher Breite errichtet werden.
- Die Abwärme von solarthermischen Kraftwerken kann zusätzlich zur Stromgewinnung für die Meerwasserentsalzung genutzt werden. Gerade Länder in Nordafrika und im Mittleren Osten, die hervorragende Standorte für die STKT sind, können so ihre Wasserversorgung verbessern

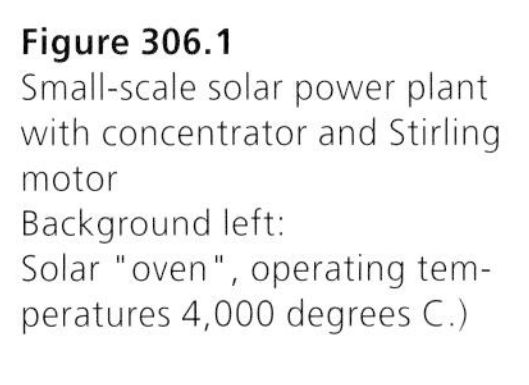

Figure 306.1
Small-scale solar power plant with concentrator and Stirling motor
Background left:
Solar "oven", operating temperatures 4,000 degrees C.)

Source: Projektentwicklung und Leitung – Schlaich Bergermann und Partner, Stuttgart, Germany

Bild 306.1
Solares Kleinkraftwerk mit Konzentrator und Stirlingmotor (Im Hintergrund ein "Solar-Ofen" mit Betriebstemperaturen bis 4.000 °C)

Quelle: Projektentwicklung und Leitung – Schlaich Bergermann und Partner, Stuttgart

Small thermal power plants

If primary electric energy requirements are smaller (< 100 MW), a type of solar power plant is being used called the small Dish-Stirling plant.

Figure 306.1 shows such a plant with Dish-Stirling engines, which are used for decentralized solar power supply. Their power is typically between 10 and 50 kW per installation, and the possibility of combining them to create larger "farms" exists. In such cases, they produce from several kW of power up to several MW, which makes them the solution for a wide range of power demands. They are the best system available to replace the expensive and environmentally harmful decentralized diesel generators of today.

Dish-Stirling systems, which concentrate the solar radiation and convert it into electricity are composed of the following important components:

- concentrator (mirror)
- tracking device
- solar heat exchanger (receiver)
- a Stirling engine with electric generator

Solare Kleinkraftwerke

Wird eine elektrische Energieleistung (primär) von deutlich weniger als 100 MW benötigt, so kommen solare Kleinkraftwerke zum Einsatz – Dish-Stirling Kleinkraftwerke.

Dish-Stirling-Anlagen, **Bild 306.1**, dienen der dezentralen solaren Stromversorgung. Ihre elektrische Leistung liegt typisch zwischen 10 und 50 kW pro Anlage mit der Möglichkeit, mehrere Anlagen zu einer „Farm" zusammenzuschalten und so einen Bedarf zwischen 10 kW und mehreren MW zu befriedigen. Dadurch eignen sich die Dish-Stirling-Kleinkraftwerke für einen weiten Einsatzbereich. Sie können die heutige umweltschädliche und teure dezentrale Energieversorgung mit Dieselaggregaten ablösen.

Dish-Stirling-Systeme konzentrieren die direkte Sonnenstrahlung und wandeln sie in elektrische Energie. Sie bestehen aus folgenden wesentlichen Komponenten:

- Konzentrator (Spiegel)
- Nachführeinrichtung
- Solarer Wärmetauscher (Receiver)
- Stirlingmotor mit elektrischen Generator

The parabolic concentrator reflects the parallel rays of the sun to a focal point, where the solar thermal heat exchanger is fixed to the Stirling engine. The receiver absorbs the concentrated solar radiation and heats up the thermal carrier medium of the Stirling engine, typically helium or hydrogen. The collected energy is transformed into rotational energy and then transformed into electricity by the generator coupled to the engine (**Figure 306.2**). Electrical power output is proportional to the solar radiation intensity, the surface area and optical performance characteristics of the mirror, and the performance parameters of the engine and the generator.

Der parabolisch gekrümmte Konzentrator reflektiert und bündelt die parallel auf ihn einfallenden Sonnenstrahlen in einem Brennpunkt. In diesem ist der solare Wärmetauscher des Stirlingmotors fixiert. Der Wärmetauscher (Receiver) absorbiert die konzentrierte Sonnenstrahlung und heizt so das Wärmeträgermedium (Helium oder Wasserstoff) des Stirlingmotors auf. Die gesammelte Wärme wird vom Stirlingmotor in Rotationsenergie umgewandelt und über einen direkt an die Kurbelwelle des Motors gekoppelten Generator in elektrischen Strom umgesetzt, **Bild 306.2**.

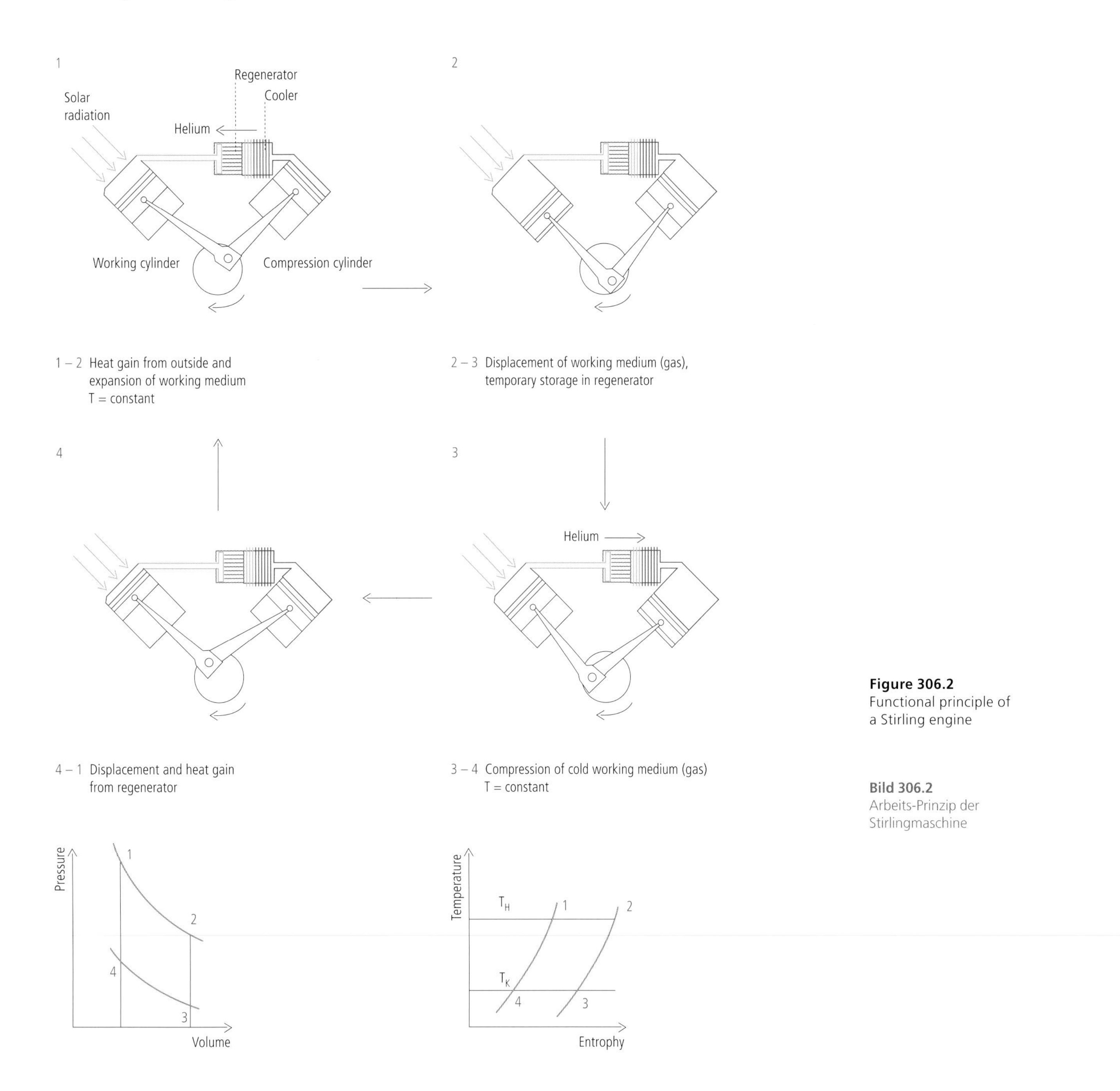

Figure 306.2
Functional principle of a Stirling engine

Bild 306.2
Arbeits-Prinzip der Stirlingmaschine

In order to reach high performance, the Stirling engines must be operated at high temperatures, and to achieve such high temperatures the concentrator needs to have a point-focusing rotational-symmetric contour. It is constructed of metallic membranes that are typically 0.23 mm thick and 1 m in width. They are welded to larger membranes with a diameter of 8.5 m. The membranes are then stretched over a cylindrical housing and bolted down or welded. The inside of the container housing is then evacuated, resulting in a precise and controlled deformation of the membrane. Additionally, the outside of the membrane is pressurized by water, resulting in a highly exact parabolic surface. As a final step, large glass mirrors 0.9 mm thick and 30 cm x 30 cm – not dissimilar to carpet tiles – are placed onto the reflector surface. The result is a torsion-resistant drum the front of which consists of a super-precise and effective parabolic mirror surface.

Because the concentrator during operation is always oriented exactly toward the sun, the system needs to be adjustable in two axles. A simple steel-tube support platform with a horizontal axle for vertical movement, plus six wheels traveling on a ring foundation, enables the necessary adjustments. Both axles are rotated by electric servo motors that transmit their forces onto a curved accentuator. A global positioning system (GPS), in conjunction with a computer, precisely calculates the position of the sun and positions the concentrator accordingly. The system status can be monitored, controlled, and diagnosed remotely from any place by a modem.

In contrast to typical Otto four-stroke or diesel engines, which are driven by internal combustion, the heat energy for the Stirling engine is provided from outside, enabling the system to convert solar-generated heat energy to electricity. In additional, based on the described principle, the engine can be operated as a solar-gas hybrid technology – during times of passing cloud cover or during the night the required heat energy may be supplied by a biogas-fired burner that is also installed on the dish receiver. Continuous operation and electricity generation can thus be secured.

In the Stirling engine, a constant volume of gas, either helium or hydrogen, is heated and cooled. An increase in temperature expands the gas, while cooling causes its contraction. The forces are then converted to a mechanical movement of pistons connected to a crankshaft to deliver mechanical energy. Because the performance of the engine increases with process temperature, combining a Stirling engine with a concentrating solar collector is ideal in that it results in very high temperatures. Also, because no internal combustion occurs, noise levels are minimal. Finally, Stirling engines are characterized by very long operating times because they do not cause combustion by-products to shorten internal engine components such as bearings and because valves are not required.

Die elektrische Leistung des Systems ist in erster Linie proportional zur Intensität der Sonnenstrahlung, zur Größe des Spiegels, zu dessen optischem Wirkungsgrad und zum Wirkungsgrad des Stirlingmotors mit Generator.

Damit der Stirlingmotor einen hohen Wirkungsgrad erreicht, muss er mit hohen Temperaturen betrieben werden. Dazu wird ein großflächiger, punktfokussierender Konzentrator mit rotationssymmetrischer parabolischer Kontur verwendet.

Der Konzentrator ist aus Metallmembranen hergestellt. Bei dieser kostengünstigen und materialsparenden Bauweise werden 0,23 mm dicke und 1 m breite Edelstahlbleche zu großen Membranen mit 8,5 m Durchmesser verschweißt und über die Vorder- und Rückseite eines zylindrischen Blechgehäuses gespannt und verschraubt bzw. verschweißt. Anschließend werden die Membranen plastisch verformt, indem das Trommelgehäuse evakuiert und die vordere Membran zusätzlich mit Wasser belastet wird. Dadurch erhält die vordere Membran eine hochgenaue parabolische Form. Dünne, nur 0,9 mm dicke, 30 x 50 cm große Glasspiegel werden dann wie Teppichfliesen auf die so verformte Blechmembran geklebt. So entsteht eine leichte, verwindungssteife Tommel, deren Vorderseite einen äußerst effektiven Parabolspiegel darstellt.

Da der Konzentrator im Betrieb stets exakt auf die Sonne ausgerichtet sein muss, ist er zweiachsig beweglich gelagert. Eine einfache Rohrkonstruktion mit einer horizontalen Lagerachse und – für die vertikale Drehung – sechs Rädern auf einem Ringfundament dient als Drehstand. Beide Achsen werden mittels Servomotoren angetrieben, deren Kräfte jeweils über eine Kette auf ein gebogenes Antriebsprofil eingeleitet werden.

Zur kontinuierlichen hochpräzisen Nachführung wird von einem PC anhand der exakten Uhrzeit eines GPS-Empfängers die Sonnenposition berechnet und der Konzentrator mittels Drehwinkelgebern positioniert. Über ein Modem und das Internet kann die Anlage fernüberwacht werden. Bedienung und Diagnose können so von einem beliebigen Ort aus erfolgen.

Im Gegensatz zu Otto- und Dieselmotoren, die durch eine innere Verbrennung angetrieben werden, erfolgt die Wärmezufuhr bei einem Stirlingmotor von außen. Damit eignet er sich vom Prinzip her besonders gut, um solar erzeugte Wärme in Strom umzuwandeln.

Deshalb kann er auch sehr vorteilhaft hybrid (solar/gasgefeuert) betrieben werden, d.h. bei fehlender Sonneneinstrahlung (Wolkendurchgang) oder während der Nachstunden erfolgt die erforderliche Wärmezufuhr über einen zusätzlich installierten und beispielsweise mit Biogas befeuerten Brenner. So ist die Stromversorgung rund um die Uhr sichergestellt.

If the gas inside of the receiver is heated up, it expands and moves the working piston, resulting in a power stroke (1 – 2). The power provided is partially used to displace the hot gas from the working cylinder into the compression cylinder (2 – 3). In this process, the gas passes a regenerator, where a large amount of the heat energy is given off and temporarily stored. Subsequent to the regenerator, a gas cooler provides further cooling of the gas (2 – 3). When the piston inside the compressing cylinder returns, the now-cool gas is compressed (3 – 4), picks up the stored heat from the regenerator, then is further heated by the receiver and pressed into the working storage device (4 – 1).

The expansion of the hot gas inside the working cylinder provides more total energy than the moving and compression of the cold gas in the compression cylinder, and this surplus energy is used by the generator and put into motion by the engine's crankshaft. Depending upon the amount of solar radiation, the generated power ranges between 3 and 10 kW. It is regulated by the pressure of the gas medium inside the engine.

The receiver typically has two functions:

- maximum absorption of radiation energy reflected by the concentrator,
- transmission of heat, with minimal losses, to the Stirling engine.

For a receiver that is operated exclusively by solar radiation, a tube-bundle heat exchanger has been developed that is placed directly on top of the cylinder heads of the engine and is passed by the working gas. Thin-walled, high-temperature-resistant tubes form an almost uninterrupted surface for the absorption of the solar radiation, and the working gas is heated up to 650 °C.

Many applications for Dish-Stirling receivers exist throughout various regions of the Earth, but they are mainly used today for the desalination of seawater. **Figure 307** depicts the current demand for desalination, **Figure 308** presents a schematic diagram of the operation by means of solar energy use, and **Figure 309** shows the conventional energy supply for a desalination process and heating and cooling energy supply. In the lower section of the image, the schematic principle of an integral solution using natural gas and solar energy is shown. Here, the majority of the cooling energy and nearly all of the total energy required for desalination is being provided by absorption chillers.

The future will involve a changed system in which the percentage of solar energy is much greater than the heat energy provided by natural gas. Evidently, the chosen system configurations are dependent upon local factors, cost, and the usage requirements.

Bei einem Stirlingmotor wird eine konstante Gasmenge (Helium oder Wasserstoff) erwärmt und wieder abgekühlt. Die bei der Erwärmung auftretende Expansion und die bei der Abkühlung erfolgende Kontraktion des Gases setzen Kolben, die über einen Kurbelbetrieb miteinander verbunden sind, in Bewegung, und liefern so mechanische Energie.

Da der Wirkungsgrad des Stirlingmotors mit steigender oberer Prozesstemperatur zunimmt, ist die Kombination mit einem konzentrierenden Solarkollektor wegen der damit erreichbaren hohen Temperaturen ideal. Aufgrund der fehlenden inneren Verbrennung ist die Geräuschentwicklung des Motors minimal.
Der Stirlingmotor verfügt über eine sehr hohe Nutzungsdauer, da es keine Verbrennungsrückstände gibt, die die Lebensdauer von Lagern etc. reduzieren.
Ein V-Zweizylinder Stirlingmotor besteht aus zwei miteinander verbundenen und mit einem Arbeitsgas (Helium oder Wasserstoff) gefüllten Zylindern (Arbeits- und Verdichtungszylinder), deren Kolben auf einen Kurbeltrieb wirken

Erwärmt man nun das Gas im Erhitzer (Receiver), zum Beispiel durch konzentrierte Sonnenstrahlung, so dehnt es sich aus, bewegt den Arbeitskolben (1 – 2) und leistet somit Arbeit. Ein Teil dieser Arbeit wird dazu verwendet, das heiße Gas aus dem Arbeitszylinder in den Verdichtungszylinder zu schieben (2 – 3). Dabei durchströmt das Gas den Regenerator und gibt dort einen großen Teil seiner Wärme ab, die hier kurzzeitig zwischengespeichert wird. Der hinter dem Regenerator angeordnete wasserdurchströmte Gaskühler sorgt für eine weitere Abkühlung des Gases (2 – 3). Wenn dann der Kolben des Verdichtungszylinders wieder zurückkehrt, wird das nun kalte Gas komprimiert (3 – 4) und strömt unter Wiederaufnahme der vorher im Regenerator gespeicherten Wärme sowie weiterer Erhitzung im Receiver zurück in den Arbeitsspeicher (4 – 1).

Insgesamt liefert die Expansion des heißen Gases im Arbeitszylinder mehr Energie, als bei der Verschiebung und Kompression des kalten Gases im Verdichtungszylinder benötigt wird. Mit dieser überschüssigen Energie kann ein Generator angetrieben werden, der direkt an die Kurbelwelle des Motors geflanscht wird.

Die abgegebene Leistung zwischen 3 und 10 kW wird je nach Sonneneinstrahlung über den Druck des Arbeitsgases in der Maschine (Zuführen bzw. Abführen von Arbeitsgas) geregelt.

Der Receiver hat grundsätzlich zwei Aufgaben:

- Soviel wie möglich von der vom Konzentrator reflektierten Strahlung zu absorbieren und
- diese in Form von Wärme bei möglichst geringen Verlusten an den Stirlingmotor weiterzugeben.

Als rein solar betriebener Receiver wurde ein Rohrbündelwärmetauscher entwickelt, der direkt auf den Zylinderköpfen des Stirlingmotors befestigt ist und vom Arbeitsmedium (Helium oder Wasserstoff) des Stirlingmotors durchströmt wird. Dünnwandige, hochtemperaturfeste Röhrchen, deren Enden an Sammlerrohre angeschweißt sind, bilden eine nahezu geschlossene Fläche, die die konzentrierte Sonnenstrahlung absorbiert und dabei das Arbeitsgas des Stirlingmotors auf ca. 650 °C erwärmt.

Ein großes Feld der solaren Nutzung ist in vielen Regionen der Welt die Aufbereitung von Frischwasser (Entsalzung) von Seewasser. **Bild 307** zeigt in etwa den momentanen Bedarf an Meerwasserentsalzung, **Bild 308** den Aufbau eines schematischen Prozessdiagramms des Betriebes unter Nutzung von Solarenergie. **Bild 309** zeigt einerseits eine konventionelle Energieversorgung zur Aufbereitung von Kälteenergie/Kühlenergie und Meerwasserentsalzung. Im unteren Teil des Bildes ist der schematische Ablauf einer integralen Lösung unter Einsatz von Gas und Solarenergie ausgewiesen. Hierbei wird ein wesentlicher Teil der Kühlenergie durch Absorptionskältemaschinen und annähernd der gesamte Energiebedarf zur Entsalzung von Wasser durch Wärmeenergie gedeckt.

Die integrale Systemlösung kann und wird sich in Zukunft zunehmend dahingehend verändern, dass der Solaranteil weitaus höher ist als der Wärmeanteil durch Gasenergie. Selbstverständlich gibt es je nach Anforderungsprofil, Kostenrahmen und Umweltangebot verschiedenartigste Lösungen, die angegangen werden können.

Figure 307
Seawater desalination, global consumption

Source:
Kernenergien 12/10/2007

Bild 307
Seewasserentsalzung, Verbräuche auf der Erde

Quelle:Kernenergien 12/10/2007

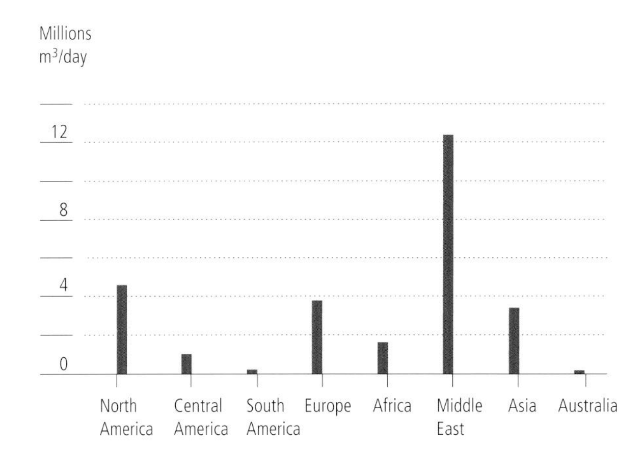

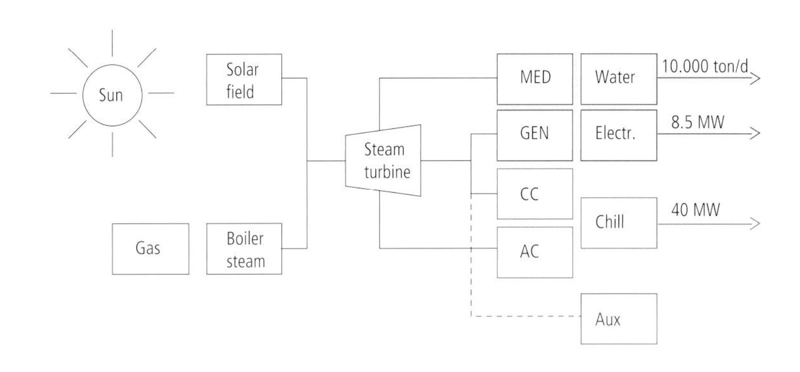

MED Seawater desalination
GEN Generator/electricity generation
CC Compressor/condenser (electric chiller)
AC Absorption chiller (thermal chiller)
AUX Auxiliary

Figure 308
Process diagram of the Jordan Solar Water Project

Source: Kernenergien 12/10/2007

Bild 308
Schema des Wirkungskreislaufes des Jordan Solar Water Projektes

Quelle: Kernenergien 12/10/2007

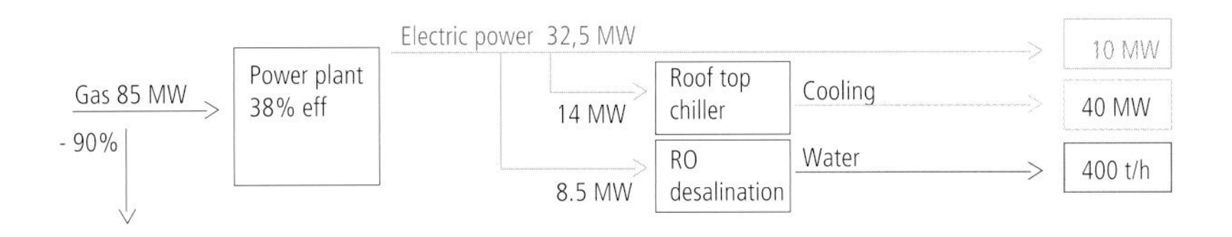

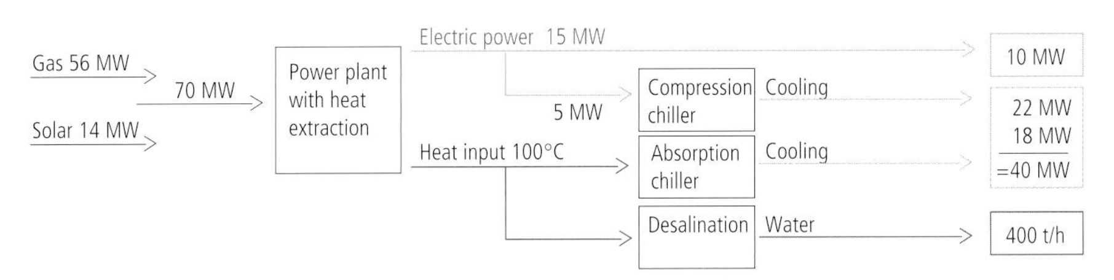

Figure 309
Comparison between conventional and hybrid power plant designs

Source: Kernenergien 12/10/2007

Bild 309
Vergleichende Betrachtung einer konventionellen- zu einer Hybrid-Lösung eines Kraftwerkes

Quelle:Kernenergien 12/10/2007

13.2.2 Solar updraft power plants

Many years ago, the engineer Jörg Schlaich from Stuttgart, Germany, developed another idea to generate large amounts of power from solar energy. His concept, which was realized in a test power plant in Spain, involves a so-called technology of solar updraft power generation in which the sun's radiation is absorbed by a large, circular, flat-plate collector field arranged around a tall, concrete updraft chimney. The air beneath the collector plates heats up, and due to the density differential between this hot air and the cooler ambient air it flows in a circular motion toward the chimney, where it rises upward. Inside the chimney, the thermal motion energy is further increased. At the bottom of the updraft chimney one or more turbines may be installed, through which the moving air is channeled. The mechanical energy of the turbine is converted to electricity in an attached generator. **Figure 310** shows the updraft power plant, and **Figure 311** depicts the diagrammatic working principle.

13.2.2 Aufwindkraftwerke

Eine weitere Möglichkeit, Energie in großem Umfang durch Solarenergie bereitzustellen, wurde von Jörg Schlaich (Stuttgart) bereits vor vielen Jahren konzipiert und in einer Demonstrationsanlage in Spanien umgesetzt. Beim Aufwindkraftwerk wird die Sonnenstrahlung von Plattenkollektoren einfacher Art, die ringförmig um einen Aufwindkamin angeordnet sind, absorbiert und die Luft unter dem Kollektordach erwärmt. Aufgrund des dabei entstehenden Dichteunterschiedes zwischen warmer Luft unter dem Kollektor und kalter Luft außerhalb des Kollektors strömt die Luft radial auf den Kamin zu und steigt in diesem auf. Innerhalb des Kamins wird zudem die thermische Antriebsenergie erhöht. Im unteren Fußpunkt der Kaminröhre können eine oder mehrere Turbinen eingebaut werden, die durch die Luftströmung angetrieben werden. Die mechanische Energie der Turbine wird in einem Generator in elektrische Energie umgesetzt. **Bild 310** zeigt ein Aufwindkraftwerk, **Bild 311** das Prinzip der Wirkung.

Figure 310
Solar thermal updraft power plant

Source (Figures 310 –313):
Low Tech-Light Tech-High Tech,
Klaus Daniels
Birkhäuser Verlag, 1998

Bild 310
Aufwindkraftwerk

Quelle (Bilder 310 –313):
Low Tech-Light Tech-High Tech,
Klaus Daniels
Birkhäuser Verlag, 1998

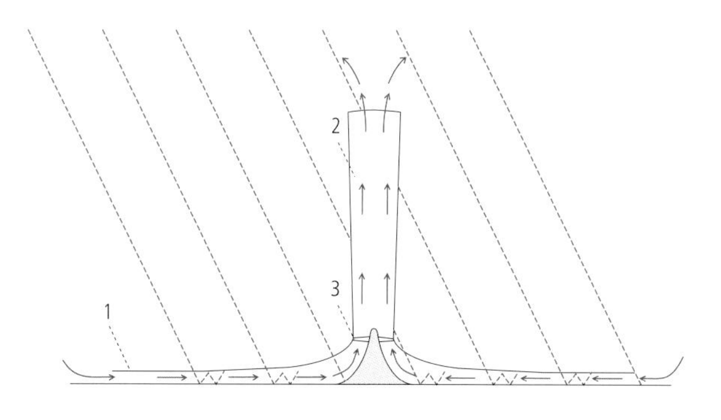

Figure 311
Principle of thermal solar thermal updraft power plant :
1 Glass roof collector
2 Chimney pipes
3 Wind turbine

Bild 311
Prinzip des Aufwindkraftwerks:
1 Glasdachkollektor
2 Kaminröhre
3 Windturbine

Depending upon collector diameter and the height of the chimney, the annual electric energy generation can be determined based on **Figure 312**.

If the entire global energy demand would need to be supplied by updraft power plants located in the nearest possible desert, the resulting space requirements can be determined using **Figure 313**. It is important to note that location for the successful implementation of solar updraft power plants would need to be characterized by a minimum insolation of about 2,000 kWh/m²a. It is not difficult to imagine solar updraft power plants being a part of the future energy generation mix, but unfortunately they will be efficient mainly in locations in Southern Europe and the global sun belts and not so much in Central Europe or similar moderate environments.

Je nach Kollektordurchmesser und Kaminhöhe werden Jahresenergieleistungen (elektrische Energie) gemäß **Bild 312** erzeugt.

Würde man den gesamten Welt-Primärenergiebedarf durch Aufwindkraftwerke in der jeweiligen nächstgelegenen Wüste mit hoher Solarzustrahlung decken wollen, so ergäbe sich ein Flächenbedarf wie er in **Bild 313** ausgewiesen ist. Wesentlich dabei ist, dass Aufwindkraftwerke in Regionen zum Einsatz kommen, die eine Energieeinstrahlung von mindestens 2.000 kWh/m²a besitzen. Wie unschwer erkennbar, gehört auch das Aufwindkraftwerk zum Szenario der Zukunft – leider nur bedingt in Mittel- und Südeuropa.

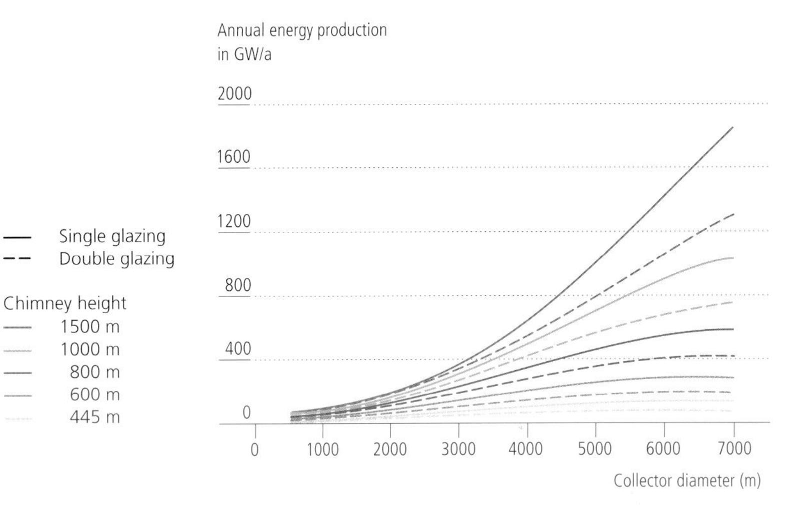

Figure 312
Annual energy production in thermal power stations (for 2300 kWh/m²a global radiation), dependent on collector diameter and chimney height

Bild 311
Jahresenergieerzeugung von Aufwindkraftwerken (bei 2.300 kWh/m²a Globalstrahlung) in Abhängigkeit vom Kollektordurchmesser und der Kaminhöhe

Figure 313
Area required to meet total world electrical primary energy requirements with thermal air tower generator located in the nearest desert area with excellent solar radiation

Bild 313
Flächenbedarf zur Deckung des gesamten Welt-Primärenergiebedarfs durch Aufwindkraftwerke in der jeweils nächstliegenden Wüste mit sehr guter Solarstrahlung

The areas indicated below symbolize areas required for using in each case one thermal air tower generator to meet the total energy demands of the world, Europe and Germany, respectively.

13.2.3 Biomass power plants

Biomass is one of the great carriers of renewable energy, and it is distinctly important because it is available at any given time – in contrast to solar and wind power. At the moment, an installed global total power capacity of 9,700 MW exists for this technology. In large power plants, biomass is currently being burned in furnace units to gain the thermal energy necessary to drive turbines and feed utility grids, but biomass power plants also in many cases burn household waste, which further increases energy generation. (**Figure 314.1**)

In addition to solid biomass, biogas and bio oils are also capable of being used, as already described in **Chapter 13.1.2**. **Figure 314.2** shows an example of a biomass power plant with the necessary components for energy generation.

13.2.3 Biomasse-Kraftwerke

Biomasse ist einer der großen Träger erneuerbarer Energien, der deshalb eine besondere Bedeutung hat, da sie jederzeit verfügbar ist – selbst dann, wenn die Solarenergie oder Windenergie kurzfristig nicht zur Verfügung steht. Zurzeit besteht eine installierte Gesamtkapazität von 9.700 MW (weltweit) durch mit Biomasse betriebene Heizkraftwerke. In Großanlagen wird die Biomasse in Großkesselanlagen verbrannt, um hieraus die notwendige Wärmeenergie zum Betreiben von Turbinen und Wärmenetzen zu erzielen, **Bild 314.1**. Neben der Biomasse wird häufig parallel eine Abfallverbrennung betrieben, wodurch die Energieerzeugung zu steigern ist.

Neben der Verbrennung von fester Biomasse lassen sich selbstverständlich auch Biogas und Bioöl zum Einsatz bringen, wie bereits in **Kapitel 13.1.2** erläutert. **Bild 314.2** zeigt beispielhaft eine Biomasse-Anlage mit den wesentlichen Elementen zum Erzeugen von Biogas.

Figure 314.1
Biomass power plant
Wood chip residues from Vermont forests and sawmills provide the bulk of the fuel used in the Burlington Electric Department's 50 MW McNeil biomass generating station.

Bild 314.1
Biomassekraftwerk zur Erzeugung von 50 MW elektrischer Energie, unter Verwendung von Holzabfällen aus dem Vermont-Waldgebiet und Sägewerken

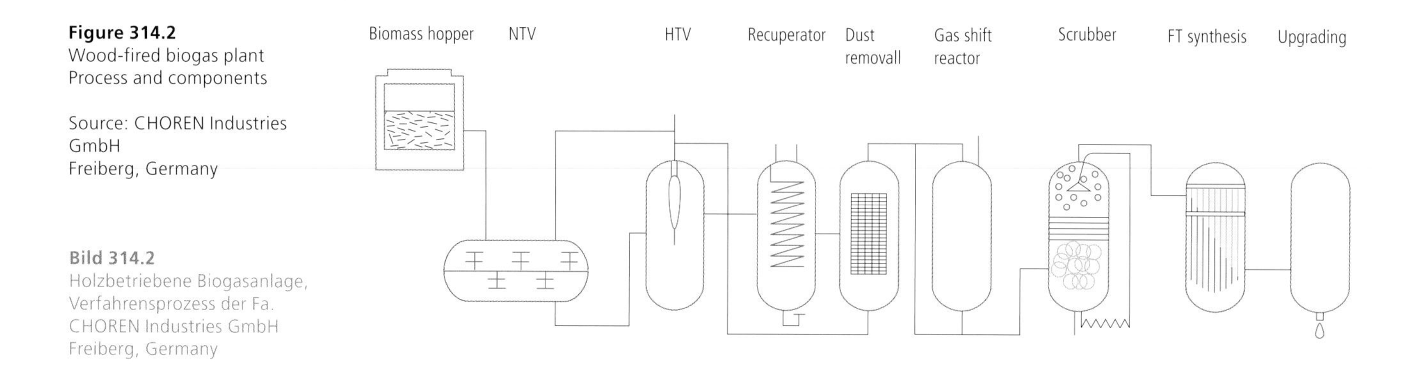

Figure 314.2
Wood-fired biogas plant
Process and components

Source: CHOREN Industries GmbH
Freiberg, Germany

Bild 314.2
Holzbetriebene Biogasanlage, Verfahrensprozess der Fa. CHOREN Industries GmbH
Freiberg, Germany

13.2.4 Wind farms

The use of wind energy has been known to humanity for centuries, yet only in our time of increased awareness of the coming energy shortage are efficient wind-wheel designs being intensely researched, developed, and implemented. The result of this effort is shown in **Figure 315**. Single units are being composed to large plants, the so-called wind farms. With 6,000 Enercon e112 wind turbines – the world's most powerful wind turbines (**Figure 316**) – 20 % of Germany's total electric energy demand could be easily supplied.

Depending upon the manufacturer and the design philosophy, the typical wind turbine consists of three-blade rotors mounted on a horizontal axis, geared or gearless. **Figure 317.1** shows the design of a wind turbine with two gears and an added generator that feeds the electric energy to the grid, and **Figure 317.2** shows a wind power plant that transmits the rotation without gears directly onto a fixed axle and into the generator. In the U.S., wind power generation has gained second rank in electricity generation, after natural gas. Electric energy generated from the wind has entered the market at competitive rates compared with generation from liquid fossil sources, but there are still two factors with regard to wind power generation that are detrimental. First, in some cases the locations of the farms and the supplied urban centers are separated by large distances, which increases transmission losses. Second, productive wind farm locations are often also scenic natural environments, creating an incompatibility with the sometimes negative visual impact of the turbines.

Power coefficient c_p*
0.7
0.6
0.5
0.4
0.3
0.2
0.1
0
Betz Law and coefficient of performance
Theoretical power coefficient for propeller type for E = ∞
1 2 3 4 5 6 7
0 2 4 6 8 10 12 14 16 18
Tip Speed Ratio λ (TSR)**

Figure 315
Typical power curves/performance for various rotor designs

Source: Daniels/author

Bild 315
Leistungskennlinien von Windrotoren unterschiedlicher Bauart

Quelle: Daniels/Autor

* Electrical power output devided by the wind energy input.
** It is the ratio between the rotational speed of the tip of a turbine blade and the actual velocity of the wind.
1 Savonius wind turbine. Type of vertical-axis wind turbine (VAWT)
2 American wind turbine
3 Dutch wind mill
4 Vertical axis rotor (Georges Darrieus machine)
5 Three-bladed rotor
6 Two-bladed (teetering) rotor
7 One-bladed rotor

13.2.4 Windfarmen

Windenergie wird schon seit Jahrhunderten genutzt, jedoch erst durch die Erkenntnis der Energieproblematik wurde in den letzten Jahrzehnten intensiv daran gearbeitet, effiziente Windräder zu produzieren und zu installieren. Das Ergebnis zum momentanen Zeitpunkt ist in **Bild 315** dargestellt. Hierzu werden zum Teil komplette Windfarmen, **Bild 315**, installiert. Mit 6.000 Windkraftanlagen des in **Bild 316** gezeigten Typs (Enercon E112) könnten 20 Prozent des deutschen Stromenergiebedarfs gedeckt werden.

Je nach Fabrikat und Firmenphilosophie werden heute im Regelfall dreiblättrige Windräder mit Horizontalachse mit und ohne Getriebe gebaut und installiert. **Bild 317.1** zeigt einen Aufbau eines Windrades mit zwei Getriebestufen und nachgeschaltetem Synchrongenerator, der seine elektrische Energie ins Netz speist, **Bild 317.2**, ein Windkraftwerk, das getriebelos seine Rotation auf einen Generator überträgt.

Figure 316
Wind power generator Enercon, Model E112. 6000 of such generators would be sufficient to supply 20 % of Germany's electricity demand.

Soure: ENERCON, Aurich, Germany

Bild 316
Windrad Enercon E 112
(mit 6.000 Anlagen diesen Typs können 20 % des deutschen Strombedarfs gedeckt werden)

Quelle: ENERCON, Aurich, Deutschland

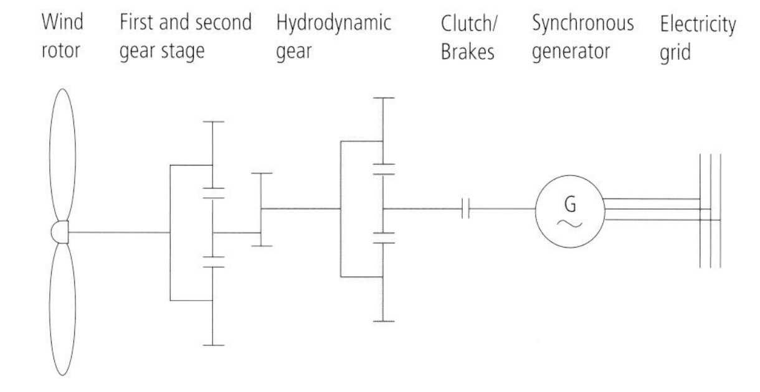

Figure 317.1
Wind power generator with intermediate transmission between rotor and generator. Hydrodynamic transmission with synchronized generator.

(Drive with variable input revolutions at rotor and constant output revolution at the generator)

Source: Voith Turbo GmbH & Co. KG Crailsheim, Germany

Bild 317.1
Windgenerator mit Getriebe zwischen Rotor und Generator Hydrodynamische Getriebe mit Synchrongenerator

(WEA-Triebstrang mit variabler Eingangsdrehzahl am Rotor und konstanter Abtriebsdrehzahl am Synchrongenerator)

Quelle: Voith Turbo GmbH & Co. KG, Crailsheim, Deutschland

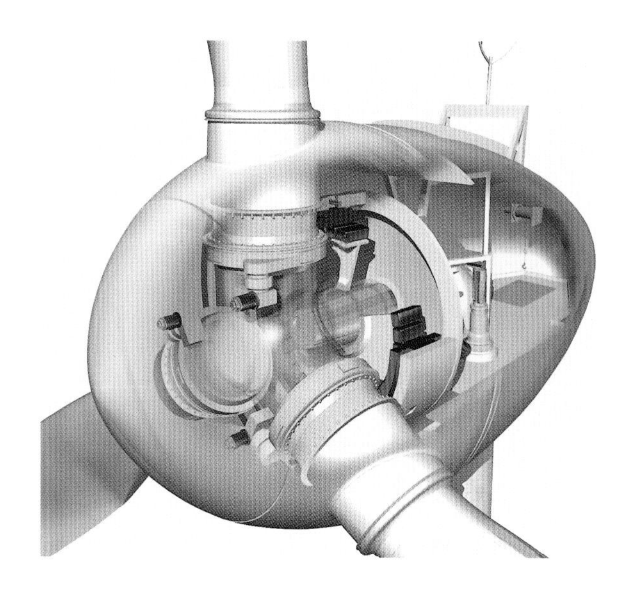

Figure 317.2
Diagrammatic view of gearless wind power generator Direct-drive model

Source:Enercon Aurich, Germany

Bild 317.2
Prinzipdarstellung Windgenerator, getriebelos/ Direktantrieb des Generators

Quelle: Enercon GmbH, Aurich, Deutschland

Wind farms may be located either on land (onshore) or at sea (offshore) to yield high productivity. When designing wind farms, either on land or offshore, the initial analysis needs to determine the annual prevailing wind directions and their frequency for the particular location, as seen in **Figure 318.1**, and the resulting average wind speeds in m/s, as shown in **Figure 318.2**. The frequency of average wind speeds is typically called the Weibull distribution, named after the Swedish mathematician Ernst W.H. Weibull (see **Figure 318.3**).

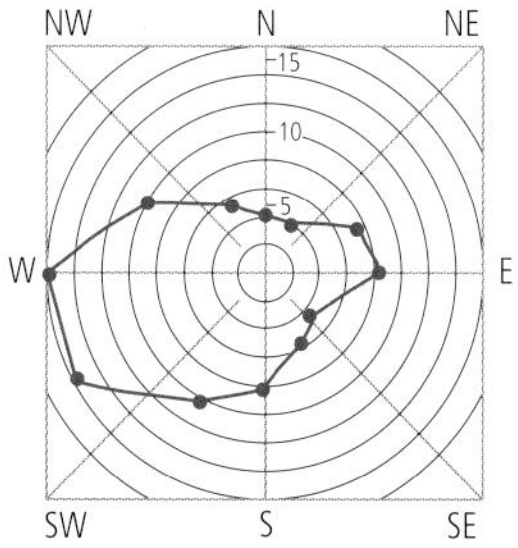

Figure 318.1
Percentage of prevailing wind directions, Central Europe. (Prevailing wind directions West and Southwest coincide with highest wind speeds)

Source: Daniels/author

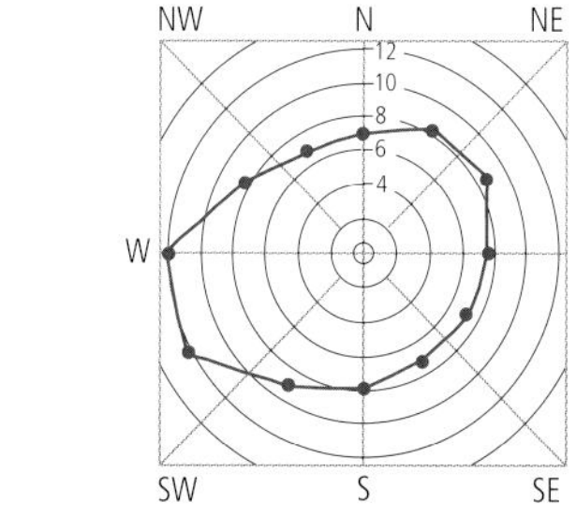

Figure 318.2
Wind diagram of Central Europe, showing typical wind directions and average wind speeds in m/s.

Source: Daniels/author

Bild 318.1
Windrichtungshäufigkeit in % eines Standorts in Mitteleuropa (Häufigste Windrichtung W – SW fällt im Regelfall mit höchsten Windgeschwindigkeiten zusammen.)

Quelle: Daniels/Autor

Bild 318.2
Darstellung von Windrichtung und mittlerer Windgeschwindigkeit (m/s), Standort Mitteleuropa

Quelle: Daniels/Autor

In den USA hat die Erzeugung von Energie durch Wind den zweiten Platz bei der Erzeugung elektrischer Energie nach Gas eingenommen. Die durch Wind erzeugte elektrische Energie steht inzwischen in einem vergleichbaren Wettbewerb zu Energie, die mit fossilen flüssigen Brennstoffen erzeugt wird. Der Windenergie stehen im Regelfall zwei Widrigkeiten entgegen. Diese sind zum Teil große Entfernungen zwischen Windfarmen und zu versorgenden Bereichen, und dass häufig Windfarmen gerade in (z.B. touristisch) interessanten Regionen entwickelt werden müssen. Die häufig großen Entfernungen zwischen Windfarm und Nutzer verteuern durch lange Leitungswege u.U. den Einsatz von Windenergie und vergrößern die Verlustleistungen.

Windfarmen werden entweder an Land (onshore) oder in der See (offshore) aufgebaut, um einen möglichst hohen Ertrag zu erzielen. Bei der Entwicklung größerer Windfarmen, onshore oder offshore, geht es anfänglich darum festzustellen, welche Häufigkeit von Windrichtungen, **Bild 318.1** und welche mittleren Windgeschwindigkeiten (m/s), **Bild 318.2**, am Standort bestehen. Die Häufigkeit mittlerer Windgeschwindigkeit wird üblicherweise als Weibull-Verteilung (nach dem schwedischen Mathematiker Ernst W.H. Weibull) bezeichnet und ist in **Bild 318.3** ausgewiesen.

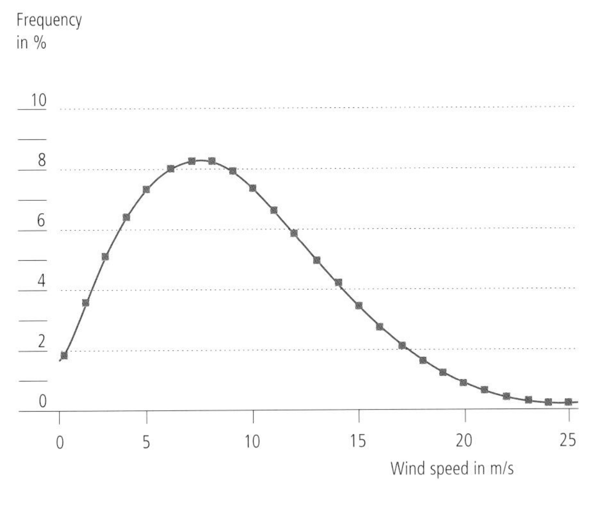

Figure 318.3
Statistical distribution of wind speeds: Weibull distribution = ratio between frequency and wind speed.

Source: Vestas Germany

Bild 318.3
Weibull-Verteilung = Verhältnis prozentualer Häufigkeit zu Windgeschwindigkeit

Quelle: Vestas Deutschland GmbH

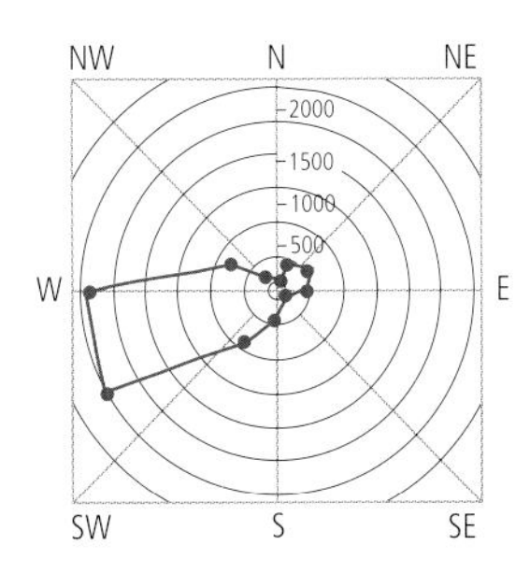

Figure 319
Wind energy generation in kWh/m²/year as a function of prevailing wind direction in a windmill-park of the baltic sea

Source: Daniels/author

Bild 319
Darstellung des Energieertrags in Abhängigkeit von der Hauptwind-Anströmung (Windenergierose inkWh/m²/Jahr) bei einem Windpark aufder Ostsee

Source: Daniels/author

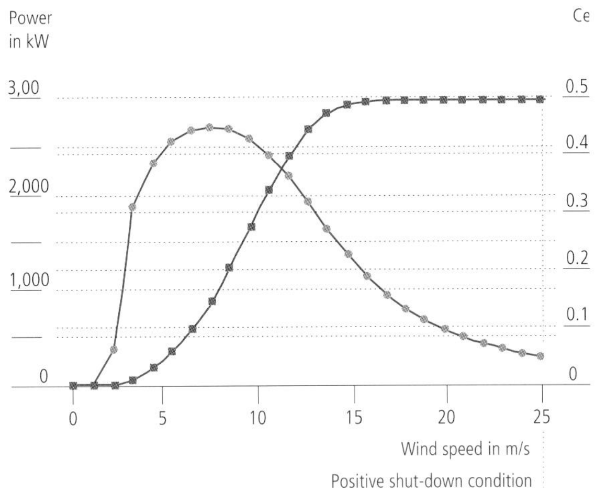

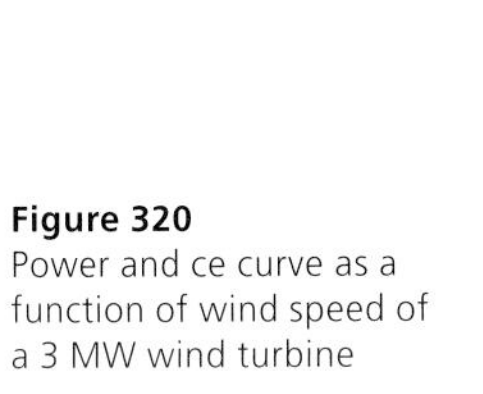

Figure 320
Power and ce curve as a function of wind speed of a 3 MW wind turbine

Source: Vestas Germany

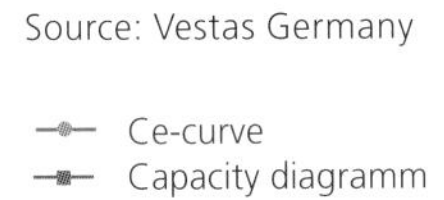

Bild 320
Leistungs- und Ce-Kennlinie in Abhängigkeit von der Windgeschwindigkeit eines 3 MW Windrades (Cp)

Quelle: Vestas Deutschland GmbH

Figure 319 shows, based on the previous diagrams, the wind energy rose, which allows the determination of primary energy yield in kWh/m²a. For each wind turbine design, a so-called power and ce-curve is given, as shown in **Figure 320**. The ce-curve is similar to the pressure coefficient, which means that at the beginning the ce-curve rises with increasing revolutions of the blades, then falls significantly. As seen in the power curve diagram in **Figure 320**, the wind turbine at wind speeds of 4 m/s first gives off minimal power; the maximum output, the nominal value of the turbine, is reached at wind speeds of 13 – 15 m/s. Above 25 m/s, the blades of the wind turbine, and consequently all turbines of a wind farm, must be stopped to prevent structural damage to the unit. Individual turbine rotors in wind farms need to be spaced at such intervals from each other that "shading" by neighboring turbines is below 10 % and the total coefficient of performance of the entire farm does not decrease below 90 %.

Extensive research has shown that even for windy locations the total average wind power performance is only around 20 – 25 %. However, wind power generation will still be of significance in the future, especially for those countries that are not situated in the world's sun belt but possess long coastlines with higher wind velocities. This is particularly the case for coastal regions above or below 50 ° to 55 ° latitude.

Bild 319 zeigt analog zu den vorigen Bildern eine Windenergierose, die den Primärertrag in kWh/m²a angibt. Für jedes Windrad bestehen eine spezifische Leistungskurve sowie eine Ce-Kennlinie, die in **Bild 320** wiedergegeben ist. Die Ce-Kennlinie entspricht in etwa dem Druckbeiwert, d.h. anfänglich steigt der Ce-Wert mit größer werdender Umdrehung, um dann deutlich abzufallen. Wie der Leistungskurve (**Bild 320**) gut zu entnehmen ist, beginnt das Windrad erst minimale Leistungen bei einer Windgeschwindigkeit von 4 m/s abzugeben. Das Maximum der Leistungskurve wird im Regelfall zwischen 13 und 15 m/s Windgeschwindigkeit erreicht – die Nennleistung. Oberhalb 25 m/s muss das Windrad oder alle Windräder eines Windparks gestoppt werden, da die Gefahr der völligen Zerstörung der Windfarm besteht. Windräder in größeren Windfarmen müssen räumlich so weit auseinandergestellt werden, dass die „Abschattung" der Windräder untereinander nicht größer als 10 Prozent wird und somit nicht der gesamte Parkwirkungsgrad unter
90 Prozent fällt.

Aufgrund von umfangreichen Messreihen lässt sich feststellen, dass selbst in windreichen Gegenden der mittlere Wirkungsgrad von Windparkanlagen zwischen 20 und 25 Prozent liegt. Gleichwohl wird der Einsatz der Windenergie in der Zukunft eine erhebliche Rolle spielen, da vor allem die Länder davon profitieren, die in weniger sonnenreichen Gegenden liegen, jedoch mit langen Küstenlinien gesegnet sind. Dies trifft insbesondere für küstenreiche Länder in den Bereichen oberhalb/unterhalb ca. 50 ° bis 55 ° Längengrad zu.

13.2.5 Geothermal energy

Except for the northern country of Iceland, geothermal power plant applications, typically ranging from 3,000 to 6,000 m in depth, are not yet implemented widely in most countries, including Germany. This is unfortunate and is mainly related to the still-inexpensive availability of fossil fuel sources. If we compare the potential of all regenerative heat and cooling for Germany, the following situation can be determined:

- solarthermal approximately 500 x 10^6 MWh
- biomass approximately 360 – 530 x 10^6 MWh
- geothermal >1,540 x 10^6 MWh

This would cover today's heat energy requirements, including 100 % supply of heat, and cooling energy, for all buildings in the country.

In geothermal power plants, two deep boreholes are drilled to capture the necessary high soil temperatures of the Earth's crust. **Figure 321** gives temperature approximations for locations in Central Europe, and as can be seen in the image, soil temperature increases by 3 K with every 100 m in depth. **Figure 322** shows various depths of so-called deep geothermal applications. If, at a given location, sufficiently high temperatures cannot be achieved even at a depth of 3,000 meters, a lesser deep borehole of around 2,000 meters may provide 60 °C warm ground water. Because this temperature range is too low for the operation of final energy consuming systems, it can be used in geothermal water-source heat pumps, which are capable of raising the temperature of the water in the supply grids significantly. **Figure 323** shows a diagram of a geothermal heat pump system using the thermal ground water with a supply temperature of 60 °C and cooling it to 20 °C. Heat pumps work on the principle of a two-stage absorption chiller (a two-stage evaporator and a two-stage absorber). Inside the evaporators, the thermal water is being cooled down by evaporation of distilled water to 20 °C. In two separate absorbers, the refrigerant vapor is then absorbed by a lithium-bromide brine solution. To achieve a closed refrigerant cycle, the refrigerant-depleted solution then returns, via a throttling device, to the absorber, where it is heated up to 150 °C and concentrated and developing water vapor that is cooled by the local district heating grid water. To increase cooling, the absorption chiller pump is equipped with two internal heat exchangers, one of which pre-heats the weak solution on its way from the absorber to the generator, while the other cools the strong solution on its way from the generator to the absorber. The heat power given off to the grid is calculated as the sum of the heat exchanger power, the cooling power of the heat pump, and the power of the necessary pumps, minus the thermal losses inside the system.

13.2.5 Geothermie

Dass geothermische Anlagen (3.000 – 6.000 m Tiefe) bisher in faktisch allen Ländern (ausgenommen Island) in so geringem Umfang wie in Deutschland zum Einsatz gekommen sind, ist nur dadurch zu erklären, dass die fossilen Brennstoffe zu kostengünstig waren. Vergleicht man die Potenziale regenerativer Wärme (und Kühlung) in Deutschland, so betragen diese in etwa:

- Solarthermie ca. 500 x 10^6 MWh
- Biomasse ca. 360 – 530 x 10^6 MWh
- Geothermie >1.540 x 10^6 MWh

Dies entspricht dem heutigen Wärmebedarf bzw. der Wärmeenergie, die zur hundertprozentigen Wärmeversorgung und Kühlung aller Gebäude notwendig wird.

Bei geothermischen Kraftwerken werden zwei tiefe Bohrungen niedergebracht, die jeweils so angelegt sein müssen, dass ausreichend hohe Temperaturen dem Erdreich entzogen werden können. **Bild 321** zeigt das grobe Temperaturprofil in der Erdkruste für Mitteleuropa. Wie das Bild darstellt, nimmt je etwa 100 m Tiefe die Erdreichtemperatur um ca. 3 K zu. **Bild 322** zeigt verschiedene Systeme der Tiefengeothermie. Kann man in der Region, die mit Wärme versorgt werden soll, selbst in größeren Tiefen keine ausreichend hohen Temperaturen vorfinden, so reicht es unter Umständen, Bohrungen bis ca. 2.000 m niederzubringen, um ca. 60-grädiges Grundwasser zu nutzen.
Da die hieraus resultierenden Temperaturen des Heizmediums zur Versorgung von Endenergieverbrauchern zu gering sind, werden geothermische Wärmepumpensysteme eingesetzt, die die Wärmeversorgungsnetze in ihrer Temperatur deutlich anheben. **Bild 323** zeigt ein Anlagenschema einer geothermischen Wärmepumpenanlage, die Thermalwasser mit einer Vorlauftemperatur von 60 °C auf 20 °C entwärmt. Die Wärmepumpentechnik entspricht der einer Absorptionswärmepumpe, zweistufig (zweistufiger Verdampfer/zweistufiger Absorber). In den Verdampfern wird Thermalwasser in zwei Stufen durch Verdampfung von destilliertem Wasser im Inneren der Wärmepumpe auf 20 °C abgekühlt. Der in den Verdampfern erzeugte Kältemitteldampf wird in getrennten Absorbern von einer konzentrierten Lithium-Bromid-Lösung absorbiert. Um einen geschlossenen Kältemittelkreislauf zu erreichen, wird die Lösung kontinuierlich in einem Austreiber bei 150 °C aufkonzentriert, wobei reiner Wasserdampf entsteht, der mit dem Fernwärmewasser gekühlt wird. Zur Verbesserung der Kältezahl ist die Absorptionswärmepumpe mit zwei internen Wärmeaustauschern ausgerüstet, in denen die Kältemittellösung auf ihrem Weg von den Absorbern zu den heißen Austreibern vorgewärmt bzw. auf dem Rückweg abgekühlt wird. Die an das Netz abgegebene Wärmeleistung errechnet sich aus der Addition der Wärmetauscherleistungen, der Kälteleistung der Wärmepumpe sowie der Antriebsleistung der Pumpen abzüglich der Wärmeverluste innerhalb des Systems.

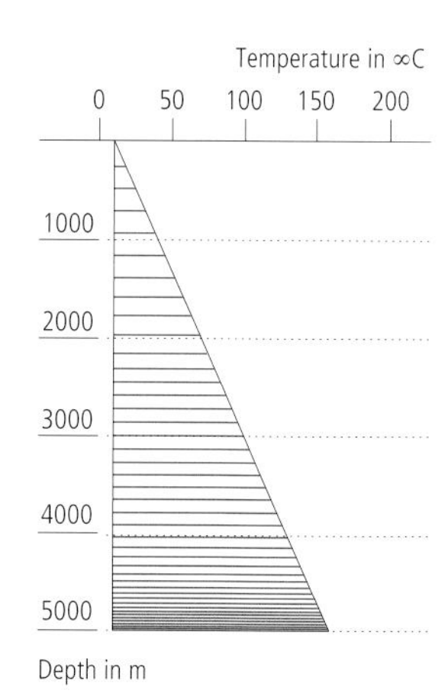

Figure 321
Temperature profile of lithosphere
In Central Europe thermal temperatures increase by 3 K per 100 m of depth. In regions with so-called "geothermal anomalies", thermal temperatures increase more rapidly.

Source:
Low Tech-Light Tech-High Tech, Klaus Daniels
Birkhäuser Verlag, 1998

Bild 321
Temperaturprofil in der Erdkruste
In Mitteleuropa nimmt die Erdtemperatur pro 100 m Tiefe durchschnittlich um 3 K zu.
In Regionen mit sogenannten „geometrischen Anomalien" steigt sie noch wesentlich schneller an.

Quelle:
Low Tech-Light Tech-High Tech, Klaus Daniels
Birkhäuser Verlag, 1998

Figure 322
Methods of heat energy generation with geothermal systems

Source: Geothermie, Unterhaching GmbH & Co KG, Munich, Germany

Bild 322
Methode der Wärmeförderung aus Geothermie, geothermische Systeme

Quelle: Geothermie, Unterhaching GmbH & Co KG, München, Germany

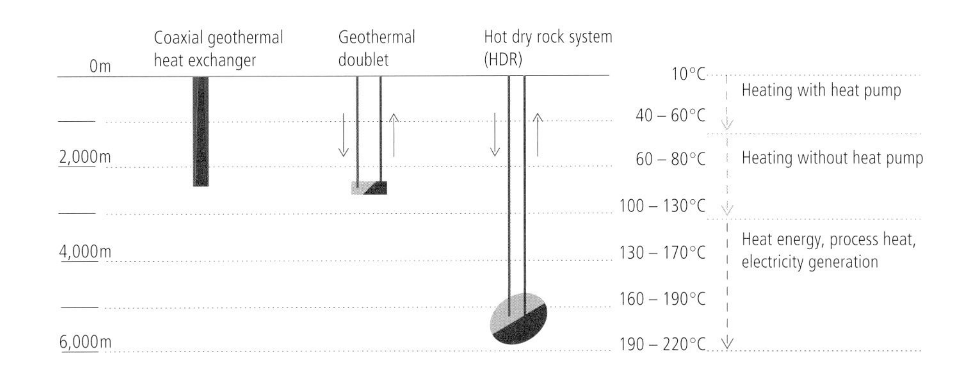

Figure 323
Diagram of geothermal heat pump installation
Entropie S.A./Rome

Bild 323
Anlagenschema einer Thermalwasser-Wärmepumpe
Entropie S.A./Rome

- Thermal water
- District heat
- Hot water
- Steam
- Condensation
- Coolant/-mixture

1 Generator
2 Condenser
3 Heat exchanger
4 Absorber
5 Evaporator

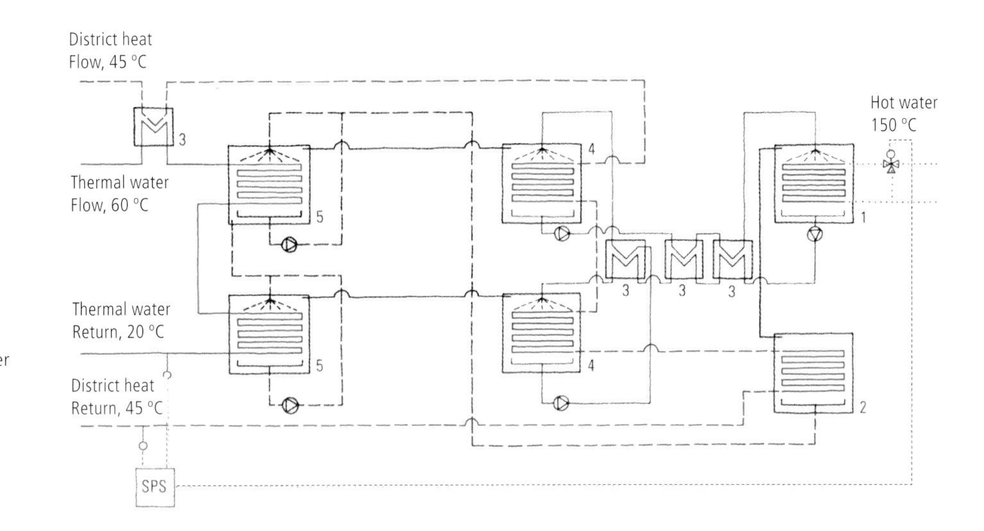

Parameters of temperatures and mass flux in a geothermal heat pump, annual readings

Parameterdarstellung der Temperaturen und Massenströme einer geothermischen Wärmepumpe gemäß Jahresdauerlinie

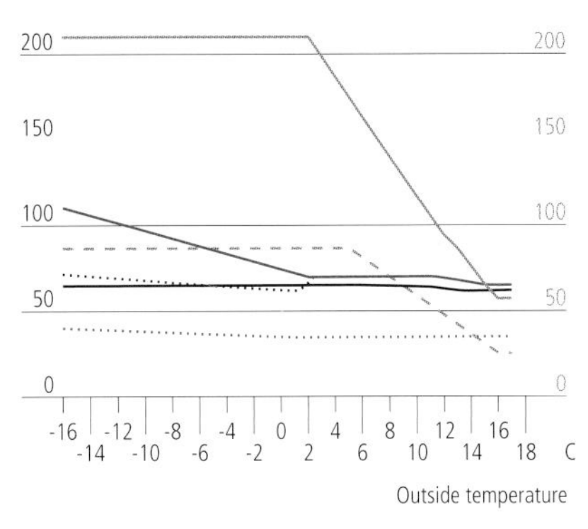

- Exit (HP)
- District heat, recirculated
- District heat, transmitted
- Inlet temperature TW to HE
- District heat volume
- Thermal water volume

Efficiency curves of geothermal heat pump installation, annual readings (cooling of thermal water to 20 ± 3 °C)

Recirculated district heat < 40 °C

Leistungskurven einer geothermischen Wärmepumpnanlage gemäß Jahresdauerlinie (Auskühlung des Thermalwassers auf 20 ± 3 °C)

Fernwärmerücklauf < 40 °C

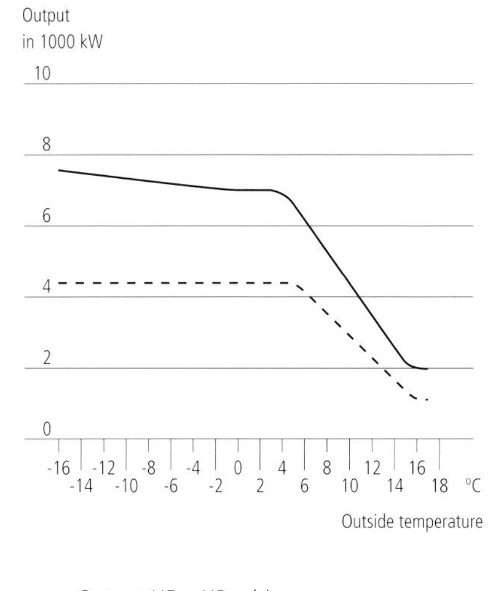

- Output HE + HP +drive
- Output from geo(thermal) heat

The power of the absorption heat pump is controlled by the value of the supply temperature of the district heating grid water, which ranges between 65 °C and 110 °C. **Figure 324** and **Table 20** provide information with regard to the planned annual cumulative power demand of a large plant and the actual operating conditions.

Die Leistungsregelung der Absorptionswärmepumpe erfolgt über die Führungsgröße Austrittstemperatur des Fernwärmewassers (min. 65 °C/max. 110 °C). **Bild 324** und **Tabelle 20** geben Auskunft über die angestrebte Jahresdauerlinie sowie die tatsächlichen Betriebszustände einer Großanlage.

		Thermal water (TW)			District heat (DH)			Heat-pump technology (HP)				Peak load boiler (PB)		
Outside temperature	Output	Incoming TW	Exit HP TW	TW mass flow	DH return	Exit HP DH	DH mass flow	Cold energy HP	Generator	Heat output HP + HE	Cooling factor	Output geo	Output PB	DH flow
[°C] t_A	[MW] Q_{Netz}	[°C] $T_{TW \cdot WT_e}$	[°C] $T_{TW \cdot WP_a}$	[kg/s] m_{TW}	[°C] T_{FWR}	[°C] $T_{FW \cdot WP_a}$	[kg/s] m_{FW}	[kW] Q_K	[kW] Q_A	[kW] Q_{WP+WT}	η_k	[kW] Q_{Geo}	[kW] Q_{SP}	[°C] T_{FWV}
-16	17,0	65,0	20	24,0	40,0	72	58,0	2.307	3.353	7.668	0,69	4.414	9.320	110
-8	13,2	65,0	20	24,0	37,5	68	58,0	2.057	3.023	7.337	0,68	4.414	5.874	92
-3	10,8	65,0	20	24,0	35,9	65	58,0	1.900	2.816	7.131	0,67	4.414	3.719	81
-1	9,9	65,0	20	24,0	35,3	64	58,0	1.837	2.733	7.047	0,67	4.414	2.861	76
6	6,6	65,0	20	22,8	35,0	70	45,2	1.703	2.556	6.613	0,67	4.157	0	70
12	3,8	63,0	20	13,1	35,0	70	26,2	989	1.615	3.778	0,61	2.262	0	70
16	2,0	62,0	20	6,3	39,9	65	19,1	600	1.102	2.000	0,54	998	0	65

- t_{Ou} Outside temperature
- Q_{Net} Output
- $T_{TW \cdot HE_i}$ Temperature thermal water heat exchanger, inlet
- $T_{TW \cdot HP_e}$ Temperature thermal water heat pump, exit
- m_{TW} Mass flow thermal water
- T_{DHR} District heat recirculated temperature
- $T_{DH \cdot HP_e}$ Temperature district heat heat pump, exit
- m_{DH} Mass flow thermal water
- Q_C Cooling output, heat pump
- Q_G Generator output, heat pump
- Q_{HP+HE} Total geothermal heat output
- η_k Cooling factor
- Q_{Geo} Output geo
- Q_{PB} Output peak boiler
- T_{DH} District heat temperature

		Thermalwasser			Fernwärme			Wärmepumpentechnik				Spitzenkessel		
Außentemperatur	Leistung	Eintritt TW	Austritt WP TW	TW-Massenstrom	FW-Rücklauf	Austritt WP FW	FW-Massenstrom	Kälte WP	Austreiber	Heizleistung WP + WT	Kältezahl	Leistung Geo	Leistung SP	FW-Vorlauf
[°C] t_A	[MW] Q_{Netz}	[°C] $T_{TW \cdot WT_e}$	[°C] $T_{TW \cdot WP_a}$	[kg/s] m_{TW}	[°C] T_{FWR}	[°C] $T_{FW \cdot WP_a}$	[kg/s] m_{FW}	[kW] Q_K	[kW] Q_A	[kW] Q_{WP+WT}	η_k	[kW] Q_{Geo}	[kW] Q_{SP}	[°C] T_{FWV}
-16	17,0	65,0	20	24,0	40,0	72	58,0	2.307	3.353	7.668	0,69	4.414	9.320	110
-8	13,2	65,0	20	24,0	37,5	68	58,0	2.057	3.023	7.337	0,68	4.414	5.874	92
-3	10,8	65,0	20	24,0	35,9	65	58,0	1.900	2.816	7.131	0,67	4.414	3.719	81
-1	9,9	65,0	20	24,0	35,3	64	58,0	1.837	2.733	7.047	0,67	4.414	2.861	76
6	6,6	65,0	20	22,8	35,0	70	45,2	1.703	2.556	6.613	0,67	4.157	0	70
12	3,8	63,0	20	13,1	35,0	70	26,2	989	1.615	3.778	0,61	2.262	0	70
16	2,0	62,0	20	6,3	39,9	65	19,1	600	1.102	2.000	0,54	998	0	65

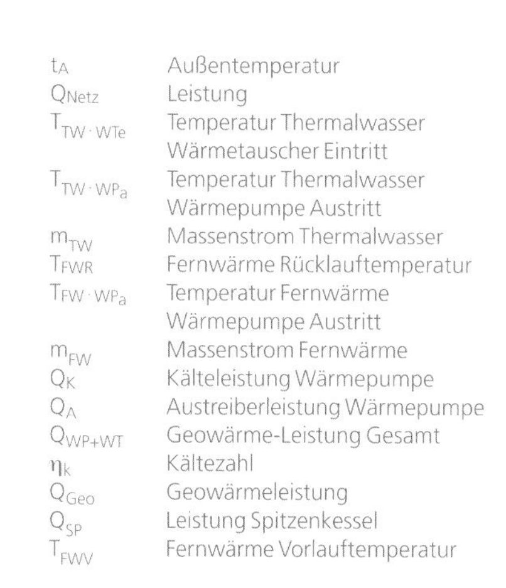

- t_A Außentemperatur
- Q_{Netz} Leistung
- $T_{TW \cdot WTe}$ Temperatur Thermalwasser Wärmetauscher Eintritt
- $T_{TW \cdot WP_a}$ Temperatur Thermalwasser Wärmepumpe Austritt
- m_{TW} Massenstrom Thermalwasser
- T_{FWR} Fernwärme Rücklauftemperatur
- $T_{FW \cdot WP_a}$ Temperatur Fernwärme Wärmepumpe Austritt
- m_{FW} Massenstrom Fernwärme
- Q_K Kälteleistung Wärmepumpe
- Q_A Austreiberleistung Wärmepumpe
- Q_{WP+WT} Geowärme-Leistung Gesamt
- η_k Kältezahl
- Q_{Geo} Geowärmeleistung
- Q_{SP} Leistung Spitzenkessel
- T_{FWV} Fernwärme Vorlauftemperatur

Table 20
Parameters of a thermal water heat pump as an example of heat pump in Figure 323 (District water return temperature < 40 °C)

Tabelle 20
Parameterliste einer Themowasser-Wärmepumpe (beispielhaft für WP gemäß Bild 323) (Rücklauftemperatur Fernwärme <40 °C)

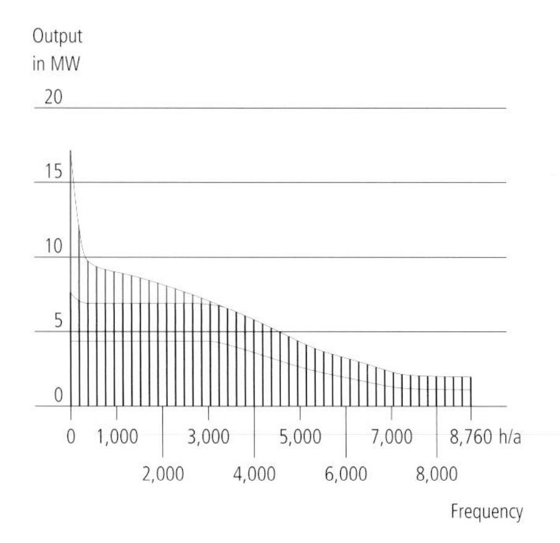

||| Output HE + Hp + generator + PB
||| Output HE + Hp + generator
||| Output HE + Hp

HE Heat exchanger (geothermal)
HP Heat pump generator hot water
PB Peak boiler (gas driven)

Figure 324
Annual output characteristics of the district heat supply in Erding (geothermal water cooled to 20 ± 3 °C, temperature of recirculated district heat < 4 °C)

Bild 324
Jahresdauerlinie Fernwärmeversorgung Erding (Auskühlung des Thermalwassers auf 20 ± 3 °C, Rücklauftemperatur Fernwärme < 4 °C)

As early as 1988, the "Atlas of Geothermal Resources in the European Union, Austria, and Switzerland," by Ralph Hänel and E. Staroste, provided volumetric heat content models for porous reservoirs assuming exploitation of geothermal energy. **Figure 325** shows the potential electricity generation for Germany as an excerpt. **Figure 326.1** shows the geological maps of Bavaria in Southern Germany and the corresponding geological cross section (**Figure 326.1**). For example, in the geographical region of the upper Jura there exists a water-carrying geological layer that extends down to great depth. There, the groundwater is heated up to at least 130 °C, which is usable for energy generation. **Figure 327** shows the geothermal heat energy potential with lines of equal energy amount extraction potential (in PJ).

Bereits 1988 wurde der Atlas der geothermischen Ressourcen in der Europäischen Gemeinschaft, Österreich und Schweiz herausgegeben (Hänel, R. und Staroste, E. „Atlas of Geothermal Resources in the European Community, Austria and Switzerland" – Hannover).

Auszugsweise sind im **Bild 325** die Potenziale zur Stromerzeugung in Deutschland dargestellt. **Bild 326.1** zeigt eine geologische Karte von Süddeutschland (Bayern) sowie den zugehörigen Schnitt, **Bild 326.2**. Im Bereich des oberen Jura und oberhalb des Grundgebirges befindet sich in dieser Region ein Malm mit einer Wasser führenden Schicht, die in große Tiefen hinabreicht. Hier erwärmt sich Wasser auf mindestens 130 °C, das zur Energieversorgung genutzt werden kann. **Bild 327** zeigt ein geothermisches Wärmepotenzial mit Linien gleicher extrahierbarer Energiemenge (in PJ).

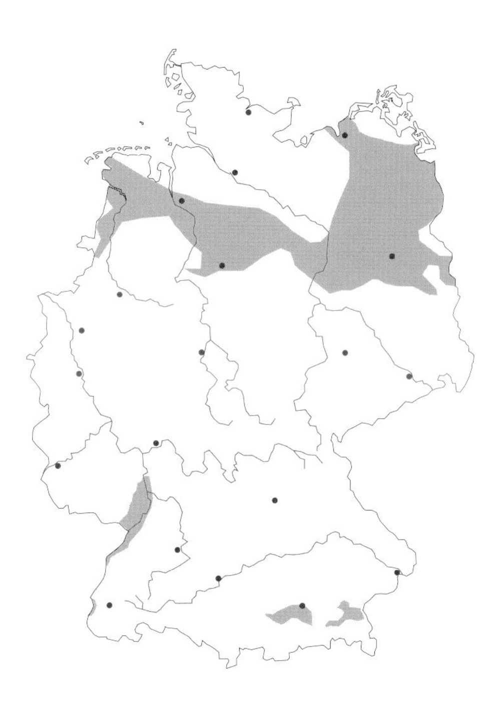

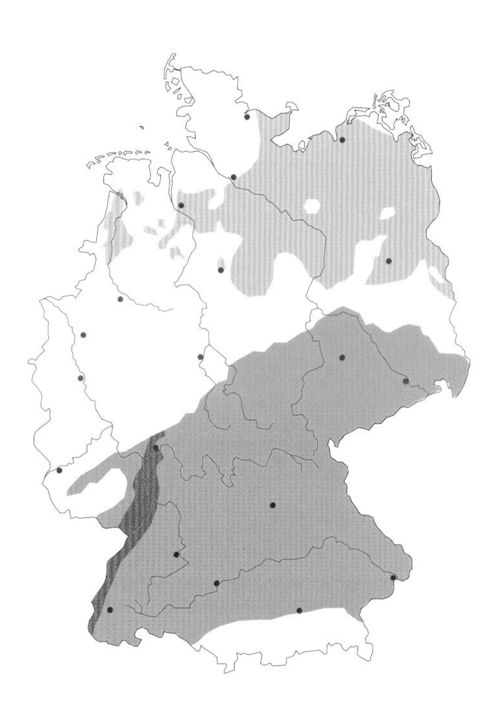

Figure 325
Potential for electricity generation with geothermal power plants in Germany.

Source: Geothermie, Unterhaching, Germany

Bild 325
Potenziale zur Stromerzeugung

Quelle: Geothermie, Unterhaching, Deutschland

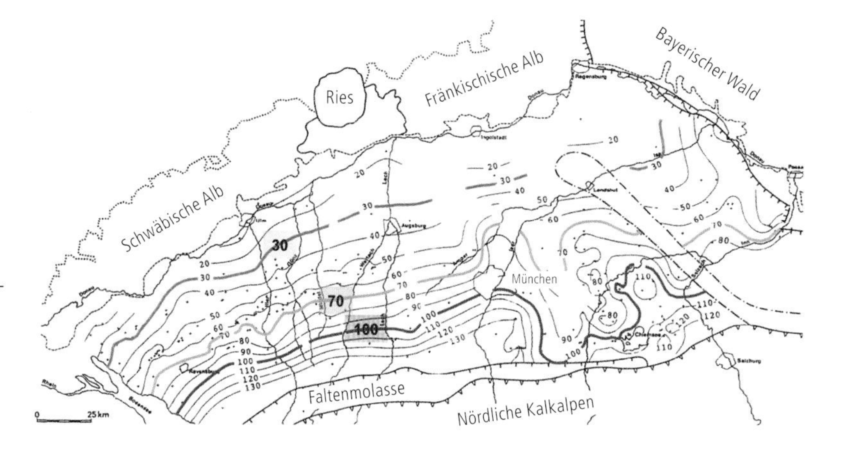

Figure 326.1
Temperature at Malm in Southern Germany, upper layer.

Source: GGA Leibniz Institute for Applied Geosciences GGA Institute

Bild 326.1
Temperatur an der Malm – in Süddeutschland, Oberkante

Quelle: GGA Leibniz Institut

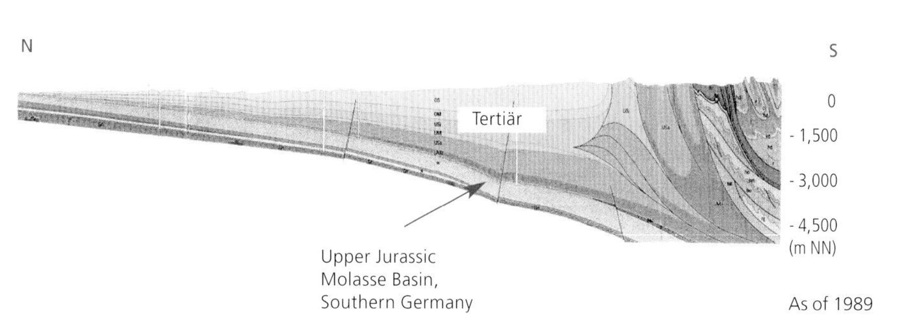

Figure 326.2
Cross section through the Bavarian molasse basin, Southern Germany

Source: GGA Leibniz Institute for Applied Geosciences GGA Institute

Bild 326.2
Schnitt durch die bayerische Molasse

Quelle: GGA Institut

In geothermal applications, extensive evaluations and assessments are carried out, especially using the method of seismic testing that result in reflections from below. For example, in the case of tests done in Unterhaching near Munich, Germany, test boreholes were drilled to a depth of 3,350 meters, where a water-carrying limestone layer, the so-called molasse basin, was discovered.

As seen in **Figure 328.1** the exemplary geothermal plant, a Kalina cycle type, is constructed in such a way that it generates both hot water and electricity from the warm groundwater. **Figure 328.2** shows the plant part that is used for electricity generation with the Kalina cycle, driven by the thermally active zone in the soil.

Nach ausgiebigen Voruntersuchungen insbesondere mit Reflexionsseismik werden Bohrungen niedergebracht, die in einer erwarteten Tiefe (z.B. ca. 3.350 m) eine wasserführende Kalksteinschicht erreicht (Molassebecken, Unterhaching/München).

Wie **Bild 328.1**. zeigt, ist die beispielhafte Geothermieanlage (Kalina-Anlage) so aufgebaut, dass das heiße Wasser aus dem Untergrund sowohl der Stromerzeugung als auch der Wärmeproduktion dient. **Bild 328.2** zeigt das eigentliche Kraftwerk zur Stromerzeugung mit Kalina-Dampfkreislauf, angetrieben aus der geothermisch aktiven Zone im Untergrund.

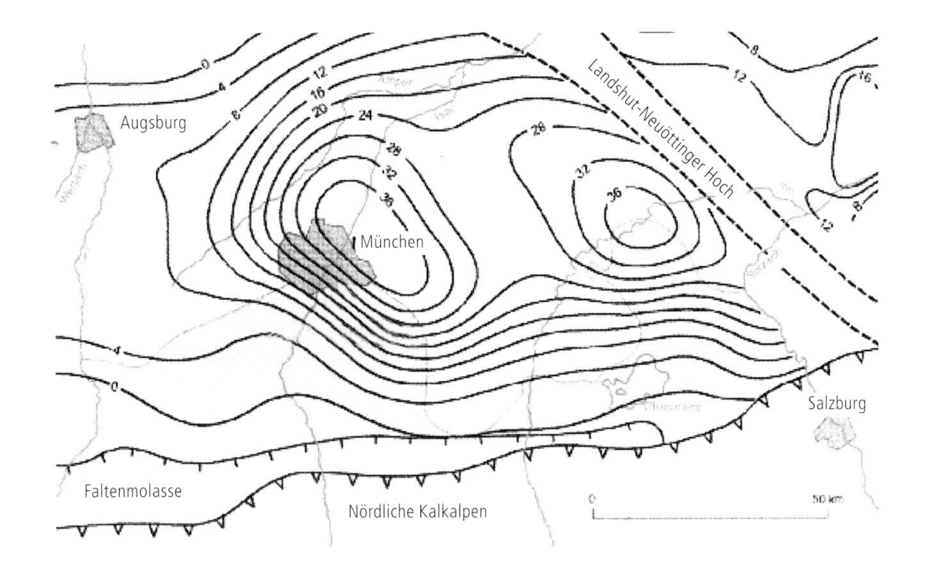

Temperature class 100 – 130 °C
–4– Lines of equal energy amount extracted in PJ, doublet operation (re-injection temperature 15 °C)

Figure 327
Map showing levels of equal thermal energy extraction capacities between Munich, Germany, and Salzburg, Austria

Source:
Leibniz Institute for Applied Geosciences
GGA Leibniz Institute

Bild 327
Karte mit Linien gleicher extrahierbarer Energiemenge im Großraum zwischen München und Salzburg, höffige Gebiete

Quelle: GGA Leibniz Institut

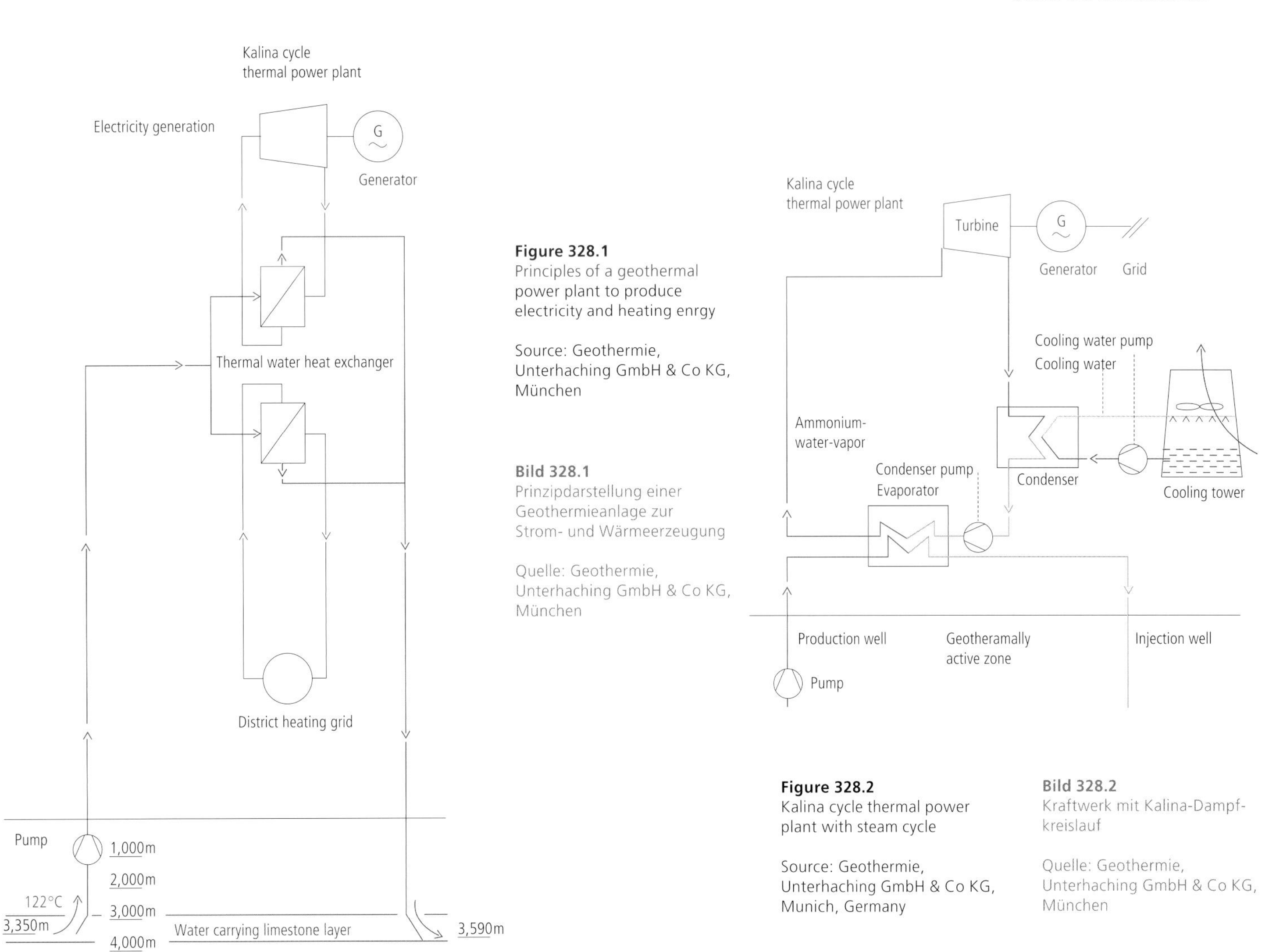

Figure 328.1
Principles of a geothermal power plant to produce electricity and heating enrgy

Source: Geothermie, Unterhaching GmbH & Co KG, München

Bild 328.1
Prinzipdarstellung einer Geothermieanlage zur Strom- und Wärmeerzeugung

Quelle: Geothermie, Unterhaching GmbH & Co KG, München

Figure 328.2
Kalina cycle thermal power plant with steam cycle

Source: Geothermie, Unterhaching GmbH & Co KG, Munich, Germany

Bild 328.2
Kraftwerk mit Kalina-Dampf-kreislauf

Quelle: Geothermie, Unterhaching GmbH & Co KG, München

The shown cooling of the turbine cycle (heat extraction from the condenser) can be rejected in summer by a cooling tower, while in winter it should be used again for warm water generation. In a Kalina cycle plant, the turbine cycle is operated by ammonium water vapor to achieve high heat transfer and to avoid at the same time isothermal evaporation. Such a process allows for high energy exploitation from the ground, even when seasonal temperature fluctuations exist. Such plants are already in operation in Japan and Iceland. **Figure 329** shows the geothermally operated heat plant, which is easily conceivable for urban environments as well.

In conclusion, borehole drilling cost as a function of depth is shown in **Figure 330** (October 2006). The data presented explain why less-deep applications are preferable. On the other hand, the hot-dry-rock (HDR) geothermal energy process, in which high-pressure water is pumped down to very hot rocks just a few kilometers below ground, may eventually be used in many parts of the world.

Die im Bild gezeigte Rückkühlung des Turbinenkreislaufs (Kondensatorwärmeentzug) kann im Sommer über ein Rückkühlwerk erfolgen und sollte im Winter wiederum der Wärmenutzung dienen. Bei der Kalina-Anlage wird der Turbinenkreislauf mit Ammoniak-Wasserdampf betrieben, um eine hohe Wärmeübertragungsfähigkeit zu erreichen und isotherme Verdampfungen zu vermeiden. Die Anpassung dieses Verfahrens erlaubt bei den geförderten Grundwassertemperaturen und bei saisonalen Temperaturunterschieden eine erhöhte Energieausbeute (vergleichbare Anlagen bestehen in Island und Japan). **Bild 329** zeigt das geothermisch betriebene Heizkraftwerk, und es lässt sich leicht vorstellen, dass dieses auch in einem städtischen Bereich noch gut darstellbar ist.

Die Bohrkosten als Funktion der Teufe sind abschliessend in **Bild 330** ausgewiesen (Stand Oktober 2006). Aus diesem Bild lässt sich leicht erkennen, warum man versucht, in nicht allzu große Tiefen vorzudringen. Andererseits ist jedoch das Hot-Dry-Rock-Verfahren bei entsprechendem Investitionseinsatz an vielen Stellen der Welt umsetzbar, und es kann davon ausgegangen werden, dass entsprechende Systemlösungen zunehmend zum Einsatz kommen.

Figure 329
Geothermal power plant

Source: GGA Leibniz Institute for Applied Geosciences GGA Leibniz Institute

Bild 329
Kraftwerksgebäude einer Geothermieanlage

Quelle: GGA Leibniz Institut

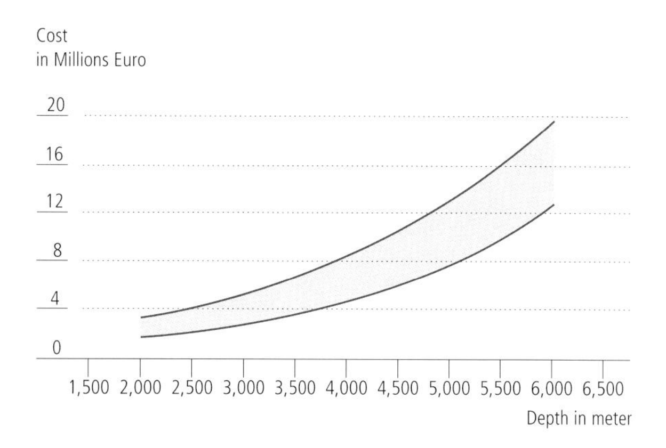

Figure 330
Cost of boreholes as a function of depth
The estimate is depending on soil condition and borehole diameter at the specific location and market availability of drilling equipment.

Source: IDEAS, Peine; Erdöl-, Erdgasfirmen; LBEG
October 2006

Bild 330
Bohrungskosten (in Mio Euro) als f (Teufe)
Grobabschätzung
- extrem gebirgs- und durchmesserabhängig (standortbezogen)
- Angebot- und Nachfragesituation von Bohranlagen

Quelle: IDEAS, Peine; Erdöl-, Erdgasfirmen; LBEG
Stand: Oktober 2006

13.2.6 Use of the energy of the oceans Tidal energy Wave energy

Tides are the result of gravitational forces of the sun and the moon onto the Earth's oceans. If tidal energy is used for energy generation, it has the advantage that low and high tides can be precisely calculated and predetermined – i.e., in contrast to solar and wind power, they are entirely independent of outside influences. The origin of the technology to use the tides for energy generation goes back to an experiment in which a bay was separated from the open sea by a dam in which wheel turbines were integrated. During high tide, the seawater flowed from the ocean side across the turbines and into the closed-off bay, and at low tide the flow was reversed, again turning the turbines. The captured kinetic energy of the water flow was then converted into electricity.

The world's largest tidal power plant is currently being constructed near Seoul, South Korea. The Sihwa Tidal power plant, which will start the generation of electricity in 2009, has a power of 254 MW, resulting in an annual power output of approximately 552 GWh/a. The plant is situated in the center of a 12-km-long dam that closes off a wide, elongated bay. Ten low-pressure turbines are driven by the high tides of the ocean (**Figure 331**), and during low tide the bay water will be released back into the ocean directly via large locks in order to improve water quality regeneration. The design was prepared by the Korea Ocean Research and Development Institute (KORDI) and executed by the South Korean industrial conglomerate Daewoo. The turbine shown in **Figure 331** contains helical rotor blades, which need to be exposed to a tidal change of at least 5 m in order to perform efficiently. In Korea, tidal power plants are currently capable of contributing 5 % to the country's total power generation.

13.2.6 Nutzung der Meeresenergie Gezeitenenergie Wellenenergie

Die Gezeiten werden von der Gravitationswirkung der Sonne und des Mondes auf das Meer ausgelöst. Der entscheidende Vorteil dabei ist, dass Ebbe und Flut berechenbare Größen und permanent vorhanden sind und im Gegensatz zu Sonnen- und Windenergie vollkommen unabhängig von äußeren Gegebenheiten zur Verfügung stehen. Ursprünglich wurde zum Einsatz eines Gezeitenkraftwerks eine Bucht durch einen Damm vom offenen Meer getrennt. In diesem Damm befinden sich auf dem Meeresgrund beidseitig ausgelegte Turbinen. Bei Flut strömt das Wasser von der Meerseite durch die Turbinen in den Staudamm. Die dabei gewonnene kinetische Energie (Bewegungsenergie) des Wasserstroms wird dabei in elektrische Energie umgewandelt. Bei Ebbe wird der große Tidenhub genutzt, um das durch den Staudamm aufgestaute Wasser von der Buchtseite zurück ins Meer laufen zu lassen – die Turbinen werden erneut angetrieben.

In Südkorea, ca. 40 km südwestlich von Seoul, wird zurzeit das größte Gezeitenkraftwerk der Welt, das Sihwa-Kraftwerk, gebaut, das ab 2009 eine elektrische Leistung von 254 MW erzeugen soll (ca. 552 GWh/a). Das Gezeitenkraftwerk liegt in der Mitte eines 12 km langen Damms, der eine langgestreckte Bucht abschließt. Zehn Niederdruckturbinen werden bei Flut vom Meerwasser angetrieben, **Bild 331**. Bei Ebbe wird das Meerwasser über Schleusen aus der Bucht direkt entlassen, um die Regeneration des Buchtwassers zu verbessern. Die Planung wurde von „KORDI" (Koreas Institut für Meeresforschung) vorbereitet und von der Fa. Daewoo weitergeführt und umgesetzt. Die in **Bild 331** gezeigte Turbine besitzt helixförmige Schaufeln, die zumindest einem Tidenhub von 5 m ausgesetzt sein müssen, um eine ausreichende Leistung zu erreichen. An den Küsten Südkoreas könnten Gezeitenkraftwerke einen Beitrag zum elektrischen Energiebedarf des Landes von 5 Prozent leisten.

Figure 331
Tidal power plant turbine with encapsulated generator

Quelle: MCE/ABB, Germany

Bild 331
Gezeiten-Turbinenanlage mit gekapseltemGenerator

Quelle: MCE/ABB, Deutschland

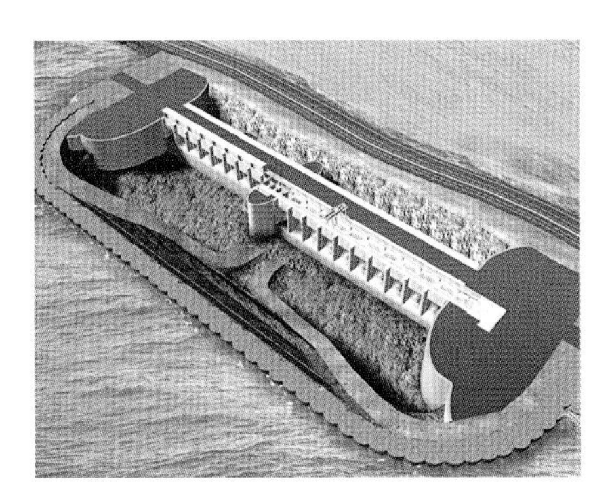

to Figure 331
TG Completition building, Eco park, 2006
Source: K water

zu Bild 331
Turbinengebäude (TG) für Gezeitenturbinenanlage, Eco Park, 2006
Quelle: K water

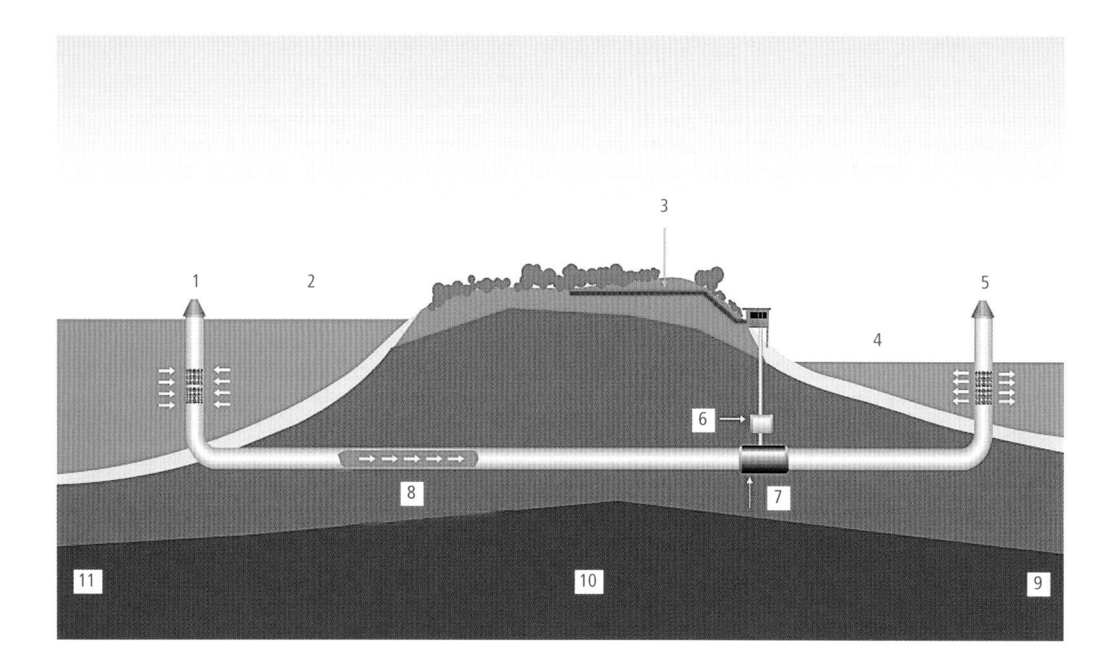

to Figure 331
Tidal power plant
Diagram

zu Bild 331
Arbeitsprinzip eines
Gezeitenkraftweres

1 Inlet/outlet
2 High tide on ocean side
3 Underground power cable
4 Low tide on bay side
5 Inlet/outlet
6 Generator
7 Turbine fitted within pipe
8 Higher level to lower level. Change of tide, situation reverses.
9 Bay
10 Peninsula or isthmus
11 Ocean

An entirely new technology for using tidal effects is power plants that are independent of tidal changes. These systems are placed directly on the ocean floor, and due to their slow rotational speed do not pose a threat to ocean flora and fauna. According to current research and assumptions by scientists, the flow energy of the ocean is capable of providing approximately 450 x 10^6 MWh of electricity per year, which is the equivalent of 40 large nuclear power plants.

The very first tidal power plant was constructed in 1967 in the estuary of the river Rance near Saint Malo, Brittany, France. It has a size of 240 MW. In 1984, another tidal power plant was opened in Annapolis Royal, a side bay of the Bay of Fundy in Nova Scotia, Canada. The 20 MW plant is mainly a research facility and generates electricity in a one-way direction during low tide. Smaller plants have been constructed in Russia and also in China, where the largest operating tidal power plant, located near the city of Jiangxia in the province of Zhejiang, was completed in 1986 and generates 3.2 MW of electricity.

Ein neuer technologischer Ansatz zur Nutzung der Gezeiten sind Turbinen, die nicht den Tidenhub, sondern auf direktem Weg die starken Strömungen am Meeresgrund nutzen. In Küstennähe sind diese besonders stark und die Bedingungen für den Bau eines Gezeitenkraftwerks damit optimal.

Ein positiver Aspekt der Technologie ist die sichtbare und ökologische Schonung des Küstengebiets, da die Turbinen auf dem Meeresgrund stehen und ihre relativ langsamen Drehgeschwindigkeiten keine negativen Auswirkungen auf den Bestand der Meerestiere und der Meeresfauna haben. Nach Expertenschätzungen kann die Strömungsenergie der Ozeane rund 450 x 10^6 MWh Strom pro Jahr liefern. Dies entspricht in etwa 40 großen Kernkraftwerken.

Das erste Gezeitenkraftwerk der Erde wurde 1967 an der Rance-Mündung in Saint Malo in der Bretagne errichtet (240 MW). 1984 wurde ein weiteres Gezeitenkraftwerk in Annapolis Royal an einer Nebenbucht der Bay of Fundy in Nova Scotia (Kanada) gebaut (20 MW). Dieses Gezeitenkraftwerk diente primär der Forschung und arbeitet im Ein-Richtungs-Betrieb (Nutzung des Ebbestroms). Weitere kleinere Gezeitenkraftwerke entstanden in Russland und China, wobei das größte chinesische Gezeitenkraftwerk bei Jiangxia (Provinz Zhejiang) 1986 fertiggestellt wurde (3,2 MW).

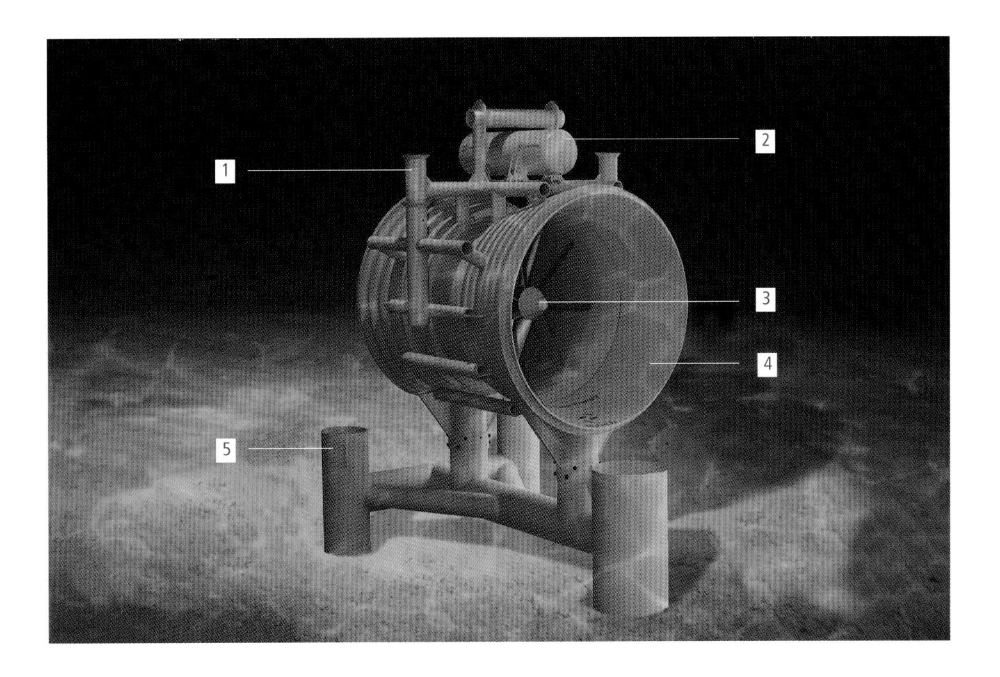

Figure 332.1
Bidirectional Rotating Tidal Turbine (RTT) of a tidal power plant
1 Megawatt unit
Duct diameter 15 meters
Duct length 19.2 meters
Turbine diameter 11.5 meters

Bild 332.1
Beispiel eines Gezeitenkraftwerkes

1 Megawatt Einheit
Durchmesser 15 m
Laenge 19,20 m
Turbinendurchmesser 11,50 m

1 Removable cassette
2 Generator modul
3 Turbine
4 Duct
5 Gravity base

On England's western coast, the currently largest tidal power plant using just the ocean's currents is being planned, with an expected power output of 8 MW. The concept involves turbines mounted on legs 10 meters tall that will be lowered to the ocean floor at a depth of 50 meters and fixated only by gravity. As shown in **Figure 332.1**, the turbines have a large cone guiding the water to a constriction where the liquid flow is accelerated. The average speed of the rotors will be just around 21 revolutions per minute, which, again, poses no environmental threat.

In the U.S., the Electric Power Research Institute (EPRI) is involved in the development of tidal energy plants, three of which three are planned for location in Canada and the west coast states of Washington and California. An additional four tidal plants are being designed for the U.S. east coast near the Canadian border.

Zurzeit geplant ist an der Westküste Englands das weltweit größte „Gezeitenkraftwerk", das nur die Meeresströmung nutzt, mit einer Leistung von 8 MW. Turbinen mit Standbeinen von ca. 10 m Höhe werden in 50 m Wassertiefe aufgestellt. Die Turbinen, **Bild 332.1**, besitzen am Anfang einen großen Einlauftrichter, der sich zur Mitte hin verengt, um den Wasserstrom innerhalb der Turbine zu beschleunigen. Die vorgesehene Drehgeschwindigkeit wird im Durchschnitt ca. 21 Umdrehungen pro Minute betragen, so dass infolge der langsamen Rotation ein sehr geringer Sog entsteht, der zu keinerlei Schäden in der Umgebung führt.

Die einzelnen Turbinen stehen, fixiert durch ihr Eigengewicht, auf dem Meeresboden.

In den USA beschäftigt sich zurzeit ebenfalls das Electric Power Research Institute (EPRI) mit der Entwicklung von Gezeitenkraftwerken, wovon drei in Kanada und an der Westküste (Staaten Washington und Kalifornien) und vier weitere an der Ostküste in der Nähe zur kanadischen Grenze liegen.

Table 21 provides an overview of the plants and presents their specific power data and cost.

	AK	WA	CA	MA	ME	NB	NS
Cross section area in m²	72,500	62,600	74,700	71,500	36,000	60,000	225,000
Power density in kWh/m²	1.6	1.7	3.2	0.95	2.9	0.7 – 2	4.5
Available power in MW	116	106	237	13.3	104	43 – 100	1,013
Extract power in MW	17.4	16	35.5	2.0	15.6	6.5 – 15	152
Unit power rated in MW	0.76	0.7	1.1	0.46	0.83	0.31	1.11
Unit speed rated in m/s	1.9	1.9	2.1	1.6	2.0	1.4	2.2
Unit avg. power yearly in MW	0.22	0.21	0.37	0.18	0.38	0.13	0.52
# of com units	66	64	40	9	12	66	250
Avg. power in MW	14.6	13.7	16.5	1.6	4.6	7.3	130
Total plant cost in Mio $	110	103	90	17	24	68	486
Yearly level O&M costs in Mio $	4.1	3.8	3.6	0.6	1.0	2.3	18
Annual energy in GWh	128	121	129	1,5	40	64	1.140
Utility gen. in US Cent/kWh	9,2 – 10,8	9,0 – 10,6	6,6 – 7,6	8,6 – 9,9	5,6 – 6,5	10,0 – 11,7	3,9 – 4,6
Muni. gen. COE US Cent/kWh	7,1 – 8,4	7,2 – 8,4	4,9 – 5,6	6,0 – 6,7	4,2 – 4,8	9,2 – 11,2	3,9 – 4,6

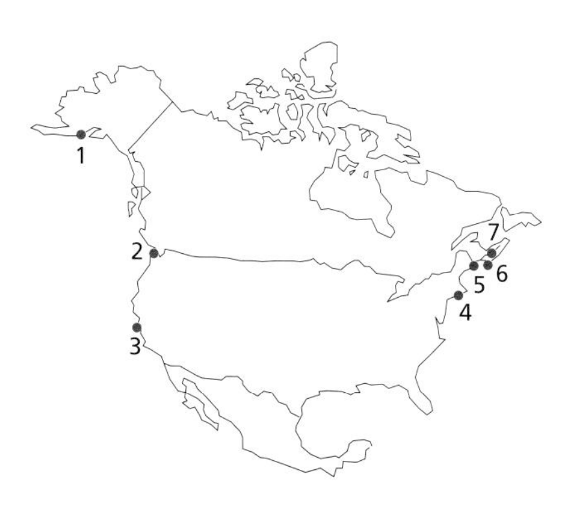

1 = AK Knik Arm, Alaska
2 = WA Tacoma Narrows, Washington
3 = CA Golden Gate, California
4 = MA Muskeget Channel, Massachusetts
5 = ME Western Passage, Maine
6 = NB Head Harbour Passage, New Brunswick, Canada
7 = NS Minas Passage, Nova Scotia, Canada

Table 21
Overview of experimental tidal power plants in the U.S. and Canada

Source: Tidal Sites Feasibility Study
Electric Power Research Institute (EPRI)

Tabelle 21 zeigt eine detaillierte Übersicht über die Anlagen mit ihren technischen Kenndaten sowie Kostenauswirkungen.

	AK	WA	CA	MA	ME	NB	NS
Querschnitt der Gesamt-Turbinen-fläche in m²	72.500	62.600	74.700	71.500	36.000	60.000	225.00
Energieleistung in kWh/m²	1,6	1,7	3,2	0,95	2,9	0,7 – 2	4,5
Erzielbare Energie-leistung in MW	116	106	237	13,3	104	43 – 100	1.013
Nutzbare Energie-leistung in MW	17,4	16	35,5	2,0	15,6	6,5 – 15	152
Energieleistung je Einheit in MW	0,76	0,7	**1,1**	0,46	**0,83**	0,31	1,11
Fließgeschwindigkeit der Einheit in m/s	1,9	1,9	**2,1**	1,6	**2,0**	1,4	2,2
Jährliche durch-schnittl. Leistung in MW	0,22	0,21	**0,37**	0,18	**0,38**	0,13	0,52
Anzahl der Einheiten	66	64	**40**	9	**12**	66	250
Durchschnittl. Gesamtleistung in MW	14,6	13,7	**16,5**	1,6	**4,6**	7,3	130
ca. Investitions-kosten in Mio $	110	103	**90**	17	**24**	68	486
Jährliche Betriebs-kosten (Betreiben, Warten, Reparieren) in Mio $	4,1	3,8	**3,6**	0,6	**1,0**	2,3	18
Durchschnittl. Energieertrag in GWh	128	121	**129**	1,5	**40**	64	1.140
Resultierende Stromkosten in US Cent/kWh	9,2 – 10,8	9,0 – 10,6	**6,6 – 7,6**	8,6 – 9,9	**5,6 – 6,5**	10,0 – 11,7	3,9 – 4,6
Derzeitige Strom-kosten in der Region in US Cent/kWh	7,1 – 8,4	7,2 – 8,4	**4,9 – 5,6**	6,0 – 6,7	**4,2 – 4,8**	9,2 – 11,2	3,9 – 4,6

1 = AK Knik Arm, Alaska
2 = WA Tacoma Narrows, Washington
3 = CA Golden Gate, California
4 = MA Muskaged Channel, Massachusetts
5 = ME Western Passage, Maine
6 = NB Head Harbor Passage, New Brunswick, Canada
7 = NS Minas Passage, Nova Scotia, Canada

Tabelle 21
Übersicht von Versuchsanlagen in den USA und Kanada

Quelle: Tidal Sites Feasibility Study
Electric Power Research Institute (EPRI)

It is also interesting that a variety of system solutions are being employed to optimize the cost-performance ratio. Different designs based on either horizontal or vertical rotor shafts are presented in **Figures 332**. Because research in this area is proceeding with great intensity, it can be said with some assurance that we will be able to see the emergence of the most effective systems as soon the next couple of years.

Weiterhin interessant ist, dass eine Vielzahl von unterschiedlichen Systemvarianten zur Anwendung kommen, um das Kosten-Nutzen-Verhältnis langfristig zu optimieren. Die verschiedenartigen Turbinen mit horizontalen und vertikalen Drehachsen sind in der **Bildserie 332** dargestellt. Wie den Bildern zu entnehmen ist, gibt es eine Vielzahl von Ansätzen, Gezeitenkraftwerke zu konzipieren – in einigen Jahren werden die besten Systeme endgültig zum Einsatz kommen, da auf dem Sektor der Gezeitenkraftwerke zurzeit sehr intensiv geforscht wird und Serienproduktionen erst in einigen Jahren endgültig einsetzen werden.

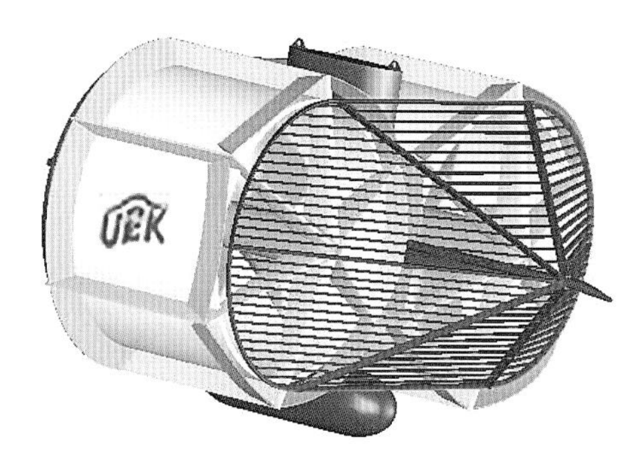

Figure 332.2
Floating double-turbine fixated by cables

Bild 332.2
Doppelturbinen-System, schwimmend – durch Trossen fixiert

	Lunar energy RTT
Axis type	H axis Duct
Diameter of the rotor	21 m dia
Rated power	2 MW

	MCT
Axis type	H axis Dual
Diameter of the rotor	18 m dia
Rated power	1,5 MW

Figure 332.3
MCT SeaGen prototype

Bild 332.3
MCT SeaGen Prototyp

Figure 332.4
Open Hydro Rim Drive Turbines

Bild 332.4
Open Hydro Rim Drive Turbinen

	Open hydro
Axis type	H axis Rim Gen
Diameter of the rotor	15 m dia
Rated power	1,5 MW

	Ver-dant
Axis type	H axis
Diameter of the rotor	5 m dia
Rated power	34 kW

Figure 332.5
Verdant Power RITE Turbine

Bild 332.5
Verdant Power RITE Turbine

to Figure 332.4
Open Hydro Rim Drive Turbine

zu Bild 332.4
Open Hydro Rim Drive Turbine

In the U.K. as well as other parts of the world, research regarding this renewable technology is also being conducted aggressively. **Figure 333** shows several regions of the U.K.'s surrounding seas with sufficiently strong currents for tidal power plants to be constructed. Those power plants will be located at water depths of 5 m to around 100 m. Unfortunately, thorough research has shown that ocean currents and their velocities are subject to notable changes. Of the currents that have been analyzed, 10 – 70 % have altered direction and speed over time, and for some tidal power plants that are oriented perpendicular to the current the resulting flow effect can only be used once – forward and back flow conditions are too unequal to generate energy twice, as was originally intended. **Figure 334** depicts the changes in flow velocities dependent upon the kinetic energy used for energy generation. It is astounding that variations in flow velocity seem to have a greater impact at a depth of 90 meters than at a depth of 65 meters.

As is the case with other renewable energy concepts, tidal power generation is a very complex topic, and much more research will be necessary to allow secure provision of energy from this significant natural source.

Auch in England wird in verschiedenen Regionen – wie bereits am Anfang beschrieben – intensiv geforscht, um zukünftig Gezeitenkraftwerke einsetzen zu können. **Bild 333** zeigt einige dieser Regionen, in Abhängigkeit von den Strömungsgeschwindigkeiten, die infolge der Gezeiten entstehen. Die Gezeitenkraftwerke liegen hier in Tiefen von 5 – 100 m. Infolge langjähriger Studien hat sich gezeigt, dass sich die Fließgeschwindigkeiten zum Teil über Jahre nicht unerheblich verändert haben (10 – 70 Prozent) und dass bei einigen Gezeitenkraftwerken die quer zu den Fließ-Stromlinien liegen der Strömungseffekt nicht doppelt, sondern nur einfach gerechnet werden kann, da offensichtlich die Hin- und Rückbewegungen des Wassers unterschiedlich ausfallen. In **Bild 334** sind die Veränderungen der Fließgeschwindigkeiten in Abhängigkeit von Energiegewinnung (kinetische Energie) dargestellt. Verblüffend dabei ist, dass in einer Tiefe von 65 m die Veränderung der Stromgeschwindigkeit nur einen geringen Einfluss hat, während der Einfluss in 90 m Tiefe deutlich stärker ist.

Wie erkennbar, ist auch das Gebiet der Energiegewinnung durch Gezeiten kein einfaches, sondern bedarf einer Vielzahl von Forschungen und Untersuchungen sowie Versuchen im Rahmen von Demonstrationsanlagen, um eine gesicherte Energiebereitstellung zu erreichen.

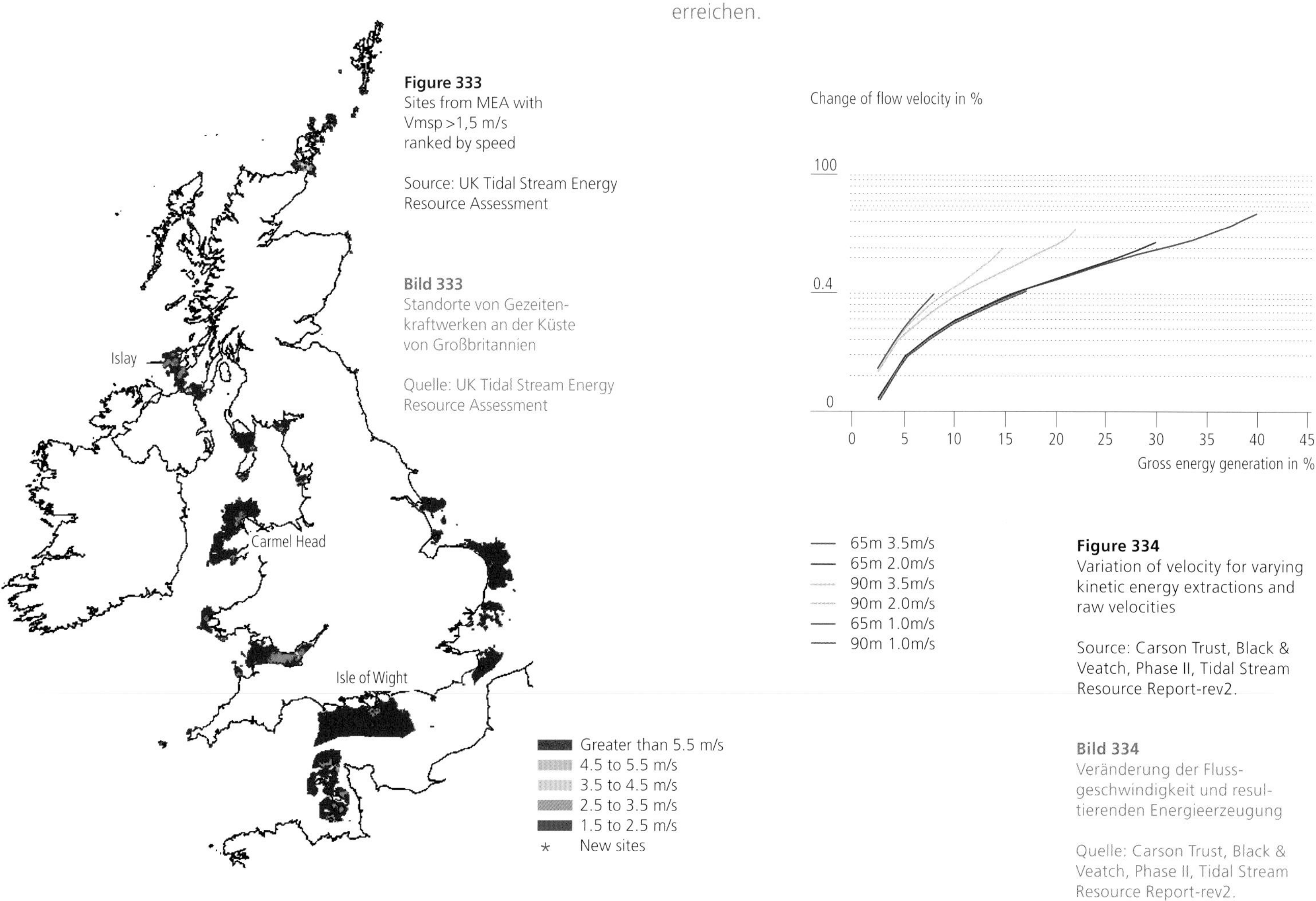

Figure 333
Sites from MEA with Vmsp >1,5 m/s ranked by speed

Source: UK Tidal Stream Energy Resource Assessment

Bild 333
Standorte von Gezeitenkraftwerken an der Küste von Großbritannien

Quelle: UK Tidal Stream Energy Resource Assessment

Figure 334
Variation of velocity for varying kinetic energy extractions and raw velocities

Source: Carson Trust, Black & Veatch, Phase II, Tidal Stream Resource Report-rev2.

Bild 334
Veränderung der Flussgeschwindigkeit und resultierenden Energieerzeugung

Quelle: Carson Trust, Black & Veatch, Phase II, Tidal Stream Resource Report-rev2.

Wave energy

For years, the company Pelamis Wave Power Limited in Edingburgh, Scotland, has been involved in research on wave energy converters and their manufacturing. Their system transforms the kinetic energy of the movement of ocean waves into electrical energy. **Figure 335** shows the system's unit, which has a length of 150 m, a diameter of 3.5 m, and a weight of around 700 tons. The device, which is semi-submerged, is composed of cylindrical sections linked by hinged joints and three power-conversion units containing hydraulic rams that pump a high-pressure fluid – agitated by the device's upward and downward motion, caused by waves – through hydraulic motors. The hydraulic presses work with velocities of 0 – 0.1 m/s and move the hydraulic liquid in the cylinders, resulting in high pressures of around 100 – 350 bar.

The conversion of pressure to kinetic energy takes place in two small turbines that are parts of the power-conversion units, which, in turn, drive two generators with a power of 125 kW each, turning at 1,500 rpm.

Wellenenergie

Seit Jahren beschäftigt sich die Firma Pelamis in Edinburgh (Schottland) mit Wellenenergiekonvertern, die über Bewegungsenergie elektrische Energie erzeugt. **Bild 335** zeigt die ca. 150 m lange und im Durchmesser 3,5 m große Gesamtanlage, die inklusive Ballast ca. 700 t wiegt. Innerhalb der Gesamtanlage sind 4 lange Teilstücke sowie 3 „Energieumwandlungs- und Sammeleinheiten" zu erkennen (power-conversion unit), die hydraulische Pressen (hydraulic rams) aufnimmt, wobei zwei beim Anheben der „power-conversion unit" und wiederum zwei beim Neigen derselben wirksam werden. Die Hydraulikpressen arbeiten mit einer Geschwindigkeit zwischen 0 – 0,1 m/s und pressen Öl durch die Bewegungen in Zylinder, wodurch ein Druck von ca. 100 – 350 bar entsteht.

Die Umwandlung je „power conversion unit" von Druck- in Bewegungsenergie erfolgt durch zwei kleine Turbinen, die wiederum zwei Generatoren mit einer Leistung von je ca. 125 kW antreiben. Die Drehgeschwindigkeit der Turbine und des Generators beträgt dabei ca. 1.500 Umdrehungen pro Minute.

Figure 335
Pelamis wave energy converter
Length approx. 150 meters, diameter 3.5 meters, weight including ballast 700 tons

Source: Pelamis Wave Power Edinburgh, Scotland. UK

Bild 335
Wellenenergiekonverter (Pelamis), (Länge ca. 150 m, Durchmesser 3,5 m, Gewicht mit Ballast ca. 700 t)

Quelle: Pelamis Wave Power Edinburgh, Schottland UK

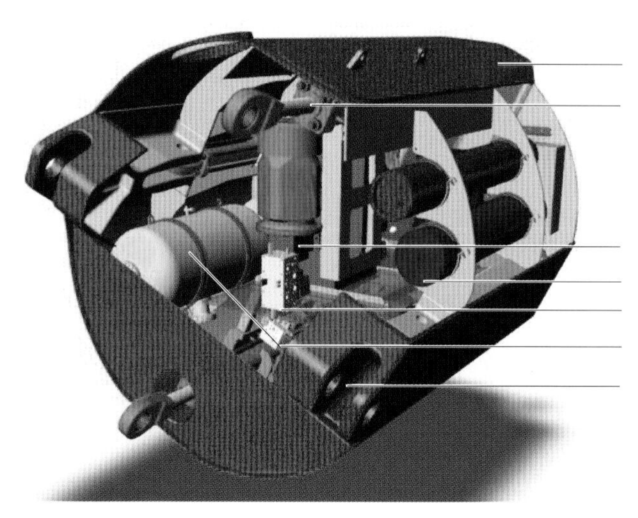

Figure 336
Intenal view of a Pelamis power-conversion module

Source:
Ocean Power Delivery LTD

Bild 336
Innenansicht eines Polamis-Energieerzeugungsmoduls

Quelle:
Ocean Power Delivery LTD

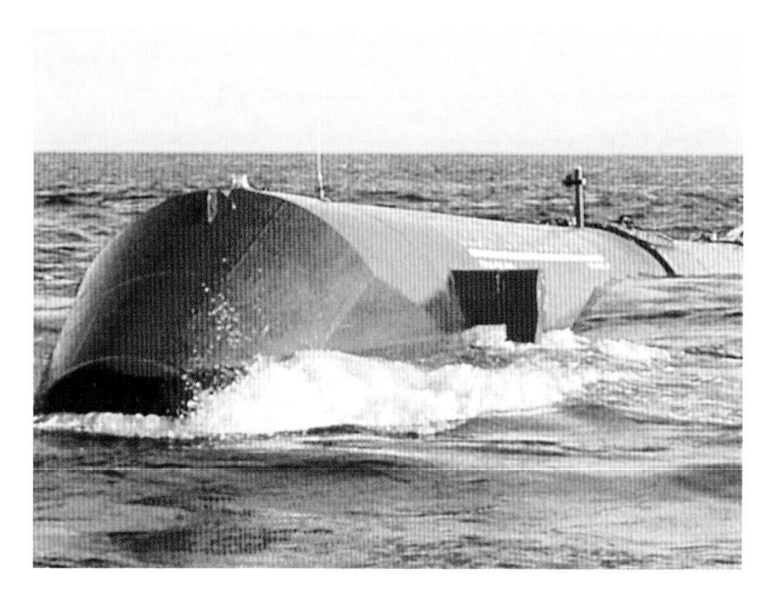

Figure 337
Wave energy converter in operation

Source:
Ocean Power Delivery LTD

Bild 337
Wellenenergiesystem in Aktion

Quelle:
Ocean Power Delivery LTD

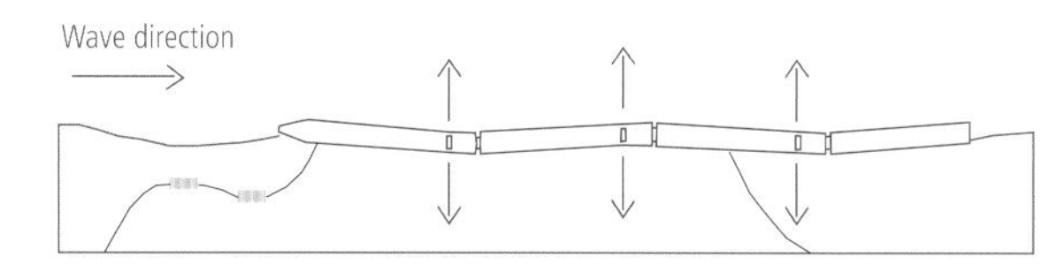

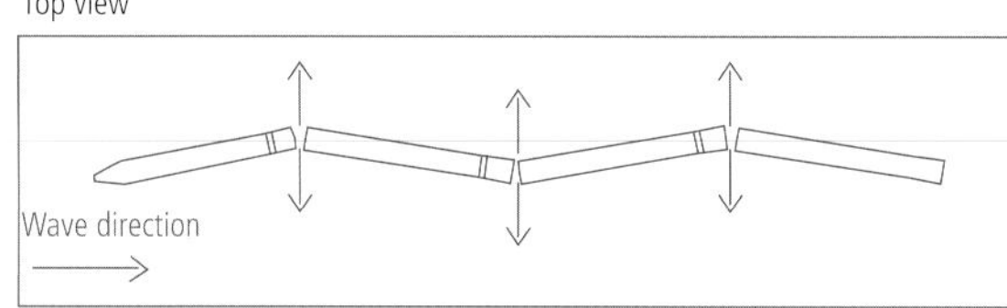

Figure 338
Operation of the wave energy converter

Data:
140 m long
3.5 m diameter
750 kW rated power
capacity factor 0.25 – 0.4

Bild 338
Betriebsweise des „Wave energy converter" mit Leistungsdaten

Länge 140 m
Durchmesser: 3,50 m
Leistung (nominal) 750 kW
Kapazitätsfaktor 0,25 – 0,4

Figure 336 shows all of the important elements of the power-conversion module as well as the hinged joints that hold the energy-generation module against the segmented main tube. **Figure 337** explains the functioning principle of the wave-power device. As can be seen, the up-and-down motion of the waves is transformed into high pressure and ultimately into electricity generation with around a 750 kW power rating, or a total of approximately 2.7 GWh/a at nominal wave energies of 55 kW/m.

The asynchronous generators inside the energy modules provide three-phase AC current with 415/690 V at 50/60 Hz, and a step-up transformer built into the system elevates the voltage to 12 kV or 33 kV, respectively.

The wave-energy converter is conceived to be attached to a slack mooring system anchored at a depth of 50 – 70 m and typically around 5 –1 0 km offshore. In such regions, the wave height is appropriate to operate the system effectively. The assumption is that, on average, every 8 sec a wave impacts the wave converter, passes by, and causes the deflection of the tube, which as a result generates energy.

Figure 338 provides an image for understanding the operation of the system.

Bild 336 zeigt alle wesentlichen Einbauelemente eines „power-conversion Moduls" sowie die Gelenke, die das Energieerzeugungsmodul gegen die Hauptrohre (vier Segmente) halten. **Bild 337** verdeutlicht die Arbeitsweise des Wellenenergiesystems. Durch Auf- und Abwärtsbewegungen der Gesamtanlage in den Wellen wird Bewegungsenergie in Druckenergie umgewandelt. Diese wiederum wird in elektrische Energie umgesetzt, so dass letztendlich eine Gesamtleistung von 750 kW entsteht und in etwa ein mittlerer elektrischer Gesamtenergiebetrag von ca. 2,7 GWh/a (nominale Wellenenergie 55 kW/m).

Die Asynchrongeneratoren innerhalb der einzelnen Energieerzeugungsteile erzeugen dreiphasigen Wechselstrom, 415/690 V bei 50/60 Hz. Ein Transformator innerhalb des Gesamtsystems hebt die Spannungsebene auf 12 kV bzw. 33 kV.

Die Wellenenergiekonverter sind darauf ausgelegt, in einem an Moorings befestigten Gesamtsystem bei einer Wassertiefe von 50 – 70 m zu arbeiten, die häufig in 5 – 10 km von der Küste entfernt anzutreffen sind. In dieser Region entstehen die notwendigen Wellenhöhen, die für den Betrieb notwendig werden. Dabei wird davon ausgegangen, dass zirka alle acht Sekunden eine Welle den Energiekonverter trifft und an diesem vorbeiläuft, wodurch er seine Nickbewegungen ausführen kann, die für die Energieproduktion notwendig sind.

Bild 338 demonstriert den Betrieb eines Wellenenergiesystems anschaulich.

13.2.7 Compressed-air power plants

Energy stored in the form of water at pumped-storage hydroelectric power plants has a tradition of more than 70 years. A precondition of the technology is that significant height differences between upper and lower reservoirs exist, resulting in sufficient pressure to operate turbines for electrical energy generation. Pumped-storage power plants pump water from a lower into an upper reservoir if surplus electricity exists, and the plant starts operation if demand peaks develop. The performance of such a system is approximately 80 %.

It is obvious that such plants can only be designed where sufficient topographical elevation differences exist, as well as the necessary spatial conditions for the placement of upper and lower reservoir lakes. However, in many parts of the world such favorable geographical and topographical pre-determinants cannot be found. It is in these places that so-called compressed-air energy storage (CAES) can be used.

The only commercial European compressed-air gas-turbine power plant has been operated near Huntdorf, Germany, since 1978. At times of lower energy demand, surplus energy is used to compress air using a motor and to pump it into deep salt caverns below ground, which are commonly found in the Northern German basin. At the location in Huntdorf two almost-identical cylindrical natural caverns exist at a depth of around 700 m. They are capable of accepting around 300,000 m^3 of compressed air at 50 – 70 bar. At times of high electricity demand, the compressed air, enriched with natural gas, drives a gas turbine. The added natural gas expands due to its temperature increase in the burner and propels a linked generator. The resulting electrical power of approximately 290 MW is given off continuously over a time period of 2 h. The power efficiency of such a CAES plant, with preheating of the compressed air by turbine exhaust gas, is around 54 %, which is a respectable value with regard to the plant's high degree of flexibility.

CAES power plants are capable of generating power to compensate for demand peaks of up to 500 MW for 15 min. However, because their operation requires the use of natural gas they cannot be called renewable energy source plants in the purest sense of the term. Yet, a reconfiguration of such plants from natural gas to biogas should be achievable in the very near future, and that is why they are included here in the group of renewable-energy power plants. **Figure 339.1** shows the entire plant configuration, and **Figure 339.2** explains the layout of the two-stage compressors used for filling the compressed-air caverns.

13.2.7 Druckluftkraftwerke

Seit mehr als 70 Jahren wird die Energiespeicherung durch Pumpspeicher-Wasser-Kraftwerke mit Erfolg eingesetzt. Voraussetzung dabei ist, dass ein großer Höhenunterschied zwischen oberem und unterem Becken besteht und somit Turbinen aus ausreichender Höhe beaufschlagt werden können, um einen nennenswerten Beitrag zur Stromproduktion zu erzielen. Pumpspeicher-Wasserkraftwerke arbeiten im umgekehrten Sinn, d.h. sie pumpen Wasser von einem unteren Becken in das obere, wenn ein überschüssiges Elektroenergieangebot besteht. Sie gehen dann kurzfristig in Betrieb, wenn elektrische Verbrauchsspitzen abzudecken sind, und erreichen dabei einen Wirkungsgrad von 80 Prozent.

Pumpspeicher-Wasserkraftwerke können selbstverständlich nur dann zum Einsatz kommen, wenn in der Region entsprechende Höhenunterschiede einerseits und das Anlegen eines Speichersees andererseits möglich ist.

In vielen Ländern der Welt bestehen keine wesentlichen Erhebungen in der Landschaft und ist das Erstellen von Stauseen möglich, so dass die Technologie der CAES-Kraftwerke zum Einsatz kommen kann (compressed air energy storage).

Seit 1978 wird in Huntorf (Deutschland) das einzige europäische Luftspeicher-Gasturbinen-Kraftwerk in Betrieb gehalten. In Schwachlastzeiten wird mit überschüssiger elektrischer Energie ein Motor betrieben, der Umgebungsluft in eine unterirdische Kaverne verpresst. Besonders geeignet sind hierfür ausgesolte Salzstöcke, die in der Norddeutschen Tiefebene häufig vorkommen. In Huntorf existieren in 700 m Tiefe zwei nahezu zylindrische Kavernen. 300.000 m^3 Luft können hier bei einem Druck von 50 – 70 bar gespeichert werden. In Zeiten hoher Stromnachfrage wird die komprimierte Luft unter Zufuhr von Erdgas in die Brennkammer einer Gasturbine eingeleitet. Die durch die Verbrennung des Erdgases erhitzte Druckluft expandiert in einer Gasturbine und treibt den zugehörigen Generator an. Hierdurch kann eine Leistung von ca. 290 MW über 2 Stunden kontinuierlich abgegeben werden.

Der Wirkungsgrad einer modernen CAES-Anlage mit Druckluftvorwärmung über das Turbinenabgas beträgt ca. 54 Prozent, was in Anbetracht der hohen Flexibilität ein noch günstiger Wert ist.

CAES-Kraftwerke können Leistungen bis zu 500 MW innerhalb von 15 Minuten ausregeln. Aufgrund des notwendigen Betriebs, d.h. der Nutzung von Erdgas, sind Luftspeicher-Gasturbinen-Kraftwerke nicht als solche zu bezeichnen, die ausschließlich durch erneuerbare Energien betrieben werden. Eine Umstellung von Erdgas auf Biogase müsste jedoch in absehbarer Zeit möglich sein, so dass Luftspeicher-Gasturbinen-Kraftwerke vollumfänglich in den Bereich der Erzeugung erneuerbarer Energien gehören. **Bild 339.1** zeigt die Gesamtanlage, **Bild 339.2** die zweistufigen Kompressoren zur Ladung der Luftspeicher.

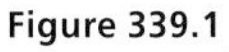

Figure 339.1
Aerial view of the compressed air energy storage power plant (CAES), Huntdorf, Germany, connected to gas turbines and generators. The image shows the covered well access areas to the subterranean storage caverns, left and right of the plant building.

Source: E.on Kraftwerke GmbH

Bild 339.1
Luftaufnahme des Luftspeicher-Gasturbinen-Kraftwerks Huntorf (D).

(Beidseits des Kraftwerkblocks sind hinten die Abdeckungen der beiden Bohrungen zu den unterirdischen Speicherkavernen erkennbar.)

Quelle: E.on Kraftwerke GmbH

Figure 339.2
Two-stage compressor of the Elsfleth-Huntorf, Germany, compressed-air storage plant. In the background the generator and the gas turbine.

Source: E.on Kraftwerke GmbH

Bild 339.2
Der zweistufige Kompressor des Luftspeicher-Gasturbinen-Kraftwerks Huntorf.
(Im Hintergrund rechts schliessen Generator und Gasturbine an.)

Quelle: E.on Kraftwerke GmbH

14 Energy storage

14 Energiespeicherung

The topic of energy storage, especially storage of renewable energy sources, is not only very important but also, in some respects, highly complex. It needs to be an integral and significant aspect of our future energy supply. **Figure 340** offers an overview of the particular operation of various storage technologies that are currently available.

Die Speicherung von Energie, insbesondere unter Nutzung von erneuerbaren Energien, ist ein besonders wichtiges und zum Teil auch schwieriges Thema, das in die gesamten Betrachtungen der zukünftigen Energiebereitstellung einbezogen werden muss. **Bild 340** gibt einen Überblick über die Betriebsbereiche verschiedener Speichertechnologien, die zurzeit angewendet werden.

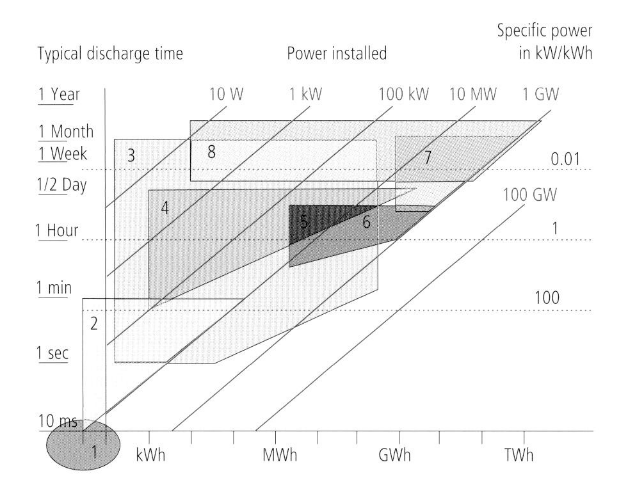

1 Capacitor
2 Super caps, flywheel
3 Batteries
4 Redox-oxidation
5 Flow batteries
6 Pump storage
7 Water reservoirs
8 Hydrogen storage

Figure 340
Operation of various storage technologies

Source: Sonnenenergie/ Jan – Feb 2008 Prof. Dirk Uwe Sauer, RWTH Aachen

Bild 340
Betriebsbereiche verschiedener Speichertechnologien

Quelle: Sonnenenergie/ Jan – Feb 2008 Prof. Dirk Uwe Sauer, RWTH Aachen

14.1 Battery storage

We are all very familiar with batteries as energy storage devices. Yet, the capability of even the largest battery system in use in buildings is not sufficient to satisfy the energy demands of entire cities. At the same time, companies such as E.on are developing very large battery systems the size of several shipping containers to store up to 4 MWh of energy. Such extremely large batteries will be necessary in the future in some applications because the electrical energy produced by wind-power farms, tidal power plants, and, in part, solar generation will need to be stored. We are at the beginning of the development of such battery storage technology, and it is planned to have operating prototypes of such mega-batteries installed by the second half of 2009.

14.1 Batterien

Batterien als Energiespeicher sind uns allen bekannt. Die Ladefähigkeit selbst üblicher größerer Batteriesysteme in Gebäuden ist jedoch nicht dazu angetan, elektrische Energie für ganze Städte zu speichern. Gleichwohl bemüht sich unter anderem zurzeit E.on, eine gigantische Batterie zu entwickeln, die in etwa die Größe von 4 Transportcontainern haben wird und ca. 4 MWh elektrische Energie speichern kann. Entsprechende Riesenbatterien werden in Zukunft zumindest zum Teil notwendig, um elektrische Energien aus Windfarmen, Wellenenergiesystemen, Gezeitenkraftwerken und zum Teil Solarenergie zu speichern. Die Speicherung elektrischer Energie steht zurzeit noch am Anfang ihrer Entwicklung, und erste große Prototypen sollen in der zweiten Jahreshälfte des Jahres 2009 in Betrieb gehen.

14.2 Hydrogen storage

After its separation, hydrogen can be stored in storage tanks with an operating pressure of up to 300 bar in the form of H_2 and O. In this respect, hydrogen is an ideal energy source that can be transported and intermittently stored (as can other gases gained through renewable processes). The cost involved, however, is still a consideration.

In principle, therefore, all gaseous fuels derived from renewable sources are excellent candidates for exploitation. Their comparatively easy storage makes them independent of the cycle of their generation, which means that they can be stored and used whenever a power demand arises.

14.3 Water storage

Water storage in the form of large reservoirs is typically combined with so-called pumped-storage hydroelectric power plants. Pumped storage is a method of storing and producing electricity to supply high peak demands by moving water between reservoirs at different elevations. As already described above, water is pumped during a power surplus condition to a reservoir at a higher altitude, and then, when power demand increases, it is run through generators back to the lower reservoir level. Conditions such as these may exist, for instance, when no alternative renewable energy sources such as wind or solar applications exist at the time of energy demand. Unfortunately, the use of large-scale water reservoirs will play only a minor role in a global future of renewable power supplies because in most regions neither large water bodies nor sufficient differences in topographic altitude exist.

14.4 Compressed air storage

As seen in **chapter 13.2.7**, compressed air power plants operate on a principle similar to that of water reservoir power plants: during times of lower demand, compressed air is generated at pressures between 50 and 70 bar, which is stored in appropriate airtight subterranean geological formations in the Earth's crust. Alternatively, such storage can be replaced by large surface tank farms if natural cavities or other useable geological storage sites are unavailable. In the future, these facilities will also play only a lesser role, similar to that of hydroelectric power generation.

14.2 Wasserstoffspeicher

Wasserstoff kann nach Zerlegung von Wasser in H_2 und O in Speichern eingelagert werden, die einen Betriebsdruck bis zu 300 bar haben können. Insofern ist Wasserstoff – bei allen bisherigen Kostenaufwendungen – ein idealer Energieträger, der sich speichern, transportieren und zwischenlagern lässt. Ähnliches gilt natürlich auch für alle anderen Gase, die, aus erneuerbaren Energien gewonnen, gespeichert werden müssen.

Somit sind grundsätzlich alle gasförmigen Brennstoffe aus erneuerbaren Energien ideal, um sie einerseits unabhängig vom Produktionszyklus und vom Verbrauch, jederzeit verfügbar zu machen – sie zu lagern und dann abzurufen, wenn sie benötigt werden.

14.3 Wasserspeicher

In Verbindung mit der Erzeugung von Energie handelt es sich bei Wasserspeichern (Stauseen) im Regelfall um solche, die in Verbindung mit einem Pumpspeicher-Wasser-Kraftwerk stehen. Wie bereits zuvor ausgewiesen, wird zu Zeiten, in denen ein Energieüberschuss besteht, Wasser von einem unteren See in einen oberen See durch Pumpen befördert, um es dann wieder vom oberen in den unteren See ablaufen zu lassen, wenn Energie benötigt wird. Dies kann zum Beispiel dann der Fall sein, wenn zum Zeitpunkt des Energieverbrauchs keine Solarenergie und Windenergie zur Verfügung stehen. Der Einsatz von großvolumigen Wasserspeichern in Form von Stauseen dürfte in der Zukunft nur eine geringe Rolle spielen, da in vielen Regionen der Welt entsprechende Möglichkeiten infolge zu geringer Höhenunterschiede und zu weniger verfügbarer Stauseen nicht vorhanden sind.

14.4 Druckluftspeicher

Wie bereits in **Kapitel 13.2.7**, Druckluft-Kraftwerke, beschrieben, wird ähnlich wie bei Wasserspeichern Druckluft zu Schwachlastzeiten erzeugt und mit 50 – 70 bar in großen natürlichen Hohlräumen der Erdrinde eingelagert. Druckluftspeicher können auch in größerem Unfang durch große Tankanlagen dargestellt werden, sofern die natürlichen Gegebenheiten am Standort z.B. natürliche Kavernen nicht hergeben. Sie dürften jedoch in der Gesamtheit der notwendigen Energiespeicherung wie die Wasserspeicher nur eine untergeordnete Rolle spielen.

14.5 Hot water storage

Hot water storage tanks are large-scale storage units in which water at temperatures of around 160 °C and at pressures of approximately 16 – 20 bar can be stored. In this case, use is typically localized because the involved volumes allow storage only for a period of around 10 operating hours.

During the years between 1970 and 1980, alternative means of thermal energy storage were introduced in the form of salt storage systems. Due to material fatigue, the life span of such storage systems has proven to be small. Together with the advances in knowledge of solar thermal power plants thermal storage of high-temperature energy will be an area of further development.

14.6 Ice storage

Much like hot water storage tanks, ice storage systems are typically used in relatively small units in or near buildings. They are in general conceived to cover the cooling energy of one day of a building, or approximately 250 – 300 kWh/m³. They operate under the principle of freezing percentages of the contained water or use brines to store cooling energy below the freezing point.

14.7 Flywheel energy storage (FES)

Under certain circumstances, electrical energy can be stored when surplus power is used for the acceleration of a flywheel with high mass to around 15,000 RPM. When used for uninterruptible power supply operations, the energy of the flywheel is given off when the regular power supply is interrupted. Flywheel storage represents a limited possibility for energy storage and is typically used in local building applications. In the past decades, such systems were mainly used to guarantee the uninterrupted power supply for computer centers. Due to their great weight, the need for maintenance, and the fact that they are only capable of bridging short supply gaps (continuous operation cycles of 30 s), flywheel storage has gone out of fashion. Nevertheless, research concerning flywheel storage continues. For example, W.R. Canders and the scientists at the Technical University Braunschweig, Germany, are working on supraconductor storage (Dyna Store) with expected power supplies of up to 2 MW for 30 s.

14.5 Heißwasserspeicher

Heißwasserspeicher sind im Regelfall großvolumige Speichertanks, in denen Heißwasser mit Betriebstemperaturen bis zu 160 °C unter Drücken von ca. 16 – 20 bar eingelagert wird. Heißwasserspeicher werden im Regelfall nur vor Ort eingesetzt, d.h. sie können aufgrund ihrer Volumina häufig nur Wärmeenergie für bis zu ca. 10 Betriebsstunden einlagern.

In den Jahren zwischen 1970 und 1980 wurden alternativ zur Speicherung von Wärmeenergie großvolumige Salzspeicher zum Einsatz gebracht, die jedoch annähernd durchweg eine zu geringe Lebensdauer infolge Materialermüdung hatten. Insofern ist die Einlagerung von hochtemperierter Wärmeenergie auf dieser Basis zurzeit ein Thema, das in Verbindung mit solarthermischen Kraftwerken wieder weiterentwickelt wird.

14.6 Eisspeicher

Eisspeicher werden wie Heißwasserspeicher im Regelfall wiederum nur in kleineren Einheiten innerhalb von Gebäuden zum Einsatz gebracht und speichern maximal den Kälteenergiebedarf eines Tages für ein Gebäude. Eisspeicher können ca. 250 – 300 kWh/m³ speichern und arbeiten entweder auf der Basis der Vereisung eines Teils von Wasser oder nutzen Salzlösungen, um Kälteenergie unter 0 °C einzulagern.

14.7 Schwungradspeicher

Elektrische Energie lässt sich bedingt auch dadurch speichern, dass bei einem elektrischen Überschuss ein Schwungrad mit großer Masse auf bis zu 15.000 Umdrehungen pro Minute gebracht wird, um bei Ausfall von Strom über das Schwungrad einen Generator zu bewegen, der wiederum elektrische Energie abgibt. Schwungradspeicher stellen nur eine eingeschränkte Möglichkeit der elektrischen Energiespeicherung dar und werden im Regelfall nur in Gebäuden eingesetzt. In den vergangenen Jahrzehnten wurden Schwungradspeicher hauptsächlich dann eingesetzt, wenn es darum ging, eine unterbrechungsfreie Stromversorgung für EDV-Einrichtungen sicherzustellen. Aufgrund ihrer hohen Gewichte und zum Teil aufwändigeren Wartung sind Schwungradspeicher aus der Mode gekommen, zumal sie im Regelfall nur als Kurzzeitspeicher anzusehen sind (Durchlaufzeiten bis 30 Sekunden). Forschungen an Schwungmassespeichern mit supraleitenden Lagern (Dyna Store) mit Leistungen bis zu 2 MW (30 s) laufen zurzeit (z.B. Uni Braunschweig, Prof. W. R. Canders).

As an alternative, ultra-capacitor technology (ultracaps) can be used for short-term storage. These storage elements have 20 times the power density of regular accumulators, and they are incapable of storing energy over longer periods.

All renewable energy sources in solid, liquid, and gaseous states are able to be stored and will eventually serve as elements in our energy supply matrix. And in the future, when the world is mainly supplied by these alternative primary energy sources, the storage technology will be very important because renewable energy supply, such as from wind or the sun, is typically non-continuous. The same is true for the generation of energy from the power of the oceans' waves or tides.

Alternativ können auch Hochleistungskondensatoren (ultra-caps) als Kurzzeitspeicher eingesetzt werden, die eine etwa 20-fach höhere Leistungsdichte aufweisen können als herkömmliche Akkus. Als Langzeitspeicher sind sie jedoch nicht geeignet.

Alle in fester, flüssiger und gasförmiger Form vorkommenden Energieträger aus erneuerbaren Energien lassen sich durchweg gut speichern und werden somit ein wesentliches Element in der Energiebereitstellung darstellen. Die Speicherung von Energien in einem Zeitalter, in dem sich die Menschheit im Wesentlichen nur aus erneuerbaren Energien versorgt, wird immer eine große Rolle spielen, da die wesentlichen Elemente der erneuerbaren Energie wie Sonne und Wind nicht immer gleichmäßig zur Verfügung stehen. Gleiches gilt auch für die Bereitstellung von Energien durch Wellenenergie und zum Teil durch Gezeitenenergien.

15 Conclusion

15 Fazit

15. Conclusion

Without doubt, it can be concluded:

- Most fossil fuels are reaching the limits of availability; thus, there is an intense need to find ways to extend those resources in order to allow our children to inhabit a world worth living in.
- Implementation of renewable energy resources could easily meet the planet's energy needs, but today such technologies need significantly greater support and practical application.
- The current apparent trend of climatic changes will continue in the future and will change living conditions in many regions on Earth drastically; scenarios once discussed by The Club of Rome, such as impending shortages of water and nutrition, are already becoming a reality for many of the world's people.
- As a result of climate change, the number and severity of natural disasters is significantly on the rise, and even the so-called developed nations will be affected to a much larger degree than in the past.
- The destructive exploitation of nature as a consequence of the unrestrained consumption of natural resources – and even, to some degree, renewable resources – needs to be ended.
- With regard to buildings, creative designs are needed to reduce energy consumption to almost zero during a structure's useful life, and we need to devise ways to disassemble structures at the end of their life spans and introduce the materials into a complete cycle of reuse.

In slightly more than one hundred years, humankind has taken possession of the world and what it has to offer in such a relentless way that in many regions the life-supporting base not only of people but also of flora and fauna has been destroyed in the process. If we are seriously concerned with the needs of our descendants, we must come to the conclusion that this destruction has to be stopped.

There is a window of opportunity in which humans need to develop many new activities and industries that will create new renewable energy sources and at the same time new jobs for many economies.

It is hard to understand why we have to be exposed to a series of dramatic crises first in order to change our ways and to react. In particular, national governments often with either non-binding regulations regarding resource usage or no policies at all will have to take the majority of the blame, not only in underdeveloped regions but also in the highly developed world.

15. Fazit

Unbestreitbar ist, dass:

- die fossilen Brennstoffe dem Ende entgegengehen und wir diese Ressourcen durch verringerten Verbrauch massiv strecken müssen, um unseren Nachkommen eine noch lebenswerte Welt zu erhalten;
- der Einsatz erneuerbarer Energien den gesamten Weltenergiebedarf ohne weiteres decken kann, jedoch heute in sehr viel größerem Umfang der Einsatz entsprechender Technologien gefördert und umgesetzt werden muss;
- zurzeit und in Zukunft ein Klimawandel vonstatten gehen wird, der unsere Lebensgrundlagen in vielen Regionen der Welt massiv negativ verändert, so dass hieraus die bereits die in den Schriften des Club of Rome dargestellten Szenarien greifen werden, wie zum Beispiel Wasserkrisen, Ernährungskrisen usw.;
- aufgrund des Klimawandels sich Umweltkatastrophen häufen und weiter zunehmen werden, so dass auch die so genannten entwickelten Länder hiervon betroffen sein werden;
- der Raubbau an der Natur durch ungebremsten Verbrauch der Rohstoffe und zum Teil erneuerbaren Energien gestoppt werden muss;
- Bauten so zu entwickeln sind, dass sie nicht nur fast keine Energie mehr verbrauchen, sondern wieder in ihre Bestandteile zerlegt werden können, um alle Ressourcen zu recyceln.

Die Menschheit hat in den letzten 100 Jahren die Welt so nachhaltig in Anspruch genommen, dass bereits viele Grundlagen nicht nur für die Menschen, sondern auch für die Pflanzen- und Tierwelt zerstört wurden. Dieser Zerstörung gilt es dringendst Einhalt zu gebieten, wollen wir unseren Nachkommen etwas hinterlassen, das für sie eine lebenswerte Grundlage bietet.

In dem Zeitfenster, das der Menschheit noch bleibt, müssen viele neue Aktivitäten und Industrien entwickelt werden, die dazu geeignet sind, nicht nur Ressourcen zu erhalten, sondern auch neue Arbeitsplätze zu schaffen.

Es ist unverständlich, warum die Menschheit erst eine Krise herbeiführen will, um dann zu reagieren. Hierbei versagt die Politik auf breiter Linie – nicht nur in den unterentwickelten, sondern in hohem Maße auch in den entwickelten Ländern, da die gesetzlichen Regularien noch viel zu wenig greifen.

In the desert of the United Arab Emirates, a walled city based on traditional Arabic urban design principles but with the inclusion of advanced technologies is planned – it will be the first carbon dioxide-neutral and energy waste-free city in the world. This six-million-square-meter new city, named Masdar, uses the urban identity of cities like Abu Dhabi yet with the sustainable urban design of a city of the future (**Figure 341**) The architects of Foster + Partners, London, in cooperation with Eta Engineers (renewable energy), Transsolar Klima Engineering, and WSP Energy (sustainability, infrastructure) are designing a city without automobiles, supplied entirely by renewable energy, and managed by digital service units. The city will be the future home of a university, the headquarters of Abu Dhabi's Future Energy Corporation, and the location for special trade zones, light industry, and an innovation center for the development of new ideas for the future of energy supply. In cooperation with the Massachusetts Institute of Technology (MIT), the Masdar Institute of Science Technology has been created, which will be a center for research dedicated to the development of future energy technologies. Housing, offices, administrative functions, a science museum, and entertainment complexes are also part of the city concept.

According to the architects and planners, the city will be connected to the existing road network of Abu Dhabi, linking it with the international airport and the downtown business district of the city. Inside the walled city, traffic nodes are never further apart than 200 meters, stimulating pedestrian traffic. Shaded pedestrian paths and narrow, self-shading street spaces are reminiscent of the traditional urban fabric of the Arabian walled city. Renewable energy is provided by wind power and photovoltaic systems, and the city is surrounded by agricultural research and production fields for the growth of biofuel plants.

He who wants to see far into the future needs to look back equally far.

Figure 341 shows the project initiated by Masdar Development.

In der Wüste der Vereinigten Arabischen Emirate soll durch die Kombination der traditionellen Planungsprinzipien der „walled city" mit fortschrittlichsten Technologien die erste CO_2- und abfallfreie Stadt entstehen. Die Millionen-Quadratmeter-Stadt Masdar korrespondiert mit der städtischen Identität Abu Dhabis, während sie gleichzeitig einen nachhaltigen urbanen Zukunftsentwurf bieten soll, **Bild 341**. Auf der Basis der Planung des Büros Foster + Partners, London (erneuerbare Energien: E.T.A., Klimaengineering: Transsolar, Nachhaltigkeit – Infrastruktur: WSP Energy), ist eine autofreie Kommune vorgesehen, die ausschließlich mit erneuerbaren Energien versorgt wird und über digital gesteuerte Serviceeinheiten verfügt. Das bestehende Programm schließt eine Universität ein, den Hauptsitz der Abu Dhabi Future Energy Company, spezielle Handelszonen, Leichtindustrie und ein Innovationszentrum für die Entwicklung neuer Ideen der Energieproduktion. Zurzeit wird in Kooperation mit dem Massachusetts Institute of Technology (MIT) das Masdar Institute of Scence Technology aufgebaut. Masdar soll einen ausgewählten Pool internationaler Bewohner beheimaten, die in progressive Energietechnologien investieren. Vorgesehen sind zudem Angestelltenwohnungen, Büros, ein Wissenschaftsmuseum sowie Entertainment-Einrichtungen.

Nach den Plänen der Architekten wird die Stadt durch ein existierendes Straßennetz, neues Schienennetz und öffentliche Verkehrsmittel mit den umgebenden Kommunen, dem Zentrum Abu Dhabis und dem internationalen Flughafen verbunden. Die Maximaldistanz zwischen den Verkehrsknotenpunkten beträgt 200 m. Schattige Fußwege und enge Straßen erzeugen in Abu Dhabis extremem Klima ein fußgängerfreundliches Umfeld und knüpfen an den kompakten Charakter traditioneller „walled cities" an. Damit die Stadt energetisch vollständig selbsterhaltend sein kann, sind auf dem umgebenden Gebiet Wind- und Photovoltaik- sowie Forschungsfelder und Plantagen zur Produktion von Biokraftstoffen vorgesehen.

Wer weit nach vorne sehen will, muss weit nach hinten blicken.

Bild 341 zeigt das von der Masdar Development initiierte Projekt.

Figure 341
Rendering of the Masdar Development, Abu Dhabi 2007
Client: Masdar-Abu Dhabi Future Energy Company
Architects: Foster + Partners
Renewable energies: ETA Transsolar

Source: archplus 184, October 2007

Bild 341
Enertopia, Abu Dhabi
Gesamtansicht des Projekts
Architektur und Städtebau: Forster + Partners
Erneuerbare Energien: ETA Transsolar

Quelle: archplus 184, Oktober 2007

The land outside of the "walled city" is planned to be used for energy generation and recreation. The concept calls for the realization of a photovoltaic field, a photovoltaic manufacturing plant, a seawater desalination plant, a wind farm, land for agricultural research, farms for the cultivation of various plant species used in biomass generation, a water reclamation plant, a recycling center, a sewage treatment plant, visitor parking, a visitor center and sport and recreational facilities.

The energy consumed in Masdar is supposed to be provided by renewable energy sources only.

To achieve this goal the following facilities are planned:

1. Large-scale utilization of photovoltaic technology. Model solution for building integrated photovoltaic power generation (BIPV) will be realized. Here all currently available solar cells are to be used: Monocrystalline, polycrystalline silicon cells, and thinfilm technology. Photovoltaic power will provide approximately half of the energy needed in Masdar, eliminating many thousands of tons of carbon emissions per year.

2. Solar-thermal power plants. It is planned to focus sunlight to raise the temperature to high levels used for the generation of steam to operate turbines to generate electricity. These solar thermal power plants will serve as the most economical method to provide solar-generated electricity.

3. It is planned to use the southwest and northeast facing corners of the site large-scale wind turbines; however, building integrated wind power (BIWP) systems are planned as "urban turbines".

4. Advanced forms of recycling, composting and combustion of waste will drastically reduce the need for landfills. Recycling and composting serves also to reduce greenhouse gas emissions.

5. Furthermore, geothermal heat pumps will be implemented. They capture the energy captured in the soil by utilizing the temperature difference between the hot desert surface and the somewhat cooler lower soil strata. Employing the geothermal technology will reduce the energy consumption for cooling by half. It is planned to integrate geothermal earth probes along all buildings.

blue: special trade zone
white: residential
red: Masdar University and headquarters of Abu Dhabi Future Energy Company
grey/purple: commercial
diagonally: city light rail system which connects the city to the airport, to Raha Beach and to the city of Abu Dhabi
yellow: light rail stops at university, headquarters and cultural centers

total area: 6.4 million m²
build area: 6.0 million m²

Land use within the "walled city":
30 % Residential
24 % Special Trade Zone
13 % Commercial
6 % University
8 % Administration
19 % Infrastructure

blau: Sonderhandelzone
weiß: Wohnen
rot: Masdar University und Hauptsitz der Abu Dhabi Future Energy Company
grau/violett: Handel
Diagonale: Stadtbahnnetz, Verbindung zum Flughafen, zur RaHa Beach und nach Abu Dhabi
gelb: Haltestationen an Universität, Headquarters und kulturellen Zonen

Grundstück: 6,4 Mio. m²,
bebaute Fläche: 6,0 Mio. m²

Flächennutzung in der "walled city":
30 % Wohnen
24 % Sonderhandelszone
13 % Handel
6 % Universität
8 % Verwaltung und Kultur
19 % Versorgungs- und Verkehrszonen

(left page)
Das Gebiet außerhalb der "walled city" ist für die Energieerzeugung und Freizeit vorgesehen. Geplant sind ein Photovoltaikfeld, eine Photovoltaikfabrik, eine Meerwasser-Entsalzungsanlage, ein Windpark, Forschungsfelder, Plantagen verschiedener Pflanzenarten zur Biokraftstoffproduktion, eine Wasseraufbereitungsanlage, Besucherparkplätze, ein Recyclingzentrum, eine Kläranlage, ein Besucherzentrum, Freizeit- und Sporteinrichtungen.

Der Energiebedarf Masdars soll ausschließlich durch erneuerbare Energien gedeckt werden. Geplant ist:

1. Die großmaßstäbliche Nutzung von Photovoltaik. Es sollen Modellprojekte innovativer gebäudeintegrieter Photovoltaik entstehen.

Alle drei Haupttypen von Solarzellen kommen dabei zur Anwendung: monokristalline und polykristalline Siliziumzellen sowie Dünnschichtzellen. Photovoltaik wird knapp die Hälfte des Energiebedarfs Masdars liefern, wodurch die Emission vieler 1.000 Tonnen Treibhausgas verhindert werden kann.

2. Der Einsatz solarthermischer Kraftwerke. Durch Fokussierung des Sonnenlichts entstehen extrem hohe Temperaturen, die zur Produktion von Wasserdampf genutzt werden, der wiederum die Turbinen für die Erzeugung von Elektrizität antreibt. Solarthermische Kraftwerke eröffnen die mit Abstand billigste Methode solarer Stromerzeugung.

3. Für die südwestlichen und nordwestlichen Ecken des Gebiets sind große Windturbinen vorgesehen, es soll aber auch einige gebäudeintegrierte „urbane Turbinen" geben.

4. Hochentwickeltes Recycling, Kompostierung und Verbrennung von gesammeltem Abfall senken drastisch den Bedarf an Deponieflächen. Recycling und Kompostierung bieten außerdem Möglichkeiten zur Treibhausgasreduzierung.

5. Der Einsatz geothermischer Wärmepumpen. Sie ziehen ihre Energie aus dem Erdreich und nutzen dabei den Temperaturunterschied zwischen der erhitzten Erdoberfläche und dem kühleren Erdreich. Es ist ein einfaches Konzept mit dem Potenzial, den Stromverbrauch zur Kühlung um mehr als die Hälfte zu senken. Vorgesehen ist die Versenkung von Erdsonden entlang der Gebäude.

Electricity from outer space

The Japanese Space Agency (Jaxa) plans on harvesting solar electricity from Earth orbit. If successful, the concept would help to tackle the problems of global warming. Also, in the U.S., the National Security Space Office (NSSO) of the Department of Defense asked the government in October of 2007 to provide funding in the amount of $10 billion for ten years to build a test satellite that would send a ray of electricity of 10 MW to Earth.

The Japanese mission, called the Space Solar Power System (SSPS), has the goal of providing a functioning solar power plant until 2030. The first step toward this technology takes place in February 2008 when a microwave system will be tested to send energy from space back to Earth. The test will bring clarification concerning the fundamental question: How can space energy be transmitted to Earth?

In a first experiment at the Taiki Aerospace Park, a transmitter antenna with a diameter of 2.4 m transmits a microwave beam to a receiver antenna, where it is being converted into electrical energy used in this test to operate a conventional electric heater. The conceived SSPS power plants would use a similar technology yet at a much larger scale: Earth-based stations with a diameter of roughly three kilometers would serve as receivers for the energy from outer space. JAXA researchers estimate that such power plants could generate up to one GW of power, enough to provide energy for 500,000 typical households (**Figure 342**).

In parallel with the microwave technology, JAXA is also researching the potential of laser focusing. Already in September 2007, the agency and the University of Osaka presented technology to convert sunlight to electricity by laser transmission, with an efficiency of 42 %. Comparing these two compelling energy sources, it becomes apparent that transmission by laser produces a much more focused beam, which means that large receiver antennas such as those required for microwave transmission would be unnecessary.

Strom aus dem All

Die japanische Weltraumbehörde Jaxa will Solarstrom im Erdorbit sammeln. Sollte es funktionieren, könnte das System dabei helfen, das Problem der globalen Erwärmung anzugehen. In den USA denkt man in eine ähnliche Richtung: Das National Security Space Office (NSSO) des Pentagon hat der US-Regierung im vergangenen Oktober empfohlen, zehn Milliarden Dollar innerhalb der nächsten zehn Jahre in einen Test-Satelliten zu investieren, der in der Lage sein soll, einen Strahl von zehn Megawatt elektrischer Energie zur Erde zu schicken.

Die japanischen Pläne laufen unter dem Namen Space Solar Power System (SSPS). Bis 2030 will die Jaxa ein funktionierendes Solarkraftwerk mit dieser Technologie bauen. Ein erster Schritt wird aber schon in wenigen Tagen unternommen: Am 20. Februar 2008 beginnen die Tests eines Mikrowellensystems zur Energieübertragung. Damit will man sich der Lösung einer fundamentalen Frage nähern: Wie soll die Energie aus dem All herunter zur Erde kommen?

Die Jaxa arbeitet parallel an einem System, das auf Laser-Bündelung basiert, und an der Mikrowellentechnologie. Im ersten Experiment im Taiki Aerospace Park wird eine Sendeantenne mit einem Durchmesser von 2,4 Metern einen Mikrowellenstrahl bis zu einer Empfangsantenne schicken. Dort wird die Mikrowellenstrahlung dann in elektrische Energie umgewandelt, die einen herkömmlichen Haushaltsheizkörper antreiben soll. Die geplanten SSPS-Kraftwerke würden eine ähnliche Technik in völlig anderen Dimensionen anwenden: Bodenstationen mit einem Durchmesser von etwa drei Kilometern sollen schließlich die Mikrowellen aus dem Orbit auffangen. Bis zu einem Gigawatt Leistung versprechen sich die Jaxa-Forscher von einem derartigen Kraftwerk – genug, um 500.000 Haushalte mit Strom zu versorgen, **Bild 342**.

Die alternative Methode, an der Jaxa-Wissenschaftler ebenfalls arbeiten, soll die Energie aus dem Erdorbit mittels Laserstrahl zur Erde transportieren. Schon im September 2007 stellten die Jaxa und die Universität von Osaka eine neue Technologie vor, mit der sich Sonnenlicht mit einem Verwertungsgrad von immerhin 42 Prozent in Laserlicht umwandeln lässt. Übertragung per Laser hätte den Vorteil, dass der Energiestrahl in viel stärker gebündelter Form auf der Erde ankäme – gigantische Auffangantennen wie bei der Mikrowellen-Übertragung würden entfallen.

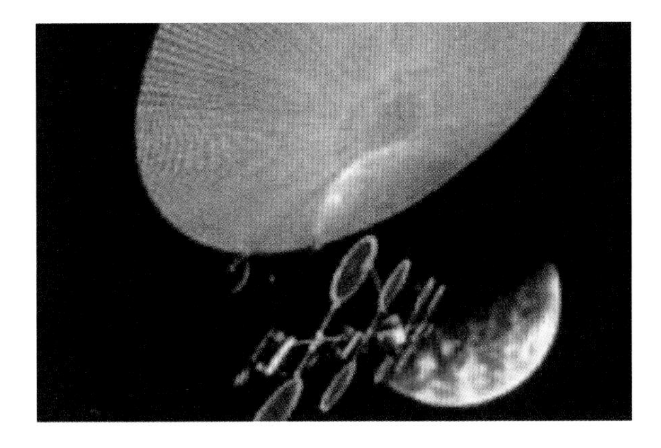

Figure 342
Solar Power Satellite (SPS) Powersat for the transmission of solar energy from outer space to Earth by microwave technolgy.

Source: Japan Space Agency Jaxa

Bild 342
Powersat (SPS, Solar power satellite) zur Übertragung von Solarenergie durch Mikrowellen aus dem All zur Erde

Quelle: Japan Space Agency Jaxa

* * *

We have begun to face the reality that, unfortunately, unlimited economic growth and development based on fossil energy sources is a delusion – unrestrained consumption will eventually cause chaotic conditions for many of us. The way in which rich, developed nations serve their needs by exploitation of the poorer countries and then use them to dump the unwanted waste of consumption is truly unsustainable. We need to not only use the resources given to us more carefully and sparingly but also forego the production of unnecessary goods so prevalent in many aspects of our economy. Industry needs to be restructured so that products are taken back and recycled at the end of their useful life – and in its own country. Products also need to include within their pricing the actual and complete cost, including the cost for their disposal. Only with the resulting increase in product price will people, especially in the developed world, start to rethink their purchase options and begin a healthy resource savings process.It is now time to continue the work of the Club of Rome, perpetuate its research, and bring it back to the conscious of our societies.

* * *

Der Traum des Wachstums ohne Ende ist ausgeträumt – der Wahn des ungehemmten Verbrauchs führt zu einem unabsehbaren Ende mit chaotischen Zuständen auf der Welt, und es kann auf Dauer nicht hingenommen werden, dass sich die wohlhabenden Länder der ärmeren Regionen dieser Welt bedienen, um ihren Wohlstandsmüll anderweitig abzuladen. Insofern wird es nicht nur dringend notwendig, weitaus sparsamer mit allen Ressourcen umzugehen wie bisher und darauf zu verzichten, unnötigen Tand und Müll zu produzieren, sondern die Wirtschaft so umzubauen, dass sie ihre eigenen Problemprodukte wieder recyceln müssen und dies im eigenen Land. Energien und Ressourcen, die verbraucht werden, müssen ihren tatsächlichen Preis haben, nämlich auch die Kosten mit einschließen, die für die Entsorgung aufzuwenden sind. Nur unter dieser Voraussetzung würden sich viele Produkte und Dienstleistungen so deutlich preislich erhöhen, dass viele Menschen – insbesondere in den entwickelten Ländern – aufgrund der hohen Preise anfangen zu sparen. Es wäre dringend sinnvoll und notwendig, die Veröffentlichungen des Club of Rome weiter fortzuschreiben, um alle neuen Erkenntnisse wieder ins Bewusstsein der Menschen zu rufen.

Company profiles
Firmenprofile

Bibliography
Literaturverzeichnis

Illustration credits
Bildnachweis

Selective index
Schlagwortverzeichnis

Imprint
Impressum

Imtech Deutschland GmbH & Co. KG
Hammer Str. 32
D-22041 Hamburg
Phone +49 (0) 40 / 6949-0
Fax +49 (0) 40 / 6949-2929
zentrale@imtech.de
www.imtech.de

Imtech Deutschland – best in technical performance
Imtech is the leading constructor and service provider for technical building services in Germany. Company's headquarter is situated in Hamburg. 4,500 employees in more than 60 branches offer technical solutions for multifunctional stadiums, airports, industrial real estates and other buildings. In 2007 Imtech Deutschland achieved the best revenue in history: an annual revenue of 970 million € and an EBIT of 43.4 million €.

Core Competences
- Technical building services
- Power plant and energy systems
- Environmental simulation / test bed engineering
- Cleanroom technology
- Contracting
- Ship and dockbuilding

Research and development
The Imtech Research and Development Centre in Hamburg is one of Europe's leading building services laboratories. Experienced engineers develop with the help of model laboratory test and computer simulations individual solutions for flow technology and air condition, smoke removal and fire protection as well as energy-oriented solutions for buildings and plants.

In cooperation with the technical university of Hamburg-Harburg and the university of Kassel Imtech developed computer-assisted programmes for energy-oriented simulations for buildings and industrial real estates. These programmes are able to simulate and optimise single HVAC systems as well as complex production units. With the help of these simulation programmes the engineers offer their clients high efficient technical solutions and guarantee individual energy and costs reductions.

Company profile
Imtech Deutschland is part of Imtech N.V., a European technical service provider situated in Gouda, Netherlands. With over 19,000 employees, Imtech achieves annual revenue of more than 3.3 billion euro. Imtech shares are listed on the Euronext Stock Exchange Amsterdam, where Imtech is included in the Midcap Index and the Next 150 index.

History
This year, Imtech Germany is looking back on 150 years of history. Founded in Hamburg on 6 July 1858 by the tradesman Rudolph Otto Meyer, the small company of the same name – Rud. Otto Meyer (ROM) – initially specialized in the construction of heating systems for greenhouses. Within just a few decades, the heating engineering firm had become the largest German provider of building services engineering. In 1997, the ROM shareholders sold their shares to the Dutch company Internatio Müller N.V., which now trades under the name of Imtech N.V.

HL-Technik

HL-Technik AG
Beratende Ingenieure
Letzigraben 89
CH-8003 Zurich
Phone +41 44 / 305 38 20
Fax +41 44 / 305 38 29
zuerich@hl-technik.ch
www.hl-technik.ch

HL-Technik AG Consulting Engineers headquartered in Zurich, Switzerland was founded in the year 1991. Executive officers are Thomas Wetter, Dipl. Techniker TS (H/S/K) and Prof. Dr.-Ing. e.h. Klaus Daniels, Professor at the Technische Universität Darmstadt and Professor Emeritus at the Eidgenössische Technische Hochschule (ETH) in Zurich.

The firm is specializing in building systems design and engineering, with an emphasis on such engineering solutions which are fully integrated and interconnected to form holistic systems. All sectors of system engineering including energy, life cycle analysis, and construction support services are part of the HL-Technik AG expertise.

The scope of engineering services covers building types of the residential, commercial, and industrial sector including buildings for the health industry.

The office provides economic feasibility studies which are based on a close client relationship and client input. HL-Technik AG develops client oriented system solutions and focuses on concepts which are sustainable.

The philosophy of the firm is based on honest service to clients which are considered to be partners in the design process resulting in excellent performance solutions. Clients, industry suppliers and own employees are treated with respect with the goal to form long lasting, positive relationships.

The office provides workshops as well as individually tailored consulting services to achieve the highest level of design quality in all practical applications.

The office hierarchy of HL-Technik AG is flat, providing the engineers with a high degree of design autonomy and responsibility. The firm is based on goal and performance evaluation. Constant improvement strategies and reviews are part of the individual design team management.

The internal organization of the company facilitates quick, direct decisions.

Response to client demand is fast and flexible. The office furthermore invests significantly in quality control and innovation.

Together with partners in the industry the office is engaged in a constant knowledge transfer, provides continuing education for employees to ascertain state-of-the-art and flexible, client-oriented results.

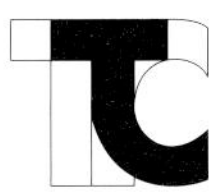

TTC Timmler Technology GmbH
Zum Wetterschacht 1
D-45659 Recklinghausen
Phone +49 (0) 23 61 / 915 96 80
Fax +49 (0) 23 61 / 915 96 89
info@ttc-technology.eu
www.ttc-technology.eu

Developing innovative solutions for new buildings and redevelopment projects in close co-operation with architects and planners

Assisting architects and planners to develop customized solutions during the planning phase is just one of the strengths of TTC Timmler Technology.
TTC supplies intelligent buildings technology for contemporary residential and work environments: LED lights, innovative air conditioning systems, design-oriented façade components and gratings for both interior and exterior applications. Our know-how and many years experience let you combine modern design, energy efficiency and economic viability. Whatever your technical requirements, we design customized solutions consisting of standard components or tailor-made components, produced to your specifications.

Kind to the environment and economically viable

People and the environment are at the heart of TTC's philosophy. We develop natural air conditioning systems that are both energy and cost efficient.

Multi-functionality
Use our know-how to enhance your design

Multi-functionality is a particularly strong point of TTC buildings technology. To name just a few examples:

– TTC Drainlight LEDs – A shining example
As with TTC gratings you can also use TTC Lighttools in our maintenance platforms to create a stunning illumination and to put your design into the »limelight«. The options Drainlight is offering are as versatile as your ideas: From Façade space lights, Power LEDs, LED light lines and tiles to wall washers – with individual designs and a wide range of materials we can deliver customized solutions for your projects. The lights can be integrated in any architectural design and create a strong feeling of depth.

– TTC Modultherm is the ideal system to noiseless air condition whole buildings cost efficiently, using the natural force of gravity.

– TTC Chilled Beams ensure an air conditioning with high comfort an very low noise in working areas. In arrangement with the architect chilled beams add themselves into the design of the ceiling.

– TTC Floorunits with different functionalities of heating, cooling and ventilation provide the free view through space high glass façades. These products combine design with functionality an energy efficiency.

– Homogeneous grating systems allow a seamless transi tion between the interior and the exterior design of a building. On the inside TTC Under Floor systems provide solutions for all your heating, cooling and ventilation requirements and on the outside they complement the TTC Façade Drainage systems.

– Filigree sun protection systems on the façade provide openness and transparency.

Schüco International KG
Karolinenstraße 1 – 15
D-33609 Bielefeld
Phone +49 (0) 521 / 783-0
Fax +49 (0) 521 / 783-451
www.schueco.com
info@schueco.com

Schüco – Sustainable solutions for building envelopes

With future oriented projects worldwide, Schüco stands for the highest quality of building envelopes. The company is known for intelligent systems and the widest range of technologies and materials. With a strong market and customer focus Schüco has always been at the forefront in providing concrete solutions to today's key challenges in terms of global warming and rising energy demand. Schüco works with the most attractive energy saving and renewable energy solutions. We combine energy efficient solutions with the highest standards in automation, security and design.

Energy^2 – Saving energy and generating energy

Schüco positions its competitive edge within a global strategy for its systems and products. The strategy is designed to protect natural resources by continuously improving the energy balance of buildings without compromising on modern architectural demands.

The mission "Energy^2 – saving energy, generating energy" forms the basis of these endeavors. Heat insulation and building automation are key tools in lowering energy consumption. The integration of photovoltaic technology, solar thermal energy and heat recovery can turn a building façade into a power plant that generates not only electricity but also heating and cooling. Schüco's energy autonomous buildings are already reality today.

Consultancy and Comprehensive Service

Technological expertise is one part of Schüco's strength, design and implementation another. Schüco places a high level of emphasis on comprehensive consultancy with all relevant actors being directly involved in the building project and covering the entire range of technical and aesthetic questions. We provide special software solutions for designers, architects, sales staff and fabricators. With the know-how and reliability of German engineering, Schüco guarantees the smooth implementation of sophisticated projects worldwide.

The Schüco project phases work like a fixed genetic code. Each construction phase represents an individual component, from design to completion. All components combined interact with one another and produce a unique solution for each building structure, based on the best cost/benefit ratio.

With Schüco's modular firm structure, key aspects such as cost, quality, innovation and individuality work hand in hand. Schüco ensures the best return on in-vestment in the field of façade construction. So when budgeting for the façade, it is important to remember its special role for the market value and the sustainability of the whole building. Schüco's CEO Dirk U. Hindrichs points out that although it usually accounts for less than 20 % of building costs, the façade represents the strongest selling tool for the investor. As European market leader and a global player in the industry, Schüco ensures that a building retains its appeal in the long term – for both investor and user.

HL-Technik Engineering Partner
GmbH
Feringastraße 10b
D-85774 Munich/Unterföhring
Phone +49 (0) 89 / 99 29 10-0
Fax +49 (0) 89 / 99 29 10-30
info@hl-technik.de
www.hl-technik.de

Excellence by experience. Inspired by vision.
Since 1969 HL-Technik founded by Prof. Dr.-Ing. e.h. Klaus Daniels is an innovative company designing building services.

If one examines future trends against the background of the major economic and ecological, social and cultural changes which are taking place, it is clear that a change in attitude has to take place which will lead to those resources which are still available being utilized as conservatively as possible, with the greatest possible protection being given to the environment.

Our aim is to create living and working environments which are comfortable and which encourage creativity in order to design environmentally friendly and energysaving buildings which are distinguished by an economical use of natural resources. This involves the passive and active use of wind energy, geothermal heat, solar energy, rainwater and the use of materials which cause as little harm as possible to the soil and air during manufacturing and in the use and disposal of water.

HL-Technik's philosophy is therefore, as part of a network of users, architects, structural and landscape planners, to help promote such developments in the planning of buildings which will be necessary in the future.
We call this **sustainable energy design**.

Integrated planning of building services
HL-Technik provides integrated services design capability that includes expert divisions which bring additional values in delivering global building solutions.

Our scope includes:
- consulting,
- overall concepts and system analysis,
- technology selection,
- optimized building services plannings,
- urban development from an ecological viewpoint.

Our range of skills also includes:
- technical advisor services,
- cost benefit analysis,
- building climate studies,
- computational fluids dynamics analysis (CFD simulations),
- aerophysics,
- system design,
- condition surveys,
- whole life costing advices.

And finally, the planning of technical building equipment:
- heating technology,
- ventilation and air conditioning technology,
- sanitary engineering,
- electro engineering,
- fire protection,
- power plants (CHP),
- conveying technology,
- daylight and lighting systems,
- information and communication technology.

Baumschlager Eberle Architekten

Having realised more than 300 buildings Baumschlager Eberle are counted among the most successful international architects. Numerous awards and credits demonstrate the high quality of the work that has been done since the office was founded.

The great acceptance of Baumschlager Eberle is a result of the concept that is able to connect immediate utility value to the users' demands and to cultural sustainability.

At six locations in Europe and Asia the teams assume special responsibility with their architectural work: For the client, the regional context and the social environment.

However, there is still enough scope to achieve aesthetical visions. This holistic approach constitutes the trust in the architecture of Baumschlager Eberle: An architecture that creates buildings that have substantially lower energy consumption values than the local regulations demand.

Offices

Baumschlager-Eberle Ziviltechniker GmbH
Lindauer Straße 31, 6911 Lochau, Austria
Phone +43 5574/430 79-0, Fax +43 5574/430 79-30
office@baumschlager-eberle.com

Baumschlager Eberle Architekturbüro
Gewerbeweg 15, 9490 Vaduz, Liechtenstein
Phone +423 23/606-46, Fax +423 23/606-49
office@be-a.net

Baumschlager Eberle P.ARC ZT GmbH
Objekt 645, 4. Stock, 1300 Vienna-Airport, Austria
Phone +43 1 7007/366-00, Fax +43 1 7007/366-01
office@p-arc.at

Baumschlager Eberle St. Gallen AG
Davidstrasse 38, 9000 St. Gallen, Switzerland
Phone +41 71/227 14-24, Fax +41 71/227 14-25
office@architectural-devices.com

Baumschlager+Eberle Architects
Floor 4, Jia No.1 Dong Bin He Road
He Ping Li East Street, Dong Cheng District
Beijing 100013, China
Phone +86 10/8 422 11-02, Fax +86 10/8 422 11-03
office@be-a.net.cn

Baumschlager & Eberle Anstalt Vaduz
Zweigniederlassung Zürich
Bäckerstrasse 40
8004 Zurich, Switzerland
Phone +41 43/322 10 00, Fax +41 43/322 10 01
office@ad-be.com
www.baumschlager-eberle.com

Bedard, Roger, Mirko Previsic, Brian Polagye, George Hagermann, Andre Casavant, Devine Tarbell and Associates, *EPRI TP-008-NA. North America Tidal In-Stream Energy Conversion Technology Feasibility Study*, Palo Alto, California, June 2006.

Beyerle, Thomas, *IMMOEBS. Newsletter 02.07, Schwerpunkt Klimawandel*, Wiesbaden, November 2007.

Daniels, Klaus, *Advanced Building Systems. A Technical Guide for Architects and Engineers*, Basel, Boston, Berlin: Birkhäuser, 2003.

Daniels, Klaus, *Low-Tech – Light-Tech – High-Tech. Building in the Information Age*, Basel, Boston, Berlin: Birkhäuser, 1998.

Daniels, Klaus, *Sonnenenergie: Beispiele praktischer Nutzung: Bericht über eine Studienreise 1975*, Karlsruhe: C. F. Müller, 1976.

Daniels, Klaus, *Sustainability and Energy Consumptions under Discussion*, Munich: HL-Technik Engineering Partner, 2007.

Daniels, Klaus, *The Technology of Ecological Building: Basic Principles and Measures, Examples and Ideas*, Basel, Boston, Berlin: Birkhäuser, 1997.

Denzler, Lukas, *Tec 21. sia. Minergiebauten,* Zurich, November, 2007.

Spiegel Special, Nr. 5, Hamburg, 2006.

Dobelmann, Jan Kai, in: *Sonnenenergie. Zeitschrift für erneuerbare Energien und Energieeffizienz,* Munich, September/October, 2007.

Energy (R)evolution, *Report Global Energy Scenario. A Sustainable World Energy Outlook*, EREC, European Renewable Energy Council, Greenpeace, 2007.

Hegger, Manfred, Matthias Fuchs, Thomas Stark, and Martin Zeumer, *Energieatlas. Nachhaltige Architektur*, Basel, Boston, Berlin: Birkhäuser, Edition Detail: Munich, 2007.

Herzog, Thomas, Roland Krippner, and Werner Lang, *Façade Construction Manual*, Basel, Boston, Berlin: Birkhäuser, 2004 (Edition Detail).

Hindrichs, Dirk U., and Winfried Heusler, *Façades – Building Envelopes for the 21st Century*, Basel, Boston, Berlin: Birkhäuser, 2nd ed., 2006.

Hindrichs, Dirk U., and Daniels, Klaus, *plusminus 20°/40° latitude. Sustainable Building Design in Tropical and Subtropical Regions*, Stuttgart, London: Edition Axel Menges, 2007.

Kraft, Sabine, Julia von Mende, and Agnes Katharina Müller, "Architektur im Klimawandel", *Archplus.* Aachen, October 2007.

Krewitt, Wolfram, Sonja Simon, Wina Graus, Sven Teske, Arthouros Zervos, and Oliver Schäfer, *"The 2°C scenario – A sustainable world energy perspective"*, *Energy Policy,* vol. 33 (2007), no. 10, pp. 4969 – 4980.

Nitsch, Joachim, *Leitstudie 2007*, *Ausbaustrategie Erneuerbare Energien. Aktualisierung und Neubewertung bis zu den Jahren 2020 und 2030 mit Ausblick bis 2050*, Berlin: Bundesministerium für Umwelt, Naturschutz und Reaktorsicherheit, 2007.

Schlaich, Jörg, *The Solar Chimney. Electricity from the Sun*, Stuttgart, London: Edition Axel Menges, 1995.

Team Germany, TU Darmstadt, *Solar Decathlon 07. Comprehensive Energy Analysis Report*, Darmstadt, August 2007.

The authors and the publisher would like to thank all those who contributed generously to this book with photographs or reproduction rights or who helped otherwise with information and expertise regarding the various topics included in the book.

The photographs in this publication are taken from individual author archives, and especially from the files of HL-Beratungs- und Beteiligungs GmbH. In rare instances, we were unable to identify authors of images despite intensive research. In these cases, we like to be notified by the respective owners of the images if possible. All drawings, schematic depictions, and tables were conceived and drawn for this book exclusively from Stefanie and Jürgen Riemer, Riemer Design, Munich.

References of illustratons (photographs, drawings, diagrams, tables) as for as known to us, are directly related to the respective captions.

outside negative pressure 213
overcast weather 301
overheating of cell 290
oxygen 293

P

Q

R

S

Z

ISBN 978-3-936681-25-3

Prepress: Reinhard Truckenmüller, Stuttgart

Printing and binding: SC (Sang Choy) International Pte Ltd, Singapore

Project management: Klaus Daniels, Munich

Translation into English: Ralph E. Hammann, University of Illinois

Editorial work: Nora Krehl von Mühlendahl, Ludwigsburg

Design and layout: Riemer Design, Munich

Cover design: Jan Riemer, Visuelle Kommunikation, Munich